PLANT MICROEVOLUTION AND CONS HUMAN-INFLUENCED ECOSYS

As human activities increasingly domesticate the Earth's ecosystems, new selection pressures are acting to produce winners and losers amongst our wildlife. With particular emphasis on plants, Briggs examines the implications of human influences on microevolutionary processes in different groups of organisms, including wild, weedy, invasive, feral and endangered species. Using case studies from around the world, he argues that Darwinian evolution is ongoing. He considers how far it is possible to conserve endangered species and threatened ecosystems through management, and questions the extent to which damaged landscapes and their plant and animal communities can be precisely recreated or restored. Many of Darwin's ideas are highlighted, including his insights into natural selection, speciation, the vulnerability of rare organisms, the impact of invasive species, and the effects of climate change on organisms. This is a thought-provoking text for students and researchers of evolution, conservation, climate change and sustainable use of resources.

Particular highlights include:

- An assessment of how neo-Darwinian concepts impact on the theory and practice of conservation in the context of climate change, alerting the reader to the implications of this novel approach
- Background information on basic elements of genetics, molecular methods, climate change, ecology and population biology, with particular reference to plants, which serves as a useful guide for students
- Case studies from many different countries, which make the book globally relevant

DAVID BRIGGS is Emeritus Fellow of Wolfson College, University of Cambridge. He completed his B.Sc. and Ph.D. at Durham University. He has served as Demonstrator in Botany, Botany School, University of Cambridge from 1961 to 1964; Lecturer in Botany, University of Glasgow from 1964 to 1974; and Lecturer in Botany, and Curator of the Herbarium, Department of Plant Sciences at Cambridge University from 1974 to 2001. For many years he was a member of Cambridge University Botanic Garden Syndicate – the committee that directs the general policy of the garden. He has a lifelong interest in conservation, evolution, genetics and taxonomy. His practical conservation experience includes being a former member of the Wicken Fen Committee of the National Trust and the Milngavie Civic Trust. He was formerly the Chair of Cam Valley Forum – an action group active in the conservation of the Cam, its flood plain and tributaries.

PLANT MICROEVOLUTION AND CONSERVATION IN HUMAN-INFLUENCED ECOSYSTEMS

DAVID BRIGGS
Wolfson College, University of Cambridge, UK

CAMBRIDGE UNIVERSITY PRESS
Cambridge, New York, Melbourne, Madrid, Cape Town, Singapore, São Paulo, Delhi, Dubai, Tokyo

Cambridge University Press
The Edinburgh Building, Cambridge CB2 8RU, UK

Published in the United States of America by Cambridge University Press, New York

www.cambridge.org
Information on this title: www.cambridge.org/9780521818353

First published 2009

Printed in the United Kingdom at the University Press, Cambridge

A catalogue record for this publication is available from the British Library

ISBN 978-0-521-81835-3 Hardback
ISBN 978-0-521-52154-3 Paperback

Contents

Preface

It is often erroneously believed that evolution is something that happened in the past. However, there is strong evidence that evolution is continuing at the present time, as plants face new selection pressures generated by human activities that destroy, damage, fragment and alter ecosystems. In a world grappling with anthropogenic climate change, such pressures are likely to increase, as human populations, presently about 6.5 billion, are projected to rise to 12.8 billion by 2050 (if fertility remains at present levels).

In human-influenced landscapes, two broad classes of plants are often recognised, based on their apparent success or relative failure. Thus, some species are 'winners' (crop plants, weeds, invasive plants etc.). Others, the endangered species, are 'losers' or 'potential losers', with extinction their likely fate. Put simply, some plant species appear to be at a selective advantage in changing ecosystems and their populations are stable or increasing, while others faced with the same selection pressures are declining and threatened with extinction. Traditionally, but with some honourable exceptions, these two facets of evolutionary change are treated as separate subjects in academic books and elsewhere. Here, the notion of winners and losers is considered as a single concept, as major insights emerge through such an approach.

Another main focus of the book is the examination of conservation efforts in the light of our understanding of evolution. Many conservationists believe that the major challenge is to persuade the general public and political leaders that conservation of biodiversity is important and resources should (indeed must) be found to secure the future of endangered species and ecosystems. However, given that a great deal of support has often been secured, the key question to confront, in the current struggle to devise successful conservation strategies and effective management, is whether such activities are likely to succeed. In essence, conservationists are attempting to ensure the long-term survival of threatened species and ecosystems by preventing or modifying the impact of deleterious selection pressures brought about by human influences. They have the belief that by re-imposing the appropriate edaphic and biotic environments long-term self-sustaining populations of endangered species may be perpetuated. How far are conservation objectives likely to succeed in the long term? If these endeavours fail, and many species become extinct, what are the likely consequences for biodiversity and human development? How will the losses or potential losses influence the future of plant evolution? It is timely to examine the theoretical and practical basis of plant conservation in a Darwinian context, especially in 2009, as this year marks the 200th anniversary of Darwin's birth and the 150th anniversary of the publication of *On the Origin of Species*.

There are a number of excellent accounts of plant evolution, conservation, environmental issues and climate change, and reference will be made to many in the text. It is not my intention to provide comprehensive coverage of these major areas of concern, but to consider, with appropriate examples, the 'interface' between microevolution and conservation in plants growing in ecosystems subject to human influence and management. Ideally, all the many life forms of biodiversity should be considered, but this is not possible within the confines of this book. The focus is on wild, introduced, invasive, feral and weedy plants, together with crop–weed and crop–wild relative interactions. The fate of rare and endangered plants will also figure large in the account.

The text deals primarily with plants. But, given that much of conservation theory and practice comes from zoological investigations, and acknowledging the manifold interactions in ecosystems, some important ideas and selected research findings from studies of animals will be briefly considered. Such studies continue to provide important models for botanical researches.

The intention in writing this book is to provide an authoritative, up-to-date text for undergraduate and postgraduate students studying evolution, conservation and aspects of climate change, while at the same time exploring the implications of recent advances to conservation practitioners. The book is also designed to appeal to the general reader with a real interest in the subject. With this diversity of readership in mind, some important background areas of genetics, landscape ecology and population biology are explored. As the text examines interconnections between complex subjects, references to important papers are provided in order that the reader can build on the framework provided.

This book is written in the same spirit as the three editions of D. Briggs and S. M. Walters, *Plant Variation and Evolution* (1969, 1984 and 1997). I have set out to provide a critical but concise account of the logical and historical development of the subject, as well as a review of current excitements and advances, while at the same time paying attention to difficulties and uncertainties. Throughout the book the aim is to engender a critical attitude of mind, reflecting my own outlook in being uncommitted and even sceptical about neat explanations and simple formulations.

Acknowledgements

I pay tribute to Max Walters, formerly Director of the University Botanic Garden, University of Cambridge, who died 11 December 2005. Together, he and I wrote the three editions of *Plant Variation and Evolution*. He was a most valued colleague and mentor, who encouraged me to write this book. I thank Max and his wife Lorna for life-long friendship and many kindnesses: they always made my family welcome with their wonderful Yorkshire teas.

I thank those teachers, colleagues and friends who gave me encouragement and provided life-changing opportunities: Ada Radford (my first biology teacher), Donald Pigott, David Valentine, Harry Godwin, Harold Whitehouse, Percy Brian, Richard West, John Burnett, Jack Harley, David Lewis and Peter Ayres.

As I accompanied them in the field or in conversations over coffee, I have discussed many issues about microevolution and conservation with a large number of friends and colleagues. To all of them I offer my thanks: John Akeroyd, Janis Antonovics, John Barrett, David Bellamy, John Birks, May Block, Margaret Bradshaw, Tony Bradshaw, Arthur Cain, Judy Cheney, David Coombe, Gigi Crompton, Quentin Cronk, Jim and Camilla Dickson, Jeff Duckett, Trevor Elkington, Harriet Gillett, Peter Grubb, Mark Gurney, John Harper, Joe Harvey, John Harvey, Peter Jack, David Kohn, Andrew Lack, Vince Lea, Elin Lemche, Roselyne Lumaret, Terry Mansfield, Hugh McAllister, Pierre Morisset, Gina Murrell, Peter Orris, Philip Oswald, John Parker, Joseph Pollard, Duncan Porter, Chris Preston, Oliver Rackham, John Raven, Tom ap Rees, Peter Sell, Alison Smith, Edmund Tanner, Andrew and Jane Theaker, John Thompson, Alex Watt, David Webb, John West and Peter Yeo.

In particular I would like to thank those I taught in the Universities of Cambridge and Glasgow: their challenging questions about microevolution and conservation stimulated me to write this book.

I am especially grateful to Joachim Kadereit, James Cullen and Suzanne Warwick who offered comments on the first drafts of this book. I very greatly value their friendship, expert advice and encouragement.

I pay special tribute to my family for encouraging me to write this book, my parents Mabel and Tom Briggs, Nancy Briggs, Jonathan Briggs, Nicholas Oates, Alastair Briggs, Françoise

Etienne, Catherine, Miranda, Ella, Judith and Adrian Howe, Norman Singer and Geoffrey Charlesworth.

Without my wife's support, tolerance and unfailing commitment, this book would never have been written. With good humour and constant encouragement, she has helped me bring this project to fruition. I thank her for all her help, especially for hours of painstaking checking and proofreading.

I am very grateful for the friendly help, advice and encouragement I have received from the staff of Cambridge University Library and the Central Science Library, and from the editors of Cambridge University Press: Denise Cheuk, Shana Coates, Annette Cooper, Alan Crowden, Rachel Eley, Jacqueline Garget, Clare Georgy, Diya Gupta, Chris Hudson, Linda Nicol, Margaret Patterson, Jonathan Ratcliffe and Tracey Sanderson. I thank her for all her help, especially for hours of painstaking checking and proofreading, with my son Alistair.

Abbreviations

AFLP	Amplified Fragment Length Polymorphism
BDFFP	Biological Dynamics of Forest Fragmentation Project
BGCI	Botanic Gardens Conservation International
BP	Before the present
CITES	Convention on International Trade in Endangered Species
cpDNA	Chloroplast DNA
EU	European Union
FAO	Food & Agriculture Organization of the United Nations
IPCC	Intergovernmental Panel on Climate Change
ISSR	Inter Simple Sequence Repeat markers
IUCN	International Union for Conservation of Nature
MVP	Minimum Viable Population
ppb	Parts per billion
PPGRI	International Plant Genetic Resources Institute
ppm	Parts per million
PVA	Population Viability Analysis
RAPD	Randomly Amplified Polymorphic DNA
RFLP	Restriction Fragment Length Polymorphism
RSPB	Royal Society for the Protection of Birds
SSSI	Site of Special Scientific Interest
STR	Short Tandem Repeat
UN-ECE	United Nations-Economic Commission for Europe
UNEP	United Nations Environmental Programme
WCMC	World Conservation Monitoring Centre

1
Introduction

I grew up in Stocksbridge, a small town in the Yorkshire Pennines between Sheffield and Manchester. Looking down into the valley, there were massive steel works, mines, coke ovens, blast furnaces, pipe works and rolling mills that polluted the air with fumes and smoke. The local river – the Little Don or Porter – was heavily contaminated close to its source. Slag and other wastes were dumped in the woodland downstream below the works. In the 1950s, our community was not alone in suffering from the effects of industrial pollution; indeed, the problem was widespread in South Yorkshire.

However, Stocksbridge had one important advantage not shared by many other towns. Looking down the valley the prospect could be depressing, but this was not the whole picture. Beyond the valley, the hills and valleys of the Peak District National Park came into view, offering some of the finest scenery in England, with farms, rough grazing, woodlands, moorlands managed for grouse shooting, and reservoirs that provided drinking water for Sheffield.

It was here, in an area of such contrasting land use, that I first considered the historical and ecological forces at work shaping the industrial landscapes and the moorland, woodlands and farmland of the National Park. In time, the scope of these reflections widened, for while recovering from some brief teenage illness, I first read Charles Darwin's *Voyage of the Beagle* (1839). Later, I was given a battered copy of *The Origin of Species* by my grandfather (The Popular Impression, 1901, from which quotations in this book have been drawn) and began to consider environmental issues in a Darwinian context. A lifetime interest has led to the writing this book.

Human influences: implications for conservation

As McNeill (2000) has shown, human beings have wrought massive changes in our world. 'To a degree unprecedented in human history, we have refashioned the earth's air, water and soil, and the biosphere of which we are a part.' Since the nineteenth century, conservationists have been attempting to prevent the loss of natural ecosystems, and in the twentieth century particular attention was directed to securing the future of endangered species, a group that might be characterised as the 'losers' or 'potential losers' in contemporary and future ecosystems. As a result of human-induced changes worldwide, it has been estimated that *c*. 10% of the world's plant species are in danger of extinction (May, Lawton & Stork, 1995). However, in the light of recent advances in our understanding of the likely impacts of climate change, this figure now looks very conservative. In recent IPCC reports (2007a–d), it is predicted that, if human activities continue in 'business as usual' mode (with no reduction in greenhouse gas emissions), by the end of the twenty-first century, much of the biosphere could be put at risk by an average global surface temperature rise of *c*. 4 °C (likely range 2.4–6.4 °C). The implications of this highly disturbing information are reported daily in the media, and there have been a number of important books highlighting the issues, and seeking solutions to mitigate such catastrophic changes. Here, the aim is to focus on climate change as one of many consequences of human activities impacting on plant microevolution and conservation.

A second major theme of this book is to highlight the fact that, not only are there 'losers' or 'potential losers', there are also 'winners' amongst the plant species in human-influenced ecosystems. Thus, McKinney and Lockwood (1999, 450) consider: 'emerging evidence shows that most species are declining as a result of human activities ("losers") and are being replaced by a much smaller number of expanding species that thrive in human-altered environments ("winners").' This point is also made by Morris and Heidinga (1997, 287). 'Humanity has induced pervasive negative impacts on the world's biodiversity. Earth is accumulating an ecological deficit that, when the accounting is complete, will be written off by mass global extinctions. But the evolutionary ledger has another side. Adversity for one group of species often represents opportunities for others.'

A third aim is to highlight the contemporary relevance of Charles Darwin's ideas on evolution, by showing that the notion of 'winners' and 'losers' is essentially a Darwinian concept. As we reach the 200th anniversary of Darwin's birth and the 150th anniversary of the publication of *On the Origin of Species* in 2009, it is timely to undertake such a review.

It is possible to consider evolution as something that happened in the past or is only of long-term concern (Stockwell, Hendry & Kinnison, 2003). But this would be a profound misreading of the situation, for, while Darwin emphasised the importance of gradual change over long-distant time scales, he was convinced that natural selection is always active, for he wrote (Darwin, 1901, 60, Darwin's emphasis):

> It may metaphorically be said that natural selection is daily and hourly scrutinising, throughout the world, the slightest variations; rejecting those that are bad, preserving and adding up all that are good; silently and insensibly working, *whenever and wherever opportunity offers* . . .

Post-Darwinian advances in our understanding of evolution confirm the ongoing ever-present nature of evolution by natural selection. This fact is recognised by some but not all conservationists. It is important to accept that the world's biota are 'controlled by evolution', for 'evolution is no mere curiosity of the natural world, but a potent process that we must understand' (Palumbi, 2001, 92, 94). Moreover, in addition to the climatic and environmental factors that have operated in geological time, a convincing case can be made that human activities present organisms with 'new' selection pressures. As we shall see, there is now a great deal of evidence for anthropogenic microevolutionary changes in human-disturbed ecosystems.

I am myself passionately interested in conservation. A fourth aim, therefore, is to consider the impact of human management within conservation areas, in collections maintained for conservation in gardens (*ex situ*), and in the wider landscape. In their attempts to secure the future of endangered landscapes, ecosystems and species, it is essential that conservationists understand the full implications of their efforts in the light of Darwin's theory of evolution in its modern form – neo-Darwinism. In particular, the possibility that management will lead to some form of domestication must be confronted. Again, this is an area of microevolution where Darwin made pioneering contributions.

A fifth aim is to highlight the uncertainties that lie ahead for ecosystems and species, in relation to the combined effects of population growth, increasing exploitation of the Earth's resources and the effects of climate change. Both the increasing human domination of the biosphere and the shrinking areas of 'natural' vegetation have major implications for the direction of future ecological research. It is highly significant that the Ecological Society of America recently called for a 'shift in [its] primary focus from the study of undisturbed ecosystems to interdisciplinary studies of human-influenced ecosystems for the betterment of human societies' (Bawa *et al.*, 2004). The original documents are available on the internet at http://esa.org/ecovisions/.

A stream of important questions flows from these preliminary observations, and these form the basis of the following chapters.

Outline of the chapters

Chapter 2 emphasises that our knowledge of microevolution in human-dominated ecosystems is illuminated by studies from very diverse fields – from archaeology and anthropology, from molecular genetics to ecology etc. In order to judge the veracity of this evidence, it is necessary to consider how it was obtained, and how reliable it might be. In essence, many advances have been made through the testing of hypotheses and models. Some important general principles will be considered about the scientific method, the status of 'facts', the possibility of 'proof' of propositions and concepts, and the public responses to researches.

Chapter 3 provides an outline of Darwin's theory of evolution by natural selection. Also, his views on domestication will be considered, as this topic is highly relevant to human-managed ecosystems. Then follows an account of post-Darwinian advances in our present understanding of these topics, paying particular attention to changes at the population level. In addition, some major concepts of population biology will be introduced relating to the establishment, survival and reproduction of natural populations. In later chapters, we examine how human activities and influences have changed, or subverted these natural processes.

A major theme of this book is the extent to which species have evolved, and are evolving, within human-influenced ecosystems, and Chapter 4 provides a brief historical review of a number of key issues. When and where did *Homo sapiens* emerge as a species, and how quickly did numbers increase across the globe? Where and when did the hunter-gatherer life-style give way to settled agriculture? How has agriculture developed, and how have many of the Earth's ecosystems – both terrestrial and marine – been transformed by ever-increasing human populations?

Chapter 4 makes the case that the Earth's ecosystems have been dramatically changed by human activities, and the time frame for these changes is counted in millennia. Drawing on evidence from various sources, Chapter 5 addresses a different, but related, issue: is there any pristine natural wilderness left? In the nineteenth century, the embryonic conservation movement was born. Such was the over-harvesting of natural resources and the destruction and pollution of many ecosystems, that attempts were made to protect surviving 'wilderness' in national parks, and, later, in nature reserves and wildlife refuges. Conservationists still employ concepts of wilderness in their fund-raising and management strategies, for instance, in the protection of the ecosystems of tropical islands and rainforests. However, recent advances in archaeology and anthropology have transformed our understanding of the extent and scale of human environmental impact in prehistoric and more recent times. Thus, many ecosystems, especially in the tropics, hitherto regarded as natural, unspoilt, untouched, pristine wilderness, are now seen as areas long impacted by human occupation and exploitation. Indeed, there is a case to be heard that parks and reserves were often far from untouched pristine natural wilderness when they were first established. Another major issue is also considered. Surveying the damaging impact of past and present human activities on the biosphere, many now consider that human well-being can only be secured if unfettered exploitation is replaced by 'sustainable development'.

Chapters 4 and 5 review our present understanding of the way humans have modified the natural ecosystems of the world, and there is abundant evidence that all ecosystems on Earth are subject to some impact from human activities. The human population has reached more than 6.5 billion. If predictions about population growth and global climate change are correct, further transformations of the biosphere are inevitable, with increasing exploitation of all natural resources.

Having provided a brief account of the 'stage' on which present microevolutionary processes are taking place, the different groups of 'vegetative' players on this stage are introduced (Chapter 6). A great deal of effort has been expended in determining the number of species that are threatened with extinction worldwide. In order to assess the scale of what many regard as an impending mass extinction (Myers & Knoll, 2001), Chapter 6 considers how far taxonomists have succeeded in naming the plants and animals of the biosphere. At first sight it would seem simple to define each of these categories, but there are many difficulties. The species concept is at the heart of our understanding of evolution, and, furthermore, both the theory and the practice of conservation are to a large extent based on concerns about species. This focus on species: how they are evolving in the contemporary landscapes, and whether it is possible to prevent the extinction of those that are threatened, raises a major issue. What are species? As we shall see, there is no single universally agreed definition.

Having presented an account of the present evolutionary stage and the plant actors on this stage, the evidence for microevolutionary change in plant populations is reviewed in a number of chapters. The general thesis will be examined that by chance, unwitting action, or by intention, human activities have increased, influenced, disturbed, distorted or reduced 'natural processes' such as gene flow, seed dispersal, plant establishment and reproduction. We examine, first in Chapter 7, how microevolutionary processes are investigated. Then we consider how plants exposed to 'new' anthropogenic selection pressures have responded to grassland management (Chapter 8), cropping in agriculture and forestry (Chapter 9), environmental pollution (Chapter 10), and introduction to new territories (Chapter 11). Chapter 12 continues the microevolutionary theme in presenting an account of how human activities have resulted in the decline of populations of many species.

As many plants are threatened with extinction, it is important to consider how quickly new species might evolve to take their place. Chapter 13 examines how contemporary species hybridisation and speciation has been influenced by human-induced breakdown of the geographical and/or ecological isolation that previously existed between species 'in the wild'. Also, the relationships between cultivated plants and weeds/wild species will also be examined, including recent studies of the interrelations between wild and weedy species and transgenic crop plants.

A major strategy for securing the future of rare and endangered species is to conserve them *ex situ* – as living plants in botanic and other gardens, or as stored seed or other structures in gene banks. Chapter 14 reviews the strengths and weaknesses of such conservation in the light of microevolutionary considerations. Several major questions have to be faced. Is it possible in the long term to maintain populations of threatened plants in their 'wild' state in gardens, or is it inevitable that some species will change genetically in cultivation? Is unwitting domestication the likely fate of 'wild' plants grown in gardens?

There is another point of interest. In order to make effective public displays of threatened plant communities, and to present conserved rare and endangered species in an authentic-looking ecological context in botanic and other gardens, many 'wild areas', such as ponds, wetlands, grasslands etc., have been created. At the same time, in many areas of the world, there is a vogue for wildlife gardens, with sowings and plantings of 'wild flowers' to give 'natural-looking' areas. Also, given that reintroduction and reinforcement of 'wild' populations of plants, in parks, reserves and increasingly outside protected areas, involves sowing and/or planting of 'wild' species, is there a danger of blurring the distinction between gardens and nature reserves? Are reserves likely to become specialist gardens and *de facto* zoos? Does reserve management, in reality, treat wildlife as a specialised crop, and, as with crop plants proper, is domestication therefore a possibility? Such moves may present difficulties in the future. If 'wild' plants are deliberately cultivated, then genetic change in the direction of domestication seems a strong possibility, and, while success in conserving the species might be achieved in the short term, if close human management in garden is diminished or withdrawn, or environmental conditions change, such stocks would be vulnerable.

Conservationists have had considerable experience of maintaining and managing national parks and nature reserves, and the rare and endangered species and threatened ecosystems they contain. Despite the claims of conservation managers that 'wildness' and 'naturalness' are being maintained in protected areas, evidence suggests that such places have frequently been subject to very active management, rather than being left to function as untouched natural wilderness areas (Chapters 15 and 16). Such management may involve the implementation of formal science-based conservation plans by professional staff. However, in other parks and reserves there may be a great deal of unofficial, often illegal, alternative 'management', inimitable to the conservation, involving hunting, logging, forest clearance for agriculture, mining etc. Conservation management is concerned not only with entire ecosystems, but also with the survival and reproduction of endangered species. In Darwinian terms, the aim is to identify threats to populations of endangered species and tailor management to counter these threats. Thus, management is about identifying and nullifying the 'adverse selection pressures' impinging on rare and endangered species, so that their populations might stabilise or increase. How far have conservationists been successful in this endeavour? Has the designation and management of protected areas succeeded in its objective of securing the immediate survival of endangered species and ecosystems? There are also questions to be faced concerning the practice of management. Should it be devised to take the form of properly designed experiments, or should 'common sense' be the guide? Are there universal general principles that can guide management, or are all conservation sites uniquely different, and, therefore, require specific local management. Often swift interventions have to be devised to prevent the loss of species and ecosystems. What are the implications of this type of crisis management?

In recent decades there have been major changes in conservation practice, with elaborate interventions, including the restoration and creation of ecosystems, and reintroductions and restocking of populations of endangered species (Chapter 16). Returning to the Darwinian theme: what are the implications for microevolution of plants in areas where such creative conservation projects have been established? How far are these endeavours likely to succeed? Indeed, by what criteria is success in ecosystem restoration to be judged? Other questions must also be faced. Is it possible to recreate original ecosystems or wilderness? Moreover, is it likely that, by careful management of planting schemes, complex communities, such as forests or species-rich grasslands, will be relatively quickly recreated?

In Chapter 17, the microevolutionary implications of size, shape, distribution, connectivity or isolation of parks and reserves will be considered in relation to the wider landscape. Does the existing network of parks and reserves provide a secure arena for the continuing evolution of plant species as members of functioning ecosystems? Many reserves and parks were designed as 'fortress' areas, from which humans are excluded except as visitors. What have been the consequences of such designs, and what would be the implications for conservation if management policies were changed, allowing local people access and control over the resources of the area, in some form of sustainable development?

Conservation management often involves short-term crisis management. However, in reality, as Frankel and Soulé (1981) point out, conservation must be a long-term effort. In one of the seminal books on conservation, they made a most important distinction between preservation and conservation. Preservation 'provides for the maintenance of individuals or groups but not for their evolutionary change' (Frankel & Soulé, 1981, 4). In contrast, conservation 'is diametrically different. Its essence is for some forms of life to remain in existence in their natural state, to continue to evolve as have their ancestors before them throughout evolutionary time' (Frankel & Soulé, 1981, 6). In practical conservation the focus is often on short-term crisis management of immediate problems. But, if the views of Frankel and Soulé are accepted, we are faced with a 'for ever time scale of concern', even if this is not explicitly stated or agreed. It is against this very challenging aim, and against this time frame that conservation efforts will be judged. Given the high costs of conservation, the practical implications of management in 'perpetuity' are daunting. This is particularly the case if separate management plans are pursued for each endangered species. Would the outcome of conservation efforts be more successful and cost effective if conservation management focused under a single management plan for the entire ecosystems in a particular area? In addition, should conservation efforts be directed largely at sites considered to be closest to wilderness, or should equal attention be paid to managing the large number of endangered plant species found in human-managed forests, grasslands, wetlands, arable lands and urban areas etc.?

The book closes with three chapters (18–20) reviewing the current concerns about the impact of global climate change in relation to conservation and microevolution. It is predicted that climate change will present plants with new selection pressures. What will be the effect on those species currently rare and endangered? How far will the species presently conserved in nature reserves and parks be put at risk by climate change? Will conservationists find themselves in the invidious position of spending a great deal of time and money trying to manage populations in their protected sites 'against' the major impacts of climate change? If charismatic plants and animals are lost from reserves and parks, will such areas be abandoned, or will they continue to be maintained for whatever wildlife they come to contain?

Considering the reaction of particular species to climate change, do species have adequate reserves of genetic variability to allow for the selection of new variants *in situ* that are more in tune with climatic change and changing interrelationships with other species? For those species with insufficient genetic variability to respond to these newly developing selection pressures in their current site, what is the probability that they will be able to migrate naturally to suitable climatic zones? If natural migration in the face of climate change is impossible, should species be artificially transferred to suitable climatic zones? Many endangered species occur outside reserves and parks. How far has it proved possible to conserve such species, and what are the prospects in the future?

To complete the review of the interface between microevolution and conservation, Chapter 21 draws together issues raised in earlier chapters. Faced with climate change, will humankind successfully adapt to limit the greenhouse gas emissions and thereby escape the most extreme impacts? Should the focus of conservation efforts be reassessed in the light of increasing anthropogenic domination of ecosystems? The concept of the 'fortress' reserve has been much criticised. Would the adoption of sustainable development management practices prove a more secure route to the successful conservation of endangered species and ecosystems? Finally, if conservation efforts falter, and there is a major extinction spasm in the coming decades, what will be the long-term effect of human domination of the biosphere on the future evolution of plants?

2
Studying change

In order to undertake a critical review of the interface between plant microevolution and conservation, it is necessary to examine published information on a very wide range of topics. Current microevolution is illuminated by ecological studies, common garden and other types of experiment, and a variety of cytological and genetic investigations that increasingly employ molecular markers. Concerning the environmental and conservation context of the success or failure of species, there are many peer-reviewed papers in academic journals and books, together with official documents and statistics produced by national governments, international agencies, interest groups and professional conservationists etc. There is also a great deal of what is called grey literature, which, for reasons of confidentiality, is not publicly available. This includes, for example, consultants' reports, on which far-reaching conservation decisions are made. There is also a great deal of information on the Internet. In total there is a substantial body of knowledge about the conservation of communities and endangered species, based not only on the experience of habitat management, but also on experiments, field observations, mapping and surveys.

Technical advances in many fields have greatly enlarged our understanding of wider environmental issues. With the increasing use of aerial photography, satellite imagery and remote sensing, it has been possible to investigate land use in a way hitherto impossible from fieldwork alone. Google Earth – http://earth.google.com – provides an unrivalled opportunity to examine in great detail the impact of human activities on ecosystems (Biever, 2005). Moreover, important advances have been made in the precise measurement of key physical and chemical properties of the air, water, soil and sea, including estimates of minute but significant concentrations of pollutants etc.

Our understanding of these complex issues has been revolutionised by investigating environmental and ecosystem change in its historical context, by examining archaeological findings, historic documents, maps, old photographs, explorers' accounts, herbarium specimens and field notes etc.

Drawing on a wide variety of historical and other sources, there are now many wide-ranging studies of environmental history (Detwyler, 1971; Goudie, 1981; Worster, 1988; Nisbet, 1991; Pontin, 1993; McNeill, 2000). Williams (2006) has studied the history of deforestation worldwide. Others have examined regional changes. For example, Whitney (1994) charts the history of change in temperate North America from 1500 to the present day; Worster (1992) reviews the ecological history of the western USA in relation to cattle ranching, irrigation and water resources; and Kirch and Hunt (1997) survey the historical ecology of Pacific islands. Major studies of historical ecology have now been carried out across the Mediterranean (Rackham & Moody, 1996; Grove & Rackham, 2001). The environmental history of the British countryside has been particularly carefully examined in the pioneer work of Hoskin (1977) *The Making of the English Landscape*, and the seminal studies of Rackham (1980, 1986). A group of other historians have investigated changes following a significant historic event. For instance, Melville (1994) has examined the changes in the Mexican landscape in the sixteenth century. Intensive sheep production was imposed by the Spanish invaders in the sixteenth century in the Valle del Mezquital region of Mexico. A densely populated region with a mosaic of different land uses, based on intensive irrigation agriculture, was converted by overgrazing to a sparsely populated mesquite desert. Major insights have also been achieved by studying small areas in detail. For example, Rackham (1975) examined the history of Hayley Wood Nature Reserve, Cambridgeshire, while Preston and Sheail (2007) have recently highlighted the transformation of the famous commons that line the riverbanks through the city in Cambridge. While these areas have a timeless quality reflecting their past use, in reality they have evolved from undrained pasture land, grazed by cattle, to areas of mixed use, especially recreation. This change has been brought about by drainage, digging out minerals and gravel, and the dumping of domestic rubbish and dredgings from the river. Widespread levelling of the ground has been accompanied by much reseeding and tree planting.

These accounts provide information into change in rural and agricultural landscapes. By examining a wide range of evidence, supplemented by careful measurement of pollutants, it has proved possible to examine the environmental impact of cities and urban industrial areas. McNeill (2000) provides a very important review of the history of air and water pollution, with reference to many case histories. For instance, the chronic pollution of Lake Michigan by the city sewage of Chicago (which included offal from the largest stockyards on the planet) was remedied by reversing the flow of two rivers, so that the pollution was diverted into the catchment of the Mississippi (Changnon & Changnon, 1996). McNeill (2000) also provides details of the revival of many European river systems, such as the Thames and the Rhine, after remedial action was taken to control pollution. He also examines, in a historical context, the growing pollution of the rivers in countries, such as India and China, that are undergoing dramatic industrial development.

Defining terms and questioning assumptions

While information from many sources provides a sound basis for judgement on questions concerning environmental history, conservation and microevolution, it is important to stress that evidence must be carefully and critically assessed. Firstly, there are differences of opinion about concepts and the definition of terms. Thus, there is much debate about 'species', 'wilderness' and 'habitat degradation' etc. For example, Barrow (1991) examines the concept of land degradation, while Grove and Rackham (2001) discuss the use of the term 'desertification'. In addition, it is important, wherever possible, to examine the assumptions on which particular studies have been undertaken. Rackham (1990, 341) draws attention to a common tendency amongst archaeologists, and others, 'to treat domestic plants and animals as a kind of artefact' and to assume 'that regions are deforested by cutting down trees; that trees and only trees protect the landscape against erosion; that there was no significant grazing or browsing before domestic animals; and that goats destroy all vegetation. All these are theories, not truisms, and need to be demonstrated, not assumed.'

Secondly, the reliability of historical documents, and the accuracy of figures and surveys etc. must be assessed, in particular the strengths and weaknesses of the methodologies used, sampling methods, the sensitivity of instruments, and the limits of critical resolution of the parameters under study. A critical attitude to published figures is clearly important. Ideally, one set of figures should be corroborated by others obtained by using different approaches. Thirdly, it is crucial to make a critical assessment of the conclusions of any investigation.

Experiments and investigations

In considering my approach to the subject matter of this book, it is important firstly to consider some general issues about the scientific method; in particular, what level of certainty flows from the results of testing hypotheses with designed experiments and other investigations. Secondly, as the notion of microevolutionary change also involves an assessment of evidence about the past, how has it proved possible to test hypotheses concerning the recent and distant past? Finally, given the major themes of this book, an assessment is also made of future trends. For instance, what will be the probable effects of predicted global climate change on species and communities of plants? Without resorting to tarot cards or Delphic oracles, how might predictions be made?

Darwin's scientific method

It was the nineteenth century philosopher Whewell who provided important insights into the philosophical background to testing ideas and concepts when he wrote: 'we can learn much about nature by seeking a "conciliance of inductions", where disparate independent lines of evidence are neatly brought together under a common explanatory umbrella' (Costa, 2003, 1030). Thus, Costa notes that the *Origin* established the reality of evolution beyond a reasonable doubt by drawing observations from palaeontology, embryology, biogeography and behaviour. He points out: 'Darwin presents each set of observations in turn and asks the reader to consider which hypothesis offers the most cogent explanation for them.'

However, Mayr (1991, 10) took a closer look at Darwin's methods of research. Firstly he notes:

> Naturalists are, on the whole, describers and particularizers, but Darwin was also a great theoretician . . . He was not only an observer but also a gifted and indefatigable experimenter whenever he dealt with a problem whose solution could be advanced by an experiment.

Darwin's method was very interesting:

> it consists of continually going back and forth between making observations, posing questions, establishing hypotheses or models, testing them by making further observations and so forth. Darwin's speculation was a well-disciplined process, used by him, as by every modern scientist, to give direction to the planning of experiments and to the collecting of further observations. I know of no forerunner of Darwin who used this method as consistently and with as much success.

Darwin stressed the importance of hypotheses in organising and directing his work. His view is clearly stated in a letter to Henry Fawcett (Darwin & Seward, 1903; quoted in Mayr, 1991, 9):

> About 30 years ago there was much talk that geologists ought to observe and not to theorize; and I well remember someone saying that at this rate a man might as well go into a gravel pit and count pebbles and describe colours. How odd it is that anyone should not see that all observation must be for or against some view if it is to be of any service!

Advances in the design of experiments

In the post-Darwinian period there have been major advances in the design and interpretation of experiments and observations, in particular the use of statistical methods to analyse experiments. R. A. Fisher, who for a number of years was responsible for statistical data analysis at Rothamsted Experimental Station, Harpenden, UK, made groundbreaking contributions to this area when he developed, and first used, analysis of variance techniques to design and analyse field experiments (Box, 1978). Fisher's role in the development of mathematical statistics is reviewed by Hald (1998), who notes that significant contributions to the theory and practice of designing experiments have been made by Yates, Kempthorne, Cochrane, Box and many others.

As a prelude to a critical evaluation of the published experiments and investigations considered in this book, the main elements of modern scientific experimentation will be briefly reviewed (see Montgomery, 1991; Sokal & Rohlf, 1993; Hicks & Turner, 1999 for more detailed accounts). Put simply: 'the experiment is the inferential tool in biological research in which conclusions are made about large groups (populations) from data obtained from smaller groups (samples)' (Zolman, 1993, 13). In the wild, many factors influence plants. By experimentation, the aim is to study in detail one or a few factors. In effect, a reductionist approach is commonly employed by scientists; a complex situation is analysed by studying its component processes in detail.

Elements of experimental design

In an important study, Gauch (2003) considers the development and philosophy of the scientific method. He emphasises the paramount importance of the designed experiment, and stresses that, while such approaches have a number of common elements, they do not involve a formulaic or standardised approach.

Taking into account previous investigations, the first step is the formulation of a hypothesis. Such hypotheses can be tested in a variety of different ways. Appreciating the most effective way to test a hypothesis is a creative rather than a mechanical process. In the design of experiments, very careful attention must be paid to the choice of critical factor(s) to test. Some groups of plants will then receive treatments with different levels of a factor (say A), while others (the control group) will receive no treatment with A, but will otherwise be identically treated as the treatment group(s). The aim is to control all the extraneous variables, reducing experimental error and providing a stable base line. Treatment conditions must be carefully devised so that there is no confounding by extraneous variables (Sokal & Rohlf, 1993). Replication must be provided in each treatment and in the control(s) to allow estimates of sample and population means, and experimental error. The problem of pseudoreplication must be avoided (Zar, 1999). Thus, if it is intended to draw conclusions about a population of plants, a sample of appropriate size, say 25, must be collected. It is not valid to make 25 measurements from a single plant. In the allocation of material to treatments, randomisation must be employed and the order of the trials be randomly determined. In the design of field and other trials, it is sometimes possible to divide an experiment into randomised blocks to make a more effective design.

The statistical analysis of the experiment must be devised as part of the design of the experiment, not as a chore to be endured after the data have been collected. Bland (1987) writes: 'More than once I have been approached by a researcher bearing a computer printout two inches thick, and asking what it all means. Sadly, too often, the answer is that another tree has died in vain' (quoted by Zolman, 1993, 6). The next step in the experiment is to test the significance of the results using appropriate statistical techniques. Usually this involves the use of computer packages, and it must be firmly stressed that *appropriate* analyses must be performed, as part of the design of the experiment, paying attention to the statistical characteristics of the data set obtained: for example whether the figures in a data set are normally distributed.

It is essential to understand what is involved in testing hypotheses. What follows draws on the excellent account of Zolman (1993, 23), which should be consulted for fuller details. We should note:

A research hypothesis . . . cannot be tested directly. The only way to provide support for the research hypothesis is to reject the *null hypothesis* (H_0). This statistical hypothesis is that the experimental treatment has no effect . . . treatment mean equals (=) control mean. Hence, an experiment cannot prove that the research hypothesis is correct but can only result in a rejection of the null hypothesis with a *specified* degree of confidence . . . If the experimental data deviates greatly from the null hypothesis, the null hypothesis is rejected and the *alternative statistical hypothesis* is asserted, which is that the treatment mean does not equal the control mean. 'Deviated greatly' is *defined arbitrarily* as the value of a statistic (e.g. t or F) which would occur on the basis of chance only a small proportion of the time, for example, 1 time in 20 ($p < 0.05$) . . . This statistical hypothesis testing, however, does not guarantee that a correct inference will be drawn. Errors of inference can be made, and there are two types. The first error, *Type I*, occurs when the null hypothesis is rejected for the sample data but is true in the population. The second error, *Type II*, occurs when the null hypothesis is accepted for the sample data but is false in the population.

In analysing the completed experiment it is critical to consider exactly what has been achieved, and avoid over-generalising. For example, in a well-thought-out experiment it is possible to derive estimates about the population from which the sample was drawn. However, generalising to all populations of the species – as is sometimes found in the literature – is unwarranted. Moreover, care must also be exercised in extrapolating in an over-simple way from the results of laboratory experiments to the situation in the field.

Quasi-experiments

Hicks and Turner (1999) have drawn attention to the fact that some experiments are of unsatisfactory design. For example, several samples selected as 'intact groups' might be given treatments with one used as a control. Such an experiment may be flawed, however, as it is not clear whether the groups were similar at the start of the experiment.

Another form of quasi-experiment is known as *ex-post-facto* research. Studies of historical records after the event may be used to try to infer whether particular factors are causally related, e.g. smoking and lung cancer. While very important information might be obtained in analysing such data sets, one must consider competing hypotheses (Hicks & Turner, 1999). We will consider this point in greater detail below.

'Proof' and falsification

The outcome of the use of scientific method is to allow decisions to be made in situations of uncertainty, but it is important to consider whether proof of a hypothesis can be achieved.

A recent review by the philosopher Cleland (2001) confronts this issue. Scientific inductivism, commonly attributed to Francis Bacon, holds that if enough confirming evidence is obtained, then a hypothesis can be accepted by the scientific community. This is the approach employed by Darwin in the *Origin*. However, there are difficulties. As Cleland (2001, 987) points out:

> Unfortunately, scientific inductivism runs afoul of the hoary problem of induction: no finite body of evidence can conclusively establish a universal generalization . . . [for] . . . a generalization is false if it has at least one counterexample. Faced with the problem of induction, many scientists embrace falsificationism, which holds that although hypotheses cannot be proved, they can be disproved.

Thus, the scientific method, as we have seen above, constructs hypotheses that are tested and then rejected (falsified) or conditionally accepted on the basis of existing evidence from experiments or empirical observations. The approach of falsification is however not without its own difficulties. Cleland notes: 'any actual experimental situation involves an enormous number of auxiliary assumptions about equipment and background conditions, not to mention the truth of other widely accepted theories'. Every scientist is aware of the possibility of the malfunctioning of equipment, the contamination of samples etc., and the strengths and weaknesses of the auxiliary theories that underpin the assumptions on which a particular experiment is based. Another point of great significance made by Cleland concerns the observation of Kuhn (1970) that many research scientists do not practice falsification. They try to salvage their hypotheses. 'They are primarily concerned with protecting their hypotheses' rather than 'ruthlessly attempting to falsify them' (Cleland, 2001, 988).

Testing hypotheses about the past

So far we have considered questions relating to the experimental study of living plants and vegetation. This book also examines historical issues, where all, or most of, the evidence comes from traces left of the past, in some cases from very distant geological, archaeological or historic times. Again, Cleland (2001) provides an important review of approaches. Firstly, it is important to stress that the principle of falsification has been accepted by many scientists whose research is aimed at elucidating the past. Cleland, considering the ideas of an eminent geologist T. C. Chamberlin (1897), writes:

> Good historical researchers focus on formulating multiple competing (versus single) hypotheses. Chamberlin's attitude towards testing of these hypotheses was falsificationalist in spirit; each hypothesis was to be independently subjected to severe tests, with the hope that some would survive. A look at the actual practices of historical researchers, however, reveals that the main emphasis is on finding positive evidence.

Technically, this is called a smoking gun, defined as a 'trace that picks out one of the competing hypotheses as providing a better causal explanation for the currently available traces than the others' (Cleland, 2001, 988).

Cleland (2001, 987) quotes Gee (1999) who recently 'attacked the scientific status of all hypotheses about the remote past': in his [Gee's] words, 'they can never be tested by experiment, and so they are unscientific . . . No Science can ever be historical.' In the review, Cleland argues strongly: 'the claim that historical science is methodologically inferior to experimental science cannot be sustained'.

Predicting the future

Prediction is fraught with difficulty. However, faced with such issues as population growth, future food production, global climate change etc., scientists have developed useful approaches. Instead of considering only one possible scenario and predicting a single outcome, a range of computer models is generated. These are based on different assumptions, and a number of possible outcomes are predicted and evaluated. In the case of climate change, such studies draw on what is known of past and present climates, and the behaviour and ecology of species and communities, in order to provide a base line for judging the likely effect of change. Prediction can in many cases be supplemented by experiments. For instance, it is clear that carbon dioxide levels in the air will rise over the coming years. In assessing the reaction of plants to this possibility, plants have been raised in growth chambers to compare the effects of ambient concentrations and the elevated level of carbon dioxide.

Weighing the evidence

Evidence from many sources illuminates our understanding of past, present and future changes. However, it is important to consider human behaviour. As we have noted above, the history of science reveals that many researchers only look for evidence that supports their own theories. Thomas (1974, 72) writing about archaeologists, but highlighting the behaviour of investigators in all fields of study, concludes: 'scholars still become rather firmly (in some cases, inextricably) wed to their "pet theories". Unfortunately, some debates in archaeology are more involved with the intransigent defense of previous ideas rather than the objective appraisal of new data.'

Concluding remarks

It cannot be too highly stressed that the results of archaeological, historical, ecological, conservation and evolutionary investigations are essentially probabilistic rather than offering proof. Carpenter (1996, 126) makes this point very clearly. The scientific method of examining both experimental evidence and historical information:

> constructs hypotheses that are then rejected (falsified) or conditionally accepted on the basis of existing evidence from experiments or empirical observations. A hypothesis may always be rejected in the future if additional information is seen to falsify it. Within this truth-seeking methodology, information is characterized by kinds and degrees of doubt, change, and availability.

Holdgate (1996, 100) has likened hypothesis testing in science to an 'evolutionary process of refinement through challenge'.

What happens, however, when scientific results reach the public domain? Caldwell (1996, 395) notes that a transformation may occur:

> To be socially credible and politically acceptable, some assumption of certainty is required in findings alleged to be scientific. These allegations in support of official policy may exceed confirmed scientific evidence or even contradict the probabilistic findings of 'good' science . . . Too often, certainty is tacitly assumed on matters that unbiased prudent judgement would regard as uncertain.

Indeed, many in the population, including conservationists, 'believe that a scientific "fact" is a hypothesis that has been proven to be true' (Jordan & Miller, 1996, 95). They continue: 'it cannot be stressed enough that no test is capable of proving a scientific hypothesis to be true'. Indeed, 'an essential function of research has been described as a search for error' (Caldwell, 1996, 402). Caution may be the habit of many in the scientific community, but this is not necessarily true of all groups of specialists, for Holdgate (1996) notes that where scientists are generally cautious about their research, many economists tend to be confident in their predictions. He also stresses that many acting in the public domain demand 'proof' from scientific experts, and use the absence of such assurances to defer action.

In some cases, a more serious situation arises. As we shall see in later chapters, climate change sceptics have employed various tactics to influence public opinion in the protection of vested interests. For example, they have suggested that the scientific community is divided on the issue, when in reality there is an emerging scientific consensus about the seriousness of the situation (IPCC, 2007a–d). There have been newspaper and other reports that politicians and others have also used a variety of means to control research, and have edited and manipulated information entering the public domain. Thus, in addition to denying the problem, some politicians and lobby groups have frustrated the efforts of the general public to understand and confront the issues. This has led to delays in trying to reach international agreement on steps to control and reduce greenhouse gas emissions (Oreskes, 2004).

Instead of introducing important evolutionary models in different chapters of the book, the next chapter has been designed to provide a connected overview of key concepts. These have emerged from the multitude of studies that have followed the publication of Darwin's theory of evolution by natural selection, which still stands at the centre of our current understanding. Also, a number of important concepts emerging from investigations of plant population biology will be introduced.

3

Key concepts in plant evolution

Our current views on evolution in plants are firmly rooted in Darwin's theory of evolution by natural selection. Before Darwin and Wallace formulated their concept of evolution, it was assumed that species had been individually created in a single act of Special Creation. They fitted perfectly their environment, and any deviations in morphology that occurred were the result of accident. Species were essentially unchanging and unchangeable 'ideal types'. Moreover, the world had been created very recently. Thus, by counting scriptural generations, Archbishop Ussher came to the conclusion that the Earth originated in 4004 BC (Mayr, 1991, 16).

The first critical appraisals of Special Creation predated Darwin's theory of evolution (Briggs & Walters, 1997), but Darwin provided the most formidable challenge to the former orthodoxy, by suggesting not only a plausible mechanism of evolution but also by assembling a wide array of evidence. The naturalist Wallace independently arrived at the concept of natural selection, but with different emphases (Sheppard, 1975). In 1858, before Darwin had published his ideas, Wallace sent him an essay on evolution by natural selection. The question of priority was resolved by Darwin's friends, who arranged a meeting at the Linnean Society of London in July 1858 at which Wallace's essay was presented and Darwin's ideas were represented by unpublished extracts from his writings. Then, over the next few months, Darwin (1901) wrote an extended account of his work: *On The Origin of Species by Means of Natural Selection*. The main strands of the concept are as follows.

A. Darwin recognised two sorts of variation in plants and animals – discontinuous variants (sports, monstrosities, saltations) and small, slight individual differences, deviations or modifications. In his view it was individual differences that were important in evolution.
B. Darwin and Wallace both acknowledged their debt to Malthus (1826) for the insight that, because of the fecundity of organisms, there would be a geometric increase [1.2.4.8.16.etc.] in numbers unless population growth was checked by natural factors. Darwin illustrates this point vividly in the *Origin* (Darwin, 1901, 47)

> There is no exception to the rule that every organic being naturally increases at so high a rate, that, if not destroyed, the earth would soon be covered by the progeny of a single pair . . . Linnaeus has calculated that if an annual plant produced only two seeds – and there is no plant so unproductive

> as this – and their seedlings next year produced two and so on, then in twenty years there would be a million plants.

It is worth noting that perhaps the most fecund organism on the planet is the giant puffball (*Lycoperdon giganteum*), a specimen of which can produce 2×10^{13} spores. If all grew in line, they would encircle the Earth at the equator fifteen times (Ridley, 2002).

However, checks to population growth occur in nature. Again, Darwin provides an apt illustration from his own experiments:

> With plants there is a vast destruction of seeds, but, from some observations which I have made, it appears that the seedlings suffer most from germinating in ground already thickly stocked with other plants. Seedlings, also, are destroyed in vast numbers by various enemies; for instance, on a piece of ground three feet by two wide, dug and cleared, and where there could be no choking from other plants, I marked all the seedlings of our native weeds as they came up, and out of 357 no less than 295 were destroyed, chiefly by slugs and insects. If turf which has long been mown . . . be let to grow, the more vigorous plants gradually kill the less vigorous . . . thus out of twenty species growing on a little plot of mown turf (three feet by four) nine species perished, from the other species being allowed to grow up freely.

C. After much examination of evidence, Darwin considered that adult populations tend to remain fairly constant; therefore checks to population increase occur. Only those individuals survive that have an inherent advantage over others.
D. The better-fitted organisms surviving this natural selection pass on their inherent advantage to their offspring.
E. Selection continues over many generations and new variants take the place of the original organisms. These later stages in the argument are best appreciated in Darwin's own words (Darwin & Wallace, 1859):

> Now, can it be doubted, from the struggle each individual has to obtain subsistence, that any minute variation in structure, habits or instincts, adapting that individual better to the new conditions, would tell upon its vigour and health? In the struggle it would have a better *chance* of surviving; and those of its offspring which inherited the variation, be it ever so slight, would also have a better *chance*. Yearly more are bred than can survive; the smallest gain in the balance, in the long run, must tell on which death shall fall, and which shall survive. Let this work of selection . . . go on for a thousand generations, who will pretend to affirm that it would produce no effect . . .

Darwin was very clear on the role of extinction in evolution (Darwin, 1901, 277):

> The theory of natural selection is grounded on the belief that each new variety and ultimately each new species, is produced and maintained by having some advantage over those with which it comes into competition; and the consequent extinction of the less-favoured forms inevitably follows.

Darwin's theory of evolution by natural selection is very far removed from an image of Mother Nature looking after her creations. As his vivid metaphors indicate, he saw the relationship of organisms in a different light – 'great battle of life'; the 'war of nature'; 'nicely balanced forces'; 'struggle for life'; 'a grain in the balance', 'which live, which die'.

Darwin also included another element in his theory of natural selection, namely sexual selection. Jones (1999, 81) makes the important point:

> sex is a marketplace for natural selection. As a merger between genetic enterprises it is, like any other business with two partners, liable to discord. To find a mate, fight off the opposition, suffer sexual congress and pregnancy, and raise a brood are all expensive and dangerous. Each involves subtle differences in the investment strategy of the parties involved . . . many of Nature's most attractive features – flowers, birdsong, mandrill bottoms – result from the rivalry amongst male cells for access to egg.

Sexual selection is therefore 'based on male competition or female choice' (Rieger, Michaelis & Green, 1968). It is responsible for the sexual dimorphism between males and females found widely in the animal kingdom.

In his theory of evolution by natural selection, Darwin saw a very strong parallel between the selection process exercised by humans in the domestication of plants and animals and natural selection in the wild. Indeed, the first chapter of the *Origin* discusses the power of 'artificial' selection, drawing attention to the conscious and unconscious elements in the process. Having discussed domestication in various animals, Darwin writes about the domestication of plants, including ornamental and crop plants (Darwin, 1901, 25):

> In plants the same gradual process of improvement, through the occasional preservation of the best individuals . . . may plainly be recognised in the increased size and beauty which we now see in the varieties of the heartsease, rose, pelargonium, dahlia, and other plants, when compared with the older varieties or with their parent-stocks.

Reviewing all the evidence concerning the evolution of domesticated plants and animals, Darwin concludes (Darwin, 1901, 30): 'The accumulative action of Selection, whether applied methodically, and quickly, or unconsciously and slowly but more efficiently, seems to have been the predominant Power.'

In *The Variation of Animals and Plants under Domestication*, Darwin (1868) published his views in greater detail. Firstly, he set domestication in its historical context (Darwin, 1905, vol. 2, 248):

> With plants, from the earliest dawn of civilisation, the best variety which was known would generally have been cultivated at each period and its seeds occasionally sown; so that there will have been some selection from an extremely remote period, but without any prefixed standard of excellence or thought of the future. We at the present day profit by a course of selection occasionally and unconsciously carried on during thousands of years.

Darwin then recognised the role of methodical selection, where 'the breeder has a distinct object in view, namely, to preserve some character which has actually appeared; or to create some improvement already pictured in his mind' (Darwin, 1905, vol. 1, 260). He also stresses the role and importance of unconscious selection. 'By this term I mean . . . the preservation by man of the most valued, and the destruction of the least valued individuals, without any conscious intention on his part of altering the breed' (Darwin, 1905, vol. 2, 242). Moreover, in further discussion Darwin concludes: 'Except that in the one case man

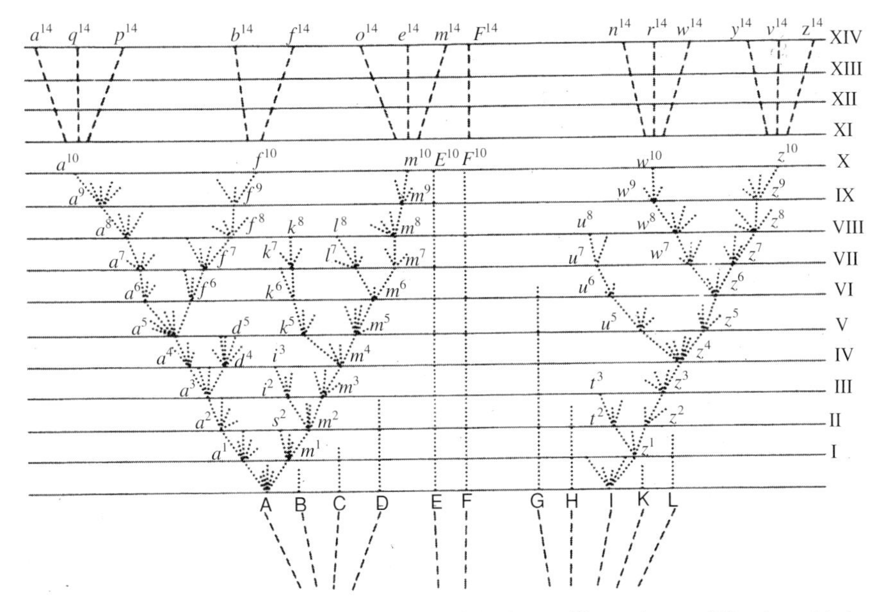

Fig. 3.1 A hypothetical evolutionary tree from Darwin's *Origin*, illustrating modification with descent through natural selection. Eleven different species are shown (A–I, K, L) of a genus at the bottom of the diagram. These diverge over time shown by horizontal lines labelled with Roman numerals and each representing 1,000 or more generations. (Darwin, 1859)

acts intentionally, and in the other case unintentionally, there is little difference between methodical and unconscious selection.' Furthermore, 'unconscious selection blends with the methodical, so that it is scarcely possible to separate them' (Darwin, 1905, vol. 2, 243).

This book is concerned with the concept of microevolution in human-dominated habitats, where management of plants is widespread, not only in common species, but also in the conservation of endangered species. Such management is often extensive and potentially never ending. Darwin points out the result of artificial and more particularly unconscious selection in plants. Could persistent management, particularly through unconscious selection, result in genetic changes akin to domestication? This possibility will remain an open question until later in the book.

Darwin's ideas about the evolution of species

Darwin also explores the origin of species, considering, by means of a diagram (Fig. 3.1) and associated text, the progress and time scales of speciation through natural selection, and exploring the part extinction plays in evolution.

A to L represent species of a genus, the spacing of which on the base line of the diagram indicates the relative morphological similarities and differences. The varying offspring of

A (and the other species) are indicated by the fan of diverging dotted lines. Horizontal lines represent each of a thousand or more generations. Only those variations 'in some way profitable will be preserved or naturally selected'. As the dotted line originating from A reaches each of the horizontal lines, derivatives marked with small numbered letters show that an increase in diversification has occurred. The process continues in similar steps for up to 10,000 generations (and at the top of the diagram in summary for up to 14,000 generations), giving rise eventually to eight distinct species from A marked by letters a^{14} to m^{14}. Likewise, species I on the base line has, by the 14,000 generation, evolved to produce six species. Descent with modification has produced a new pattern of variation. Darwin suggests that the six descendants of I could now form two sub-genera or even genera, while the eight descendants of A will be in very distinct genera or even sub-families. In contrast, Darwin also recognised the possibility that species might remain unaltered or change only to a slight degree. Hence, over 14,000 generations F has not evolved into a number of derivatives.

Darwin devised the diagram to stress his belief that evolution was a slow gradual process and involved not only descent with modification, but also extinction. Thus, parent species and intermediate stages became extinct. However, he did recognise that 'both child and parent' might survive together if they did not come into competition. Which variants are 'winners' and which are 'losers' is the theme of a number of eloquent passages in the *Origin* and Darwin makes it clear that, in his opinion, rarer forms, with smaller populations, are the most vulnerable (see Chapter 12).

Post-Darwinian contributions to our understanding of evolution

The theory of evolution by natural selection is at the heart of modern biology and there is a huge literature presenting post-Darwinian advances. Here, to provide the background to material in later chapters, a number of key concepts are outlined that have a bearing on a central themes of this book – microevolution and conservation in human-influenced habitats.

Cyto-genetic studies

Since the publication of Darwin's work there has been enormous progress in our understanding of the mechanisms of heredity. Darwin's theory of natural selection is based on the principle that those organisms surviving and leaving more progeny than others will generally inherit the characteristics of their parents. When Darwin was writing the *Origin*, some biologists favoured a form of blending inheritance and he developed his own version of this called 'pangenesis' in his book *Variation of Animals and Plants under Domestication* (Darwin, 1868). He postulated that gemules carrying genetic information circulated in the organism, but experiments by Galton (1871) involving cross transfusion of the blood of rabbits of different colour failed to confirm this theory. Darwin, himself, also carried out some crossing experiments with plants, but he did not succeed in unravelling the principles of heredity, although some of his experiments revealed patterns of genetic segregation (Whitehouse, 1973).

The crucial experiments founding the science of genetics were carried out by Mendel (1866). The significance of these investigations was not appreciated by biologists until the beginning of the twentieth century. In crossing experiments between pure breeding stocks, followed by the careful analysis of the characteristics of progenies, Mendel demonstrated the inheritance of seven genetic traits in peas. He explained patterns of segregation in crosses in terms of particulate hereditary factors, later called genes by Johannsen (1909). In their behaviour, Mendel's factors could be dominant or recessive. Two copies of each factor were present in most cells, with the gametes having a single factor.

In the early twentieth century, the realisation that genes were carried on chromosomes began to emerge, and the chromosome theory of inheritance was proposed by Sutton (1903). This theory was confirmed in experiments with *Drosophila melanogaster* (Morgan, 1910; Bridges, 1914, 1916). Over the next few decades cytological observations and crossing experiments made it possible to devise maps representing the linear order of the different genes on specific chromosomes (see Whitehouse, 1973). By the 1940s, it was accepted that specific genetic information was carried in the genes present as linear arrays on chromosomes that were located in the nucleus. However, the identity of the hereditary material was a mystery: and there was speculation that only protein molecules were sufficiently diverse chemically to carry the necessary information (Schrödinger, 1944).

Chemical basis of hereditary information

Although deoxyribonucleic acid (DNA) was first identified by Miescher in 1869, it was only in the mid twentieth century that it became clear that this is the primary molecule of life, specifying in its structure the sequence of hereditary information that provides the instructions by which cells grow, divide, differentiate and reproduce. In its properties and mutability, DNA is also at the heart of evolutionary processes.

In 1953, Watson and Crick, studying the results of investigations by Wilkins and Franklin and others, deduced that DNA has a double helix structure of two complementary polynucleotide chains held together by hydrogen bonds. Key components of the structure are four bases. Through specific hybridising bonds, adenine (A) always pairs with thymine (T) [A-T] and guanine (G) with cytosine (C) [G-C]. On replication, the DNA double helix opens, and the highly specific bonding of the bases [A-T, and G-C] allows self-copying (Fig. 3.2). The base sequence and, hence, the hereditary information in the original strand is maintained in replicated strands generally without changes, but sometimes mutational changes in base sequence occur (see below).

DNA is the key component in the system regulating biochemical activity (Fig. 3.3). Of prime importance is the transcription of the information in DNA into another nucleic acid molecule ribonucleic acid (RNA). RNA transcripts (messenger RNA, mRNA) are produced

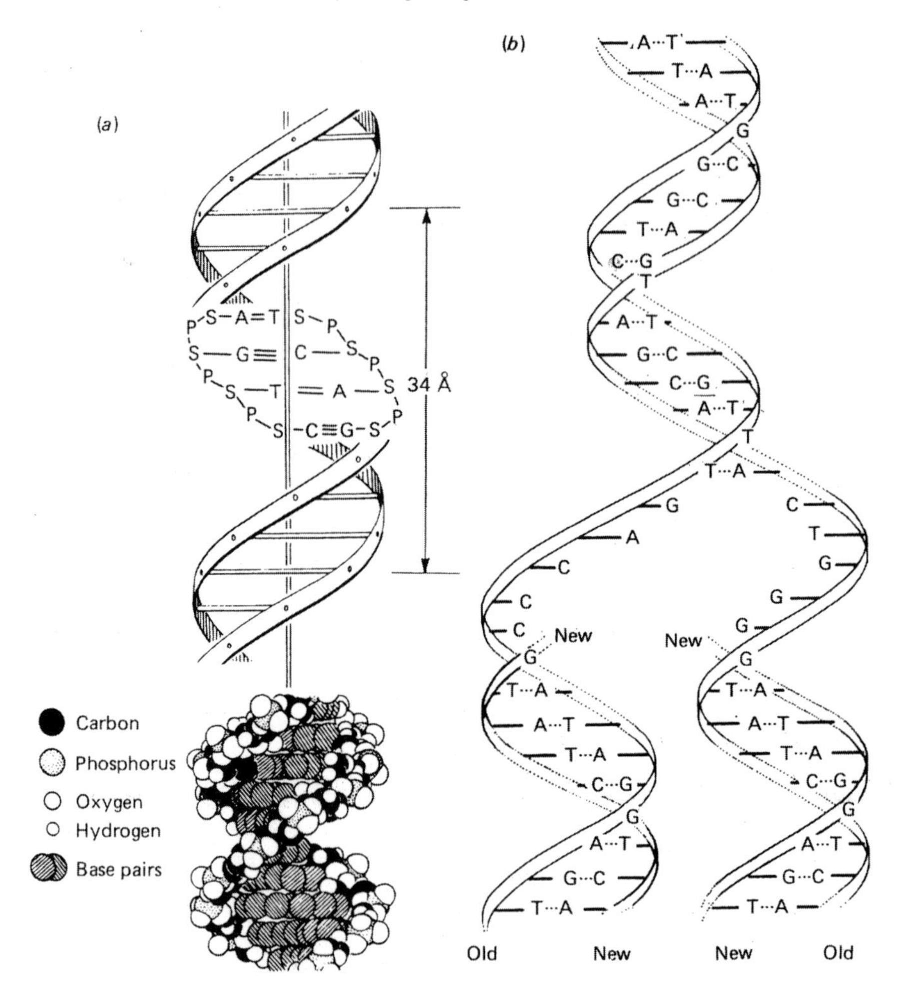

Fig. 3.2 The structure and mode of replication of DNA. (a) Structure of DNA, showing the complementary paired strands linked together by hydrogen bonds. Adenine (A) and thymine (T) always pair together, as do guanine (G) and cytosine (C). The backbone of the DNA molecule consists of the phosphate (P) and deoxyribose sugar (S) groups. Thus, the DNA molecule is built up from many basic units called nucleotides, each of which consists of a base linked to a molecule of deoxyribose plus a phosphate group. (From Briggs & Walters, 1997; after Berry, 1977) (b) Replication of DNA as suggested by Watson and Crick. The complementary strands are separated to produce a replication fork and each forms the template for the synthesis of a complementary daughter strand. (From Watson, J. D. (1965), *Molecular Biology of the Gene*. Reprinted by permission of Addison Wesley Longman Publishers, Inc.)

by replication of the same base sequence as the DNA coding strand, except that the base thymine is replaced by uracil in RNA. The nucleotide sequence in the mRNA is then translated into a sequence of amino acids constituting the polypeptide chain of a protein. The sequence of the bases carrying the genetic code for amino acid sequence is read in triplets (see Fig. 3.4).

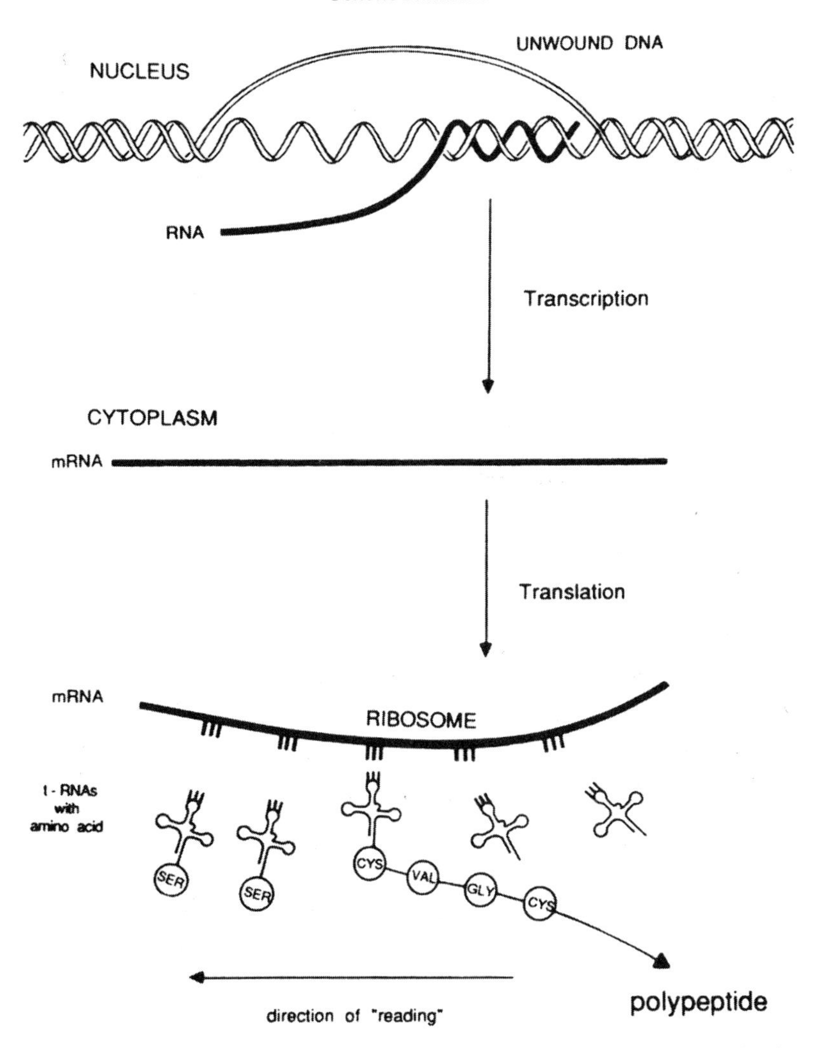

Fig. 3.3 The transcription and translation of genetic information in the production of polypeptides. The information encoded in the DNA is transcribed in the mRNA. The translation of this information occurs outside the nuclear membrane. The assembly of amino acids into polypeptides is achieved as each three base codon in the mRNA in turn forms a complementary 'pair' with the anti-codon of the appropriate tRNA. There are 20 such tRNAs each carrying a different amino acid. (Reprinted by permission of John Wiley & Sons, Inc. from Crawford (1990), *Plant Molecular Systematics*, © 1990)

Genetic mutation

Although the DNA molecule can be accurately replicated without change, by chance, mutation may occur. Many naturally occurring oxidising chemicals and radiation sources, such as X-rays and ultraviolet light, act as mutagenic agents (Sniegowski, 2002).

Considering the molecular changes involved in mutations, the quantity of DNA may increase and/or there may be changes in the base sequence(s). Relative to genetics of the

Second position

First position	U		C		A				Third position
U	UUU	Phe	UCU	Ser	UAU	Tyr	UGU	Cys	U
	UUC		UCC		UAC		UGC		C
	UUA	Leu	UCA		UAA	Stop	UGA	Stop	A
	UUG		UCG		UAG		UGG	Trp	G
C	CUU	Leu	CCU	Pro	CAU	His	CGU	Arg	U
	CUC		CCC		CAC		CGC		C
	CUA		CCA		CAA	Gln	CGA		A
	CUG		CCG		CAG		CGG		G
A	AUU	Ile	ACU	Thr	AAU	Asn	AGU	Ser	U
	AUC		ACC		AAC		AGC		C
	AUA		ACA		AAA	Lys	AGA	Arg	A
	AUG	Met	ACG		AAG		AGG		G
G	GUU	Val	GCU	Ala	GAU	Asp	GGU	Gly	U
	GUC		GCC		GAC		GGC		C
	GUA		GCA		GAA	Glu	GGA		A
	GUG		GCG		GAG		GGG		G

Fig. 3.4 The genetic code. The triplet codons given are those of the RNA, which provides the template for the production of the polypeptides from amino acids. The amino acids are given the standard three letter abbreviations Met (Methionine), Val (Valine) etc. Note that almost all the amino acids are specified by more than one triplet. Stop codons mark the end of a sequence specifying a particular protein. (From Kendrew, 1994. Reproduced with permission of Blackwell Publishing)

parental plant, a mutated DNA sequence may result in a changed genotype, and, as the organism develops, the interaction of genotype and environment may lead to a changed appearance (a different phenotype). The investigation of the base sequences in the DNA of mutants reveals that mutation may involve a modification of one or more nucleotide sequences. In some mutants it has been found that many DNA sequences may be changed representing one or a block of genes, or indeed mutation may involve changes to whole chromosomes. Some of these changes involving point mutations, inversion or deletion of sequences are illustrated in Fig. 3.5.

Chromosome changes

Sometimes mutations involve major changes to the chromosomes, which contain a linear DNA molecule (Cooke, 1994; McVean, 2002). The role of the chromosome, through the *a* Original sequence

. . ATG GTG CTC AGC ATA GCT TAT AGC . . .
. . Met Val Leu Ser Ile Ala Tyr Ser . . .

b Point mutation (missense)

. . ATG GTG **G**TC AGC ATA GCT TAT AGC . . .
. . Met Val **Phe** Ser Ile Ala Tyr Ser . . .

c Insertion leading to frameshift with premature termination

. . ATG GTG CTC AGC ATA GCT TA**T** TAG C . . .
. . Met Val Leu Ser Ile Ala Tyr **STOP**

d Insertion leading to frameshift with altered amino-acid sequence and premature termination

. . ATG GTG C[**GA TAT CTC TGT GT**]T CAG CAT AGC TTA TAG C . . .
. . Met Val **Arg Tyr Leu Cys Val Gln His Ser Leu STOP**

e Deletion leading to frameshift

. . ATG GT[G CTC AGC ATA G]CT TAT AGC . . .
. . ATG **GTC TTA TAG C** . . .
. . Met **Asp Leu STOP**

f Silent mutation no change

. . ATG GTG CT**A** AGC ATA GCT TAT AGC . . .
. . Met Val Leu Ser Ile Ala Tyr Ser . . .

Fig. 3.5 The diagram shows the effect of different types of mutation on an original sequence of triplets of DNA (top line) each coding for a different amino acid (second line). The amino acids are given the standard three letter abbreviations Met (Methionine), Val (Valine) etc. (a) Original sequence. (b) Point mutation resulting from the substitution of one base by another. (c–e) Mutation in the reading frame of the triplets caused by the insertion or deletion of bases. (f) Because of the redundancy of the code, a base change may lead to a silent mutation as it does not result in a change in the amino acid sequence. (From Kendrew, 1994. Reproduced with permission of Blackwell Publishing)

process of mitosis, is to deliver a copy of the replicated DNA to daughter cells. Our understanding of chromosome structure has greatly increased in recent years. Evidence suggests that there is compartmentalisation of the DNA. For example, some chromosomes may be partly or wholly of heterochromatin, which consists of repeated DNA sequences containing few if any genes. Faulty division of the chromosomes may lead to cells with fewer or more chromosomes than normal, so called aneuploidy.

In comparing related variants, species etc. it may be revealing to study the karyotype, i.e. a display of metaphase chromosomes set out to show the number, shape, size and the consistent chromosome banding patterns obtained by using certain fluorescent stains, e.g. giemsa (G-banding) and quinacrine (Q-banding).

Meiosis is a process in which homologous chromosomes pair, chiasma formation takes place allowing reciprocal exchanges between homologous pairs, and gametes are produced with the haploid (n) number of chromosomes. Fusion between two gametes restores the diploid ($2n$) chromosome number. Studies of hybrids between related plants stocks may reveal, through their pairing behaviour at meiosis, that major changes in chromosome structure have taken place involving deletions, inversions and translocations (Fig. 3.6).

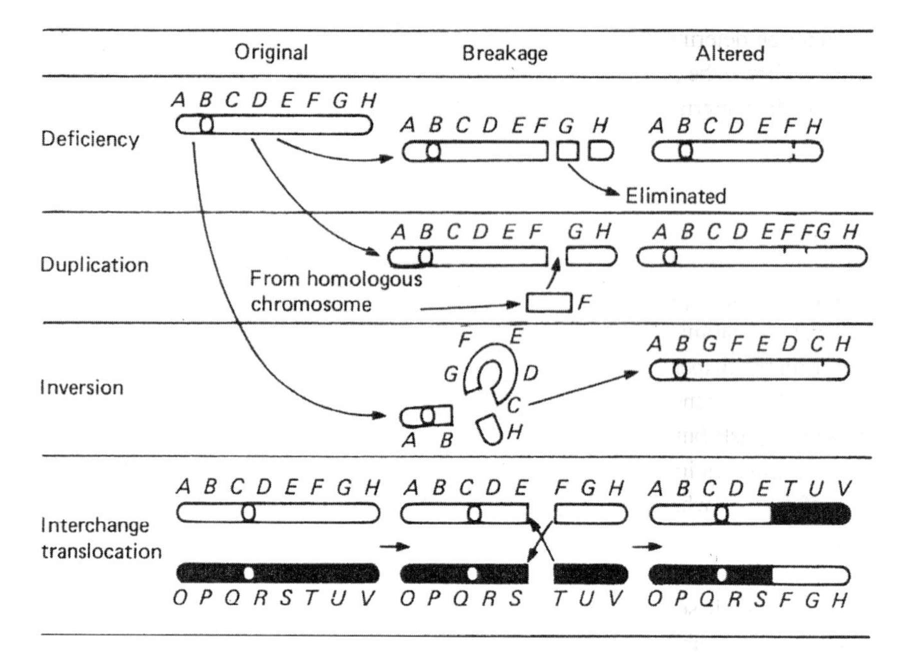

Fig. 3.6 Diagram to show how chromosome breakage and reunioning can give rise to the four principal changes that chromosomes undergo. (From Stebbins, 1966. Reproduced by permission of R. L. Stebbins)

Furthermore, gametes with unreduced chromosome number may be formed and yield polyploid plants with more than two complete genomes, i.e. basic haploid sets of chromosomes. The evolutionary significance of such plants will be considered below.

Microevolution in plant populations

A number of other post-Darwinian advances have offered new insights into microevolution, which refers to 'genetic change within the lifetimes of populations and species' (Pellmyr, 2002, 731). He also noted that the term macroevolution is employed 'to describe patterns of diversification at and above the species level, governing issues such as the origin of major evolutionary novelties and rates of species formation in different groups of organisms' and stressed 'the processes that generate microevolution can also account for macroevolution'.

Populations

For most purposes the term population refers to the number of organisms of a species at a site at a particular time. However, the population unit of outstanding significance in the study of variation is quite different, being 'an interbreeding group of individuals sharing a common gene pool' (Dobzansky, 1935). Population geneticists have called such groups Mendelian or panmictic populations. These groups are model systems, the exact limits of which cannot be determined in the wild. The best that can be achieved is to examine the source and estimate the probable destination of pollen successful in achieving fertilisation, and examine the pattern and the dispersal of seeds and fruits.

Gene flow

The concept of gene flow refers to the introduction of new alleles into a population from another population. In plants, this may involve the movement of immigrant pollen carried by wind, water or animals, or the incorporation in the population of immigrant individuals resulting from the dispersal of seeds/fruits.

Until the 1970s, gene flow in natural populations was not properly examined. Pollen is very widely distributed in air, and fruits and seeds appear to be very well adapted to dispersal. Observations supported the view that gene flow might be widespread and effective. However, when pollen and gene flow was carefully investigated, it was concluded that gene flow in agricultural and wild plants was more restricted (Levin & Kerster, 1974). In these experiments the movement of dyed pollen and the behaviour of genes was studied. As we shall see later, our understanding of gene flow in populations of plants in human-modified habitats has been greatly increased by employing isozyme and DNA markers.

Gene flow and population structure

Studies of gene flow have provided a number of important models of plant populations. First, as gene flow is limited, it has been suggested that a plant species in a particular geographical area is represented by a multitude of sub-populations rather than a single or several large area-wide populations. This conclusion emerges from many theoretical and experimental investigations that studied the subdivision of populations. Tests for subdivision make use of F statistics, devised by Wright (1951), that partition total genetic diversity to reveal variation within and amongst sub-populations. It is beyond the scope of this book to provide a full account of the theoretical framework and means of estimating F statistics (see Nei, 1987; Excoffier, Smouse & Quattro, 1992; Holsinger, Lewis & Dey, 2002; Beebee & Rowe, 2004). Goodman (1997) and Beerli and Felsenstein (1999) provide full information on computer programmes available for calculating F statistics from genetic markers.

Population structure in fragmented populations is intimately related to gene flow. At one extreme there is the 'open' population, which is connected to others by active gene flow of pollen and/or seed fruits. At the other, there is the 'closed' population, where gene flow is absent, resulting in self-contained isolated population units. Reflecting the complexities in the 'wild', a family of models has been devised to explore the implications of migration from a mainland population to islands of different size and shape. Another group of models explores migration as a stepping stone process, where gene flow takes place only between adjacent populations.

In addition, the concept of the 'metapopulation' has been developed by Levins (1969) for spatially structured sets of populations 'that experience local extinction and recolonization' (Whitlock & Michalakis, 2002, 725). Thus, species of interest may have been recorded at many sites, but not all are occupied at any one time. Over time the situation is dynamic: local extinction is followed by recolonisation from a nearby occupied site. As we shall see below, metapopulation models are of particular concern to conservationists, as they grapple with the fact that human activities have disrupted and changed previously continuous natural habitats, leaving only small isolated fragments of vegetation, many of which contain rare and endangered species.

Genetic markers can be employed to 'infer' the movement of pollen grains and seed (and sometimes vegetative propagules) within and between populations. It is also possible to estimate migration from F statistics, as this quantity is related to population size and level of gene flow (Sork *et al.*, 1999). However, it should be emphasised that these estimates of migration give an indication of historical evolutionary rates of gene flow reflecting common ancestry (Steinberg & Jordon, 1998). It must be emphasised very strongly that they may not be an indication of current gene flow.

Effects of chance in populations

Natural selection has a very profound effect on population variability. However, this is not the only influence to consider, for chance may also be a major factor. Populations of plants and animals, particularly small populations, are subject to a process called genetic drift, in which major changes in allele frequencies can arise by chance (Wright, 1931). Such a process may be visualised by considering a large sample of balls of various colours. A small sample is drawn at random from this population. The frequencies of the different colours in the smaller sample might be unrepresentative of the original frequencies. If the small sample is then expanded to the original number, while maintaining the new colour frequency, and a further small sample taken, another accidental shift in frequencies might occur. Repeated sampling could lead to complete loss of some colours and to the fixation or near-fixation of others. This simple model reveals the effects of sampling error on allele frequencies, and it is clear that genetic changes will occur in small populations of plants and animals.

The effects of chance are often highly significant when large populations are reduced to small following the so-called bottleneck effect (Fig. 3.7). As a consequence of the effects of chance through genetic drift, gene frequencies may change, particular alleles may become more frequent, but, and this is a point of particular significance, some alleles may be lost. As we shall see, the potential loss of variability in declining populations of rare and endangered species is a central concern for conservationists. However, drift is a problematic area to investigate in the field. Its study involves the examination of changes in supposedly non-adaptive selectively neutral traits. It is difficult to be sure that any trait is indeed non-adaptive and, in truth, the effects of drift cannot be separated from those of natural selection.

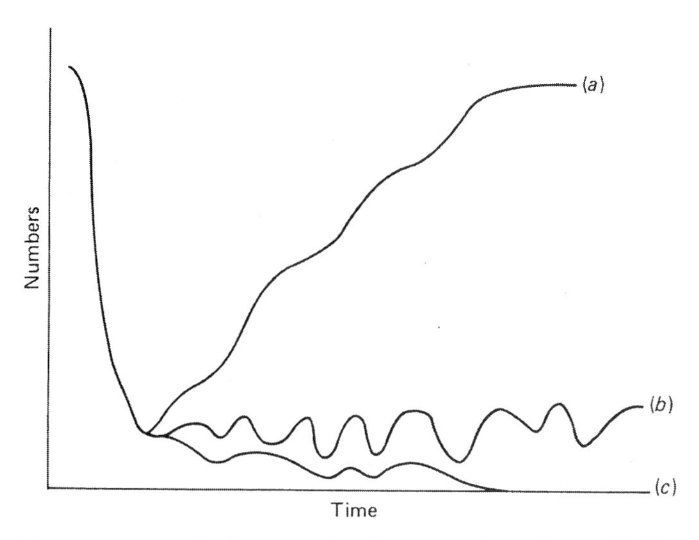

Fig. 3.7 The effect of a severe reduction in numbers in a population – the so-called bottleneck effect. (a) Recovery of numbers: the new population may not differ phenotypically from the original, but some genetic variability may be lost. (b) No recovery in numbers. (c) Extinction. (From Briggs & Walters, 1997. Reproduced with permission of Cambridge University Press)

Another important example of the operation of chance concerns the so-called founder effect, in which one or a few individuals initiate a new population. In cases of long-distance disposal, daughter colonies might be geographically distant from the parent population. Such new populations may lack the genetic variability of their source populations.

Stability of natural populations

In his concept of evolution by natural selection, Darwin took the view that adult populations were often stable in their numbers. There is now abundant evidence, for both plants and animals, that natural populations are not stable, indeed they may show great fluctuation in size in relation to climatic factors, successional changes, incidence of pests and diseases, introduced organisms etc. (Harper, 1977). This point, although briefly stated here, is of very great importance; natural and human-influenced populations may suffer severe reductions in numbers, and individual populations may or may not recover from such bottleneck effects.

Animal and plant interactions

In the *Origin* Darwin provided a number of examples of the interrelatedness of plants and animals. For example, he set out a famous series of interactions based on his own studies of clovers (Darwin, 1901, 53):

> Visits of bees are necessary for the fertilisation of some kinds of clover . . . protected heads [of *Trifolium pratense*] produced not a single seed . . . humble-bees alone visit red clover, as other bees cannot reach the nectar . . . the number of humble bees in any district depends in a great measure upon the number of field mice, which destroy their combs and nests . . . now, the number of mice is largely dependent, as every one knows, on the number of cats . . . Hence it is quite credible that the presence of a feline animal in large numbers in a district might determine, through the intervention first of mice and then bees, the frequency of certain flowers in that district.

Darwin's friend, Thomas Henry Huxley (1906, 244), considered that 'old maids' should be added to the interaction as 'they keep the cats'.

Ecologists are fascinated by interactions between organisms, studying in great detail the food chains and food webs found in the wild, the interactions of plant/animal relationships involved in pollination, seed and fruit dispersal, and the effects of pests, diseases and predators, symbionts and mutualisms of every kind. As we see in later chapters, these manifold interactions are important parts of the functioning communities. Some of the interactions between the members are opportunistic, but other intimate relationships are seen as the product of co-evolution. Thompson (2002, 178) characterises co-evolution as follows:

> . . . process of reciprocal evolution between interacting species driven by natural selection. These include some of the remarkable reciprocal adaptations found in many plants and their pollinators, predators and their prey, and parasites and their hosts. By linking species together, co-evolution has shaped the fundamental structure of life on earth . . . A major result of the co-evolutionary process is that most species become specialized to interact with only a small number of other species.

Plants are different from animals

Many theoretical and practical advances in our understanding of microevolution and conservation have come from the study of animals. It must be stressed that plants are very different from animals. Land plants, once they are established, are generally rooted in soil. If the environment becomes inappropriate, hostile, polluted etc., they are unable to move elsewhere. Plants cannot escape pests and predators by moving: many invest in defence mechanisms, which take the form of physical structures – prickles, stinging hairs etc. – and/or chemical defences.

Plants species are of many different life forms and growth habits. They may be annual, biennial, perennial, short- or long-lived. In general, plants have a greater phenotypic variability than animals, which are held in tight bounds in their development. Thus, a two-headed dog is a greater rarity than a tree with two trunks. Many plants are clonal in their growth, as they expand vegetatively from a single colonising individual. Unless mutation occurs, such plants are genetically identical. There are species-specific differences in the extent and consequences of vegetative growth. In some herbaceous species, radial growth takes place by underground (e.g. rhizomes) or above-ground structures (e.g. stolons, runners). Even large patches may operate as single connected individual plants. In other cases, vegetative connections break down and groups of separate individuals, or indeed lineages of individuals are formed. Some tree and shrub species are also capable of producing large clonal patches, e.g. certain *Ulmus* species. Many water plants are clonal: typically the developing individual eventually 'falls apart', and daughter plants, carried on water currents, continue their separate existence.

Clonal propagation under experimental conditions provides a means of studying the remarkable phenotypic plasticity exhibited by many plants. Thus, a number of individuals of the same genotype can be raised from a single plant and replicate samples may be submitted to different growing conditions. Such experiments have revealed that vegetative characteristics – height, leaf length etc. – are more phenotypically plastic than floral characteristics. Also, there may be remarkable differences in the size and number of seeds/fruits produced by different phenotypic variants of the same genotype raised in very different conditions.

In reproduction, pollen is carried between plants by inanimate agencies of wind or water, or by animals. The products of sexual reproduction – seeds/fruits – are dispersed by the same agencies. Pollination and dispersal of fruits and seeds by animals may involve their attraction to plants by 'rewards' or by 'deceit' of apparent reward.

In the higher animals, there are separate sexes – male and female. In contrast, relatively few species of flowering plants have separate sexes. Most plants are hermaphrodite, with both male and female structures. Such plants have a wide array of floral structures and temporal/spatial mechanisms that encourage or enforce outbreeding (crossing between genetically different individuals) rather than inbreeding (selfing or crossing between closely related individuals). Repeated inbreeding, especially in normally outbred species, may result in inbreeding depression. Continued selfing produces more uniform lineages resulting, some generations later, in weak plants of reduced vigour or fertility. Crossing unrelated plants can restore plant vigour and fertility in a process known as heterosis.

Plants exhibit a wide array of breeding systems and it is important to stress that the random mating (assumed in many population genetics models) may not occur. Three different modes of reproduction are found in plants (Richards, 1997).

A. Some plants are self-sterile by virtue of their genetic self-incompatibility systems. Thus, not all individuals are necessarily fully fertile with all others in the population. The self-incompatibility alleles are decisive in the self-incompatibility reaction that arises from an interaction of gene products of pollen and of style preventing free access of the pollen tube to the ovule.
B. Some other species are self-fertile, sometimes producing seed within closed structures, e.g. cleistogamous flowers.
C. In other groups of plants, reproduction is achieved by apomixis, i.e. without fertilisation, the sexual processes being wholly or partly lost. In the so-called agamospermous plants, normal seed is set, but no sexual fusion has taken place. Unless mutation occurs, the offspring have the same genotype as the parent plant. In other cases, there may be no reproduction by seed: plants may reproduce vegetatively, for example, by means of bulbils etc. Some authorities refer to this form of reproduction as vegetative apomixis.

Studies of plants in the field have revealed that many species reproduce in more than one of the three modes just outlined. Risking a generalisation, it is clear that, typically, lineages of plants have a capacity to generate variability as well as reproducing their own genotype more or less faithfully. For example, the following combinations have been reported: separate sexes (obligate outbreeding) + vegetative reproduction; outbreeding + occasional self-fertilisation; mixed reproduction involving selfing and outcrossing in different proportions; and facultative apomixis (where seed production involves both apomictic and sexual reproduction).

Finally, there are a number of other major differences between plants and animals. Firstly, some species of plants, including many weeds of arable land, have a very large persistent seed bank of viable seed in the soil. Thompson and Grime (1979) and Vighi and Funari (1995) draw the distinction between a transient and a persistent seed bank. In a transient seed bank, none of the seed remains in the soil for more than a year, with 'new' seed added at the end of each cycle of reproduction. In contrast, persistent seed banks contain seed that has survived in the soil for two or more years before germinating. Secondly, some species have a considerable 'bud-bank' under the soil in the form of dormant meristems in bulbs and corms, and buds on rhizomes etc. (Harper, 1977). Thirdly, interspecific hybridisation is more common in plants, and there is also a much greater incidence of polyploidy (where species have three, four or more chromosome sets, instead of two as are found in diploids) than in higher animals (Briggs & Walters, 1997).

Different modes of natural selection

Turning now to developments in our understanding of microevolution, evolutionary biologists have found it useful to define a number of different modes of natural selection, and two or indeed all three may operate simultaneously (Fig. 3.8). Different researchers have coined different 'terms': synonyms are given in brackets. What follows is based on Rieger *et al.* (1968, 398), and follows the terminology of Mather (1953).

A. In stabilising (centripetal, normalising) selection a particular genetic variant in a population is at a selective advantage from generation to generation, with the elimination of peripheral variants. 'In stable environments and populations which have already achieved a high state of adaptation, a certain range of genotypes of proven fitness is preserved from generation to generation. Stabilizing selection does not result in evolutionary changes, but rather maintains an existing state of adaptation.'

B. Where there is a progressive change in the environment, directional (progressive, linear, dynamic) selection may favour a particular variant, resulting in a systematic shift in gene frequencies, leading to a changed (or changing) state of adaptation.

C. In disruptive (centrifugal) selection there is simultaneously selection for more than one optimum in a population occupying a heterogeneous environment. For example, at a site where there are two very distinct habitat types adjacent to each other, as a result of disruptive selection on a polymorphic population, each habitat is occupied by a different optimally adapted genotype.

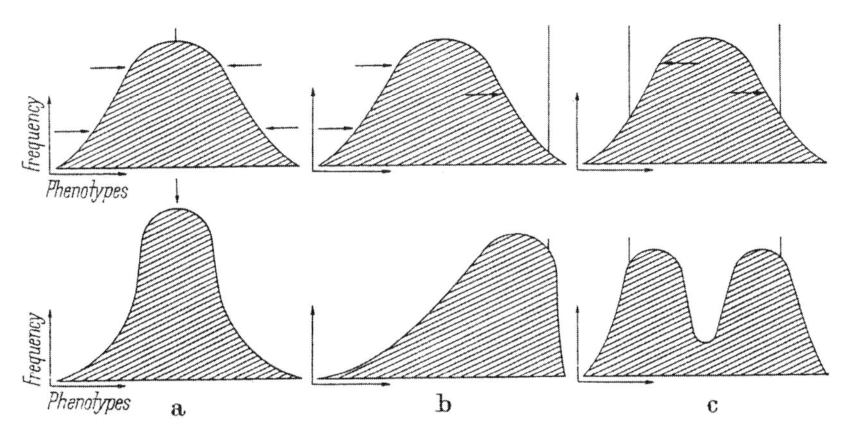

Fig. 3.8 Modes of selection within populations: (a) stabilizing selection; (b) directional selection; and (c) disruptive selection (the horizontal arrows show the direction of selection; the lower curves represent the new patterns of variation after response of the population to the new selection pressure. (After Mather, 1948, from Rieger *et al.*, 1968, with kind permission from Springer Science and Business Media)

r- and K-selection

It is important to appreciate the dynamic nature of ecosystems and note that they are subject to successional change, from bare ground and open water to forest etc. In considering the way that plants allocate their resources of energy and time in open and closed habitats, two different responses have been characterised, both of which involve the major types of selection outlined above (Harper, 1977, 760).

r-selection

In open, disturbed habitats a great deal of bare ground is available to be colonised by plants. Such areas are common in human-dominated landscapes, and range from arable fields to waste dumps and overgrazed areas. Harper concludes:

> During phases in which a population increases after a crash or after a new invasion, the individuals are by definition free from interference from neighbours and of their own species. During the phase of population increase the pre-eminent force selecting the individuals that contribute most descendants to subsequent generations is their fecundity and the effect of fecundity is enhanced by precocity (provided that generation follows hard on the heels of generation).

It is likely that there will also be ‘selection for the allocation of a large fraction of resources to seed production’ and ‘there is no premium on attributes that confer competitive ability’.

K-selection

The situation begins to change, however, when habitats begin to fill with plants. Harper (1977, 760) notes:

An increasing population may reach densities at which individuals begin to interfere with each other. The 'struggle for existence' that ensues contributes a special element to the process of natural selection, for the chance of an individual leaving descendants now depends on the share of needed resources that can be captured from competitors. Precocity and high fecundity are now relatively unimportant compared to factors that confer aggressiveness, e.g. height, a perennial or quickly renewable canopy, larger but fewer seeds. In dense communities both inter- and intraspecific encounters favour individuals with longer life, longer juvenile periods spent gaining competitive ascendancy.

Woody or clonal life-styles are at a selective advantage. As competition increases, 'the struggle ceases to favour the fecund, short-lived and instead favours the selfish and conservative growth strategies'.

Fitness

It is important at this point to define the terms associated with natural selection. Firstly, selective advantage may be defined as 'an advantage in the competition for survival between one genotype as compared with another showing "selective disadvantage" and producing fewer viable offspring' (Rieger *et al.*, 1968, 409). Fitness gives 'a measure of the advantage of a particular genotype over others in a population, which is reflected by that genotype's higher frequency in the next generation' (Kendrew, 1994, 377). Jones (1999, 82) notes: 'Evolution is an examination with two papers. To succeed demands a pass in both. The first involves staying alive long enough to have a chance to breed, while the mark in the second depends on the number of progeny.' However, Harper (1977, 649) stresses the important point that fitness is a measure not of numbers of offspring but the number of descendants. Thus, he writes:

Natural selection operates on individuals in a population and selectively favours those individuals and those genotypes that contribute most descendants to subsequent generations. This process leads remorselessly to an increase in the population of those forms that contribute more descendants than their neighbours. The critical word in this argument is descendants – the contribution of parents to subsequent generations. *It does not follow that an organism that produces a large number of progeny will also leave a large number of descendants.*

To obtain a quantitative measure of selection, the selection coefficient (s) may be calculated. It is:

The proportional reduction in the gametic contribution of a particular genotype compared with a standard genotype, usually the most favoured. The contribution of the favoured genotype is taken to be 1, and the contribution of the genotype selected against is then $(1 - s)$. This expresses the fitness of one genotype compared with the other. (Rieger *et al.*, 1968, 400)

Neutral theory of selection

An area of natural selection emerging in the post-Darwinian period relates to what happens at the molecular level. Studies have radically changed our perception of the selection, and the variability of populations and its significance (Kimura, 1983). Enzymes are the products of DNA transcription and translation, and electrophoretic studies of the mobility of enzymes in tissue extracts showed that several forms of the 'same' enzyme could be found in a population and high levels of genetic polymorphism were often detected in populations. Such observations were unexpected. Classical views on the action of natural selection assumed that new alleles of a gene could occur by mutation. There might be a degree of variability, while selection was taking place, but it was usually assumed that the most-fit allele would be selected. However, Kimura concluded that most variability within species and most amino acid changes were neutral with respect to selection. Different alleles (that had given rise to the different enzyme variants studied in electrophoresis) were considered to have little or no effect on individual reproductive success. The situation has been illustrated as follows. Two amino acids spaced a certain distance apart are the essential elements in the metabolically active site of an enzyme. However, in terms of fitness, which amino acids lie between these does not appear to be critical. Thus, a spacer sequence is essential to the activity of a protein, but in terms of selection, it is not critical which amino acids fill the gap (Alvarez-Valin, 2002).

Studies of the variability of DNA extracted from different populations and a wide range of species have offered support for the neutral theory. The implications of the theory are that only a small portion of the genetic code would appear to code for functional molecules. Also, while there are some regulatory sequences, the rest of the DNA appears to be without discernible function, and in the genomes of a wide range of organisms there are repeated DNA sequences, clustered or dispersed, of parasitic or selfish DNA. Thus, the model of neutral evolution considers that the majority of genetic differences between individuals or between species are 'due to the accidental accumulation of mutant DNA sequences that have little or no effect on individual reproductive success' (Dover, 1998). As neutral mutations accumulate at a constant rate depending only on the constancy of the mutation rate, the accumulation process has been likened to a 'molecular clock'. While the neutral theory has contributed much to evolutionary debate, and there is agreement that random effects and natural selection are important in evolution, 'there is disagreement, however, on the relative contribution of each force' (Alvarez-Valin, 2002, 821).

Recently, critical analysis of the neutral theory of molecular evolution has intensified. It is now being proposed that apparent 'junk DNA' may not be non-coding and functionless, but may play an active role in cellular activities (Greally, 2007). It is beyond the scope of this book to examine these researches, which are emerging from investigations of *Drosophila* (see, for example, Kondrashov, 2005), and the ENCODE pilot project studying the human genome (Birney *et al.*, 2007).

Another point of considerable interest to evolutionists emerges from studies of molecular biology. There are many duplicated sections of DNA, and a mutation lethal in a single copy may be neutral if a second unaffected copy is present. However, while one of the copies of an essential portion of the DNA sequences must remain 'unchanged' for survival, the existence of copies of this sequence offers the possibility that duplicated sections might, in their subsequent evolution, come to have new functions. Over time, divergence of descendant lineages may take place. Such descent with modification is what Darwin predicted.

Post-Darwinian models of speciation

As we have seen, Darwin visualised speciation as a gradual rather than an abrupt process. He considered that new species could arise with the extinction of the parental and intermediate generations. However, if parent and parental species did not compete with each other, both could live in the same area. Moreover, new species could also evolve on islands from long-range dispersal from a mainland source. For example, on the voyage of the *Beagle*, Darwin collected mocking birds (*Mimus*) species. The celebrated ornithologist John Gould examined Darwin's material and reported that on each of three Galapagos islands there was a different species. Darwin realised that a single South American mainland species had given rise to three daughter species (Mayr, 1991, 5, 22).

Darwin's notion that a derivative species could sometimes both arise and co-exist in the same area as its parent – i.e. sympatically – was the source of a bitter controversy with the naturalist Moritz Wagner (1868, 1889) who insisted that geographic isolation was necessary for speciation. 'Unfortunately Wagner beclouded the issue by also insisting that natural selection could not operate unless the population was isolated' (Mayr, 1991, 33).

Speciation of geographically isolated sub-populations derived from a common source is believed to be the predominant form of speciation in many organisms (Levin, 2000). In this type of speciation (also known as allopatric speciation) geographical isolation reduces or prevents interbreeding. Separate populations evolve independently and gradually diverge through the process of 'race' formation, and, later, achieve species-hood through the evolution of isolating mechanisms, a concept devised by Dobzansky in the 1930s, and popularised by the zoologist Mayr (1942, 1963).

There are a number of key points in this process. In cultivation experiments from the eighteenth century onwards, evidence was obtained that many tree species have geographically distinct races (see Briggs & Walters, 1997). Moreover, in cultivation experiments Turesson (1930 and earlier papers cited therein) detected races in many widespread European herbaceous plants. He called these ecotypes; climatic ecotypes being distinguished in samples taken from widely different areas and edaphic ecotypes found on populations growing on different soil types. These classic studies stimulated others to examine the constancy, or otherwise, of morphologically and physiologically distinct regional variants of trees (provenance trials) and herbaceous plants in common garden experiments. Justifiably the most famous are the intensive transplant and cytotaxonomic studies of Clausen and associates, working in California on races in a number of species, for example *Potentilla glandulosa* and *Achillea* species (see Clausen, 1951). These experiments established the importance of race formation, called ecotypic differentiation by Turesson. The degree to which races are always clear and distinct has intrigued many researchers. Perhaps only in certain situations are very distinct races evident. In other cases, where populations occur over a very wide range of climatic and edaphic conditions, much more complex patterns have been detected (Briggs & Walters, 1997).

Gradual speciation

In the model of geographic speciation, long-continued isolation allows independent evolution to take place and, if the derivatives A1 and A2 from a common ancestor (A) come to occupy the same territory, they may be found to have diverged genetically to the point where they are reproductively isolated, and two derivative species have been formed from a common ancestor. (We note at this point the fact that we are considering the biological species concept here. There are different species concepts and these will be examined in Chapter 6.)

The model predicts that, as a consequence of genetic changes evolving in isolation, pre-zygotic barriers to crossing may develop and these become evident when long-isolated populations, with a common ancestor, are brought together. Interbreeding does not take place because derivatives have pre-mating (pre-zygotic) barriers to crossing, as they may become sexually mature at different times of day or different seasons, and/or may have come to occupy different ecological niches, and/or they may have co-evolved with different pollinators and have different flower types. Post-zygotic isolating mechanisms may also be detected when two derivatives, long isolated geographically, come together. As a consequence of genetic changes, often involving chromosome restructuring, hybrids may not be formed. Or they may be produced, but exhibit hybrid invariability or weakness either immediately, or in F_1, F_2 or later generations. Generally, biological species are separated by more than one isolating mechanism. Moreover, 'barriers to crossing are not encountered simultaneously, but may be seen as a series of resistances which have to be overcome if crossing is to be effected' (Briggs & Walters, 1997, 261).

Speciation and founder effects

In 1942, Mayr suggested that another factor was important in geographic speciation. He called this factor the Founder Principle and coupled it with the notion of the co-adapted gene complex, with alleles selected for their ability to work successfully together. A small sub-population (one or two or few individuals) might be accidentally isolated and would, by sampling error, have only a small fraction of the genetic variation of the parental population. The new small population might then increase in numbers and a new co-adapted gene complex could be built up from the subset of founding genes augmented by mutation. Experiments and observations confirm the importance of long-range dispersal of one or a small number of founders, especially in the development of island floras (Levin, 2000). Advances in molecular techniques may make it possible to test the notion of co-adapted complexes as visualised by Mayr.

Considering the present status of the concept of isolating mechanisms, it is important to note that some biologists find the notion of 'mechanism' misleading, in that it lumps together many different by-products of natural selection and/or drift, and perhaps implies that the formation of species is qualitatively different from the formation of races and subspecies. Experimentalists are still very active in this area of research, but have chosen to look in detail at mate choice, hybrid inviability etc.

The process of geographical speciation suggests that isolated derivatives from a common ancestor *will* in the course of time develop into new species. However, this view is incorrect. During periods of environmental changes, such as glaciations or by human agency, incipient species can be brought together at any stage in their evolution and such differentiation, as has occurred, might be lost. Thus, races of a species may cross freely when they come into contact after a period of isolation. In contrast, derivative populations long isolated might have developed partial isolating mechanisms. In contact, these isolating mechanisms may be reinforced or may break down (Turelli, Barton & Coyne, 2001).

Reinforcement is an intriguing possibility that has been insufficiently studied. A simple model may be used to illustrate what might occur in plants. From a common ancestor (A) two long-isolated derivatives (A1 and A2) have a common pollinator, but different overlapping flowering times, and hybrids between them are found to be infertile. Thus, A1 flowers from April to May, while A2 is in flower from May to June. Individuals of A1 flowering in April and A2 flowering in June will not produce A1/A2 or A2/A1 hybrids and have the capacity to produce fully fertile offspring. In contrast, crosses A1 × A2 and A2 × A1 produced in the period of overlapping flowering will give infertile offspring. Thus, in the next generation only the progeny of early flowering variants of A1 and later flowering variants of A2 will be represented. Effectively, an imperfect pre-zygotic isolating mechanism between A1 and A2 has started to be reinforced when the two derivatives came to occupy the same territory. This theoretical model has been tested experimentally and received support in investigations, for example with maize (Paterniani, 1969).

Introgressive hybridisation

While reinforcement may occur when derivatives from a common ancestor come together after a period of geographic isolation, in other cases, where an incomplete isolating mechanism has developed and where the derivatives A1 and A2 are faced with an ecologically complex situation, such 'genetic isolation' as exists between them may be broken down, temporarily or perhaps longer-term, through hybridisation. Anderson (1949), reflecting on the possibilities, concluded that where the habitat is hybridised (i.e. where major upheavals in ecosystems are occurring or have taken place, and a wide range of different habitats are being/have been formed), as might occur as a consequence of post-glacial changes or by human activities, not only the parental but also the hybrid derivatives might be at a selective advantage in some part of the complex ecological system. Continued hybridisation might occur to produce a hybrid swarm and, through backcrossing, genes from one derivative might be transferred to the other by a process he called introgressive hybridisation. In the case of reinforcement, the outcome is the co-existence of two derivatives as separate species: upgrade evolution continues. In the case of hybrid swarms and introgression, a temporary or long-lasting breakdown of isolating mechanisms might occur: the speciation processes, which had produced a degree of differentiation between derivatives, are subverted by 'down grade' events.

Sympatric speciation

Darwin's model was based on his conviction that evolution of new species was a slow process based on small genetic changes. There is now abundant evidence that abrupt sympatric speciation occurs widely in plants through the process of polyploidy. Some years ago, Goldblatt (1980) estimated that at least 70% of monocotyledons are polyploids, the figure for dicotyledons being even higher at 70–80% (Lewis, 1980). This conclusion was based on chromosome numbers, which in many genera are simple multiples of a minimum or basic number. However, in some genera there are no representatives with low multiples of the base number, extant species having only high polyploid chromosome numbers. These are considered to be of ancient polyploid origin. For instance, the fern *Ophioglossum reticulatum* has $2n = 1440$, which is believed to be 96-ploid. Very high chromosome numbers occur in some flowering plants: the dicotyledonous species *Sedum suaveolens* has $2n = c. 640$, *c.* 80-ploid, while the monocotyledon *Voaniola gerardii* has $2n = c. 596$, which is *c.* 50-ploid. Many other polyploid groups have every step in a polyploid series 'occupied' and some of these are likely to be of more recent origin. Probably, many evolved during the turbulent climatic upheavals of the Quaternary period. Only those species survived that adapted to changing situations *in situ*, or were able to 'migrate' to areas that had an appropriate climatic envelope for that species. As plant distributions changed in response to repeated climate perturbations, new polyploids were produced when species evolving originally in allopatry were brought into sympatry.

Evidence from recent molecular studies of gene sequences has provided many new insights into polyploidy, transforming our understanding of the frequency and importance of polyploidy in plant evolution (Paterson *et al.*, 2005). Providing a commentary on these investigations, Soltis (2005) notes: 'one of the most startling results of recent genomics studies is the near ubiquity of polyploidy. The small genome of *Arabidopsis* may have been derived from rounds of polyploidization (Vision, Brown & Tanksley, 2000; Bowers *et al.*, 2003)... *Oryza* has been shown to have an anciently duplicated genome (Wang *et al.*, 2005).' These findings force us to 'accept that probably all angiosperms – and maybe all plants – are polyploid to some extent'.

Cytogenetics of polyploidy

In a diploid plant there are two genomes – two homologous chromosome sets – arising from the fertilisation of haploid male and female gametes; namely, A+A gametes fuse to give a AA diploid, which in turn produces gametes A. By somatic misdivision, or more usually by the fusion of a reduced (A) and an unreduced gamete (AA) an autotripolyploid (AAA)

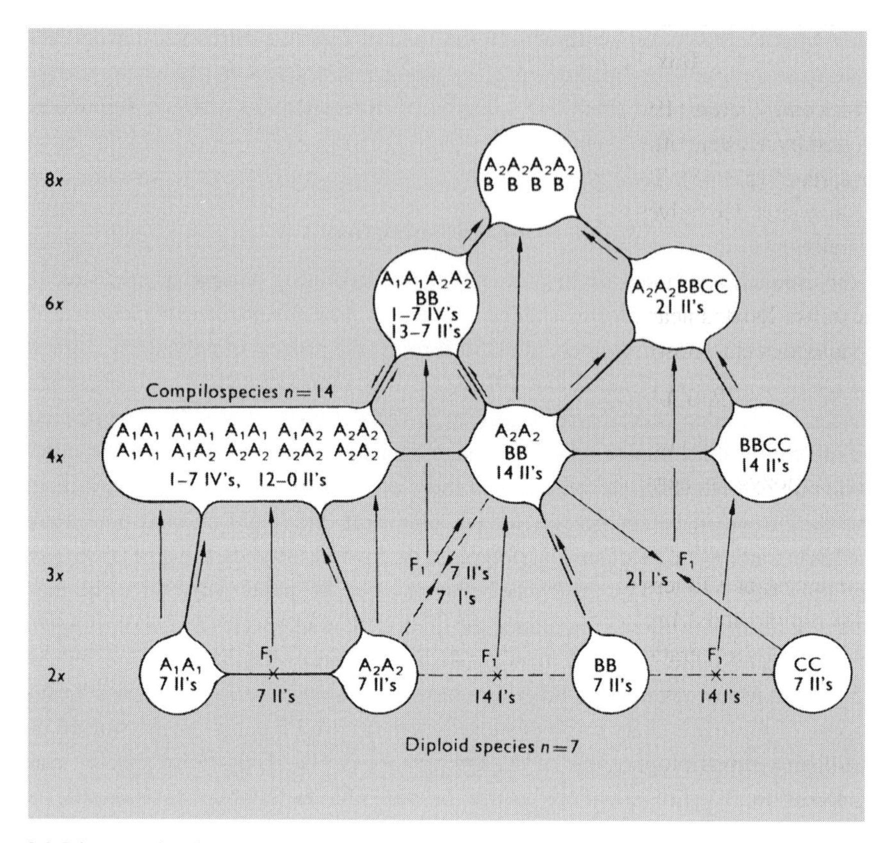

Fig. 3.9 Diagram showing a typical polyploid complex, and the various ways in which it can evolve through hybridisation and polyploidy. Genomes are indicated by letters AA, BB etc. and how chromosomes associate at meiosis is indicated by Roman numerals, i.e. unpaired chromosomes – univalents: I; paired chromosomes – bivalents: II; associations of four chromosomes – quadrivalents: IV. (From Stebbins, 1971. Reproduced by permission of R. L. Stebbins)

is produced. An autotetraploid with four (AAAA) or more homologous chromosome sets may also arise by the fusion of unreduced gametes.

Following interspecific hybridisation, allopolyploids may be produced. These are polyploids that have a doubling of chromosome sets (or higher multiple) from different species or genera. Thus, diploids of genomic constitution AA and BB may produce a polyploid derivative AABB. For the purposes of further illustration, Fig. 3.9 shows the origin of a number of auto- and allopolyploids from a series of diploid parents.

Polyploid derivatives are indeed new species, as they are reproductively isolated from their diploid parental species. For example, diploids species AA and BB may produce a fertile polyploid species AABB that is capable of regular meiotic pairing of the chromosomes (chromosomes of the A genome have pairing partners, as do the chromosomes of the B genome). Crossing AABB to AA produces a triploid AAB, while crossing AABB

Box 3.1 Polyploids and inbreeding depression

Briggs and Walters (1997) discuss this point. They note: 'In the heterozygous diploid, the dominant allele often shelters recessive alleles that are deleterious in the homozygous state' (p. 143). They continue (p. 341): 'repeated self-fertilisation may lead to homozygous derivatives and inbreeding depression . . . In a diploid plant of genotype *Aa* selfing produces progeny in the familiar Mendelian ratio 1*AA*:2*Aa*;1*aa* and 50% of the progeny are homozygous. In a tetraploid plant of genotype *AAaa*, in which alleles are located near the centromere on different chromosomes and where the four homologous chromosomes separate at random in pairs, the ratio on selfing is:

1/36 *AAAA*: 8/36 *AAAa*: 18/36 *AAaa*: 8/36 *Aaaa*: 1/36 *aaaa*.

Segregation follows Mendelian principles but with a different ratio, yielding only 1/18 homozygous derivatives; 94.4% are heterozygotes of various genotypes. Models may be constructed in which selfing proceeds for several generations and where genotype frequencies are not influenced by selection. Thus, to reduce the percentage of heterozygotes to less than 1% from a population initially wholly heterozygous (*Aa* in the diploid and *AAaa* in the autotetraploid) will take 7 generations in the diploid but it will take 27 generations for the tetraploid.'

with BB gives the triploid ABB. These triploids are infertile, because there are three sets of homologous chromosomes in a triploid. Chromosome pairing at meiosis in triploids is highly irregular, as there are three homologous chromosomes rather than the normal pairs in diploids. Figure 3.9 shows the genomic relationships, expressed through chromosome pairing, in members of a polyploid complex.

The success of polyploids

Soltis and Soltis (2000) review the attributes of polyploids that contribute to their success. They conclude:

> Based on a limited number of studies . . . (i) Polyploids, both individuals and populations, generally maintain higher levels of heterozygosity than do their diploid progenitors . . . (ii) Polyploids exhibit less inbreeding depression than their diploid parents and can therefore tolerate higher levels of selfing [Box 3.1 considers why this should be so] . . . (iii) Most polyploid species are polyphyletic, having been formed recurrently from genetically different diploid parents . . . (iv) Genome rearrangement may be a common attribute of polyploids . . . (v) Several groups of plants may be ancient polyploids.

In addition to speciation by polyploidy, there is evidence, which we examine below, that sympatric speciation can occur without change in chromosome number.

Evidence from the fossil record

Many post-Darwinian advances in our understanding of extinction and speciation have come through studies of the fossils. In the fossil record there is evidence for mass extinction events. Such events are germane to our concerns in this book, for many conservationists consider that we are at present in another mass extinction event, as human activities, ever more destructive of the environment, are driving a huge number of species, both animal and plant, towards extinction, with many more under threat. Also, there have been active debates about whether speciation is a gradual continuous process, as envisaged by Darwin, or whether there are periods of rapid speciation interspersed by long periods of little change. If human impacts are to be properly assessed, it is essential to understand the background rates of natural speciation and extinction, and whether they are continuous and gradual or episodic.

Mass extinctions

Considering the evidence for mass extinctions in the fossil record, it is important to make it plain that the evidence comes mostly from investigations of animal groups. Thus, Raup (1991) reports researches on the extinction rate in marine invertebrate groups in the fossil record. For much of the time, there was a low rate of extinction, but there was evidence for five major mass extinction events at *c*. 443, *c*. 364, *c*. 248, *c*. 206 and *c*. 65 million years ago (MYA: figures from Willis & McElwain, 2002, 233), with correlated perturbations in vertebrate and plant groups (but see below). Each episode of mass extinction was different in detail. The most severe event was at the Permian–Triassic boundary, *c*. 248 MYA, when many animals were lost from the fossil record, including ‘up to 90% of all durable skeletonized marine invertebrates and 54% of all marine families’ (Willis & McElwain, 2002, 233).

The most recent mass event was at the boundary of the Cretaceous and Tertiary (65 MYA), which included the extinction of the non-avian dinosaurs. Archibald (1997) notes that there are as many as 80 theories concerning the extinction of the dinosaurs, including the frivolous (Martians hunted them to extinction) and the currently untestable (they died of a plague). Other hypotheses link extinction to the extreme volcanism of the period, marine transgressions and/or decreases in atmospheric oxygen. Alvarez *et al.* (1980), however, proposed that the extinction followed the impact of a huge asteroid on the Earth’s surface. A crater of appropriate size and age has now been detected at Chicxulub, on what is now the Yucatan Peninsula in Mexico. Several ‘extinction scenarios’ have been proposed (Hallam, 2002). On impact, the asteroid caused extensive wildfires, and huge quantities of dust blocked solar energy from reaching the Earth’s surface and waters, thereby reducing photosynthesis. Initial rapid cooling was perhaps later followed by the release of carbon dioxide from sedimentary carbonates, resulting in the trapping of heat in the atmosphere – by the so-called greenhouse effect. It is also possible that sulphur dioxide expelled from gypsum deposits gave rise to sulphuric acid in the atmosphere resulting in damaging acid rain.

While many scientists have accepted the impact theory of mass extinction 65 MYA, there is still fierce debate about details of environmental change following asteroid impact, and what part other factors such as the extreme volcanism of the period and marine incursions might have played (see Glut, 2000).

In reviewing the cause of the other four mass extinctions, it has been suggested that in each case an extraterrestrial body landed on the Earth's surface (see Raup, 1991; Glen, 1994). However, there is evidence that major impacts by asteroids did not always lead to mass extinctions and there would appear, as yet, to be no evidence of an impact at the time of the Permian mass extinction event, but much evidence of extraordinary volcanic activity. The causes and extent of mass extinction events continues to be debated by both scientists and non-scientists (Hallam, 2002, and references cited therein). One major issue, generally overlooked in these debates, is whether there were mass extinctions in plant groups, coincident with those in animal groups. Willis and McElwain (2002, 236) have recently examined the evidence in detail and conclude that 'there are no major peaks of extinction in the plant fossil record comparable to those of the faunal mass extinction events'.

Punctuated equilibrium

Turning now to the speed and periodicity of speciation, Gould and Eldredge (1977, 1993) challenged the 'phyletic gradualism' of Darwin, proposing instead the 'punctuated equilibrium' model, in which there were periods of rapid change in the number of species followed by long periods of little or no change (stasis).

A recent review of 58 studies concluded that 'paleontological evidence overwhelmingly supports a view that speciation is sometimes gradual and sometimes punctuated and that no one mode characterizes this very complicated process in the history of life' (Erwin & Anstey, 1995). However, the review did not consider speciation in plants.

Examining the fossil record for plants, Willis and McElwain (2002, 288) came to the conclusion:

> plant evolution appears to be concentrated in distinct intervals in geological time rather than evenly spread throughout. Similarly, increases in morphological complexity and total plant species diversity have occurred during relatively short intervals of geological time, interspersed by longer periods of relative stasis (in terms of major evolutionary innovation) and plateaus in species diversity. The peaks of plant originations, seen in the plant fossil record, do not, however, appear to match those seen in the animal record, which tend to follow mass extinction events. Furthermore, the peaks of plant extinction do not coincide with at least four of the 'big five' mass extinction events.

What are the mechanisms driving change in plants? One major factor is likely to be continental drift.

Continental drift

In 1915, Wegener proposed that continental land masses do not have a fixed position, but have moved over time. While some accepted Wegener's views, others were not convinced. However, in the 1960s, the mood of earth scientists moved from scorn to acceptance of his ideas, in the development of the theory of plate tectonics. Evidence emerged of mid-ocean ridges in the Atlantic, Pacific and Indian Oceans, and plausible mechanisms were established of continental movement by the spreading of the sea floor.

Geologists have discovered that at about 300 MYA the continents had coalesced to produce a super continent Pangea, and at about 200 MYA rifting started to divide this land mass and the continents, in the course of millions of years, and masses arrived at their present configuration (Hallam, 1973). Essentially, a single land mass was broken into separate continents (each fragment with its own evolving biota). Over millions of years these fragments drifted so far apart as to prevent natural gene flow, and different species and ecosystems evolved in geographical isolation. During continental drift there were periods of great tectonic activity when carbon dioxide levels were increased. The theory of plate tectonics also accounts for the development of a ring of volcanic islands on the edge of some continents and elsewhere. Advances have also been made in our understanding of mountain building processes, and changes in sea level leading to the development of oceanic islands and islands of uplifted regions in many areas of the world. Following rare long-distance dispersal from continental sources, these islands and mountain peaks have been colonised and have been sites of independent speciation in geographical isolation (Nunn, 1994; Vitousek, Loope & Adersen, 1995).

Orbital variations

While continental drift is one of the keys to understanding patterns of speciation in plants, it has been claimed that climatic changes associated with orbital variations are also extremely important (Bennett, 1997). Willis and McElwain (2002, 289) reveal: 'Over geological time, calculations indicate that variations in the Earth's axial tilt, precession, and changes in the eccentricity of the elliptical orbit around the sun have produced differences both in the amount of solar radiation received by the Earth and also in the latitudinal and seasonal distribution of this radiation.' These variations, known as Milankovitch cycles, have resulted in large 'astronomical forcing of the climate at intervals of approximately 400, 100, 41 and 23–19 thousand years', leading to numerous cold/warm stages, including 'the glacial-interglacial cycles of the Quaternary period (the last 1.8 million years), and the pacing effect of these Milankovitch cycles has been proposed as the pulsing mechanism behind punctuated equlibrium'. During cold stages in the cycle, vegetation retreated to isolated refugia in favourable climate zones, and speciation may have occurred in situations of geographical isolation. In the warmer interglacials, some species are likely to have adapted and remained in the refuges, or migrated to reoccupy some or all of their former territory, or they may have failed to migrate and become extinct. In the migration process, plants of genetically related lineages, isolated in different refugia in the cold period, may have come together again as distinct species. Alternatively, if differentiation to full species status was incomplete, such differentiation as had occurred could have been lost in hybridisation, or partially differentiated lineages might have attained species status as partial barriers were reinforced (see above). In summary, it has been suggested that continual drift, and its associated tectonic changes, and orbital forcing may have resulted in episodes of speciation rather than the universal gradual evolution pictured by Darwin (1859). It is very difficult to devise critical tests of these important ideas. What must be stressed, however, in the gradualism versus abrupt change debate, is that present evidence supports Darwin's notion that microevolutionary processes are continually ongoing, and, therefore, with ever-changing circumstances, gradual change is always occurring.

Conclusions

The ideas of Darwin and Wallace, first set out in the middle of the nineteenth century, form the centrepiece of our current views on evolution. As we have seen above, the results of genetic, cytological and molecular biological investigations have greatly enriched our understanding. Both abrupt and gradual processes are involved. Chance events are of great importance. Plant evolution has not produced one or a few highly adaptable and phenotypically plastic species that can colonise and grow in every habitat from the tropics to the poles. Instead, the plant and animal kingdoms are highly speciose – evolution has produced millions of species, some with wider tolerances than others.

Secondly, it cannot be assumed that the evolutionary concepts outlined above somehow stand outside the human realm. While it is true that many insights have emerged from a multitude of observations and experiments on living wild species and from the fossil record, as we see in later chapters, many elements of our understanding have come from studies of human-managed species and habitats. Thus, our evolutionary concepts owe much, for instance, to the study of cultivated plants, examination of selection in polluted areas, interspecific hybridisation as a result of human breakdown of geographical and ecological barriers etc.

A major post-Darwinian insight is that human activities and influences on microevolution are not restricted to domestication alone, as is perhaps implied in Darwin's presentation of selection. With regard to plants, he classifies selection into Natural and Artificial. In making the distinction, clearly he does not see human actions as part of natural selection. He treats human activities as a special case. In considering the origin of varieties of crop, horticultural and cultivated garden plants, he draws attention to both deliberate and unconscious elements of selection. In his introduction to *Variation of Plants and Animals under Domestication*, Darwin concludes, with regard to domestication: 'Man, therefore, may be said to have been trying an experiment on a gigantic scale' (Darwin,1868, quoted from 1905 edition, Vol. I, 3). Darwin does not mention the consequences of other human activities and whether they might constitute selective forces.

As we shall see in the following chapters, the sphere of human influences on microevolution extends beyond strict domestication of crop and garden plants, and I suggest that the scale of 'the human experiment', as Darwin visualised it, is immeasurably enlarged, for human activities increasingly present plants and animals with many 'unintentional' and 'accidental' selection pressures, e.g. in relation to medical interventions (antibiotic resistance in bacteria), agricultural practices (in the use of pesticides, herbicides, fertilisers etc.) and animal and plant responses to pollution. It is helpful at this point to introduce a number of animal examples to reinforce the developing argument. For instance, mortality in fish populations is often selective, because fishing gear tends to catch some individuals more readily than others. Indeed, in some cases the gear is designed to catch the largest fish. Thus, gill-nets act as filters that retain the large fish while letting the smaller fish escape. Bell (1997, 300) notes: 'in lakes where gill-nets were set, whitefish (*Coregonus*) tended to become longer and thinner at a given age, and reproduced earlier in life and at smaller sizes'. Nets are in effect acting as a form of size-selective predation. In another celebrated case of microevolution, a *Drosophila melanogaster* population from inside the wine cellar of the Chateau Tahbilk, Victoria, Australia, was found to have greater larval ethanol tolerance than a neighbouring population from outside the cellar (details in McKechnie & Geer, 1993). Geneticists across the world now frequent wine cellars to investigate the physiological and evolutionary mechanisms at work in alcohol dehydrogenase polymorphism. Warfarin is used in the control of rats, mice, feral pigs and vampire bats. In 1950, the poison was first used on populations of brown rat, *Rattus norvegicus*, in the UK. However, in 1958, genetically determined resistance to warfarin was detected in the Scottish lowlands, and later, in 1960, in sites on the Welsh/English border (Boyle, 1960). By 1972, resistance had been found in 12 areas in Britain. Warfarin resistance has also been detected in rats on the European mainland (Denmark) and in the USA. Population studies have revealed variability in resistance. For example, at Forden, mid-Wales, rats living in barns showed variability in resistance and this was linked to the periodic (rather than continuous) attacks on the rat populations with warfarin. In contrast, in adjacent hedgerows the rats were rarely if ever treated and, thus, there was great heterogeneity in the selection, and populations close together had very different levels of phenotypic resistance. Turning to a final example, resistance to an insecticide was detected in 1908, when it was discovered that the plant scale (*Aspidiotus perniciosus*) was discovered to be resistant to lime sulphur. In the twentieth century, cheap synthetic insecticides have increasingly been employed, the first of which was DDT, introduced for practical use in 1941. Later, organo-chlorine and organo-phosphorus insecticides were developed. Resistance to these insecticides has developed in many insects (see Bishop & Cook, 1981). It has been found that resistance is genetically determined – sometimes a single major gene. In a minority of cases resistance is dominant. Some insects have proved to be resistant to more than one insecticide; resistance can develop independently in separated sites; and in response to a particular pesticide, different resistance mechanisms have been found in different populations of the same species. In addition, when the use of an ineffective insecticide is discontinued, it has been found that resistant genotypes to this once-effective insecticide are at a selective disadvantage relative to susceptible genotypes. Thus, when selection is relaxed the frequency of resistant genotypes is reduced. A further example of microevolution in action is provided by changes in the House Mouse brought to the remote Gough Island, South Atlantic, on ships. Now numbering more than 700,000, mice have adapted to a carnivorous diet of live chicks of a range of seabird species and are now double the size of their ancestors (Wanless *et al.*, 2009).

Finally, recognising the dramatic transformations found increasingly across the world, it is important to stress that a number of environmentalists have extended the concept of domestication to include anthropogenic effects on ecosystems and across landscapes (Kareiva *et al.*, 2007). For example, Western (2001, 5461) refers to 'human domestication of ecosystems' and the 'expansion and intensification of domesticated landscapes'. In the next two chapters, these ideas are further explored, in examining the accelerating impacts of human activities on the biosphere.

4
The origin and extent of human-influenced ecosystems

Human-influenced ecosystems are the arena for contemporary microevolution, and the increasing spread and intensity of human activities are major concerns for conservationists. It would be possible to assume that the historical and contemporary effects of human influences are so widely understood and appreciated that only a few paragraphs are needed at this point in the book. However, such an approach would fail to set the scene adequately for what follows. In a brief historical review, this chapter examines several major issues: (a) the environmental consequences of inexorable rise of human populations; (b) how changes in agriculture and technology have led to abrupt and gradual environmental transitions; and (c) how human activities have resulted in the widespread introduction of plants and animals into new areas across the globe. This chapter also considers the rise of public concern about the profound human impacts on the world's vegetation, animal resources and soil, leading to the development, in the post-Second World War period, of increased environmental awareness, focusing on human population growth (now at over 6 billion and projected to rise), habitat destruction, over-harvesting of natural resources, the extinction of species, the impact of pollution and the dramatic consequences of the widespread introduction of alien species across the world.

The origin of humans

Weber (2005) notes that only about 10% of the population of the USA fully accept a Darwinian explanation of the origin of humankind, many believing that the world and its plants and animals are of very recent origin. Avise (2001, 202) stresses both the immensity of the time scale of evolutionary events, and the comparatively recent evolution of humans. He writes that, if the whole of geological history is compressed into a single calendar year, life on Earth began on 9 October. The rise and fall of the dinosaurs lasted a mere four days from 24 to 28 December. Humans did not appear amongst a fantastic array of evolved biodiversity until 10.30 pm on the final day of the year, 31 December, with the birth of Christ occurring at 11.59 plus 56 seconds. The dramatic rise of human populations has occurred within the last second of time.

Concerning the evolution of humans, in 1871, Darwin published his views after an exhaustive review of the evidence then available (Darwin, 1989, 631–2). He concluded:

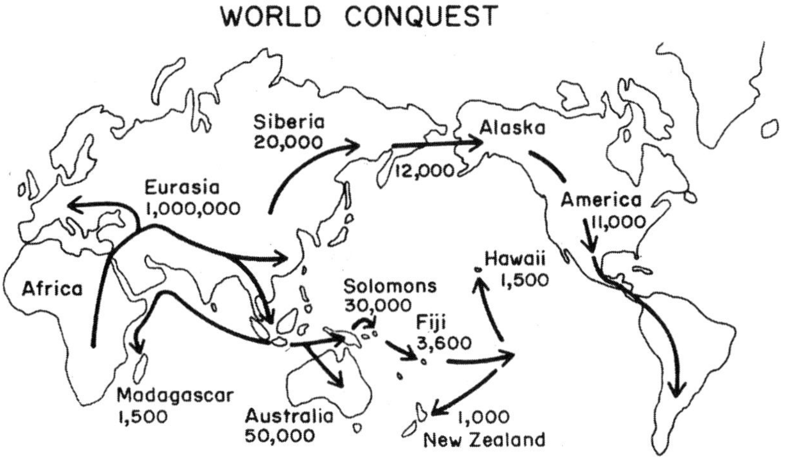

Fig. 4.1 Diagram, based on current archaeological evidence, of the spread of mankind across the world from African origins. Numbers stand for estimated number of years before the present. (From *The Rise and Fall of the Third Chimpanzee* by Jared Diamond, 1992, published by Hutchinson. Reproduced by permission of Random House Group, Ltd.)

> . . . man is descended from some less organized form. The grounds upon which this conclusion rests will never be shaken, for the close similarity between man and the lower animals is embryonic development, as well as in innumerable points of structure and constitution . . . Now when viewed by the light of our knowledge of the whole organic world, their meaning is unmistakable. The great principle of evolution stands up clear and firm . . . He who is not content to look, like a savage, at the phenomena of nature as disconnected, cannot any longer believe that man is the work of a separate act of creation . . . the close resemblance of the embryo of man to that, for instance, of a dog – the construction of his skull, limbs and whole frame on the same plan with that of other mammals . . . and a crowd of analogous facts – all point in the plainest manner to the conclusion that man is the co-descendant with other mammals of a common progenitor . . . . man is descended from a hairy, tailed quadruped, probably arboreal in its habits, and an inhabitant of the Old world.

Diverse lines of evidence, including fossil remains, now support Darwin's views on the early history of humans, suggesting that anatomically modern humans arose in Africa over 100,000 years ago and by 40,000 years ago were fully modern in anatomy, behaviour and language (Diamond, 1992). Modern humans have spread to all corners of the Earth (Fig. 4.1) and from what were initially very small numbers, populations have reached more than 6 billion. They have had such profound effects on the environment that much of the Earth's surface has been transformed.

Human uses of plants

In a highly original book, Evans (1998) considers the history of human population growth in relation to the use of plants, particularly for food. The difficulty of estimating population size in prehistoric and indeed historic human populations is acknowledged. Best estimates are offered. I acknowledge the use of this very helpful framework and supporting ideas in this brief account of the transformation of the Earth. Here, the concern is to reveal the historical context of the present arena of plant microevolution. Thomas (1956), Detwyler (1971), Holdgate (1979), Goudie (1981), Nisbet (1991) and McNeill (2000) may be consulted for more information on the historic roots and major issues in environmentalism.

Reaching 5 million: hunter-gatherers

Before the advent of agriculture, humans, then probably represented by small, scattered populations, obtained food as hunter-gatherers. While some insights have been obtained by archaeologists, a good deal has also been discovered from historic and recent field studies of populations practicing this life-style. Thus, in the last 200 years, since European settlement of Australia, important information has been obtained from studies of aboriginal peoples, who have lived as nomadic hunter-gatherers since they entered Australia at least 40–50,000 years ago (Diamond, 1992). Revealing insights have also been obtained from studies of the !Kung bushmen of the Kalahari of Africa (Lee, 1984), the North American Indians and inhabitants of tropical forests (Harlan, 1995). However, a note of caution must be entered here. It is important to stress that the present-day life-styles of native peoples may provide an imperfect model of the behaviour of ancestral foragers, for many have partially or completely abandoned such a way of life.

It is clear that contemporary hunter-gatherers have a comprehensive knowledge of the flora and fauna of the territory in which they live and that they eat a wide range of animals and plants (O'Dea, 1991; Harlan, 1995). There has been much debate about the efficiency of the foraging life-style. Evidence, both from the studies of present-day hunter-gatherers, and in experiments by botanists in Turkey, Mexico, Africa etc., suggests that, in some circumstances, it is possible to collect a substantial quantity of food quite quickly, with far less expenditure of energy than that used in cultivating plants (Harlan, 1975; Evans, 1998). Another important point must be stressed. While it is clear that the Australian aboriginal peoples have neither domesticated nor cultivated plants, they have manipulated plant populations in different parts of the country to a significant extent. For example, tubers of yams (*Dioscorea*) are collected, and the 'tops' are replanted to allow the plants to grow again and provide a further harvest at a later date (see Harris, 1996). Also, it is important to confront the growing evidence that, in prehistoric times, hunter-gatherers had a profound effect on plant and animal communities, including probably hunting some animals to extinction (see Chapter 5). Across Australia and in other parts of the world vegetation has been decisively influenced by the repeated use of fire by aboriginal peoples (Russell-Smith *et al.*, 2007), the purpose of which was to ambush animals, promote new growth of grasses and facilitate the germination of scattered seeds of food plants (Evans, 1998). Hunter-gatherers of Australia also dug ditches to increase the supply of fish (Walters, 1989), and diverted water to flood forests in the dry season (Campbell, 1965). On the Great Plains of North America, hunter-gatherers also used fire, made dams, irrigated areas and sowed seeds (Steward, 1934).

Reaching 50 million: the beginnings of agriculture

In the period from 8000 to 2000 BC, when human populations rose to *c.* 50 million, dramatic changes occurred, as the hunter-gatherer life-style increasingly gave way, independently in different parts of the world, to settled agriculture with cultivated crops and domesticated animals. Pontin (1993) refers to this as the First Great Transition in human societies.

Different crops were developed in different areas of the world. The availability of suitable wild plants was an obvious precondition, and more or less all the staple crops we now use were domesticated in this period. Major advances in our understanding of the origin and location of domesticated crops have been achieved by the close study (often involving microscopy) of radio-carbon-dated plant remains. Recently, archaeologists have investigated both specific and intraspecific differences. For example, the starch grains of maize and manioc are highly distinctive. Moreover, the starch grains of wild chilli are 5–6 micrometres long: in contrast, domesticated variants have grains 20 micrometres long.

In the 1950s, the domestication of crops was thought to have occurred in only two areas, the Near East and the Americas (Schwanitz, 1966). Now, there is evidence that plant domestication took place in at least ten widely separated centres of origin, including West Africa, China, South India and New Guinea. As plant remains rarely survive in tropical areas, it is unclear whether some plants were also domesticated there. Overall, archaeologists now conclude that Old World crops were fully domesticated a little later than once thought, while those of the New World are a little older (Balter, 2007). For example, squash and maize were once thought to have been domesticated around 5,000 years ago. Now, there is evidence from investigations in Mexico that the squash was domesticated nearly 10,000 years ago (Dillehay *et al.*, 2007), while domesticated maize originated 9,000–8,000 years ago (Matsuoka *et al.*, 2002).

Some have viewed the development of agriculture as an abrupt step. For example, Childe (1925, 1952) postulated a 'Neolithic Revolution' in the Near East, the stimulus for which was the dramatic climate changes at the end of the last Ice Age some 11,500 years ago. It has been suggested that a rise in carbon dioxide in the late Pleistocene might have removed a limiting factor in plant productivity and this was a spur to the successful establishment of agriculture (Sage, 1995). However, other archaeologists consider that the origin of domesticated crop plants was a more gradual process. Indeed, some scholars reject the idea of a rigid distinction between the activities of hunter-gatherers and agriculturists. Tudge (1998, 4) provocatively suggests that from at least 40,000 years ago humans were managing their environment to such an extent that they can properly be called 'proto-farmers', and that, at the time of the Neolithic Revolution, farming was already 'an ancient and established set of crafts'.

The process of domestication of plants

Darwin viewed domestication as a process rather than an event. Human selection, either consciously or unconsciously, occurred, leading to a genetic shift making them 'better adapted to the environment created by cultivation, but at the same time ill-adapted to their original habitat in the wild' (Ladizinsky, 1998, 7). As a broad generalisation (but stressing that there are many exceptions) this better adaptation of crop species often involves the same characteristics in different species. Gepts (2004) explores the so-called domestication syndrome, which involves retention of seeds on the crop; loss of dormancy in some species; different, often shorter and more compact growth; increase in the size and properties of the organs harvested; changes in the breeding system towards more selfing; and loss or reduction in the levels of toxic compounds associated with plant defence in the wild.

Concerning the genetics of individual traits, many are inherited in a simple Mendelian fashion, but others are polygenically controlled. Reviewing the evidence, Gepts (2004, 29) confirms 'the predominace of recessive mutations among domestication alleles'. However, dominant alleles are also found. An example is provided in barley. As part of the dispersal mechanism of the grains, in wild barley the spike of grains breaks down into units. In contrast, in cultivated barley, as a consequence of a dominant mutation in one of either of two genes, the spike remains intact at maturity and the seeds are not dispersed (Nilan, 1964). This genetic change makes for a more efficient harvesting of the crop. Concerning the genetic architecture in crop plants, it has often been found that important traits are linked on a small number of chromosomes. Pernès (1983) suggests that the 'linkage of domestication genes' is of great significance 'because it would maintain the cohesion of some essential elements of the domestication syndrome' in a situation where hybridisation with the wild progenitor species was highly probable (quoted from Gepts, 2004, 25).

Regarding the context of domestication, some scholars have suggested that the first crop plants originated from weedy species naturally growing around human habitations and refuse dumps, rather than from plants deliberately cultivated (see Abbo *et al.*, 2005). There has been discussion about whether humans, once embarked on crop husbandry, provided strong consistent selection towards domestication through the repeated use of techniques of soil preparation, sowing, harvesting and seed selection leading quickly to genetic isolation between crop and wild relatives. Archaeological evidence suggests that mixtures of wild and proto-domesticated variants of the same species were often grown together and this has led some to propose that the domestication process might have been longer, as weaker, perhaps intermittent, selection may have occurred. Thus, in years of poor harvests and food shortages, the selection pressures associated with harvest may have been relaxed as the crop was collected by hand. Moreover, in the year following, if all the previous year's crop had been consumed, wild seed may have been sown, rather than 'proto-domesticated' seed. These observations suggest the importance of the cultural context in which domestication was taking place, a viewpoint strongly presented by Rindos (1989).

Following domestication of cultivars of food plants, increasing amounts of labour were expended in developing field systems (often with irrigation), and advances were made in food storage. Thus, there was an increasing reliance on cultivars for food with a decreased dependence on the collecting of wild foods. A sedentary life-style based on agriculture provided the resources for the development of the first civilisations in the Middle East, Mexico etc., which were characterised by the increasing exploitation of natural resources to service the development of major settlements, communications etc. This was achieved through further deforestation, and the management of water resources to expand agricultural production etc.

Turning to another important question, archaeologists have long considered which came first, animal rearing or plant cultivation. The evidence, by which such questions might be resolved, comes from the study of bones, plant remains (such as carbonised seeds) and other artefacts found in dated deposits. For both plants and animals there are often very considerable difficulties in separating the 'first' cultivated plants and 'first' domesticated animals from their human-exploited wild ancestors. Ladizinsky (1998) takes the view that some sort of animal husbandry probably started long before animals were domesticated, and even perhaps before plant cultivation. Thus, he envisages the possibility that wild animals were kept alive until they were needed for food. This view is shared by Bohrer (1972), who considered that, in pre-agricultural times, humans harvested wild plant material not only as food for themselves, but also collected and stored huge amounts of fodder for captive animals. Thus, opportunistic and planned capture of prey by hunter-gatherers gave way to the management of the animals, through ecosystem manipulations. In the early stages of domestication some 'tamed' animals might have been given the status of pets. But the process of domestication was only properly begun when breeding populations of animals were isolated out of range of their wild relatives. Such stock-raising led to genetic isolation from their wild ancestors, many of which were exterminated by human actions. Evans (1998) reviews the available evidence and concludes that sheep and goats were domesticated in the Fertile Crescent at about the same time as wheat and barley. Animal domestication contributed hugely to settled life by providing meat, milk, hides, horn, manure, and beasts of burden and traction (Harris, 1989, 1996).

Reaching half a billion: the spread of agriculture from 2000 BC to AD 1500

In this period, agriculture spread from its centres of origin. In the case of the Near East – the so-called Fertile Crescent – agriculture was adopted along the coasts of the Mediterranean to Spain and via the Danube into Central Europe, reaching Britain about 5,000 years ago. Other diffusion paths saw the establishment of agriculture eastwards to India and beyond, and southwards to North Africa. The speed of diffusion was quite slow. It has been estimated that it 'took about 2,500 years – the equivalent of roughly 100 human generations – for early farming to "travel" from Greece to Scandinavia' (Paine, 2002, 20). The enlarging horizons of agriculture were associated with local and regional deforestation and other major ecological changes.

Two models of the spread of farming practices have been proposed. These may have been passed between groups, in cultural change. Alternatively, farmers themselves may have migrated in the expansion of agriculture. A great deal of archaeological research has recently been carried out involving precise dating of seeds etc. with accelerator mass spectrometry (AMS), genetic investigations on the geography of human genetic traits and, more recently, the study of genetic markers on the human Y chromosome (offering insights into the history of genetic movement, as the Y chromosome passes in inheritance only along the male line). According to Paine (2002), evidence suggests that the spread of agriculture involved not only cultural diffusion, as knowledge of farming practices spread, but also the migration of farming peoples themselves (Ammerman, 2002; Paine, 2002). Further research is in progress on this complex area.

With regard to the use of tools, the earliest cultivation was carried out with digging sticks and hoes. Developments in metal working and animal husbandry provided the means for more efficient cultivation of the soil. The development of the ox-drawn heavy plough in the sixth century AD represented a very considerable technical advance. In the face of increasing populations, the construction of cultivation terraces on the hills began. Later, in the medieval period in Europe, there were further major developments that contributed to food production. The horse-drawn wheeled plough, with mouldboards that turned a furrow, proved to be a major invention, as it allowed the cultivation of heavy clay soils. A wider range of crop plants was also exploited, e.g. chickpeas, beans, lentils, and the harnessing of wind and waterpower increased the efficiency of grinding corn etc.

The introduction and spread of agriculture allowed settled life through the development of cities. Later, by means of long-range wars and military expansion, large empires were created (e.g. Persian, Roman, Arab, Mogul), and the first 'market economies' (e.g. Venice) were established, providing trade across the known world. All these developments provided opportunities for the transfer of alien species, both deliberately as ornamentals etc., and accidentally as contaminants of crop seed etc. (Mooney & Hobbs, 2000).

The first billion humans: from subsistence to commercial farming 1500–1825

In this period, agriculture in Europe began to change from subsistence farming in a feudal system of common fields, and later to commercial farming. New developments saw advances in seed sowing based on seed drilling (rather than scattering seeds). This change allowed much more efficient weed control. Also, the first steps in systematic plant breeding were made, and new crop rotations were introduced employing clover and rye grass. Following the expansion of Islam in the eighth century, new crops were introduced to Europe, e.g. figs, limes, lemons and oranges.

From *c.* 1500 onwards Europeans began to establish colonial territories overseas. In the mid-latitude regions of the globe, colonies were established in the Americas, Africa and Australia with the purpose of settlement, and the exploitation of the land, and its biological and mineral resources. In contrast, in areas of less congenial climatic conditions such as the tropics, colonies of exploitation were established. Here, there were limited opportunities for establishing complete European societies and tropical crops were grown under European management, often with imported slave labour (Christopher, 1984).

As Europeans colonised territories overseas, there was increased opportunity for transfer of crop plants (e.g. maize, tomatoes, potatoes etc. from the New World to the Old), accompanied by the unwitting introduction of weeds, diseases and alien species of animals and plants into new territories. Increasingly, the barriers that had kept separate the different biological realms on distant land masses were broken down. Whole floras and faunas that had evolved in geographic isolation were brought together by human activities (Mooney & Hobbs, 2000).

Populations rising to 2 billion: the expanding frontiers of agriculture 1825–1927

This period saw the rise of the industrial economies of Europe and North America – the second great transition in human history (Pontin, 1991, 265). With populations rising in the period to 2 billion, there was an increasing demand for agricultural produce of all kinds.

Following the early experiments in the fifteenth and sixteenth centuries on West African Islands (Madeira, Canary Islands etc.), very extensive plantation industries were later established in the West Indies, Central America, West and North Africa, and the Far East etc.). Commercial pastoralism developed in such countries as South Africa to produce wool, hides, skins, and such fashion items as ostrich feathers. Further intensification of stock rearing, for example, in North and South America, Australia, New Zealand and Africa was facilitated by technological developments (e.g. the invention of barbed wire, advances in the refrigeration and chilling meat). Spectacular landscape transformations took place in other grassland regions, e.g. USA, Russia, Canada, South Africa and Australia, with the introduction of extensive cereal production.

Throughout this period, and in spite of the introduction of quarantine regulations in many countries, the pace of the breakdown of geographical isolation between the species of different continental and island ecosystems accelerated. For instance, European grasses were exported to refurbish pastures overseas, and Australian trees were transferred to lands with low yielding forests (Christopher, 1984). In many parts of the world, for example the USA and Australia, acclimatisation societies were established (Lever, 1992). Species were deliberately brought from overseas to determine whether they offered potential as ornamentals, crops or medicinal plants, or were possible subjects for commercial ventures in forestry, land reclamation, pasturage etc. As we shall see later, botanic gardens played a key role in the transfer of plants. For example, rubber from the New World was introduced via Kew Gardens into Asia, forming the basis of an extensive plantation industry (Brockway, 1979).

Many ornamental species were also widely introduced through the activities of botanic gardens and zoos. In the absence of familiar 'targets' overseas, a range of animal species was introduced in distant lands to offer the sportsman (and woman) something to shoot, chase and kill (for example fox and rabbit in Australia, and many species of deer in New Zealand) (Elton, 1958). Some releases were eccentric in the extreme. The European starling, now widespread in the USA, was released in the 1890s by Eugene Schieffelin in Central Park, New York, with the aim of introducing all the birds mentioned in the plays of Shakespeare (Lever, 1992).

Concerning the models used by colonists to fashion their new territories abroad, Christopher (1984, 4) notes that in Africa: 'The settlers, as in the Americas and Australia, transformed the landscape by creating a new one as close to their image of Europe as possible.'

Thus, in the development of outposts of Empire many plants species were introduced from Europe. Darwin, writing in December 1835 (Darwin, 1986, 396) of his visit to Waimate, New Zealand, while on the voyage of the *Beagle* wrote with great feeling:

> After having passed over so many miles of an uninhabited useless country, the sudden appearance of an English farm-house, and its well-dressed fields, placed there as if by an enchanter's wand, was exceedingly pleasing . . . fine crops of barley and wheat in full ear were standing; and in another part, fields of potatoes and clover . . . there were large gardens, with every fruit and vegetable which England produces; and many belonging to a warmer clime. I may instance, asparagus, kidney beans, cucumbers, rhubarb, apples, pears, figs, peaches, apricots, grapes, olives, gooseberries, currants, hops, gorse for fences, and English oaks; also many different kinds of flowers . . . When I looked at this whole scene, I thought it admirable. It was not merely that England was vividly brought before my mind, yet, as evening drew to a close, the domestic sounds, the fields of corn, the distant country with its trees now appearing like pasture land, all might well be mistaken for some part of it.

The 'transplantation' of elements of the English landscape abroad was also attempted in the hill stations of India. These settlements allowed members of the administration to seek relief from summer heat on the plains by moving to the relative coolness of the mountains. In these 'holiday resorts' the design of houses and bungalows was reminiscent of England, and they were given names of English towns ('Richmond Hill', 'York Villa', 'Windsor Bungalow') or of its plants ('Oak Lodge', 'The Elms', 'Violet Dale'). There was active planting of a wide variety of imported 'familiar' species, as well as the rich variety of Himalayan flowering shrubs (Christopher, 1984, 156).

In a period of major industrial and urban growth in the nineteenth and early twentieth centuries (which resulted in an increase in environmental problems associated with the burning of fossil fuels, and industrial processes of every kind), agricultural advances of great importance were made with the development of artificial fertilisers. In addition, progress was made in the chemical control of pests and diseases (Evans, 1998). The first commercial use of biological control was successfully tested at this time. In 1889, cottony-cushion scale (*Icerya purchasi*) in Citrus was brought under control by releasing the Australian ladybird Vedalia beetle (*Rodolia cardinalis*); and in Australia the highly invasive prickly pear (*Opuntia*) was controlled by introducing the Argentinean insect *Cactoblastis cactorum*. Evans (1998) also stresses the role in increased food production of the new farmlands developed in Russia and the Americas. This period saw the beginnings of a clearer understanding of the evolution and genetics of domestication following the work of Darwin and Mendel.

From 2 to over 6.5 billion: the rise of industrial food production 1927 to the present day

In a period when mankind began to exploit nuclear energy, agricultural production was increased by further mechanisation, and intensive use of agrochemicals of all kinds, especially fertilisers and pesticides (Blaxter & Robertson, 1995). Arable and pasturelands were more thoroughly exploited, and further agricultural lands were developed, especially at the expense of tropical rainforest. In many parts of the world, arid lands were planted with exotic species in afforestation schemes (di Castri, 1989). There was also another burst of domestication. 'By 2000 years ago, an estimated 90% of the species presently cultivated on land had been domesticated' (Duarte, Marbá & Holmer, 2007, 382). In contrast, many 'new' domestications have taken place in the twentieth century with the rise of aquaculture of a range of organisms in both freshwater and marine environments.

In only 15 years, from 1961 to 1975, the population rose from 3 to 4 billion. There was an increasing internationalisation of research efforts in agriculture (Evans, 1998) that led, amongst other advances, to the so-called 'Green Revolution'. By using dwarf varieties and appropriate fertilisers there was a dramatic increase in food production in such countries as the Philippines. The human population increased from 4 to 5 billion in only 10 years, 1977–1986, and, in a further 12-year period – 1986–1998/9 – the world's human population increased by an additional billion. This period saw attempts, in the developed world, to devise ever more efficient agricultural systems using carefully formulated combinations of pesticides, fertilisers, irrigation water and biological control (Evans, 1998). There was also increasing globalisation of markets with the continued rise of multinational companies. Throughout the developed and developing worlds the technological and organisational changes associated with the drive towards highly 'efficient' productive mechanised industrial agriculture produced higher food yields, but these have come at a cost. Ecosystems throughout the world are increasingly threatened. For example, in Britain there have been major losses of downland, wetlands, moorland etc. with significantly reduced populations of farmland birds etc. (Shoard, 1980, 1997). Tattersall and Manley (2003, x) also note 'the polarisation of agriculture into an arable east and a pastoral west, a doubling of the number of sheep on the land, larger field sizes, and the removal of hundreds of thousands of miles of hedges'. In addition, the 'quality' of the food produced by 'mechanised' agriculture has become, for some, a major issue. These concerns have resulted in an increase in 'organic' food production, and the search for ways of sustaining wildlife populations within modern agricultural landscapes.

What is the present extent of human-modified ecosystems?

It is clear from a mountain vantage point or from an aircraft crossing the Alps, or the Sahara, that not all the Earth is equally susceptible to human occupation and exploitation. Nicholson (1970) draws attention to the fact that water covers about two-thirds of the planet, only one-third is 'solid'. Moreover, about 10% of the land surface is ice, and another 10% is covered by the major deserts of the world. To this unusable 20% must be added a further 10% of inhospitable mountainous and rocky areas that carry no soil or vegetation.

It has proved possible by examining aerial and ground surveys, and more recently by satellite imaging, to estimate with high accuracy the extent to which the Earth has been modified by human activities, although it is important to acknowledge the difficulty in defining categories, for example for 'degraded', 'eroded' land (Grove & Rackham, 2001).

Vitousek *et al.* (1997) estimated that perhaps as much as 50% of the Earth's land surface has been transformed or degraded, and that humans utilise *c.* 39–50% of the biological productivity of the planet. They emphasise:

Land transformation encompasses a wide variety of activities that vary substantially in their intensity and consequences. At one extreme, 10 to 15% of the Earth's land surface is occupied with row crop agriculture or by urban-industrial areas, and another 6 to 8% has been converted to pastureland; these systems are wholly changed by human activity. At the other extreme, every terrestrial ecosystem is affected by increased atmospheric carbon dioxide (CO_2), and most ecosystems have a history of hunting and other low-intensity resource extraction. Between these extremes lie grasslands and semiarid ecosystems that are grazed (and sometimes degraded) by domestic animals, and forests and woodlands from which wood products have been harvested; together these represent the majority of Earth's vegetated surface.

The effects of domestic animals should not be underestimated. From 1950 to 1997, the 'number of cattle, sheep, goats, camels, horse and pigs in the world increased . . . from 2.3 to more than 4 .0 billion', and 'poultry production grew from 3 to more than 11 billion over the same period' (FAO estimates quoted in Pond & Pond, 2002, 37).

Natural forests and other pristine habitats are often partially destroyed or damaged by storms, natural fires etc., but, in the course of time, such areas have the potential to regenerate. In contrast, human activities may so modify or destroy natural ecosystems that regeneration is impossible. Sometimes the scars of human exploitation remain in the landscape for centuries, but it is important to acknowledge the power of recolonisation by plants: much of the human landscape that is not deliberately planted or managed is covered by secondary vegetation that is different in species composition from the original natural vegetation.

The scale of alteration of natural habitats can be illustrated by the increasing human dominance of forested ecosystems, now facilitated by the use of chain saws, bulldozers and specialist heavy machinery for transporting timber, and the cultivation and management of trees as a crop.

Of the *c.* 3.54 billion hectares of forested lands (about a third of the Earth's land surface) about 150 million hectares are plantations and another 500 million hectares are classified as actively managed for goods and services. However, this is a considerable underestimate of the area of forest affected (and often dominated) by human activity as it excludes large areas affected by indigenous gardening, hunting and gathering, and indirect management such as changed fire regimes . . . Earth's forested estate has shrunk by about a third (2 billion ha) since the rise of agriculture-based civilisations and continues to be eroded at dramatic rates. Harvesting for wood and fuel is currently about 5 billion m^3 annually and is increasing by about 1.5% (7 million m^3) per year. In addition, some 10 million hectares of new land, cleared largely from forest, is needed each year to support the increase in world population at current levels of nutritional and agricultural yields. Estimates suggest that forest clearing averaged over 13 million hectares per year from 1980 to 1995, which was only partly compensated for by about 1.3 million hectares per year of new plantations. (Noble & Dirzo, 1997)

In many regions there is great pressure to increase agricultural production, and, particularly in tropical and sub-tropical areas, yet more forest is being removed to make way for crops. Using satellite imagery, Miettinen, Langer and Siegert (2007) have mapped 600,000 ha of forests in Borneo burnt in preparation for agriculture.

In many parts of the world the vegetation is strongly influenced by the deliberate use of fire by humans. Thus, burning has been used as a tool for deforestation, and periodic controlled burning has also been used in the traditional management of grassland, rangelands and heathlands. In addition, fire has also been used as an important weapon in warfare. In denying resources to the enemy, the motto has always been '*incendit et vastavit*' – burn and lay waste. Accidental fires have also profoundly influenced ecosystems. In his classic book on the role of fire in America, Pyne (1982) cites many examples of devastating individual fires. For instance, in 1942, huge fires broke out during the construction of the Alaskan highway: a single blaze consumed an area of forest the size of Ireland. However, this blaze was dwarfed in 1998 by fires in the Russian Far East (Gawthrop, 1999), which destroyed forest greater than the area of Switzerland. Not only are very large individual fires important in anthropogenic ecosystem change, many areas are repeatedly damaged by fire, e.g. in areas of 'Mediterranean' climate in Europe, South Africa, California and Australia. For example, in 2007 in the hottest summer for 50 years, catastrophic fires occurred in Greece. More than 10% of the forests were burnt (Smith, 2007). It is claimed that some fires were the result of arson. Many nature reserves were affected, including Mount Taygetos, which had just begun to recover from devastating fires in 1998. Some endangered species of animals and plants have been affected. Two major concerns are being voiced. If there is very heavy rainfall in the winter, the environmental consequences of the fires could be compounded by erosion. In addition, there are very strong political pressures to allow many coastal-forested areas to be developed as tourist resorts. In the long term, forests recover from fire, but, given the present devastated state of these areas, will it prove more difficult politically to resist development pressures?

Humans also influence ecosystems through fire suppression policies, which are directed towards the rapid detection and speedy extinguishing of fires, be they natural from lightning strikes, accidental or deliberately started by arsonists. In such situations huge fuel loads build up in ecosystems, and spectacular fires may then occur, e.g. at Yellowstone National Park in 1988 (see Chapter 15).

In the transformation of the land to agriculture, not only forests have been cleared, but also wetlands have been drained. Coastal areas have also been reclaimed for agriculture by the construction of sea walls and other defences. In addition, in many parts of the world, rivers have been diverted and dams built to provide water for domestic purposes, irrigation and power generation. Overall, it has been estimated that 'more than half the runoff water that is fresh and reasonably accessible' is put to human use (Vitousek *et al.*, 1997). And about two-thirds of the world's rivers are regulated in their flow, with about 36,000 dams worldwide (Humborg *et al.*, 1997). Only 2% of rivers in the USA run naturally. Many rivers are used so intensively that very little water reaches the sea, e.g. the Colorado, Ganges, Nile (Abramovitz, 1996). This has resulted from what Gleick (2003) calls the 'hard path' approach to water management, with an infrastructure of dams, levees, flood defences, aqueducts, canals, pipelines, water and sewage treatment works, and defences such as walls, dykes and barriers etc. erected against flooding by rivers and the sea. In addition, huge quantities of water – both renewable and fossil – are drawn from underground (Gornitz, Rosenzweig & Hillel, 1997). Such management has a profound effect on freshwater aquatic and wetland systems.

Human geomorphic activities

Human changes to the surface of the Earth have occurred on such a massive scale that geographers have made a special study of 'human landforms' (Gill, 1996). Some of these are agricultural in origin and fashioned by human toil, such as the spectacular terracing to be found in Asia, South America and the Mediterranean. Irrigation schemes on a vast scale have produced very striking effects, e.g. there has been a huge reduction in the water area of the Aral Sea in Russia. Dramatic landscape changes have also been wrought in the development of cities, the building of motorways, canals, railroads, airports, docks, harbours, marinas etc.

Now, increasingly, massive purpose-built earthmoving, drilling and mining machines are transforming the Earth's surface. Exploitation of underground resources – water, gas, oil – can also lead to landscape changes through subsidence. The exploitation of mineral resources through the mining and processing of coal, ores, stone, gravel, sand etc. has resulted in massive open-cast and deep mines, and pits harvesting gravel, clay, sand etc. Sometimes, the landscape is scarred with huge spoil heaps, but, increasingly, an attempt is made to restore the original topography sometimes by landfill with household rubbish, or the wastes from industrial or energy plants, capped with topsoil. However, there is growing concern about the policy of 'filling holes' with toxic and other wastes, as discarded materials are potentially recyclable.

The rise of concern about the environment

In the remainder of this chapter we consider, briefly, the rise of environmentalism, a movement that subjects our increasing exploitation of the Earth's resources to the most critical scrutiny, highlighting 'environmentally damaging' activities and promoting the view that human use of the biosphere must become more sustainable. Evidence that *particular* human activities had dramatic effects on the environment is available from the writings of the earliest civilisations through to the present day. What was new, in the post-1945 period, was the emergence of a *comprehensive* overview based on a study of the cumulative effects of different aspects of human activities. These concerns were reflected in the establishment of many international organisations, such as Greenpeace (begun under another name in Vancouver in 1969), and the founding of political parties – for example, Die Grünen (The Greens) in Germany. The international dimension of many environmental problems became increasingly clear and after effective campaigning by environmentalists, as we shall see below, legislation was enacted to counter many major ecological problems.

It is helpful at this point to examine some of the major concerns of environmentalists.

Conservation through protection

Environmentalism may have its origins in the late eighteenth century ideas developed in the Romantic Movement. Edmund Burke, Thomas Carlyle, Coleridge, Shelley and Wordsworth were key figures in Britain, while in the USA, Alexander Wilson, Emerson and Thoreau were highly influential. These thinkers rejected the materialism of the age. They spoke against the increasing domination of nature by humans. George Perkins Marsh is another figure of great significance to the early conservation movement, through his book *Man and Nature: or, Physical Geography as Modified by Human Action* (1864). The first fruits of growing environmental awareness came in 1872, when the majestic lands at Yellowstone were designated as the first national park in the USA.

The aim of those championing Yellowstone and other parks that were established in many countries, was, at first, essentially preservationist. Such areas were designated to safeguard wilderness areas containing sublime scenery. In other areas, there was a concern to protect wildlife, particularly game. In all these concerns, it has been argued that parks were devised as playgrounds for the wealthy, e.g. the game parks of Africa and India, where gentlemen hunted for sport (see Chapter 15).

From the end of the nineteenth century, new organisations arose with different environmental agendas. Nature reserves were established to protect threatened species, and, in Europe and across the developed world, there were pressures to safeguard areas of countryside and allow increased access to those living in urban areas. With improving transport systems, tourism increased significantly in the twentieth century, especially in coastal and mountain areas containing significant biodiversity and renowned landscapes (Lichtenberger, 1988). Thus, many scenic and biodiverse areas have been significantly altered or destroyed by the building of hotels and facilities for tourism and summer and winter sports. Extreme sports are becoming popular, with the increase in motorcycle and speedboat racing, river rafting, abseiling, off-piste skiing etc. The affluent, displaced from many of their former haunts by mass tourism, are seeking out destinations in remote parts of the world, formerly only accessible to expeditions. Many overseas visits are now sold under the label of ecotourism, where some benefit to conservation or local development at or near the area is promised. Many of those supporting and managing areas of high biodiversity, and the wider countryside outside reserves, are motivated by a very strong desire to protect cherished landscapes and wildlife. However, they are faced with a difficult task. Given the recent increases in the use of the countryside for leisure pursuits, as well as mass tourism to more distant destinations, many conservationists are pointing to the dangers of large numbers of visitors having access to ecologically fragile sites, particularly to national parks, reserves and scenic areas of high conservation significance. For while many cherish the landscape they visit, others destroy the tranquillity and beauty of such areas, and disturb or destroy the wildlife of the area by arson, illegal tipping of wastes and rubbish, illegal hunting, fouling by dogs, using off-road vehicles etc.

The wise use of resources

In the late nineteenth century there was increasing evidence that natural products were being over-harvested, e.g. timber extraction, whaling, hunting for big game, fur-bearing animals and birds to supply the fashion trade etc. Reflecting on the colonial period, Christopher (1984, 88) concludes that the 'commercial exploitation of Africa had the elements of a robber economy'.

It was against this background that at the end of the nineteenth century a second and different strand of conservation began to emerge in the USA. The so-called Progressive Era Conservation concept was particularly associated with the name of Pinchot, who was in charge of the Forest Service. Meine (2001, 888) summarises the main elements of the approach as follows:

> The forests were not regarded as 'reserves' to be 'locked up', but 'as lands to be worked' for the greatest good of the greatest number for the longest time... In practice, this meant that the forests were to be managed by a trained, professional workforce; that scientific principles were to guide the efficient and sustained exploitation and processing of forest resources; and that the wealth derived from the forests was to be equitably distributed for the common good... the core of utilitarian conservation was the concept of sustained yield.

The ideas have been extended well beyond the use and management of forest. Thus, Meine notes that the famous conservationist Aldo Leopold later (1933) observed, 'under the Progressive conservation banner wild life, forests, ranges and water power were conceived... to be renewable organic resources, which might last for ever if they were harvested scientifically, and not faster than they reproduced' (Meine, 2001, 889).

The ideas of the progressive era have been highly influential, not only where resources are under state control but also where they are in private ownership. However, over-harvesting continues in many parts of the world. Taking the figures assembled by the Food and Agricultural Commission in 1994, 22% of recognised marine fisheries were over-exploited or already depleted and 44% more were at their limit of exploitation (Vitousek *et al.*, 1997). Typical forestry practice in the tropics results in the over-harvesting of mahogany and other prime tropical woods. In many areas, only damaged 'island' fragments remain in a sea of agricultural land.

In the 1980s, there was a renewed concern to 'search for sustainable use' of ecosystems (Milner-Gulland & Mace, 1998). The concept of sustainable development was formulated. It was adopted by the World Conservation Strategy (IUCN/UNEP/WWF) in 1980 and was accepted by many conservationists. 'By referring to two concepts that are universally popular, it became a phrase that few could disagree with – perhaps we could have our cake and eat it?' (Milner-Gulland & Mace 1998, 7). Meffe and Carroll (1994, 564) write of the 'need for humans to live on the income from nature's capital rather than on the capital itself'. Considering the wise use of resources, Milner-Gulland and Mace (1998, 349) consider: 'the ideal of sustainable development is that the exploitation of these resources can be managed at a level that has relatively little ecological impact, whilst having great social and economic benefit'. In Chapter 21 we consider how far the notion of sustainable development has been helpful.

Environmental concerns about pollution

Human communities produce large quantities of wastes and detritus. Gases and particulate matter enter the air, water and soil, together with volumes of sewage, household rubbish, farmland effluents etc. Also, there are enormous amounts of industrial and commercial gases and wastes associated with the extraction, refining and use of metals, oil, gas, coal, nuclear power etc. and their use in transportation, generating gas, electricity and nuclear power, and in manufacturing a wide range of consumer goods.

Many pollutants are naturally occurring substances. For instance, plants may be exposed to heavy metals arising from naturally exposed metal-bearing veins of rocks. However, human activities have produced 'higher than normal quantities' of certain atmospheric gases, smoke, organic compounds, compounds of nitrogen, phosphorus etc., heavy metals from mining, smelting etc. with major impacts on natural ecosystems. In addition, chemical species 'new' to the biosphere have been devised, e.g. pesticides.

During the 1950s and 1960s, concerns about environmental pollution were heightened by the discovery of major pollution in the Great Lakes Region of North America and mercury poisoning at Minamata Bay in Japan. The damage caused by the oil spills from stricken oil tankers or platforms – the Torrey Canyon off the British coast and the Union Oil Company off Santa Barbara – were given worldwide publicity. In the same period, Rachael Carson (1962) published her famous book *Silent Spring* that highlighted the dangers posed by environmental pollution. The environmental movements in the developed world began to mobilise public opinion about the dangers of pollution, leading eventually to political action at international, state and local levels.

Many steps have been taken to measure, monitor and control pollution in the developed world, and these have been reviewed in a number of sources (e.g. Holdgate, 1979; Bell & Treshow, 2002). For instance, advances in our understanding of anthropogenic atmospheric pollution have come from measurements made from instruments carried on satellites and from ground stations and ships at sea. It is abundantly clear that pollution is a global phenomenon and that emissions produced in one country affect populations elsewhere. As a consequence, there have been major efforts to control pollution in the developed world through legislation.

In Britain, the first control laws were enacted in 1863 following agitation to control pollution from coal burning at alkali works (Ashby & Anderson, 1981). Many more pieces of legislation followed in Europe and North America. From the 1960s onwards, there was mounting pressure for the effective control of pollution and many international agreements were negotiated – the control of dumping at sea (1971), regulation of discharges via rivers (1974), bans on the release of substances depleting the ozone layer (1987) etc. In the developed world, the control of pollution is covered by many pieces of legislation. In general, the pollution burden has been reduced rather than eliminated, but it is important to note that, in many areas of the developing world, almost unrestrained pollution occurs.

As we shall see in later chapters, plants and animals have responded to these new or intensified anthropogenic selective forces, including pollution. Here we examine some of the different types of environmental pollution.

Atmospheric pollution

Air is polluted by many compounds, for example, ozone (O_3), sulphur dioxide (SO_2), nitrogen oxides including nitrogen dioxide (NO_2), hydrogen fluoride (HF), peroxyacetyl nitrate (PAN), non-methane hydrocarbons (NMHC) or volatile organic compounds (VOC), metals such as lead (Pb), cadmium (Cd) and mercury (Hg) and a wide range of organic compounds and particulates (Frangmeier *et al.*, 2002). Pollutants may act individually and in combination, and may also react to form new compounds, causing adverse or toxic effects to humans, plants and animals. Atmospheric pollution is of the utmost complexity, as ecosystems may be exposed to mixtures of gases and particulate matter, and pollutants may act simultaneously or sequentially. There may also be seasonal effects. Plants weakened by pollutants may be more susceptible to frost damage, and disease organisms may obtain entry via lesions caused by pollutants.

In the 1970s, environmentalists in Europe and North America called attention to the phenomenon of 'acid rain', acting downwind of industrial areas, power stations, cities and urban centres. Investigations revealed that nitrogen oxides are formed from the exhaust gases emitted by traffic, and arise from many industrial sources. Acid rain was formed when these oxides of nitrogen and sulphur dioxide produced by fuel combustion were converted in the atmosphere into nitric acid (HNO_3) and sulphuric acid (H_2SO_4) and related compounds that return to earth in rain. Also, it was discovered that pollution by 'dry' deposition of acid-forming substances such as ammonium sulphate ($NH_4)_2SO_4$) was also significant in some areas.

Acid rain is carried in the atmosphere to remote areas with few local pollution sources of their own, sometimes from one country to another (e.g. USA to Canada, and industrial Europe to Scandinavia), causing acidification of lakes, and damage or loss of fish stocks. The problem was most acute where underlying geology yielded lake waters that were naturally acid. It was also claimed that acid rain damaged forests, causing crown thinning and death, as for instance in those forests downwind of areas in Central Europe where high sulphur containing coal is used as fuel.

However, while it has been shown in experiments and observations that many pollutants damage trees, and may make trees more susceptible to frost, diseases etc., it cannot simply be deduced that the increased acidity of the rain alone is the direct cause of any tree damage. Acid rain may interact with other factors. For instance, acid rain may lower the pH of the soil, and adverse plant responses may result from the mobilisation of aluminium and magnesium compounds. Acidification of groundwaters and soil may also increase the solubility of toxic compounds of cadmium, copper etc. Field studies have revealed the death or poor performance of trees in some sites. Relating this damage to acid rain alone, in the absence of experimentation, may be premature. Other explanations must be investigated in particular cases, e.g. various tree diseases may cause damage to foliage/or roots. Also, insect damage and the effect of drought etc. should be considered. Moreover, the demography of tree populations may be an issue – a particular forest could be dominated by old senescing trees.

In an attempt to limit the damaging effects of acid rain, legal controls have been negotiated on the emissions of sulphur dioxide in the USA and Europe, but emissions of nitrogen oxides continue. Unregulated high levels of air pollution occur in many parts of the world e.g. the Asia-Pacific area (Shah *et al.*, 2000).

Another major component of air pollution is the gas ozone (O_3) that is formed by the reaction of pollutant nitrogen oxides and hydrocarbons. This gas is toxic to humans, animals and plants, and because it has a long lifetime in the atmosphere (1–2 weeks in summer, 1–2 months in winter), it can be transported from one continent to another all year long. The concentration of surface ozone was *c.* 15–25 parts per billion by volume (ppbv) in 1860. The current figure may exceed 40–50 ppbv. There is evidence that levels of pollution are already high enough in many areas to jeopardise agriculture and natural ecosystems (Akimoto, 2003).

The greenhouse effect and global climate change

Reflecting the concerns of environmentalists on the effect of human activities on the biosphere, this issue has come to be seen as one of the biggest threats to the future of mankind and Earth's biodiversity. Here, preliminary comments are presented about this potentially catastrophic problem. Chapters 18–20 explore our current understanding of climate change in some detail.

There is increasing evidence that atmospheric pollution is altering the global climate (IPCC, 2007a-d). This effect comes not from our generation of heat, but by our alteration of the flows of energy in the atmosphere through changes to its composition. Heat is received on Earth from the Sun. Some of this energy is reflected back from clouds and the Earth's surface, but some is trapped by the so-called greenhouse gases (water vapour, carbon dioxide, methane and nitrous oxide) and warms the planet. The rest escapes to space.

Since the beginning of the industrial revolution there have been dramatic increases in greenhouse gases resulting from the burning of fossil fuels, deforestation (accounting for *c.* 20% of greenhouse gas emissions) and other human activities. These gases have long lifetimes and produce an accelerating greenhouse effect in the atmosphere by their accumulation over decades. For example, carbon dioxide has increased from 280 parts per million by volume (ppmv) to more than 379 ppmv today, with half that increase arising since 1965. As we see in later chapters, there have also been dramatic increases in other greenhouse gases. While some date the rise of greenhouse gases to the period of the industrial revolution, a new hypothesis suggests that Stone Age farmers' activities, 8,000 years ago, produced significant amounts of carbon dioxide from the clearing of forests by burning and from about 5,000 years ago elevated quantities of methane were produced through the irrigation of increasing areas of land in the cultivation of rice (Ruddiman, 2005).

Climate has varied in the past, but looking to the future, the magnitude of the change driven by anthropogenically produced greenhouse gases will depend upon whether emissions are reduced from current levels. Estimates of global warming reveal that, if no action is taken to control emissions and 'business as usual' continues, average global surface temperatures could increase by 4 °C (likely range 2.4 to 6.4 °C) by 2100. Thus, on present evidence the rate of human-induced climate change is projected to increase well beyond the effects of naturally occurring processes, certainly those prevailing over the last 650,000 years. The likely outcome is more extreme climatic events, with frequent heat waves, changing rainfall patterns, declining water balances, droughts, extreme precipitation events, and related impacts – such as wild fires, heat stress, vegetation changes and sea level rise (Hannah *et al.*, 2002; Akimoto, 2003; IPCC, 2007a–d). It is also possible that the frequency of tropical cyclones and other storms might increase (Webster *et al.*, 2005). Unless greenhouse gas emissions are stabilised, and thereafter reduced, it is predicted that there will be catastrophic effects on human societies and the Earth's biodiversity.

While some sceptics remain unconvinced about the seriousness of the challenge posed by climate change, recent IPCC reports have revealed a growing consensus amongst the scientific community about the seriousness of the issues. Recently, interest in environmental issues has markedly increased. The media are full of pieces about human impacts on the environment; with climate change a main issue of the day. In many parts of the world, national and local politicians are attempting to translate these concerns into practical action. But, as yet, it has proved impossible to reach full binding international agreements on the control of emissions of greenhouse gases that are contributing to climate change. It remains to be seen whether effective timely action will be taken.

Essentially the control of anthropogenic greenhouse gases involves reduction in pollutants of the atmosphere. While for the purposes of travel by air, 'air space' is seen as a national possession – in reality the atmosphere is a shared resource held in common by all humanity living in every country on the planet. Pollutants produced locally are transported and mixed in the atmosphere globally. While the capacity of the atmosphere to be a sink for pollutants would appear to be infinite, this is not so.

While the principles of sustainable development can be applied to resources held locally by private individuals or regionally under state control, there are particular, often intractable, difficulties in the case of resources 'held in common' – the atmosphere, fisheries, the oceans outside territorial limits etc.

The source of these difficulties was illuminated by Hardin (1968) who published his influential paper 'The Tragedy of the Commons' with fisheries, water use etc. particularly in mind. In the words of Phillips and Mighall (2000, 238) he considered:

> ... if humans have free access to common areas or resources these will be ruined because, more often than not, individuals ... maximise their own benefit ... without considering any potential damage that may result from misuse ... most people will place their own short-term needs before the long-term preservation of that resource or the interests of other people, and that most users will exploit the resource as much as possible because they think that if they do not somebody else will.

Many have considered how resources held in common might be controlled and regulated. With regard to fisheries, water resources etc. it has been proposed that those using the resources should enter into binding agreements to share and sustain the resources. However, obtaining and enforcing 'agreements' is often very difficult, but not impossible (Dietz, Ostrom & Stern, 2003).

A much bigger challenge is offered by the control of atmospheric pollution, especially of greenhouse gases. It is cheaper for manufacturers, local populations etc. to dispose of their pollution directly into the atmosphere rather than remove the pollutants at source or minimise the level of emissions. However, essentially, the Earth's atmosphere is a resource held in common by all the nations on Earth and the effective control of greenhouse emissions that are predicted to lead to global climate change can only come through effective action following international agreement. These issues are considered in greater detail in later chapters.

Heavy metal pollution

Pollution of the environment by heavy metals has a long history, but because many mining areas have continued to be exploited to the present day, it is often difficult to determine the extent of ancient mining and quarrying (Scarre, 1996). However, archaeological studies have revealed prehistoric copper mines in Southern Europe, Wales, Argentina etc. (Sherratt, 1996). Heavy metal pollution is evident in different areas, namely where the ores are mined, smelted and processed, where products are used in the human environment and, finally, where they are discarded.

Compounds of a range of metals (cadmium, copper, cobalt, zinc, manganese, mercury, lead, nickel, chromium) and the metalloids arsenic and selenium are major pollutants in human-influenced ecosystems. However, it is important to note that some (for instance, copper and zinc) are micronutrients essential for normal plant growth. It is only when elevated levels of these metal ions are present in the environment that they become toxic pollutants.

Nitrogen (N) compounds

These are released as pollutants from human settlements in the form of sewage, animal wastes and remains. They are also produced in deliberate and accidental burning of biomass, fuels etc. Agriculture is a major source of nitrogen pollution. To secure the increased agricultural yields associated with mechanised farming, large quantities of nitrogen-containing fertilisers are applied to farmland. While some of the applied nitrogen is taken up by crops and harvested, some is leached out from the soil, enters streams and rivers and is carried into lakes and wetlands and into near-shore coastal systems with profound effects on these ecosystems (Scavia & Bricker, 2006; Schindler & Donahue, 2006). Overall, human activities have more than doubled the pre-industrial rate of supply of nitrogen compounds to terrestrial ecosystems. By 2050, it is estimated that nitrogen input could reach as high as three to four time the pre-industrial rate.

Nitrogen-containing compounds also impact natural and semi-natural ecosystems through air pollution. Ammonia is volatilised in great quantities from the wastes created by the intensive agricultural husbandry of farm animals. Atmospheric nitrogen deposition from all sources in non-forested areas of Europe and North America was of the order of 1–3 kg N ha^{-1} yr^{-1} in early 1900, and can now be as high as 20–60 kg N ha^{-1} yr^{-1} (Bell & Treshow, 2002, 201).

Phosphate (P) compounds

These are also very widely used in fertilisers. It has been estimated that the current use of such compounds globally doubles the levels found in natural ecosystems, and phosphate use could double again by 2050. Phosphate compounds are less soluble that those of nitrogen. Nonetheless, widespread dispersal of these compounds occurs into terrestrial and aquatic ecosystems through leaching and water pollution, spray drift, and atmospheric transport of various pollutants including dust.

Pollution by N and P compounds can be short range, but it can also be carried far and wide to have profound effects on many terrestrial ecosystems (e.g. forests, bogs, heathland etc.), especially those that have soils naturally low in N and P compounds. For example, in a study of grasslands in the UK, long-term increase in nitrogen deposition is associated with loss of species richness (Stevens *et al.*, 2004). There are major changes too in aquatic ecosystems, where increases in anthropogenic N and P contribute to increasing frequency of damaging algal blooms, the toxins from which may kill many aquatic organisms.

Pesticides

Pests and diseases cause enormous losses in food production, with losses as much as 40% globally and 70% regionally. Since the Second Word War, thousands of synthetic chemical pesticides have been developed including insecticides (30%), fungicides (19%), herbicides (45%), with other formulations acting as nematicides and arachnicides (6%). Currently, more than 5 million tons of pesticides are used worldwide (Matson *et al.*, 1997). They are designed as 'biocides' to kill certain target plants or animals. However, even when they are used with care, they impact on other non-target plants and animals through spray drift and run-off from treated areas. Pesticides differ in their toxicity to humans and wildlife, but they may have drastic effects, especially in the case of persistent pesticides, which may enter food chains and be concentrated in fish and birds in the higher trophic levels (Holdgate, 1979).

Organic pollutants from various sources

Extensive monitoring of the river basins of the European Union has revealed the presence of many organic compounds originating from a wide variety of human activities, e.g. phenolic compounds, dioxins, synthetic oestrogens and androgens. Such chemicals have direct and indirect effects on plants, e.g. some toxicants are known to have mutagenic effects, while ecosystem disturbance may influence plants indirectly through the endocrine disruption in animals etc. (Brack *et al.*, 2007).

Concerns about the loss of biodiversity

From the 1950s onwards, there has been mounting concern about the habitat loss and threats to biodiversity, and legislation to create additional national parks and wilderness areas, establish nature reserves, protect endangered species etc. was enacted in many countries. In the 1970s, the first moves were made to make a scientific assessment of the number of extinct and endangered species of plants and animals, and to devise means of ensuring the survival of such species. In 1975, recognising the damaging trends in the utilisation of wild species of animals and plants, the World Conservation Union secured international agreement on the Convention on Trade in Endangered Species (CITES), which provided a legal framework to protect exotic species through restricting trade.

Confronting the adverse effects of introduced organisms

In the transformation of the Earth, plants, domesticated and wild animals, crop and wild plants, and ornamentals have been deliberately carried to different parts of the world. Other animals and plants, disease organisms and pests have been accidentally transported from their original native habitats to other continents and to islands. Thus, organisms that evolved in isolation in different parts of the world have been introduced to new territories. Some, but not all, introductions have become invasive and have had profound effects on native ecosystems. We shall be considering introduced organisms in detail in later chapters. To emphasise the importance of introductions, two examples may be cited here. Introduced plant species have taken over most of the woodland savannahs and grasslands of California, and coastal habitats contain many introduced exotic species (Bossard, Randall & Hoshovsky, 2000). A virus disease of ruminants, rindepest, which was introduced to Eritrea in north-east Africa in 1887, illustrates the devastating effects that may come from the introduction of a disease organism, for as it moved south to the Cape in 1900, over 90% of the African cattle population died. Antelope, giraffe, buffalo and wildebeest were also infected, leading to starvation of native carnivores and profound changes to the ecosystems and human communities of southern Africa (Bell, 1987; MacKenzie, 1987).

Soil erosion and salination

Traditional management in arable areas results in exposed soil being subjected to wind and rain. Likewise, animal trampling and over-grazing may result in the exposure of grassland soils to the elements. In such situations, serious soil erosion may occur, with the topsoil being blown or washed away (e.g. over-grazing by sheep in montane grasslands of New Zealand has resulted in serious erosion with gulley and canyon formation (McCaskill, 1973). Eroded material may be deposited locally, but in many instances it is carried downstream in rivers and deposited far away, modifying the geomorphology of deltas, estuaries or coastlines.

In classical times, deforestation and agricultural development gave rise to widespread soil erosion, being responsible, for instance, for the abandonment of villages in Jordan as early as 6000 BC (Pontin, 1993). Reflecting on the Greek landscape a century before the birth of Christ, Plato noted: 'What now remains compared with what then existed is like the skeleton of a sick man, all the fat and soft earth having wasted away, and only the bare framework of the land being left' (quoted by Pontin, 1993, 76). In the development of agriculture in China, erosion was a serious problem in many areas. For example, the Yellow River was so named because of the abundant waterborne sediments it contained that were released following deforestation and erosion from farmland. Spectacular soil erosion occurred in the 1930s, when the Great Plains region of the USA was affected by severe droughts. Agriculture had destroyed the natural grass cover of the region and soils were blown away in spectacular dust storms. Soil was deposited as far away as Chicago and was detected on ships out in the Atlantic.

The growing of crops in seasonally hot dry areas is limited by climatic constraints. This was clear to early agriculturalists in the Near East. To grow crops through the dry season required the storage of water and irrigation. However, after initial success in this region of low rainfall and exceedingly high summer temperatures (as high as 40 °C), crop yields declined as evaporation resulted in accumulation of salt (often containing also boron, selenium and other toxic compounds) at the surface of the soil (salinisation). Wheat proved to be salt sensitive and was replaced by the more salt-tolerant barley. Eventually, the land became useless for cropping and new land had to be cultivated, where it could be found. Salinisation was one of the major factors that contributed to the decline of ancient civilisations in Mesopotamia and other sites in the Near East (Jacobsen & Adams, 1958; Pontin, 1993). In studying a number of case histories in detail, Diamond (2005) has analysed the factors contributing to the decline and fall of a number of societies. He concludes that environmental crises, including issues of soil fertility, have contributed significantly to disintegration of some societies. Given the challenges of higher temperatures and altered rainfall patterns that are predicted to lie ahead under climate change, it is important to note that salinisation remains a very major issue for farmers in many parts of the world, including the USA, India, Turkey, Australia, Russia etc.

Maintaining agricultural production in the longer term has proved a major challenge to all farmers. Some of the strategies currently important were discovered in classical times. For example, the works of Roman authors such as Cato and Columella contained first-hand advice on the importance of animal and green manures, and the benefits of crop rotations involving fallow periods (Seymour & Girardet, 1986). Today, different strategies are available from which to devise a management regime to minimise soil erosion and salinisation, including fallow periods, planting areas with trees, green manuring, contour ploughing, mosaics of different cropping, drip irrigation, cultivation of crops with minimal ploughing and direct drilling of stubbles etc. However, looking ahead to the likely effects of climate change, the age-old problems of maintaining soil fertility, and controlling erosion and salinisation, will be a major challenge with changing temperature and rainfall patterns and a greater frequency of extreme weather events.

The inexorable rise of human populations

It is self-evident that since humans first evolved there has been a relentless rise of their populations resulting in the transformation of the Earth's ecosystems (Evans, 1998; Cohen, 1995, 2003). The environmental movement has proposed that there is a clear relationship between population size and environmental impact, and such a linkage is widely acknowledged. However, this proposition is difficult to examine statistically, as many other interacting factors are involved. Looking at the figures for population growth, it is important to emphasise once again a point mentioned earlier: all demographic statistics about the past, present and future are estimates.

Summarising the trends, Cohen (2003, 1172) concludes:

> Earth's population grew about 10-fold from 600 million people in 1700 to 6.3 billion in 2003 . . . It took from the beginning of time until about 1927 to put the first 2 billion people on the planet; less than 50 years to add the next 2 billion people (by 1974); and just 25 years to add the next 2 billion (by 1999). The population doubled in the most recent 40 years. Never before the second half of the 20th century has any person lived through a doubling of global population. Now some have lived through a tripling.

Concerning the relationship between population size and the available food is a matter of great debate, to which Malthus (1766–1834) made such important contributions. More recently, Boserup (1965) has provided valuable insights. In essence, the debate ponders the precise relationship between the supply of food and population size – the extent to which increase in food supply *promotes* population increase, and, in contrast, the degree to which the pressure of increasing population *forces* humans to devise ever-more ingenious ways of increasing the food supply by agricultural developments. Of course food supply is only one of the many variables governing population growth and size. Demographers are also concerned with the effects of migration, famine, disease, warfare, religious beliefs, and social and sexual behaviour etc.

Demographic projections for the next 40–50 years

Environmentalists have played a major role in highlighting the potential for major increase in the human population in the future, which would have major consequences for the biosphere. A major area of uncertainty about future population size concerns the possibility of a 'demographic transition' in countries currently experiencing major population growth. Mace (2002, 235) characterises a demographic transition as follows: '[It] refers to a phenomenon observed throughout the world, in which societies with high fertility and high mortality change to having lower fertility. This transition to having fewer children is generally (but not always) associated with increasing living standards and decreasing mortality risks and represents a drastic cultural change.' Clearly, given the possibility of such a change across the world and other imponderables, it is impossible to predict with any certainty the size of human populations in the future. However: 'were fertility to remain at present levels the population would grow to 12.8 billion by 2050, more than double the present size' (Cohen, 2003, 1172).

Conservationists have been influenced by the concept of the demographic transition. Looking to the future, it is proposed that if this transition, detected in many societies, becomes more widespread, human populations would fall and there would be less pressure on the environment. With this concept in mind, reserves, zoos and botanic gardens are seen as arks that would secure the future of biodiversity through a temporary demographic winter (Soulé *et al.*, 1986). As human populations declined, the conserved biodiversity would be able to emerge from its safe havens to reclaim lost territory, either by natural spread or through active restoration and reintroduction.

Looking to the future, population growth has been estimated on a range of assumptions with high, medium and low variants, together with models based on 'business as usual' (Young, 1998; Cohen, 2003). They give little comfort to the notion of a temporary demographic winter. 'According to the medium variant, the world's population is expected to grow from 6.3 billion today to 8.9 billion by 2050. . . . The anticipated increase by 2050 of 2.6 billion over today's population exceeds the total population of the world in 1950, which was 2.5 billion' (Cohen, 2003, 1172).

Ecological footprints

Environmentalists have attempted to quantify human impacts on the biosphere by calculating 'ecological footprints'. Early models defined footprints in terms of 'the total area of land and water ecosystems required to produce the resources that the population consumes, and to assimilate the wastes that the population generates' (Rees, 2001). Such analyses reveal that the average citizen of developed countries, such as the USA, Europe, Australia and Japan, requires 5–10 ha of biophysically productive land and water to support their consumer life-styles. In contrast, in China the ecological footprint is less than 1.5 ha and for Bangladesh is 0.5 ha. Clearly such figures will change with circumstances. For instance, China is in the midst of massive industrial development and this will lead to an increased average footprint. The methodology and calculation of ecological footprints is open to criticism, but serves to illustrate the environmental cost of different life-styles (Sanderson *et al.*, 2002). Rees (2001) provides a full analysis of the methodology, sample calculations and critique of the strengths and some obvious weaknesses of the concept of ecological footprints. There have been many developments in footprint calculation. Recently, in view of the growing concerns about climate change (see below) 'carbon footprints' have been calculated estimating the amount of carbon dioxide emitted per year for each person. It has been calculated that on average each American emits *c.* 20 tonnes per year, while the figures for Australia are 18 tonnes, UK 9, Switzerland and Sweden 6, China 3, India 1 and Ethiopia 0.1 tonne (figures from http://environment.independent.co.uk/climate change).

Considering the inequalities in the distribution of resources implied by different footprint figures, global statistics conceal vastly different stories in different parts of the world. About 1.2 billion people live in the economically rich, more developed regions: Europe, North America, Australia, New Zealand and Japan. The remaining *c.* 4.9 billion live in economically poor, less developed regions (Cohen, 2003). Balmford *et al.* (2002, 953), quoting from the UN Human Development Report of 2001, reveal the shocking statistic that more than 1.2 billion people are existing on less than 1 US$ per day. With the possibility of further huge increases in population in the years ahead, mankind will face mounting difficulties, especially as there are, at present, more than 800 million undernourished people in the developing world, some of whom are starving. One billion people lack safe drinking water, 2.4 billion have no effective sanitation, and 2–5 million die annually, and hundreds of millions people suffer, from water-borne diseases (Gleick, 2003).

Estimates of population growth and the necessity to increase food output must be seen against predictions about agricultural production. Pimentel and Pimentel (2002, 260) report: 'more than 99% of food comes from the land, less than 1% is obtained from the oceans'. They continue: 'throughout the world, fertile cropland is being lost from production at an alarming rate . . . general over-use of the land' being responsible for the 'loss of about 30% of the world's cropland during the past 40 years'. The 'most eroded and unproductive land is being replaced with cleared forest land and/or marginal land. Indeed the urgent need for more cropland accounts for more than 60 to 90% of the world's deforestation . . . Per capita world cropland is declining, currently standing at only 0.27 ha per capita. This is only about 50% of the 0.5 ha per capita that is considered the minimum land area required for the production of a diverse diet similar to that of the United States and Europe . . . Other countries have even less land. For example, China now has only 0.08 ha available per capita, about 15% of the accepted minimum.'

Turning to questions of future food production, Young (1998) reviews the 65 estimates of future food resources in relation to population growth. Some of these estimates offer reassurance. For instance, increases of agricultural efficiency might raise cereal output by as much as 15 times. In addition, if post-harvest loss and waste could be avoided, the world could support 10–11 billion people. Indeed, some have calculated that the Earth could support 40 billion people. However, others consider that a reduction in population growth is necessary to bring into balance food and population.

Conclusions

The world's population has grown dramatically and most analysts forecast further massive increases. While some humans live in an affluent society, very many suffer from disease, malnutrition and a lack of clean water. Even with surpluses of food in some areas, in other parts of the globe, populations face starvation. Concerning the question of population growth and food supplies, as we have seen above, many models have been published. However, predictions may quickly become out-of-date, as they have not taken into account all the emerging interacting factors influencing agricultural production, for example, the implications of the development and wider use of genetically modified crops. At the time of writing, rising food prices and shortages are being reported in many developing countries. These have been linked to poor harvests (due to adverse growing conditions in some regions), civil unrest, the rising costs of fuel and agricultural chemicals, the purchasing of more meat and milk by emerging economies of India, China etc., the diversion of food crops to make biofuels etc.

It is clear from this brief review that the ecosystems of the Earth have been transformed by human activities, often in an unsustainable fashion, and further transformations are inevitable as an increasing population seeks to exploit and manage the Earth and its resources, against a background of global climate change. Given the efforts of environmentalists to encourage the wise use of resources, will it prove possible to manage these in a more sustainable fashion? Can effective control of greenhouse gases be achieved? We return to these questions in later chapters.

Given the scale of human transformation of the Earth's surface, have any natural, pristine ecosystems survived? Is wilderness still to be found on Earth? This important issue is the subject of the next chapter together with an examination of further concepts illuminating the role of humankind in the biosphere.

5

Consequences of human influences on the biosphere

In the last chapter the origin, scale and accelerating impact of the rising tide of humanity were reviewed together with an account of the origins of environmentalism, a movement that subjects our increasing exploitation of the Earth's resources to the most critical scrutiny with the hope that human use will become more sustainable. This chapter introduces a number of other concepts and ideas that are helpful in understanding the setting for the contemporary microevolutionary drama.

Cultural landscapes

Humans are social beings and it is recognised by anthropologists, archaeologists and others that different regions of the world have their own characteristic 'cultural landscapes', each of which reflects the different civilisation, customs and artistic achievements of the people who produced it and live within it. Today, there is an astonishing array of cultural landscapes ranging from the dwindling lands of the last remaining hunter-gatherers, through the multitude of different agricultural landscapes to the cityscapes and urban industrial landscapes of modern civilisations.

Cultural landscapes first began to evolve as humans migrated to distant parts of the globe as hunter-gatherers, and these were superseded by the landscapes of the settled agriculturist. From these early beginnings, a wide diversity of cultural landscapes has developed. Many of the observable differences relate to the crop plants grown, and current agricultural practices related to the climate, soils etc. of different regions. Grigg (1974) considers the characteristics and distributions of different agricultural systems, which include shifting agriculture, wet-rice cultivation, pastoral nomadism, Mediterranean agriculture, the mixed farming of western Europe and North America, intensive dairying, ranching in arid regions, plantation crops and large-scale grain production. It is important to stress that each of these different major types of agriculture has its own locally evolving variants.

For example, there are different forms of shifting cultivation in Asia, South America and West Africa (Grigg, 1974). Typically, there is a short period of cultivation of cleared plots in forest. These areas are abandoned to long fallow periods allowing bushes and trees to invade, and settlements are moved to new areas. Some years later, the farmers return and the land is re-cleared for cultivation. In other types of shifting cultivation, the cultural landscape contains longstanding village settlements surrounded by many widely scattered fields. Rather than having permanent cultivation of any particular plot and rotation of crops, there is a rotation in the use of fields. After a period of cropping each plot is allowed several years of natural fallow in which trees and shrubs invade, and then the field is cleared and cultivated once again. These shifting agricultural systems allow soil fertility to be restored after a period of cropping and provide a means of controlling weeds. Each of the other types of cultivation systems has its variants.

A major point to stress is that, if the history of different cultural landscapes is investigated, in some cases there is evidence for gradual change over considerable periods, while at others there have been dramatic abrupt events, as, for example, when the peoples of one region have been conquered by another, with the victors imposing their own culture on the vanquished. In other areas, settlements may have flourished for a while, but then they were abruptly abandoned. Historians may point to the significance of single factors. For instance, in the fourteenth century the first outbreak of the Black Death resulted in the death of more than 33% of the population in some parts of Europe, and some villages were reduced in size or were abandoned because there were too few to till the soil. In other cases, several interacting factors may be implicated in the failure or abandonment of settlements, including coastal erosion, retreat from areas of poorer or exhausted soils, pestilence, climate change, crop failures, soil erosion, eviction of farmers on arable land to make way for flocks of sheep, removal of villages to allow the creation or extension of the great parks surrounding country houses etc. (Glasscock, 1992; Davison, 1996). The evolution of cultural landscapes, through abrupt or gradual change and abandonment, may leave very obvious traces – ruins, landmarks and artefacts of different kinds (Fig. 5.1). In other situations, remains may be few, or buried, and await the attentions of archaeologists.

Regarding cultural landscapes, Atkins, Simmons and Roberts (1998, 219) conclude:

> In ancient and complex societies . . . the landscape is a composition of hundreds or thousands of years of small additions and modifications, and sometimes large-scale changes, overlaid on top of each other. It is, therefore, a cultural artefact or construction and acts as a record, not only of people's physical efforts in clearing the wood or draining the marsh, but also of their ways of thinking, the technology they used, and even their social and political structures.

Thus, a landscape will reflect not only current influences but also, in varying degrees, it may contain elements reflecting the past. A metaphor for landscape change is the 'palimpsest'. This was a medieval and early modern European document written on vellum – a prepared animal skin. The vellum was so expensive that it was reused, even though the properties of the skin made it impossible to remove all traces of writing from previous uses. The analogy of the palimpsest is helpful. However, it is clear that modern technology, such as deep ploughing with heavy machinery, is capable of eradicating surface evidence of 'yesterday's landscape' so effectively that no visible traces may remain (Atkins *et al.*, 1998).

Humans have always climbed to vantage points on high trees, and on hills and mountains to survey the landscapes below. Now, many have had the privilege of larger panoramic

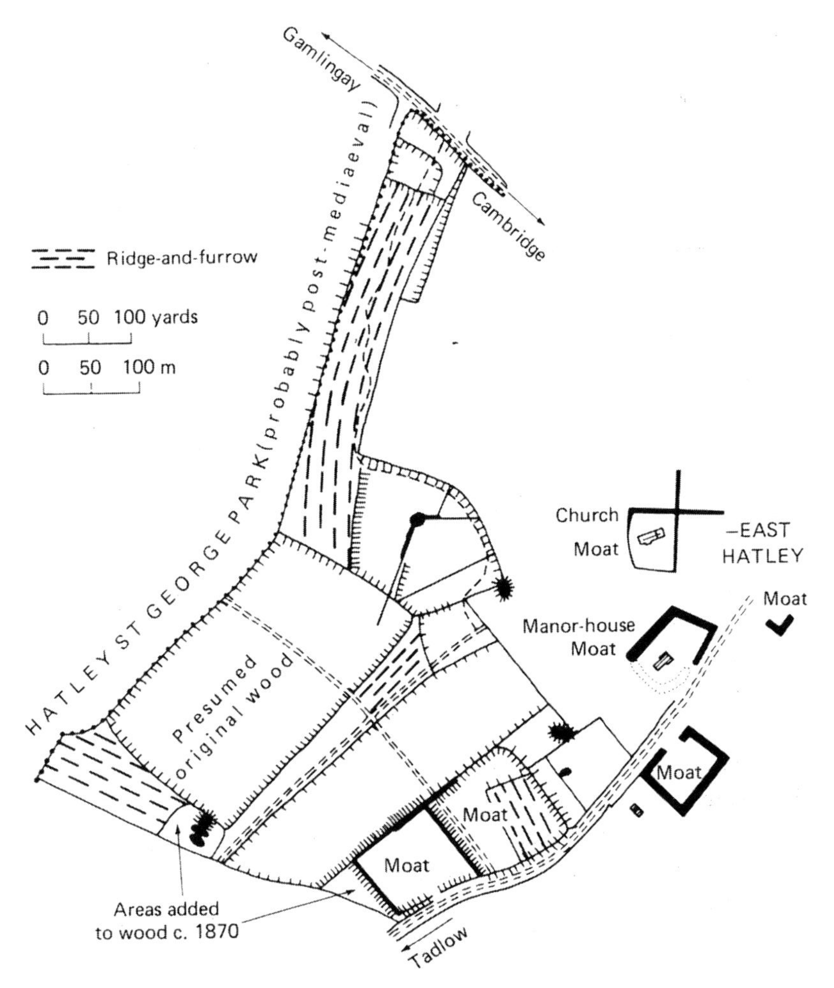

Fig. 5.1 Historical ecology of Buff Wood, Cambridgeshire, England. This small wood of *c.* 16 ha is situated on a calcareous boulder clay plateau west of Cambridge. Its history is complex; there is an original core of presumed ancient woodland surrounded by areas that have been allowed to revert to woodland from agricultural use or after the decline of human settlement in various periods from 1350 onwards. This conclusion is supported by the presence of distinctive features within the present-day wood. The map shows the presence in secondary woodland of earthworks and moats, and also patterns of visible ridge and furrow, indicating former plough land. (Reproduced by permission from Rackham, 1980. For further insights into the history of Buff Wood see second revised edition, Rackham (2003))

views from aircraft, hot air balloons, gliders etc. Therefore, it is easy to confirm the presence of wide expanses of cultural landscape, and reflect on the fact that the development of these areas has been at the expense of natural communities. As cultural landscapes now predominate across the world, the question arises as to whether any wilderness has survived.

Is there any wilderness left?

In the course of the nineteenth and early twentieth centuries, agricultural and urban development in the USA proceeded to such an extent that steps were taken to protect the last remaining 'wilderness' in national parks, and these efforts have continued through the Wilderness Act of 1964 (Woods, 2001).

The definitions of wilderness, employed by the 1872 National Park and 1964 Wilderness Acts, have proved highly influential, not only for conservation in the USA, but also for the wider conservation movement. The 1964 wording (quoted from Woods, 2001, 350) reads:

> A wilderness, in contrast to those areas where man and his works dominate the landscape, is hereby recognised as an area where the earth and its community of life are untrammeled by man, where man himself is a visitor who does not remain. An area of wilderness is further defined to mean... undeveloped Federal land retaining its primeval character and influence, without permanent improvements or human habitation, which is protected and managed so as to preserve its natural conditions and which (1) generally appears to have been affected primarily by the forces of nature, with the imprint of man's work substantially unnoticeable; (2) has outstanding opportunities for solitude or a primitive and unconfined type of recreation; (3) has at least five thousand acres of land or is of sufficient size to make practical its preservation, and use in an unimpaired condition; and (4) may also contain ecological, geological, or other features of scientific, scenic, or historical value.

In an important review of wilderness, on which this account is partly based, Woods (2001) notes that, in addition to many national parks, sanctuaries and reserves, more than 600 wilderness areas have now been declared in the USA and, through the activities of the World Wilderness Congress, tracts of wilderness have been designated all over the world, e.g. Australia, Canada, New Zealand, Zimbabwe.

The notion of wilderness is often used by those selling holidays, and safaris and ecotourist expeditions to distant places. For instance, Scotland is often presented as 'a pristine Highland wilderness' (Warren, 2002, 3). And, in their fund raising, conservation organisations frequently refer to wilderness under threat from human destruction. The concept is also widely employed in publicity material and conservation literature, as for example in *Wilderness: Earth's Last Wild Places* (Mittermeier *et al.*, 2003) published by Conservation International. This book describes and illustrates 37 wilderness areas in tropical rainforest, wetlands, deserts and arctic tundra. These areas are defined as having 70% of their original vegetation cover, and are at least 10,000 square kilometres in area. Surprisingly, not all the wildernesses they described are uninhabited: they include some areas with fewer than five people per square kilometre.

The concept of wilderness raises a number of very interesting questions. Critics of the thinking behind the Wilderness Act of the USA point to the fact that human influences are so pervasive and extensive that 'if we define wilderness as untrammeled, unimpacted, uninhabited, non-human nature' (Woods, 2001, 355), then there is no such thing as wilderness.

Wilderness concepts

Given these different opinions, it is helpful to look more closely at the concept of wilderness. At the outset it is important to note that in many contexts the term is often ill-defined, as are other similar concepts – the 'pristine environment', the 'virgin forest', the 'natural world', the 'undamaged habitat', 'wild nature' etc.

Turning to the origin of the notion of 'wilderness', Oelschlaeger (1991) argues that the concept emerged with the beginnings of agriculture, when the distinction was made between the 'domestic' and the 'wild', and these terms were applied to animals and plants and territories. 'The wild was un-tame, lying beyond the tended fields and managed woods' (Adams, 2004, 102).

Wilderness and the 'Pristine Myth'

These notions of wilderness were taken to lands overseas. According to Nash (1982), when the first Europeans arrived in America they considered themselves to have arrived in a condition of wilderness, 'empty of civilisation, populated by wild beasts and savage men' (Nash, 1982). Also, colonial observers 'saw the landscapes of America and Australia and in due course Africa as open, under-used and largely un-settled. They seemed to lack effective human occupation, because they lacked visible improvement... The Australian outback was considered a wilderness in the original European sense, to be settled and civilised by European farmers and their introduced livestock and technologies' (Adams, 2004, 103).

An essay by Denevan (1992, 369), published as part of the quincentenary of the arrival of Columbus, reviews in detail what he calls the 'Pristine Myth', which asserts: 'In 1492 the Americas were a sparsely populated wilderness, a world of barely perceptible human disturbance.' Denevan considers that the myth 'is to a large extent the invention of nineteenth-century romanticist and primitivist writers such as W. H. Hudson, Cooper, Thoreau, Longfellow and Parkman, and painters such as Catlin and Church'.

There is evidence from many sources, however, that the Native American landscape of the early sixteenth century was 'humanised landscape' almost everywhere. Populations were large. Denevan estimates that at the time of the arrival of Columbus in the Americas, there were *c.* 54 million inhabitants (*c.* 4 million in North America; almost 20 million in Mexico and Central America, *c.* 3 million in the Caribbean and *c.* 24 million in South America). As a consequence of human activities: 'forest composition had been modified, grasslands had been created, wild life disrupted and erosion was severe in places. Earthworks, roads, fields, and settlements were ubiquitous' (Denevan, 1992, 345).

Moving to a longer perspective, by about fifteen thousand years ago, humans had colonised all but the harshest territories on most of the planet and archaeological studies in different parts of the world are revealing that the impact of humans in prehistoric times was probably very considerable. For instance, some authorities consider that humans were responsible for the extinction of certain prehistoric animals.

Megafaunal extinctions

As long ago as 1875, Alfred Russel Wallace made the observation: 'We live in a zoologically impoverished world, from which all the hugest, and fiercest, and strangest forms have recently disappeared . . . not in one place only, but over half the surface of the globe' (Wallace, 1876, 150). At first he thought that the extinctions were brought about by the ice age glaciations, but it became clear that many of the large animals had survived the glacial epoch only to become extinct at a later date, and Wallace became convinced that extinction was due to 'human agency' (Wallace, 1911).

Archaeological studies of fossil bones of large extinct mammals, including radio carbon dating, have confirmed that at the end of the Quaternary ice age the continents of the world lost 85 genera of large mammals. Geologists and archaeologists have long tried to determine the cause(s) of this megafaunal extinction spasm, and research continues (see Martin & Klein, 1984, for a review of the evidence). In summary, while 'regional extinctions often coincide with changes in climate . . . attempts at a synthetic treatment that incorporates the intensity, timing and character of late Pleistocene extinction into a world-wide model related to climatic changes are more difficult'. On a global scale the pattern of late Pleistocene extinction appears to follow the prehistoric migrations and activities of humans much more closely than global climate change (Martin, 1984). The notion that humans were responsible for prehistoric overkill is supported by coincidence of human migrations and the megafaunal extinctions, and for some, this is the critical clue to what happened in the past. Thus, according to Martin, New World megafaunal extinction occurred as humans entered and spread across the continent after crossing the Bering Strait (at that time a land bridge) around 11,000 BP. The extinction of birds and mammals in Madagascar and the giant moas in New Zealand occurred later at the same time as the arrival of humans.

The extinction process has been visualised by some as a rapid 'blitzkrieg', but recent studies of radio-carbon-dated bones has revealed that the demise of a flightless sea duck (*Chendytes lawi*) in California was a protracted affair (Jones *et al.*, 2008). Also, some archaeologists are speculating that the Pleistocene extinctions may have come about not only through over-hunting, but also because human activities destroyed important habitats (Miller *et al.*, 2005; Grayson, 2008).

Cultural landscapes and fire

There is growing evidence that hunter-gatherers in North America, Australia and other parts of the world had a profound effect on natural ecosystems through their use of fire. For instance, in Australia the vegetation has been greatly influenced by aboriginal 'firestick farming' practices, with fires set to ambush animals, promote new growth of grasses and facilitate the germination of scattered seeds of food plants (Evans, 1998). Repeated burning of aboriginal fires has had a major impact on the vast savannah lands of Australia, and presents complex problems for those attempting to devise fire management strategies to conserve these ecosystems, especially in densely settled areas (Russell-Smith *et al.*, 2007).

Krech (1999) has reviewed the evidence of the Indian uses of fire in North America. They burned forest 'to drive and encircle animals' in the course of hunting; 'to promote the growth of grasses and other forage favored by large animals'; 'to increase the production of berries, seeds, nuts and other gathered food'; 'to drive off unwanted strangers and enemies'; 'to signal to each other'; 'to ease travel' by removing the undergrowth in forests; and 'to kill grasshoppers, which they gathered up and ate with relish'. Sauer (1950, 1958, 1975) has argued that the great grasslands of the New World are of anthropogenic rather than climatic origin. Thus, it is alleged that the eastern prairies of the USA were maintained by 5,000 years of near annual burning of the grasslands by Indians. It became clear that forest can grow on the prairie lands, for in the nineteenth century when a fire suppression policy was initiated, many grasslands, in Wisconsin, Illinois, Kansas and elsewhere, were invaded by forest. Debate about the role of humans in the evolution and maintenance of the major grassland areas of the world continues (Anderson, 2006; Behling *et al.*, 2007).

How 'virgin' is virgin rainforest?

Denevan (1992, 373) also considers the impact of humans in American forests and notes: 'tropical rainforest has long had a reputation for being pristine, whether in 1492 or 1992. There is, however, increasing evidence that the forests of Amazonia and elsewhere are largely anthropogenic in form and composition.' At the ninth Pacific Science Congress in 1957, Sauer (1958, 105) challenged the views of the celebrated tropical botanist Paul Richards who believed that, until recently, the tropical forests had been largely uninhabited, and that prehistoric people had 'no more influence on vegetation than any of the other animal inhabitants'. Sauer concluded: 'Indian burning, swiddens and manipulation of composition had extensively modified the tropical forest'. Support for this interpretation comes from Balée (1987, 1989) and Uhl *et al.* (1990), who regard the forests of Amazonia as 'cultural' and 'anthropogenic', concluding: 'large portions of Amazonia forests appear to exhibit the effects of past human interference'. The view that the rainforest has been subject to many anthropogenic influences has been strengthened by recent archaeological research. Nonetheless, the term 'wild' nature is often used for ecosystems of high biodiversity and conservation concern, such as the biodiverse supposedly 'virgin' tropical rainforest. The underlying assumption is still made that such areas are still pristine or relatively undisturbed by human activity (e.g. Balmford *et al.*, 2002).

However, having reviewed the archaeological evidence, Willis, Gillson and Brncic (2004, 402) reach a different conclusion. There is increasing evidence that: 'So-called "virgin" rainforest blocks might not be as pristine as originally thought, and have in fact undergone substantial prehistoric modification'. Thus, large areas of what are often taken to be undisturbed rainforest in the Amazon, Congo and Indo-Malay regions have all been influenced by extensive prehistoric human activities, as part of former cultural landscapes.

Looking at these areas in turn, in the Amazon there are up to 50,000 ha of fertile 'terra preta' soils. The formation of these soils is now attributed to prehistoric burning and agricultural activities (*c.* 2,500 years ago). In the Upper Xingu regions of Brazil there are prehistoric settlements (dated *c.* AD 1250–1600) indicating intensive management and development of the landscape, closed forest being replaced with agricultural land and parkland. Depopulation, in the period AD 1600–1700, resulted in extensive re-growth of forests, giving the impression of wilderness.

A similar picture emerges from multi-disciplinary studies of the lowland Congo basin in Africa. Extensive finds of stone tools, oil palm nuts, subsoil charcoal horizons, pottery fragments and banana phytoliths (silica fragments allowing the positive identification) point to extensive forest clearance, habitation and cultivation practices dating from *c.* 3,000 to *c.* 1,600 years ago, when there was a dramatic decline in populations, allowing the forest to recolonise. Archaeological studies in western central Africa have revealed the existence of iron-working furnaces dating from *c.* 650 BC, an activity 'that would have had a serious impact on the forest through the extraction of wood for charcoal production and smelting' (Willis *et al.*, 2004). Many forest types are now found in the Congo. Often characterised as 'virgin', these forests are now interpreted as being the result of secondary succession (van Gemerden *et al.*, 2003).

It would appear that prehistoric modification of Indo-Malayan rainforests took place at a very early date (Willis *et al.*, 2004). Various lines of evidence point to the existence of cultural landscapes in Papua New Guinea *c.* 7,000 years ago, and the lowland rainforests of Thailand were apparently managed as early as 8,000 years ago. Evidence from ecology, archaeology and oral history suggests that on the Solomon Islands there was settlement and development until about 150 year ago, when populations declined and migration occurred, allowing extensive forest regeneration with many trees species now thought to be typical of secondary successions. Willis *et al.* (2004) conclude that for many tropical forests it is no longer 'acceptable to suggest that land loss caused by previous human activities was too small to have had a major impact'. The presence of apparently 'virgin' forest attests to the resilience of forest ecosystems, the capacity for forest regeneration, and various lines of evidence point to a likely time frame for such regeneration of secondary forest after human population movements or population crashes.

Wild areas in Europe

In 1970–71, Netting (1981) made a thorough study of land use in the Swiss Alps, by examining in detail the village of Törbel, Vispertal, at 1,500 m. He points out that travellers armed with images of 'wild' untouched natural landscapes must accept that, with the exception of the high peaks, all the elements of the Swiss landscape have been intensively managed and decisively influenced by generations of villagers, principally in the tilling of the soil, grazing of stock and the mowing of hay. Close study revealed the intricate and all-pervasive actions of the villagers, who made intensive use, both near and far, of the natural resources available in the area where they were living.

The fieldwork revealed that the major part of the village land was used for hay: low-lying meadows were regularly manured and supplied with water through a series of irrigation channels producing two cuts of hay, plus some grazing by tethered animals in the spring and autumn. Grassland too steep or stony provided periodic grazing, with goats on the most rugged precipitous outcrops. Wherever the slope allowed, hay was cut too from alpine meadows. Summer grazing was also provided for cattle taken up to high altitude, where small groups of men would make cheese from the milk. In protected locations near the village fruit trees were grown to produce apples, hazelnuts, plums and walnuts. Wild berries, pine nuts (from stone pines) and many medicinal herbs were harvested. In the past, hemp and flax were grown for household textiles. Within terraced walls, vegetable gardens were patiently tilled and fertilised until everything but the cliffs appears 'ordered and domesticated'. Every field, gully and habitation has its own place name. Woods covered a great deal of the slopes. These anchor the soil, reduce the danger of avalanches and protect the watershed. Field studies revealed that larch and fir stands had been thinned and protected. Trees marked for felling were allotted by the drawing of lots. Fallen branches and pine needles were collected for fuel. Windfalls and dead standing-timber were auctioned by the forest steward on Sunday after mass. Netting describes a community that was to a great extent self-sufficient. Only a vineyard was lacking within the village borders! And this deficiency was made good by having plots of vines lower down the Vispa valley. The picture that emerges is of a cultural landscape well adapted to alpine regions, with a wealth of hay barns in various parts of the village, and chambers and cellars in the houses in which to store hay, grain, foods and drink for winter consumption, and as an insurance against bad years.

The notion that wild untouched areas might still occur *between* segments of cultural landscapes has also been the subject of much study in Europe. There is evidence that many supposed natural areas, containing endangered species of plants and animals, have been exploited by humans. For example, reflecting on the vegetation history of Scandinavia, Faegri (1988, 1) concluded: 'with some small and doubtful exceptions all vegetation types were created or modified by man'. Faegri emphasises a profound shift in emphasis in our ecological understanding: 'the "natural landscapes" of preceding generations are now understood for what they really are: relics of earlier types of land-use'.

Considering the situation in Scotland, where the 'Highland Clearances' of local populations were carried out to allow an expansion of sheep farming, Hunter (2000, 6) has pointed out: 'Highland wilderness, as celebrated by today's conservationists, is not so much natural as the wholly artificial result of the nineteenth century landlords having excluded humanity from ecosystems of which people have been a part for ten millennia.' Indeed, as a result of intensive studies of a wide range of evidence, Bennett (1995, 36) concludes that 'the whole Scottish landscape is anthropogenic' (see Hunter, 1995).

Thus, throughout most of the developed world the concept of islands or blocks of agricultural land in a matrix of natural vegetation has had to be abandoned. Evidence suggests that there are 'gradients of human impact'. In considering any particular tract of supposedly pristine vegetation, first impressions may be deceiving, as evidence from the researches of archaeologists and historical ecologists may reveal significant human influence on the ecosystem.

The myth of oceans as wilderness

Wilderness concepts have been applied not only to areas on land, but also to the seas. Jackson (2001, 5411) provides a critical appraisal in concluding:

> The persistent myth of the oceans as wilderness blinded ecologists to the massive loss of marine ecological diversity caused by over-fishing and human inputs from the land over past centuries. Until the 1980s, coral reefs, kelp forests, and other coastal habitats were discussed in scientific journals and textbooks as 'natural' or 'pristine' communities with little or no reference to the pervasive absence of large vertebrates or the widespread effects of pollution.

The recent publication of a map, based on 17 global data sets for 20 marine ecosystems, reveals the increasing impact of human activities (Halpern *et al.*, 2008).

The myth and its implications

Returning to the origins of the Pristine Myth in the Americas, Denevan (1992, 379) concludes that early observers were 'unaware of human impacts that may be obvious to scholars today'. In addition, it is clear that many of the eyewitness accounts of wilderness in America and elsewhere were written not at the time of the first European colonisation but later, in the period 1750–1850, when the vegetation in the New World was recovering after dramatic reduction in human numbers following the devastating effects, post-1492, of introduction by adventurers of Old World diseases such as smallpox, measles etc.

These new insights into the myth of pristine wilderness have important implications for conservation. As Denevan (1992) points out, the landscape of the Americas bears an imprint of Indian cultures and then, after 1492, 'the familiar Euro-American landscape was grafted, rather than created anew'. Furthermore, he stresses that much of what is at the present day protected or proposed to be protected from human disturbance is far from being pristine, having a long history of environmental modification by human activities. Thus, and this is a point of great importance, conservationists in many parts of the world are attempting to conserve what they take to be wilderness, but increasingly evidence suggests that the territories involved are or were cultural landscapes.

The first conservationists

Recent studies of the activities of indigenous peoples in prehistoric time also allow us to examine another idea widespread in the conservation literature, namely that native peoples were the first conservationists. For example, Sale (1990) makes the claim: 'New World peoples lived in harmony with nature and refrained deliberately from altering their environments, to a degree that they were somehow able to maintain an idyllic ecological equilibrium. Europeans, by contrast, had a ruthless land ethic, were driven only by materialistic goals' (quoted in Butzer, 1992, 347).

The concept of 'The Ecological Indian' is examined in detail by Krech (1999, 212). He considers available evidence concerning the environmental impacts of North American native peoples, with regard to pleistocene extinctions; uses of fire, hunting of deer, beaver and buffalo (by driving them over cliffs etc.), and land degradation under agriculture. He concludes:

> Native peoples clearly possessed vast knowledge of their environment . . . prior to the twentieth century, the evidence for Western-style conservation in the absence of Western influence is mixed. On the one hand, native peoples understood full well that certain actions would have certain results: for example, if they set fire in grasslands at certain times, they would produce excellent habitat for buffalos one season or one year later. Acting on their knowledge, they knowingly promoted the perpetuation of plant and animal species favored in the diet. Inasmuch as they left available, through these actions, species of plants and animals, habitats, or ecosystems for others who came after them, Indians were 'conservationists.' On the other hand, at the buffalo jump, in the many uses of fire, in the commodity hunt for beaver pelts and deerskins, and in other ways, many indigenous peoples were not conservationists.

Man and nature

Returning to the topic of wilderness and its protection, it is important to consider another issue. The notion of preserving wilderness areas by excluding humans provokes many questions about the relationship between man and nature. As we saw earlier in Chapter 4, Darwin concluded that humans were the product of evolution, a species amongst species, part of nature. Extending this line of reasoning, it could be claimed that as humans are part of the natural world, logically anything humans do is natural, including the conversion of wilderness into cultural landscape through the development of agricultural land, the management of grasslands and forests, the building of towns and cities.

However, many conservationists do not subscribe to this view. They make a distinction, indeed, see a separation, between the natural world and human activities. Such a view has a long history, for in the Judaeo-Christian tradition, as related in Genesis, man was created into a preformed world of nature, conveying the idea that humans are separate from nature, are at the centre of things, with every thing created to serve their purposes (Gruen & Jamieson, 1994, 15). Man had a precise place in the order of creation and a destiny: 'Be fruitful and multiply, and fill the earth and subdue it' (Genesis: Chapter 1).

The words 'nature' and 'natural' are complex with different usages. Williams (1980, 67) confronts this issue:

> Some people when they see a word think the first thing to do is to define it. Dictionaries are produced . . . but while it may be possible to do this more or less satisfactorily, with certain simple names of things and effects, it is not only impossible but also irrelevant in the case of more complicated ideas. What matters in them is not the proper meaning but the history and complexity of meanings: the conscious and unconsciously different uses.

Thus, the word 'natural' has many concrete and abstract meanings.

The dualistic thinking that separates nature and humans is commonplace. Very often the distinction is drawn and separation made between people/culture and nature. Such anthropocentric/homocentric views are particularly associated with the Enlightenment and Romantic movements of the eighteenth and nineteenth centuries (Williams, 1976, 188). Nature was contrasted with what had been made by man. And again: 'nature has meant the "countryside" and the "unspoilt places", plants and creatures other than man . . . '. 'Nature is what man has not made.' The distinction between nature and culture was widely drawn, particularly as great landscaped parks were being laid out in England and landowners were rearranging and improving on nature in their huge estates.

Environmentalists too often see humans in opposition to nature rather than a part of it. Thus, in the nineteenth century, Marsh (1864), in decrying damaging human activities writes of man as the 'disturber of natures harmonies', and many in the twentieth century speak of 'man's alienation from nature' (e.g. Pontin, 1993).

In classifying vegetation, plant ecologists have employed a number of categories: natural, semi-natural and deliberately created cultivated landscapes. Yet again a distinction is being made between the natural and the man-made. For example, Tansley (1945, 1) writes: 'all vegetation which "has come by itself", but under conditions determined by human agency, may be called "semi-natural", as contrasted with that which is entirely natural, unaffected by man, and on the other hand with sown crops and plantations which he has deliberately created'. Tansley also uses the terms 'half-wild' and 'wild' in describing British vegetation (Tansley, 1945, 1). Rackham (2001, 675) also uses the term semi-natural denoting: 'Vegetation that owes its character to human activity, but in which the plants are wild, not sown or planted.'

Human activities as viewed by environmentalists

Those concerned with conservation of species often focus on the damaging effects of human activities through what Diamond (1989) calls, the 'Evil Quartet' of habitat destruction and fragmentation, overkill and over-harvesting, and the impact of introduced species, leading to 'chains of extinctions'. However, reflecting on this view, Western (2001) considers that this type of analysis does not sufficiently examine human motives. 'Did we create anthropogenic environments intentionally or not? Do they fulfil human goals? Ecologists are quick to judge the result without looking at the cause, implying that we destroy nature without thought to the outcome. But is our behaviour really that aberrant?'

Human activities: the concept of niche construction

Viewed from a utilitarian perspective, humans have purposely and successfully converted a 'wilderness' of natural ecosystems into productive 'humanscapes' that meet our needs. Restating this proposition in evolutionary terms, it is possible (and I would argue necessary) to view human activities as evolutionary strategies to ensure survival and reproductive success. For example, Laland (2002, 821) notes that traditionally:

Table 5.1 *Some characteristics of intentionally modified ecosystems*

High natural resource extraction
Short food chains
Food web simplification
Habitat homogeneity
Landscape homogeneity
Heavy use of herbicides, pesticides and insecticides
Large importation of non-solar energy
Large importation of nutrient supplements
Convergent soil characteristics
Modified hydrological cycles
Reduced biotic and physical disturbance regimes
Global mobility of people, goods and services

From Western (2001). Reproduced with permission © 2001, National Academy of Sciences, USA.

adaptation typically is regarded as a process by which natural selection molds organisms to fit a pre-established environmental 'template' . . . yet, to varying degrees, organisms choose their own habitats, choose and consume resources, generate detritus, construct important components of their own environments (e.g. nests, holes, burrows, paths, webs, pupal cases, dams and chemical environments), destroy other components and construct environments for their offspring. *Thus, organisms not only adapt to their environments but in part also construct them.* [my emphasis]

Such a process has been called niche construction (Laland, Odling-Smee & Feldman, 1999; Odling-Smee, Laland & Feldman, 2003), and is exemplified by the activities of ants, bees, spiders, wasps, termites etc.

Clearly, humans also construct their niches, but compared with ants and termites, their efforts are on a truly monumental, indeed sometimes grandiose, scale. Laland (2002, 822) concludes: 'few species have modified their selective environment to the same extent as humans', noting that the capacity for changing the niche has been 'further amplified' by human cultural development. Thus:

Human innovation and technology have had an enormous impact on the environment; they have made many new resources available via both agriculture and industry; they have influenced human population size and structure via hygiene, medicine and birth control; and they have resulted in the degradation of large areas of the environment. These are all potential sources of modified natural selection pressures. Cultural processes that precipitate niche construction might be expected to have played a critical role in human evolution for many thousands, perhaps millions, of years.

Western takes the view that it is helpful, in considering human impacts on the environment, to distinguish intended from unintended effects, while acknowledging the difficulty of categorising human activities into two such groups (Tables 5.1 & 5.2).

Table 5.2 *Some ecosystem side effects of human activity*

Habitat and species loss (including conservation areas)
Truncated ecological gradients
Reduced ecotones
Low alpha diversity
Loss of soil fauna
Simplified predator–prey, herbivore–carnivore and host–parasite networks
Low internal regulation of ecosystems due to loss of keystone agents
Side effects of fertilisers, pesticides, insecticides and herbicides
Invasive non-indigenous species, especially weeds and pests
Proliferation of resistant strains of organism
New and virile infectious diseases
Genetic loss of wild and domestic species
Over-harvesting of renewable natural resources
High soil surface exposure and elevated albedo
Accelerated erosion
Nutrient leaching and eutrophication
Pollution from domestic and commercial wastes
Ecological impact of toxins and carcinogenic emissions
Atmospheric and water pollution
Global changes in lithosphere, hydrosphere, atmosphere and climate

From Western (2001). Reproduced with permission © 2001, National Academy of Sciences, USA.

Considering the intended effects, Western (2001, 5459) notes:

> The most universal and ancient features of our 'humanscapes' arise from a conscious strategy to improve food supplies, provisions, safety and comfort – or perhaps to create landscapes we prefer, given our savanna ancestry. The domestication of species, the creation of open fields, the raising of crops, and the building of shelters and settlements are the most obvious of intentional human activities, each practised for millennia.

All of these endeavours 'are deliberate strategies to boost production and reproduction. As an evolutionary strategy, our success at commandeering resources and transforming the landscape to meet our needs has been phenomenal.' Considering Western's notion of intentional modifications, we can see that there is a very wide array of ways in which humankind benefits from plants and animals and their biological processes. In making a judgement it is important to take account of not only obvious immediate benefits, but also the variety of essential services provided by ecosystems (Balmford *et al.*, 2002). Benefits range from 'sea food, timber, biomass fuels and the precursors of many industrial and pharmaceutical products' (Daily & Dasgupta, 2001). But, there are also many very important and complex non-extractive services, e.g. natural pest control, renewal of soil fertility, watershed protection and water purification, coastal and river stability, pollination of our crops, etc.

(Daily & Dasgupta, 2001). While further research is necessary to establish how plant communities might influence or control climate locally, regionally and globally, it is clear that they may provide a crucial 'service' by moderating regional and local weather. For instance, in the Amazon, 'transpiration of plants in the morning contributes moisture to the atmosphere that then falls in thunderstorms in the afternoon, damping both moisture loss and surface temperature rise'. Thus, '50% of the mean annual precipitation is recycled via evapotranspiration by the forest itself. Amazon deforestation could so dramatically reduce precipitation that the forest might be unable to re-establish itself following large-scale destruction' (Daily & Dasgupta, 2001, 353). The services of plants and animal communities also include 'life-fulfilling functions encompassing aesthetic beauty and cultural, intellectual and spiritual values derived from nature' (Daily & Dasgupta, 2001, 357).

Returning to Western's (2001) review of the effects of humans on the biosphere, he also lists many unintended side effects (Table 5.2) that lead to a multiplicity of consequences at every spatial scale, for instance the loss of wild habitats and species, soil erosion and sedimentation leading to land degradation, and pollution of air, land and water.

Ecosystems: natural and human influenced

The interrelationships of humans and the biosphere may be viewed in yet another context. In closing this chapter, it is important to examine the development of the ecosystem concept. Odum (2001) has considered the history of advances in our understanding of the interrelations of organisms and their environment, and notes that it was the ecologist Tansley (1935), who introduced the term ecosystem, for functional units of the natural world encompassing the complex dynamic relationships between all the organisms in an area under study, in relation to their physical and abiotic environments (Virginia & Wall, 2001). In Fig. 5.2a such an ecosystem is represented diagrammatically as a box, within which solar energy is transformed into organic matter by autotrophs (plants and certain bacteria). Symbols indicate, in a simplified fashion, the food relationships linking autotrophs and heterotrophs (feeders on others, including humans, i.e. dependent on organic carbon from an external source), together with the flows of energy, cycling of materials and storage within the ecosystem. Ecosystems are to be visualised as convenient study areas. Thus, for different purposes the ecologist might study the large or small areas in detail, e.g. the Greater Yellowstone Ecosystem, which includes the Yellowstone National Park, the nearby Grand Teton National Park, as well as adjacent areas of National Forest (Fig. 5.3). Alternatively, on a completely different scale, an ecologist could examine the ecosystem of a single lake in Yellowstone.

In the early part of this book we have discussed how much humans have transformed the Earth. Natural areas are being lost: over very wide areas human-dominated landscapes prevail. The recognition of the importance of human activities has led, some would say belatedly, to the recognition of the widespread occurrence of what Odum calls human techno-ecosystems (Fig. 5.2b). Thus, domesticated ecosystems (agricultural land, managed forestry and urban industrial areas etc.) are major land uses of the planet. These highly managed/ordered ecosystems are interspersed by others that are on degraded, abandoned or derelict areas. All ecosystems, both terrestrial and aquatic, are impacted to various degrees by human pollution and wastes. Moreover, new energy sources – solar power, fossil fuels and atomic technologies – provide the energy by which technological conversions drive our fuel-powered cities, industries and agriculture. These new relationships are explored by Odum (2001), who visualises natural and human techno-ecosystems co-existing side by side. However, such a view can be challenged. Ecologists have travelled far and wide in an attempt to find truly natural areas, however, there are few, if any, entirely natural ecosystems free from any human influence. And as the impact of human activity increases

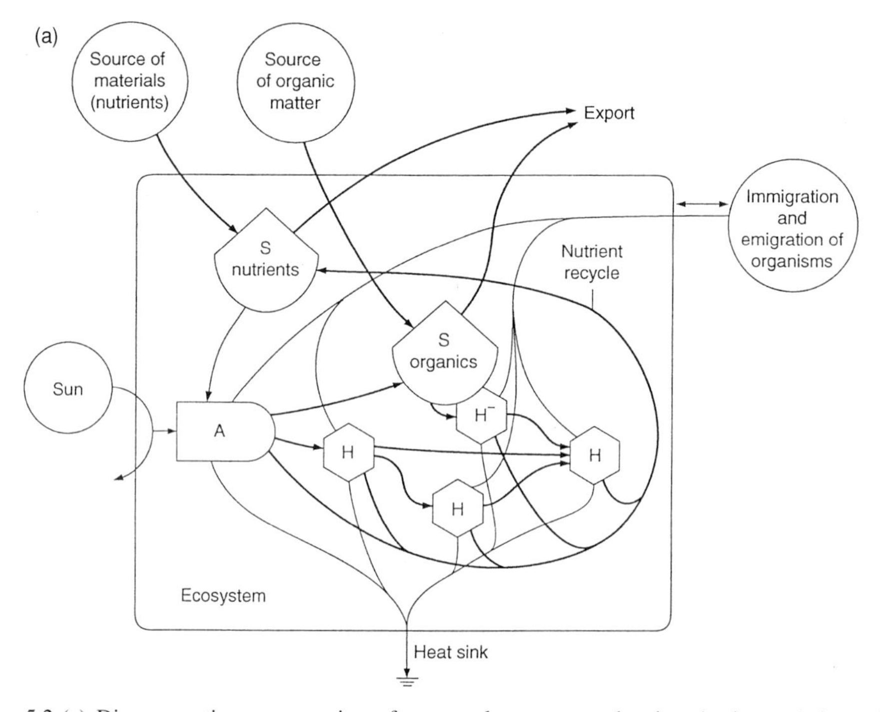

Fig. 5.2 (a) Diagrammatic representation of a natural ecosystem showing the internal dynamics involving energy flow, material cycles and storage (S), as well as food webs of autotrophs (A) and heterotrophs (H). Natural ecosystems are solar powered. Man is regarded as one of the heterotrophic elements of the system and not specifically identified. (b) A human-dominated techno-ecosystem, which differs from the natural in its use of fossil fuels, nuclear power etc. as sources of energy. A substantial amount of pollution and waste is generated, especially where humans live in concentrated populations such as cities (from Odum, 2001. Reproduced with permission © Elsevier). (c) A diagrammatic representation of a complete self-regulating ecosystem such as a woodland, meadow or lake as envisaged by Ellenberg (1988). Here, the role of 'man' in ecosystems is highlighted. Through their food preferences, humans profoundly alter food chains. And in acting as a 'superorganic factor', they can 'consciously or unwittingly influence any part of the ecosystem'. (Reproduced by permission of Ellenberg, 1988, and Cambridge University Press)

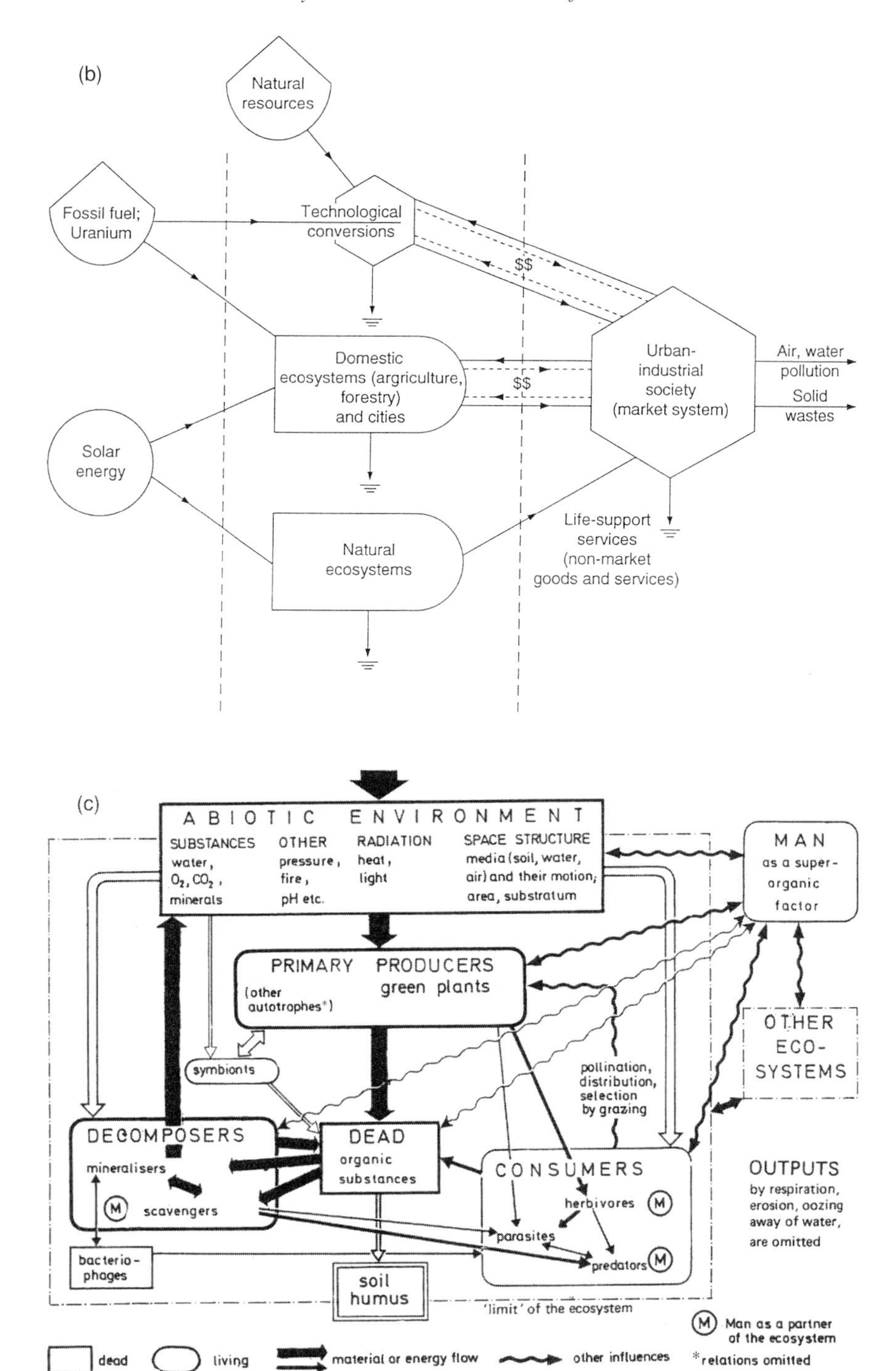
(b)
Natural resources
Fossil fuel; Uranium
Technological conversions
$$
Domestic ecosystems (argriculture, forestry) and cities
$$
Urban-industrial society (market system)
Air, water pollution
Solid wastes
Solar energy
Natural ecosystems
Life-support services (non-market goods and services)
(c)
ABIOTIC ENVIRONMENT
SUBSTANCES water, O2, CO2, minerals
OTHER pressure, fire, pH etc.
RADIATION heat, light
SPACE STRUCTURE media (soil, water, air) and their motion; area, substratum
MAN as a super-organic factor
PRIMARY PRODUCERS green plants
(other autotrophes*)
symbionts
OTHER ECO-SYSTEMS
pollination, distribution, selection by grazing
DECOMPOSERS
mineralisers
scavengers
DEAD organic substances
CONSUMERS
herbivores
parasites
predators
OUTPUTS by respiration, erosion, oozing away of water, are omitted
bacterio-phages
soil humus
'limit' of the ecosystem
Man as a partner of the ecosystem
dead
living
material or energy flow
other influences
*relations omitted

Fig. 5.2 (*cont.*)

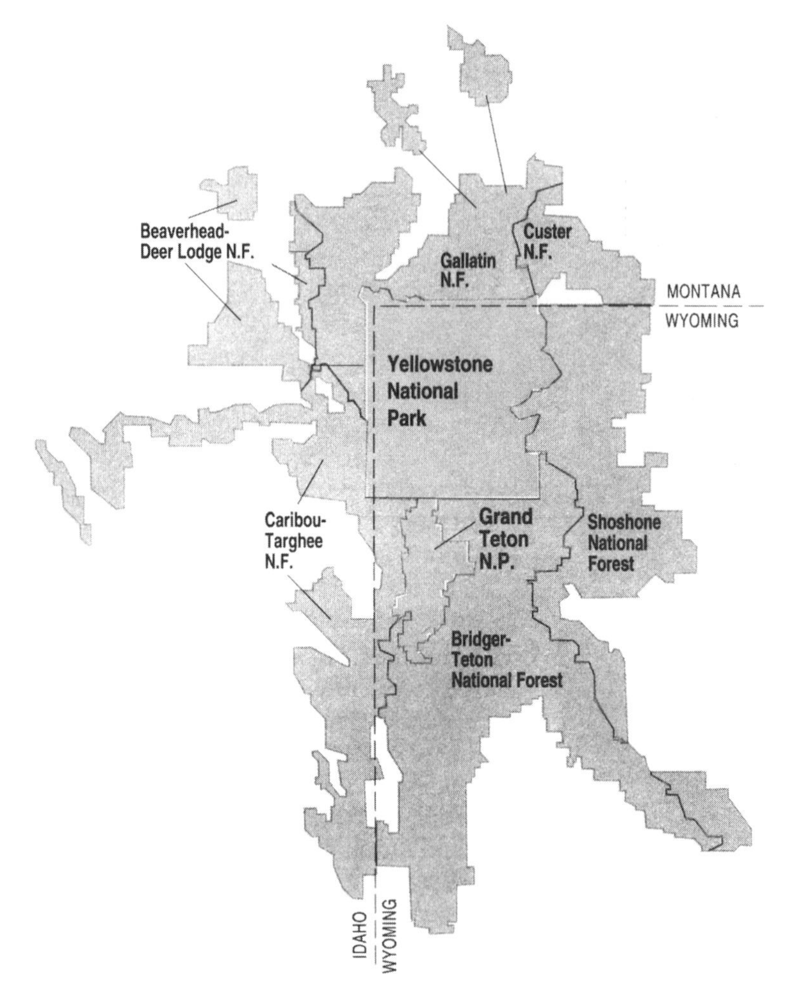

Fig. 5.3 The Greater Yellowstone Ecosystem. (Reproduced by permission of the Division of Interpretation, Yellowstone National Park, Anon., 2004)

in the future, it is likely that even more ecosystems will be decisively and deliberately exploited, domesticated and managed. Moreover, even in relatively unpopulated areas, plant and animal communities are being unwittingly impacted by anthropogenic effects, such as pollution. Unless the levels of greenhouse gases are first contained and then reduced, it is predicted that there will be increasing climate change (involving altered temperatures, rainfall patterns and more extreme weather events). Thus, for most, perhaps all, study areas, it is important to acknowledge explicitly the many human influences impacting on the ecosystem. For example, Fig. 5.2c illustrates in diagrammatic fashion an appropriate model for a cultural landscape of forest in Central Europe.

Conclusions

Firstly, human activities have transformed much of the Earth's surface and influenced every part of the globe, and a range of models have been devised by different groups of scholars to illuminate these developments – the cultural landscape, the constructed niche, and the human techno-ecosystem. Many conservationists stress the separateness of nature and human activities, while others see humans as an evolved and evolving part of nature.

Secondly, through the activities of anthropologists, archaeologists and others, advances in our knowledge of the past have undermined the notion that many so-called wilderness areas are 'pristine'. However, it has to be recognised that myths are highly potent ideas not easily displaced by scientific advances (see Callicott & Nelson, 1998). Stott (2001) concludes that there are many myths associated with tropical rainforests, which have become 'an icon for all "Green" movements, for environmentalists, for Deep Ecologists, and for New Age folk throughout Europe and North America'. They are frequently presented as 'primeval', very ancient (thousands or millions of years old), and 'undisturbed', and represent a 'veritable cathedral of the wild'. 'To disturb the ancient harmony of this last surviving Eden was the ultimate human sin.' However, Stott points out that most 'tropical forests are less than 12,000 years old' and many have been cut and burnt by humans. He sees the notion of the 'tropical rainforest' as a Northern neo-colonial myth that has been exported worldwide through empire and education. In his view, 'it is a fine example of the prevalence, persistence and perils of *idées fixes* in environmental history and debate'.

Given these observations, it is important to consider the question: what determines our perception that something is natural or pristine? Jackson (2001, 5411) concludes: 'Our concept of what is natural today is based on personal experience at the expense of historical perspective.' Furthermore, 'we see the world through a model of our own creation that organizes and filters understanding'. Jackson notes that this theme is explored in a famous painting by Magritte called *La Condition Humaine*. Schama (1995, 12), describing the picture, reveals:'a painting has been superimposed over the view it depicts so that the two are continuous and indistinguishable'. Quoting the words of Magritte, Schama continues:

> 'We see it as being outside ourselves even though it is only a mental representation of what we experience on the inside'. What lies beyond the windowpane of our apprehension, says Magritte, needs a design before we can properly discern its form, let alone derive pleasure from its perception. And it is culture, convention, and cognition that makes that design.

Thus, different observers may come to different conclusions about the origin and significance of particular landscape features. For instance, Fairhead and Leach (1998, 2) studied the forest–savannah mosaics in West Africa. Their research revealed that the villagers regarded forest patches not as 'relics of destruction', but as areas 'formed by themselves or their ancestors in the savanna'. Others, especially colonial administors, however, arrived at a different reading of the landscape, regarding forest fragments as relics and savannahs as derived. Forested islands were therefore evidence of the wilderness being destroyed?

In considering the supposed 'wilderness' status of particular tracts of vegetation, a cautious evidence-based approach would seem appropriate. Rackham (2001, 675) admirably reflects this viewpoint in stating:

Some of the world's biodiversity is associated with 'virgin forest' and other ecosystems that are supposed to have escaped human interference until now. This category shrinks as archaeological research reveals more of the extent and persuasiveness of past human activities. The landscapes of most of the world's land surface result from long and complex interactions between human activities and natural processes.

Thirdly, Vitousek *et al.* (1997, 499), considering the implications of human activities, stress our increasing management of the biosphere, and conclude:

Humanity's dominance of Earth means that we cannot escape responsibility for managing the planet. Our activities are causing rapid, novel and substantial changes to Earth's ecosystems. Maintaining populations, species and ecosystems in the face of these changes and maintaining the flow of goods and services they provide humanity, will require *active management for the foreseeable future*. There is no clearer illustration of the extent of human dominance of Earth than the fact that maintaining diversity of 'wild' species and the functioning of 'wild' ecosystems will require increasing human involvement. [my emphasis]

Finally, while some have taken the view that efforts to protect wilderness are bound to fail as wild areas disappear (Woods, 2001), others have argued that efforts should be made to *restore* wilderness conditions to many landscapes (Callicott, 1991; Nelson, 1996; Callicott & Nelson, 1998). Later in the book, in considering a range of restoration projects in 'wilderness' and cultural landscapes, we examine how far such an endeavour has succeeded.

Chapters 4 and 5 have provided a brief sketch of the changing stage on which the present microevolution of plants and animals is being played out. In the next chapter, focusing largely on the plant component of ecosystems, but stressing the complexity of plant–animal relationships, we examine the different groups of plants that provide the 'actors' in the current evolutionary drama. How plants are classified into convenient groupings – weeds, invasive, endangered species etc. – reveals their 'role' in human societies. And this determines how they are treated and what human-induced deliberate and unwitting selection pressures are placed upon them.

6
Categories

The world has been transformed by human activities, and mankind is centre stage in the evolutionary drama now being played out in the world's ecosystems. In this chapter, different groups of plant species in the evolutionary play are introduced – native, wild, crop, weedy, introduced, invasive, feral and endangered species. On the face of it there would appear to be little difficulty in assigning plants to species, and deciding into which category particular species fall. But as we shall see, in confronting the following major questions, there are many complications. How are species to be defined? How many species of plants are there in the world? How are native and introduced species distinguished? What do conservationists mean by wildlife? How many species are cultivated, weeds or invasive? In response to different kinds of environmental change and disturbance, how many species are presently at risk of extinction, and are current extinction rates greater than those of the past? Is there any agreement on the numbers of species that are endangered? Is there any evidence that conservationists might have exaggerated the situation?

Species

The concept of the species is at the heart of any consideration of evolution, and both the theory and the practice of conservation are to a large extent based on these concerns. Different groups of plants – wild, cultivated etc. – are widely discussed in terms of the *numbers of species* they contain. In considering these estimates, however, a major concern must be faced. Stated bluntly, Rojas (1992) argues that there is 'no agreement on what species are, how they should be delimited, or what they represent'.

Early ideas about species

Plants are so important in human life that, from the earliest times, systems of naming plants developed in different parts of the world, and it is from European folk taxonomies that scientific naming, description and classification has developed (Briggs & Walters, 1997). Plant taxonomy was well developed at the time of Linnaeus, and, thus, many of the elements of the current taxonomic system were devised in the period of a belief in Special Creation and the fixity of species. There is a continuity of approach between these early endeavours and present-day taxonomic studies. Herbarium specimens are the key element in naming, describing and classifying plants, supplemented with information from cultivated material and plants collected in the wild. Species are defined on morphological criteria, based on the premise that the members of a species share common characteristics, and that there are morphological discontinuities between species.

In the post-Darwinian period, there has been a profound change of view. Species are not fixed, but have evolved and are evolving. Biologists continue to be fascinated by patterns and processes in microevolution. Investigations of evolutionary relationships have provided insights into ancestry and descent, and it has become clear that patterns of variation in some groups are complex. From these studies, many different sources of information are now available to those attempting to name species. Thus, for certain well-investigated groups, the taxonomist can draw on fossil evidence, ecological and geographical information, the results of crossing experiments and natural hybridisation, reports of chromosome numbers, biochemical data, and, increasingly, on the results of investigations employing molecular methods. There is considerable excitement at the prospect of finding genetic markers that could act as 'bar codes' to distinguish between species. Research reveals that the matK gene, located in the chloroplast, might be a suitable marker for bar-coding. (For reports of this research and full details of earlier literature on the subject see Lahaye *et al.* (2008)). Of course, this additional information is only available for a relatively small number of groups of plants, for much of the world's flora has still not been properly examined.

Species concepts

In their investigations of processes of evolution different botanists have defined species in relation to their own different purposes. As we saw in Chapter 3, to those concerned with breeding behaviour the Biological Species is characterised as: 'groups of actually or potentially interbreeding natural populations which are reproductively isolated from other such groups' (Mayr, 1942). A number of other species definitions are found in the literature. From an ecological perspective the species has been defined as: 'a lineage (or closely related set of lineages) which occupy an adaptive zone minimally different from that of other lineages in its range, and which evolves separately from all other lineages outside its range' (van Valen, 1976). To evolutionists, the Evolutionary Species is: 'an independent lineage that has its own unitary evolutionary role (ecological tolerance) and evolutionary tendency' (Simpson, 1961; Wiley, 1981). There are many other species definitions, including the Cohesion, the Genealogical, and the Phenetic Species concepts.

With regard to the diversity of species definitions advanced by those interested in evolutionary processes, Levin (2000, 9) writes:

> Whereas the processes of evolution are universal, the products are highly idiosyncratic owing to their different ancestries, potentialities, and evolutionary histories . . . we must keep in mind that species concepts are just concepts . . . Powerful as they may be in organizing diversity into distinctive

packages, any attempt to neatly fit biological diversity into a single species concept is likely to be futile.

Thus, botanists working as taxonomists in herbaria and others studying living plants in the wild, or in experiment, approach the species question from different perspectives. It is apparent from these considerations that species are 'equivalent only by designation and must therefore be regarded to a considerable degree as convenient categories to which a name can be attached' (Stace, 1980).

We arrive now at a point of serious disagreement. Some botanists – naïve realists – take the view that species are 'out there' in nature waiting to be recognised. Other botanists take a contrary view. For them, in a very real sense, species 'are made' by the process of looking (Briggs & Walters, 1997). We learn to recognise and circumscribe species in the context of our previous experience. Species are not objects waiting to be described, but a category employed by us in making useful classifications. Thus, the question for the naïve realist is: how many species are there in a particular region or group? In contrast, others would consider that the appropriate question should be: 'how many species is it convenient to recognise' in a particular set of material (Gilmour & Walters, 1963 and references cited therein). Clearly, botanists of different philosophical outlook could recognise different numbers of species within the same group of specimens.

In considering the vexed 'species' question, it is instructive to examine the views of Darwin (1901, 399), as they were set out in the concluding chapter of the *Origin*.

When the views advanced by me in this volume, and by Mr. Wallace, or when analogous views on the origin of species are generally admitted, we can dimly foresee that there will be a considerable revolution in natural history. Systematists will be able to pursue their labours as at present; but they will not be incessantly haunted by the shadowy doubt whether this or that form be a true species. This, I feel sure and I speak after experience, will be no slight relief. The endless dispute whether or not some fifty species of British brambles are good species will cease. Systematists will only have to decide (not that this will be easy) whether any form be sufficiently constant and distinct from other forms, to be capable of definition; and if definable, whether the differences be sufficiently important to deserve a specific name. This latter point will become a far more essential consideration than it is at present; for differences, however slight, between any two forms, if not blended by intermediate gradations, are looked at by most naturalists as sufficient to raise both forms to the rank of species. Hereafter we shall be compelled to acknowledge that the only distinction between species and well-marked varieties is, that the latter are known or believed, to be connected at the present day by intermediate gradations whereas species were formerly thus connected. Hence, without rejecting the consideration of the present existence of intermediate gradations between any two forms, we shall be led to weigh more carefully and to value higher the actual amount of difference between them. It is quite possible that forms now generally acknowledged to be merely varieties may hereafter be thought worthy of specific names; and in this case scientific and common language will come into accordance. In short, we shall have to treat species in the same manner as those naturalists treat genera, who admit that genera are merely artificial combinations made for convenience. This may not be a cheering prospect; but we shall at least be freed from the vain search for the undiscovered and undiscoverable essence of the term species.

Numbers of species in different categories

Not only do biologists have differing views about what constitutes a species, there are also concerns about how the practice of taxonomy might influence the numbers of species recognised in specific regions. This point may be illustrated by considering *Flora Europaea* (Tutin *et al.*, 1964–1980). Thus, Walters (1995, 365) writes: 'Early estimates of the number of species to be treated in full in the *Flora* produced a figure of 16–17,000.' This figure, synthesised from the numbers of species recognised in accounts of individual countries within Europe, proved to be an overestimate. The final total of species included in the *Flora* is *c*. 11,300. Not all the species named in European floras were accepted in *Flora Europaea*. Walters explains:

> Throughout the nineteenth and early twentieth centuries many Floras of individual nation-states of Europe were produced by botanists, whose main concern was too narrowly parochial. This restricted view . . . expressed itself in two directions; in the first place, their attempts to equate the native species of their own country with those of adjacent territories were often ineffective and, even more seriously, led to an over-enthusiastic 'splitting' of widespread and variable species, so that the newly-recognized taxon could be described as endemic to the country concerned. To a surprising extent, nationalistic zeal was able to distort the taxonomic picture.

A good example is provided by *Dianthus polonicus*, described from the Polish Carpathians by Zapałowicz in 1911. In *Flora Europaea* this taxon is now seen as an infraspecific variant within the widespread and variable species *Dianthus carthusianorum*.

The example quoted from Walters illustrates the process of 'lumping' together taxa previously thought to be separate. (A taxon (plural taxa) is defined as a classificatory unit of any rank.) However, there are many cases where, with increasing knowledge, the number of species has been increased by splitting, especially in problematic 'critical groups', e.g. *Rubus*, *Alchemilla*, *Sorbus*, *Taraxacum* etc. In early taxonomic treatments of these groups a few 'species' were recognised, but later, as a result of close study, such species were often split into a number of endemic 'micro-species'. Here, taxonomists are operating as 'splitters' rather than lumpers. Clearly, the degree to which taxonomists 'lump' or 'split' the material will determine the total number of 'species' in a region.

Conservation uses of the term 'species'

'Species' and their fate are of central concern to conservationists (Rojas, 1992). Loss of biodiversity is most often expressed in terms of the numbers of 'species' in danger of extinction, or may focus on threats to an individual 'species'. The management of reserves and other protected areas is to an extent based on the endangered 'species' they contain. Furthermore, such concepts as areas of species richness, biodiversity hotspots, centres of endemism, and Pleistocene refugia (regions where species survived the last ice age) are all based on 'species' distribution and numbers. In interpreting the significance of these concepts in conservation, it is important to acknowledge the complexities and uncertainties

Table 6.1 *Plant species diversity: known and estimated*

Taxon	Known diversity	Estimated diversity	% known
Chlorophyta[a]	14,200 – 16,250	34,000 – 124,000	11 – 48
Rhodophyta[a]	2,500 – 6,000	5,500 – 20,000	12.5 – 100
Chromista[a]	13,400 – 14,100	118,200 – 134,700	10 – 12
Protozoa[a]	2,650 – 3,050	5,500 – 13,000	20 – 55
Bryophyta[b]	16,500 – 17,000[c]	20,000 – 25,000[a]	68 – 85
Pteridophyta[d]	10,500 – 11,300[c]	12,000 – 15,000[e]	70 – 94
Gymnospermae	766[d]	835[ef]	92
Angiospermae	220,000 – 231,000	275,000 – 290,000	76 – 84
Dicotyledons	170,000 – 178,000[cdg]	210,000 – 220,000[ef]	77 – 85
Monocotyledons	50,000 – 53,000[cdg]	65,000 – 70,000[d]	71 – 82

[a] John (1994)
[b] Dr A. Touw (Rijksherbarium/Hortus Botanicus, Leiden University (RHHB), pers. comm.)
[c] Wilson (1988)
[d] World Conservation Monitoring Centre (WCMC) (1992)
[e] (also) based on extrapolation of Roos estimate of species richness in Malesia (Roos, 1993)
[f] Woodland (1991)
[g] Heywood (1978)
From Roos (2000). Reproduced with permission of the Linnean Society of London

of species concepts in plants. Bearing in mind the lack of a universal definition of plant species, it is appropriate to try to determine the number of plant species on Earth.

How many plant species are there in the world?

By the eighteenth century, many plant species had been named and a number of classifications proposed. However, information was scattered, and the difficulties of identification were becoming acute. 'This was the position which faced Linnaeus (1707–78), a Swede of tremendous energy and enthusiasm whose mission it was to record the works of the Creator – and who got up before dawn to achieve it' (Davis & Heywood, 1963, 16). In 1737, Linnaeus described all the genera of plants accepted by him, and, in 1753, in his *Species Plantarum*, he set out details, in the form of diagnoses, of nearly 6,000 species in 1,000 genera. As more areas of the world have been explored, botanists have described an increasing number of species.

Roos (2000, 57) provides a very useful summary of the number of species currently recognised in each of the major groups of plants, and estimates of the size of the task

Table 6.2 *Plant species diversity (tropics versus temperate areas)*

Taxon	Known diversity	Estimated diversity	% known
Algae	36,000	165,000 – 290,000	12 – 22
Cormophyta	252,000	308,000 – 330,000	76 – 82
Tropics[a,b]	162,000	208,000 – 230,000[c]	70 – 78
Temperate[a,b]	90,000	100,000	90
TOTAL	290,000	475,000 – 620,000	47 – 61
TROPICS	> 2/3 of the world's species diversity and *c*. 1/3 of the species described (1.5×10^6 in total) are tropical		

[a] Koopowitz & Kaye (1990)
[b] Raven (1988)
[c] Roos (1993)
From Roos (2000). Reproduced with permission of the Linnean Society of London

Table 6.3 *Characteristics of an ideal weed*

1. Germination requirements fulfilled in many environments
2. Discontinuous germination (internally controlled) and great longevity of seed
3. Rapid growth through vegetative phase to flowering
4. Continuous seed production for as long as growing conditions permit
5. Self-compatible but not completely autogamous or apomictic
6. When cross pollinated, unspecialised visitors or wind utilised
7. Very high seed output in favourable environmental circumstances
8. Produces some seed in wide range of environmental conditions; tolerant and plastic
9. Has adaptations for short- and long-distance dispersal
10. If a perennial, has vigorous vegetative reproduction or regeneration from fragments
11. If a perennial, has brittleness, so not easily drawn from ground
12. Has ability to compete interspecifically by special means (rosette, choking growth, allelochemics)

After Baker (1974) from Holzner & Numata (1982). Reproduced by permission from *Annual Review of Ecology and Systematics* © 1974 by Annual Reviews, www.annualreviews.org

ahead (Tables 6.1 & 6.2). Comparison of the estimated number of species in temperate and tropical areas is also provided. Evidence suggests that more than two-thirds of the world's plant species are found in tropical ecosystems, yet perhaps only about one-third of the species in the tropics have been named and described. Considering the Cormophytes (flowering plants, conifers, ferns etc.), about 90% of the species diversity in temperate floras has been documented, and much of the information is available in local and regional

Floras. In contrast, Roos (2000) estimates that only between 70 and 78% of the tropical species have been described, and notes that the existing taxonomic information on tropical plants 'is difficult to retrieve', as it is scattered in numerous publications, few of which provide detailed 'syntheses' for particular areas or groups.

While this book is primarily about plants, it is important to consider how many species of organisms are to be found in the ecosystems of the Earth. By employing different approaches, several estimates have been made, including ratios of known to unknown faunas and floras (Stork, 1997, 47). For instance, in well-known animal groups (mammals, birds etc.) there are twice as many tropical as temperate species. If this is true for other organisms, then, as about 1.5 million species have been described and two-thirds of these are temperate, globally there might be 3 million species. Estimates of the number of fungi have also been made. Stork (1997, 49) cites the work of Hawkesworth (1991), who discovered that in intensively studied areas the ratio of vascular plant to fungal species was in the range 1:1.4 to 1:6.0. Assuming that there are of the order of 270,000 species of vascular plants worldwide, 'Hawkesworth argued that this would give a conservative estimate of about 1.5 million species of fungi (including allowances for species in unstudied substrata)'. Others have tried to estimate the number of species worldwide by studying the number of organisms of various size classes in well-known ecosystems. Then, by using these ratios, zoologists have estimated the number of species in regions where the species of larger animals are well known, but the smaller organisms are relatively less examined (May, 1990). Further approaches have involved the sampling of organisms in particular places, and then the number of species globally has been estimated by extrapolation. The most famous of these studies is the investigation by Erwin (1983) of the diversity of beetles in the canopies of *Luehea seemannii* trees in Brazil. Collecting devices were set up below the trees and knock-down insecticides were blown into the canopies using a fogging machine. The total number of host-specific canopy beetles and arthropods was then calculated. Given that there are about 50,000 species of tropical trees, each with its own species-specific insects, and making some allowance for the number of arthropod species on the forest floor, by extrapolation, Erwin calculated that there might be as many as 30 million species of tropical arthropods worldwide. It is clear that there are huge numbers of insect species in tropical forests, and this has been confirmed in other intensive sampling surveys in different parts of the world, e.g. Sulawesi (Stork, 1997). The assumptions behind the '30 million' insects claim have been critically examined and many zoologists have reservations about the estimates obtained by extrapolation (see May, 1990; Stork, 1997). On present evidence, Stork (1997, 61) concludes:

> There seems little case to be made for estimates of 30 million or more global species, a more probable total is 5–15 million. If a single figure is to be selected, then that proposed by Hammond (1992) of 12.5 million species would seem reasonable at this stage. However, upward or downward revisions of this number could easily occur.

Discussing two major projects, Stork (2007) makes the case that large-scale sampling efforts will provide improved understanding.

About '1.8–2.0 million names have been used by taxonomists'. But 'in reality this probably equates to 1.4–1.6 million species' taking into account the problems of practical taxonomy in locating type specimens etc. (Stork, 1997, 43). It is clear, therefore, that a great deal of taxonomic work will be necessary to describe and name the Earth's biodiversity. While the mammalian and avian faunas of the world are well studied, an enormous amount of research is needed on the taxonomy of insects, fungi and micro-organisms in terrestrial ecosystems, and algae, annelids, molluscs and arthropods in the marine environments. Unfortunately, there are too few specialist taxonomists, especially in tropical countries (Gaston & May, 1992). Given the threats to many ecosystems, it is imperative that taxonomic research is better funded in the future. Otherwise, many species will become extinct before they are even discovered, described and named. Having explored some of the complexities of the species concept in plants, we turn to other categories that are important to the issues raised in this book.

Native and introduced species

When writing floras, botanists almost always try to determine whether plant species are native or introduced. If the natural components of ecosystems of a region are to be recognised and the natural distribution of a species determined, the distinction between native and introduced species is crucial. Moreover, such information provides a base line for the study of ecological and distributional changes over time. It also permits the recognition of the subset of native species that are rare or becoming rare and endangered, for which conservation measures are often devised to prevent their extinction. Conservationists are clear that particular attention should be paid to securing the future of endangered species that are narrowly endemic, i.e. those species that are native to a limited geographical area, and do not occur naturally elsewhere. Native species are those that have colonised an area by natural means of dispersal. These are distinguished from other species that have been introduced either accidentally or deliberately by human activities. However, it is clear that conservationists are not only concerned with native species; management is in place to safeguard species that are almost certainly introduced as agriculture spread from the Near East across Europe, e.g. *Adonis aestivalis*, *Adonis flammea* and *Agrostemma githago* in Central Europe (Scherer-Lorenzen *et al.*, 2000, 352) (Table 6.3).

Lines of evidence concerning status

In considering status, botanists take account of the findings of geologists and archaeologists. In addition, written and other records of various types have proved important. Mack (2000) has presented a very important review of the different classes of evidence available to understand the initiation and progress of invasions by introduced animals, plants and pathogens. Here in Box 6.1, we extend Mack's analysis: the same type of evidence provides valuable clues as to which species are native and which are in decline and possible candidates for extinction.

Box 6.1 Lines of evidence for determining the 'status' of different species

1 Early observations and concerns

- On the voyage of the *Beagle*, Darwin (1839, 454–5) commented on native and introduced floras.
- In 1865, the great environmentalist Olmsted produced a report on the 'newly set-aside' Yosemite Valley, noting, 'unless actions were taken, its vegetation likely would be diminished by common weeds from Europe' (quoted in Bossard *et al.*, 2000, 15).
- Bean (1976, 360–3) gives an account of the nineteenth century exploitation of the stands of Coastal Redwood (*Sequoia sempervirens*) and Big-tree (*Sequoiadendron giganteum*) in California. In an address to the Royal Institution in 1878 Hooker set out the concerns of many botanists, when he wrote: 'The doom of these noble groves is sealed. No less than five saw-mills have recently been established in the most luxuriant of them, and one of these mills alone cut in 1875 two million feet of Big-tree lumber... The devastation of the Californian forest is proceeding at a rate which is utterly incredible, except to an eye-witness.' Fortunately, Hooker's predictions were not realised. Following public protest, groves of these trees were offered protection, and are now conserved in protected areas.
- In Britain, from the 1840s onwards, a number of botanists expressed concern at the rapacity with which collectors went about their work (see Allen, 1980, 1987).
- In an assessment of the modern-day botanical collecting, it is recognised that some species may be driven into decline or even extinction (Norton *et al.*, 1994).

2 Herbarium specimens

Plant material – dried, mounted and labelled – can provide a verifiable record of the existence of a species at a particular place and at a particular date. Herbaria usually have material collected at different dates, and specimens can be used to prepare dot and other types of map, revealing the arrival and spread of an invasive species, and the distribution and decline of many rare species. Mack (2000) makes it clear that the interpretation of herbarium records involves certain assumptions. 1. The earliest dated herbarium specimen approximates to the first appearance of a plant in a new locality? If collecting is thorough and frequent, then this may be a realistic assumption, but, in areas little visited by botanists, herbarium records may not indicate with any accuracy the date of the first arrival of a species in a new territory. 2. Inconspicuous species may easily be overlooked, and some plants may not be obvious at certain times of year (Rich & Woodruff, 1992). 3. Once a species has arrived in an area it persists there, with the possibility of an expanding range. 4. It cannot be assumed that because there are no herbarium records that a particular species never existed at a particular site.

3 Independent dated evidence

Mack (2000) notes that survey records, for instance, for land-use changes such as railways, sometimes included plant lists. Also, special surveys were sometimes initiated to assess the importance of newly expanding weed populations.

- Towards the end of the nineteenth century, the threat of the weed Russian thistle (*Salsola iberica*) in North and South Dakota was so great that surveys were made to determine the distribution and severity of the infestation (Dewey, 1894).
- Local floras are very informative about dating the appearance and spread of introduced plants. For example, Mack (1984) has studied the information in four floras – published in 1892, 1901, 1914 and 1928 – in his studies of introduced plants of the Columbia Plateau of southern Washington.
- In Britain, there is a tradition of writing local county floras. For most counties successive floras provide comprehensive lists of all the species found in a defined and largely fixed territorial area, presented in a historical context. The decline or extinction of any species within the county is documented, as is the arrival of new introduced species, some of which later become invasive.
- McCollin, Moore and Sparks (2000) have analysed the comprehensive archival botanical information in two *Floras* of Northamptonshire – Druce (1930) drawing on the records of the Natural History Society from 1876 onwards, and Gent and Wilson (1995). About 100 species recorded by Druce in 1930 had become extinct by 1995. Also, many species not listed by Druce had been introduced at some time, but had failed to establish themselves. A number of species had become successful invaders, e.g. *Impatiens capesis* and *I. glandulifera* along waterways; *Pilosella aurantiaca* ssp. *aurantiaca* along railway banks; *Conium maculatum*, *Lactuca serriola*, and *Heracleum mantegazzianum* along roadsides; *Veronica filiformis* on recreation grounds and church yards; and the 'salt-spray' species *Cochlearia danica*, *Puccinellia maritima* and *Aster tripolium* along those roads where salt is spread in winter to combat ice and snow. Overall, since the Second World War, there has been a marked increase in those species associated with higher soil nitrogen, reflecting the increasing use of chemical fertilisers (McCollin *et al.*, 2000).
- Middlesex was a small county containing the greater part of Central London. The *Flora of Middlesex* (Trimen & Thiselton-Dyer, 1869) contains many early records. Kent (1975) discovered that 78 native and naturalised species have become extinct during the intervening century, and many species had seriously declined in frequency. In this case, species losses were the result of the relentless growth of London. Kent also records many additions to the county list including 100 introduced species, e.g. the North American species *Bromus carinatus* escaped *c.* 1919 from the Royal Botanic Gardens, Kew; and *Epilobium adenocaulon*, thought to have been introduced with fodder in the First World War. Some introductions have increased in frequency, e.g. *Senecio squalidus* (originally from Sicily) escaped from the Oxford Botanic Garden and spread along railway lines of the Great Western railway into central London, and later to many other parts of Britain.
- An on-line *Historical Flora of Cambridgeshire* has been prepared by Crompton and associates, which documents the published and unpublished historical records for the changing flora of Cambridgeshire dating back to 1538 and includes information from the first county flora – John Ray's *Catalogus Plantarum circa Cantabrigiam nascentium* of 1660. The database

includes information in later floras of the county – published in 1785, 1786, 1793, 1802, 1820, 1860, 1939) and Perring, Sell & Walters (1964), together with material in published and unpublished sources and information on herbarium specimens (www.cambridgeshireflora.com).

- Dirnböck *et al.* (2003) employed a number of sources of historical information in their assessment of threat to the native vegetation posed by invasive species on Robinson Crusoe Island, in Juan Fernandez Archipelago, Chile. They conclude that many of the 124 endemic plant species on this famous island, which is in a Chilean National Park, are threatened by invasions of introduced plants, in particular *Acaena argentea* (first report of introduction 1864), *Aristotelia chilensis* (1864), *Rubus ulmifolius* (1927) and *Ugni molinae* (1892).

4 Dot maps and other grid-based surveys

Walters (1957) reports the pioneering work of Hoffmann, Professor at Giessen, near Frankfurt, Germany, who produced what is possibly the first dot map showing the local distribution of *Prunella grandiflora* and *Dianthus carthusianorum* in the Kissingen area, and possibly the first use of grid methods in studying the distribution of *Sambucus ebulus*. Later developments of grid-based surveys has led to the production of atlases of the distribution of the plants of many regions e.g. Great Britain and Ireland (Perring & Walters, 1962), Germany (Haeupler & Schoenfelder, 1989), and the Netherlands (Van Der Meijden, Plate & Weeda, 1989). Such enterprises are of inestimable value as dated base-line accounts.

- Important limitations in the use of cell-by-cell assessments of the changing fortunes of native and introduced plants species. 1. Grid squares are not all visited with equal frequency or intensity. 2. Some species are more evident at particular times of year (Rich & Woodruff, 1992). Some of these limitations, once recognised, can be overcome by the sort of deliberate searches that have characterised recent studies of the British flora, and which have produced such impressive and thorough data sets.
- The monumental *New Atlas of the British and Irish Flora* (Preston, Pearman & Dines, 2002) provides a wealth of historically based information from systematic surveys carried out at different dates, revealing many changes in the distribution of native and introduced elements of the flora. Thus, a number of species have increased their range, e.g. the introduced species *Buddleja davidii* and *Lysimachia punctata*, while others have seriously declined, e.g. *Ranunculus arvensis* and *Scandix pecten-veneris*.
- Deliberate searches have played a major role in determining the status of particular species. *Berberis vulgaris* is the intermediate host for stem rust of wheat (*Puccinia graminis*) in the USA. A federal survey begun in 1918 was carried out to inspect more than 1,800,000 farms and destroy any *Berberis* plants they found.
- Searches have also resulted in the refinding of species thought to be extinct. The optimistically named Santa Barbara Island Live-forever (*Dudleya traskiae*), found only on a small island off the coast of California, was believed to be extinct (last sighted 1968). However, in 1975, after reduction in the population of introduced hares, several plants were discovered (Lucas & Synge, 1978). A year later, another population was detected on a cliff inaccessible to hares.

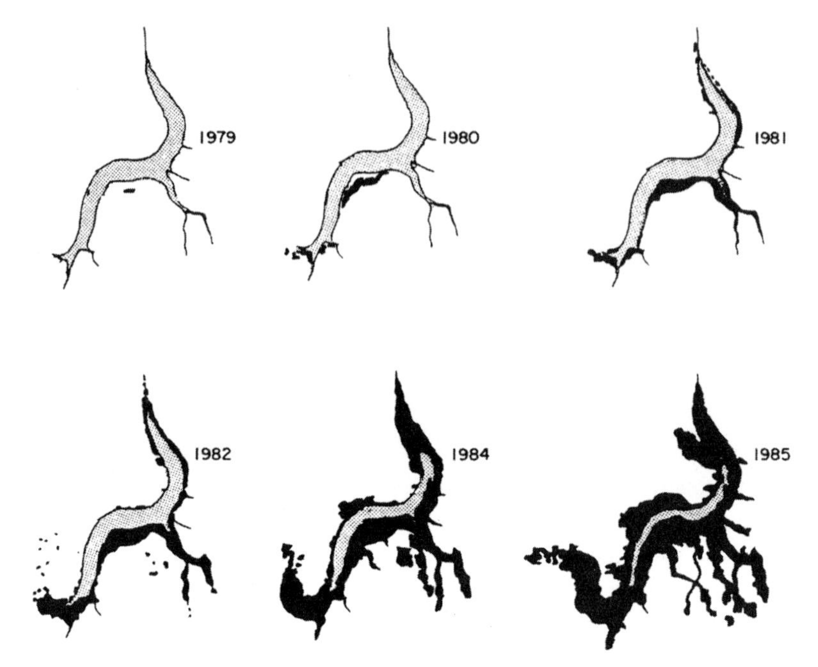

Fig. 6.1 Spread of *Mimosa pigra* in the Adelaide River floodplain, Northern Territory, Australia: evidence from aerial photographs. (Reproduced by permission of the British Ecological Society, from Lonsdale, 1993)

5 Indirect methods

The use of remote sensing technology (aerial photography and satellite imaging) offers many possibilities, especially where provisional conclusions are checked by fieldwork (see Mack, 2000, 158).

- The use of aerial photography is illustrated by studies of the dated advance of colonies of the introduced invasive species *Mimosa pigra* in the remote Northern Territory of Australia (Fig. 6.1).
- Using SPOT satellite imagery, supplemented by ground surveys, a pilot study from 1994 was made of water hyacinth infestations on Lake Victoria (Anon., 1995).
- In many cases it may not be possible to distinguish between native and introduced species using those satellite-imaging techniques currently available. The spectral signature of invasive *Melaleuca quinquenervia*, deliberately introduced to the Florida Everglades from Australia, does not differ significantly from that of native plants.
- But developments in aerial photography – the use of infrared techniques – have made it possible to estimate the extent of the infestations by the four principal invaders of the Florida Everglades (Australian Paperback (*Melaleuca*), Brazilian Pepper (*Schinus terebinthifolius*), Castuarina (*Casuarina equisetifolia*) and Latherleaf (*Colubrina asiatica*)) (Welch, Remillard & Doren, 1995).

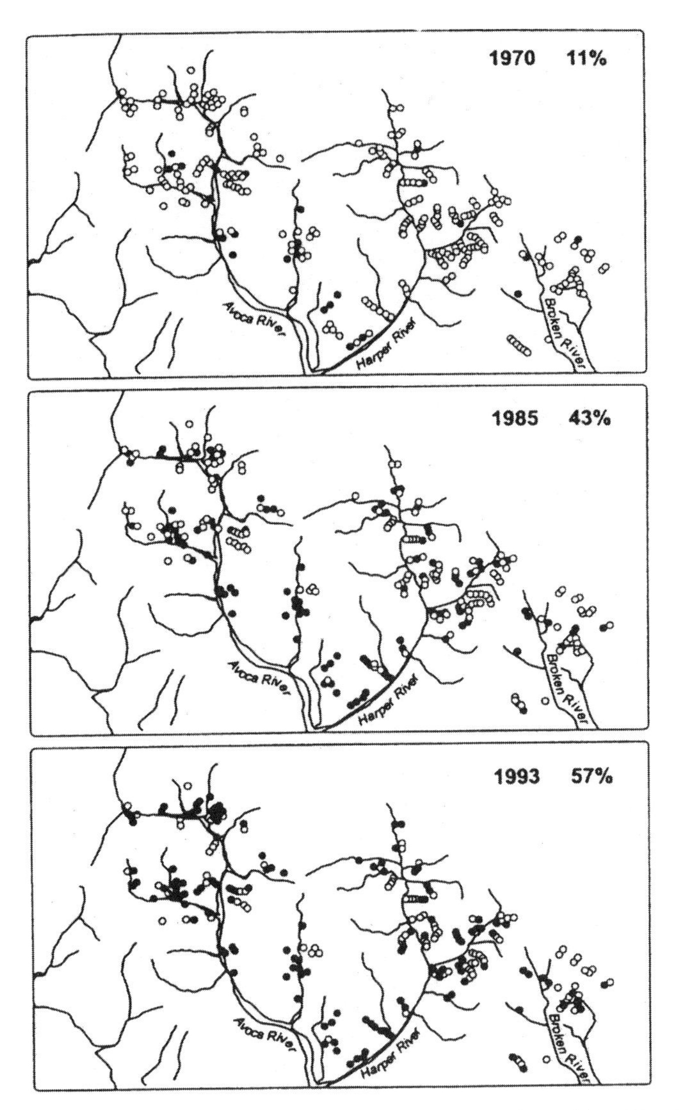

Fig. 6.2 The course of invasion in the eastern ranges of the Southern Alps, New Zealand by *Hieracium lepidulum*. Evidence from studies of permanent plots recorded 1970, 1985 and 1993: black circles indicate the presence of *Hieracium*; white circles the species was not present. (Reproduced by permission of the Ecological Society of America from Wiser *et al.*, 1998)

6 Permanent plots

The repeated examination of permanently marked plots has the potential to reveal information about the competitive interactions of endangered species and invasive taxa in relation to the other components of the ecosystem (see review by White, 1985, which refers to the pioneering studies of A. S. Watt). Mack (2000) stresses that permanent plots will provide valuable information about the effects of global climate change.

- Anable, McClaren and Ruyle (1992) used 75 plots to examine the spread from 1959 onwards of Lehmann Love Grass (*Eragrostis lehmanniana*) in Arizona. This grass was deliberately introduced into an area of 200 hectares at the Santa Rita Experimental Range, and has now spread to 85% of marked plots across 20,200 hectares.
- New Zealand has a network of plots, including more than 3,000 in grasslands. The invasive behaviour of *Hieracium lepidulum* in the Southern Alps is clearly documented (Fig. 6.2).

Assessing the evidence: criteria for native and introduced status

In 1985, Webb produced a very helpful review of the criteria for presuming native status. The paper deals with the plants of Britain and Ireland, but the general principles it presents may be applied worldwide. Webb considers the agents of natural dispersal – animals, wind, water etc. – but recognises that humans have become agents for dispersal on quite a different scale.

> While he was still a hunter or food-gatherer he may well have shifted some seeds from one place to another, but only in the same manner as a bear or an ape. But as soon as he began to herd flocks and to till the ground his impact on plant-geography suddenly increased enormously, and he ceased to be in any ordinary sense a part of nature but became a phenomenon *sui generis* – the only one of its kind. (Webb, 1985, 231)

Webb examines eight criteria for presuming native status.

Fossil evidence

In some well-studied regions of the world, fossil pollen, leaves, fruits etc. have been discovered, and positive identifications have established the native status of some species. For example, there is clear evidence in Britain that certain weedy species, previously regarded as introduced, are indeed native, e.g. *Aethusa cynapium*, *Solanum nigrum*, *Sonchus oleraceus*. In the post-glacial period it is possible that many of the species we now call weeds were plants of naturally open ground, such as 'landslides, moraines, beaches, dunes and similar habitats', and it was from these sites that they invaded the first agricultural sites and assumed the status of weeds (Salisbury, 1964, 24).

While there are cases where the assignment of a species to native status can be made with confidence, the status of others is uncertain, as it is not possible to identify all fossil plant remains to a specific level. Also, it must be stressed that where there are no fossil remains, the status of species cannot be determined.

Historical evidence

Although there are no direct records of plants introduced by prehistoric movements of humans into new territories, early records and herbarium specimens are available as evidence for recent migrations. For example, as we shall see in Chapter 11, it has proved possible to estimate the date of arrival and chart the progressive spread of some introduced species that have become invasive and threaten the native floras of North America, Australia, New Zealand etc.

Habitat

If a species only grows in man-made habitats then it is likely to be an introduction; if, on the other hand, the species only grows in natural habitat then it is likely to be a native plant. However, some introduced plants have invaded semi-natural areas, for example *Rhododendron ponticum* in Britain. Moreover, the use of this criterion begs an important question: What is a natural habitat? As we have seen above, most, if not all, ecosystems have been influenced to some degree by human activities.

Geographical distribution

If a species is accepted as introduced in an 'adjacent' territory, then a very good case has to be made for accepting the same species as native in adjoining territory.

Frequency of naturalisation

A claim for native status in a particular area must be examined in the light of the plant's behaviour and ecology in other parts of its range.

Genetic diversity

Webb suggests that patterns of genetic variation might be helpful in determining whether a species is native or introduced. Generally, a new population is founded by one or a few individuals, and, therefore, recent immigrant populations may be genetically depauperate (see Chapter 11). In contrast, native species might be more variable. However, as we shall see in later chapters, lack of variability in itself may not be indicative of status, for many populations of rare and endangered species, presumed to be native, lack variability through processes of genetic erosion.

Reproductive pattern

It might be assumed that native species, especially on islands, would reproduce by seed. In contrast, some introduced species are known to reproduce entirely by vegetative means, e.g. the self-incompatible species *Veronica filiformis*, introduced from the Caucasus found on many Cambridge lawns, does not reproduce by seed. Clonal material of single genotype spreads between gardens, as fragments are carried on lawn mowers, and in lawn mowings etc. (Lehmann, 1944). However, not all vegetatively reproducing species are aliens, e.g. the native montane species *Polygonum viviparum* and *Festuca vivipara* reproduce by vivipary (detachable bulbils or shootlets).

Possible means of introduction

If it is alleged that a species is introduced, then some plausible means of introduction should be presented.

Webb concludes that in Britain far more introduced species (often called aliens) are represented as natives than vice versa. An additional criterion for determining status was suggested by Preston (1986) and concerns the relationship of the species to oligophagous insects. For example, Coombe (1956) demonstrated that the presence of 'species-specific' insects on *Impatiens noli-tangere* can be used to distinguish sites where the species is probably native (NorthWales and north-west England), from sites in southern England where it is a garden escape. However, it is clear that introduced plants can be adopted by insects from related native species, as is the case in *Impatiens parviflora*. It must also be borne in mind that aliens introduced into new territory as vegetative plants may bring with them their native insect associates, e.g. *Azolla* and *Carpobrotus* species in Britain.

Three different groups of introduced plants have been distinguished in Europe. Those that were established before *c*. AD 1500 – the archaeophytes – have often attained the more or less continuous distributions typical of native species. Those plants first introduced after 1500 or present only as casuals (see below) before 1500 are called neophytes by Preston *et al*. (2002). Many neophytes are introductions from the Near East and the New World, reflecting the colonisation of new lands by Europeans post AD 1500. While many introduced plants form successful long-term colonising populations, it has proved useful to recognise a third group – the casuals. These are introduced species that fail to persist in the wild for more than approximately five years. Only by repeated reintroduction do they remain as long-term members of ecosystems. In the present account we have followed Preston *et al*. as to dates and definitions. Heywood (1989, 33) draws attention to different usage of the terms archaeophyte and neophyte. In the MedChecklist archaeophytes are defined as plants introduced 'before the end of the fifteenth century' (Greuter, Burdet & Long, 1986), while Webb (1985) proposes 'about AD 1550' as the dividing line. Some authors regard archaeophytes as part of the native flora, for example the MedChecklist. There are differences in usage of the term neophyte. For instance, in considering the adventive flora of Israel, Dafni and Heller (1982) reserve the term neophytes for recent introductions.

It is clear that it may be difficult or impossible to decide the status of a species – native or introduced. Moreover, it is important to face a further complication. The conclusion that a species is native to a particular territory may be supported by firm evidence, but it could be erroneous to conclude that *all* the populations of that species in a particular territory are necessarily of pure 'native' origin. Plants and/or seeds of many species are widely introduced across the world. For example, stocks of many native British trees, shrubs and herbs have been imported from Europe and planted in Britain. With regard to weeds, Salisbury (1964, 28) emphasises the same point about introduced species:

> It is highly probable that . . . there have been repeated replenishments, perhaps reintroductions, and these may well have brought strains physiologically if not morphologically different from those already present . . . So the weed of the present, though belonging to the same species, must not be assumed to be identical with the weed of the past.

To determine the situation for a particular species, each case must be investigated individually. The interaction of stocks of different status must also be considered, for while some pure non-native populations may occur, hybridisation between native and introduced stocks might take place where they grow together.

Wildlife

A major focus of conservation efforts worldwide is to safeguard the wildlife of the area. Which species of plants and animals are to be given the status of 'wild'?

In 1945, Tansley published his famous book: *Our Heritage of Wild Nature: A Plea for Organized Nature Conservation*. In post-war Britain his concern was for the future of native vegetation, rural beauty, and in particular for the conservation of wildlife, a category that he carefully defined. His definition of wildlife, although devised in a British context, is of wider significance. Firstly, he noted: 'Many people mean by wild life only the larger wild animals, the birds, and sometimes fishes.' But Tansley's definition was more inclusive (Tansley, 1945, 4): 'By "wild life" is here meant all kinds of native animals and plants which maintain themselves without human assistance.' Wild plants are therefore those that are self-sufficient: i.e. able to disperse, establish and reproduce on their own, as distinct from plants that are deliberately sown, cultivated or distributed by human agency. Tansley did not limit the term 'wildlife' to native plants and animals, for he includes 'those introduced species which are securely established and reproduce themselves freely in the country'. Thus, many species that have been deliberately or accidentally introduced in Britain are included in wildlife, if they pass the test of reproductive independence.

The notion that wildlife is a subset of creatures that maintains itself without human assistance is a keystone of Tansley's definition. However, as human activities have placed much wildlife at risk of extinction, increasingly conservation efforts are needed to encourage, protect and indeed ensure their survival. Thus, through management of national parks, nature reserves, gardens and in the wider agricultural and urban landscapes, humans often deliberately manipulate/regulate the numbers in wild populations, sometime even providing legal protection. As we shall see in later chapters, without this human protection and management some wildlife would be unable to survive.

Wild and cultivated plants

Considering Tansley's definition of 'wildlife', it is important to recognise that such has been the use of plants for millennia that it is problematic to categorise plants into the watertight categories – wild, cultivated or domesticated (Harlan, 1975). For instance, research has established that in prehistory a very wide range of plants were used for food (Renfrew, 1973), with other plants providing further valuable resources, e.g. fibres, medicinal plants etc. (Dimbleby, 1967).

Considering the contemporary world, Solbrig (1994, 18) concludes:

> The number of species of plants and animals on which human society depends is very narrow. Twenty species of plant and five species of animal account for over 90% of all human sustenance and international commerce in foodstuffs. Three cereal plants, wheat, rice and maize, account for 49% of human calorie intake. If we increase the list to 100 species, we cover 98% of important economic plants and animals, and if we enlarge the list to a thousand species we include essentially most cultivated and useful plants and animals, excepting some ornamentals as well as species used in folk medicine.

In making independent estimates, Heywood and Stuart (1992, 103) consider that '5000 plant species are cultivated by human kind, and at least twice as many are harvested from the wild'. Other estimates have been published, for instance, Vietmeyer (1995) considers that as many as 20,000 plant species have been used for food.

While no published listings or databases claim to provide a complete list of species cultivated as ornamentals, two sources reveal the enormous number of species grown in gardens. Griffiths (1994) lists 60,000 taxa in an *Index of Garden Plants*, while *The European Garden Flora: A Manual for the Identification of Plants Cultivated in Europe, Both Out-of-doors and Under Glass* provides information about more than 17,000 species (Dr J. C. Cullen, personal communication). In addition, to this list of cultivated species must be added an enormous number of cultivars developed by plant breeding, hybridisation etc. For instance, *c.* 55,000 named cultivars are commercially available in the UK (*The Plant Finder*: Anon., 1997). This figure is unlikely to be an accurate reflection of the total number of cultivars: for example, 20,000 cultivars of *Rhododendron* have been registered with the Royal Horticultural Society, UK (Dr J. C. Cullen, personal communication).

In practice, it is abundantly clear that the attempt to divide plants simply into 'wild' and 'crop' often presents major difficulties, for the term 'crop' is used for all those plants that are harvested without implying that the plants concerned are domesticated. Harlan (1975) provides a number of interesting cases of the intermediate states between wild and cultivated plants. In West Africa the oil palm (*Elaeis guineensis*) is a very valuable tree. It occurs in wild stands in areas of shifting cultivation associated with slash and burn agriculture. In slashing the bush prior to the burning process, the palm is spared and thick stands develop. Here is a species that has never been deliberately planted from seed that yields important harvested products. Selection is practised in the palm stands. Variants with thin shells around the oily kernel are harvested for palm oil, while trees with a very thick kernel wall, from which it is more difficult to extract the oil, are tapped for sap to produce palm wine, a process that finally kills the tree, reducing the frequency of thick kernel types in the population.

In Melanesia there are wild and domesticated variants of the sago palm *(Metroxylon sagu)*, a tree with an edible starchy pith. Human selection is applied in wild populations – thorny variants are selectively cut back when young, leaving non-thorny trees in more or less pure stands (Harlan, 1975).

Throughout the world, many 'wild' species are regularly harvested as a crop. Others are used as 'emergency', starvation or famine foods (Etkin, 1994). Indeed, in Britain 'it is difficult to discount any species as never having been used for some purpose' (Raybould, 1995).

Feral plants

Complex situations have also sometimes arisen when cultivated plants have reverted to a 'wild feral' state outside their agricultural environment (Raybould, 1995). For instance, in Yora, Peru, 'wild bananas' have been found. It has been discovered that these are feral populations probably established on riverbanks from rootstocks washed downstream from cultivated areas (Hill & Kaplan, 1989).

Raybould (1995) describes several different types of feral crop in Britain. In some cases such populations are short-lived, e.g. from seeds of cereal crops scattered in other areas. However, in other cases, persistent populations may be developed, but it is not clear whether they are fully self-sustaining, e.g. oilseed rape. In many cases it is not possible to distinguish between domesticated plants and non-domesticated wild variants of the same species, e.g. forage grasses and legumes, and, therefore, it is not clear how far feral populations develop. Also, it may be difficult to distinguish between feral and native populations. Raybould (1995) discusses the case of populations of wild cabbage (*Brassica oleracea*) found on sea cliffs in Britain. Such populations could be native to the British Isles, but perhaps they might be longstanding feral populations arising from cultivated cabbage first brought to the British Isles by the Romans. Finally, Raybould (1995) observes that many ornamental species introduced and grown in gardens have developed feral populations in the British countryside, e.g. rhododendron (*Rhododendron ponticum*), Japanese knotweed (*Fallopia japonica*), Indian balsam (*Impatiens glandulifera*) and giant hogweed (*Heracleum mantegazzianum*).

Worldwide, feral plants are important components of many ecosystems. A famous example is provided by the feral form of the cultivated globe artichoke. On the voyage of the *Beagle*, Darwin reports that he found enormous populations on the pampas of Argentina on his overland journey from Bahia Blanca to Buenos Aires: 'Very many, probably several hundred square miles are covered by one mass of these prickly plants and are impenetrable by man nor beast. Over the undulating plains where these great beds occur, nothing else can now live.' It also occurs in California, where it was introduced in the 1860s. Reporting that by the 1930s the species had covered more than 150,000 acres of rangeland, Kelly (2000, 141) also provides a useful commentary to Darwin's observations. Cultivation of edible artichokes appears to have spread from the Naples region into the broader Mediterranean in the fifteenth century. The globe artichoke (*Cynara scolymus*), grown by dividing plants vegetatively, produces large spineless edible flower-heads. Plants raised from seed of the edible artichoke revert to a wild form, which has large spines. Emigrants from the Mediterranean region carried the edible plant to many countries, where feral populations escaped cultivation, e.g. on the pampas in Argentina.

Weeds

Harlan (1975, 85) points out that the term weed involves a value judgement: 'A weed is a herbaceous plant, not valued for use or beauty, growing wild and rank, and regarded as cumbering the ground or hindering the growth of superior vegetation' (Murray *et al.*, 1961: *Oxford English Dictionary*). For others, the aim has been to define weeds in biological terms. Harlan and de Wet (1965) regarded a weed as 'a generally unwanted organism that thrives on habitats disturbed by man'. Radosevich, Holt and Ghersa (1997) cite the views of Zimmerman (1976) who uses the term 'weed' to describe plants that '(1) colonize disturbed habitats, (2) are not members of the original community, (3) are locally abundant and (4) are economically of little value [or are costly to control]'.

In a critique of the definitions of weeds, Holzner (1982, 4) notes that some ecologists regard weeds as 'colonising plants with a special ability to take advantage of human disturbance of [the] environment'. While this obviously applies to such areas as arable land, it does not adequately cover the wide range of weeds that have been described in plantations, managed forests, grassland and aquatic habitats etc. Therefore, Holzner (1982, 5) promote a wide and more inclusive definition: 'weeds are plants adapted to man-made habitats and interfering there with human activities'.

Concerning the number of weed species, Holm *et al.* (1977a, b) concluded that less than 250 species were universally recognised as troublesome weeds. But it is clear that, on a local scale, a large number of other species often grow where they are not wanted.

It has proved useful to recognise several categories of weeds (King, 1966; Holzner, 1982, 6). Agrestals are weeds found on many types of cultivated ground, e.g. tilled, arable land supporting root, vegetable or cereal crops. They are also found in orchards, gardens, and plantations of such crops as sugar cane, tea and coffee. Different agrestal weeds are at a selective advantage in different types of cultivated land. For those areas subject to frequent cultivation, annual species may predominate, but there may also be some very serious perennial weeds. In climatic zones where there is a cold season, the timing of the cultivation processes – ploughing, sowing, pesticide and fertiliser treatments, harvest practices etc. – has a major influence on the species composition.

In the management of plantation crops and orchard crops, cultivation may be less frequent, and perennial and even woody plants may be serious weeds. In some countries, the distinction between orchard/forest/grazing and arable land is very clear, but in others there may be patches of arable or grazing land within orchards/plantations, or complex polycultures of species of many life-forms including trees. For example, there are a variety of different practices in the Amazon. Miller and Nair (2006) describe systems 'ranging from deliberate planting of trees in home gardens and fields to the management of volunteer seedlings of both cultivated and wild species. These practices result in various configurations of agroforestry systems, such as home gardens, tree/crop combinations in fields; orchards of mixed fruit trees and enriched fallows.'

Ruderals are weeds of human-disturbed areas such as rubbish dumps, wasteland roadsides and railway lines. This highly heterogenous group of human-made habitats may be characterised by severe episodic disturbance, and may involve the dumping of plant, animal, agricultural and industrial wastes. In so far as these sites occur within a matrix of agricultural land, ruderal habitats may act as reservoirs of agrestal weeds. In newly disturbed sites, annual pioneer species may predominate, but, if areas are left undisturbed for a period, early and later successional perennial plants including shrubs and later trees may become established.

There are also very characteristic 'weeds' of various grasslands (grazing lands, pastures, meadows and lawns etc.). Foresters too have to contend with weeds. Thus, annual weeds occur in nursery areas of tree seedlings, but, following transplantation, perennial, shrubby, climbing and woody weed species are frequent in older plantations. In some parts of the world, there are parasitic species of weeds. Economically important weeds are also found in water bodies, rivers and canals. The species composition of contemporary aquatic communities is highly influenced by the scale and composition of any water contamination that might occur, e.g. pollution by industrial and agricultural chemicals. Many species, including introductions, have been able to spread widely in river systems that have been linked by canals dug for irrigation and transport.

Invasive plants

Invasive species – animals, plants, micro-organisms – may modify natural and semi-natural habitats to such an extent that they endanger native species and ecosystems. It has been estimated that invasive species now 'dominate 3% of the Earth's ice free surface' (Mack, 1985). 'Vast land and waterscapes, in certain regions, are completely dominated by alien species, such as the star thistle (*Centaurea solstitialis*) in the range lands of California, cheat grass (*Bromus tectorum*) in the intermountain regions of western United States, and water hyacinth (*Eichhornia crassipes*) in many tropical lakes and rivers' (Mooney & Cleland, 2001, 5446). Cronk and Fuller (1995, 1–2) define invasive plants as aliens 'spreading naturally (without the direct assistance of people) in natural or semi-natural habitats, to produce a significant change in terms of composition, structure or ecosystem processes'. In this context they define natural or semi-natural as 'communities of plants and animals with some conservation significance, either where direct human disturbance is minimal or where human disturbance serves to encourage communities of wild species of interest to conservation'. Thus, Cronk and Fuller try to make a clear distinction between 'invasive' plants and the weeds of agricultural and other highly human-disturbed areas. However, it is obvious from the literature that others are less precise in their use of the term 'invasive'. For example, Mooney and Hobbs (2000, xiii) focus on alien species 'that are exacting a toll on ecosystem diversity or ecosystem processes or "services"'. They draw attention to the variety of terms used by those publishing in this area by noting: 'the invaders go by many names – exotics, aliens, pests, weeds, introduced species, non-indigenous species'. It is important to note, however, that not all invasive species are introduced. In human-disturbed habitats native species may, in some situations, become invasive. A good example is provided by bracken in various parts of the world, including Britain (Green, 2003).

There are no comprehensive worldwide lists of invasive species. Daehler (1998) assembled a 'global data set' of 381 species gathered from 16 published papers and books, supplemented by species noted on websites of 'local resource management agencies'. The sample included 'invasives' from a wide range of habitats and tried 'to sample globe evenly'.

Given the focus of this book, it is important to note that invasive plants have entered many national parks, e.g. Kakadu National Park, Northern Australia (Cowie & Werner, 1993), Teide National Park, Tenerife (Dickson, Rodriguez & Machado, 1987); Kruger National Park (Macdonald, 1988); highlands of the Galápagos Islands (Macdonald, Ortiz & Lawesson 1988); and Yosemite National Park, California (Underwood, Klinger & Moore, 2004).

Concerning other accounts of invasive plants of particular territories, Mooney and Drake (1986) examine the biological invasions of North America and Hawaii. The wide-ranging survey of Mooney and Hobbs (2000) contains chapters reviewing what is known about the invasive plants of South Africa, Germany, New Zealand and Chile. Another extremely valuable source provides details of 78 invasive species that are damaging the wild lands of California (Bossard *et al.*, 2000). Cronk and Fuller (1995) have provided an illuminating set of case histories, and some useful lists, of selected invasive plants from around the world.

While it is appropriate to pay attention to the numbers of invasive species, it is important to make the point that individual invasive species have had a profound effect on the vegetation of various territories. For instance, Heywood (1989) emphasises the disproportionate effects of *Casuarina littorea* in the Bahamas (Correll, 1982), and *Andropogon pertusus* in Jamaica (Adams, 1972). In Egypt, there has been a huge expansion of *Parthenium hysterophorus*, a Texan species apparently accidentally imported, in 1960, as an impurity in a widely sown batch of grass seed (Boulos & el-Hadidi, 1984).

Endangered species

There have been several approaches to the problem of determining how many species are endangered. One of the most important derives from the observation made by many biologists that small islands have fewer species than large islands. The general rules for the distribution of species diversity on islands were explored by MacArthur and Wilson (1967) in their investigations of island biogeography. Conservationists have become interested in the properties of 'island' habitats because human activities have resulted in a reduction and fragmentation of many natural ecosystems. For example, where once there were uninterrupted large tracts of natural forest, now, often, there are only 'small islands' of forest in a sea of agricultural land. Intuitively, it might be expected that massive loss of forest habitat could result in species extinction.

At the heart of the island biogeography model is the species–area relationship, which May *et al.* (1995, 13) characterise as follows:

> The species–area relation is an empirical rule, based on a variety of studies of how the number of species, S, of a particular taxonomic group (beetles, birds, vascular plants etc.), found on individual islands within an archipelago depends on the area of the island, A. The islands may be real islands in the sea, or virtual islands, such as freshwater lakes or isolated mountain-tops. Very often, log-log plots of S against A show a straight line, so that the relation can be written: $S = cA^{z}$. Here, c is a constant, and the parameter z has values in the rough range $z \sim 0.2$–0.3 (MacArthur & Wilson, 1967; Diamond & May, 1976). This rough rule is often expressed by saying that if the area of suitable habitat is reduced to 10% of its pristine value, the number of species will be halved.

Wilson (1992, 280) employed the species–area relationship to determine the loss of species from tropical forests. For the calculations it is necessary to estimate, firstly, the number of species in tropical forests, and secondly, the rate of deforestation (Cleuren, 2001). Given the high estimates for the number of species proposed by Erwin and others, in selecting the figure of 10 million, Wilson could be seen as making a conservative estimate. There are various estimates of forest loss. Using the figure of 1.8% per annum, Wilson concluded that as many as 27,000 species could be going extinct per annum.

How does this figure compare with 'normal background' extinction rates? Given that the typical lifespan for a species could be of the order of 1 million years (Wilson, 1992, 132–141), it would follow that, if there were 10 million species in tropical forests, this would give a background extinction rate of 10 species per year. Wilson's estimate of 27,000 species becoming extinct per year is an alarmingly high figure. Predictions were also made by other conservationists, and they sometimes arrived at even higher estimates of the rate of species extinction. For example, Myers (1979, 31) speculated: 'it is likely that during the last quarter of this century we shall witness an extinction spasm accounting for 1 million species'.

However, Mann (1991) considers that the 'doom laden prophecies of mass extinction are based on assumptions that have modest scientific support and are wide open to question'. What are the assumptions behind calculations of extinction rate using the species–area relations? Many of the points raised by Mann (1991) have been reviewed in depth by Reid (1992), Heywood and Stuart (1992), Pimm (1998) and May *et al.* (1995).

To determine the extinction rate it is necessary to have accurate estimates of three parameters: the rate of habitat loss, the shape of the species–area curve and the absolute numbers of species found in the ecosystem under review. There are difficulties in making accurate estimates of all three.

Loss of habitat

Most of the calculations attempting to estimate species extinction in tropical rainforests employ estimates of the loss of this biodiverse habitat, which is threatened by logging, conversion to agricultural use etc. (Whitmore & Sayer, 1992). Until the advent of satellite remote sensing technology, it was impossible to determine objectively the forest cover of a large region at a particular point in time. Now, many estimates of the rate of forest destruction have become available using remote sensing and some of the problems in making estimates have now clearly emerged. There are many types of forest and, furthermore, the distinction between closed forests and open forests is not, and forgive the pun, clear cut. Also, it is difficult to separate natural from man-made disturbance. For centuries, shifting-cultivation agriculture has been practised in tropical forests, as described, for example, by Ellenberg (1979). Recently, population pressures in some parts of the tropics have resulted in a shortening of the shifting agricultural cycle: referred to as 'slash and burn'. This more frequent disturbance has been especially condemned by conservationists as being particularly destructive of biodiversity. In considering human impacts on forests, we must face the important question raised in Chapter 5. How much of the forest is pristine, in the sense of being unaffected by human activities?

Current human 'assaults' on the forests result in different outcomes. Deforestation is the clearance of forest and the conversion of the land to other uses, usually agricultural. However, cleared land may be used for crops for a time, and then is abandoned to forest development. Alternatively, primary forests may be logged and the area left to develop into secondary forest, which may differ from the primary forest in structure, and species composition. The wide range of land use has to be faced in interpreting satellite images (see Whitmore & Sayer, 1992).

Realisation of the complexity of tropical ecosystems raises another issue. The insights obtained through the study of island biogeography were obtained through the study of islands of different size surrounded by water, a hostile environment to terrestrial plants and animals. Are forest fragments equivalent to islands? Clearly, the situation is more complex (Mann, 1991). Fragments of old-growth forest may be surrounded by 'hostile' agricultural developments, but the environment may not be totally inimitable to forest plants, for some trees may survive or regenerate within the agricultural matrix. In other cases, islands of old-growth forest may remain in contact or close to large tracts of regenerating secondary forest.

Species–area curves

The species–area relation indicates that increase in area is associated with an increase in species. However, what happens if the sample area is made wider and wider? Perhaps the upper part of the species–area curve levels off. If so, it may be that a large area could be lost without substantial species losses (Mann, 1991).

Number of species in tropical ecosystems

As we have seen earlier in this chapter, nobody knows how many species there are in the world, and, while it is clear that tropical ecosystems are very biodiverse, we can only guess at the number of species they contain. Mann (1991) concludes: 'those who prophesy the end of half the world's species find themselves in the awkward position of predicting the imminent demise of huge numbers of species nobody has ever seen'.

The taxonomic community becomes aware of the possibility of large-scale extinctions

Observations of the threats and extinction of endemic and other plant species are scattered through the taxonomic and other botanical literature, but from 1968 to 1974, Dr Ronald Melville of the Royal Botanic Gardens, Kew, began to gather together this information on threatened plant species. The first *Red Data Book* on plants (Volume 5: Angiospermae) was issued by Melville in 1970–1. In 1974, building on this initial far-sighted study, a network of specialist research workers was set up as the Threatened Plants Committee by the International Union for Conservation of Nature and Natural Resources (IUCN). They produced Red Data Sheets of information about endangered species. Information was collected for different regions, leading to the publishing of *Red Data Books* and the setting up of a database of information at the World Conservation Monitoring Centre (WCMC), Cambridge, UK. Species are placed in one of a number of categories (wording quoted from Lucas & Synge, 1978; Table 6.4).

The current extinction rate and prospects for the future

Smith *et al.* (1993) have examined the accumulated records of extinct and threatened species of animals and plants (dating back to *c.* 1600) that are held by the WCMC. Although there are fossil remains of some species, we have no knowledge of how many species have become extinct in prehistoric and historic times before records were kept. In this period, human activities caused massive habitat changes, as for example in the deforestation of the Mediterranean region several thousand years ago.

They examined the number of recorded animal and plant extinctions by geographical region. Considering the likely total number of species worldwide, the numbers are low. For example, there are only 21 recorded extinctions (2 animals, 19 plants) in the whole of South America, which includes the large species-rich Amazon basin and its tropical rainforest. This figure may reflect the relatively recent assault on the forests and/or the lack of records of extinctions. There is a huge difference between the number of recorded extinctions on island and continental areas. Probably this figure reflects the massive exposure to extinction forces on many islands. However, the results are perhaps skewed by the intensity of investigation on islands such as Hawaii, and the lack of detailed knowledge of other areas. Figures for other areas are also subject to distortion. Forty-five plant species have become extinct in

Table 6.4 *IUCN categories. The categories listed below have been used by conservationists for many years for a whole range of purposes, including the preparation of red data books etc. After much debate, a new listing has been published (Anon., 1994). The aim is to define a number of categories – Extinct, Extinct in the wild, Critically endangered, Endangered, Vulnerable, and Lower risk (including Conservation dependent, Near threatened, and Least concern) – in a more objective fashion by introducing some quantitative measures. These include estimates of present and future: (a) population size; (b) area occupied; and (c) numbers of populations. Estimates of future risk are to be made within a specified time frame. It remains to be seen how far these new categories prove workable given our imperfect knowledge of endangered species and the complexities of clonal growth, seed banks etc. found in many plants.*

EX: Extinct species

E: Endangered species

Taxa in danger of extinction and whose survival is unlikely if the causal factors continue operating. Included here are taxa whose numbers have been reduced to a critical level or whose habitats are so drastically reduced that they are deemed to be in danger of extinction.

V: Vulnerable species

Taxa believed to move into the Endangered category in the near future if the causal factors continue operating. Included are taxa of which most or all the populations are decreasing, or with populations seriously depleted, or those with still abundant populations but under threat from several factors.

R: Rare

Taxa with small populations that are not at present Endangered or Vulnerable, but are at risk.

I: Indeterminate

Species which are Endangered, Vulnerable or Rare but where there is not enough information to say which of these categories is appropriate.

nt: Species which are not now rare or/and threatened

From Briggs and Walters (1997). Reproduced by permission of Cambridge University Press

the sub-Saharan region. But all the recorded extinctions are in South Africa (Smith *et al.*, 1993).

Considering the future extinction rate, Smith *et al.* (1993) stress the differential taxonomic and geographic coverage by field workers. More is known about birds and mammals, and certain groups, such as the gymnosperms and palms, have been more intensively studied than many other groups. Therefore, it is not clear whether totals reflect 'differential research effort', or the different threats posed in different regions by different levels of economic development. Taking the figures at their face value, it would appear that the percentage of plant species under threat in the 1990s was estimated at 9%. Measured against a very low figure for the background rate extinction (see above) this figure is depressingly high. However, Smith *et al.* (1993) consider that the figure of 9% may be an underestimate.

The time frame of the extinction process

Heywood and Stuart (1992) have drawn attention to a very important point about the extinction process, which arises from the cautious conclusions drawn by Reid (1992) from his studies of extinction rate in species growing in tropical forest ecosystems employing the species–area relations. Reid concluded: 'If forest loss continues to accelerate, by the year 2040 some 17–35% of tropical forest species could be committed to eventual extinction when equilibrium numbers are reached.' Heywood and Stuart (1992) approve of Reid's very careful use of the phrase 'committed to eventual extinction'. Clearly, a species could become extinct in an instant, if its entire habitat is totally and irrevocably destroyed. However, in other cases the extinction process may be very long. Heywood and Stuart (1992, 102) point out that there are 25,000 plant species in the Mediterranean region of which about a quarter are endemic. Despite the many pressures dating back thousands of year, few species are reported to be extinct, but large numbers are endangered. It may be supposed that these endangered species have started along the road to extinction. And indeed they may be 'committed to eventual extinction'. But how far is it possible to reverse the process? Are some species so far along the track to extinction that essentially their fate is sealed? We return to these issues in later chapters.

Assessing the threat of extinction

Heywood and Stuart (1992) are critical of the use of the equilibrium theory of island biogeography for estimating the threat of extinction, and consider: 'prediction of extinction rates is an almost impossible task' and 'feel that we cannot attach any great degree of confidence to any predictions of species extinction rates' (Heywood & Stuart, 1992, 108). Moreover, the theory does not predict which species are likely to be at risk, and, therefore, is of limited use in devising management of populations and ecosystems.

They also stress another extremely important point. Whatever the scientific accuracy or probability of these predictions concerning species extinction, 'they have served the important purpose of advocacy with respect to the environment'. The possibility of widespread plant and animal extinctions in tropical rainforests is very well publicised, and has led to many conservation initiatives. However, some have asked whether environmentalists are 'crying wolf' on the issue of imminent mass extinction (Mann, 1991). After all, as Mann points out:

> In a frightening example of environmental degradation, the island [of Puerto Rico] one of the few tropical places where long term biological records have been kept, was almost completely stripped of virgin forest at the turn of the century. Yet it did not suffer mass extinctions. Even birds lost only 7 of 60 species – a painful, even unacceptable, total, but not an eco-catastrophe. Now, 90 years later Puerto Rico is thickly covered with trees.

Reviewing the practical issues arising from difficulties of making estimates of extinction rates, Heywood and Stuart (1992) conclude that conservationists should make careful and cautious statements that are 'scientifically defensible, since too large margins of error could undermine the credibility of conservation case in the political arena'.

Assessing the risk of species extinction under climate change

So far, in this review of endangered species of plants, historic initiatives have been examined. In essence, the data sets on individual species held at the WCMC have been compiled by regional experts in different countries, who have examined the different factors causing endangerment, and decide the level of risk species by species, country by country, using agreed criteria. These early analyses give information for individual countries or territories rather than for species ranges or biomes. They take into account the effects of such major factors as habitat fragmentation and loss, over-harvesting, pollution and introduced species etc.

However, even more complex and difficult times lie ahead, for, if the likely effects of another hitherto neglected factor, namely climate change, are included in the analyses, the problems encountered in assessing the risks to species are increased by orders of magnitude. As we shall see in later chapters, the degree of climate change depends upon whether emissions of greenhouse gases continue to increase, are stabilised (or reduced), and, crucially, at what level. Is the world facing a 2-degree rise in average global temperature, or are we destined to live in an even hotter world? The use of models with different assumptions leads to a range of predictions for climate in a particular region, and, therefore, a range of general predictions about the fate of species.

It is important to grasp the scale of the threat to species and ecosystems posed by climate change, which is predicted to lead to higher temperatures, changed rainfall patterns and an increase in the frequency of extreme weather events etc. In the light of these changes, many commentators are predicting that in the future increasing numbers of plant and animal species will face a very greatly increased threat of extinction (Lovelock, 2006; Meyer, 2006; Lynas, 2007). Given these uncertainties, it is clear, therefore, that it is impossible to form a settled view of how many species are becoming endangered and will be further threatened in the future; the figure depends upon whether effective international action is taken to control greenhouse gases, and the interaction of changing climate with other major factors, population levels, future use of resources etc. It is to be hoped that the WCMC will be supported in their work of collecting and interpreting information, as they have played an extremely important role in focusing attention on the risks to biodiversity, and have provided 'base-line' data sets for judging the effectiveness of conservation measures and predicting future trends.

Threats to cultivated plants and forest trees

The threats to biodiversity in wild plants are high on the agenda of conservationists. However, there are other pressing issues concerning the loss of biodiversity and genetic variability. To provide a fuller picture, several concerns are briefly noted here. The development and widespread planting of advanced agricultural and horticultural cultivars has led to the loss of some of the older cultivars, landraces and threats to the wild relatives of crop plants (Colwell, 1994). Also, changes in fashion have led to the neglect and possible loss of the older cultivars of ornamental plants. Furthermore, the widespread planting of elite stocks of plantation crops and forestry trees is often associated with a loss of variability in the species concerned. Considerable progress has been made in securing the future of some of these stocks by the establishment of seed banks (see Chapter 14).

Conclusions

It is clear from this brief account that the categories used to examine micro-evolution and conservation are far from simple. We are faced with a complex wild (native)–weed–crop–feral continuum, and there is mounting evidence that species do not fit neatly into a single category. Plant species – key actors in the evolutionary drama – may play many roles. For instance, in some cases cultivated plants have been abandoned and have become weeds. Crabgrass (*Digitaria sanguinalis*) was once cultivated in Europe, but later became a weed (Harlan, 1975). Plants of *Beta vulgaris*, in different circumstances, can be classified as wild, a crop and a weed. Indeed, as we shall see in a later chapter, in many areas both crop beet and weedy beet plants co-exist.

Dudley (2000, 53) provides another excellent example of multiple 'roles'. The giant reed (*Arundo donax*) is considered by some authorities to have originated in the Mediterranean, but there is evidence that it came from the Indian subcontinent. 'Giant reed was brought to North America quite early, being abundant by 1820 in the Los Angeles River, where it was harvested for roofing material and fodder.' The plant is now grown in commercial plantations to provide reeds for woodwind instruments. Not only is the plant grown as a crop, it is also an ornamental plant in gardens. In addition, it is often planted as part of erosion control. To add to the complexity, invasive populations are found in many wetlands, river channels and flood plain areas in California, the 'result of escapes and displacement of plants from managed habitats'. As the plant does not set viable achenes in North America, invasive populations result from the rooting of rhizomes and fragmentation following vigorous vegetative growth. Thus, we have the paradoxical situation that in many places *Arundo* is being carefully grown as a valued crop, while at other sites it has been planted as an ornamental or for erosion control. At the same time, at other locations, huge areas are being sprayed by herbicides to control feral populations.

There are other examples of the blurred crop/weed categories. Harlan (1975, 68) writes:

> One man's weed is another man's crop. The wild oat may be a serious pest to the Californian wheat grower, but to the cattleman of the coast and foothill ranges, it may be a most important forage. Johnsongrass may be a hated weed to the Texas cotton farmer, but a valuable hay crop to his neighbour . . . a man may fight *Cynodon dactylon* with a passion in one field, deliberately plant an improved variety in another field, and nurture still another variety with tender care [as a lawn] about his house.

Not only do different groups make different value judgements about weeds, attitudes may change with time. In the pursuit of high crop yields farmers from time immemorial have tried to exert maximum control of weeds in their fields. Now, by employing all the techniques of modern intensive agriculture – selective herbicides, immediate cultivation of stubble fields after cropping, early (often autumn) instead of spring sowing etc. – there are fewer longstanding stubble fields. Thus, in many areas of Europe, there is less food available for wild farm birds in the winter months, and there have been remarkable declines in numbers of farmland species. With the aim of providing additional feeding for birds in the UK, the Royal Society for the Protection of Birds (RSPB) is carrying out experiments on an experimental farm to study the effects of leaving unsprayed strips and patches. Because these areas have not been sprayed with herbicide, they support larger populations of weeds, the seeds of which will provide a source of food for farmland birds. We have arrived at the somewhat paradoxical situation that in the future the conservation of birds may involve the encouragement of certain weeds, as a valuable sub-crop within the farmland ecosystems. Clearly, such imaginative steps are necessary as evidence suggests that the expansion of agriculture across the world has been accompanied by major changes in the avifauna, with many species lost or declining. Worryingly, Teyssedre and Couvet (2007) estimate that 27–44% of bird species could be lost following agricultural expansion from the Neolithic to 2050.

On the basis of changes sketched in the early chapters of this book, it may be accepted that human activities are impacting to a greater or lesser degree on all the ecosystems of the world. Certain plants have increased in geographical area, numbers of populations and numbers of individuals (weeds, invasive species). In contrast, other species have become restricted and threatened with extinction. The major hypothesis of this book is that these changes in status are the result of microevolution in ecosystems being increasingly influenced by human activities. In the following chapters this proposition will be critically examined.

7

Investigating microevolution in plants in anthropogenic ecosystems

Earlier chapters have been designed to serve particular purposes. Chapter 3 provides a connected overview of the concepts underlying our current understanding of the long history of plant evolution. Chapter 4 reveals that while on a geological time scale, humankind arrived on the scene relatively recently, human impacts on the Earth's ecosystems have been very dramatic and are destined to increase. Examples of the effects of human activities on ecosystems are all around us, and it could be claimed that such interactions are so obvious they hardly need to be documented. However, as we have seen above, evidence suggests that the cumulative impact of human activities on the biosphere, both directly and at a distance, have often been underestimated. Also, the full complexity of the interactions of plants and humans is becoming evident, resulting in the recognition of many different categories. For the convenience of devising helpful groups, a species may often be classified in different categories according to the circumstances (Chapter 6).

In total, there have been many thousands of published papers of different facets of plant evolution. Some have investigated the early evolution of plants and animals, others have examined the patterns and processes that have generated, through time, the millions of species that live on Earth. The aim of this book is to focus on plant evolution in relation to human activities, and, therefore, only a subset of plant evolutionary case histories will be considered here, namely those that have been specifically designed to investigate human-influenced microevolution, including selection, domestication, gene flow, hybridisation, speciation etc.

Natural selection

An excellent starting point is provided by a consideration of Darwin's views on whether contemporary evolution in progress might be detected. Reflecting on the situation in the 1930s, Robson and Richards (1936, 182) emphasise that when he published the *Origin* Darwin did not offer any evidence 'that a selective process has ever been detected in nature'. Natural selection 'is suggested and assumed: its actual occurrence is nowhere demonstrated. Stated briefly, the argument is as follows: selection has plainly "worked" in domesticated races, analogous results and appropriate processes and conditions are found in nature. In short, the proof is based on circumstantial rather than direct evidence, the mainstay of the case is the analogy between Artificial and Natural Selection.'

Did Darwin have a view on whether natural selection might be detectable? According to Sheppard (1975, 22), Professor E. B. Ford reports: 'Charles Darwin, in conversation with his son Leonard, said that if data were properly collected, they might reveal, "perhaps in no more than fifty years", the progress of evolutionary change.'

The high esteem in which Darwin's theory of evolution is held by many might suggest that it is unnecessary to provide any experimental or other demonstration of natural selection. However, in order to convince a sceptic, it is clear that properly devised observations and experiments should be carried out to test this most basic concept that underpins all our ideas about evolution.

Studies of wild populations: some early experiments

Surprisingly, it was several decades after the publication of the *Origin* that the first attempts were made to detect natural selection in the wild. Robson and Richards (1936) reviewed all the available evidence from the period 1898–1932, including the studies of Kellogg (1907) and Plate (1913). Early studies of selection were concerned with animal populations, and examined cases of protective and warning coloration, mimicry etc.

It is essential to consider the context in which Robson and Richards (1936) wrote their review. Some biologists were convinced Darwinians. However, others were supporters of the 'mutation' theory of speciation, which proposed that speciation was abrupt, followed by periods of stability (De Vries, 1905; Bateson, 1913). This concept gained support from the sudden appearance of new variants, for instance in the plant genus *Oenothera*. As we have seen in Chapter 3, it is now clear that the evolution of species involves both gradual and abrupt events, but, in the 1930s, Darwin's view that speciation was always a gradual process had been seriously challenged (Huxley, 1942).

It was against this sceptical background that Robson and Richards (1936) reviewed all 19 case histories of the supposed demonstration of natural selection in wild populations.

Robson and Richards (1936, 310) concluded that there were severe problems in the sampling methods used, and that the design of the experiments was generally poor. Moreover, there had been 'a pathetic trust in observation *per se*' and 'wholly inadequate data have sometimes been brought forward in support of the adaptive origin of certain examples of mimicry, protective coloration etc.' (pp. 187–8). Overall the review concludes: 'the direct evidence for the occurrence of natural selection is very meagre and carries little conviction' (p. 310).

Conveniently, 50 years later, another review of the field was published, this time covering both animals and plants (Endler, 1986). By this later date, major advances had been made in studying natural selection in the wild (see, for example, Ford, 1971; Sheppard, 1975; Berry, 1977; Roughgarden, 1979; Bishop & Cook, 1981; Bradshaw & McNeilly, 1981). Such advances were made against the background of a sound theoretical framework provided by

Fisher (1929), Haldane (1932), Wright (1931) and others (see Provine, 1986, 1987) that provided the foundations for modern neo-Darwinian views.

How might natural selection in human-influenced ecosystems be studied?

As we have seen in earlier chapters, the ecosystem concept envisages the interrelations between species, as they respond to each other in food webs and to climatic and edaphic factors etc. Consider a natural forest with its hundreds of interacting populations of different species. Trying to frame these interactions in selectionist terms, by considering relative fitness of different individuals, presents us with a challenge beyond our capacity to measure or, indeed, fully comprehend. In earlier chapters, the proposition has been advanced that human activities have produced additional or changed selective forces. Thus, to the intricacies already acknowledged, human activities provide a further dimension of complexity in potentially altering the relative fitness of one intraspecific variant relative to another, or one species relative to another. It is immediately and abundantly clear that there are very many ways that humans can alter ecosystems, and many possible interactions thereafter.

Realising these complexities, it is clearly impossible to make an analysis of whole ecosystems in terms of Darwinian fitness. The solution to the problem has been to adopt a reductionist approach, examining the responses of one or a few species/intraspecific variants, in investigating situations where ecosystems are impacted by very *extreme* human-induced habitat factors, involving changes in land use or environmental pollution. In such situations, a 'tolerant' species/variant/genotype survives and reproduces in an extremely demanding environment, while in the same environment a 'susceptible' species/variant etc. may die or barely survive, with little or no reproduction. Thus, through studies of genetic polymorphism in extreme situations, it has proved possible to estimate differences in fitness, speed of change, responses to habitat change etc. (Bradshaw, 1971).

Many advances in our understanding of microevolution have come from studies of animal populations. At appropriate points below, several highly influential studies on responses to pollution, pesticides and hunting are very briefly considered, as they provided a stimulus to botanical studies. For instance, with the finding of pesticide resistant populations in animals, it was predicted that such strains of plants would arise following the introduction of herbicides.

Methods of detecting natural selection used in plant studies

As we shall see below, the study of those plant populations with appropriate genetic polymorphic traits has proved the most revealing. Ten methods of detecting natural selection – both direct and indirect – are considered by Endler (1986) in his review of both animal and plant studies. Here, basing our account on this excellent review, we consider the three approaches that have been used to studies of plant populations subject to human influence.

A. By observations and experiments of various kinds, selection has been studied, or inferred, following gradual, rapid, temporary or continuous acute or chronic disturbance produced by human activities, e.g. effects of haymaking, grazing, weed control of various kinds including the use of herbicides. Human cultural landscapes are characterised by mosaics of land use, e.g. grazing/arable/managed woodlands etc. and in such areas disruptive selection has been examined in species/intraspecific variants found in two or more components of the patchwork landscape.
B. Genetic demography or cohort analysis. In this approach, the fate of parents and their offspring with different traits is studied over a considerable period by examining 'marked' individuals. The aim is to investigate establishment, survivorship, fertility, mating and fecundity etc. By this means differences in fitness of individuals of different trait may be estimated.
C. Comparison of age classes or life-history stages may reveal the action of natural selection, e.g. the comparison of juvenile versus adult; breeding versus non-breeding individuals etc.

Important techniques for studying selection

Garden trials, tolerance tests

Very valuable information has been obtained by growing plants from different sources in common garden trials to examine morphological traits and life cycle characteristics. Water cultures, or simple test situations, have been employed to measure the tolerance/sensitivity of individuals to some important environmental contaminant or pollutant, such as a herbicide. There are a number of issues to consider in designing and interpreting garden experiments. How far are the residual differences maintained in cultivation genetically based? This question can only be elucidated by crossing experiments carried out in tandem with the cultivation trials. In addition, it must be acknowledged that some of the differences detected in cultivation of diverse stock from seed may arise from maternal effects (Roach & Wulff, 1987; Mousseau & Fox, 1998). Such effects may arise from the fact that, in many species, seed and fruits have considerable food reserves. As some seed-bearing plants might be diseased, malnourished or damaged, the seed they produce may be provided with fewer resources than those of well-grown disease-free stocks. As the early growth of seedlings and plants may be determined, to a high degree, by the food resources in the seed, differences between plants at an early stage in growth may be due to maternal effects. It is often assumed that in fertile soils maternal effects may disappear as plants mature, but this may not be correct.

Reciprocal transplant experiments

Of particular significance in the study of natural selection is the reciprocal transplant experiment. This method of studying patterns of variability has a long history (Briggs & Walters, 1997). Suppose that there are two very different habitats (a and b) in the same region or side by side. Preliminary tolerance tests on a species found in both habitats may suggest that each of two variants (these could be species or intraspecific variants, A and B) is best suited to its own native habitat (A in a and B in b). To test this hypothesis plants of A and B are collected from the wild. They are grown in a common environment to minimise carry-over effects (see below). Then, a replicated full set of material of the variants A and B is planted out in the two different habitats a and b. The growth and reproduction of the plants is then closely monitored, giving comparative data of the behaviour of both variants A and B in environment a, and, likewise, both A and B in environment b. By this means it is possible to test the hypothesis that each variant has higher Darwinian fitness in its native habitat.

Experiments of this kind must be designed with great care. Some means of marking plants must be devised or stocks with different genetic markers can be employed (e.g. *Trifolium*, Harper, 1977). To provide material for experiment, many plant species can be clone-propagated. Daughter individuals, barring mutation, will be of the same genotype as the parent, and can then be used to examine the effect of different environments on genetically identical stock. It is very important, however, to generate uniform clonal material from parental material by selecting equivalent vegetative fragments of equal size, vigour and food reserves. If fragments differ in these parameters, then differential growth under the experimental conditions may not reflect genetic differences but be influenced by unequal provisioning. To minimise possible 'carry-over' effects, diverse stocks are often grown for a time in a common environment before the cloning process.

Turning to other issues, in the development of crop trials it was discovered that, however carefully plots are selected for experiments, differences in fertility, drainage etc. occur. In order to counter these differences, experiments are now set out in randomised plots in properly designed field trials (see Chapter 2). Instead of ignoring soil differences, the aim is to take them into account and *estimate* the effect of differences. Furthermore, as part of a designed experiment, an appropriate statistical analysis should be devised.

In the course of garden trials the experimental material may be attacked by insects, fungi etc., and the experimenter must decide whether to try to control any infestations. If so, all the stocks in the experiment must be treated in the same way. As diverse stocks in an experiment may include material resistant to disease or pest damage, the decision is often taken not to use pesticides. However, in such cases, there is a risk that the whole experiment may be lost or so badly damaged as to provide no information with which to test the hypothesis under study.

A further point should be stressed. It is generally assumed in science that, if an experiment is repeated, the same result will be obtained. However, this is not the case in garden trials and experiments performed out-of-doors. Year by year differences in rainfall, frost, temperatures etc. occur and different results are possible in different years (see Warwick & Briggs, 1979, 530).

The use of molecular markers in microevolutionary studies

Very valuable insights, unobtainable by other means, have been achieved through the use of molecular techniques. As the examples in the following chapters show, such approaches have transformed our understanding of patterns and processes in human-disturbed habitats by providing information on gene flow, founder effects, hybridisation, speciation etc. Moreover, there is enormous potential for further advances.

Table 7.1 *Characteristics of different molecular methods for assessing genetic diversity*

Method	Source	Non-invasive sampling	Cost	Development time[a]	Inheritance
Electrophoresis	Blood, kidney, liver, leaves	No	Low	None	Co-dominant
Microsatellites	DNA	Yes	Moderate	Considerable	Co-dominant
DNA fingerprints	DNA	No	Moderate	Limited	Dominant
RAPD[b]	DNA	Yes	Low–moderate	Limited	Dominant
AFLP	DNA	Yes	Moderate–high	Limited	Dominant
RFLP	DNA	No	Moderate	Limited	Co-dominant
SSCP	DNA	Yes	Moderate	Moderate	Co-dominant
DNA sequencing	DNA	Yes	High	None	Co-dominant
SNP	DNA	Yes	Moderate–high	Considerable	Co-dominant

[a] Indication of the time taken to develop the technique so that genotyping can be carried out in studies of threatened species.
[b] There are sometimes problems of repeatability with RAPDs (see text for further details). All other methods are highly repeatable.
From Frankham, Ballon & Briscoe (2002). Reproduced with permission of Cambridge University Press

Different molecular markers have been used in the study of plants. It is beyond the scope of this book to provide detailed information. Here, a number of important points are outlined. For further information, including protocols etc., the reader should consult Avise (1994), Karp, Isaac and Ingram (1998) and Baker (2000). Table 7.1 summarises the characteristics of different molecular methods employed in assessing genetic diversity (from Frankham *et al.*, 2002).

1. The collection of plant material for molecular studies has received great attention. Hyam (1998) may be consulted for a review of the different approaches to collecting field samples of plants for DNA analysis. Many researchers use the method of Chase and Hillis (1991) in which the collected samples are dried immediately in silica gel and stored in a refrigerator before use. When the samples are rich in polysaccharides and/or certain secondary metabolites such as polyphenols, terpenes, resins etc. special extraction techniques are necessary (Rueda, Linacero & Vázquez, 1998). Zhang and Hewett (1998) review the methods of extracting DNA from preserved materials.
2. Molecular studies make use of gel electrophoresis techniques in which enzymes and DNA fragments may be separated by the use of polyacrilamide and agarose gels, at a specific pH, in the presence of an electric field. Generally, larger molecules travel slower than smaller molecules. Patterns of DNA fragments are generally located on the gel by staining, although, in certain techniques, radioactive labelling has been employed.
3. Microevolutionary studies have often investigated isozymes, which are 'structurally different molecular forms of an enzyme system with qualitatively the same catalytic function'

(Müller-Starck, 1998). These variants originate from base changes in the DNA sequence producing alterations in the amino acid sequence of an enzyme. Such changes may yield derivatives with changed net charge and/or differences in their spatial structure and, thus, they may have different electrophoretic mobility. Allelic variations at a particular enzyme gene locus are often revealed in electrophoresis of samples from different individuals/populations etc.: the different alleles being represented by molecules with different mobility. In cases where isozymes have been shown to be different alleles of the same gene, they are known as allozymes. Electrophoretic studies are carried out using ground, fresh plant material in buffer on starch or polyacrilamide gels. After electrophoresis, the location of the enzymes is revealed by providing a substrate/stain that takes advantage of the precise substrate-specificity of the enzymes. The study of allozymes has many advantages. Markers are expressed co-dominantly, and, as homozygous and heterozygous genotypes can be distinguished precisely, it is possible to measure heterozygosity, genetic diversity and genetic differentiation. Isozymes have their limitations. Essentially, a non-random sample of the genetic variability is being studied; only a small number of water-soluble enzymes are being examined. Water-insoluble and enzymes bound to structures are not sampled by the methods usually employed. Moreover, only nucleotide substitutions in the DNA that affect the mobility of the enzymes are revealed, as two different enzyme variants that yielded a band at the same position cannot be differentiated.

4. DNA may be cut into fragments by employing one or more of a wide range of restriction endonuclease enzymes obtained from bacteria. Each different enzyme cleaves double-stranded DNA at a different recognition site characterised by specific sequences of nucleotides (Baker, 2000). This behaviour is the basis for the study of Restriction Fragment-Length Polymorphism (RFLP), an approach employed in many studies. Figure 7.1 shows how a single base change within a restriction site can result in polymorphism, and how this is visualised on an agarose gel. RFLP techniques are of relatively modest cost, requiring only moderate expenditure of time to perfect the techniques for any particular species and inheritance is co-dominant (Frankham *et al.*, 2002).
5. Protocols for studying molecular markers exploit other properties of DNA. For example, in the presence of primers (small selected pieces of DNA that attach to specific target sites on single stranded DNA), the enzyme DNA polymerase and appropriate nucleotides, DNA replication will take place. Figure 7.2 shows how amplification of a gene or targeted DNA segments may be achieved by use of the Polymerase Chain Reaction (PCR) that employs thermostable Taq DNA polymerase obtained from *Thermus aquaticus*, a bacterium that lives in hot springs (Birt & Baker, 2000). Amplification through the PCR techniques has been exploited in the development of the Random Amplified Polymorphic DNA technique (RAPD), which typically employs 'randomly selected 10-base primers' without prior knowledge of any specific primer sites in the genome. An individual primer of such length is able to hybridise with several sites within the target DNA and provides the spaced 'marked' double-stranded sections between which the DNA copying can occur in the PCR reaction. The result of many amplification cycles is the production of DNA fragments of different length that can be separated by electrophoresis. RAPD techniques are in the low to moderate cost range, with limited development times (Frankham *et al.*, 2002). The method is not appropriate for population genetic studies as the RAPD markers are dominant. Thus, homozygotes and heterozygotes cannot be distinguished. Furthermore, there is a major concern about the reproducibility of RAPD profiles. In a study of the problem, DNA from two *Populus* × *euamericana* clones was circulated to nine laboratories together with two primers, chemicals all

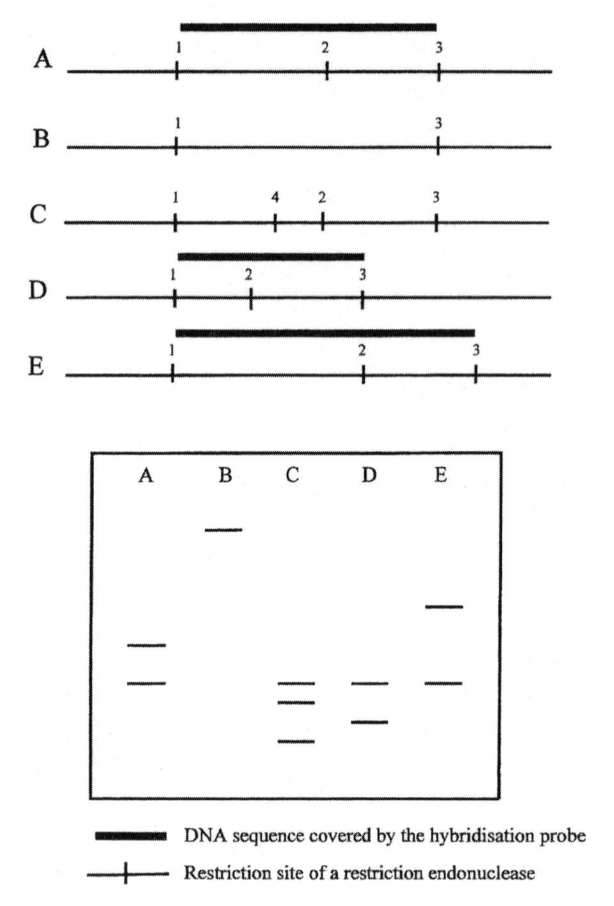

Fig. 7.1 RFLP analysis. Top: A chromosome segment digested with a restriction enzyme. The numbers indicate different cleavage sites. The bar marks the segment covered by the hybridisation probe. In comparison with A, B represents a polymorphism caused by the loss of one cleavage site (2); C a polymorphism caused by the creation of a new cleavage site (4). D and E show RFLPs originating from deletion (D) or insertion (E) of DNA fragments between two cleavage sites (1 + 2). Bottom: Scheme of an exposure obtained when genomic DNA is digested as indicated above and then blotted after electrophoresis to a membrane, which is hybridised by a probe represented by the bar. (Reproduced from Karp *et al.*, 1998, by permission of Chapman & Hall)

drawn from a common batch and detailed agreed protocols (Jones *et al.*, 1998). As a result of these studies, it was concluded that the method was not sufficiently reproducible between laboratories. Indeed, the contributors concluded that even within a single laboratory changes in chemical, equipment or personnel might make it difficult to produce reproducible results. None the less, many workers have used the technique because of its relative cheapness.

6. In further developments, the Amplified Fragment Length Polymorphism (AFLP) technique was devised. The procedures involved in this approach are represented in Fig. 7.3. Essentially, the method consists of an automated process exploiting the heat-stable properties of Taq polymerase in 'selectively amplifying a subset of restriction fragments from a complex mixture of DNA

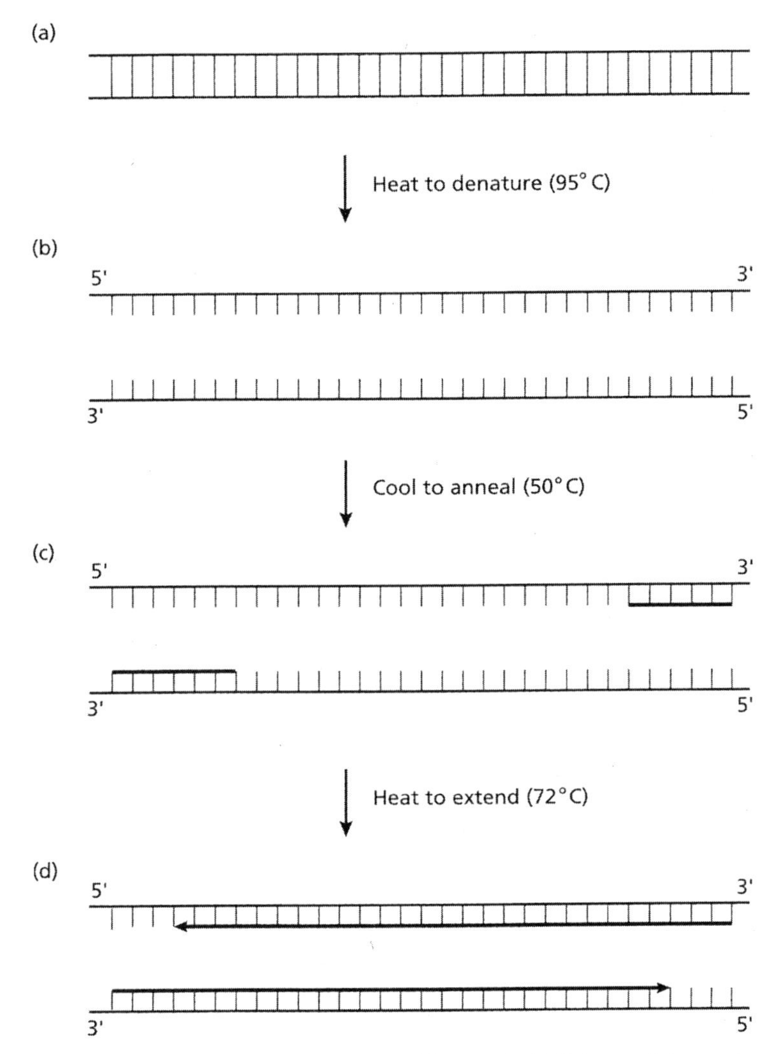

Fig. 7.2 The polymerase chain rection (PCR). Double-stranded DNA containing regions complementary in base sequence to oligonucleotide primers is heated to a temperature sufficient to cause dissociation of the two strands. Cooling of the resulting single-stranded DNA then permits annealing of the primers to their respective binding sites on alternate strands. The temperature is then elevated to the activity optimum of *Taq* polymerase, which then extends the annealing primers. Repeated temperature cycling results in a geometric increase in amplified products. (Reproduced from Baker, 2000, by permission of Blackwell Scientific)

fragments obtained after digestion of genomic DNA with restriction endonucleases' (Matthes, Daly & Edwards, 1998). AFLP techniques are in the moderate to high cost range, require only limited time to develop for a species of interest and inheritance is co-dominant (Frankham *et al.*, 2002). The method has the advantage that only very small amounts of DNA are necessary for the PCR reaction. The reproducibility of the AFLP method has been studied in a collaborative project

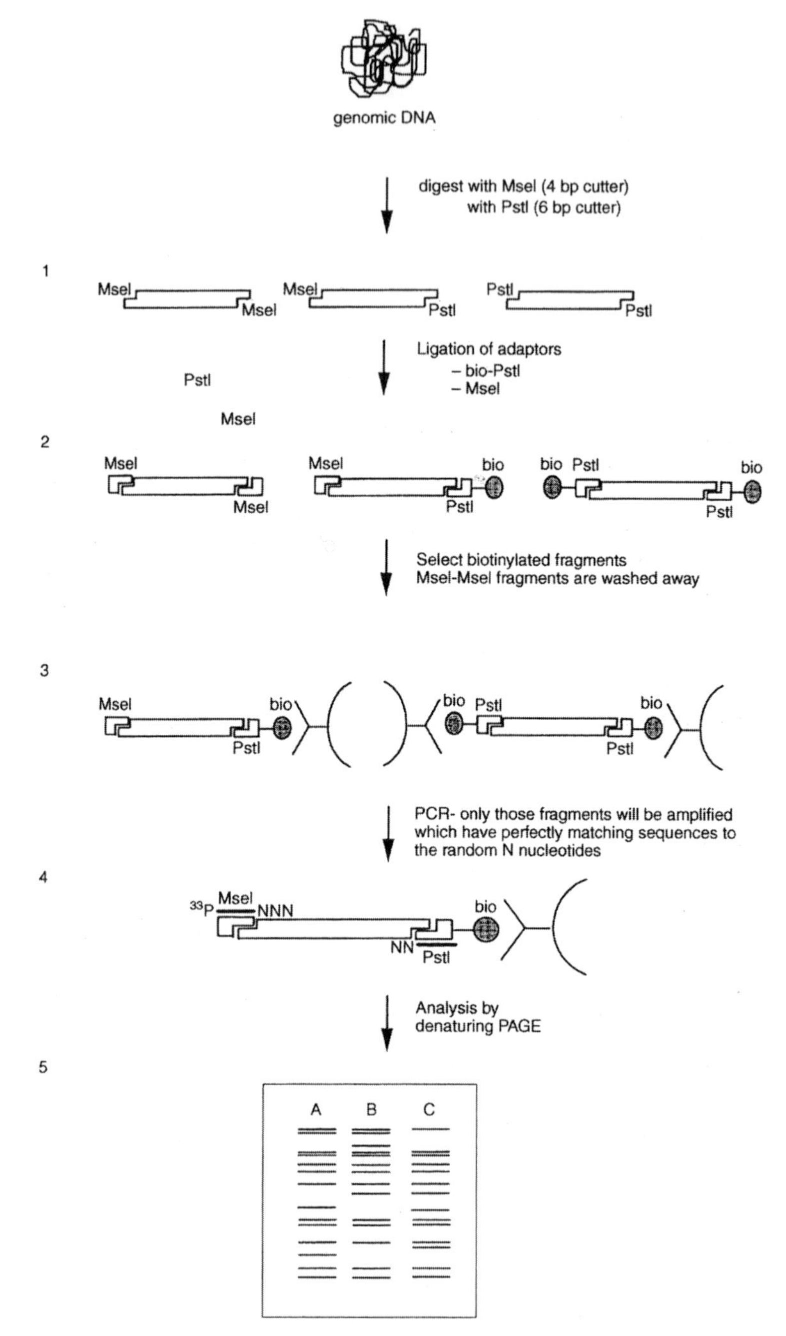

Fig. 7.3 Schematic representation of the AFLP procedure. In the AFLP technique, DNA fragments are obtained after digestion of genomic DNA with restriction endonucleases and adaptors are ligated to the ends (stage 1). PCR is then used to amplify selectively a subset of restriction fragments from the complex mixture (steps 2 to 4). Polymorphisms are detected as differences in the length of the amplified fragments after polyacrylamide gel electrophoresis (PAGE). (Reproduced from Karp *et al.*, 1998, by permission of Chapman & Hall)

involving seven laboratories working to exactly the same protocol with sub-samples of the same DNA sample, and with the different laboratories drawing from the same batches of chemicals. As a result of this test it was concluded that the AFLP method had high reproducibility (Jones *et al.*, 1997). Moreover, very small amounts of material are necessary: the DNA from a single hair can be amplified. However, a serious potential problem with the PCR methods must be faced. Contamination must be strictly avoided, as contaminant DNA is copied along with the target DNA.

7. Various aspects of the molecular biology of plants have been investigated. Patterns of repeated 10–100 base-pair (bp) units have been detected in some studies – the so-called minisatellites (Scribner & Pearce, 2000, 236). Also, microsatellites have been investigated, consisting of arrays of 1–10 bp sequences in short tandem repeats (STRs), or simple sequence repeats (SSRs) (Scribner & Pearce, 2000, 236). How these repeat sequences, sometime present in high numbers, have been employed in genetic fingerprinting for population genetics and conservation is reviewed in Bruford and Saccheri (1998) and Baker (2000). The use of machines, such as those employed in the Human Genome project, have also made it possible to determine the DNA sequences of individual genes and also of entire genomes, e.g. *Arabidopsis*, grapevine, rice etc.
8. Chloroplasts contain DNA as a 'circular genome present in multiple copies with about 100 genes' (Dyer, 2002). Chloroplast DNA is generally maternally inherited as the female gametes are much larger and contributes entirely or almost entirely the chloroplast in the zygote. Plant cells also contain mitochondria. These organelles also contain DNA and patterns of inheritance are predominantly maternal (Randi, 2000).

Assessment of microevolutionary studies

In the ideal situation, all the elements of pattern and process in microevolution etc. will have been investigated by appropriate historical studies and well-designed experiments and observations. However, as we shall see in later chapters, published studies report investigations at different stages of resolution. Incomplete partial studies are still valuable, in pointing to profitable new areas of research, and providing a valuable collection of case histories, which help to build an overall picture. In many situations, selectionist interpretations of the effects of environmental change may be advanced, but without any experimental investigations. These situations should be seen for what they are, hypotheses awaiting rigorous testing. It is important, therefore, if possible, to go beyond the collection of a series of *just-so stories* to the stage of proper investigation. As we shall see, in some cases comprehensive critical studies have been carried out. However, it has to be acknowledged that only a few microevolutionary situations have been thoroughly investigated. In designing and interpreting one's own and published experiments, there are many other important questions to consider.

Sampling

It is essential to collect appropriate samples (of adequate number) for investigation and experiment. Sampling may be random, spaced or deliberate. An appropriate sampling strategy must be designed to test the hypothesis under study (Briggs & Walters, 1997).

Genetic differences

Evolution by natural selection is concerned with genetic differences between individuals. Crossing experiments and the analysis of progenies provide the means of investigating the genetic basis of important morphological and behavioural traits. In cases of apparent selection etc. it is essential to establish that any differences that are detected are genetically based, by carrying out the appropriate crosses. However, while in some cases comparatively simple Mendelian situations have been discovered, in others more complex polygenic systems have been detected that may be more difficult to study. In many cases the heritability of significant traits has been established, but not the precise mechanism of genetic control.

Ecological context

It is clearly of crucial importance in conservation to have a detailed understanding of the ecology of the ecosystem under study. Thus, the temporal and spatial distribution and biological activity of significant factors in the environment must be studied. If it is alleged that a particular factor is important, this should be demonstrated. For example, in testing the hypothesis that copper contamination is creating a new selection pressure in a polluted site, it should be established that copper is present in toxic quantities, and in a form available to the plant in question. Tolerance tests may then be carried out to see if plants growing at the site are more copper tolerant than those in uncontaminated sites. Such tests might examine the effect of copper in comparatively simple water culture experiments, as well as investigating the responses of potentially tolerant and sensitive plants to natural and contaminated soils.

Estimating fitness

On the face of it, the estimation of fitness and the examination of what happens over time would appear to be relatively simple to study, but in reality such investigations in plants are fraught with difficulty (see Christiansen, 1984, on which this account is based).

Determination of fitness requires the study of both survival and reproduction, together with an estimate of the number of descendants arising from these progeny. Survival is a character that can be measured, provided that it is possible to determine the limits of an individual plant. This may be simple in the case of trees and herbs with an upright growth habit. But many individual plants, including trees and shrubs as well as herbaceous species, have a clonal growth habit: creeping above the soil, or with underground parts extending the individual out from the point of establishment. Later, these patches may fragment into clonal daughter colonies, which, in their turn, continue to expand. In such cases, the measurement of reproduction is much more complex.

As we have seen above, different species of plants have different breeding systems. While some are self-fertile, others are genetically self-sterile. Yet more have apomictic reproduction – either obligate or facultative. In outcrossing, several individuals could be involved in supplying pollen to a seed parent. Moreover, any particular seed parent, in its turn, could be providing pollen to several other plants. In a 'complete' estimate it would be necessary to determine these reproductive interrelationships, and also understand the genetics of the trait differences between the plants whose fitness is to be determined. Another complicating factor is the fact that many plants have overlapping generations, which considerably increases the complexity of any estimations of lifetime fitness. In addition to numbers of progeny, studies of fitness require information on the number of descendants, and this requirement increases the difficulties. Given these issues, it is not surprising that many studies do not attempt to estimate 'full' fitness through studying the numbers and performance of gametes, zygotes, developing individuals, descendants etc. Instead, *components of fitness* are estimated. Even making these estimates is far from simple. For example, it may not be possible to count the seeds maturing over a period on a very large plant. Most investigations involve estimation of numbers of seeds/fruits and this must be carefully carried out as it involves sampling. One particular problem results from density effects in plants in crowded populations (Begon, 1984). Sometimes one or a few individuals dominate the populations. Therefore, estimates of individual fitness will differ markedly from estimates of average fitness of populations.

Speed of change

As we shall see, in investigations of plants, documented evidence has proved invaluable in judging the speed of microevolutionary change and the dating of events. For example, dated botanical records and herbarium specimens have provided insights into the history of introduced species. On a longer time frame, several lines of evidence have revealed the scale, detail and successive episodes of landscape change, e.g. evidence from archaeology; analysis of pollen, seeds and other plant remains in peat and other deposits examined in Quaternary studies; radiocarbon dating of plant remains; place names; maps; documents; landmarks; tree rings; satellite, aerial and other photographic images; field records and information on herbarium sheets; long-term monitoring of fixed quadrats in various types of vegetation etc.

The relationship of microevolution to conservation

In the next few chapters we examine the evidence for microevolution in anthropogenic ecosystems, and comment on many issues of major conservation interest. It is important to stress that many conservationists have yet to acknowledge that insights gained from the study of microevolution have the potential to contribute to the effective conservation of rare and endangered species. Microevolutionary studies concentrate on the population as the key unit of study. In contrast, many naturalists focus on the species as the significant 'unit' of conservation, the ultimate objective being effective species management in reserves and national parks.

Conserving species: typological approaches

Conservationists often appear to accept what might be called a 'typological view' of species. The basic concept of typology is that species may be viewed as having 'an invariant, generalised or idealised pattern shared by all members of the group [i.e. the ground plan]' (Davis & Heywood, 1963, 9–10). Historically, at the species level, 'typology found its expression in a denial of variability, or at least of its importance'. The fact that some intraspecific variation occurred was accepted by early typologists, but it was usually thought to be accidental or unimportant, the effect of climate or habitat upon the ground plan of the species created by God: an imperfection of nature. 'The deficiencies of eighteenth century herbaria are largely attributable to this narrow typological concept – any specimen was thought to represent the species adequately.'

Rojas (1992) considers that some conservationists have a 'typological' outlook. In considering the requirements of a species in conservation, they act on generalised information on the species as a whole, rather than recognise that species are usually genetically variable entities, often consisting of a collection of differently adapted local populations.

Changing goals: preservation vs. conservation of evolutionary potential

It is my contention that conservationists should consider their endeavours in the light of what we know about species and evolution. The neo-Darwinian insight of fundamental importance is that evolution occurs in populations within the context of ecosystems. In so far as species continue to exist and evolve, they do so as individuals in populations. If species fail, and become extinct, this process takes place in populations. The concern that conservationists have for 'species' extinction should always lead to a consideration of what is happening at the population level, taking into account the characteristics of interacting species and other aspects of the ecosystem in which they live. The aim should be to try to understand the extinction process, and thereby to gain understanding of how it might be countered by conservation action. A further point is crucial. If an endangered species is to survive in the longer term, it is essential to move away from the notion of preservation and grapple with the much more difficult concept of how to maintain the evolutionary potential of species. If a species is to have a future, it is essential to maintain its ability to evolve in self-sustaining populations. The intention of conservation management, therefore, must be to maintain a capacity for evolution in response to habitat change in self-sustaining populations. As we see below, if endangered species are to survive, some evolutionary changes will need to take place in their populations in response to human activities, especially in relation to anthropogenically induced climate change. There is also an additional requirement. The maintenance of the evolutionary potential of a species must include the survival of co-evolved organisms associated with the endangered plant species, for example, the animal–plant interactions involved in pollination and in fruit dispersal, and the fungal–plant mycorrhizal associations.

Conclusions

By employing a range of techniques in the examination of populations of plants growing in extreme environments, it has proved possible, through decades of research, to investigate aspects of microevolution in response to anthropogenic impacts. The following chapters examine the evidence for such microevolution in areas with different human land use, together with comments on the importance of the various findings to plant conservation. The material is arranged as follows: areas managed as grasslands (Chapter 8), sites where cropping is practiced in arable land and forests (Chapter 9), plant populations in polluted zones (Chapter 10), regions influenced by introduced organisms (Chapter 11), and populations in decline (Chapter 12).

In many cases, molecular research has added greatly to our understanding. The potential for even greater insights has recently been emphasised. Kreitman and Rienzo (2004, 300) explore the possibility that 'natural selection is expected to leave distinctive signatures on patterns of neutral variation that are tightly linked to a site carrying an advantageous mutation'. Molecular investigations of the model plant *Arabidopsis thaliana* have transformed our understanding of genetic architecture of a species. In addition, comparison of different ecotypes, in the regularly self-fertilising species, has revealed that they have different patterns of markers detected through 'fine-scale genotyping of their genome sequence' (Shindo, Berasconi & Hardtke, 2007, 1043). These patterns of DNA markers reveal the 'signature of natural selection', and the stage is set for detailed evolutionary genetic investigations comparing the fitness of different ecotypes of this species by employing the traditional approaches of garden trials, tolerance tests, reciprocal transplants etc. For further discussion of the possibilities in this rapidly developing area, see for example, Edelist *et al.* (2006) studying microsatellite signatures of ecological separation in sunflowers; Harris and Meyer (2006) and Mattiangelli *et al.* (2006) investigating human adaptation; studies of *Drosophila* by Kreitman and Rienzo (2004); and the review by Ellegren and Sheldon (2008).

8

Plant microevolution in managed grassland ecosystems

Herbivory in anthropogenic ecosystems

Microevolutionary insights have come from investigations of a number of different kinds of managed grassland (hay, pasture, lawns and sports turf), and also from arable lands and managed forests. While, at first sight, management regimes are very different, they all involve selective herbivory.

Managed grasslands, rangelands etc.

In such habitats, humans take a harvest directly from the sward, either through haymaking or fodder collection, or, indirectly, through the management of domestic, semi-wild and wild animals. From the plant's eye view, harvesting by cutting and grazing are quite different. Harper (1977) makes this point very clearly, when he notes that, in clipping or mowing, some portion of the sward is more or less evenly removed. In contrast, grazing is patchy, and animals differ in their behaviour. 'The cow rolls her tongue round a bunch of grass and pulls . . . sheep bite leaves between the incisors of the lower jaw and a pad on the upper jaw whereas the rabbit cuts leaves with teeth on both jaws.' In addition, Harper makes it plain that the un-eaten portion of the sward is impacted by grazing animals, in their deposition of urine and dung and their tramping of the vegetation. Thus, 'grazing animals frequently sit, lie, scratch and paw on pasture in addition to walking, running and jumping on it' (Harper, 1977, 449).

Not only do herbivores graze in different ways, it is important to look closely at the selectivity of eating behaviour. All plants, in wild and managed settings, are targets for herbivores, and there is now abundant evidence that, in the course of evolution, species have evolved defence and deterrence mechanisms (Pollard, 1992). These take different forms, with plants having distasteful or poisonous chemical compounds in their tissues, or external defences such as toughened leaves, spines, prickles, stings, glandular hairs etc. Herbivores include animals of different sizes – from insects to elephants. Investigations have revealed, for example, that some defences are only successful against insects or molluscs, they do not deter grazing by larger animals. Both these points are illustrated by investigations of populations of *Trifolium repens* in managed pasture in Wales, UK, that have been found to be genetically polymorphic in their ability to produce HCN – some genotypes (HCN+) produce HCN from their crushed leaves, while other genotypes (HCN−) lack this ability (Harper, 1983). Also, *Trifolium* is genetically polymorphic in leaf markings (genotypes range from those with unmarked leaves to variants with different inverted v-shaped white or coloured marks on the leaves). Studies of sheep grazing have revealed that when animals are faced with a variety of leaf mark morphs, they eat more of the unmarked and the commoner leaf marking variants – an example of apostatic selection. There was no evidence that sheep were avoiding HCN+ genotypes. Moreover, HCN production was not associated with lack of markings. In contrast, experimental studies in controlled conditions and reciprocal transplant experiments in the field with various invertebrates reveal that selective predation occurs – cyanogenic (HCN+) plants being less eaten than acyanogenic (HCN−) variants.

Selective grazing by domestic stock has a major impact on plant community composition and structure. Evidence suggests that where overgrazing is occurring there may be a dramatic increase in unpalatable species. For example, species with distasteful foliage or stinging hairs and spines may come to predominate in areas where there is dramatic overgrazing by feral rabbits escaping from former rabbit warrens (Rackham, 1986). Another example is informative. In the Stalinist era vast herds of sheep were driven along the Russian-Georgia Military Road that runs through sub-alpine meadows in the Caucasus Mountains of the Republic of Georgia. The vegetation of the region was massively overgrazed and populations of the spiny *Cirsium obvalatum* and the distasteful *Veratrum lobelianum* increased greatly. The presence of unpalatable species in overgrazed areas has, however, a surprising side effect. Stands of unpalatable species harbour within them many palatable species, whose size and reproductive ability and fitness are increased by the protection offered by having an unpalatable neighbour (Callaway, Kikvidze & Kikodze, 2000).

Arable lands

Arable agriculture also involves selective herbivory not only of the crop but also weeds, the control of which involves burial by ploughing, damage or death by burning, digging, weeding, rolling, hoeing, clipping, mulching and herbicide applications, and curtailment or interruption of life cycles by cultivation strategies such as fallow periods. It is important to note that sometimes weeding processes involve above-ground 'herbivory' only, and that plants may grow again from intact below-ground parts. In other cases, the aim, if not the outcome, is to destroy the weed plants completely.

Forests

The microevolutionary patterns and processes imposed by humans on forested areas will be considered briefly in the next chapter. Here, it should be stressed that exploitation of trees may also be considered as a form of selective herbivory, with the potential to affect species differentially. Anthropogenic interventions impact on trees at all stages of their development, from seeds/fruits, seedlings and saplings to mature trees. Management of forested resources involves many interventions in which trees are selectively burnt, weeded out, grazed, ringed, cut down, pollarded, coppiced, stripped of branches, poisoned, or replaced by others etc.

For convenience, separate accounts are provided of the effects of the management of grazing lands (this chapter) and arable/forestry (Chapter 9). However, it is important to emphasise the interconnectedness of former rural practices in the developed world (some of which are still current in the developing world). For instance, domestic animals are sometimes allowed to graze the stubbles on arable land. Moreover, domestic animals will eat leafy twigs, and traditionally branches were collected as fodder from forests in Switzerland (Ellenberg, 1988) and Scandinavia (Emanuelsson, 1988). Also, grazing and forest management were combined in various types of wood pasture (Rackham, 1987). Furthermore, in Central Europe and elsewhere, it was traditional practice to collect turves from grassland and heathland and use these as bedding for animals, before the material, mixed with dung, was finally spread on arable land to improve soil fertility (Ellenberg, 1988).

The antiquity of grassland management for pasture and hay

Grazing lands have been managed by humans for millennia, e.g., hunter-gatherers used fire in various ecosystems to control populations of wild animals. Later, as we saw in Chapter 4, human intervention in ecosystems increased in providing grazing for domestic and fodder for animals. Many complex management systems have been developed in different parts of the world (Prins, 1998; Bakker & Londo, 1998). In addition, in many parts of Europe, transhumance was practised, stock being driven long distances between summer and winter grazing areas (Rinschede, 1988). Such practices were probably first developed in the Neolithic period (Poschlod & WallisDeVries, 2002), and later became widespread in alpine regions of the Pyrenees and Central Europe (Fig. 8.1). For instance, from the fifteenth to the nineteenth centuries it is recorded that very large sheep flocks grazed on summer pastures at higher altitudes in the Franconian Jura Mountains. For the winter, animals were driven several hundred kilometres to climatically milder areas in the river valleys of the Danube and the Rhine. In addition, before refrigeration of meat was possible, it was not unusual for stock to be moved long distances. For example, there are records of animals being driven from south-west Germany more than 600 km for slaughter in Paris (Poschlod & WallisDeVries, 2002).

Not only did management involve grazing areas, in addition, fodder, especially hay, was collected to enable animals to survive the winter or seasonal dry periods. The English word for hay is of Saxon origin, and there are many place names, containing the elements 'maed' or 'hamm', suggesting the presence of hay meadows. However, haymaking in northern Europe has a long and continuous history probably dating back at least to the Bronze Age

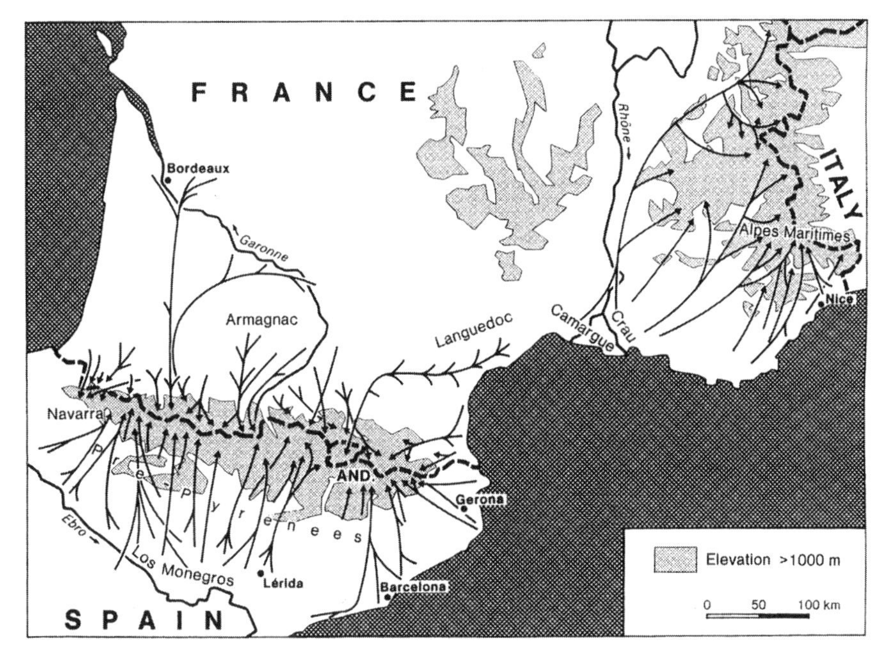

Fig. 8.1 Traditional transhumance routes of sheep in the Alps and Pyrenees. (From Rinschede, 1988. Reproduced by permission of Rowman & Littlefield Publishers)

(*c*. 1000 BC). This conclusion comes from the analysis of plant material (leaves, seeds, pollen etc.) preserved in anaerobic deposits of different ages, and, in later periods, from the study of documents, maps etc. From the earliest sites, such as lake sediments in Switzerland, grassland species have been identified, but it is often impossible to determine to what extent meadow and pasture were differentiated in the prehistoric period. In northern Europe, better and more convincing evidence has come from the Roman era, e.g. from the analysis of the contents of Roman wells in Britain, the species composition of charred hay discovered in excavations from the Roman stables at Dormagen in Rhineland, and the finding of the principle tool for cutting hay – the scythe – at other sites.

It is important in considering winners and losers in a microevolutionary perspective to realise that parcels of grassland have been subjected to different longstanding treatments – grazing and haymaking. In considering landuse in Britain, Rackham (1987, 332) makes it clear why this should be so.

> The division of grassland into meadow and pasture has to do with feeding animals round the year. Until the tractor was invented every parish, and almost every farm, had both ploughland and livestock. The livestock could feed on pasture for much of the year, but from January to April there was a hungry period when grass was not growing. The horses and oxen that drew the ploughs were working hard at this time and had to be fed well, not merely kept alive. Hence the practice of storing hay. The best grassland, on which the crop grew thickest was therefore reserved for hay.

The provision of adequate hay for animals presented a formidable challenge in many areas. For instance, European settlers in the north-eastern USA faced long and severe winters of 4–5 months duration, and through this period it has been calculated a cow would consume approximately 2 tons of hay, the produce of one to two acres of meadow land (Whitney, 1994, 250). In Britain, haymaking was carried out in late June or early July, partly timed so that the farmer could complete haymaking before the first of the grain crops was ripe (Greig, 1988; Feltwell, 1992). Many different types of meadows were developed in Europe. Hay was so important that, as a general rule, animals were prevented from grazing hay meadows before cutting. Moreover, it was traditional practice until the twentieth century in many areas of Europe to enrich existing hay fields by sowing seed from hay barns (Poschlod & WallisDeVries, 2002). In addition, many special systems were devised to increase hay yield. For instance, in the Alps and in Italy, irrigation systems were often constructed to counter the effects of seasonal drought and bring nutrients to the meadowland. In England, water meadows with their elaborate irrigation channels represent a 'supreme technical achievement of English farming' (Rackham, 1987, 338). The management of water meadows was quite complex. Rackham notes: 'the usual downland practice was to irrigate the meadows in winter; to turn off the water in March and let the sheep (by then near lambing) pasture the grass; to take the sheep off in late April and let the grass grow up to hay for two months'.

Many traditionally managed meadows and pasturelands have been 'improved' by ploughing and reseeding, and by fertiliser treatment. In cropping for silage production, grasslands are now more frequently mown. Other meadows have been converted to arable land etc. Economic considerations have brought an end to many of the transhumance practices. For instance, In south-west Germany most shepherds have now given up the practice of moving stock to winter grazing areas. Stables have been built near the summer pastures (Dolek & Geyer, 2002). As traditional transhumance shepherding practices have become uneconomic, afforestation has occurred in many areas, and some of the traditional grazing and hay meadows have been abandoned. As a consequence of these changes many species-rich grasslands of high conservation value have been lost or damaged (WallisDeVries, Poschlod & Willems, 2002). As Poschlod and WallisDeVries (2002, 368) observe: 'remnant areas of calcareous grassland are now embedded in a hostile matrix of intensively managed farmlands, forests and roads, and the original tight dovetailing of pasture-lands and sheep tracts has been fragmented into more or less isolated patches'. In addition, many alpine grassland ecosystems have been changed in the transition to a mixed landscape economy, increasingly dependent upon winter sports and recreation (Allan, Knapp & Stadel, 1988).

However, in many areas of the Alps, change has been resisted and conservation management is directed towards maintaining traditional practices. As Rackham (1987, 340) has observed: 'from Dauphiné to Slovenia mixed meadows and pastures are the rule and not the exception, brilliant with rampion and oxeye, scabious and cranesbill, and many umbellifers and orchids'. He also makes the critical point: 'These plants are not weeds: Alpine farmers, who are at least as skilled stockmen as we, regard hay containing fewer than a dozen plant species as unfit for bovine consumption.' Such traditionally managed areas have provided study sites for very revealing microevolutionary investigations.

Early studies of grazing and hay ecotypes in garden trials

Gregor and Sansome (1927) and Stapledon (1928) collected material of *Dactylis glomerata* from many different habitat types in Britain, including hay meadows, pastures, wasteland, cliffs etc. Plants were transplanted in spaced, weed-free gardens to see if the distinctive growth habitat and behaviour of the material in the field was maintained in cultivation. As plants were grown singly, persistent differences in growth habit could be fully appreciated. There were six main growth-forms: lax hay, dense hay, tussocks, cups, spreading pasture and dense pasture. Stapledon found that in old grazing pastures there was a considerable proportion of prostrate forms, while hay meadows had taller types usually earlier flowering.

Following clone-transplant experiments with one or sometimes several gardens, various biotic and seasonal ecotypes have been described in sexual species. In addition, biotic ecotypes have also been discovered in apomictic plants (e.g. *Alchemilla* studies by Turesson, 1943). Building on these early observations, Bradshaw (1963a, b, 1964) and Walters (1970, 1986) investigated dwarf alchemillas in garden experiments. They discovered that collections of *A. minima* from northern England remained dwarf and much-branched in cultivation and postulated that the species was at a selective advantage in sheep-grazed pasture, a habitat where other taller species adapted to hay meadows – *A. xanthochlora*, *A. glabra* – would be unable to flower. Cultivation experiments also revealed that the dwarf habit of *A. faeroensis* (var. *pumila)* from the Faeroes persisted in cultivation. Such variants are considered to be adaptive in intensively sheep-grazed grasslands (Walters, 1970).

While these experiments are interpretable as revealing that selection has occurred in human-managed habitats, more direct evidence has come from other sources.

The finding of unplanned dateable 'experiments'

An early study in Southern Maryland, USA, is very informative (Kemp, 1937). As in many studies of microevolution, this example shows the value of appreciating the finding of 'unplanned experiments' that provide important insights into patterns and processes. An area was reseeded with grasses and legumes, and the owner then divided the ground with one area used as a hay field, protected from livestock. The other portion of the field was heavily grazed. Three years later, Kemp collected samples of *Poa pratensis*, *Trifolium repens* and *Dactylis glomerata* from the two plots and these were transplanted in an experimental garden. In cultivation, a high proportion of the samples from the 'hay' maintained their upright growth habit. In contrast, many of the samples from the heavily grazed area proved to be of dwarf procumbent growth. It was inferred that the different management of the subplots had exerted high and different selection pressures on the plants all originating from the same seed sample, and that such differences had arisen in only three years.

Artificial selection experiments

Evidence of the power and speed of selection has also come from a range of artificial selection experiments, in which plants from a common mixed seed stock have been submitted to a range of conditions (see Briggs & Walters, 1997 for examples).

For example, seed of a mixture of ecotypic variants of *Lolium perenne* and *Phleum pratense* was sown in experimental plots, which were thereafter managed for four years as hay or grazing (Sonneveld, 1955). The plots were subsequently sampled by taking rooted tillers and these were grown in an experimental garden. It was discovered that early flowering variants predominated in haymaking plots, while late flowering variants were predominant under frequent grazing.

In another experiment, Charles (1966) prepared a seed mixture of *Lolium* cultivars in which early flowering and late flowering variants were present in equal numbers. Two varieties of clover were also included in the mix. Sub-samples of this seed mix were sown at ten sites at different elevations on a range of soil types, and managed in different ways: seven with different grazing regimes and three managed for hay (followed by grazing). The possibility of contamination of the sites by wild seed was considered and rejected by Charles. Samples of the plants surviving in each plot were collected in various years and grown in a common garden experiment to determine the percentage of early flowering *Lolium* plants. By the second year, statistically significantly higher levels of early flowering types were detected in two of the three hay treatments.

Experimental investigations of selection

While common garden trials have provided much valuable information, they could be criticised for studying material in an over-simple context, in which typically there is no competition from other plants. Thus, many studies report the behaviour of spaced single plants in weed-free plots at a single site. More direct insights into selection processes have come from experiments in which seeds or plants are grown at several sites within human-managed communities (see Bennington & McGraw, 1995 and references therein).

Seasonal ecotypes

Seasonal races – early (aestival) and late (autumnal) flowering types – have been described in such genera as *Campanula*, *Euphrasia*, *Galium*, *Melampyrum*, *Odontites* and *Rhinanthus*. Some of these are hemi-parasitic plants that derive their food resources partially from other species through root attachment. Sometimes the variants are given specific rank, while in other cases intraspecific variants have been named (Sterneck, 1895; Wettstein, 1900). Wettstein considered that such variants had arisen in historic times in man-made habitats, the regular annual mowing of the grass to make hay being a potent selection pressure. Others took a different view. For example, Soó (1927) considered that the ecotypes had arisen prior to the influence of human activities. Krause (1944) agreed, suggesting that alpine management of hay meadows and grazed pastures had not been in place for a long enough period (*c*. 400 years in some places) for the evolution of seasonal races.

Zopfi (1993a, b, 1995) has re-examined the situation in studies of *Rhinanthus alectorolophus* and *R. glacialis*, both annual, highly polymorphic hemiparasites that grow in different types of grassland in the Alps. Investigations of field collections of *R. alectorolophus* from

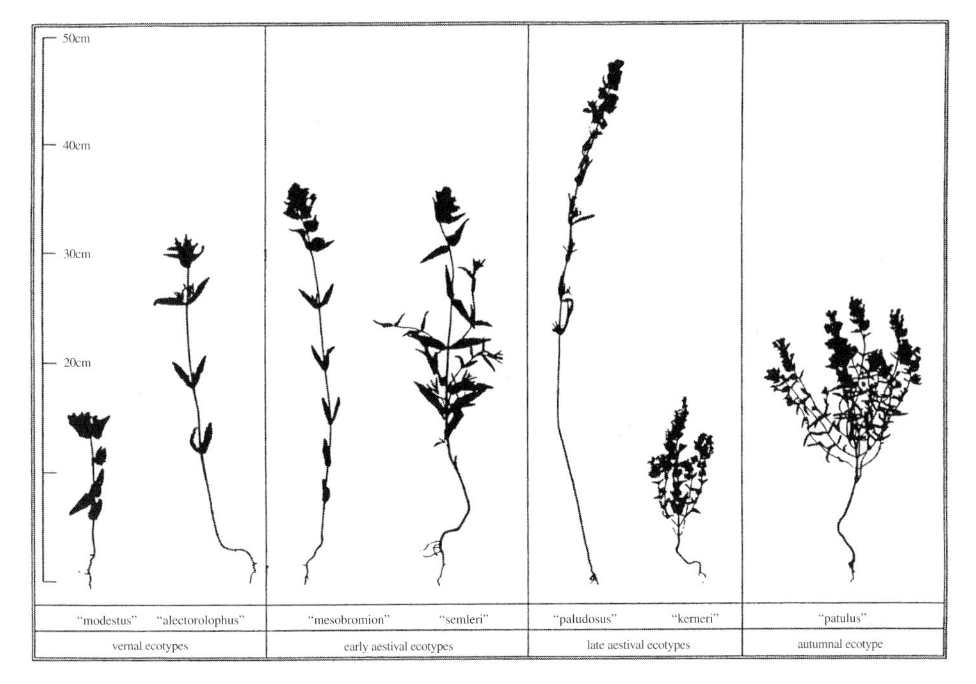

Fig. 8.2 Typical specimens of the vernal (spring), aestival (summer) and autumn flowering grassland ecotypes of *Rhinanthus alectorolophus*. (From Zopfi, 1993a. Reproduced from *Flora* by permission of Elsevier)

eastern Switzerland revealed that there were seven distinctive phenetic groups (Fig. 8.2). Seeds of these seven types were sown in grassland vegetation of different types at four sites. In addition, seeds were sown in pots with different hosts, and also in pots with no host species in Zürich Botanic Garden. These studies provided information on the 'constancy' of key characters, in particular the flowering time in relation to the number of internodes. Internode number proved to be independent of environmental conditions, and there was a near-perfect correlation between number of internodes and flowering time. Thus, ecotypes differed chiefly in the length of the vegetative phase: a difference of one extra internode corresponded to about four days 'delay' in flowering. The persistence of major differences in flowering behaviour in plants raised from seed, grown under different conditions, provides strong evidence for the heritability of such traits.

In interpreting these results Zopfi described seven different ecotypes in differently managed grasslands (Fig. 8.2). These ecotypes fall into four groups.

4–8 internodes

Very early flowering variants are found in alpine areas where the growing season is short and where hay is cut early. These variants are seen as a result of directional selection for short vegetative phase and quick reproduction in this annual species imposed by the timing of hay cropping.

8–13 internodes

These later flowering plants are found in meadows at lower altitude that are mown two weeks later. Grassland management allows a longer preflowering period of growth.

11–16 internodes

These variants are found in autumn-mown litter fields, in which there is high competition for light from accompanying species. A particular ecotype of *R. alectorolophus* is found in grasslands grazed for a period in the early summer, when its seedlings are small. Later flowering time can be seen as adaptively significant, as plants mature after grazing has ceased.

15 plus internodes

This ecotype occurs in areas intensively grazed by cattle in early summer, when *Rhinanthus* is protected by its rosette-like habit. Later, after the traditional grazing period is over, plants set seed in the autumn.

Considering the origin of the different ecotypes detected in *R. alectorolophus*, it seems possible that such variants might have existed in pre-managed natural habitats. However, the extent and variety of grasslands in pre-human ecosystems is unclear. But, it seems unlikely that the precise conditions of grassland management were 'replicated' in natural vegetation. While grazing by domestic stock may mimic to some extent the grazing by natural herbivores, hay cropping produces a 'man-made' habitat, without exact parallel in the wild.

Zopfi's elegant studies are best interpreted as revealing the presence of seasonal biotic races in *R. alectorolophus*, precisely and finely adapted to the various human-managed grassland types. Of the selection pressures imposed by haymaking, the date of the cutting is most important for, as the harvest is essentially herbivory of all or most of the above-ground parts, this sets the limit to growth and reproduction of plants in the sward. Other factors of great importance are the responses to light in plants growing in closed swards, the co-ordination of the growth of *Rhinanthus* with the developmental rhythms of the host plants, and the effects of grazing by cattle. Considering all the different grassland types, Zopfi considers that there has been very strong directional selection for length of vegetative period at the differently managed grassland areas. But, given the importance of flowering time in relation to grassland management, there is strong stabilising selection within sites.

In the past, some authorities have thought that there has been insufficient time for the evolution of seasonal ecotypes in human-managed grasslands. Archaeological studies have provided a probable time frame. Evidence suggests that the first man-managed extended pastures and meadows in the area studied by Zopfi may be not more than 1,000 years old.

Furthermore, other grasslands, for instance, fertilised meadows cut twice, may be no more than 200 years old (Zoller, 1954). However, Zopfi points out that selection experiments in other species have shown that early and late variants can be selected from a variable stock in a few generations (e.g. in *Capsella* (Steinmeyer, Wöhrmann & Hurka, 1985) and in *Lolium perenne* (Breese & Tyler, 1986)). Overall, *R. alectorolophus* is a species that shows great variability in flowering behaviour. Considering the genetics of flowering time, Zopfi suspects that perhaps one or only a few genes are involved. It seems possible, therefore, that the ecotypes could have arisen quite quickly. Moreover, the presence of colonies of *R. alectorolophus* ecotypes in many different isolated alpine valleys suggests that they may have arisen independently several times.

Turning briefly to investigations of *Rhinanthus glacialis*, 14 Swiss populations were studied from seven different habitat types (Zopfi, 1995). As in the earlier study of *R. alectorolophus*, seeds of different variants were sown into natural populations, and plants were also grown with and without hosts in Zürich Botanic Garden. Early flowering variants of *R. glacialis* were discovered in montane hay meadows, with later variants in managed grasslands at lower altitudes. The study also considered the concept of trade-offs (Stearns, 1992). 'A trade-off exists where a benefit realised through a change in one trait is linked to a cost paid through a change in another.' Following comparative morphological studies of all the variants, Zopfi suggests that the adaptively significant benefit of early flowering and fruiting in alpine hay meadows is bought at the cost of reduced plant height, fewer but longer internodes, less branches, fewer flowers and fewer seeds.

Zopfi also considers the evolution of the seasonal ecotypes in *R. glacialis*. He postulates that natural ecotypic differentiation occurred in the post-glacial period, resulting in different climatic ecotypes at different altitudes, and different edaphic ecotypes in acid and limestone grasslands. Subsequently, human management of grasslands has resulted in recent ecotypic differentiation reflecting the date of haymaking, litter cutting, and the degree to which the areas are later grazed.

Comparative studies of pasture and hay ecotypes

Studies of *Plantago lanceolata*, which is found both in grazed areas and in hay fields, have provided further insights into patterns and processes in grasslands. A number of uniform garden trials have revealed the presence of distinct ecotypic variants that come true from seed indicating the genetic basis of differentiation in this variable species. Demography studies have been carried out by van Groenendael (1986) using permanent quadrats set out in two highly contrasted habitat types – a hay field and a pasture. The deliberate choice of extremely different sites is a characteristic of many well-designed experiments. The open pasture habitat in which r-selection forces would be expected to predominate: in contrast, K-selective forces would be expected in the dense swards of the hay meadows.

The investigation revealed that ecotypic differentiation involves not only morphological differences, but also life-history characteristics related to the two very different habitats of origin, namely, a dry dune grassland and a hay meadow. The dune area has been common grazing land for cattle and horses for at least three centuries. The open vegetation was subject to trampling and grazing. Van Groenendael (1986) discovered that the *Plantago lanceolata* populations in the area were of an ecotypic variant with small flat many-leaved decumbent rosettes (often with side rosettes), a high number of small globular inflorescences on short ascending scapes with light seed. Demographic studies revealed that seeds showed innate dormancy and, as a result, there was a considerable seed bank in the soil. The main germination period was in the spring. In this habitat, there was equal risk of mortality in juveniles and adult plants. Rosettes were short-lived and flowered in their second year.

The contrasting study area was a hay field with a high water table and tall vegetation traditionally mown, once per year, at the beginning of July. The ecotypic variant of *Plantago lanceolata* in this habitat had an erect rosette (with no side rosettes), a few tall erect leaves, and few long erect scapes bearing larger inflorescences and producing heavy seeds. Demographic studies revealed that seeds germinate readily in the autumn without forming a seed bank. Juveniles were at greater risk of mortality than adults and several years elapsed before first flowering. Adult plants, in contrast to those in the grazed area, were long-lived and flowered repeatedly, successfully producing seed.

The experiments of van Groenendael (1986) provided a test of the concept of co-adapted traits (Stearns, 1976), which is based on the realisation that different habitats pose different 'problems' of survival, growth and reproduction. The finding of different co-adapted life-history traits in grazing and hay ecotypes of *Plantago lanceolata* provides strong support for Stearn's model, with different variants arising in response to the different selective forces active in contrasting human-managed habitats.

Insights into the operation of selective forces involved, particularly the effect of haymaking were obtained by studies carried out by van Tienderen and van der Toorn (1991a, b). Three sites in the Netherlands, each with its own ecotypic variant of *Plantago*, were selected for study. (a) Late mown hay field on peaty soil at Bruuk; (b) early mown hay field on clay soil at Heteren; and (c) pasture grazed by cattle on sandy soil at Junne.

In order to examine the survival and performance of the different ecotypic variants in their own 'native' and 'alien' sites, reciprocal transplant experiments were carried out in which seeds of the three variants were sown at each site. In addition, cloned material was carefully produced from each site and this material was transplanted, with appropriate replication at each site. The design of the experiment allowed an assessment of the performance of all three populations tested side by side in all three contrasting habitats, providing a native and two alien contrasts at each site. The traditional grassland management appropriate to each site was continued during the experiment.

As expected, some plants died during the experiment, and there was variability in the degree to which plants that survive reproduced. This behaviour presented some difficulties in formal statistical analysis, as the data set was unbalanced. From a biological perspective, clear-cut results were obtained. The three populations were very well differentiated in vegetative characters, time of onset of flowering etc., confirming the findings of van

Table 8.1 *Relative performance under field conditions of clonal transplants of* Plantago lanceolata *from an early mown hay meadow (Heteren) and a grazed pasture (Junne) reciprocally transplanted into hay and pasture. For comparative purposes the performance of plants transplanted into their 'native' sites is set at 100%. 'Alien' transplants perform less well than plants in their native site. The comparative reproductive failure of the pasture variant transplanted in the hay field is particularly significant. It is late flowering and fruiting, and is damaged by early season haymaking before it reaches reproductive maturity.*

Habitats	Hay cut as usual		Pasture grazed by cattle	
Source of cloned material	pasture	hay	pasture	hay
Vegetative survival	70	100	100	90
Success in flowering	52	100	100	60
Success in producing at least one ripe fruiting spike	50	100	100	30
Total seed yield	3	100	100	40

Data sets taken from experiments of van Tienderen & van der Toorn (1991a,b); from Briggs & Walters (1997), reproduced with permission of Cambridge University Press

Groenendael. In particular, plants from the early mown hay field flowered earlier that those from the late mown field, with plants from the pasture being the last to flower.

Comparative measures of performance in the clone transplant experiment are shown in Table 8.1 with native performance set at 100%. Overall, native populations performed much better than alien transplants. Transplants are presented with a series of hurdles to surmount – survival, vegetative growth and reproduction. In terms of survival, alien plants had higher scores than those for the more exacting tests of successful flowering and reproduction, where scores were much lower. At the final 'hurdle' aliens performed poorly against native plants in terms of seed output – a crucial component of fitness. Thus, seed yield was reduced in all alien populations in relation to the native population by 40–95%.

Studying reproduction in more detail reveals what happens to the three populations at haymaking. In the early mown hay field, native *Plantago* plants flower earliest and most spikes had time to mature and produce seeds. However, the variant from the late mown hay meadow was much less successful: flowering was just beginning at the time of hay cut in the early mown meadow. As a consequence, most of the seeds were not fully mature at the date of the hay cut. This hay ecotype is adapted to the later cutting date operating in its native habitat. Plants from grazed areas did not succeed in flowering before the hay cut: they flowered later in the season.

The results of this study parallel those for *Rhinanthus*, where more than one hay ecotype was detected. Thus, in *Plantago* there is clear evidence of two different hay ecotypes, each adapted to the haymaking management of its native habitat. There may of course be further hay ecotypes in the species adapted to different cutting dates.

Park Grass Experiment: *Anthoxanthum odoratum*

If hay crops are taken from a site repeatedly, there is a danger that soil fertility will fall and yields of hay decline. Traditionally, hay meadows were sited in areas on flood plains where nutrients were replenished in winter floods. Alternatively, meadows were irrigated with nutrient-rich waters, as in the Alps, and fertilised by animal dung. However, in the mid nineteenth century, agriculturalists became interested in whether the yield of hay could be increased by the use of artificial chemical fertilisers.

The classic Park Grass Experiment at Rothamsted, UK, was set up by Lawes and Gilbert in 1856, to examine the effect of fertilisers on hay crops. The site chosen had been grassland for at least a century. Separate plots were marked out to receive different treatments, comparing the effect of artificial and natural fertilisers on hay production by weighing crop samples and determining species composition.

It was soon evident that artificial fertilisers did increase yield. Fortunately, the experiment was continued. One effect of fertiliser treatment, particularly in plots treated with ammonium sulphate, was the acidification of the soil (originally probably pH 5.7). To counter this growing acidity, in 1903, regular liming was applied to the southern half of most of the plots. In 1965, further subdivision of the experiment was made to give three subplots each with a different lime treatment: a fourth subplot provided the unlimed 'control'. Over the course of time species composition has changed. The unfertilised plots are the least productive in terms of yield of hay, but richest in the number of species. In fertiliser-treated plots, yields are greater, but different and fewer species are found. As we shall see in a later chapter, this finding is of exceptional importance to those concerned with the management for conservation of species-rich grasslands.

Looking at the experimental site – now maintained for more than 140 years – there are striking visual differences, especially before harvest, between plots and subplots, in productivity, height of vegetation and species composition. Areas given different treatment are clearly defined with crisp boundaries between plots and subplots, the 'transition' between treatments occupying 30 cm or less. This indicates that there has been little sideways movement of applied nutrients on this level site.

The Park Grass Experiment, with its mosaic of differently treated plots, has provided a serendipitous opportunity to investigate microevolution. *Anthoxanthum odoratum* was selected for study, as this species is widely distributed, occurring both on limed and unlimed plots (Snaydon, 1976). Samples were collected and grown in a uniform garden trial and morphological differences were detected. Overall, plants from tall sward plots grew taller in the garden than those from short vegetation plots. It was concluded that such correlated differences were adaptive.

In addition, the response of plants to different soil nutrients has been examined (Fig. 8.3). Populations from limed plots (pH 7) were affected more by the calcium concentration

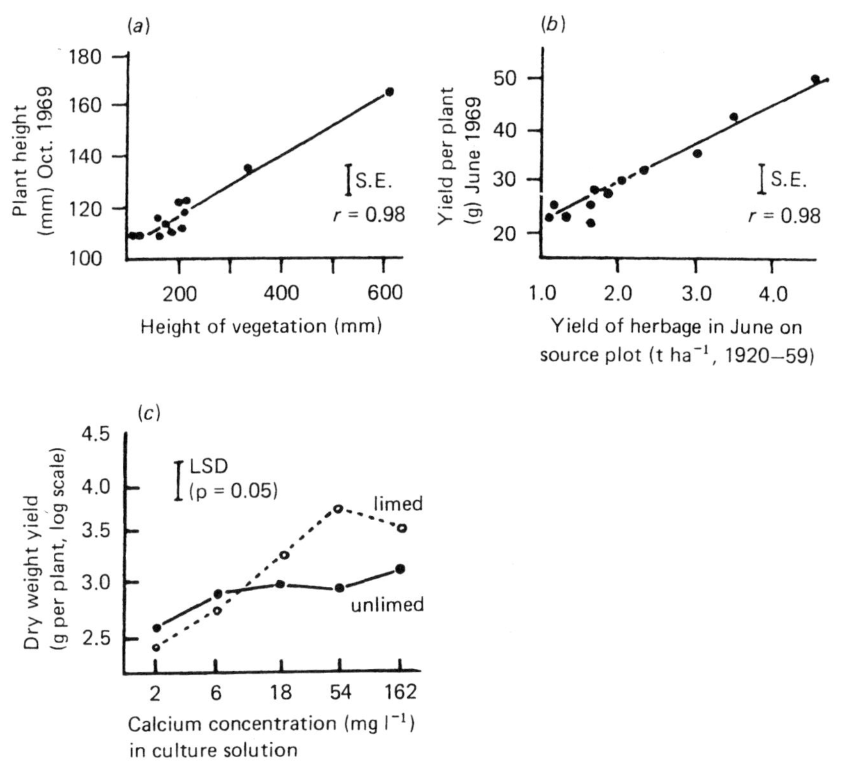

Fig. 8.3 Graphs summarising some of the studies of *Anthoxanthum odoratum* collected from Rothamsted Park Grass Experiment and grown in garden trials. (Reproduced from Snaydon, 1976, with permission of the author) (a) Relation between plant height and height of vegetation in source plots. (b) Relation between plant weight and yield in source plots. (c) Effect of calcium on dry weight of plants from limed and unlimed plots. (Snaydon, 1970, 1976; Snaydon & Davies, 1972; Davies & Snaydon, 1973a. See also Davies, 1975 – responses to potassium and magnesium; Davies & Snaydon, 1974 – response to phosphate.)

in the culture solution than plants from unlimed plots (pH 4). The growth of the plants was correlated with the pH ($r = 0.9$) and calcium status ($r = 0.8$) of their source plots (Davies & Snaydon, 1973a, b). Experiments on the responses to different soil nutrient factors proved very revealing. The soils with the Park Grass Experiment have come to have different levels of phosphate. In a sand-culture experiment with material of *Anthoxanthum* from plots with contrasting phosphate status, the response of plants to applied phosphate was closely correlated ($r = 0.70$) with the extractable phosphate content of its source soil. Over a wider range of applied phosphate concentrations, the response of plants was more closely correlated with the pH of the soils from which they came (Davies & Snaydon, 1974).

Acid soils have a higher soluble-aluminium content than calcareous soils. It is of microevolutionary significance that plants of *Anthoxanthum* sampled from unlimed (more acid) plots proved to be more tolerant of high aluminium concentrations than those from limed plots (Davies & Snaydon, 1973a, b). Growth in different concentrations of applied potassium was also examined (Davies, 1975). Considering the results of all these experiments, there is clear evidence that the response of the *Anthoxanthum* samples closely matched the sward and soil conditions of the source plots. As the different soil types had arisen from differently treated subplots of a once uniform area, they deduced that the disruptive selection had occurred in the mosaic plots within the experiment.

Investigations also provided some insights into gene flow in the Park Grass Experiment. These were obtained by using one of the standard approaches of examining gene flow and selection outlined above – namely comparing the variability in adult plants and their seedling progenies. Plants and progenies collected across the boundary between plots differed in height, yield and other characteristics, suggesting that cross fertilisation between the plants growing in different plots occurred, but was only important within about 2 m of the boundary.

In addition, reciprocal transplant experiments were carried out (Davies & Snaydon, 1976). Cloned material from different parts of the Park Grass Experiment was grown in an experimental garden to minimise carry-over effects. Then, cloned material was transplanted back into the Park Grass Experiment. The experimental design allowed the measurement of survival and performance of the native and alien plants in each of a number of plots. It was discovered that plants grew best in their native plots. Relative fitness of alien plants proved to be only 0.59 on the limed subplots and 0.75 on the unlimed subplots.

Reflecting on their results Davies and Snaydon (1974, 705) concluded: 'Each population seems to have evolved specific and complex ecological adaptation to the environment of its source plot.' Overall, 'the morphological and physiological differences between populations are closely correlated with differences in analogous environmental variables on their source plots'. Reflecting on the key decision to add lime to some of the subplots in 1906, they concluded that the differences they detected in *Anthoxanthum* in different parts of the experiment had 'evolved within 65 years and over distances of less than 30 m' (Davies & Snaydon, 1974, 705).

This result gives us some purchase on the question of the speed of evolutionary change in human-managed areas. The recent change in the liming treatments in 1965 provided an ideal opportunity to test whether microevolutionary change could be even more rapid. The new regime provided the possibility of comparing material from subplots that had received no lime for 100 years with adjacent subplots, which had been similarly treated until they received an addition of lime in 1965. Samples were collected and grown in a spaced garden trial. Plants were studied for 3 years and 21 attributes were measured (including morphology, yield, disease resistance etc.). It was concluded that considerable genetic change had occurred after only 6 years between the plants growing on never-limed and newly limed subplots (Snaydon, 1976).

This highly imaginative use of a classic 'agricultural' experiment has provided many insights into pattern and process in *Anthoxanthum* in a human-managed patchy environment (Snaydon, 1978). It is to be expected that the selection pressures experienced by *Anthoxanthum* will have impinged on other species found within the experiment. In preliminary investigations, differences have been detected in other species collected from the

Fig. 8.4 Budding's lawn mower, patented about 1830. Three advertisements for the mower illustrate 'the machine's rapid social progress, as the operators change from a rustic, to an elegant composed gentleman, who might be beautifying his suburban villa'. (From Thacker, 1979)

Park Grass Experiment, namely *Lolium perenne* (Goodman, 1969) and *Holcus lanatus* and *Dactylis glomerata* (Remison, 1976). Also, adaptively significant differences may be found in *Taraxacum officinale* agg. Richards recognised 12 microspecies on the plots. Taller species occurred in the tallest hay and those with flattest rosettes occurred in the plots with the shortest vegetation (see Thurston, Williams & Johnston, 1976).

The results of the Park Grass Experiment and other microevolutionary investigations point to a critical issue for the conservation management of species-rich meadows. Throughout Europe, many hay meadows have been lost, as management has changed, through ploughing, reseeding and regular fertiliser addition allowing multi-cropping for silage etc. This management leads to high-yielding but species-poor swards. If the remaining species-rich traditional hay meadows, with their distinctive and beautiful plants, butterflies etc. are to be successfully conserved, each must be managed as it had been in the past, especially in relation to the date of mowing, manuring, irrigation etc., rather than given some universal 'standard' hay treatment, with a date of cutting 'sometime in the summer' (Feltwell, 1992).

Selection in lawns and golf courses

The herbage cut from hay meadows is a most valued crop. However, there are large areas of grassland across the world that are carefully managed and mown, but where the 'crop' is discarded or used as mulch. What is valued is the remaining turf – the green sward of gardens, parks and sports fields. It has been estimated that there are 31 million acres of lawns in the USA. Worldwide there are also large numbers of sports fields. For example, there are more than 31,000 full-length golf courses worldwide, with about 23,000 in the USA and 6,000 in Europe.

Grassy areas have been a part of garden design for centuries. In Britain, lawns in the modern sense probably date back to the Tudor period. Until the invention of the mowing machine by Budding (see Fig. 8.4) and the devising of horse-drawn machines, lawns were cut with scythes or kept short by grazing. Grassy areas may have evolved *in situ* from grazing land by persistent management, but traditionally lawns were often laid with cut turf from the close swards of grazed coastal grasslands, or, more recently have been sown with a seed mixture of selected low-growing hard-wearing cultivars of grasses. However, increasingly turf is now grown as a commercial sown crop. Turf is lifted, transported and set out as lawn on prepared sites. Thus, the lawn has evolved from its role as a grand setting for the stately home, to a desirable element in gardens, generally. However, in reality, lawns range from the scruffy and neglected to the ancient and celebrated. But even the most ancient lawns cannot rival the centuries of consistent land use represented by some hay meadows.

The aim of most lawn-keepers is to have a weed-free sward. However, weeds are found in many lawns, which in their most manicured form represent a most extreme environment for plant growth. The vegetation is cut repeatedly, to give a closed sward of height as low as 0.5 cm. The aim is to avoid bare areas and lawns are often patched or reseeded. The species composition is manipulated by herbicide treatments, influenced by addition of fertilisers, and affected by irrigation practices.

Temperature-controlled germination

The investigation by Wu, Till-Bottraud and Torres (1987) has revealed genetic differentiation in golf course populations of *Poa annua* on Davis Municipal Golf Course, California. The study site was a golf course about 20 years old. In this area, new golf courses become seriously invaded by *P. annua* within five or six years. The species is predominantly inbreeding, but some outcrossing has been detected in experiments. As a consequence of human activities, especially contract-mowing by machine, gene flow by seed is likely, within and between golf courses.

In California, rainfall is concentrated in the period November to March. Summers are dry with almost no rain between May and September, and parts of the golf course have to be irrigated. *P. annua* occurs as a weed on all three of the types of grassland that make up a golf course.

A. Greens are mown to give a short 0.6 cm sward that is watered daily or more frequently in dry weather, to maintain the playing surface.
B. Fairways, mown twice a week to a height of 2.5 cm, are watered only two or three times per week.
C. The rough is occasionally mown, and is not irrigated.

In an experiment by Wu *et al.* (1987) seed samples were collected from plants in all three grassland types. Germination experiments revealed that the three populations differed in their responses to high temperature (25 °C). The mean germinability of the seed from the rough was uniformly low and less than 30%, with a mean of 5%. In contrast, the germination of seed from the greens was higher than 60%, with a wide range of behaviour amongst the fairway population. The heritability of this behaviour was established by testing progenies produced by polycrossing separately isolated groups of plants representing each of the three populations – greens, fairways and rough.

The patterns of genetic differentiation in germination behaviour discovered in the experiment are explicable in relation to the turf irrigation treatment. Thus, in the rough, seedlings are unable to survive the dry summer period, as they receive no watering. Germination is controlled by temperature: high temperature inhibits the process, and lower temperature triggers germination at a period when rainfall is predictable and when there is a high chance of plants completing their life cycle. Thus, in the rough the weed is basically annual, the result of selection in alternating wet and dry seasons. In contrast, golf greens receive frequent irrigation, and seeds germinating there, even in the high temperatures during summer, are guaranteed favourable moisture conditions. Such sites support populations with a high incidence of temperature-independent germination. However, it is unclear why selection is not sufficiently intense to produce populations entirely composed of individuals with this type of germination behaviour. Perhaps, the responses of the different variants to other selection pressures – mowing/competition/herbicide – may be complicating factors. Finally, the populations on fairways contained a mixture of *P. annua* variants, and this is perhaps to be expected in what is an intermediately watered heterogeneous environment.

This elegant investigation offers insights into microevolutionary processes in a habitat with no counterpart in nature. Evidence suggests that differentiation can occur quickly. Furthermore, independent selection takes place at each of the greens, which are separate 'islands' within the golf course.

The ecological relationships between the unsown species *P. annua* and other species (in particular *Agrostis stolonifera*) found on the greens of a golf course in Melbourne, Australia have been examined by Lush (1988a, b).

Disruptive selection in *Poa annua*

Differentiation has been studied in *P. annua* growing on a bowling green, adjacent to the University Botanic Garden, Cambridge (Warwick & Briggs, 1978a, b). The green was established in 1928, with new turf being laid in 1938. The sward is maintained at a height of 0.5 cm. Around the bowling green there are flowerbeds. *P. annua* is present in both bowling green turf and in the flowerbeds.

There are two different variants of *P. annua*: (a) quick-flowering short-lived plants with erect growth habit, and (b) prostrate and semi-prostrate perennial plants that are slower to come into flower and fruit. In order to determine the distribution of the variants within the study area, plants growing on the bowling green and flowerbeds were sampled on three occasions, transplanted into pots of soils, and grown for seven months to determine their growth habitat. All the bowling green plants had prostrate habit at the time of collection. In cultivation, 95–98% retained their prostrate habit, but 2–6% grew erect, revealing that they were phenocopies (i.e. they were non-hereditary modifications that mimicked the genetic prostrate variants). Samples of plants from the flowerbeds contained both variants in the ratio prostrates to erects *c.* 1:2. A study of esterase isozymes revealed that there were no bands that distinguished prostrates and erects. But there was heterogeneity within groups, prostrate and erect categories being represented by a number of genotypes. Segregation was also detected within family lines.

Poa annua is predominantly self-fertilising, and to examine the morphology and life-cycle characteristics of the material from the bowling green and flowerbeds, inflorescences of erect and prostrate plants were bagged to produce selfed seed. Family lines were grown. Progenies from the different seed parents proved to be highly homogeneous, with erects and prostrates maintaining their distinct growth habit. There were highly significant differences in the length of the vegetative phase. Thus, erect plants flowered early, means 45–63 days (with a grand mean 55 days from sowing to anthesis), while prostrate plants flowered later, means 70–89 days (grand mean 80 days).

Observation suggests that regular defoliation is a major selective force in lawns. Cloned material of both prostrates and erects was grown in pots and, to simulate the mown environment, plants were clipped to a height of 2 cm each week. The clippings were retained and dried to estimate cumulative loss of dry weight. Control plants, in which there was no clipping, simulated the flowerbed environment. Prostrate plants grew and flowered more successfully than erect plants under clipping. Coefficients of selection against erect plants under clipping ranged from 0.53 to 0.68. In contrast, erect plants were at a selective advantage in control conditions, with the coefficient of selection against the prostrate plants being 0.77.

In addition to regular mowing, bowling greens are also subject to trampling pressures. To simulate the effects of trampling, prostrate plants from the bowling green and erect plants from the flowerbeds were grown in pots and they were subjected to light and heavy trampling using a mechanical foot (Warwick, 1980). This consisted of a metal weight calculated to give a pressure equivalent to that of the average human (70 kg), when dropped onto the plants from a standard height of 5 cm. It was discovered that genetically determined prostrate variants of *P. annua* were better able to withstand trampling damage than erect variants. Coefficients of selection acting against erect forms under trampling were particularly high for reproductive characters (0.79). In considering whether the experimental situation exactly simulated trampling on the lawns, Warwick concluded that while the mechanical foot may have reflected adequately the trampling pressures exerted by the human foot, the simulation did not reproduce the rotary twisting associated with human trampling.

The patterns of adult variability in *P. annua* detected at the site support the notion that disruptive selection occurs in the bowling green–flowerbed environment, genetically prostrate plants being at a selective advantage in the lawns, and also that a small number of plants have sufficient phenotypic plasticity to produce the prostrate phenotype as a phenocopy. In contrast, in the weeded environment of the flowerbeds it might be expected that erect plants with their quicker development from germination to fruiting would be at a selective advantage. If selection pressures are intense at each site, it would be expected that prostrates might produce most seed on the lawns and erects most seed in the flowerbeds. However, gene flow must also be considered.

The investigations of *P. annua* at the bowling green site also examined the seed bank in the soils of the bowling green and flowerbed. These seed banks reflect the seed produced *in situ* and also seed arriving from elsewhere by seed dispersal. At each site the seed bank populations of erect and prostrate variants were estimated by growing seed present in soil samples. It was discovered that potential populations in the soil of bowling green and flowerbed were both more variable than the actual adult populations. This evidence offers strong support to the model of disruptive selection. Concerning the patterns of variability represented within sites, it is significant that there was a higher frequency of erect plants in the soil at the edges of the lawn than in the centre. Perhaps, seed of *P. annua* is introduced as a contaminant when the worn edges of the lawn are reseeded at the end of the playing season. In addition, seed could also be carried onto the lawn on footwear, and would be more likely to be deposited on the edge of the lawn rather than the centre. Initially, it was difficult to understand why there were a high number of prostrate plants in the flowerbed seed bank. But a study of the gardening practices revealed that a mulch of mowings from the playing surface, presumably containing prostrate seed, was added periodically to the flowerbeds. Compost made from lawn mowings was also added.

Comparative studies of lawns and other grassland types

Cambridge is a city of lawns, the most famous of which are the long-established college lawns of King's, Christ's and St John's Colleges and the Botanic Garden etc. Despite the determined efforts of generations of gardeners to maintain short turf in courts and gardens, some weeds survive in lawns. As we have seen, mowing and trampling present plants with extreme selection pressures. Are the weeds in college lawns ecotypically distinct?

Samples of five weed species were collected from the lawns, and investigated in comparative cultivation over a two-year period, with samples of the same species from other grassland types: heavily grazed areas; seasonally mown or grazed grasslands; and areas not known to be grazed or mown (Warwick & Briggs, 1979, 1980a, b, c). There were very marked differences in phenotype of the plants at point of sampling. Some variants retained distinctness, but others, in their growth responses, indicated the importance of phenotypic plasticity. Firstly, the distinctive variants will be considered.

Lawns

It might be supposed that dwarf or prostrate plants might be at a selective advantage in lawns, and some dwarf or prostrate variants of *Bellis perennis* (King's College), *Prunella vulgaris* (St John's College), *Achillea millefolium* (King's, St John's) and *Plantago major* (King's, St John's and Botanic Garden) retained their characters in cultivation.

Grazed areas

Port Meadow, Oxford, has a very long and continuous history of grazing by cattle and horses (Baker, 1937), and it might be supposed that selection would favour prostrate or dwarf habit in such sites. In cultivation, samples of *Plantago major* were distinctive in their prostrate growth habit and short scapes. Likewise, dwarf plants of *Achillea millefolium* were also found.

Oxford hay meadows

As we have seen above, early flowering variants are at a selective advantage in hay meadows. Such variants were detected in *Prunella vulgaris* and *Plantago lanceolata* in cultivation of samples from long-established, traditionally managed hay meadows at Yarnton and Pixey Meads.

Areas not known to be regularly grazed or mown

No dwarf or prostrate variants were detected in cultivation from samples taken from this group of habitats.

The presence of distinctive variants at sites subject to years of consistent lawn, grazing or hay management suggests the importance of selection. However, the responses of many of the samples in garden trials indicate the crucial role of phenotypic plasticity, particularly in *Achillea millefolium*, *Bellis perennis*, *Prunella vulgaris* and *Plantago lanceolata*. For *Prunella*, our results were similar to those of Nelson (1965) who found that some prostrate plants growing in lawns were genotypically prostrate, while others proved to be 'phenocopies', growing taller in cultivation.

Reciprocal transplant experiments

Garden trails and tolerance tests (clipping and simulated trampling) have provided a great deal of evidence about lawn weeds. However, these investigations were essentially simulations of the effects of grazing and trampling. How would plants from lawns and contrasting grassland types fare in real turf? Such experiments were carried out with *Plantago major* (transplanted seedlings), and clonally propagated material of *Bellis perennis* and *Achillea millefolium*.

Plantago major

First, in a clipping experiment simulating the mowing in the lawn environment, genetically prostrate plants from lawns were able to produce thousands of seeds, but the erect-growing plants originating from roadsides were severely damaged, and lost all their reproductive structures. How far were these responses a reflection of what would happen in mown and tall plots of actual turf?

In a reciprocal transplant experiment employing mown and tall-sward plots, lawn plants in the mown treatment again produced much seed. However, the behaviour, in turf, of plants originating from roadsides was different from that under clipping, as some plants from this habitat were able to produce a little seed. This response would appear to be related to the flexibility of the leaves and scapes and the cushioning effect of neighbouring species in the 'real' lawn. In the tall-sward plot, roadside plants had a greater reproductive output than lawn plants. Selection against roadside plants under clipping and in the mown plot proved to be high – with estimates in the range 0.78 to 1.0. In the control of clipping experiment and in tall grass plots coefficients of selection against lawn plants were also high (different estimates 0.74 to 0.77). Thus, the results of these experiments with *Plantago* gave similar results to those of the investigations of *P. annua* outlined above. As the clipping, trampling and reciprocal transplant experiments revealed, prostrate lawn variants of *Plantago* were better able to withstand the lawn environment than erects from grasslands not subject to extreme lawn management.

The performance of Bellis *plants in a transplant experiment*

In an experiment, plants from lawns and contrasting grassland types were planted into a lawn that was regularly mown to a height of 2–3 cm. There was no evidence that all the *Bellis* plants sampled from lawns were equally well adapted, but four very successful individual plants with very short scapes (flowering and fruiting structures held close to the ground) were detected: three from very well-maintained longstanding lawns (Botanic Garden, and two from King's College) and one from Port Meadow, Oxford. There are a number of possible explanations as to why apparently 'less well-adapted *Bellis* plants' may survive in lawns. Lawn surfaces may be sufficiently uneven to allow plants to escape decapitation. Also, even though successful establishment by seed may be low in the closed turf of the lawn, some plants do establish opportunistically, and, in the absence of herbicide use or hand weeding, survive as vegetatively vigorous clonally reproducing patches, even if flowering and fruiting structures are frequently removed by mowing.

The responses of Achillea *in mown and tall plots*

In a clone transplant experiment employing mown and tall-sward plots, there was no evidence that plants originating from sampled lawns and grazed sites (that were notably dwarf in standard cultivation) were more successful than others in vegetative spread and reproduction. Moreover, individuals originating from non-mown sites grew successfully in the lawn plots. Therefore, there is no support for the notion of distinctive lawn ecotypes in this species: many genotypes are able to survive the rigours of repeated mowing. Fruiting is rare in well-kept lawns, but clonal reproduction is important. The rate of individual clonal spread of different genotypes was followed in the experiment. Plants originating from lawns were not statistically significantly better able to reproduce by this means than other genotypes from different habitat types. There was no evidence in the experiments for tall grassland ecotypes.

The importance of phenotypic plasticity in lawn weeds

The transplant experiments to a large extent confirm the results of the garden trials. While there is evidence for lawn races in some species, in others, colonisation may be seen as opportunistic, and survival, often by clonal growth, involves considerable phenotypic plasticity. Considering the microevolutionary significance of plasticity, Bradshaw (1965) makes the important point: If plasticity requires less energy, or the flexibility derived from plasticity gives as great a fitness as that associated with genetic adaptation involving limited modifiability, then there will be no selection acting to produce genetic differentiation within the species.

Gene flow in grasslands by movement of hay

In addition to the results of investigations of gene flow reported above, studies of populations of *Crocus scepusiensis* have provided important insights. Flower colour is variable, with four different variants (morphs) – violet or white flowers with orange or white stigmas. Over a 5-year period, a survey of the distribution of these colour variants was carried out in the Gorce Mountains, of the Western Carpathians, Poland (Rafinski, 1979). *Crocus* is apparently restricted to man-made grasslands in glades within beech or spruce forest that are (or were) hay meadows or pastures. Populations vary widely in size, from small discrete patches to very large populations with hundreds of thousands of plants. Surveys revealed that while the frequency of morphs does not change year on year, there were dramatic differences of frequency of stigma colour in populations separated by small distances, such as a belt of trees or a stream. But, in other cases, patterns of variability did not appear to be associated with any topographical or ecological factor.

Turning to the microevolutionary significance of colour variation, it is possible that it is adaptively significant in insect pollination. However, Rafinski takes the view that the variability in stigma colour might result from non-selective processes. Reproduction occurs by seed, and by vegetative propagation through adventitious corms. Seeds are large and with no apparent means for dispersal. Rafinski considers that founder effects, associated with seed dispersed in the human management of the grasslands of the area, are a major influence on population variability. Thus, hay containing crocus seed is collected from woodland glades at the time of fruiting of the crocus, and transported along forest trails. Some of the crop falls by the wayside and small isolated populations of crocus become established in grassy areas. Which colour variant establishes at a particular site is a matter of chance. Other factors are also important. For instance, some crocus populations are released from snow cover early in the season, while in other places with a different microclimate, snow lingers and plants flower later. Such differences in flowering behaviour could influence levels of gene flow between colour variants in this self-compatible species. While our knowledge of the variability of flower colour is as yet incomplete, in particular the genetic basis for polymorphism, overall Rafinski's study reveals the likely importance of founder effects on population variability in this species and highlights the role of human activities in the traditional management of grasslands in the landscape.

Seed banks

A number of studies have been made of the buried persistent seed bank of viable seeds in the soil (Harper, 1977; Thompson, Bakker & Bekker, 1997). For example, grasslands managed for hay may have 38,000 seeds/m^2 (Chippindale & Milton, 1948). Recently, investigations have revealed a decline in the total density of seeds in the persistent soil seed banks in grasslands and heathlands that have been abandoned or modified (Bossuyt & Hermy, 2003). As we shall see later, the potential loss of soil seed banks has profound implications for conservationists setting out on habitat restoration projects, and planning the management of rare and endangered species.

Gene flow by dispersal of seeds in animal husbandry

Turning to another issue, evidence suggests that there has been a decline in species richness in semi-natural grassland and heathland habitats, and this has been largely attributed to habitat fragmentation (Brunn & Fritzbøger, 2002). But changes in land use and abandonment of former agricultural practices are also important. Thus, there have been dramatic changes in livestock husbandry over the last 200 years resulting in decreased chances of seed dispersal in agricultural landscapes. For example, seed dispersal by the movement of livestock in Denmark has changed on three spatial scales in the last 200 years (Brunn & Fritzbøger, 2002).

A. At a scale of 1–10 km. In the medieval common field system there were large areas of sparsely fenced land. Domestic animals moved around the agricultural landscape for most of the year, on pastures and heathland, and also on stubbles and fallow land. Some of the grazing was on 'commons', the areas of which could be uniform or include intermingled patches of alder carr, coppice and high forest, as well as heathland, dry and wet grasslands. Hay, heather, straw, leaves and twigs were collected and fed to livestock. The manure produced by the animals was spread on village fields. The common field systems, which were formerly widespread in Europe, were swept away or modified in the enclosure movement resulting in fenced farmland and forests, a process which began in Denmark around 1760 and was more or less completed by 1810. The nineteenth century also saw the introduction of stall-feeding of cattle.

B. In the pre-enclosure period, there was considerable movement of livestock between landscapes on a scale of 10–50 km. This involved the driving of animals to distant areas of pasture, commons, heathlands, woodlands and salt marshes. 'Forest grazing' was particularly important in the raising of pigs in the sixteenth to eighteenth centuries; large flocks were taken some distance to oak and beech forests to feed on the acorns and beech mast before the Christmas slaughter. The enclosure movement restricted grazing in forests, but mast grazing for pigs was allowed.

C. From the late medieval period onwards, there was an increasing export of live beef cattle and horses to markets hundreds of kilometres away in Germany and the Low Countries. Animals driven perhaps 10 km per day were sometime fed en route. After 1847, with the rise of railways, there was a decline in long-distance droving of oxen and horses.

In considering the gene flow by seed dispersal associated with livestock mobility, evidence suggests that in Denmark, as in many other parts of Europe, livestock movement on all three spatial scales has declined dramatically with the development of intensive animal husbandry and stock-free intensive arable farms. This decline is likely to have had major consequences for the dispersal of seeds and fruits of many species, for it has been noted by Brunn and Fritzbøger (2002) that 'a large proportion, more than two thirds, of Danish plant species growing in (semi)- natural habitats can survive the passage through cattle or sheep intestines or can attach to mammal fur of the same animals'. Also, on intensively managed farms, populations of wild birds and mammals are often in decline, and this has probably led to a reduction in dispersal of seeds and fruits by 'wildlife'.

In Denmark the effects of the loss of mobility of domesticated stock on seed dispersal may be somewhat counterbalanced by increases in deer populations. However, reflecting on current animal husbandry practices, Brunn and Fritzbøger (2002, 431) conclude: 'the freedom of movement of livestock has decreased dramatically'. Now, farm animals are transported in trucks. Continuing their analysis they point out: 'seeds transported in mud attached to animal hoofs and fur are also capable of attachment to footwear and vehicles, and the potential seed dispersal distance associated with motor traffic is probably orders of magnitude greater than that of animals'. However, they conclude: 'the dramatic rise of motor traffic during this century is not likely to replace livestock as dispersers of plants from semi natural habitats. The plants dispersed by cars are mainly weedy species found on road verges' (see also Hodkinson & Thompson, 1997).

Concluding remarks

Experiments have revealed strong evidence for the evolution of habitat-specific genetic variants of plants in different types of anthropogenic grasslands. Where very distinctly different grassland management occurs in sites adjacent to each other, as at Park Grass, and in bowling green lawns and golf courses, there is evidence for disruptive selection leading to locally adapted populations. Studies have also examined aspects of gene flow and estimated components of fitness. In several investigations, the speed of evolution has been estimated, with very quick responses in some cases. In studies of grassland ecosystems, there is evidence that both ecotypic differentiation and phenotypic plasticity are important components of the adaptive responses of plants. Other grassland species have been studied in relation to land use: for example, *Agrostis tenuis* (Bradshaw, 1959a, b; 1960); *Cynosurus cristatus* (Lodge, 1964); *Dactylis glomerata* (van Dijk, 1955); *Euphrasia* spp. (Karsson, 1976, 1984); *Gentianella germanica* (Zopfi, 1991); *Heracleum sphondylium*

(Jaeger, 1963); *Lolium perenne* (Gregor & Sansome, 1927); *Lotus corniculatus* (van Keuren & Davis, 1968). *Odontites lanceolata* (Bollinger, 1989); *Phleum pratense* (van Dijk, 1955); *Potentilla erecta* (Watson, 1970); *Rhinanthus serotinus* (Ter Borg, 1972); and *Sesleria albicans* (Reisch & Poschlod, 2003).

As agriculture has evolved, movement of stock for feeding and on-the-hoof to market has declined in many areas, and the local and regional dispersal of fruits and seeds (of species able to withstand transit through the digestive system of cattle, sheep, pigs etc.) has been greatly reduced. While this facet of human-influenced dispersal has declined in many areas, widespread dispersal of plants and animals to new territories has come from the increasing globalisation of trade. Microevolutionary investigations of the patterns and processes of plant introduction are considered in some detail in Chapter 13, together with an appraisal of the consequences for conservation.

In the next chapter we examine patterns and processes of microevolution in arable and forested cultural landscapes.

9

Harvesting crops: arable and forestry

In *The Variation of Animals and Plants under Domestication*, Darwin (1905, vol. 390) considered an issue raised by Loiseleur-Deslongchamps in his book *Les Cereales*. As cereal crops have evolved under domestication, perhaps the weeds infecting the fields in which these crops have been grown have also changed. It is interesting to quote Darwin's reaction to this point and his cautious conclusion.

Loiseleur-Deslongchamps has argued that, if our cereal plants have been greatly modified by cultivation, the weeds which habitually grow mingled with them would have been equally modified. But this argument shows how completely the principle of selection has been overlooked. That such weeds have not varied, or at least do not now vary in any extreme degree, is the opinion of Mr H C Watson and Professor Asa Gray, as they inform me; but who will pretend to say that they do not vary as much as the individual plants of the same sub-variety of wheat? We have already seen that pure varieties of wheat, cultivated in the same field, offer many slight variations, which can be selected and separately propagated; and that occasionally more strongly pronounced variations appear, which as Mr Shirreff has proved, are well worthy of extensive cultivation. Not until equal attention be paid to the variability and selection of weeds, can the argument from their constancy under unintentional culture be of any value.

There is now abundant evidence that weeds have co-evolved with crops in arable fields (Barrett, 1983). Considering this issue, Wulff (1943, 106) makes two general points.

Of exceptional importance is the circumstance that the very same changes in biological peculiarities and morphological structure . . . in plants purposely cultivated may also be observed in those weeds that constantly accompany definite crops, thanks to which they are involuntarily cultivated by man. [Moreover,] certain of such weeds, without man himself being aware of the process, may gradually be transformed into direct objects of his cultivation.

In general, the management of the arable environment aims to provide a uniform set of growing conditions for the crop, with appropriate weed control, maintenance of soil fertility and adequate water supply. In attempting to achieve these ends, there have been many changes in agriculture. In most areas, shifting cultivation has given way to long-term settled agriculture. Peasant agriculture has employed, and in many places continues to rely on, traditional irrigation, terracing, manuring, fallow and crop rotations etc., together with hand weeding, and labour-intensive threshing and winnowing of the crops. But, increasingly, these traditional approaches have been swept away with mechanisation and the use of fertilisers, herbicides, new crop varieties, new and improved methods of threshing, seed cleaning, area irrigation schemes, contour ploughing, minimum tillage management systems to control soil erosion, and focused pesticide use in integrated farm management etc. In addition, there have been 'counter movements' against modern agricultural methods. For example, organic farming practices that do not employ artificial fertilisers and chemical pesticides are increasing, and also efforts are being made to reverse the adverse effects that modern farming methods have on wildlife. It is proposed that the evolving management of arable land presents weeds with many selection pressures. How have weed species responded?

Arable weedy populations: general purpose genotypes or specialist races?

Given the formidable colonising ability of weedy species and their broad ecological tolerance, two models have been proposed that are not mutually exclusive. Selection may favour 'general purpose genotypes' that would have high Darwinian fitness in many situations. In contrast, it is possible that weedy species contain specialist races – sometimes called agroecotypes – that have co-evolved with particular crops. Three sets of case histories investigating microevolution in arable land will be examined – crop mimicry, evolution of life histories including seed dormancy, and responses to herbicides.

Some species contain not only cultivars, but also con-specific wild and/or weedy variants, with which they are interfertile. Microevolution in these groups is therefore influenced by the frequency and consequences of intraspecific hybridisation. In addition, crossing between related species may influence the evolution of crop–wild-weed complexes. While some reference to hybridisation between crop and weed will be introduced in this chapter, detailed consideration of this topic is reserved for Chapter 13.

Investigations of selection processes in arable weeds may be divided into two broad categories. In some cases, the investigations have studied or inferred the Darwinian fitness of *intraspecific* variants of genetically polymorphic weed species. Often the morphology and behaviour of cultivated, wild and weedy variants has been compared. In other investigations, it has proved possible to examine the changing fortunes of different arable weed species.

1. Crop mimicry
   The more a weed resembles the crop in its ecological requirements, growth habit, life history, flowering and fruiting behaviour, the more difficult it is to control. The evidence for co-evolution is largely circumstantial being based on historical, taxonomic and distributional studies, supplemented with some genetical and cultivation experiments. For the most part the selective processes contributing to patterns of variability are those of traditional farming methods. While these persist in many regions of the world, in others they have been abandoned or modified by more modern agriculture.
2. Life-history variation and dormancy
   These have been studied experimentally in comparative investigations of intraspecific variation in widespread species, populations of which grow not only as weeds in agricultural areas, but, also, in more natural habitats.

3. Herbicide resistance

   Weed control has increased in efficiency, as hand weeding with simple tools in broadcast crops has given way to modern intensive mechanised methods and the widespread use of herbicides, in crops sown in rows. Some of the most comprehensive and 'convincing' demonstrations of unwitting Darwinian selection resulting from human activities have come from the study of the differential effects of herbicides on species and intraspecific variants of weed species.

Crop mimicry

Vegetative mimicry

The practice of removing weeds from crops is as old as agriculture itself (Vavilov, 1951; Barrett, 1983). Hand weeding, the oldest form of control, was less effective in broadcast crops than in crops grown in rows. Row agriculture became possible when Jethro Tull perfected the seed drill (1733), which allowed the weeds growing between the rows to be controlled by hoeing. Salisbury (1964) suggests that this new husbandry began the decline of the grass darnel (*Lolium temulentum*), which was once a common weed of arable land but is now rare in Britain (see Radosevich *et al.*, 1997, 21).

Effective weeding requires the weed to be identified amongst the crop. In many cases this presents no problems, but some weed species are vegetative mimics of the crops in which they grow. The vegetative resemblance of weed seedlings to the crop may be so close that it is impossible to distinguish weed from crop. As a consequence, early weed control – by hand, or by hoe – is ineffective, and the weed grows on to flowering and fruiting, when it may be visibly different from the crop. At this late stage, weed control by hand or by tools is by then not attempted, as it causes too much loss of yield in the crop.

Seed mimicry

Many weed seeds may be inadvertently harvested with the crop. To remove these impurities, a number of traditional winnowing techniques were developed. The relatively heavy grain of the crop was separated from the lighter weed seeds and chaff by tossing the mixture into the wind or into a forced air current. However, some weeds have seeds that possess similar winnowing characteristics to those of the crop, and these 'mimic seeds' are therefore likely be at a selective advantage as they remain with the grain, and may be sown with the 'seed corn ' planted to produce the next crop. Thus, the farmer, through the winnowing process is operating as an unwitting selective force. Crop mimicry does not necessarily produce weed seeds that closely resemble those of the crop grains; selection favours weed seeds of such shape, size, density and weight that the winnowing process fails to separate them from the grain. In many regions of the world, winnowing of crops has been superseded by more effective seed cleaning techniques, involving sieving etc. To remove or reduce the incidence of the seeds of dodder (*Cuscuta* spp.) in alfalfa, the crop is mixed with iron filings, which attach to the rough reticulate seed coat of the dodder, and may be separated from the smooth seeded alfalfa by the use of magnets (Radosevich *et al.*, 1997, 123).

As we shall see below, there are a number of case histories where weed species well adapted under traditional winnowing have become endangered or extinct under modern seed cleaning regimes. Former 'winners' in the traditional arable landscape have become 'losers'. Crop mimicry has been examined in a number of crops raised by traditional farming methods. Flax cultivation has been abandoned in many areas of Europe and this has led to regional extinction of the specialised weed flora (e.g. see Kornás, 1961).

Crop mimicry in flax

Zinger (1909) and Sinskaia and Beztuzheva (1931) studied crop mimicry in the genus *Camelina*, and the value of their largely morphological investigations has been greatly increased by the genetical investigations of Tedin (1925). The following account summarises their extensive findings, supplemented with material from the account of crop mimicry in Stebbins (1950). Where appropriate, I use the past tense, for it seems highly likely that the traditional cultivation techniques have been abandoned at all the sites studied more than 70 years ago.

Camelina sativa subsp. *linicola* is a weed variant found exclusively in flax fields. Such plants closely resemble flax in their vegetative and reproductive characteristics, with large seeds that simulate those of flax and fruits that dehisce with difficulty, so that the seeds were harvested with the flax crop. It is suggested that *C. sativa*, a summer annual, evolved from the related winter annual *C. microcarpa* (Fig. 9.1). While both *C. sativa* and *C. microcarpa* are weedy plants, there is archaeological evidence suggesting that both may have been cultivated by early man for their oil-bearing seed.

The flax variant of *C. sativa* subsp. *linicola* evolved from *C. sativa* as a result of responses to the selective forces imposed by the flax crop, which grows tall and straight, casting a dense shade beneath. Thus, selection favoured weedy variants of *C. sativa* that mimicked the flax growth-form of straight unbranched stems. Initially, such a response was likely to have been by phenotypic plasticity, phenocopies of the 'flax' growth-form being produced. Evidence for this proposal comes from the demonstration that *C. sativa* produces 'flax-like' phenocopies when grown with flax. However, as *C. sativa* subsp. *linicola* retained its 'flax-like' growth habit when grown apart from flax, it is clearly a genetically distinct adapted variant.

C. sativa subsp. *linicola* also mimics flax in reproductive behaviour, flowering and fruiting at the same time as the particular flax crop with which it grew. Thus, in the north of Russia, '*linicola*' fruited early, as did the flax crop grown for its fibre, with flax and its mimic fruiting progressively later in southern areas, where some of the flax was grown for oil. There is also another adaptation of great importance. The seedpods in '*linicola*' failed to dehisce. Synchrony of seeding and indehiscent seedpods ensured that the weed was harvested with the flax. In traditional practice, the harvested flax seed was winnowed to remove weed seeds and foreign material. The flax mimic seed proved to have the same seed size/weight relations as that of the crop, and, therefore, while the winnowing process effectively removed some weed species from the flax, it failed to remove those of '*linicola*',

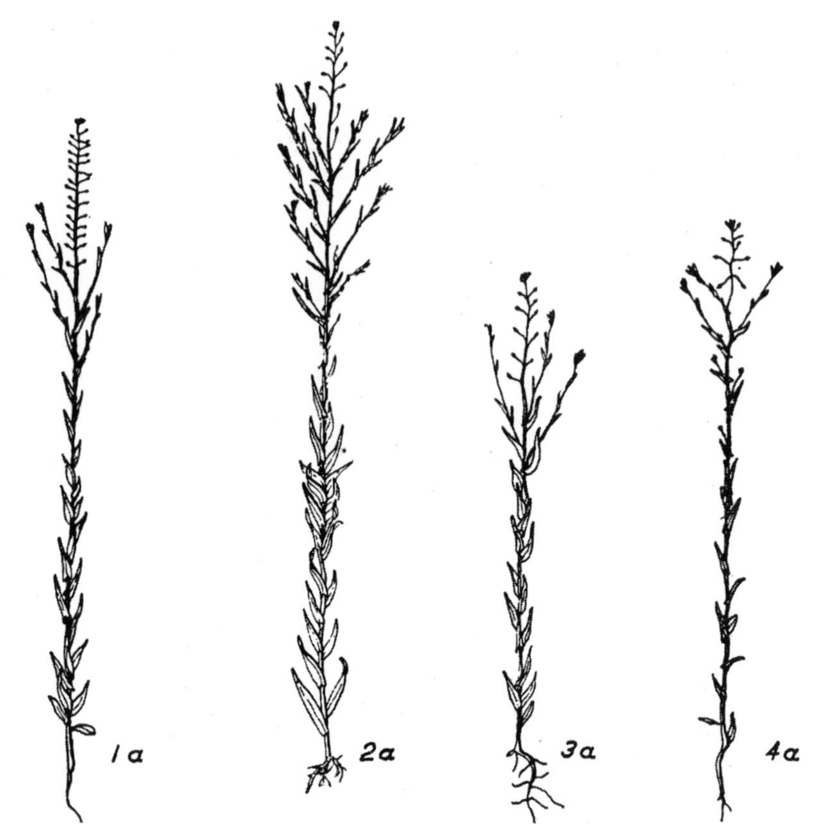

Fig. 9.1 Growth habit of four species of *Camelina*: 1a *C. microcarpa*; 2a *C. pilosa*; 3a *C. sativa*; 4a *C. sativa* subsp. *linicola*. (From Stebbins, 1950)

as its seeds were blown to the same distance as the flax. Regionally, flax differed in seed size and it is of great interest that there were different variants of '*linicola*', matching the flax with which they were associated. Such findings reveal the precision of the mimicry of '*linicola*' to the flax, and, furthermore, studies of the genetics of large-scale geographical patterns of seedpod shape in relation to seed size suggested that different variants of weed were likely to have had an independent origin.

Crop mimicry in the genus does not lead only to a single outcome. Further studies revealed that some ancient cultivars of flax had dehiscent seed capsules and these were harvested before maturity to extract oil. In these flax fields, dehiscent weedy variants of *Camelina* were found, the seeds of which were not mixed with those of flax harvest, but fell in the fields.

A number of other weed species are associated with flax (*Agrostemma linicola*, *Eruca vesicaria*, *Lepidium sativum*, *Lolium remotum*, *Polygonum lapathifolium*, *Silene gallica*, *S. cretica*, *S. linicola*, *Sinapsis alba* and *Spergula arvensis*). *Cuscuta epilinum* has small seeds but a double seeded variant with larger seeds grows with flax (Wickler, 1968). Some of these weeds are mimetic of flax in vegetative and seed characters, but the extent of mimicry is unclear.

Crop mimicry in wheat

Archaeological evidence suggests that Corn Cockle (*Agrostemma githago*), a species that probably originated in the eastern Mediterranean, has been associated with man for at least 4,000 years, as a seed contaminant of cereals (Firbank, 1988). In some deposits, seeds of corn cockle are common, e.g. in Roman grain stores, cesspits and rubbish dumps. The species is not only a weed in cereals, where it occurs as a large-seeded variant (var. *macrospermum)*, but, as we have noted above, it is also found as a weed of flax in the former USSR (var. *linicola* – with small almost smooth seeds).

The association of corn cockle as a contaminant of wheat and rye has been examined in some detail. It is reported that young plants are 'surprisingly similar to growing cereals', providing perhaps yet another example of vegetative mimicry of crop and weed (Firbank, 1988). Fruiting of corn cockle occurred at the time of grain harvest. Capsules dry out and open, but some seeds are too large to leave the capsule or remain firmly attached within it. This seed retention is almost certainly the result of unwitting selection by man (Firbank, 1988). Thus, many seeds were harvested with the grain. Because the size/shape/weight relationship of corn cockle seeds matched that of cereal, it proved impossible with former threshing and winnowing methods to clean the seed, and corn cockle remained as an impurity, to grow again, when seeds were sown to raise another cereal crop. As long as it was impossible to separate the seeds of the weed from those of the grain, it was impossible to control the *Argostemma* as a weed.

However, in the early part of the twentieth century, refined seed cleaning methods were introduced and populations of corn cockle rapidly declined in Europe during the period 1910–60, because, in general, there is no persistent seed bank of corn cockle seed in the soil (Svensson & Wigren, 1983). Thus, this once abundant weed of cereal crops is now a rare plant. It is important to note that the decline in the species predated the use of herbicides (Thompson, 1973a, b).

Allium vineale is another species exhibiting crop mimicry, but in this case the mimicry of cereal grains is by aerial bulbils. In the UK, reproduction in this species is almost entirely by vegetative means, offset bulbs being produced underground. Also aerial bulbils may be produced (Richens, 1947, 209). Individuals are found in which the 'flowering' stalks carry heads containing flowers alone, mixtures of flowers and aerial bulbils, or bulbils alone. Richens reports that large numbers of aerial bulbils may be produced (up to 300) per head and: 'these are roughly the size of wheat grains and ripen at the same time, so that when the plant is present in a cereal crop these bulbils are readily disseminated in grain used as seed'. Scott (1944) confirms that, in cereal crops, bulbils ripen at the same time as the crop and are borne at about the same level above the ground. In addition, they are approximately the same size and density as the grain. This weed is difficult to control, even by seed cleaning methods, as large numbers of the bulbils are produced underground and may remain dormant for 2–6 years.

Nigella arvensis ssp. *arvensis* is another species whose distribution and abundance has been influenced by crop mimicry and modern seed cleaning methods. Recorded as early as the ninth century in Switzerland, it was formerly a widely distributed European weed in arable fields, especially cereals (Strid, 1971). While there are limitations in the use of records based on herbarium specimens, general trends in changing distribution have emerged from the study of dated material. The species began to decline in frequency in the period 1890–1929, especially in northern and Atlantic parts of its range, and it is now only common in the Balkan Peninsula. Strid suggests that three factors have contributed to the decline. *Nigella* seeds were formerly a regular contaminant of cereal crops, but such contamination has declined with improved threshing and seed cleaning methods. In addition, *N. arvensis* fruits in late summer to early autumn, and, with the trend toward early harvesting of cereal crops, there is a risk of the weed being cut down at cereal harvest before it fruits. Finally, the use of herbicides has increased the pressure on the species and the combined effect of all factors has led to the extinction of many populations. Perhaps, *Nigella* will eventually become restricted to the east Mediterranean in what is likely to have been its original native area of distribution.

Crop mimicry in rice

Barnyard grass (*Echinochloa crus-galli*) is a widespread and variable weed, in which a number of rice mimics have been reported in both flooded and upland (non-flooded) rice cultivation. Traditionally, rice was (and still is) grown in small plots in intensively hand-weeded cultivation. The barnyard rice mimic has a similar growth form to rice in its upright growth of tillers and leaves: other weedy variants of barnyard grass have drooping leaves and a lax, decumbent growth habit. Weeding represents a powerful selective force, selection favouring variants that are indistinguishable from the growing rice, both during the transplantation phase of rice cultivation and its subsequent growth in the fields. Such variants are likely to have arisen under ‘primitive’ labour-intensive agriculture in Asia.

Comparative cultivation of weedy and mimic barnyard grass variants reveals that the rice mimic has heavier seeds, weaker dormancy, more synchronous germination, larger and more vigorous seedlings, and is able to establish under anaerobic conditions in flooded soils. At reproduction, anthesis in the mimic usually coincides with that of the rice crop. In traditional systems, it is likely that some barnyard seed is harvested with the rice crop, other seeds falling in the fields. Rice is now grown intensively, for example in California. Hand weeding is not practised: a major selection pressure has now been removed. However, mimic variants of barnyard grass occur and continue to be spread by seed.

In a study by Ling Hwa and Morishima (1997), 24 rice strains were examined that grew as weeds in and around arable fields. One group, with low seed shedding, were of the crop mimic type and survived by being sown and harvested together with the crop. As low seed shedding is typical of rice cultivars, it is proposed that these weedy rice variants may be remnants of old cultivars. Other weedy variants shed their seeds, as in wild rice, and have strong seed dormancy. This group are perhaps derived from crosses between wild and cultivated variants, or may represent hybrid derivatives between distantly related cultivars. A third group of weedy forms, from the lower Yangtze Valley, are possibly the product of natural hybridisation between cultivars and a wild rice that may now be extinct. With intensive modern cultivation methods populations of weedy rice are decreasing.

Crop mimicry in maize

Teosinte (*Zea mexicana*) is a variable species, adapted not only to natural habitats, but also to fields of maize in Central America and Mexico, where it is a crop mimic. Teosinte is fully interfertile with maize and hybridisation has contributed to the precision of mimicry. For instance, where maize cultivars are grown that have reddish pigmentation in leaves and stems, the weed has developed these colours. In attempting to control the mimetic weed in maize, weedy variants are cut from the crop, but as they are often fed to cattle, seed may be returned to the fields in manure (see Wilkes, 1977).

Life-history variation

There are a number of genetically polymorphic species that have weedy variants in arable land and different ecotypic variants in more natural habitats. Comparative cultivation of the different variants can be very revealing, providing insights into the adaptive significance of various life-history and morphological traits in different habitats.

Considering the weedy variants, evidence suggests that selection favours those individuals that develop and reproduce in synchrony with the germination, growth and harvesting of the crop. As we have shown in the case of crop mimicry, weed seeds contaminate the harvested seed and are sown with this seed to infest the crop in its next cropping cycles. Non-mimetic weeds may also have high fitness, if their seeds are shed into the soil just before or at harvesting. In contrast, as we shall see below, very different selection pressures operate on populations growing in ruderal and semi-natural habitats.

Growth strategies in relation to land use

Comparative cultivation experiments with weedy and non-weedy ecotypes of common plants have often revealed major differences in life history (Cavers & Harper, 1964, 1966, 1967a, b; Harper, 1977). A typical example is provided by studies of *Rumex crispus*, where cultivation experiments with transplants and seedling progenies in common garden trials have provided valuable insights into morphological differentiation, and life-history variation (Akeroyd & Briggs, 1983a, b). *Rumex crispus* is a genetically very variable species, with morphologically distinct inland variants that grow in ruderal habitats, and as weeds of cultivation. In addition, the species also grows in semi-natural habitats on seashores, especially shingle beaches, and on tidal mud. Experiments have revealed that there are three morphologically distinct variants that come true from seed and are therefore likely to be genetically distinct. Many inland plants behave as annuals, some, but not all, dying after flowering in their first year. In contrast, plants from coastal areas and from riverine tidal mud (e.g. from the banks of the River Wye at Tintern Abbey, Monmouthshire) only flower in their second (or later) year of growth. Variants differ, too, in the timing of flowering; plants from inland and tidal mud being early flowering, while those from coastal areas flower later. The adaptive significance of flowering behaviour of the three variants is explicable in terms of the differences in the selection pressures in the different habitat types. Annual, early flowering variants are at a selective advantage in or near agricultural sites, where they are often subjected to weeding pressures, while longer-lived and perennial variants would appear to be at a selective advantage in semi-natural habitats that present formidable problems of seedling establishment and where there is no weed control. Baker (1974) detected a similar pattern in *Picris echioides*. Wild ancestral perennial decumbent self-incompatible variants occur on cliffs and at the edges of salt marshes. In contrast, taller annual self-compatible plants occur in wheat fields (Baker, 1954).

Speed of development in relation to weeding pressures

Human activities include not only the cultivation of monocultures of crop and vegetable species, but also very large areas set out as ornamental gardens. In the traditional formal garden, plants are grown in weed-free displays. Weeding, in this context, can be seen as a form of extreme herbivory, where the whole plants are removed. While some gardens may appear to be weed-free, close examination reveals that seasonally some weeds survive and produce seeds. In such gardens, where intense weeding is practised, selection should favour variants able to reproduce quickly.

Evidence for such a possibility is provided by a number of studies (Briggs, Hodkinson & Block, 1991) . In cultivation experiments with *Capsella bursa-pastoris*, Sørensen (1954) discovered that plants from the garden of the Royal Agricultural College, Rolighedsvej, Denmark, were quicker to come into flower than those from seed collected at other sites. Imam and Allard (1965), studying variation in *Avena fatua* in California, found that plants raised from seed collected in areas subject to weeding flowered more quickly than those where there was less or no weed control. In a cultivation trial, Sobey (1987) discovered that plants raised from seed of *Stellaria media* from a garden in Belfast, Northern Ireland, flowered 11 weeks earlier than material from a seabird colony at Hackley Head, Aberdeenshire, Scotland. These findings confirmed the preliminary investigations of Vegte (1978) studying arable and coastal variants of *S. media* in the Netherlands. Cultivation trials of populations of the annual weed *Polygonum aviculare* sampled from weeded and less disturbed and trampled sites were carried out by Meerts (1995). He discovered that material collected from weeded areas had a 'shorter lifespan, an earlier flowering date and a higher biomass allocated to reproduction compared with genotypes from less disturbed areas'. Also, plants from trampled areas had a distinctive growth habit – smaller internodes and shorter shoots. However, the notion of tightly controlled morphology and behaviour was rejected. Meerts concluded that the species complex had 'evolved a "dual" adaptive strategy i.e. a combination of genetic polymorphism and high phenotypic plasticity'.

Speed of development has been intensively studied in *Senecio vulgaris*. Kadereit and Briggs (1985) discovered that stocks raised from achenes (fruits) collected in Cambridge University Botanic Garden, while showing a degree of variability, developed more quickly to first fruiting than plants from areas subject to less weeding pressure (arable fields, waste land etc.), or to no weeding (coastal shingle). Briggs and Block (1992) made a further investigation in the Botanic Garden and discovered that most of the progenies raised from a sample of seed parents (18/20) appeared to be acting as inbred lines for rate of development. To examine how far rate of development might vary between lines, an experiment was set up in which six inbred lines were grown out-of-doors in fertile and infertile soils, and in a glasshouse on fertile soil. It is clear from this experiment that the speed of development and the rank order of lines to first fruiting is not fixed. Thus, relative fitness may vary seasonally and in soils of different fertility. This behaviour, in a species able to reproduce at all seasons of the year, could be a major factor in the maintenance of population variability within the garden. However, other factors are also important in influencing the length of life cycle; namely, any spatial and seasonal variation in the efficiency of weeding, and the effects of predators, pests and diseases. For instance, it has been discovered that various fungal pathogens influence the length of the life cycle (*Erysiphe fischeri* (Ben-Kalio & Clarke, 1979; Harry & Clarke, 1986); *Botrytis cinerea* and *Puccinia lagenophorae* (Paul & Ayres, 1986 and references cited therein)).

Extending the investigation into life-history variation in *S. vulgaris*, Theaker and Briggs (1992, 1993) carried out cultivation trials on seed samples from a number of other sites. It was discovered that plants from very well-weeded botanic gardens – Oxford, Cambridge and Kew – were all precocious in their development. However, populations were not homogeneous within gardens: a degree of variability in speed of development was detected. In contrast, in the same experiments, material from coastal sites in Eastern England had a slower rate of development, and there was no evidence of population variation, suggesting that genetic drift of founder effects might be very important, or that selection favoured a very narrow range of variants. Furthermore, at Shingle Street, Suffolk, there was evidence of sharp local differentiation. In cultivation, as in the wild, plants from the shingle beach were quite different from those of adjacent farmland, in having a dwarf habit, a high concentration of internodes at the base of the stem, profuse adventitious roots on the stem, highly serrated and slightly glaucous leaves, and hairy achenes (Theaker, 1990). In contrast, plants grown from a population from a field margin 50 m away were taller, quicker to develop to fruiting, and had fewer adventitious roots and less hairy achenes.

In experiments with inbred family lines of *S. vulgaris* grown on a range of soil types, Theaker (1990) discovered significant maternal effects on time taken to maturity, but that these were only likely to be significant in plants growing in very infertile soils. In addition, the experiments confirmed the earlier finding that the speed of development was not fixed, and that there was a very high degree of morphological plasticity in the species. Under favourable conditions an individual could produce hundreds of achenes. In contrast, on very poor soil, phenotypically dwarf plants would produce only a single capitulum with very few achenes.

The studies outlined above have investigated variability in *S. vulgaris* (var. *vulgaris*) that has rayless capitula. A distinct morphological variant of *S. vulgaris* having capitula with short ray florets (ssp. *denticulatus*) has also been investigated (Kadereit, 1984). This variant is a winter annual plant with an Atlantic–Mediterranean montane distribution. Strictly coastal outside the Mediterranean, ssp. *denticulatus* has strong seed dormancy and requires a substantially longer period to complete its life cycle than *S. vulgaris* var. *vulgaris*, which typically flowers and fruits all year round. It has been proposed that ruderal and weedy variants of var. *vulgaris* have evolved from forms similar to *denticulatus* (Kadereit, 1984). However, the situation may be more complex, as preliminary studies by Theaker (1990) have established that some coastal variants of *S. vulgaris* (taxonomically var. *vulgaris* because they lack ray on their capitula) are also characterised by seed dormancy and winter annual habit.

Vernalisation and the winter and summer annual habit

Our understanding of the underlying genetic basis of different types of annual life cycles has come from the study of *Arabidopsis*. Jones (1971a, b, c) investigated the speed of development in populations of *A. thaliana* from different habitat types. Samples were grown from gardens that were subject to the selection pressures of regular weeding. Also, material was grown from semi-natural areas that are not subject to weeding pressures, and from areas of waste ground. This latter habitat – disused railway tracks – was formerly subject to weed control, but at the time of the study was being invaded by ruderal plants. The behaviour of the material reflected what appear to be the major differences in selection pressures in the areas. Plants raised from seed collected from heavily weeded garden sites developed much more quickly from germination to fruiting and had little or no vernalisation requirement. Thus, potentially more than one generation was possible in each growing season. In contrast, samples from semi-natural areas behaved as winter annuals requiring a long cold (vernalisation) period before flowering in the spring. Material from presently disused railway tracks proved to have a mixture of variants.

Arabidopsis thaliana has been the subject of many investigations of the molecular-genetic basis of physiological behaviour. With regard to vernalisation, experiments have revealed that two genes, Frigida (FRI) and Flowering Locus C (FLC), act synergistically in determining vernalisation requirement. Investigations of a range of summer-annual ecotypes reveal that they have independently evolved from winter annuals following different mutations of the FRI locus. Further investigations have shown that rapid cycling summer-annual variants have also evolved independently following mutations at the FLC locus (Michaels *et al.*, 2003). Considering the adaptive significance of the different life cycles discovered in populations of *A. thaliana*, evidence suggests that the winter-annual variants are at a selective advantage in northern latitudes. Variants without vernalisation requirement would appear to be at a selective advantage in areas of milder climate. However, this is not the whole story, as a study of a range of samples reveals that there is no correlation between flowering time and winter temperatures or flowering and altitude. It seems likely that flowering early without vernalisation requirement may be of advantage where winters are so severe that they prevent germination until the spring. In addition, and this is important in the context of the evolution of arable weed floras, the summer-annual life history is clearly of adaptive significance in agricultural/horticultural conditions, where intense selection pressures favouring quick reproduction are imposed by weed control, harvest dates, and crop husbandry, especially in areas where there are summer droughts.

Timing of maturity in relation to crop harvest

The timing of the crop harvesting, both of hay (see above) and arable crops, presents weeds with strong selection pressures. With regard to arable harvesting, Salisbury (1964) noted in preliminary studies that, following the introduction of the reaper, dwarf forms of *Aethusa* and *Torilis* were at a selective advantage in fields of cereals. Before harvest, these variants grew initially below the height at which the crop was cut, only later did they flower and fruit in the open stubbles. Taller variants of these species are present in other rural habitats.

Studies of the important weed Yellow foxtail (*Setaria lutescens*) from several states in the USA have provided another example of variation in growth habit in relation to cropping of alfalfa. Schoner, Norris and Chilcote (1978) discovered a prostrate variant in California (*c.* 50 cm) with taller variants (*c.* 85–115 cm) in other states. Interestingly, in California the crop is cut more frequently (on a 21–28 day cycle), and therefore it is proposed that selection favoured a relatively prostrate growth habit.

Dormancy

In many species there is another crucial element in the life cycle of weeds. Many weed species of arable crops have dormant seeds, but it is important to stress that some species exhibit genetic variability for the trait. Thus, comparative studies of weedy and non-weedy variants have detected major differences in dormancy and germination behaviour of adaptive significance (see Harper, 1977).

The investigations of Cavers and Harper (1966) on *Rumex crispus* may serve as an example. They carried out investigations in which seeds were sown at different sites, and discovered that inland plants in the agricultural and disturbed ruderal habitats possessed polymorphic seed dormancy, resulting in intermittent seed germination. Thus, a reservoir of viable seed was maintained in the soil seed bank at inland sites over the two years of the experiment. Such behaviour is adaptively significant, enabling *Rumex* seeds in soils under fallow, or neglect, to germinate in response to another cycle of crop cultivation or disturbance. In contrast, the seed of the long-lived coastal plants sown in its native habitat germinated in the first few months, suggesting that different selective forces are operating in these semi-natural habitats.

Avena fatua is another species with variants having dormant or non-dormant seeds. Jana and Thai (1987) carried out an experiment in which experimental plots were sown with *A. fatua* containing equal frequencies of true-breeding dormant and non-dormant variants. From 1978 to 1985 plots were subjected either to continuous cropping or to two-year rotation consisting of one-year land cultivation followed by a year of summer fallow. They discovered that the frequency of dormant lines increased substantially in the regime that included the summer fallow regime.

Seed production and soil seed banks

Many arable weeds combine high seed production with considerable longevity of seed in soil (Radosevich *et al.*, 1997, 124–5). For instance, *Verbascum thapsus* may produce *c.* 223,200 seeds per plant, combined with a potential longevity of 100 years. Estimates for *Rumex crispus* are 29,500 seeds per plant with 80 years longevity in the soil. In agricultural land there is often a very large persistent bank of viable dormant seeds in the soil. Light only penetrates 1–2 mm in soil and, for many small-seeded species of weeds in the soil seed bank, dormancy can be induced by even a shallow burial. Furthermore, many weed species require light for germination. Thus, when the seeds of many weed species are brought to the soil surface – other factors such as temperature and moisture levels being favourable – light triggers germination. Taking a long evolutionary perspective, this behaviour is regarded as of selective advantage for small-seeded plants that may not survive germination in the lower soil (Pons, 1991). However, the phenomenon is of particular significance in more recent microevolution, as a persistent soil seed bank and light-stimulated germination could be seen as one of the many evolutionary 'strategies' found in common weedy species growing successfully, not only in areas where human activities lead to regular disturbance of the soil profile in arable agriculture, but also in ruderal sites, waste ground, rubbish dumps etc.

Studies of soils on the Broadbalk field at Rothamsted have established the scale of the soil seed bank in arable areas. This long-term experiment has been sown to winter wheat every year since 1843. By germinating the seeds in the soil, it was estimated that the soil seed bank contained up to *c.* 34,000 seeds/m^2 of 47 different species. Remarkably, two-thirds of the whole seed population was of *Papaver* species. The experiment offers some measure of the effectiveness or otherwise of weed control by certain agricultural practices. Major differences were detected between the plots given different cultivation treatments. Thus, a fallow period might be expected to reduce weed populations and the rain of weed seeds entering the seed bank. However, in the Broadbalk experiment, after plots had been given two years of fallow, the total seed content of the seed bank was reduced by only 6%. There were also reductions in the seed bank of individual species, but these were not dramatic, the number of seeds of *Scandix* in the seed bank fell by only 5% from its previous density (Harper, 1977).

Very intensive treatment may be necessary to deplete a seed bank. For example, repeated tillage can reduce the number of viable seeds in the seed bank (Chancellor, 1985). But, five months of weekly or bi-weekly harrowing was necessary to eradicate *Cyperus rotundus* (purple nut sedge) from farmland in Alabama, as this species produces underground tubers as well as seeds (Smith & Mayton, 1938).

Herbicide resistance

It has been estimated that in 1830 the cultivation and harvest of an acre of cereal crop required 58 hours: in contrast, 150 years later the task could be accomplished in only 2 hours (Kirby, 1980). This amazing transformation was achieved with the mechanisation of agriculture, together with the widespread use of herbicides. Shortly after herbicides were introduced it was predicted that herbicide resistance would evolve in weeds (Abel, 1954; Harper, 1956). This prediction was based on studies of the evolution of pesticide resistance in insects and fungi. Relatively quickly, the use of herbicides resulted in the evolution of resistance in some species of weeds. Studies of this phenomenon have provided one of the best understood and most convincing examples of microevolutionary change in human-managed environments.

The first organic herbicides produced on an industrial scale were developed in the 1930s. Dinitro-ortho-cresol (DNOC) was manufactured in France, and later 2-(sec-butyl)-4,6-dinitrophenol (DINOSEB) was discovered to be an effective weed killer in cereal crops in California. Later, in Britain, investigations of plant hormones led to the development of 2-methyl-4-chlorophenoxyacetic acid (MCPA) and 2,4-dichlorophenoxyacetic acid (2,4-D). Many herbicides were developed from the 1950s onwards, and now there are in excess of 250 phytotoxic chemicals based on *c.* 20 families of organic structures (Vighi & Funari, 1995, 54).

Herbicides, in a multitude of different formulations, are available to attack weeds at every point in the cropping cycle, from the pre-crop sowing (or pre-planting) period, to the post-sowing stages, where the herbicides may act pre- or post-crop emergence. (For a review of the range of herbicide and their properties see Zimdahl (1999).) Some herbicides are active in contact with foliage or seedlings etc., while in others are absorbed and act as they are translocated within the plant. In addition, some herbicides have a short life; others (residual herbicides) retain their activity in the soil for longer periods. An arsenal of different types is available to the grower. At high enough concentrations herbicides damage and kill all plants, and the herbicide concentrations employed in field applications are designed to kill weeds and have as little adverse effect on the crop as possible. Treatments are often temporally separated from the crop cycle. But in many cases weed control is carried out in the presence of the growing crop and in this case shields are often used on spraying equipment to direct the herbicide onto the weeds and away from the crop canopy. Spray jets are designed to minimise the 'drift' of herbicide from treated areas into adjacent vegetation. In addition, commercially formulated 'safeners' have been devised that protect the crop either by enhancing detoxification or by the 'competitive antagonism of herbicide and safeners at a common target site' (Vighi & Funari, 1995, 62). Herbicide spray drift into adjacent semi-natural vegetation is a serious conservation issue, not only from ground-based applications but also where crops are grown in very large fields, and herbicides are sprayed from airplanes or helicopters.

The economic impact of weeds is not easy to quantify, but various estimates have been made. For instance, the value of total losses in US agricultural crop production has been estimated at $7,468 billion (US) (Charadattan & DeLoach, 1988). Astonishingly, *Amaranthus* species were responsible for more than 11.9% of this figure, with *Setaria* species close behind at 9.8%. Most herbicide is used in field crop production, but, in the USA, more than 23% of herbicide is employed in 'weed control' on industrial sites, parking areas, roadsides, railway tracks, and sports and amenity areas. In many of these areas, total weed control has been attempted and, because they are cheap and long lasting, triazines have often been the herbicides of choice (see Powles & Shaner, 2001, 15). (Recently, the use of triazines has declined, as there are concerns about the safety of this group of chemicals.) In the management of forests, wetlands, aquatic habitats, rangelands, grasslands and coastal communities a range of herbicide sprays are commonly applied. They are also widely used, often controversially, in conservation management, for example in the control of invasive species (Vighi & Funari, 1995, 62).

Incidence of herbicide resistance

Microevolutionary insights into winners and losers have come from the study of herbicide resistance. Resistance to the herbicide simazine, first introduced in 1956, was discovered in *Senecio vulgaris* at sites in Washington State in 1968 (Ryan, 1970). By 1993, resistance to triazines was reported in 57 weed species, including both monocotyledons and dicotyledons. In addition, 64 species had biotypes resistant to one or more of 14 other herbicides (Radosevich *et al.*, 1997, 94). Now, more than 250 cases of herbicide resistance have been reported (see the International Survey of Herbicide-Resistant Weeds: www.weedscience.org for up-to-date listings, which include six cases of resistance to the extremely widely used herbicide glyphosate) (Baucom & Mauricio, 2004). While the number of species exhibiting some form of resistance is increasing, the overall number relative to the world's flora is very small indeed. Furthermore, many of the 76 species identified as the world's worst weeds have not so far developed herbicide resistance (Heap & LeBaron, 2001, 16). Herbicide-resistant weeds have been detected not only in arable crops, orchards, pastures, plantations and forests, but also in non-agricultural sites such as roadsides, railways and industrial areas (Heap & LeBaron, 2001, 7). Herbicide resistance in weeds is very much a phenomenon associated with agriculture in the developed world. In the developing world, economic conditions and cheap labour have limited the use of herbicides, but resistant weeds have been reported. Given the increasing use of herbicides, more cases of resistance are to be expected, especially as cost often severely restricts the range of herbicides that can be employed, and a single cheap herbicide, such as atrazine, is therefore repeatedly used in weed control.

There is no uniform definition of the terms resistance and tolerance, which are often used interchangeably. Resistance (the preferred term for many experts), in contrast to susceptibility, 'denotes the evolved capacity of a previously herbicide susceptible weed population to withstand a herbicide and complete its lifecycle when the herbicide is used at its normal rate in an agricultural situation' (Heap & LeBaron, 2001, 2). Resistance and tolerance are useful categories, but Cousens and Mortimer (1995) make the important point that there is a continuum of response to herbicides, from sensitivity and non-tolerance through to tolerance and resistance. Some species, including crops, show a degree of natural tolerance to particular herbicides. Such tolerance is inherent and enables some plants to survive and reproduce after herbicide treatment. This ability has not involved selection or genetic manipulation, plants being naturally tolerant and able to sustain an amount of damage without any corresponding loss of fitness. An example is the tolerance by maize cultivars of triazine herbicides, applications of which can be used to control weeds.

Studies of herbicide-tolerant plants have detected two other important categories of response, cross-resistance, 'where a single resistance mechanism confers resistance to several herbicides', and multiple resistance, where 'two or more resistance mechanisms are present' (see Powles & Shaner, 2001, 3).

Studies of the genetics of herbicide resistance have revealed a number of different situations (Table 9.1). In many cases a single dominant or semi-dominant allele is involved, but in others resistance is controlled by multigenic systems. Resistance to triazine herbicides takes yet another form: it is generally maternally inherited (but see below).

The underlying physiological mechanisms of herbicide resistance have been intensively investigated, and may involve enhanced metabolic detoxification or sequestration of the herbicide or the modification of the biochemical/metabolic binding of herbicides (see Radosevich *et al.*, 1997; Cobb & Kirkwood, 2000). For example, the mono-oxygenase enzyme in wheat can detoxify several herbicides that have been used to control weeds that lack this capacity (Vighi & Funari, 1995, 60). However, resistance has evolved in *Alopecurus myosuroides*, a major weed of wheat fields. Analysis of the physiological basis of resistance has revealed that there has been an abrupt rise in the detoxification capacity in the species. This has been interpreted as a subtle example of co-evolution, in the form of biochemical mimicry of the level of detoxifying enzyme found in wheat. As a consequence, in England, the grass has become cross-resistant to most of the herbicides used to control weeds in wheat, i.e. 'a single resistance mechanism confers resistance to several pesticides' (quoted from Heap & LeBaron, 2001, 2).

An example of resistance involving changes in binding sites is provided by the case of the triazine group of herbicides, e.g. atrazine, simazine etc. Triazine resistance is due to changes in the herbicide-binding domain on the D1 protein associated with photosystem II of photosynthesis (Trebst, 1996). Molecular and biochemical analysis has revealed that resistance usually involves a point mutation in the chloroplast gene resulting in the substitution of amino acid Ser 264 to Gly in the D1 protein of the chloroplast *psbA* gene (Preston & Mallory-Smith, 2001). The presence of the mutated gene in resistant plants reduces photosynthesis electron transport by roughly 1,000-fold compared to wild-type chloroplasts. As the chloroplast genes are generally maternally inherited in plants, so too is

Table 9.1 *Inheritance of herbicide resistance in weeds*

(a) Mendelian inheritance in selected weed populations

Herbicide	Weed	Number of genes
Atrazine	*Abutilon theophrasti*	1 semi-dominant
Chlorotoluron	*Alopecurus myosuroides*	2 additive
Diclofop	*Lolium multiflorum*	1 semi-dominant
Fenoxoprop	*Avena sterilis*	1 semi-dominant
Fluazifop	*Avena sterilis*	1 semi-dominant
Haloxyfop	*Lolium rigidum*	1 semi-dominant
Metsulfuron	*Lactuca serriola*	1 semi-dominant
Paraquat	*Arctotheca calendula*	1 semi-dominant
	Conyza bonariensis	1 dominant
	Conyza philadelphicus	1 dominant
	Erigeron canadensis	1 dominant
	Hordeum glaucum	1 semi-dominant
	Hordeum leporinum	1 semi-dominant
Trifluralin	*Setaria viridis*	1 recessive

(b) Quantitative inheritance in weed and wild populations

Herbicide	Weed	Heritability
Barban	*Avena fatua*	0 to 0.63
Glyphosate	*Convolvulus arvensis*	Additive
Simazine	*Senecio vulgaris*	0.22

Adapted from Cousens & Mortimer, 1995; after Darmency, 1994. Reproduced by permission of Cambridge University Press

triazine resistance. Thus, as chloroplast genes are transmitted in the egg but not the pollen in many species, gene flow of herbicide resistance is possible through seed dispersal but not via pollen (Vighi & Funari, 1995, 59). However, triazine resistance is not universally maternally inherited. Occasional paternal transmission via pollen has been detected in triazine resistant *Poa annua* (Darmency & Gasquez, 1981), and in *Abutilon theophrasti* it has been discovered that atrazine resistance is controlled by a single partially dominant nuclear gene (Anderson & Gronwald, 1987).

Speed with which resistance develops

Table 9.2 presents information on the date of introduction of widely used herbicides, and the time taken for resistance to appear. Resistant biotypes arise as a consequence of random mutation (Vighi & Funari, 1995, 5) and resistance to a particular herbicide typically arises where weed control has involved the continuous use of one particular herbicide for a 5–20 year period (Cousens & Mortimer, 1995). In an attempt to lower the selection

Table 9.2 *(a) Increase in the number of organisms resistant to pesticides. (After Holt & LeBaron, 1990: from Cousens & Mortimer, 1995. Reproduced by permission of Cambridge University Press): arthropods (circle); plant pathogens (triangle); and weeds (square).*

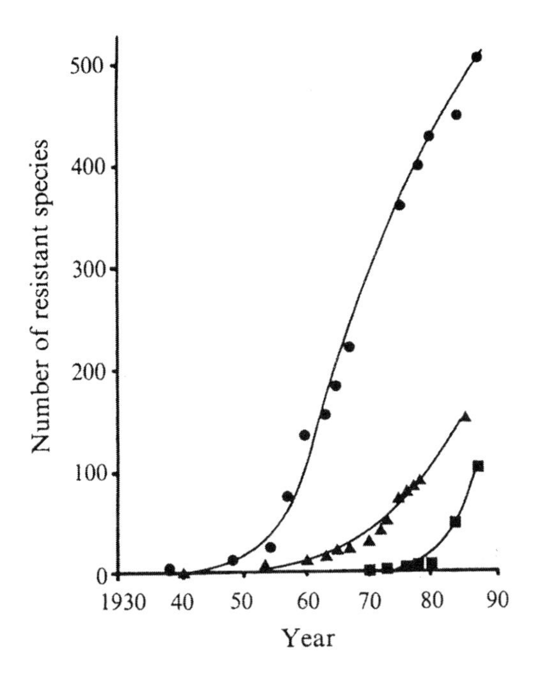

(b) Estimated number of years for natural section of herbicide resistance to appear in some weed species

Species	Chemical selection agent	Years for resistance to be recognised
Kochia scoparia	Sulfonylureas	3–5
Avena fatua	Diclofop methyl	4–6
Lolium multiflorum	Diclofop methyl	7
Lolium rigidum	Diclofop methyl	4
Senecio vulgaris	Simazine	10
Alopecurus myosuroides	Chlorotoluron	10
Setaria viridis	Trifluralin	15
Avena fatua	Triallate	18–20
Carduus nutans	2,4-D or MCPA	20
Hordeum leporinum	Paraquat/diquat	25

Maxwell & Mortimer (1994) © 1994. Adapted from *Herbicide Resistance in Plants*, S. B. Powles. Reproduced by permission of Taylor and Francis Group, LLC, a division of Informa PLC pressures in favour of resistance, agriculturalists recommend the rotational or simultaneous application of a range of herbicides with different biochemical mechanisms of action, together with a judicious use of crop rotations and mechanical weed control. In this way it is hoped to delay the appearance of resistance to particular herbicides and also cross and multiple resistance, where weeds are resistant to more than one herbicide (Vighi & Funari, 1995, 61).

Fitness: costs and benefits

Given that herbicide-resistant variants of many common weeds have appeared, it might be considered that, once evolved, these would persist in agricultural sites in perpetuity. However, what happens in populations is related to the 'costs and benefits' of herbicide tolerance to the plants concerned. In land treated with herbicide A, resistant (R) variants are clearly at a selective advantage, being able to grow and reproduce in a situation where sensitive (S) variants, receiving effective field doses of the herbicide, will be killed. Thus, the herbicide-tolerant variants have greater fitness. However, it has been discovered that costs are involved in resistance and these can be estimated by examining the relative fitness of R and S plants in unselecting (herbicide-free) habitats. In such circumstances the growth and vigour of the S plants of *Senecio vulgaris* exceed that of triazine-resistant (R) plants (Weaver & Warwick, 1983). Studying the same species, Gressel (1991) concludes that when in competition with S, the fitness of the R individuals may be reduced by 10–50%. Regarding this difference in fitness, Cousens and Mortimer (1995) note that herbicide-resistant plants may have 'a less efficient electron transport system, arising from the alteration of the herbicide binding site' on the thylakoid membrane of the chloroplast. The 'acquisition of resistance lowers photosynthetic potential and thus reduces both vigour and overall ecological fitness'.

Resistance has developed to the broad-spectrum non-selective, translocated weed killer glyphosate ('Roundup'), marketed since 1971 for controlling invasive perennial weeds. This herbicide is becoming the predominant herbicide in managed systems (Baucom & Mauricio, 2004) and the cost/benefits relationships involved in herbicide resistance have been investigated in *Ipomoea purpurea*. Studies of maternal lines raised from seed collected from a range of seed parents, in populations in Georgia, USA, reveal that they have begun to develop genetic resistance to glyphosate. The 'fitness costs' involved in tolerance were estimated. Comparison of tolerant and non-tolerant plants revealed that, in the absence of glyphosate, the tolerant lines produced 35% fewer seeds. These costs are such that, in the absence of the herbicide, it could be supposed that resistant variants would be at a selective disadvantage and, in time, would be eliminated from the population. However, considering generalised models of the population dynamics of weed populations in which resistance has appeared several factors are important.

A. While it is generally accepted that herbicide resistance comes with a cost, it is not clear whether this is always the case. For instance, there have been reports in a number of species that R and S variants do not differ in fitness (Holt, Powles & Holtum, 1993; Warwick & Black, 1994;

Jasieniuk, Brule-Babel & Morrison, 1996). It is possible to argue that these estimates may not be fully realistic as they are based on components of fitness during part of the life cycle, often in non-competitive situations. Furthermore, estimates of fitness are generally made with 'wild' R and S stocks that almost certainly differ in genetic background. In addition, fitness is not a static quantity: R and S plants may hybridise within a population; in addition, gene flow may bring new genotypes into the interbreeding population. While it would be possible to obtain more exact estimates of the effect of herbicide resistance on fitness by comparison of isogenic lines of weeds that differ only in alleles, which confer resistance/sensitivity, such plants are not generally available.

B. Taking into account the life history of the weed, is every cohort of R and S equally exposed to the herbicide(s)? Sometimes herbicide applications are restricted to the growing season of the crop. However, weeds may be able to reproduce at other times of year, when herbicides are not applied or when the effect of the last application has diminished or been totally lost (Cousens & Mortimer, 1995).
C. Are fields uniformly treated or are there sub-sites within fields where sub-optimum doses of relevant herbicide are received by the weeds?
D. Considering the population dynamics of R and S genotypes to a particular herbicide, it is very important to consider whether other herbicides are being used in the area and, if so, with what frequency?
E. How frequently do new independent sites of herbicide resistance arise? Will long-continued use of a particular herbicide on very large populations inevitably result in the evolution of new independent sites of herbicide resistance? Or is the evolution of resistance very infrequent and newly discovered populations are the result of gene flow from other sites? In a review of the literature, Powles and Shaner (2001) conclude that resistance to triazines has been found at widely scattered sites, and the majority represent independent separate evolutionary events, rather than the spread of resistant seeds. However, evolved resistance to triazines occurs in populations in car parks, and along roadsides, pathways and railways, where spread of resistance is likely to have been by seed dispersal on vehicles.

In other cases too, evidence suggests that independent events are highly likely. Thus, in 1987, resistance to sulfonylurea in *Kochia scoparia* appeared in six different states of the USA, and also in Canada. Given the more or less synchronous appearance of the resistant variants and the fact that the sites of origin were widely separated with no common agricultural connection, it seems highly likely that herbicide resistance had independent polytopic origins. Likewise, in the 1980s, strains of *Alopecurus myosuroides* resistant to substituted urea herbicides appeared in Germany, Spain and the UK. In this species, natural dispersal is limited and grain contamination is unlikely, therefore, polytopic origin is the likely explanation. Recent research has provided new evidence of the evolution of herbicide resistance in *Alopecurus myosuroides*. This species is an exceedingly abundant weed in cereal crops in France. Since the late 1980s, ACCase-inhibiting herbicides have been employed to control this weed, but resistant variants of the grass have appeared, as a dominant trait conferred by mutations at the acetyl-CoA carboxylase (ACCase) gene. Gene sequencing studies have detected seven different alleles of this gene that have arisen as point mutations, and five distinct amino acid substitutions have been identified. Using molecular techniques, it has proved possible to determine the broad patterns of distribution of these different resistant variants across the whole of France and, in a parallel study, the detailed distribution of the herbicide resistance alleles in the Côte d'Or region. In total, 13,151 plants were genotyped across 243 fields. Pollen dispersal in the species has been examined (Chauvel, 1991). It was discovered that more than 70% of the pollen is dispersed less than 1 m. However, gene flow may extend to 60 m from donor plants. It would appear that long-range dispersal of resistance alleles by natural pollen flow is rare. Moreover, self-dispersal of seeds is also very limited: most seeds being dispersed 50–60 cm from the seed-bearing plant (Colbach & Sache, 2001). Given the gene flow biology of the species and the locally distinct patterns of allele distribution, Menchari *et al.* (2006) conclude that resistance has developed polytopically by independent mutants. They also conclude that as self-dispersal by gene flow is limited, the patterns of distribution of resistance alleles found in different farms or groups of farms point to the importance of random transfer of seeds by combine harvesters and tractors at a scale of *c.* 8.3–22 km. Machinery is operated in a territorial fashion over such distances, and it is proposed that identical mutants, detected at distances beyond the range of use of particular sets of machinery, are likely to be independent mutational events. These findings offer very important insights into microevolution, providing clear evidence of the significance of farm-scale operations that result in mosaics of different herbicide regimes, leading to the polytopic origin of resistant mutants and the random dispersal of such mutants by machinery, whose range of use has territorial limits set by land ownership and management arrangements.

Another line of evidence concerning polytopy comes from studies of *Lolium rigidum* in Australia. Patterns of resistance in this species are very complex, including multiple and cross resistances, suggesting that different populations have different mechanisms of resistance to the herbicide. These findings support the hypothesis of multiple independent polytopic origins of resistance.

F. Is there a persistent seed bank in the soil? In the case of glyphosate-resistant *I. purpurea*, viable seed may last for at least 7 years in the soil, and this could have important buffering effects influencing the balance of resistant/susceptible variants in the population (Baucom & Mauricio, 2004).

Withdrawal of herbicide treatment

What happens to the R genotypes when the herbicide to which they are resistant is withdrawn? Grignac (1978) studied the decline in metoxuron resistance in *Poa annua* populations no longer subjected to the selective pressure of herbicide application, and detected a more or less rapid return to low level of resistance.

Changes in populations of triazine-resistant *Senecio vulgaris* have been studied in Cambridge, UK. Since the 1980s, contractors working for the Cambridge City Council have used herbicides to keep the city tidy by controlling annual and perennial weeds in city streets, under railings and street furniture. Triazines (atrazine/simazine) were widely used for their high residual activity that may persist in the soil for longer than a year (www.extoxnet.orst.edu). In 1988, plants of *S. vulgaris* resistant to simazine were detected at several sites in the city, including two oil depots. With regard to these two commercial areas it is significant that triazine-resistant *S. vulgaris* had been found at about the same time at the oil refinery at Harwich, from which oil supplies are carried by road to Cambridge. Another population of triazine-resistant plants was detected by Hobson's Brook, under ornamental railings on the western edge of the Cambridge University Botanic Garden.

Following concerns about the pesticide contamination of the nearby Hobson's Brook, the City Council discontinued the use of triazine herbicides in *c.* 1988, relying instead on applications of glyphosate. However, in 1991, triazine-resistant variants of *S. vulgaris* were detected in studies of the railings site (Briggs *et al.*, 1992). Additional investigations carried out by Mount (1992) confirmed the presence of resistant plants at this site. But, by 1999, no resistant variants were detected at the site in tests of plants growing from soil samples, and seedlings raised from adult and seedling plants found growing at the site (Scott, 1999).

These findings suggest that, in the absence of triazine applications, selection acts against the R variants, favouring S plants. The time scale for the disappearance of the R plants reflects the very short period in which the achenes of *S. vulgaris* are viable in the soil seed bank (Grime *et al.*, 1989). These findings agree with the model of Conrad and Radosevich (1979), who predicted that, on the basis of fitness estimates, populations of *S. vulgaris* could switch from being 98% resistant to more or less wholly susceptible in 9 or 10 generations. The population by Hobson's Brook was 100% resistant in 1991. Given that *S. vulgaris* could readily pass through two or three life cycles per year, there has been sufficient time for population change. However, triazine resistance had not entirely disappeared from the area around the Botanic Garden. In the 1999 survey, one resistant variant was detected at a previously 'triazine treated' site in nearby Panton Street (Scott, 1999).

Effects of herbicide treatments: winners and losers

Studies of herbicide resistance in individual weed species have provided very convincing examples of 'natural selection in action'. Thus, if triazine-R and -S achenes of *Senecio vulgaris* are germinated in the presence of atrazine/simazine, both sets of seedlings will grow successfully to the cotyledon stage. However, while R seedlings will continue to grow and produce new leaves, S seedlings, in the presence of the herbicide, are unable to photosynthesise properly, and they die when all the food reserves in the cotyledons are exhausted.

How have weed populations changed under herbicide treatments?

Selection can be seen to operate in experimentally devised situations, and, as herbicides are effective in weed control, it is logical to accept the overwhelming evidence that natural selection is operating in response to herbicide use in the agricultural landscape.

How have weed floras responded to chemical weed control? Changes in weed floras have been studied, but, given the complexity of agricultural practices and the variability of weeds, the precise effect of particular herbicides is not always clear. However, some general trends have been noted.

A. Given that different weed species are differentially sensitive to herbicides, repeated treatments might lead to an overall reduction in the diversity of species (see Ashton & Crafts, 1981; Ross & Lembi, 1985; Devine *et al.*, 1993).
B. Where annual weeds have been intensively controlled by herbicide applications, the weed flora contains a higher proportion of perennial species.

C. There may be a shift in the relative abundance of dicotyledonous weeds relative to monocotyledons. For example, the repeated use of 2,5-D to control dicotyledonous weeds in cereal crops in Britain has led to a decline in many susceptible broad-leaved weeds and an increase in frequency of *Avena* species and blackgrass (*Alopecurus myosuroides*) (Fryer & Chancellor, 1979). Similar findings have been reported in Germany (Bachthaler, 1967).

D. Weedy species taxonomically related to the crop are much more difficult to control by herbicide applications than non-related species. Thus, a taxonomic shift in the spectrum of weeds at a site may occur, with weeds of the same family as the crop increasing relative to non-related weeds (Radosevich *et al.*, 1997). For example, graminoid panicoid weeds (*Sorghum halepense*, *Digitaria sanguinalis*) have increased in importance in atrazine-treated corn, while festucoid grass weeds (*Alopecurus mysuroides* and *Avena* spp.) have increased in abundance in wheat treated with auxin-type herbicides (Vighi & Funari, 1995, 59). Another example is provided by use of the herbicide trifluralin to control grass weeds in tomatoes and potatoes. Both crops are members of the Solanaceae, and they are relatively unharmed by this weed killer. However, while grasses and other weeds are successfully controlled, repeated use of the herbicide has resulted in an increase in Solanaceous weeds.

Selection pressures associated with the development of modern agricultural practices

Agricultural practices are constantly changing, and it is important to appreciate how changes in modern agro-ecosystems are imposing significant new selection pressures on weed species. Clements *et al.* (2004) and Murphy and Lemerle (2006) have considered this question in relation to new cropping methods. Reflecting on a wide range of situations, they report that significant selection pressures have been inferred from observing the outcome of particular agronomic practices on changes in weed species frequency, revealing winners and losers.

- Many crops are now sown at higher densities. This provides selective advantage to climbing weeds such as bindweed (*Polygonum convolvulus*) and cleavers (*Galium aparine*), at the expense of rosette species such as sowthistle (*Sonchus asper*) (Håkansson, 1983).
- Fallow periods are now employed as part of some cropping systems. In addition, experimental and field trials have revealed that weeds may be managed by cutting, chopping, the use of mechanical devices such as brush hoes, flame weeding powered by propane gas, freezing with liquid nitrogen or carbon dioxide and microwaving the soil etc. (Lampkin, 1990). As a practical farm practice, pasture phase fallows with grazing animals are sometimes incorporated in crop rotations (Liebman, Mohler & Staver, 2001).
- To reduce the effects of erosion, particularly in dry climates, and help in weed control, minimal soil disturbance agriculture is being practised with stubbles retained in the system. In such systems, residues of a number of species release chemicals toxic to some other plants, influencing, in the shorter term, which species are winners and which losers (Liebman *et al.*, 2001).
- With the increasing adoption of conservation tillage (no-till or reduced frequency tillage) in North America and elsewhere, there is a greater ground cover of weeds and an increase in biennial weed species, such as horseweed (*Conyza canadensis*), bull thistle (*Cirsium vulgare*) and yarrow (*Achillea millefolium*). Also, in the absence of regular ploughing there is shallower burial of seeds

in the soil. These changes present new selection pressures. Regular ploughing presents an open crop-seeded habitat for colonising weed floras. In contrast, conservation tillage on the same land allows the development of more closed vegetation, in which intraspecific interactions are increased. It will be interesting to examine microevolutionary changes in weed populations faced with such dramatic changes in cultivation (Murphy *et al.*, 2006).

- Fertiliser applications. At Rothamsted Experimental Station, long-term experiments have revealed that, under high nitrogen applications, *Stellaria media* would appear to be at a selective advantage, while *Medicago lupulina* and *Equisetum arvense* were commoner on low nitrogen plots (Moss *et al.*, 2004). In other studies, crops with high nutrient supply tend to grow at increased density, favouring such weed species as *Stellaria media* and *Chenopodium album* that have some capacity to tolerate shade. In such situations *P. convolvulus* and *Galium aparine* may be at a selective advantage, as they are able to climb into better-lit upper crop canopy.
- Mechanisation of harvesting. Small-seeded weed species may be at a selective advantage with the combined harvesting, as their seeds are ejected with the chaff behind the machine. In contrast, the larger seeds of other weed species may be carried off with the crops. Machinery has been devised to catch the chaff (and weed seeds it contains) that was formerly discharged behind the machines.

Differential responses to farming practices

Given the complexity of cropping systems, Murphy and Lemerle (2006) stress that it is difficult to isolate the effect of any particular 'selective' factor. Greater insights will only be available if observations and monitoring lead to more detailed investigations. To take account of the effects of current agricultural practices, many variables must be considered, including: site history; past and current crops; the weed flora of particular sites, especially species represented in the seed bank; and the tillage systems, fertilisers and weed management regimes being employed. In addition, soil factors and climatic variable must be taken into consideration.

Winners and losers in forested areas

Arable and grazing areas are not the only ecosystems that provide crops. Forested areas across the world have been exploited to provide a variety of products, hunting grounds for plants, animals, fungal species etc. There are many different ways in which the microevolution in forested ecosystems has been influenced by human activities. Many are highly site specific. Here, in a brief account, based on the admirable review of Ledig (1992), issues and hypotheses relating to winners and loser are confronted.

Deforestation

In a region, deforestation to provide agricultural land has a profound effect on which species/variants survive. For example, in 1900, conifer and hardwood forests covered 40% of Ethiopia, but by 1985 only 2.7% of native forests remained.

Exploitation

Forests are also routinely 'damaged' by many human activities – from warfare to harvesting. For example, Thucydides records: 'Demosthenes set fire to the forest of Sphacteria to harry the Spartans during the Peloponnesian wars in 435 BC' (Ledig, 1992). In the Vietnam War, US forces sprayed forests (about 10% of the land area of the country) with herbicides. In the exploitation of forests, local extinction of particular tree species may occur. For example, Pitch pine (*Pinus rigida*) was so extensively used as fuel wood on the island of Nantucket that the species became extinct. In other cases, species survive, but patterns of harvesting determine which are winners and which are losers. At one end of the spectrum of harvesting is the selective removal of large specimens of commercially important trees, while, at the other extreme, forests may be effectively clear-cut. Thus, in the first wave of exploitation of White pine (*Pinus strobes*) in north-eastern USA in the seventeenth century, selected trees were harvested to provide ship's masts. In contrast, clear-cutting of forests is a widespread practice across the globe that may leave behind only small useless crooked diseased trees and radical disturbance to the ground flora. Additionally, there may be dramatic changes to the flora, if the deforested ground is burnt. Looking to what happens after harvesting, patterns of winners and losers will be dictated by species biology, the location of seed parent trees, and subsequent land use, especially whether secondary woodland is allowed to develop. Where remaining commercially valueless trees are left, these could be the source of seed colonising the area, or seed may arrive from adjacent woodland areas or isolated trees. Regarding the structure of secondary woodland, it is important to note that while some woody species have little or no capacity to recover, others are not killed by felling but are able to re-grow from the stumps or underground parts (Figs. 9.2. & 9.3).

Fragmentation

Deforestation and exploitation often lead to the fragmentation of forests, and such changes may have profound effects on the microevolution of the tree species and all the other associated species in the ecosystem. Fragmentation, especially if it is associated with reduced population size, can lead to loss of genetic variability through the combined effects of loss or restriction of gene flow, inbreeding effects and genetic drift. However, if very large genetically isolated fragments remain, there is the potential for each to evolve in isolation, leading to population differentiation in the longer term.

Demographic changes

The transition from old-growth forest, with its diversity of age classes, to managed secondary forest may lead to marked demographic effects. Selective felling may alter the seed rain in the forest. Mass establishment of cohorts of plants may follow clear-felling leading to even-age stands, in which the proportions of the different species may be radically different from the pre-harvest ecosystem. Overall, there may be reduction in the average age of trees

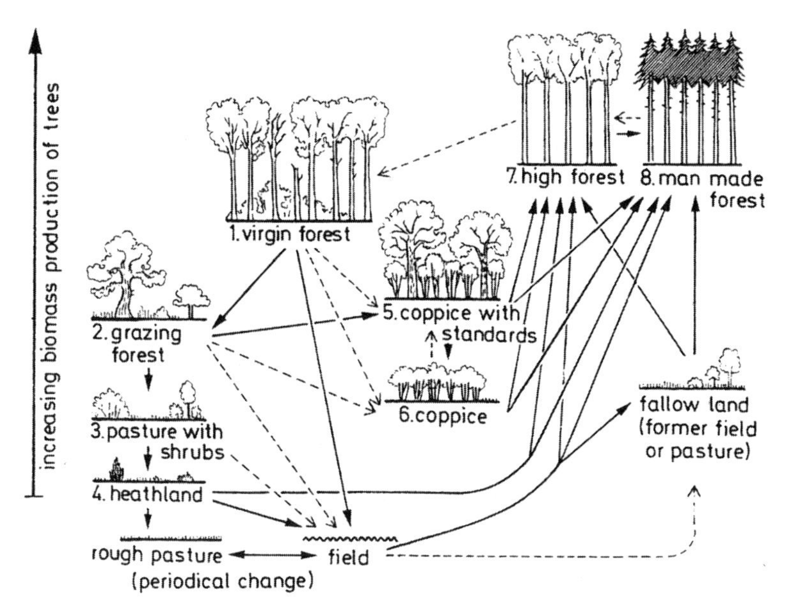

Fig. 9.2 Transformation of the primeval forest through grazing, farming and forestry in Central Europe. (From Ellenberg, 1988. Reproduced by permission of Cambridge University Press). Troup (1952) and Watkins (1998) discuss the history of different regional management of European forests. For a historical account of the exploitation and management of the forests of the USA see Williams (1989).

(Figs. 9.2, 9.3, 9.4). However, each situation must be considered carefully. Coppiced woodlands are repeatedly harvested, and the size of the coppice stools indicates that individuals may be of great age.

Habitat alteration

Changes in habitat may encourage hybridisation between species that were formerly ecologically isolated. Thus, after three centuries of logging and extensive fires in the Maritime provinces of Canada, extensive intercrossing has occurred between the red (extensively logged) and the black spruce in bogs and river valleys. Fire control has also had major effects on forest structure and dynamics. Some species of pines are fire-adapted. Their cones are sealed with resin and seeds are only released when the cones are subjected to the heat from fires. Some species, such as pitch pine, are genetically polymorphic; with variants producing either sealed (serotinous) or unsealed cones. In human-influenced forests, modelling predicts that the proportions of the different variants in secondary communities will be influenced by the frequency of fire. Unsealed variants increase in frequency under fire prevention regimes, while sealed variants will flourish if fires are quite frequent.

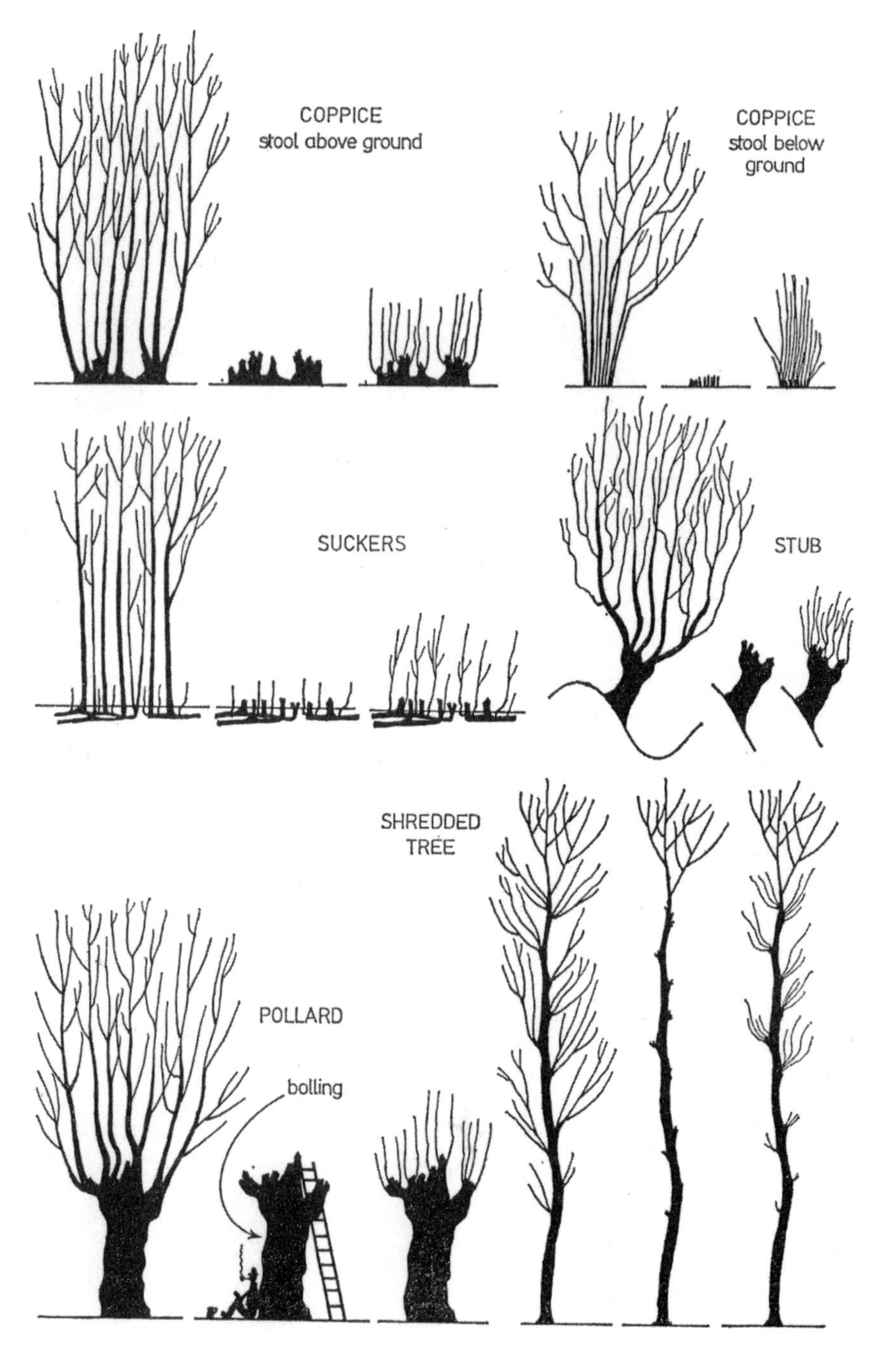

Fig. 9.3 Different ways of managing wood-producing trees in the British landscape. For each method, the tree or group of trees is shown just before cutting, just after cutting, and one year after cutting. All are drawn to the same scale. New growth from managed trees and shrubs is relished by large herbivores. Therefore, in traditional management, coppiced woodland was protected from grazing domestic and other stock by ditches etc. The system of pollarding ensured re-growth above the browse level of domestic stock and wild herbivores, and therefore provided the means to combine grazing land with woodland production through the development of wood pasture systems. Shredding trees provided fodder for stock. (From Rackham, 1986, who provides further details of the woodland management systems. Reproduced with permission from Rackham, O. (1986), *Trees and Woodland in the British Landscape*. Published by J. M. Dent, a division of Orion Publishing Group.)

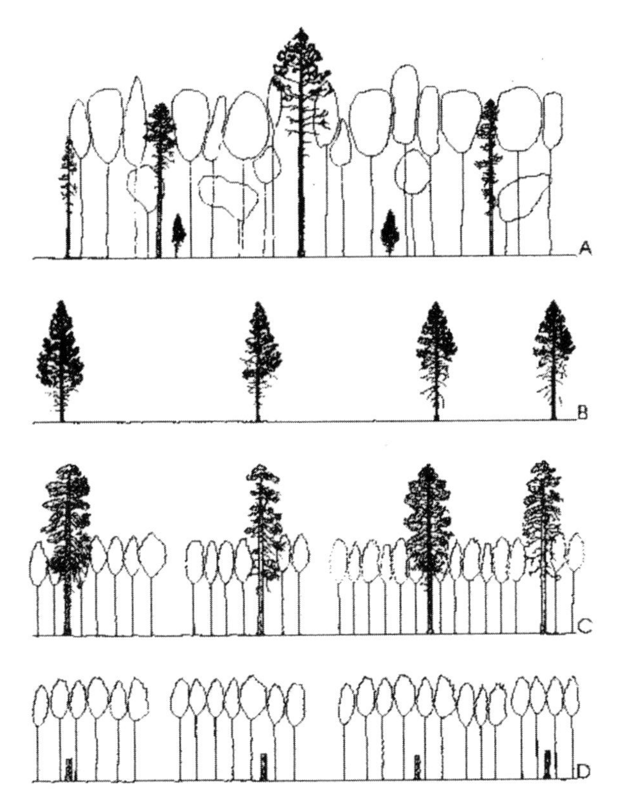

Fig. 9.4 Human activities have a major influence on the genetic structure of woodlands. A model of changes in eastern white pine populations in New England. (A) Pre-colonial forest with scattered pines. (B) Conversion of forest to agriculture with pine remaining in fencerows and woodlots. (C) After the abandonment of agricultural land the progeny of widely spaced trees filled the gaps. (D) Management of resulting woodland has involved the removal of senile trees, including the seed parents of the old-field colonists. (From Ledig, 1992. Reproduced with permission of Blackwell Publishing)

Environmental deterioration

Ledig (1992) recognises multiple stresses that determine which tree species will be winners and losers. Here, we note factors discussed in later chapters, namely the effects of atmospheric pollutants, and heavy metal contamination, introduced herbivores, pests and diseases.

Introduction of new tree species and variants

In the management of forests, native or introduced species are often planted and in some cases introduced species may cross with native trees. Ecotypic differentiation is well documented in trees, and the use of non-local stocks of native species may have implications for microevolution and conservation. For example, in the early management of the Redwood

National Park in the USA, seeds of *Sequoia sempervirens*, Douglas Fir and Sitka Spruce (*Picea sitchensis*) from other areas were scattered in aerial seeding.

Domestication

Ledig considers that many forest tree species, especially those cultivated in plantations, are effectively being domesticated. He sees a continuum of situations from the wild to the domesticated, in which directed selection is taking place for desired traits. Also, he points to many situations of unwitting 'selection' in the collecting of seeds, the cultivation practices employed in seedling nurseries and the thinning and harvesting regimes of the maturing plantations. In addition, there is clear evidence that some woody and ground flora species fail to survive under particular forest management regimes (e.g. fern species in some coppiced woods in the UK): others are regarded as weeds and removed by hand or by herbicide treatment.

Testing hypotheses

Many ideas are set out in Ledig's review. But, until comparatively recently, progress in testing hypotheses was slow, as the only means of investigation involved crossing experiments and the study of progenies. However, the development of molecular tools has made it possible to carry out critical investigations on a shorter time frame. Here, to illustrate the range of studies, are a few examples. For more detailed reviews see Finkeldy and Ziehe (2004), Degan *et al.* (2006), Hosius *et al.* (2006) and Mathiasen, Rovere and Premoli (2007).

Effects of fragmentation

Detailed studies have been carried out on samples drawn from remnant populations of the southern Florida slash pine (*Pinus elliottii* var. *densa*). Fragmentation of populations has resulted from logging, clearing for development and agriculture, and hurricane damage. Population structure has been examined using microsatellite markers (Williams *et al.*, 2007). They found no evidence for genetic erosion in any of the fragments, but emphasised that the time frame for possible alteration to the genetic structure of populations must be considered, as the species is long-lived.

Effects of logging on population variation

In studies of an insect-pollinated tree in the Amazonian rainforests (*Carapa guianensis*), investigations using microsatellite markers revealed no short-term genetic impact of selective logging (Cloutier *et al.*, 2007). Another informative study was carried out on *Scaphium macropodum* populations in Malaysia using RAPD markers (Lee *et al.*, 2002). Genetic variation was examined in two 'control' unlogged areas and in two plots where the species was regenerating in a logged area. On the assumption that all plots were genetically identical before logging, there was clear evidence of substantial genetic erosion in one plot of regenerated forest relative to unlogged areas. In studies of *Acer saccharum*, genetic variation was examined in populations in old-growth forest and in an area recolonised since logging (Baucom, Estill & Cruzan, 2005). Decreased genetic variability was detected in logged areas, and it was concluded.: 'even one extensive bout of logging can alter the level and distribution of genetic variation in this forest species'.

Pollen flow

Studies of the consequences of logging on pollen flow have been investigated in African Mahogany by employing microsatellite DNA markers to examine paternity of progenies (Lourmas *et al.*, 2007). Evidence for long-distance pollen flow was detected with a low rate of selfing (below 2%). The main factor limiting regeneration is the number of seed-bearing trees in and around logged areas.

Seed gene flow

By employing microsatellite markers, parent–seedling matches were carried out in Spanish populations of *Pinus pinaster*, which is currently found in small, fragmented, scattered populations (Gonzales-Martinez *et al.*, 2002). These paternity studies revealed 'there was an excess of parent–offspring matches in a radius of 15 m from the parent trees'. These findings highlight the restricted seed flow in this relatively heavy-seeded species. Patterns of variability in the progeny were different from those in the adult trees. 'Animal-mediated secondary dispersal' and mortality in the severe Mediterranean climate were identified as important factors.

Forest management

In Australian forests, two different forestry practices are employed in managing regeneration of such species as *Eucalyptus consideniana* – clear-felling with aerial sowing of seeds, and leaving selected trees in the logged areas to act as seed parents (the seed tree system). Samples of saplings regenerating under the two regimes were investigated using RFLP and microsatellite markers (Glaubitz, Murrell & Moran, 2003). The results suggested that genetic erosion is more likely under the seed tree system.

Coda

As we have seen in the last two chapters, human activities lead to complex mosaics of rural land and urban use involving arable, lawns, gardens, grazing lands, forested landscapes etc. There is increasing evidence that the management regimes in these different 'new' ecosystems exert deliberate or unwitting selection pressures resulting in winners and losers amongst plant species. It is important to stress, however, that within the rural and urban landscapes there are areas of temporarily neglected or abandoned land, waste dumps etc., and some species, excluded by intensive management from certain areas, may find a refuge there.

In this chapter, the focus has been mainly on rural ecosystems. Recent research draws attention to the possibility of rapid plant evolution in the city. In Montpellier, France, there are many street trees, each of which is surrounded by a small patch of earth, which is so often disturbed that few perennial weeds grow there. These patches are invaded by *Crepis sancta*, an annual weedy species, which produces two types of fruit within the capitulum (fruiting head) – central fruits, with a feathery pappus, are light dispersing structures, while peripheral fruits are heavier and non-dispersing. Studies of the dispersal biology of the species have been carried out by growing samples – from both rural and urban populations – in a common environment (Cheptou *et al.*, 2008). The results of these studies reveal that the city populations have a higher proportion of the non-dispersing fruits than their rural counterparts. Essentially, the street habitat for the *Crepis* consists of a series of very small islands. Dispersal of fruits and seeds always incurs a 'cost'. The persistence of *Crepis* through several seasons in the island systems could be threatened if widespread dispersal of the fruit occurs. Thus, evidence is consistent with the hypothesis that selection favours lower dispersal in the urban areas. It is estimated that this rapid evolution of fruit dispersal may have arisen over *c*. 5–12 generations of selection.

Arable and grazed communities have been more extensively investigated than forested areas, and there is evidence for co-evolution of crop and weed. However, co-evolutionary explanations of crop–weed interrelations should be made with care. In the case of hay meadows, it would be possible to see grasses as the crop. But what is the status of the other species in the floristically rich swards? Are other species merely co-evolving 'weeds', responding to the date of hay harvest etc.? Such an interpretation would be over-simplistic, as many of the non-grass members of the hay sward are part of the crop. As we have seen above, hay with accompanying forbs is prized by alpine farmers. It would be more realistic, in this case, to consider the entire hay meadow ecosystem as a co-evolving whole, responding from site to site in response to specific human management. Extending this argument to situations not yet fully investigated, models of co-evolution of ecosystems could with profit be explored in the case of other elements of the rural landscape.

With regard to the plant species in rural landscapes, it is possible to designate winners and losers. However, in the course of time, the status of species may change. Likewise, there might be alterations in the status of intraspecific variants. Agricultural and rural practices are always developing: species/intraspecific variants that are winners in one context may quickly become losers. This is very clear, as we have seen, in the case of weeds whose seeds mimic those of a crop. The fate of these species was sealed with the development of new seed cleaning techniques.

There is currently little support for the notion that adaptation to the complexities of the grazing and arable landscape involves general-purpose genotypes (but further consideration of this concept appears below). In contrast, there is an impressive body of research supporting the widespread occurrence, in human-made ecosystems, of intraspecific genetic races in many plant species – the so-called biotic ecotypes. These findings stress the importance of the population concept. Case studies reveal site-specific histories and unique genetic patterns and processes. While only a sample of species has been studied in any detail, it is worth stressing that, as the processes of microevolution are universal, it is highly likely that ecotypic differentiation is also a ubiquitous phenomenon. Thus, concerning a species of interest or concern, it is important to stress again the dangers in typological thinking, in which members of a single population could be taken to represent the species as a whole. Furthermore, where reintroductions are being considered (see Chapter 16) great care must be exercised in the choice of stock. This follows, for example, from the discovery that hay meadow management is site specific, leading to the evolution of an array of different intraspecific ecotypes. Thus, for most species, there is no single 'hay' ecotype that would be appropriate for the restoration of hay meadows at different altitudes under different management.

Chapters 8 and 9 have considered patterns and processes of microevolution in areas subject to different human management – hay/pasture; weeded/unweeded areas etc. As we have seen, there is increasing evidence that selection favours different variants in specific habitats. Such adaptations involve life-history traits related to the length of the vegetative phase/flowering time etc. In some cases evidence has been found in crossing experiments that such traits are genetically determined. For instance, Comes and Kadereit (1996) made crosses between *Senecio vulgaris* var. *vulgaris* × *S. vulgaris* ssp. *denticulatus*. Var. *vulgaris* shows precocious flowering relative to *denticulatus*. The analysis of F_2 phenotypes revealed patterns of segregation in flowering behaviour that could be explained by differences in a single major gene and the genes controlling speed of development may be homologous to genes identified in *Arabidopsis* (early flowering genes: elf 1–3).

In the last ten years, major progress has been made in the study of flowering behaviour in *Arabidopsis thaliana* raised under controlled laboratory conditions in which manipulation of environmental factors has been exactly controlled. It is beyond the scope of this account to report on the findings of these experiments – Roux *et al.* (2006) and Jaeger, Graf and Wigge (2006) provide helpful reviews. Here, two points are emphasised from Roux *et al.* (2006). *Arabidopsis* proves to have 'a complex genetic network that can detect environmental and internal signals' and microevolutionary change to early flowering may involve 'a limited number of genes'. The degree to which the findings on *Arabidopsis* are applicable to other species is under review. The sequencing of many of the genes from *Arabidopsis* (micro-arrays) has provided the means of studying gene sequences and flowering behaviour in other species. For instance, molecular studies of ecotypic variants of *Capsella bursa-pastoris* (a close relative of *Arabidopsis*) are revealing the genetic architecture of adaptive flowering behaviour in regional and biotic ecotypes (Slotte *et al.*, 2007).

Major studies have revealed how winners and losers emerge in polluted ecosystems and these investigations are considered in the next chapter.

10

Pollution and microevolutionary change

As we have seen in earlier chapters, Darwin did not provide direct evidence for natural selection in his *Origin of Species*. In the post-Darwinian period, many patterns of variation in plants explicable in terms of selection were described, but the demonstration of natural selection in wild populations remained elusive. Then, in the 1950s, detailed studies of air pollution and sites contaminated with heavy metals began in earnest, providing a number of thoroughly investigated and convincing cases of natural selection in action in animals and plants (Antonovics, Bradshaw & Turner, 1971; Taylor, Pitelka & Clegg, 1991; Macnair, 1981, 1990, 1997; Shaw, 2001).

Bell and Treshow (2002) discuss the long history of concern about the effects of pollution. For example, the diarist John Evelyn published the celebrated account *Fumifugium: Or the Inconvenience of the Aer and Smoake of London Dissipated* in 1661. As the industrial revolution gathered pace in the nineteenth century, there was a marked deterioration in air quality in towns and cities. As urban growth accelerated and industrialisation developed in Europe and North America, trees and other vegetation were often killed or damaged, especially in towns and cities, and near smelters, factories and industrial installations. Lichens proved to be particularly sensitive indicators of the growing problem of air pollution (Bates, 2002).

Here, our concern is to examine the proposition that pollution exerts potent selective pressures on plant populations. One of the earliest papers on the subject was published by Dunn (1959). He reported that smog had damaged some, but not all, of the plants grown in his experiment on variation in Californian populations of *Lupinus bicolor* set out in a garden at Highland Park in Los Angeles. Eight populations from different parts of the state were investigated; one from vacant lots in the city. At this time important advances were made in the understanding of the origin, chemistry and destructive power (to plants) of smog (Bell & Treshow, 2002). Reflecting on his results in the light of these developments, and drawing on the results of growing subsets of his material in growth chambers at the Earhart Plant Research Laboratory at the California Institute of Technology, Dunn concluded that within *Lupinus bicolor* some variants were susceptible and others were resistant to smog. The local Los Angeles population 'was more resistant to smog and was able to produce flowers and seeds more readily than any of the other subspecies from other parts of California', several of which 'were killed outright, suggesting very strongly that smog has been a subtle selective agent over a rather extended period of time'. He continued: 'I therefore suggest that the various gaseous fumes of large cities may well be active as selective agents in the local plant population.'

Mutation may occur as a consequence of human activities. Significant levels of radiation have been released in both civil and military uses of radioactive substances and from accidents at nuclear power stations, the most infamous of which was the explosion at Chernobyl in the Ukraine in April 1986. In addition, some fungicides, pesticides and other chemicals found as air and/or water pollutants have been shown to cause genetic or chromosomal damage (Sharma & Panneerselvam, 1990). Such behaviour is often detected and measured using bioassay tests. For example, studies of herring gulls in the Great Lakes Region 'have demonstrated elevated mutation rates near steel mills' and these findings have been confirmed and extended by comparing mutation rates in laboratory mice kept in cages at a control rural site and a polluted site down-wind of steel mills in Hamilton Harbor (Somers *et al.*, 2002).

Tests involving animals raise many ethical issues. In a useful review, Grant (1994) makes a strong case for using plants in bioassays, citing a range of investigations employing *Allium cepa*, *Arabidopsis thaliana*, *Tradescantia*, *Vicia faba* and a number of other species. For example, *Tradescantia* clones have been selected that are heterozygous for flower colour with blue dominant and pink recessive. The stamen assay investigates somatic mutation involving the examination of the staminal hair cells to detect any 'pink' mutated cells. The effects of chemical and radiation pollution – in soil, water or air – may be tested by comparing the mutation rate in staminal hairs in cloned material set out in control unpolluted sites with material grown in areas subject to pollution. By this means it was discovered that the ambient air around a chloroprene rubber plant in Armenia had genotoxic properties (Arutyunyan *et al.*, 1999). Also, cytological bioassays of meiosis in flower buds of *Tradescantia* clones, set out in control and polluted sites, provides the means of studying whether pollution of various kinds leads to increased chromosome mutation in the germline, detectable through the production of chromosome aberrations, in particular the formation of micronuclei. Investigations of radiation-induced mutation have often employed bioassay techniques. For example, complex germline mutation has been discovered in wheat plants grown for one season in the heavily contaminated exclusion zone around the damaged Chernobyl Nuclear Power Station (Kovalchuk *et al.*, 2003).

Reductionist approaches in the study of the effects of pollution

Dunn's experiments provide one of the earliest studies of the complex effects of the cocktail of pollutants that constitute smog (ozone, peroxyacyl nitrates etc.), and investigations of mutation have often examined situations where many pollutants are impacting ecosystems.

Turning to later developments in the study of microevolution in polluted areas, researchers have frequently taken the 'reductionist route' in studying the biological effects of individual gaseous pollutants (sulphur dioxide, ozone, oxides of nitrogen, particulates etc.).

Investigations have taken advantage of our increasing knowledge of the patterns of pollution in both time and space, made possible by the development of sensitive monitoring equipment (Bell & Treshow, 2002).

To test the responses of plants to such highly reactive pollutants as sulphur dioxide, environmentally controlled tunnels have been constructed. Horsman *et al.* (1979a) give details of a suite of replicate fan-driven tunnel systems in which air is drawn through an illuminated compartment containing experimental plants. Replicate tunnels provided an opportunity to test the effect of different concentrations of the gas against controls of ambient air (with its burden of SO_2) drawn from outside the laboratory or 'clean air' produced by passing air through activated charcoal. Experiments in this field have often involved short-term high levels of SO_2 (acute) or chronic fumigations involving lower doses extending over longer periods. Later, open top and other types of fumigation chambers have been developed (see Bell & Treshow, 2002).

Effects of sulphur dioxide pollution

Multiple sources

A commercial cultivar of *Lolium perenne* (S23) was discovered to be more sensitive to SO_2 pollution than the wild populations of the same species collected from a heavily polluted site at Helmshore, East Lancashire (Bell & Clough, 1973; Bell & Mudd, 1976). Experiments were also carried out on *Lolium perenne* populations from polluted sites in Liverpool, using control populations of the same species sampled from less polluted areas on the Wirral (Horsman *et al.*, 1979a, b). Plants were subjected to 2600 $\mu g\ SO_2\ m^{-3}$ for two weeks (acute) or 650 $\mu g\ SO_2\ m^{-3}$ for eight weeks (chronic treatment). The material from Liverpool proved to be more SO_2 tolerant than the control.

Pollution from 'point sources'

Several studies have investigated resistance/sensitivity in populations collected near point sources of industrial SO_2 pollution with appropriate controls originating from 'clean sites'. Taylor and Murdy (1975) discovered that plants raised from seed of populations of *Geranium carolinianum* growing within 700 m of a 31-year-old coal-fired power station in Georgia, USA, were more tolerant than populations in non-polluted sites. In another study, tolerant populations of the annual *Lepidium virginicum* were detected growing near a heavily polluting smelter at Copper Basin, Tennessee (Murdy, 1979). It is of particular interest that, following fumigation, pollution-sensitive populations from 'clean sites' had higher seed abortion than tolerant plants.

In another series of investigations, greater tolerance was discovered in a number of species growing near to a factory at Askern than in material collected some distance away in an area of low contamination (Wilson & Bell, 1986). From this point source, SO_2 pollution was produced in the manufacture of smokeless fuels.

Table 10.1 *Speed of development of sulphur dioxide tolerance*

Species	Location	Ambient SO_2 (ppb) regime	Time (years)
Lolium perenne	Helmshore, UK	30–60 (winter)	<150
Festuca rubra			
Dactylis glomerata			
Holcus lanatus			
Phleum bertolonii			<25
Lolium perenne	Manchester, UK	44–67 (annual)	<17
Festuca rubra	Askern, UK	46–53 (annual)	<50
Dactylis glomerata			
Holcus lanatus			
Trifolium repens			
Festuca rubra	London, UK	75–250 (annual)	<34
Agrostis capillaris			<22
Lolium perenne	Manchester, UK	44–67 (annual)	3.5 (acute)
Poa pratensis			4.5 (chronic)
Phleum pratense			4.5 (acute)
Lolium perenne	Liverpool, UK	75–250 (winter)	200?
Lepidium virginicum	Copper Basin, TN, USA	>500 3 h max. for 25% of days	<75
Geranium carolinianum	Newnan, GA, USA	Unspecified	<31
Bromus rubens	Nipomo mesa, CA, USA	900 (mean max. daytime concentration)	<25

Adapted from Bell, Ashmore & Wilson (1991). Reproduced with permission from Springer Verlag

Studies of the rate of development of resistance

Table 10.1 summarises the information from different investigations, including studies of grass species in lawns of different ages (dated *c.* 1770, 1875, 1946, 1958 and 1977) in the grounds of The Temple, The Law Courts, in Central London (Wilson & Bell, 1985).

Examining the evidence accumulated from different studies, it was concluded that concentrations of SO_2 'as low as 30 to 45 ppb are sufficiently deleterious to act as a selective force for resistance' (Bell *et al.*, 1991). But they note: 'in most cases it is difficult to determine with any accuracy the time scale over which resistance has evolved'.

However, studies of an experiment in Philips Park in Manchester revealed that resistance could develop over a short period (Ayazloo & Bell, 1981). This experiment was set out in 1975 by the Sports Turf Research Institute in the UK. The aim was to determine which species/cultivar of grasses performed best as sports turf, in an area known to be subject to high levels of SO_2 pollution. Ayazloo and Bell (1981) sampled plants from this experiment in their studies of SO_2 tolerance, and Table 10.2 sets out the details of their findings on five species. They were able to sample the experiment plots annually from 1978 to 1982 and, because the grasses are all perennials, living material was maintained in cultivation 'representing different years' for comparative tests of tolerance.

If selection for sulphur dioxide had indeed occurred in the Phillips Park experiment, it is likely that the population would change over time. It proved possible to test this proposition because Ayazloo and Bell were able to grow samples representing the base line unselected population by raising plants from the original seed samples. Table 10.2 shows the percentage leaf injury after acute fumigations with SO_2 for the samples of different species taken in different years. To follow the course of any changes in the populations, samples for particular years were tested against the original seed population.

There were statistically significant increases in tolerance in *Lolium perenne* in 1979/1980 but not in 1981/1982. In *Poa pratensis* (1980) and *Phleum pratense* (1981) the samples of that year showed statistically increased tolerance, but this effect did not persist beyond the year in question. It was concluded that there was strong evidence for the development of resistance to SO_2 injury in three species in a period as short as four to five years 'where mean ambient annual concentrations were generally 40 to 60 ppb' (Bell *et al.*, 1991, 46). Furthermore, they concluded: 'the resistance to acute injury apparently disappeared in all three species by 1982'. Examination of the pollution records for the area suggested an explanation for the loss of resistance. 'Local air pollution monitoring data reveals that this coincided with a marked fall in mean SO_2 concentration at Philips Park to 23 to 30 ppb.' Therefore, selection pressures of SO_2 were significantly reduced, and 'less resistant individuals were at a competitive advantage in that they have been demonstrated as invariably showing faster growth than resistant individuals when grown in clean air' (Bell, 1985). These changes in SO_2 tolerance in plants were devised to examine consequences of the passing of the Clean Air Act 1970 (Bell *et al.*, 1991; Bell & Treshow, 2002). They were influenced by the famous studies of changes in the frequency of melanic moths in industrial areas, as smoke levels were reduced (see Box 10.1; Majerus, 1998).

Ozone pollution

Before examining the evidence for microevolutionary changes in response to ozone pollution, a number of major points must be stressed. What follows draws on the excellent review of Ashmore (2002). Ozone is not emitted directly into the atmosphere as a pollutant but, as studies of the chemistry of smog have revealed, the highly toxic gas is the product of the complex reactions, some photochemical, between hydrocarbons and nitrogen oxides, produced as pollutants principally in the exhaust products of cars and other internal combustion engines. Atomic oxygen is produced, which combines with oxygen to give ozone.

Table 10.2 *Percentage leaf injury after acute fumigations in samples taken in different years from the Philips Park (P.P.) experiment, compared with plants raised from the original seed (O.S.) See text for details.*

Species	Population	1976: Injury (%)	1978: Injury (%)	1979		1980		1981		1982	
				SO_2 ($\mu g\ m^{-3}$)	Injury (%)	SO_2 ($\mu g\ m^{-3}$)	Injury (%)	SO_2 ($\mu g\ m^{-3}$)	Injury (%)	SO_2 ($\mu g\ m^{-3}$)	Injury (%)
Lolium perenne	O.S.	20.5	23.5	4184	9.7	4568	42.8	5038	4.8	2749	5.5
	P.P.	18.5	20.1		4.3[b]		33.0[a]		4.9		4.8
Lolium multiflorum	O.S.	50.7	48.0	4087	20.0	4503	56.0	4762	7.3	5098	23.6
	P.P.	52.5	44.7		14.9		64.1		12.1		18.8
Poa pratense	O.S.	6.1	9.4	4032	3.4	5317	44.5	4149	12.7	Not fumigated	
	P.P.	5.6	8.8		4.4		35.0[a]		10.7		
Festuca rubra	O.S.	4.5	4.4	5139	4.1	5816	24.5	3949	9.9	3040	17.6
	P.P.	4.4	3.9		8.1		20.7		9.7		23.2
Phleum pratense	O.S.	59.8	67.0	3924	5.8	5052	67.2	4641	4.6	3409	29.8
	P.P.	61.4	62.5		9.0		54.3[a]		0.8[a]		35.5

[a] $P < 0.05$; [b] $P < 0.01$

From Wilson & Bell (1985). Reproduced with permission of *New Phytologist*

Box 10.1 Industrial melanism in the Peppered Moth (*Biston betularia*) (Account drawn from Majurus, 1998)

The Peppered Moth has two distinct forms. Forma *typica* has a recessive allele conferring white wings with black speckling. In contrast, forma *carbonaria*, with the dominant allele, is almost completely black. This night-flying moth rests during the day on tree trunks and branches, where it is subject to predation by birds. Forma *typica*, because of its cryptic coloration, is at a selective advantage on the lichen-rich surfaces in non-polluted areas. In contrast, the melanic f. *carbonaria* variant is at a selective advantage in polluted areas where lichens have been killed by pollution (in particular by sulphur dioxide) and the trees are blackened by soot. The level of predation of the moth depends, therefore, on the effectiveness of its crypsis against predation by birds. Against a background of initial scepticism concerning the alleged predation of the moths by birds and the role of crypsis, Kettlewell and Tinbergen provided important insights, when they filmed the 'selective' eating behaviour of wild birds on moths of both variants introduced into polluted and non-polluted test environments in the field. In addition, the alleged crypsis of the moth has been investigated in detail and appears to be effective as the bird 'sees it' (see Majerus, 1998 for details). Interestingly the visual acuity of birds is different from humans, and so it is not possible for the human eye to judge exactly the level of crypsis conferred by 'protective coloration' against different backgrounds. One very important question still has to be resolved. Protective coloration in crypsis will only be effective if the insect rests against an appropriate colour/ textural background. Thus, there may be linkage between the gene for wing colour and behavioural traits. Many investigations have examined the genetic variability of populations in different areas (including the presence and genetics of intermediate variants, forma '*insularia*'), and the fact that gene frequencies would appears to be influenced not only by selective predation but also by migration. Studies of *Biston* are remarkable for they record, through continuous monitoring, the high frequency of the melanic variant in urban and industrial areas *before* and its slow decline in frequency *after* a major perturbation in the environment – namely the effect of anti-pollution legislation in the 1950s, the aim of which was to reduce atmospheric pollution. Thus, in sample areas the frequency of the melanic variant was *c*. 90% in 1959. In the same sites this figure had fallen to less than 20% by 1995.

In many areas of the world, ozone is now reaching high enough concentrations to injure and depress the yield of many cultivars of crops. The gas can travel very large distances in the atmosphere, and, if concentrations of the gas reach critical levels, susceptible plants may be injured over wide areas. Thus, ozone is an invisible significant pollutant gas active not only in towns and cities, but also in distant agricultural, grassland and forested ecosystems (Davison & Barnes, 1998).

Ozone damage to plants may be invisible, or visible symptoms of damage may occur, such as yellowing, stippling, flecking or blotching of the leaves. Injury may be so severe that plants may suffer premature senescence or die. For instance, in the USA and Europe, there are reports of damage to crops and trees by episodes of ozone pollution (see Ashmore, 2002). There are also reports of crop damage elsewhere, for instance, in Taiwan, Punjab, Mexico City and Egypt (Marshall, 2002). In many places where ozone damage has not yet been reported, rural levels of ozone may be high enough to cause damage, e.g. in parts of Asia and South America. There is now an enormous literature on the effects of ozone on forest trees in Europe and North America. In North America, even forests in national parks may show ozone injury symptoms, as for example in the Great Smoky Mountains National Park, Tennessee, with 47% of the *Prunus serotina* trees damaged (Chappelka *et al.*, 1997).

Accurate measurement of ambient ozone is now possible with sensitive instruments. In addition, cultivars of radish and tobacco have been used to provide semi-quantitative monitoring of the degree of ozone pollution through their extreme sensitivity (e.g. tobacco Bel-W3) or lesser sensitivity (e.g. Bel-B and Bel-C; Heggestad & Middleton, 1959). Also, fumigation chambers have been designed to allow the replicated testing of the sensitivity of plants to ozone (Reiling & Davison, 1992, 1994, 1995).

The evolution of ozone resistance

Resistance to ozone has been examined in investigations of *Populus tremuloides*. Cloned material of *Populus* from several national parks was grown in a field trial in which the plants were exposed to high levels of ozone (Berrang, Karnosky & Bennett, 1988, 1991). Clones differed in their response: some showed no injury, while others had visible leaf damage. The damage proved greatest in those clones that came from parks subject to lower ozone exposure.

Very extensive investigations have been carried out on *Plantago major*. In preliminary studies, this perennial species was shown to be an appropriate test species in its sensitivity to ozone (Reiling & Davison, 1992a; 1994). To examine patterns of variability across the UK, seed was collected from 27 populations of *Plantago major* subject to different degrees of ozone pollution (Reiling & Davison, 1992b). The relative resistance of young plants was studied using two replicate environmentally controlled growth chambers in which plants were exposed to 70 nl O_3 1–1 for 7 h d^{-1} over two weeks. Two additional replicate chambers acting as controls were supplied with charcoal filtered air ($<$10 nl O_3 1^{-1}). Figure 10.1 shows the resistance found in the different populations (expressed as mean relative growth rate of treatment/control $\times$ 100) plotted against the exposure to ozone at the sample sites experienced by the populations in the year preceding the collection of the seeds. (Figures for exposure to ozone from the Warren Springs Laboratory UK Monitoring Network.) The regression line of resistance to exposure is significant ($P < 0.001$); the most resistant plants were found in southern Britain, which regularly experiences concentrations above the UN-ECE critical levels for crops (Nebel & Fuhrer, 1994). Less resistant populations were detected in the less sunny, cooler regions of northern England and Scotland.

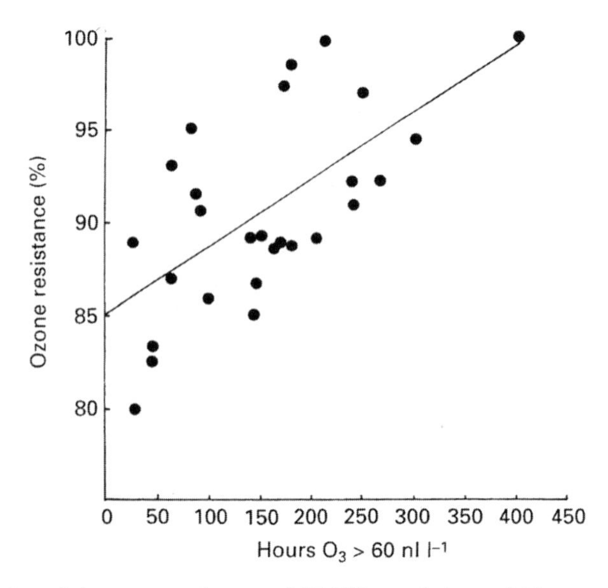

Fig. 10.1 Regression of the ozone resistance of 27 UK populations of *Plantago major* against an ozone exposure index for each collection site, the number of times the hourly mean O_3 concentrations exceeded 60 nl l^{-1} in 1989, the year before the seeds were collected. $r = 0.631$, $P < 0.001$. (Reiling & Davison, 1992b. Reproduced with permission of *New Phytologist*)

In a companion study, ozone resistance in *Plantago* populations from different parts of Europe subject to different levels of ozone were investigated (Lyons, Barnes & Davison, 1997). This experiment, together with the earlier UK survey, supported the view that current ambient ozone levels in the UK and other parts of Europe were high enough to promote the evolution of resistance. But there are some unanswered questions.

In order to understand the microevolutionary significance of these results, other investigations were carried out. Three populations from the Yorkshire/Derbyshire area were sampled and investigated from 1985 to 1991. In this period, ozone pollution was changing. In the mid 1980s, concentrations were relatively low at these sites, but in 1989 and 1990 pollution became more severe, 'the number of hours when concentrations exceeded 60 nl l^{-1} was 3–4 times greater than in the proceeding two years' (Davison & Reiling, 1995, 337).

The ozone resistance in the samples taken at different dates from the three populations was examined and in two of the three populations there was a statistically significant increase. Did these changes reflect selection within populations or were these results a reflection of the arrival of new more resistant plants arriving from elsewhere?

In order to judge the significance of these findings a number of additional studies were carried out. Studies employing molecular markers (RAPD and inter-SSR markers) revealed that 'individuals from later years represented a subset of the genetic variation present at that location in earlier years'. This supports the view that the change in ozone resistance was not due to the migration of more resistant genotypes, and 'suggests that the change was probably the result of selection of genotypes already present in local populations (selection *in situ*)' (Wolff, Morgan-Richardson & Davison, 2000, 501).

A critical question remained: was resistance to ozone in *Plantago* a heritable trait that could be selected in the field? Studying material from Lullington Heath (resistant) and a sensitive population (Bush), artificial selection was applied to each population, attempting to select a more resistant and more sensitive line (Whitfield, Davison & Ashenden, 1997). Two statistically significant responses to the selection occurred. Thus, it proved possible to select a significantly more resistant line from the Bush population and a line with decreased resistance in the Lullington Heath population. These results demonstrate that ozone resistance is heritable.

Resistance to heavy metals

Very important insights into microevolution have come from detailed studies of heavy metal tolerance in plants. In the history of science there are many cases where a crucial early investigation in a new field was not fully appreciated. This is true of studies of heavy metal tolerance. Initial experiments were carried out by Prat (1934), who showed that a population of *Melandrium silvestre* from a copper mine grew better on an artificially contaminated soil than plants from an uncontaminated site. He suggested that mine populations had evolved tolerant races. The phenomenon was 'rediscovered' by Bradshaw in the 1950s, when he investigated metal tolerance in populations of *Agrostis capillaris* (*A. tenuis*) growing on mine spoil (see Bradshaw, 1952, 1976).

It is very interesting to consider the context of the early studies of heavy metal tolerance by Bradshaw and co-workers. Firstly, in the 1950s, the major aim of many plant ecologists was to try to understand patterns and processes in natural vegetation. It was unusual, therefore, to embark on the close study of the vegetation of mines, abandoned smelters and active and derelict industrial areas. However, for the study of microevolution such areas have major advantages. Metal-contaminated mine sites are often extremely contaminated severe environments in which only a few species can survive. Often mine debris is acidic, with very low organic matter, and plant-available heavy metal toxicity is at a maximum. Another advantage is that heavy metal contamination is long lasting, even in derelict mine sites, and many of the selected sites have sharply defined boundaries between contaminated and non-contaminated areas. It is possible, therefore, to examine the critical microevolutionary processes of gene flow and selection etc. Also, as many of the study sites were in areas comparatively recently developed, the detailed history of mines, smelters and industrial sites can be determined, making it possible to estimate the speed of evolutionary change. In pursuing different lines of research on microevolution, hundreds of research papers have been produced, together with many important reviews of the subject (Antonovics *et al.*, 1971; Bradshaw & McNeilly, 1981; Baker, 1987; Shaw, 1990; Macnair, 1993). The aim here is to explore the concept of 'winners and losers' in heavy metal contaminated sites.

Naturally occurring areas with high heavy metal content

Firstly, it is important to stress that heavy metals occur naturally in the bedrock of certain areas or in surface materials transported to other regions. Inputs of heavy metals into ecosystems arise naturally in volcanic activity, and through dust and leaching of heavy metals from rocks into soils and water bodies. Metal-enriched areas often occur as islands with sharp gradients in edaphic characteristics at their boundaries. In the longer term, heavy metal outcrops are often colonised by endemic species or ecotypes as part of unique vegetation types that are generally of low biodiversity. Some of these plants hyperaccumulate heavy metals, and characteristic site-specific endemic variants and floras have been described from four different rock types.

A. Ultramafic (serpentine) floras found in many parts of the world, growing on soils rich in chromium and nickel that are deficient in molybdenum, nitrogen and phosphorus (Brooks, 1998).
B. Areas of copper/lead/zinc sulphide mineralisation found in Europe, USA and Zaire.
C. Selenium-rich sites in the USA and Australia.
D. Areas rich in copper and cobalt, found for example in Zaire (Brooks & Malaisse, 1985).

Molecular studies are beginning to reveal the evolutionary history of natural colonisation of extreme soil types. For example, studies of cpDNA variation in *Arabidopsis halleri* in Central Europe revealed that heavy metal tolerant variants have evolved independently several times from non-tolerant variants (Pauwels *et al.*, 2005).

In addition to these naturally occurring sites, human mining activities have resulted in an increasing, sometimes chronic, contamination of ecosystems by toxic metals from several sources. Often such contamination is a cumulative process, occurring in areas with no a-priori experience of metal pollution, providing opportunities for evolution *de novo*. The levels of heavy metals in such areas, while still presenting a significant stress to plants, may be somewhat lower than those found in soils with a 'natural' burden of heavy metals.

Sources of anthropogenic heavy metal pollution

1. Burning of forests and fossil fuels contributes to atmospheric pollution by heavy metals.
2. Mining, smelting, refining, storing and working of metals produces dust and airborne contamination, and leaves behind a legacy of contaminated mine tailings, slag heaps, toxic drainage from mines and various industrial wastes. As a result of modern open-cast mining techniques, some of the highly specialised endemic floras found on natural outcrops, including hyperaccumulators, are threatened with extinction, e.g. the copper/cobalt flora of Shaba province in south-eastern Zaire (Brooks & Malaisse, 1985). Ores, refined and scrap metals and finished metal products are shipped across the world leading to contamination of sites at a distance from the original mining workings.
3. The beginnings of metallurgy reach back to about 8000 BC, but it is often difficult to date the earliest mines, as metal deposits have often been reworked many times (Muhly, 1997). Archaeological evidence suggests that the first metal used in antiquity in the Near East was copper. Deposits of native metal occur in some places and, as early as *c.* 8000 BC in south-eastern Anatolia, copper was hammered into beads, hooks and pins. The finding of copper slag at ÇatalHöyük in Central Anatolia, dated *c.* 6000 BC, reveals that smelting was practised. A lead bracelet of

the early seventh millennium found in northern Mesopotamia could only have been made from smelted ore. Later, the preparation of copper/tin alloys, to produce bronze, came into common use *c.* 3000 BC. Evidence from archaeological investigations reveals the expansion of prehistoric copper metallurgy into the Aegean, the Balkans and south-eastern Europe around the second half of the fifth millennium BC. The distribution of hyperaccumulators has provided evidence of the location of: (a) long-abandoned Roman and pre-Roman mines in Germany, and in England (e.g. Derbyshire and the Mendip Hills); (b) pre-colonial copper workings in Zaire; and (c) ancient copper mines in China (see Brooks, 1998).

4. Heavy metal contamination of ecosystems may occur through atmospheric and biotic inputs from a multitude of sources, including human uses of metals and their compounds (Friedland, 1990; Seaward & Richardson, 1990). For instance, galvanised zinc is used to coat steel products to prevent corrosion. Copper compounds are employed as pesticides in hop fields, vineyards and apple orchards (Seaward & Richardson, 1990; Ernst, 1998), and contamination becomes evident with change of crop.
5. Heavy metals may be present in significant quantities in some commercial agricultural products added to soils, e.g. fertilisers, crushed limestone, slag, sewage and municipal effluents. Irrigation with contaminated water can also lead to contamination of soils with heavy metals, e.g. agricultural soils in Japan irrigated with copper-mine drainage water. Coal-fired power stations produce fly ash that can be contaminated with heavy metals. Burning domestic and industrial wastes also leads to air pollution by heavy metals, and residual ash contains metal compounds (Seaward & Richardson, 1990).
6. Contamination also occurs as metal-containing compounds and commercial products are dumped in landfill and other sites. Even the attempt to find a suitable use in 'recycling' contaminated material can produce new centres of pollution; for instance, in the use of smelter waste as road building material in the Netherlands (Dueck, Endedijk & Klein-Ikkink, 1987).

Heavy metals in soils

What happens when heavy metals enter soils depends upon the interaction of many factors in soil chemistry (see Friedland, 1990). In the present context, it is important to stress a number of critical issues. Heavy metals are more soluble in acidic than in basic soils, where they may be present as insoluble compounds that are 'less available' to plant roots. Also, toxic metals may be adsorbed on mineral fractions of the soils. In addition, heavy metals have a strong affinity with organic materials, complexes being formed through the processes of chelation, complexation and adsorption (Friedland, 1990). Thus, in soils with well-defined surface organic horizons, trace metals may concentrate in the upper layers of the soil.

Given the complexity of soil chemistry, it is important to be able to estimate not only the total level of contamination by different metals but also the fraction of the heavy metal burden that is 'available' to plant roots. In estimating different fractions of the heavy metal content, a number of procedures have been devised. The following extractants, employed sequentially, are believed to provide estimates of heavy metals in different fractions: H_2O – water-soluble; KNO_3 – exchangeable; $Na_4P_2O_7$ – organically bound; HNO_3 – sulphides; $HNO_3 + H_2O_2$ – residual (see Miller, McFee & Kelly, 1983).

Table 10.3 *Chemical analyses of some mine spoils and normal soils from the British Isles*

Site	pH	Pb	Zn	Cu	N	P	K
Minera, Clwyd	7.3	14,000	34,000	625	164	97	1,960
Y Fan, Powys	4.5	42,400	6,700	376	122	245	3,400
Parys Mountain, Gwynedd	3.6	327	124	2,060	88	141	2,670
Goginan, Dyfed	5.4	16,800	2,700	134	120	103	458
Ecton, Staffordshire	7.2	29,900	20,200	15,400	110	116	825
Snailbeach, Salop	7.2	20,900	20,500	25	100	110	1,780
Darley, Derbyshire	7.3	6,000	4,600	80	32	93	1,550
Normal soils	4.5–6.5	2–200	10–300	100–200	200–2,000	200–3,000	500–3,500

Values are total content of air dried soils, as $\mu g\ g^{-1}$
From Bradshaw & McNeilly (1981). Reproduced with permission of Cambridge University Press

Definitions

According to Baker (1987), heavy metals operate as stress factors in the environment, causing a range of physiological reactions including reduction of vigour and in the extreme case plants are unable to survive and grow. He carefully defines important terms. *Sensitivity* 'describes the effects of a stress which result in injury or death of the plant'. *Resistance* 'refers to the reaction of a plant to heavy metal stress in such a way that it can survive and reproduce and in doing so contribute to the next generation'. Resistance to heavy metals can be achieved by either of two strategies: *avoidance*, by which a plant is protected externally from the influences of the stress, and *tolerance*, by which a plant survives the effects of internal stresses. Tolerance is 'conferred by the possession of specific physiological mechanisms which collectively enable it to function normally even in the presence of high concentrations of potentially toxic elements'. Investigations confirm that the presence of the physiological mechanisms conferring tolerance in particular genotypes are heritable attributes not shared by sensitive genotypes.

Testing for metal tolerance

Around the world, surveys have revealed that at newly established mining and industrial sites, where pollution of normal soils by heavy metals becomes chronic, most of the existing vegetation is damaged and dies (Bradshaw & McNeilly, 1981). Some plants of a limited range of species may survive or invade the contaminated area from adjacent populations. Table 10.3 illustrates the level of contamination found at a range of contaminated and uncontaminated sites.

The evolution of tolerance can be reproduced on a small scale with sowings of seed onto contaminated soil. For most species, seedlings die but in certain species, although mortality is still great, a few may survive and grow to maturity, revealing clear differences in Darwinian fitness. A tolerance test, carried out in the manner outlined above, shows that these individuals have greater tolerance to the metal(s) in the test soil than the base population. Bradshaw, McNeilly and Putwain (1990) see such screening experiments as providing 'a very simple evolution "do-it-your self" kit'.

While some studies of metal tolerance have been carried out using mine and normal soils, contaminated soils frequently contain several different toxic metals. In order to understand patterns of tolerance/susceptibility to the suite of metals encountered by plants at particular sites, the responses of plants to each metal in turn must be examined. These responses have been usually assessed by calculating the Tolerance Index in a hydroponic root elongation test devised by Wilkins (1957, 1960, 1978), in which root growth is measured in a solution containing heavy metal and in an uncontaminated control.

Tolerance index = growth in metal solution/growth in control solution.

The use of simple test solutions is understandable, as the use of hydroponic solutions containing the full range of plant nutrients leads to the precipitation of heavy metals as phosphates, sulphates etc. Thus, in many investigations a single concentration of heavy metal ion as nitrate or sulphate has been employed in a solution of calcium nitrate at a concentration of 0.5 or 1 g per litre. The control solution contains calcium nitrate alone at the same concentration. However, realising that such test solutions are nutritionally deficient and, for instance, could affect the growth of seedlings, some researchers have chosen to study the effect of heavy metals using a more or less complete nutrient solution with the omission of phosphates and sulphates to prevent precipitation (see Baker, 1987). Other have persisted with the use of simple test solutions, but have examined growth in a range of concentrations, the results of which have been examined by more sophisticated techniques such as regression or probit analyses.

Another aspect of the test must be examined. In calculating tolerance indices some researches have used the parallel method, where there is simultaneous measurement of growth of replicate individuals in metal and control solutions. Such an approach requires the cloning of the plants to be tested. Others have employed a sequential method with less replication, growth in control solution being measured first, and then elongation in a metal-containing solution. However, such an approach makes the assumption that growth rate is constant, but this may not be so.

A number of other tests have been devised for metal tolerance. For instance, pollen tube growth in test solutions has been measured. In *Silene dioica*, *S. alba* and *Mimulus guttatus* there was good agreement between indices of zinc and copper tolerance calculated from pollen and root growth indices using the traditional methods (Searcy & Mulcahy, 1985a, b, c).

Genetics of tolerance

Heritable differences in tolerance have been detected in studies of a number of species (Macnair, 2000). These have been investigated using the index of tolerance models and also by studying growth in simple solutions of heavy metal salts. Many early studies were of grasses. While these plants may be vegetatively propagated and therefore had the advantage of providing adequate replication in experiments, each flower produces only one grain in a cross. In contrast, other species – e.g. *Silene vulgaris* and *Mimulus* spp. – produce more seed per crossed flower providing larger families with which to study genetic segregation.

In crosses of tolerant and sensitive plants, continuous distributions in tolerance were often detected and in early reviews of heavy metal tolerance these were interpreted as indicating polygenic inheritance. More recently, a number of cases have been published in which one or a few major genes are involved in tolerance for arsenic, cadmium, copper and manganese, and there are instances where minor genes act as modifiers (Macnair, 1993, 547).

The soils around mines and other sites are often contaminated with more than one toxic metal. Figure 10.2 shows the results of investigations of metal presence and tolerance in a number of populations of *Silene vulgaris*. Interpreting these results Macnair (1993, 550) concludes:

> While populations generally show tolerance to metals present in excess, they also exhibit a number of cases where a high level of tolerance to a particular metal could not be explained by elevated levels of that metal. For instance, the Imsbach population has a very low level of Zn in the soil, but has high Zn tolerance; all the mine populations seem to show some Cd tolerance, even though the concentrations of this metal are low; many populations also show an increase in Co and Ni tolerance. This suggests that selection for Cu tolerance also increases tolerance of Zn, Cd, Co and Ni, while selection for Zn, Cd and Pb tolerance also leads to tolerance to Co and Ni.

Much more research is needed in these areas to determine whether independent mechanisms of tolerance occur for the various metals, or whether tolerance to one metal naturally confers tolerance to another or others.

Origin of tolerant variants

Firstly, metal tolerant variants may have long existed on naturally occurring outcrops of rocks, and it is the descendents of these variants that are found colonising mine spoil heaps etc. However, this explanation is not applicable to secondarily contaminated sites that had normal soils in the past.

Secondly, a number of researchers have suggested that metal-resistant plants may be dispersed by the activities of miners and/or by the transport of ore (see Ernst, 1990). For example, the New World copper moss *Scopelophila cataractae* in South Wales was discovered on slag heaps. This species is restricted to substrates with higher than average copper content (Corley & Perry, 1985).

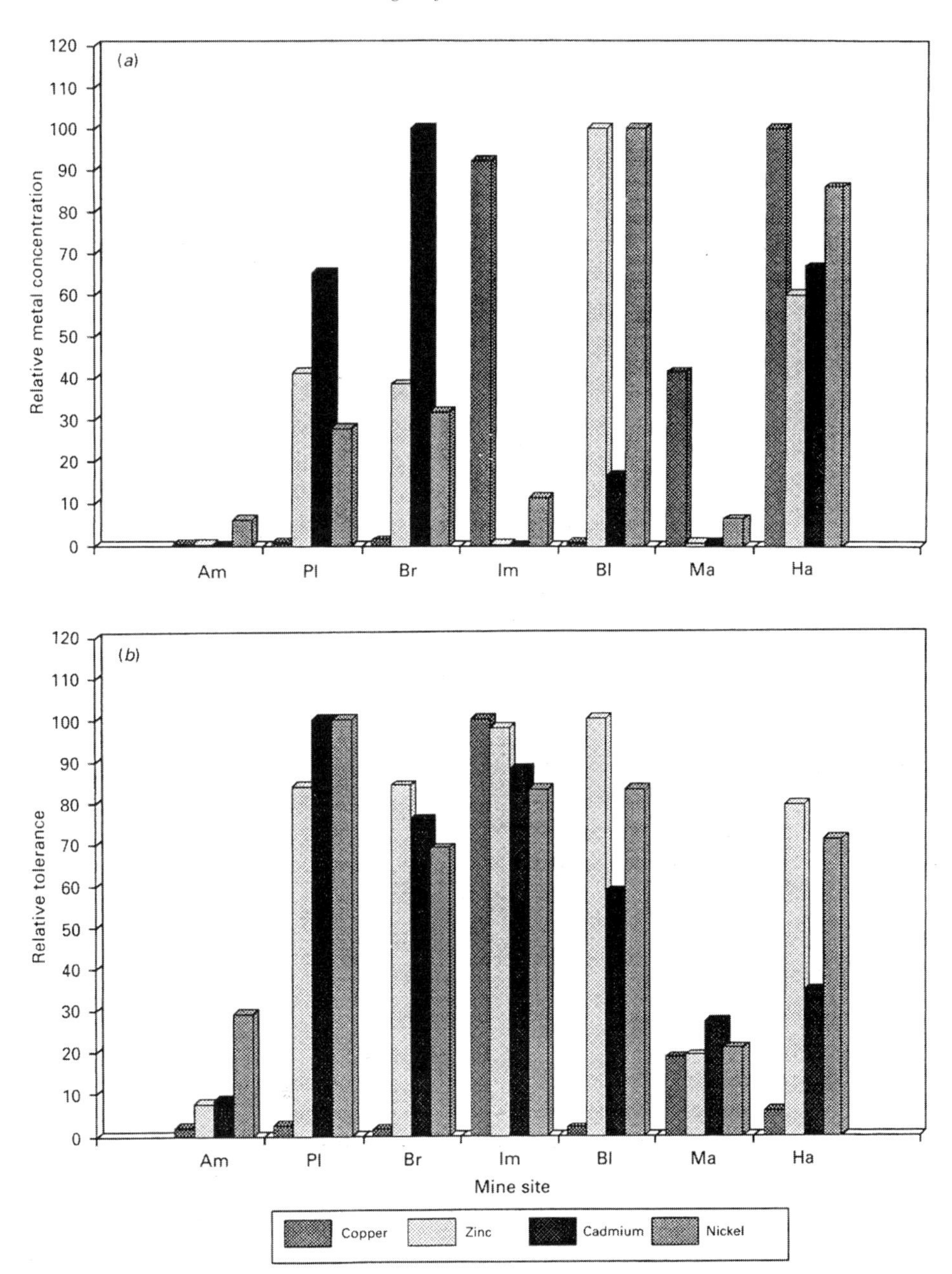

Fig. 10.2 Metal tolerance in *Silene vulgaris*: (a) relative available metal ion concentrations at study sites; (b) relative metal tolerance for six European populations of *Silene*. In both graphs, and for each metal, the population with the highest value has been given the value of 100; all other populations have been expressed relative to this. Am, Amsterdam (normal control); Br, Breinig; I, Imsbach; Bl, Blankenrode; Ma, Marsberg; Ha, Harlingrode. (From Macnair, 1993. Diagram from data in Schat & ten Bookum, 1992. Reproduced by permission of *New Phytologist* ©)

Thirdly, heavy metal tolerant variants may be present in 'normal' populations. A number of investigations have explored and confirmed this possibility (Gartside & McNeilly, 1974; Walley *et al.,* 1974; Wu *et al.*, 1975). For example, 2000 seeds of each of a number of populations of *Agrostis stolonifera* from mine sites not contaminated by copper were sown on copper contaminated soil (Wu *et al.*, 1975). Although most plants died, there were five survivors of moderate copper tolerance. Gartside and McNeilly (1974) screened populations of eight species and found small numbers of tolerant plants in two species that have been found on copper mines.

As heavy metal tolerant populations of the same species often occur at widely separated sites it is possible that each has an independent origin. Support for this model comes from studies of the grass *Deschampsia cespitosa* that has colonised heavy metal rich soils in the mining regions of Sudbury and Cobalt in Canada (Bush & Barrett, 1993). Samples from these areas, together with others from uncontaminated sites, were examined in studies of isozyme variation (nine enzyme systems – 19 putative isozyme loci). Bush and Barrett concluded: 'the results corroborated the prediction that colonization of contaminated habitats reduces levels of genetic variability, particularly where populations are recently established'. In addition, 'Cobalt and Sudbury populations were clearly differentiated by unique alleles at a number of enzyme systems, providing evidence for the independent origin of metal-tolerant populations in the two mining regions.'

Further evidence is provided by the investigations of Schat, Vooijs and Kuiper (1996), who studied the genetics of heavy metal tolerance in *Silene vulgaris* in five populations originating from metal-rich sites in Germany and Ireland, together with a non-tolerant population from Amsterdam. Evidence from crossing experiments suggested that there were two distinct major gene loci for zinc tolerance amongst the five tolerant populations. Also, as the German and Irish populations had common major genes for tolerance, it was concluded that such widely geographically separated populations must have resulted 'from independent parallel evolution in local non-tolerant ancestral populations'.

Restraints on the evolution of tolerance

Investigations have revealed that species vary in their basic susceptibility to metal contamination (Baker & Proctor, 1990). Some species would appear to be constitutively tolerant, while others are generally more sensitive. Such a conclusion may be drawn from the fact: 'at least in solution culture, the tolerance index of normal populations of different species at low levels of metal may differ considerably' (Macnair, 1997, 6). However, in judging the responses of species to heavy metals, it is always important to establish whether phytotoxic levels are present in any particular ecosystem (Macnair, 1997, 6). The development of metal tolerance will depend upon two factors – the degree to which species are constitutively tolerant and the frequency of tolerant individuals in normal populations.

In addition, it is important to stress that tolerance genes may not always be at low frequency. In studies of populations of *Holcus lanatus* in mining areas of south-west Britain, Meharg, Cumbes and Macnair (1993) discovered high levels of arsenic with tolerant plants at high frequency (92–100%). Furthermore, the survey also revealed that populations on non-contaminated soils might have a high frequency of arsenic tolerance (average 45.3%). Extending the geographic area revealed that arsenic tolerance was also common in populations from southern Britain. It is not yet clear why this should be so.

Baker *et al.* (1986) have studied another phenomenon of importance: there is evidence that metal tolerance may be inducible. This conclusion was drawn from investigations of cadmium tolerance in *Holcus lanatus*, *Agrostis capillaris*, *Deschampsia cespitosa* and *Festuca rubra* from aerially polluted and mine sites, and involved the investigation of whether metal tolerance was a stable character. Could tolerance be changed by growing tolerant plants on normal soils? Investigations revealed that there was a 13% reduction in mean tolerance in clonal plants grown for a period on normal soil in a glasshouse. *Holcus lanatus* proved capable of the largest 'phenotypic adjustment'. In addition, it was discovered that a degree of cadmium tolerance could be partially induced in *H. lanatus* from uncontaminated sites. While the physiological basis of these phenotypic adjustments is unclear, if they occur in the wild, they may be very important in the early colonisation of contaminated sites.

Gene flow and selection

The fact that islands of contaminated mine workings occur within pastureland has provided an opportunity to examine the microevolutionary processes of selection and gene flow by pollen. In the classic studies of *Agrostis capillaris* (*tenuis*) at the Drws-y-Coed copper mine, McNeilly (1968) examined variation in adult plants and their progenies in two transects from the mine and out into the adjacent pasture (Fig. 10.3). One transect extended down the narrow glacial valley in the direction of the prevailing wind, while the other transect was set at right angles to it. Adult plants from the mine had a high index of tolerance: in contrast the tolerance of pasture plants was much lower. With regard to the progenies, as a result of gene flow from plants in the adjacent pasture, progenies grown from seed collected on the mine had a wider tolerance than the adults, This difference is explicable in terms of the high selection for tolerance operating on the mine.

Studies of the tolerance in progenies from pasture plants revealed that they had a higher tolerance than the adults on which they developed, revealing that genes for copper tolerance were distributed by windborne pollen from the mine, fertilising seed parents growing in the adjacent pasture. Off the mine, adult plants were less tolerant than their progenies. Thus, there was evidence of selection against metal tolerance in the pasture. These patterns are explicable in terms of wide gene flow at reproduction, with a pronounced effect of the prevailing wind, together with strong disruptive selection for increased tolerance on the mine and against tolerance in the pasture. The driver for disruptive selection becomes clear when it was discovered that there are 'costs' associated with tolerance. Because they are able to grow on copper-contaminated soils, metal-tolerant variants have higher fitness than non-tolerant plants on highly contaminated sites; indeed non-tolerant plants may not survive. However, metal-tolerant races are slower growing and less competitive than non-tolerant plants on normal soils (Cook, LeFèbvre & McNeilly, 1972;

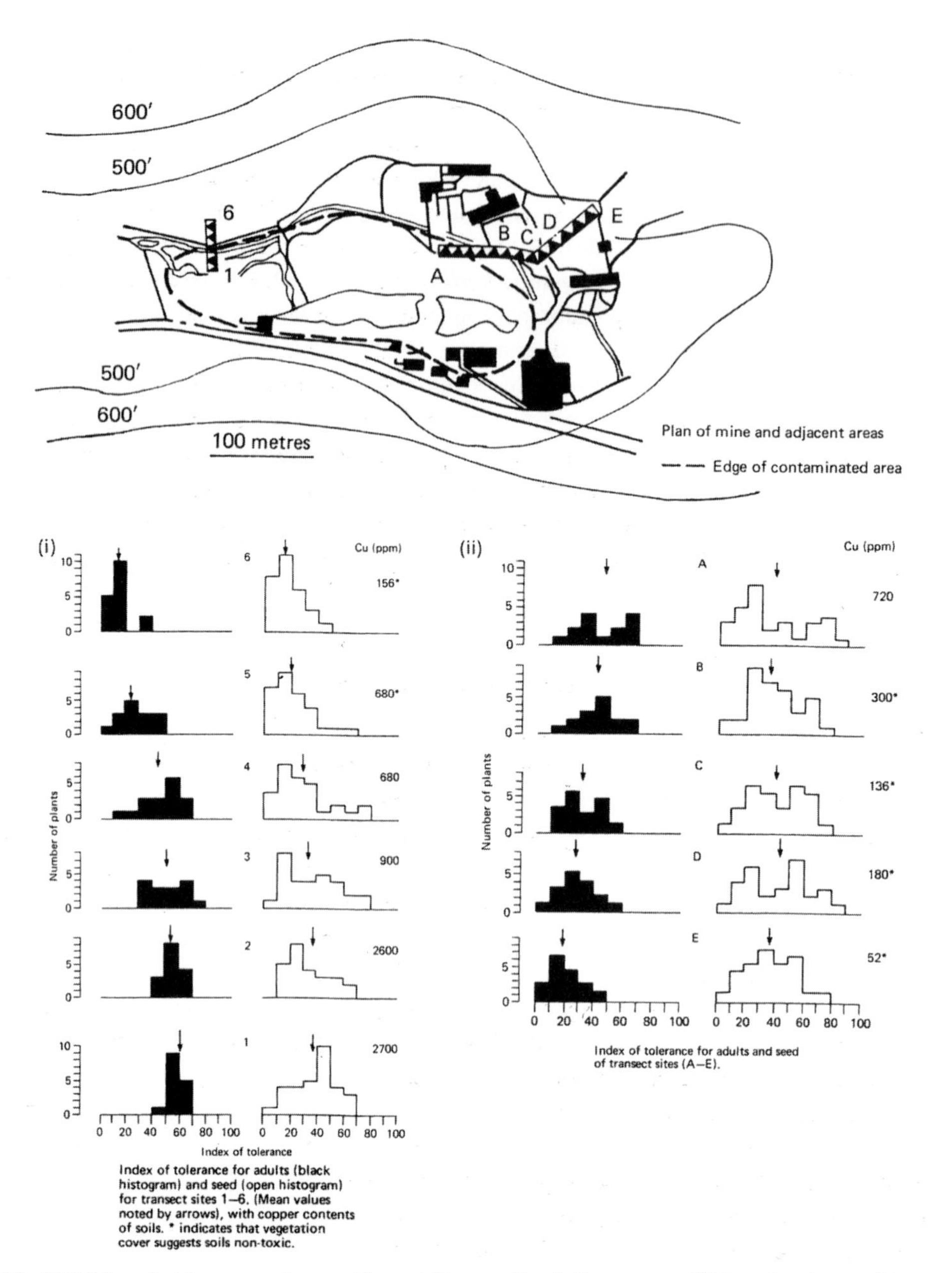

Fig 10.3 Map of old copper mine workings at Drws-y-Coed, Caernarvon, Wales, showing positions of transects sampled by McNeilly (1968). In his studies of *Agrostis capillaris* (*A. tenuis*), using a water culture technique, an index of copper tolerance for adult plants and seed produced by different adults was determined for material from two transects: (i) sites 1–6; and (ii) sites A–E. Adults from the mine proved to be more copper tolerant than plants from non-contaminated pasture adjacent to the mine. Studies of the seedlings, produced from wild collected seed, revealed a wider spectrum of variation than the adult plants. This pattern was particularly clear in transect A–E, where evidence of considerable gene flow of copper-tolerance genes downwind of the mine was discovered in the progeny of copper-sensitive plants. This experiment is consistent with the view that strong natural selection occurs on the variable products of sexual reproduction. The only seedlings to survive to adulthood are likely to be copper tolerant on the contaminated areas and non-tolerant variants on pasture areas. The latter have been shown to be better competitors in uncontaminated soils than copper-tolerant plants. (From *Heredity*, McNeilly, 1968. Reproduced with author's permission)

Hickey & McNeilly, 1975). Thus, tolerant plants exhibit reduced fitness when growing on normal soils. 'The "cost" of metal tolerance may be viewed as an energy expenditure for operation of the mechanisms of tolerance, and may result in slower growth rates and lower biomass production by comparison with their non-tolerant counterparts' (Wu, 1990, 277). Further research is necessary to understand how these costs arise in physiological terms. It is beyond the scope of this book to provide a review of the physiological basis of metal tolerance, including the role played by mycorrhizal fungi. The proposed mechanisms of tolerance include complexation and compartmentalisation of metals at the cellular and tissue level, and delayed translocation from roots to shoots may be important (Ernst, 1998; Broadley *et al.*, 2007). Many species colonising metal-contaminated soils have mycorrhizal associations in their roots. The role that such associations play in metal tolerance is receiving increasing attention (see Meharg, 2003; Gohre & Paszkowski, 2006). A further area of interest relating to the physiology of tolerance concerns the capacity of some metal-tolerant variants or species to hyperaccumulate very high levels of metals from the soils. It has been demonstrated in trials that such plants can be used to remove heavy metals from industrial and urban contaminated soils (Schnoor *et al.*, 1995; Brooks, 1998). Extending these studies, there is a growing interest in employing plants to 'degrade, assimilate or detoxify' a wide range of chemical pollutants including metals, hydrocarbons, pesticides, solvents and explosives through phytoremediation of contaminated areas (Susaria, Medina & McCutcheon, 2002).

Speed of development of heavy metal tolerance: lines of evidence

A number of investigations have examined the speed with which microevolutionary change occurs.

Galvanised netting

To exclude rabbits from the permanent quadrats in his long-term experiments on Lakenheath Warren, Suffolk, Dr A. S. Watt, University of Cambridge, erected galvanised (zinc coated) netting in 1936 (renewed 1958). Realising that zinc is leached by rainwater from galvanised structures, Snaydon discovered that the area under the fencing had toxic levels of zinc and populations of *Festuca ovina* and *Agrostis canina* growing beneath the fence were significantly more tolerant of zinc than neighbouring populations on uncontaminated sites. As there are no natural sources of zinc contamination in the vicinity and the localised source of zinc was introduced to the site at a known date, it is clear that zinc-tolerant populations have developed *de novo* over a period of *c*. 25 years (for details see Bradshaw, McNeilly & Gregory, 1965, 334).

Copper refinery

Around 1900 a new copper refinery was opened at Prescot, near Liverpool, resulting in aerial copper pollution in a previously uncontaminated area. Few species survived in the most contaminated areas around the factory and attempts to establish lawns were not entirely successful. Detailed studies of the site have provided insights into the build-up over 70 years of copper contamination and the rapid evolution of copper tolerance in *A. stolonifera*, a species found in three lawns of different age.

About 15 years before the study, the oldest lawn was constructed by laying turf (grown elsewhere) on contaminated soil. This new sward was subject to aerial copper pollution from the refining processes and, when the investigation was carried out, contamination of the new turf had built up to *c*. 2,600 ppm in 14 years. The mean copper tolerance of plants of *A. stolonifera* in the turf was 42%, but there was a range of values indicating a degree of variability in the population. It is likely that the maximum level of copper tolerance has yet to develop in the population, as typical values for copper tolerance in mine plants is of the order of 70%, with a figure of *c*. 5% for copper-sensitive plants in uncontaminated pasture populations near mines. The closed sward of the turf of this old lawn was made up of only two species – *A. tenuis* and scattered plants of *A. stolonifera*. As the lawn was mown regularly it seems likely that reproduction by seed would be greatly curtailed or prevented, any recruitment of 'new' individuals into the sward would come by gene flow of seeds from adjacent populations.

A second lawn – 7 years old – had developed following repeated sowings of commercial grass seed onto pre-existing contaminated soil with 4,800 ppm copper. Only *A. stolonifera* was present in the sward, which had large amounts of bare ground. Mean copper tolerance of plants in this lawn was 32%. It is postulated that from the base population of grass sown, only copper-tolerant variants were able to survive the seedling stage on the toxic soil. Fifteen clones were studied in further detail and it was discovered that there was a statistically significant correlation between of size of the clone in the lawn and its index of copper tolerance: the larger the clone, the more tolerant the plant. This finding suggests that selection was a two-stage process, with selection acting not only at the seedling but also at the adult stage.

A population of *A. stolonifera* that was growing in the rough boundary grassland near the refinery was also examined. Here, the sward, which consisted of a variety of grasses and herbs, had been contaminated for a period of 4 years following the opening of a new refinery. In this period, copper contamination rose to 1,900 ppm. Mean copper tolerance in the *A. stolonifera* was 21%. Copper tolerance was in the process of evolving in this population, but had yet to reach the level of that in the older lawns.

The evolution of copper tolerance at Prescot occurred within a 70-year period. Studies of vegetation around a new zinc and cadmium smelter at Datteln, in Germany, has revealed an even more rapid evolution, where zinc tolerance was detected in *A. capillaris* exposed to intense pollution for only a 5-year period (Ernst, 1990 and references cited therein).

Electricity pylons

Steel electricity pylons that criss-cross modern landscapes are zinc coated (galvanised) to prevent corrosion. Rainwater leaches the zinc from the structures, and there is a build-up of zinc in the soil. There are no deleterious effects on the plants growing under pylons on soils of high pH, but, in contrast, where geology, soil structure and chemistry lead to acid

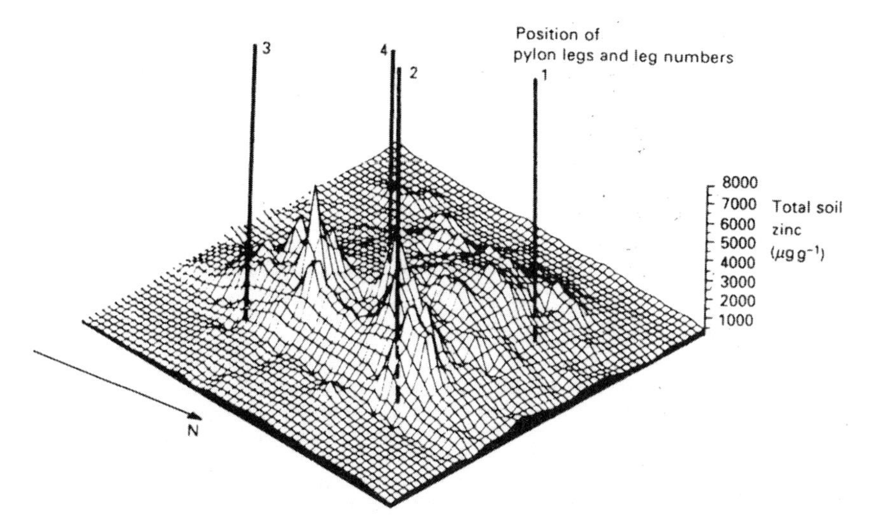

Fig. 10.4 Total soil zinc concentration in 399 soil samples from under an electricity pylon in North Wales. Data smoothed by extending sampling scale ×2.5; intermediate values obtained by averaging. (Reproduced by permission of *New Phytologist* © from Al-Hiyaly, McNeilly & Bradshaw, 1988)

soils, zinc contamination damages existing vegetation (Harris, 1946). Does the continual build-up of zinc contamination present an increasingly extreme habitat imposing selection pressures for the evolution of zinc tolerance? This possibility has been examined in studies of *Agrostis capillaris* from a series of nine pylons on acid soils in North Wales. Pylons represent a unique series of replicated islands of contamination in the midst of normal soils and vegetation (Al-Hiyaly *et al.*, 1988).

Firstly, it was confirmed that the area beneath the pylons was highly contaminated with zinc. Patterns of contamination were high near the pylon legs, but, in a thorough study of a single pylon, the distribution of contamination was skewed indicating an effect of the prevailing wind and downward slope (Fig. 10.4).

Acetic acid extractable zinc was determined to provide some measure of the plant availability of the toxic metal, and high values were obtained for the soils beneath the pylons. By growing seedlings of a normal commercial variety of *A. capillaris* on soils from under the pylons, the toxicity of the soil was confirmed. There was a marked reduction in dry weight of 12-week old plants, relative to growth on a control soil, sampled beyond the limits of the pylon.

Zinc tolerance of plants of *A. capillaris* was determined in samples collected under the pylon, and also for control plants collected at least 50 m away. Tolerance was examined by measuring root growth in simple solutions of the heavy metal plus control with no metal. It was discovered that some of the plants growing under the pylons had zinc tolerance equivalent to that found in plants on zinc mines. Others had lower values or differed little from the zinc tolerance of control plants collected at least 50 m away from the pylons.

The evolution of zinc tolerance in *A. capillaris* under pylons can be seen as a cumulative process. Water-borne zinc from the superstructure of the pylon accumulates in the soil over time, but does not yield a uniform pattern. Evidence suggests that in less than 30 years the zinc contamination has imposed selection pressures sufficient to initiate the evolution of zinc tolerance in populations. However, relative to mines, the evolution of zinc tolerance is at an early stage. The evolutionary outcome at the series of effectively replicated situations did not yield identical end points. The spectrum of variation was different at each pylon. The presence of individuals with a variety of tolerance is not unexpected, as there was a variable level of zinc contamination in the soils beneath each pylon.

Other species growing around pylons

In a further study metal tolerance was examined in four other species found growing under the pylons – *A. stolonifera*, *Anthoxanthum odoratum*, *Deschampsia cespitosa* and *Festuca ovina*. Zinc tolerance was assessed in plants collected from under pylons and in controls from adjacent uncontaminated soils. Some degree of zinc tolerance was found for all four species.

Considering the patterns of variation in the species studied, there were several important findings.

A. The grasslands surrounding the pylon sites contained many species, but only a small subset grew under the pylons.
B. Despite the fact that *Anthoxanthum* and *Festuca ovina* were common constituents of the vegetation surrounding virtually all the pylons, zinc-tolerant individuals of these species only occurred at some, but not all the study sites.
C. Where *D. cespitosa* was absent from beneath pylons, this seemed to be due to the absence of the species in the surrounding vegetation.
D. *A. capillaris* was abundant in nearly all the grasslands surrounding the pylons studied and was present beneath most, but not all the pylons.

Al-Hiyaly *et al.* (1988, 1990, 1993) conclude from their investigations that the genes necessary for the evolution of tolerance were not universally present within the species in the surrounding vegetation. On the basis of these and other studies, Bradshaw (1984a, b) formulated the important concept of genostasis. Evidence suggests that natural selection operates in response to metal pollution, but whether tolerant populations arise depends upon whether the appropriate genetic variation is available in the populations affected. If genes for metal tolerance are not present in populations of the species in the area surrounding the pylons, then 'further evolution is held up by lack of appropriate variability. There will be genetically determined condition of stability, which can be called *genostasis*. Metal tolerance provides excellent evidence for this. There is no sign that the species excluded from metal-contaminated sites are in the process of evolving tolerance. They appear to be stuck as they are. Genostasis is the only logical explanation for all situations where no evolution is occurring' (Bradshaw *et al.*, 1990, 531).

The hypothesis of genostasis was examined in a further set of experiments on *A. capillaris*. Five pylon sites were selected: under four of the pylons (group A) some degree of zinc tolerance was discovered. In contrast, there was no zinc tolerance under a fifth pylon (ZK-180). Samples of seeds were collected from uncontaminated pastures from around each pylon and these were kept separate and grown and tested to see if they contained any zinc-tolerant seedlings. There was significantly higher zinc tolerance in group A than in that adjacent to pylon ZK-180. The next part of the experiment, using four of the five populations, examined whether selection and breeding from the best-rooted seedlings would result in increased zinc tolerance. At all times the material from each site was kept separate and, by isolating the stocks at flowering, crossing occurred only between plants of the same population, the so-called polycross method. Two cycles of selection and breeding were carried out. Marked increases in tolerance were detected in three of the four populations, but there was no response to selection in the fourth population ZK-180. It is concluded that the lack of zinc tolerance under the pylon ZK-180 is due to the lack of appropriate genetic variation within the population adjacent to the pylon. This is in marked contrast to the other three sites where some level of zinc tolerance revealed itself in the artificial selection and breeding experiments. The experiment provides (i) clear evidence for genetic constraints limiting evolution (Bradshaw, 1984a, b) and (ii) highlights the stochastic nature of the microevolutionary process related to randomness in the patterns of genetic variability.

Conclusions

Significant progress has been made in exploring the microevolutionary patterns and processes associated with pollution. These findings provide important insights complementing and extending those obtained in the study of grasslands, arable and forested ecosystems. There is clear evidence of rapid directional selection as pollution increases at sites of SO_2 and O_3 pollution and in areas of heavy metal contamination around mine sites, smelters, pylons etc. In the case of SO_2 pollution at Philips Park, the effect of reversal of selection pressures has been detected, as legislative steps were taken in the UK to reduce atmospheric emissions. Disruptive selection and gene flow have been explored where mine and pasture occur in close juxtaposition. Very revealing studies of replicated sites of pollution – mines and pylons – have provided clear evidence for independent evolution at different sites, in those species and populations where the appropriate local genetic variability exists. However, not all species/populations were able to respond. As contamination increased around the smelter at Prescot, fewer species were found in the lawns. In the most contaminated site only one species – represented only by metal-tolerant genotypes – was able to grow.

This chapter has concentrated on air and heavy metal pollution. For completeness, other cases of microevolutionary responses to pollution are briefly noted here. In many countries subjected to severe winter weather, a mixture of rock salt (sodium chloride) and grit is spread to de-ice the road surface. A number of ecological effects of this practice have been noted. Seasonally, species growing at the edge of roads may be very badly damaged by salt spray. Also, the vegetated road margins may be invaded by coastal species (Scott & Davison, 1985) and, in common species, salt-tolerant variants may be selected (Briggs, 1978; Kiang, 1982). Studies of roadside habitats also identified another selective force. In the post-war period, additives including lead tetra-ethyl etc. were used in petrol as antiknock agents, and exhaust fumes containing lead compounds contributed to air pollution in towns and cities and along the marginal vegetated strips along busy highways. (In recognising the toxicity of lead compounds, many regulatory authorities have banned the use of leaded petrol in many states.) In a period when the use of leaded petrol was widespread, investigations revealed that roadside populations of *Plantago lanceolata* and *Cynodon dactylon* were more lead tolerant than populations some distance from the roadway (Wu & Antonovics, 1976).

Turning to the wider implications of studies of microevolution, it is clear that the responses of plant populations to new human-induced extreme situations, beyond those that that can be accommodated by phenotypic plasticity and developmental flexibility, will depend upon whether populations contain appropriate reserves of genetic variation. Those species able to respond, by having the genetic potential to exploit the new circumstances, will be winners, while others – lacking the necessary variability – are likely to fail. On the face of it, it might be supposed that studies of plants growing below electricity pylons would offer no general principles that might underpin effective conservation. But this is not the case. As we see below, conservationists must fully understand the implications of microevolution studies if there is to be hope for endangered species in a changing world.

11

Introduced plants

While there are examples of remarkable natural long-range dispersal of plants (Ridley, 1930), human-mediated trans-global movement of a wide range of organisms has resulted in an extraordinary dispersal of alien species. A very large number of species have been, and are being, accidentally or deliberately transported vast distances to new regions across the world, well beyond the limits of natural dispersal. Some of these have proved to be winners: others have been less successful.

There are a number of important reviews of the introduced species, including invasive species (Gibbs & Meischke, 1985; Drake *et al*., 1989; Williamson, 1996; Mooney & Hobbs, 2000; Sandlund, Schei & Viken, 2001; Sax, Stachowicz & Gains, 2006). These sources provide information on a wide range of issues, including the economic damage wrought by introductions, and provide detailed information on the attempts to control invasions. The control of introduced plant species by pesticides or biological means will be discussed in some detail in Chapter 15.

Here, our main focus is on plants and considers microevolutionary issues relating to success or failure of introduced taxa. However, reference will also be made to introduced animals, fungi and micro-organisms, as these influence the native plants in many ecosystems. The following questions are considered: by what means have plant species been introduced to new territories, and how many of these have become established in functioning populations? What factors are important in determining whether or not species are successful as introductions, and what factors determine whether introductions become invasive? Is there any evidence that introduced species evolve in their new territories? Taking a wider perspective, how have ecosystems been altered and modified?

As an introduction to this important subject, it is instructive to examine a case history that fascinated Darwin, as this will help define some of the key issues to be examined in this chapter. In search of information, Darwin was in correspondence with many of the active naturalists of his day, and a number of letters reveal his intense interest in introduced plants, especially those that had become invasive. Darwin's theory of evolution by natural selection is based on the premise that checks to the natural increase in species occur. Darwin was therefore extremely intrigued by examples of naturalised species that were increasing very rapidly. Were these introduced species somehow released from the natural checks affecting native species? Also, he asked the important microevolutionary questions:

was the introduced species eaten or controlled by anything and was the increase in one species at the expense of another, as he predicted in his theory of evolution by natural selection?

Darwin's active interest in the subject is revealed in his letters about the introduction and spread of the aquatic Canadian Pondweed (*Elodea canadensis*) in the River Cam at Cambridge, UK (Preston, 2002). *Elodea* was detected in 1842 at Duns Castle, Berwickshire, and was also found, in 1847, in ponds associated with the canal system at Foxton Locks near Market Harborough, Leicestershire. Initially, it was considered to be an overlooked native species and was named *Anacharis alsinastrum* by Babington (1848), but further investigation revealed that it had been accidentally introduced from Canada. Foxton canal system is at the heart of the waterways of England, and *Elodea* spread widely through the interconnecting canal systems. Regarding the infestation of the River Cam, published records indicate that material from Foxton was grown in the Botanic Garden, Cambridge, and in 1848 the Curator of the Garden placed a piece of the plant in the Hobson's Conduit, a waterway that runs along the boundary of the garden. Fed by water from springs arising in the Gog Magog Hills, this artificial watercourse was dug to provide the town's water supply, eventually out-flowing into the River Cam. By 1851, the plant had spread from the conduit into the river in small quantities, but by 1852 plants had spread for many miles, so that *Elodea* 'blocked narrow water courses, and the mouths of docks and sluices, impeded barges travelling on the Cam through Cambridge, obstructed rowers and swimmers, interfered with fishermen and impeded drainage' (Preston, 2002). As *Elodea* is dioecious, and only female plants were detected in the early invasion of the species, all this prodigious population growth was achieved through vegetative spread (Simpson, 1984). In a letter to John Stevens Henslow, dated 26 March 1855, Darwin thanks his former teacher at Cambridge for sending a living specimen of *Anacharis* [*Elodea*], which Darwin grew and studied (Darwin, 1985). Then, on 2 April 1860, Darwin sent another letter with queries about the course of the infestation on the Cam (Darwin, 1985). My dear Henslow:

> I have since you were here found a statement by Babington that this plant [*Elodea*] is now not so common as it was at first introduction. I shd. very much like to have this confirmed, and what creatures destroy it. – *Lastly, what plants in any particular spot or pond it has nearly or quite exterminated by taking their places*? [Darwin's italics]

Henslow took his students on a field excursion to examine the situation, and wrote to Darwin to report his findings: 'I found the *Elodea* had made great progress since I saw it a year or two ago.' With regard to Darwin's final question, Henslow concluded: 'So far as mere recollection guides me . . . it seems to have diminished the quantity of *Ranunculus aquatilis* var. *fluitans* which used to abound in the stream at the part we visited' (Darwin, 1985).

Later observations suggest that *Elodea* populations in the Cam 'declined after an initial explosive expansion' (Preston, 2002), and this pattern of boom and decline has been noted in introductions of the species to waters elsewhere in Britain and overseas, e.g. New Zealand (Cook & Urmi-König, 1985).

The 'introduction process'

As we saw in Chapter 6, for introductions that took place in prehistoric periods, there are no precise details of date and number of plants/seeds etc. However, with the invention of writing, some remarkably early records have survived. In 1500 BC incense trees were imported by Queen Hatshepsut to Egypt from the Land of Punt (Hodge & Erlanson, 1956). As a result of more recent movements of human populations, for instance those involved in European colonisation of territories overseas, much more direct evidence is available on the introduction process, and in a discussion of alien plants, with particular reference to North America, Mack and Lonsdale (2001) defined three overlapping non-mutually exclusive phases of introduction.

1 The accidental phase

From 1500 onwards, immigrants travelling to North America brought with them their own crop seed, which contained weed seeds, as seed cleaning was poor. Seeds were also found in imported bedding, hay and fodder and attached to animal fur. Evidence of early introductions sometimes comes from unlikely sources. For example, there are plant remains of several introduced European species (e.g. *Erodium* spp., *Centaurea melitensis*, *Brassica nigra*) within the structure of adobe bricks used to make the first buildings of Spanish settlements of Mexico and California in the 1700s (Hendry, 1931). It is also suspected that many species were introduced in dry and wet ship's ballast. Sometimes an opportunity arises to make a proper examination of ballast and see what plant propagules it contains. For instance, Nelson (1917) examined a ballast dump at Linnton, Oregon, and found 93 plant species not known elsewhere in the state. In historic times, as at the present day, accidental dispersal of plants occurs through trade in raw materials, such as logs, wool etc. and through the export and import of vegetables, seeds and fruits, and minerals, ores and coal etc. Turning to a more contemporary example of accidental dispersal, it is believed that Witchweed (*Striga*) entered North America on 'returning military equipment' (OTA, 1993).

2 The utilitarian phase

After the initial colonisation of new territories, Mack and Lonsdale (2001) suggest that settlers often deliberately introduced species that were deemed to be useful. In the 'new territories' colonists were sometimes reluctant to eat the foods of indigenous peoples, and, therefore, the seeds of many food and medicinal plants were imported. In addition, the native grasses were often regarded as poor forage for introduced domestic stock, and potential forage species from overseas were introduced. Thus, the Spanish and Portuguese colonisers introduced African grasses to South America (e.g. *Brachiaria mutica*, *Melinus minutiflora* and *Panicum maximum*) (Parsons, 1970).

In a later period, the Australian government was involved in the introduction of many species thought to be good candidates for increasing forage production, and a range of legumes was grown to increase nitrogen levels in the soil. Many of these introductions have since become weeds and caused economic damage. According to Lonsdale (1994): 'Of the 463 non-indigenous species introduced between 1947 and 1985 only 21 have subsequently been recommended as useful as forage in Australia.' But 60 of these introductions 'have been naturalised as weeds, including 17 of the 21 "useful" species'(Lonsdale, 1994 quoted in Mack & Lonsdale, 2001, 98). They make the further point: 'In Australia (and probably elsewhere) the potential for plant naturalisation was increased either by inexplicable decisions to evaluate even unpalatable species as forage (e.g. *Calopogonium mucunoides*) or simply by abandoning trial plots without destroying the experimental plants.'

Turning to another group of introductions, quick-growing woody plants and trees have been planted to combat perceived deficiencies in the supplies or properties of the local species. Large-scale tree planting is a very ancient practice. For instance, timber and crop producing trees were planted in the Mediterranean region as long ago as 255 BC (Richardson, 1999, 238). As early as the fourteenth century, conifers were planted to stabilise coastal dune systems in Portugal, and this practice was later expanded to many coastal sites in Europe. To combat a shortage of firewood, the seventeenth-century Dutch colonists of the Cape area of South Africa introduced non-native species such as *Acacia saligna*, *Hakea suaveolens* and *Pinus pinaster*, and these species have become seriously invasive (McNeely, 1999). Also, in the colonial period, plants were introduced to act as hedges. For instance, *Lantana camara* was brought to India from America via Sri Lanka to serve as a hedge plant. It has proved to be highly invasive in India as elsewhere (Cronk & Fuller, 1995).

There is a long history of tree planting in Europe, leading in the twentieth century to very large-scale agroforestry employing alien species, in many parts of the world. Richardson and Higgins (1999) note the remarkable development of plantations in Chile, Brazil, Argentina, Australia and New Zealand, many employing *Pinus* species particularly *P. radiata*. Large-scale planting of other species is also found in temperate regions, involving species of *Abies*, *Eucalyptus*, *Fagus*, *Larix*, *Quercus*, *Picea*, *Populus* and *Pseudotsuga* etc. Many of these developments have used conventional 'plantation' methods of planting and management. However, it is important to stress that tree planting is widespread across the globe, employing hundreds of tree and shrub species in a multitudes of different schemes with different aims – to beautify gardens; to provide food products, timber, firewood and fodder; to act as shelter belts, shade trees or windbreaks; to combat and promote stabilisation and rehabilitation of derelict and damaged lands; and as part of multi-cropping schemes with crop plants. Leguminous trees and shrubs are widely introduced to improve soil fertility through their ability to fix nitrogen (Hughes & Styles, 1989).

While tree and shrub plantings provide valuable products and services, in many cases the species have invaded outside these man-made ecosystems and become seriously invasive in natural and semi-natural habitats. Thus, many of the dominant tree genera used in commercial forestry have become alien invaders (e.g. species of *Acacia*, *Pinus*, *Eucalyptus*). Richardson (1999, 244) notes, of the more than 2,000 or so species used in forestry: '135 species (*c*. 7%) were weedy'. Also, of the 653 species listed in the *Invasive Woody Plant*

Database, '34 (18%) were introduced for forestry and another 49 species (27%) for amenity purposes' (Binggeli, 1996).

A very good example of the fate of utilitarian plantings is that of *Tamarix* species (salt cedars), which were deliberately planted as windbreaks and to stabilise stream sides etc. in the dry lands of south-western USA. These species have now become seriously invasive and, by the late 1980s, more than 1.5 million acres of riparian habitat were dominated by salt cedars (DiTomaso, 1998).

It is important to stress that public policy and perceptions about introduced species may change. For instance, *Melaleuca quinquenervia* (Australian Paperbark) and *Schinus terebinthifolius* (Brazilian Pepper) were deliberately introduced in the 1900s to help with 'drying out the Everglades' of Florida as part of the conversion of wetlands into agricultural land. Indeed, seeds of *Melaleuca* were scattered from an airplane in 1936 (see Cronk & Fuller, 1995). While these efforts might be judged a success by some, they are now seen in a different light by others, who are seeking to protect these Florida wetlands. Thus, the aim of much contemporary management is to conserve these world famous natural wetlands, and managers are now faced with the almost insuperable task of controlling the two introduced dangerously invasive woody plants. The scale of the task is enormous, as Brazilian Pepper infests 800,000 acres, while 500,000 acres are dominated by the Paperbark (Cox, 1999, 116).

While some of the deliberate movements of plants around the world have involved commercial stocks of plants, it is important to stress that in many cases botanic gardens have often been involved in receiving, cultivating and redistributing plants to colonial territories (Cronk & Fuller, 2001). For example, the Royal Botanic Garden at Kew was closely associated with a large and effective network of interchanges of plants between gardens throughout the British Empire and Commonwealth (McCracken, 1997).

3 The aesthetic phase

Once settlers were established in their new territory, Mack and Lonsdale (2001) propose that a new phase in the importation of plants began. Having secured the plants essential to survival, colonists turned their attention to ornamental plants. Evidence suggests that they wished to surround themselves with the familiar plants of their homelands and these became increasingly available through the horticultural trade. Research has revealed that, from the late eighteenth century onwards, settlers establishing themselves in new colonial territories in different parts of the world bought their seeds of crop, vegetable and medicinal plants from the catalogues of European seed merchants (Mack, 1991). Responding to an increasing demand for ornamental species in the nineteenth century, British, German and Japanese seed merchants offered an increasing number of plant species and cultivars in their catalogues (Mack & Lonsdale, 2001). Many species introduced as ornamentals overseas have become established 'in the wild'. For example, many exotic plants have escaped from gardens into coastal native forests in northern New Zealand (Sullivan, Timmins & Williams, 2005). In addition, in the nineteenth century, there was a growing horticultural market in

Europe for seed of plants originally collected in the territories overseas. This flourishing domestic trade of 'new' ornamentals, based on horticultural catalogues, was an important factor in the introduction of a wide range of species that later became invasive in Britain and other parts of Europe (Dehnen-Schmutz *et al.*, 2007).

The aesthetic phase of plant introduction continues to this day. An astonishing number of ornamental plants are grown in gardens. For example, more than 10,000 taxa are listed as garden plants in New Zealand (Bryant, 1998). Remarkably, many species now known to have invasive potential are still being offered to the gardener as packets of seeds. For instance, recently Mack reports that the very well-known invasive *Schinus terebinthifolius* was available at seed merchants in the USA, together with such potentially detrimental weedy species as *Verbascum thapsus* and *Tanacetum vulgare* (Mack & Lonsdale, 2001).

Establishment: founder effects, genetic drift and multiple introductions

As a consequence of human activities, plants are introduced into new territories, sometimes to different continents. Models predict if one or a small number of individuals found the new colony, a single generation bottleneck occurs, giving a new population that will be genetically depauperate relative the parental population(s) from which it came (Frankham *et al.*, 2002). Such founder effects result in the loss of genetic diversity, as some genetic variation may be totally unrepresented in the new population.

Whether a species establishes in a situation where there are bottleneck and founder effects depends to a great extent on the breeding system of plants. Baker (1955) made an important contribution in formulating what has come to be known as Baker's Law (Stebbins, 1957). Baker suggested that in self-compatible species a single propagule would be sufficient to initiate a sexually reproducing colony in a new habitat. However, with self-incompatible plants a minimum of two propagules must arrive at the same place and more or less at the same time. The presence of an appropriate pollinator is also essential. Baker's Law has proved to be an important model for considering the establishment of naturalising species.

In addition, issues relating to the initial colonisation and stochastic factors are likely to be very important in the early development of naturalising colonies. Thus, if the new population remains small through several generations, perhaps with several bottleneck effects, genetic drift may occur. At reproduction the allele compositions and relative frequencies represented in the parental gametes might not be preserved in the offspring, as 'each offspring receives one allele, selected at random, from each parent. Just by chance, some alleles, especially rare ones, may not be passed on to the offspring and may be lost' (Frankham *et al.*, 2002, 178).

In interpreting surveys of potential founder situations, especially where the event took place some generations ago, it is important to acknowledge the importance of other factors in influencing variability. In small founding populations of plants naturalising in new territories, matings between relatives are inevitable, and, in the longer term, inbreeding could lead to loss of variability through reduced survival, and reproductive failure (Frankham *et al.*, 2002). In addition, natural selection may be a very important factor in the establishment and subsequent development of naturalising populations. A number of careful studies of naturalising populations have been carried out, but in some other cases only small numbers of samples have been studied, and a fuller picture will emerge when more is known about the breeding behaviour of the species.

Establishment in naturalised populations: founder effects

Founder effects have been described in a number of studies of colonising populations in different areas (see review of Bossdorf *et al.*, 2005). A wide range of species have been examined, e.g. *Abutilon theophrasti* (Warwick & Black, 1986); *Avena barbata* and *Bromus mollis* (Clegg & Brown, 1983); *Chondrilla juncea* (Burdon, Marshall & Groves, 1980); *Emex spinosa* (Marshall & Weiss, 1982); *Phragmites australis* (Saltonstall, 2003); *Setaria* spp. (Wang, Wendel & Dekker, 1995); *Sorghum halepense* (Warwick, Thompson & Black, 1984); *Striga asiatica* (Werth, Riopel & Gillespie, 1984); and *Xanthium strumarium* (Moran & Marshall, 1978).

It is instructive to examine a number of case histories of single and multiple founder effects, the first of which involves the very well-known self-compatible insectivorous pitcher plant (*Sarracenia purpurea*), which occurs in low nutrient boggy habitats along the Atlantic seaboard of North America. Schwaegerle and Schaal (1979) investigated genetic variation of a population on Cranberry Island Bog, Licking County, Ohio. Here, in the 1830s, an artificial water body – Buckeye Lake – was created, in which a 17-acre floating *Sphagnum*-dominated island developed. In 1912, Freda Detmers, a graduate student of Ohio State University, introduced a single specimen of pitcher plant, of unknown origin, onto the island. In the 1970s, the population of pitcher plant had grown through *c.* 15 generations to a population of over one hundred thousand plants. Schwaegerle and Schaal (1979) studied the genetic variability of this and several other populations by examining eight isozyme systems. They concluded that the Cranberry Island population:

> exhibits a reduced genetic variability when compared with the species as a whole. Only one polymorphic locus was found where the mean number for all populations of *S. purpurea* in this study is 2.5 loci. Similarly, average heterozygosity per individual for all loci at Cranberry Island, 0.042, is 50% below the species mean of 0.089. The unique history of the Cranberry Island population, established by a single founding individual, leads us to attribute this reduction in genetic variability to an extreme founder effect.

However, the level of variability was not significantly below several other populations examined in their investigations: four of the ten other populations exhibited reduced variability. Schwaegerle and Schaal (1979) suggest that these populations might also have been influenced by founder effects, as a consequence of the complex post-glacial colonising history of the species in the region. In addition, genetic drift might also have been very important, especially if populations had at some point in their development been very small and geographically isolated.

Sarracenia purpurea is a celebrated plant and its absence from the European flora has been remedied by bringing the plant not only to gardens but also by planting in more natural habitats. For instance, a thriving population was established at Termonbarry Bog, Roscommon, Ireland, in 1906, using an unknown quantity of seeds and rootstocks of Canadian origin. From the 1930s to 1960s, new populations were established at other sites (see Taggart, McNally & Sharp (1989) for details). In six cases (including Coolatore, Co. Westmeath), plants were taken directly from the original site at Termonbarry, but, at a seventh site, plants from Cooltore were transferred to Woodfield, Co. Offlay. Some of these transfers were prompted by a concern for the future of the population at the Termonbarry site, where commercial peat cutting has reduced the site from 35 to *c.* 10 hectares. Taggart *et al.* (1989) carried out a survey of genetic variability of plants at Termonbarry and at five of the seven sites, where plants were surviving in the late 1970s. Twenty-five isozyme systems were examined, of which 14 provided useful information. Overall genetic variability was low and within the range of North American populations. Regarding the variability in the satellite colonies, Taggart *et al.* (1989) drew attention to the fact that four of the extant populations had been established by transferring very small numbers of plants (only 2–4). They concluded that these transfers represented severe founder events as the derived populations had a reduced number of polymorphic loci and in some cases alleles present in the Termonbarry population were absent from the satellite colonies.

Seeds of *S. purpurea* were also taken to Switzerland in the nineteenth century. A colony, source unknown, was established around 1900 at the high altitude Tenasses peat bog. Historical evidence suggests that around 1950 perhaps a single individual from this colony was transferred about 30 km away to a lowland acidic marsh near Lausanne (currently with dense vegetation). Parisod, Trippi and Galland (2005) have made a detailed study of the two populations using RAPD molecular markers to examine the genetic variability in parental (currently >25,000 plants) and derivative (*c.* 120 individuals) populations. They discovered that while the two populations had some markers in common, they were clearly well differentiated. It is likely that, as a consequence of a bottleneck, allele frequencies were changed through a founder effect. However, they draw attention to the likely importance of selection forces, as the two habitats are very different.

Giant Hogweed, *Heracleum mantegazzianum*, which grows up to 3.5 m in height, was introduced into British gardens from the Caucasus area. It has naturalised in many areas especially in waste places near rivers. By using microsatellite DNA markers, Walker, Hulme and Hoelzel (2003) detected considerable population genetic variability, with populations from different river systems being more variable than those within the same catchment area. It seems likely that there was a large initial introduction of seed into the country or there were multiple introductions. Recent studies in the Czech Republic have shown the value of dated aerial photographs in assessing the history of regional spread of the species (Müllerova *et al.*, 2005). Other investigations have revealed further insights into the invasion by giant *Heracleum* species. It has now become clear that at least *three* giant invasive *Heracleum* species were introduced into Europe in the nineteenth century from south-west Asia. Population variability and the genetic relationships between species have been investigated by Jahodova *et al.* (2007). Reviewing all the available information, they conclude: ‘the majority of the invading populations were not affected by a genetic bottleneck’ and that, in their subsequent rapid evolution, genetic drift and species hybridisation played a role in some populations.

Establishment of introduced species: founder effects in clonally propagating species

Since its introduction as a garden plant, Japanese Knotweed (*Fallopia japonica*) has become a serious weed in Britain, despite the fact that no male fertile plants have been discovered. Hollingsworth and Bailey (2000) studied the genetic variability of the species. In Britain, 150 samples of this species were collected, together with 16 samples of this introduced species from Europe and the USA. DNA extracts were analysed using the RAPDs technique. All the samples produced an identical profile. While this does not provide proof of complete genetic identity, it provides strong evidence that the plants naturalising in different areas of Europe and the USA came from a common source by vegetative propagation. The absence of the male plant in naturalising populations in Britain is a further example of the importance of founder effects in introduced species.

Establishment: a case of successive founder effects

The native range of *Rubus alceifolius* stretches from northern Vietnam to Java. It has been introduced to Madagascar, and the islands in the Indian Ocean. In some areas it is a seriously invasive weed, e.g. La Réunion. It has recently been introduced to Queensland, Australia. By investigating the genetic variability in the different parts of its range it has been possible to deduce the probable sequence of introductions (Amsellen *et al.*, 2000). Examination of DNA samples using AFLP molecular markers revealed very great variability of the species in its native range. In contrast, except in Madagascar, genetic variability is negligible. The reproductive biology of the species remains to be clarified; in particular the extent of vegetative and sexual reproduction, and whether the species is apomictic. The evidence so far available suggests:

> Each population sampled in the other Indian Ocean islands (Mayotte, La Réunion, Mauritius) was characterised by a single different genotype of *R. alceifolius* for the markers studied, and closely related to individuals from Madagascar. Queensland populations also included only a single genotype, identical to that found in Mauritius. These results suggest that *R. alceifolius* was first introduced into Madagascar, perhaps on multiple occasions, and that Madagascan individuals were the immediate source of plants that colonized other areas of introduction. Successive nested founder events appear to have resulted in cumulative reduction in genetic diversity.

While the precise dates of introduction are not known, it has been possible to a remarkable degree to deduce the microevolutionary history of naturalised populations of this species. The means of dispersal of the species is not known. The authors conclude that propagules could have been carried by birds but dispersal was ‘perhaps more likely by humans’.

Establishment: variation in native and introduced ranges

While populations of some introduced species exhibit reduced variation relative to those presumed to be ancestral, in other cases, introductions may exhibit comparable levels of variation to those in their native range, e.g. *Apera spica-venti* in Canada (Warwick, Thompson & Black, 1987); *Avena barbata* in California (Clegg & Allard, 1972); *Bromus mollis* in Australia (Brown & Marshall, 1981), *Echium plantagineum* (Brown & Burdon, 1983); *Pueraria lobata* in south-eastern USA (Pappert, Hamrick & Dovovan, 2000); *Trifolium hirtum* in California (Jain & Martins, 1979) and *Trifolium subterraneum* in Australia (Brown & Marshall, 1981). It is instructive to examine a number of case histories.

Establishment: insights from the use of molecular markers

Historical records are not sufficient in themselves to build up a full picture of the introduction process, but they are the starting point for molecular studies that have provided remarkable insights. By this means it has proved possible to identify novel genotypes unique to parts of the native range and detect these same genotypes in the introduced range. Thus, molecular methodologies, including the study of isozymes, provide a powerful retrospective means of testing models of immigration, based on historical information and herbarium records.

A recent study of *Alliaria petiolata* has provided many insights into the genetic variability of a self-fertilising introduced species (Durka *et al.*, 2005, 1696–7). Native to forest edges and moist woodland in Europe, the plant was introduced to North America in the nineteenth century, as a flavouring in cooking and for medicinal purposes. *Alliaria* is now a serious invader of woodland habitats. Twenty-six introduced populations were sampled from Wisconsin eastwards to New York, and from Tennessee northwards to Ontario, Canada. Twenty-seven populations were sampled in the 'native' range from Sweden to Italy, and from Britain to the Czech Republic. The European material was grouped into six regional areas: Britain, northern, central, south-central, eastern and southern Europe. DNA samples were prepared and genetic variability was studied using microsatellite molecular markers. It was discovered that 'overall, introduced populations were genetically less diverse' than those in Europe. 'However considerable variability was present and when compared to the probable source regions, no bottleneck was evident . . . The high allelic diversity in the introduced range strongly suggests multiple introductions.' Evolutionists have long been interested in trying to unravel the geographic origin(s) of colonising species. From which parts of the native range did the colonising species originate? Patterns of microsatellite markers in the introduced range suggest that three European regions – British Isles, northern and central Europe – were most likely to be the source.

A recent study of *Lepidium draba*, an invasive species across North America, has also revealed a number of insights into the colonisation process (Gaskin, Zhang & Bon, 2005). The species, which is reported to be self-sterile (Mulligan & Frankton, 1962), is native from the Balkans eastwards to southern Russia, western and central Asia. Gaskin *et al.* (2005) examined variability in highly polymorphic chloroplast DNA (cpDNA), which is maternally inherited and, therefore, provides a marker for the movement of seeds only (not pollen). The investigation involved the study of a specific section of the cpDNA genome (the intergenic region between trn S and trn G genes). Plants representative of Asia (186), Europe (188), and the USA (360) were examined and 41 different variants (haplotypes) were detected, allowing regional comparisons of variability. The results revealed that 66% of allelic richness found in the native area was retained in the introductions to the USA, a finding that suggests either multiple introductions or a few very diverse introductions. It seems highly likely that *Lepidium* was carried to the USA as a seed contaminant in alfalfa or clover seed imported from Europe and central Asia (e.g. Brown & Crosby, 1906). This view is supported by the findings of Hillman and Henry (1928), who discovered that 40% of seed batches of alfalfa from Turkistan contained *Lepidium*. The cpDNA analysis also provided information about patterns of variability within the USA. A low level of regional structuring was detected, suggesting that considerable dispersal has occurred since the first arrival of the species. At a local level, mixtures of variants were often found. As *Lepidium* seeds have no obvious natural means of dispersal, it is likely that the plant has been distributed by agricultural operations. Some of the cpDNA variants detected in the USA are not matched by those found in the native range, suggesting that they might be rare in Eurasia. A major issue of general concern is raised by this study, namely the concept of the 'native range'. Such has been the human-induced movement of plants that it is often difficult, perhaps impossible, to determine the natural native range. This is very well illustrated by the case of *Lepidium*. Is the species native to southern Europe or was it introduced when alfalfa was brought from the east by farmers, *c*. 2,000 BP?

Capsella bursa-pastoris is one of the most frequent and widespread species on Earth. The colonising history in North America has been investigated by examining variability in three multilocus allozymes systems in samples taken from Europe (9,000 plants from 592 populations) and North America (2,700 plants from 88 populations from California eastwards to New York; from Missouri and Virginia to British Columbia and Alaska) (Neuffer & Hurka, 1999). Little is known about the very early history of the introduction of this weed to North America, but inferences from DNA have been very revealing. Overall, North America has a smaller number of genotypes than European populations (by *c*. 50%) and, on the basis of the distribution of genotypes, it seems likely that a minimum of at least 20 independent introductions took place in the colonisation of North America. Regarding the geographical routes of introduction, genotypes with distinctive allozyme markers proved to be characteristic of the Mediterranean region, being particularly common in the Iberian Peninsular. Of major significance is the fact that the same variants were also detected in California and former Spanish colonies in South America. This pattern of variability supports the hypothesis that Spanish colonisation and introduction of weeds in grain, fodder etc. resulted in a south to north wave of colonisation by *Capsella* in California. Moreover, the evidence from allozyme distribution in the more temperate regions of North America offers support to the proposition that there was another broad wave of colonisation, this time from east to west across North America, reflecting the gradual colonisation westwards by other immigrants principally from other European nations. Neuffer and Linde

(1999) have made further investigations of *Capsella* in California involving the study of isozymes and RAPD markers, the analysis of crossing experiments, and cultivation trials to study such characteristics as leaf shape and flowering behaviour. Patterns of co-segregation were also studied. The conclusions of the earlier study were largely confirmed. Southern populations of early flowering variants have a particular multilocus isozyme genotype (MMG) that is shared with presumed ancestral populations in Spain. This commonality reflects the importance of introductions in the colonial period. Northern populations are of later flowering variants in which the MMG genotype is absent, reflecting their introduction from northern Europe via the eastern USA. Concerning the colonisation history, Neuffer and Linde consider the Gold Rush of 1848 to be another crucial period in the colonisation of the species. Very large numbers of immigrants arrived in California from many parts of the world. Some came via the eastern USA and others through the port of San Francisco, after travelling round Cape Horn or overland from Panama. In this interpretation, the current geographical distribution of the different genotypes does not show the limits of waves of invasion, but rather the effect of natural selection on a widely distributed array of preadapted genotypes. Furthermore, as the predominantly self-fertilising breeding system of the species and differences in flowering time permits very little hybridisation between early and late genotypes, these different variants persist.

Establishment: tracing the spread of Bromus tectorum

Another valuable case history is provided by the invasive self-fertilising weedy species *Bromus tectorum*. This introduced species was reported in Pennsylvania in 1790. It was discovered in British Columbia in 1889 and by 1900 it had established in the intermountain west. *Bromus* was probably introduced and spread in 'discarded packing material, livestock bed straw discharged at railway sidings, adulterated and contaminated grain seed, transported livestock and even the occasional deliberate sowing of the grass as forage' (Novak & Mack, 2001, 116). Now, *B. tectorum* is highly invasive in North America particularly in the intermountain west, where it is the dominant plant in at least 200,000 km^2. The plant has been widely introduced across the world including Australia, the British Isles, Hawaii, Japan, New Zealand and South America.

In a study of allozymes, Novak and Mack (2001) examined genetic variation in 164 populations (5,513 plants) collected from the native ranges in Eurasia and North Africa and the introduced ranges in Argentina, the Canary Islands, Chile, Hawaii and New Zealand. In addition, 94 populations (3,254 plants) were investigated from different geographical areas of North America. These results supplemented and extended those obtained from a smaller preliminary study. Electrophoretic studies were carried out on 15 enzymes encoded by 25 putative genetic loci. It was discovered that some marker genes have a restricted distribution in native range of *B. tectorum*, and that these markers can also be detected in the introduced range. Thus, by studying the distribution of genetic markers, it is possible to examine the genetic relationships and make deductions about migration routes.

Bromus populations from the Canary Islands proved particularly interesting, as these islands were a staging post for ships travelling from Europe and North Africa to North

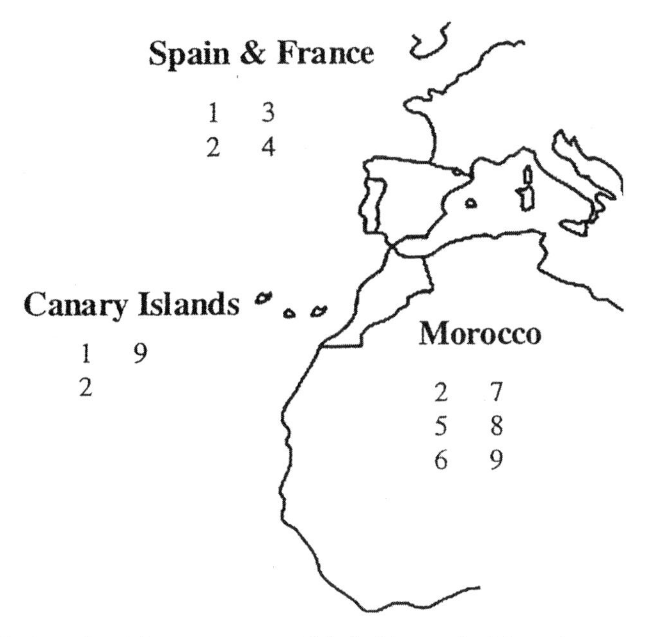

Fig. 11.1 Distribution of multilocus genotypes (labelled by number) in native range populations of *Bromus tectorum* in Spain, France and Morocco, and the introduced range in the Canary Islands. The distribution of these markers suggests that multiple introductions may have occurred from populations in Europe and North Africa. (Reproduced by permission from Novak & Mack, 2001 © American Institute of Biological Sciences)

America. Evidence suggests that there have been multiple introductions of *B. tectorum* to these islands from Europe and North Africa (Fig. 11.1). Considering introductions further afield, valuable evidence of worldwide introduction routes had been provided by the three multilocus enzyme systems.

With regard to the invasion of North America, the evidence suggests there were at least seven independent introductions into western North America and several introductions in the east (Fig. 11.2 & Fig. 11.3). There are too few samples to deduce post-introduction pathways, but Novak and Mack consider that the gradual east to west movement of settlers was very important, especially in the development of overland transport including railways. In some cases the same genetic markers were detected across the USA, suggesting the possibility of separate introductions or movement from east to west. The data set revealed that there was high variability in the introduced range. Also, some unique markers were found only in the west; these observations make it highly likely that there were several independent direct introductions. In addition, novel recombinant genotypes were detected, pointing to crossing between different variants. Novak and Mack make it clear that they regard the maps as provisional and likely to be modified as more data become available. Further surveys would be very valuable. For example, which variants have colonised the Yukon region of North America? Are the populations now found in Canada separate

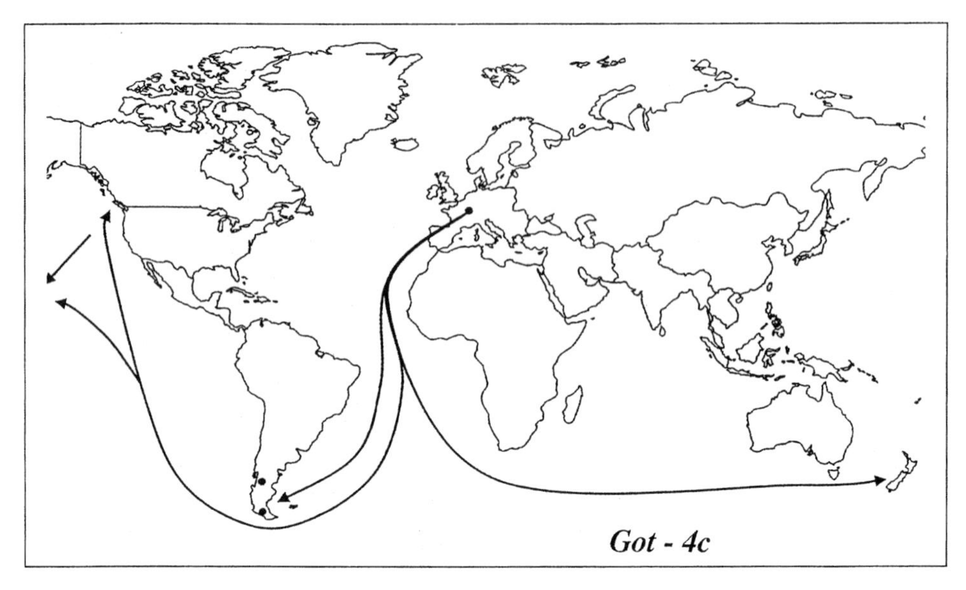

Fig. 11.2 Distribution of populations with the Got-4c marker genotype. Source populations apparently occur in central Europe, and arrows indicate the possible routes of dispersal from this native range to western North America, Argentina, Hawaii and New Zealand. (Reproduced by permission from Novak & Mack, 2001 © American Institute of Biological Sciences)

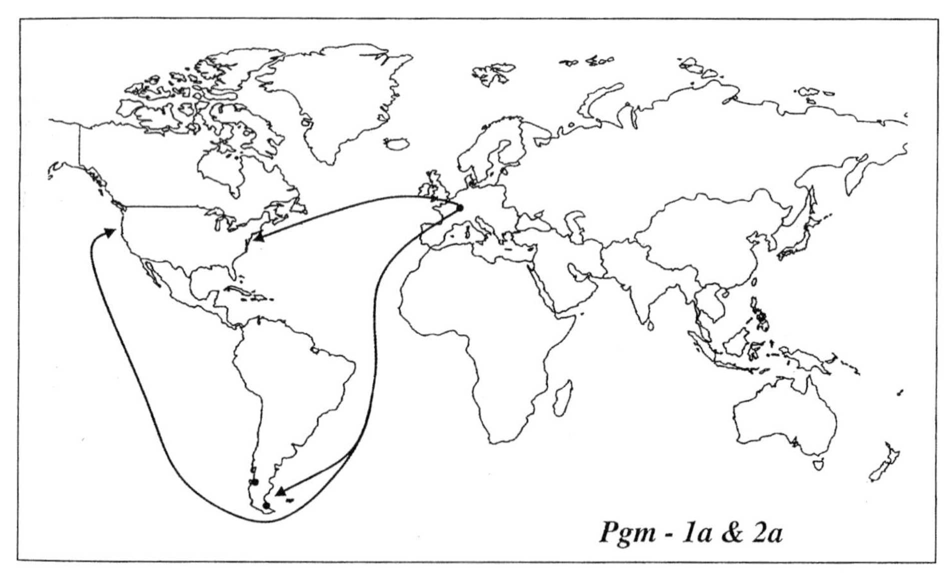

Fig. 11.3 Distribution of populations with the Pgm-1a and Pgm-2a genotypes. Source populations apparently occur in central Europe, and arrows indicate the possible routes of dispersal from this native range to eastern and western North America and Argentina. (Reproduced by permission from Novak & Mack, 2001 © American Institute of Biological Sciences)

introductions or invaders from the USA? Could further surveys disentangle the overland routes of the migration of *B. tectorum* in the USA?

Establishment: what proportion of introduced species becomes invasive?

Bromus tectorum is a celebrated example of an important class of introduced species, namely those that have become invasive. At this point, in the discussion of winners and losers, it is important to consider what proportion of introduced species become 'invasive', for as we shall see, this group of species are potentially exerting a disproportionately large effect in endangering species and ecosystems, and becoming major factors in the microevolutionary processes worldwide. Williamson and Fitter (1996) considered this issue and, as a broad generalisation, have formulated the 10s rule, which proposes a series of three nested 10% categories. Thus, of the species introduced into a territory by natural or human-assisted dispersal, approximately 10% will survive, *c.* 10% of these will establish self-reproducing populations and *c.* 10% of this final category will prove to be invasive. Thus, out of 1,000 introduced species, 100 will establish, but only one will prove to be seriously invasive. The 10s rule is intended to provide an approximate guide to performance, and receives some support from surveys (Williamson, 1996). For example, more than 13,000 species of plants have been introduced to the islands of Hawaii, about 900 of which have established and 100 become serious pests. Taking the figures for Hawaii Volcanoes National Park, of 475 exotic plant introductions, 53 (11%) are judged problem species (Williamson, 1996). While the 10s rule is a helpful indicator, it is often difficult to test in practice because of the different usage of the world 'invasive'. Some authors restrict the use of the term to economically important aliens that cause economic damage etc. Others do not make a very rigid distinction between introduced and invasive taxa.

Plant introductions: winners and losers

What determines the success or failure of the initial introduction? Moreover, what factors are important in the subsequent dispersal and establishment of the plant, and what role does the reproductive biology of the plant play in the success or failure of establishment? These questions are examined in turn; important issues concerning the role of interspecific hybridisation in the invasion process will be deferred until Chapter 13.

Some species are successfully introduced but many fail to establish

While many species may be successfully introduced into a territory, not all will establish and produce viable long-term populations (Sakai *et al.*, 2001). So far in this chapter, the effects of chance (founder effects and genetic drift) have been emphasised. Natural selection is also of crucial importance. Indirect evidence for selection comes from the study of wool aliens on Tweedside, Scotland, and at Port Juvenal, Montpellier, France (Hayward & Druce,

1919). In the rangelands of Australia, South Africa, Argentina and elsewhere, viable seeds of many species become entangled in the wool of the grazing sheep. Early in the twentieth century, the practice was to import raw unwashed fleeces and, in the processing, seeds of many species contained in the wool were released. Observant naturalists, hunting for plants that grew to a size where identification was possible, recorded 348 wool alien species on Tweedside, while 526 appeared at Port Juvenal.

Concerning the list from Tweedside, these figures may not reflect the full richness of the flora of seeds in the wool. The washing process was particularly harsh and some would be damaged or destroyed. In order to remove the species with burs and barbed fruits/seeds, the wool was first washed in soap and alkali at 110–120 °F and rinsed in warm water. Then, to carbonise the burs etc., the wool was treated with sulphuric acid and dried at a temperature of 180 °F. Finally, the wool was passed through rollers to crush the burs, which could then be winnowed away from the wool. Then, following the *laissez faire* attitudes of the day, wastewater containing many seeds was flushed into the rivers Gala and Tweed. Many wool aliens were detected downstream, winter floodwaters depositing the seeds on shingle banks along the river.

It is of considerable interest that some species were able to establish. For example, the sub-tropical species *Bidens pilosa* is abundant as a wool alien. This species, most probably a native of South America, is a serious weed on grazing land in Australia, New Zealand, Argentina and the South African Cape. Observations and experiments revealed that seeds of the species were found in quantity in imported wool. However, while the species germinated and grew in Scotland, plants were sensitive to frost and were therefore 'too tender to flower'.

Another contaminant of wool, *Xanthium spinosum* (one of the worst weeds of Australian sheep country), is also frost sensitive. In contrast, other species survived the winter, e.g. *Senecio lanatus* from Australia and New Zealand. However, only a few species produced persisting populations. A good example is *Rumex Brownii* from Australia, which was recorded as 'quite naturalised on Tweedside' and was 'abundant in successive years in the same places'. In similar fashion, while many species were introduced to the Port Juvenal area in France, only a few became established in the local flora.

Turning from the particular to the general, the successful establishment of introduced plants can only succeed if newly arrived seeds/plants encounter not only a congenial climate but also appropriate ecological conditions. The concept of safe sites, introduced by Harper (1977), has proved influential. For an introduced species to germinate and establish, the species must encounter the appropriate climatic, physical and biological conditions. Thus, on Tweedside, wool alien species from sub-tropical areas were unable to survive the Scottish winter. Taking another example, many Mediterranean species have been introduced to overseas territories. The places where these introductions have succeeded – California, Chile, South Africa and south-west Australia – all have similar 'Mediterranean' climates, characterised by hot dry summers and cool wetter winters (Fox, 1989).

It is important to stress that the climate of 'home range' may not be a reliable guide to the current or potential range of a species in its introduced area. Mack (1996) cites a number of examples. Water Hyacinth (*Eichhornia crassipes*) is a species of tropical origin, but as an introduced species it is found to latitude 37° N in northern California (Mack, 1996, 111). Other species in which there is a poor fit between the climatic tolerance in their native range and their introduced range include *Sorghum halepense* introduced into Canada (Warwick *et al.*, 1984) and *Andropogon scoparius* in Hawaii (Sorenson, 1991). A particularly striking example is provided by *Pinus radiata*, which has a small natural distribution in coastal areas of California, but has become an important invasive species over a much wider climatic range in New Zealand, Australia and South Africa (Richardson & Bond, 1991). Such observations prompt an important general question: how far does the current geographical distribution of a given species reflect its climatic tolerances? How far are species constrained by competition from other species?

A number of investigations have been carried out to see if a species can survive and flourish beyond its present range. For instance, *Lactuca serriola* is a widely naturalised annual in Britain. Seeds were sown in plots in south-western and northern Britain, well outside the current range of the species (Prince & Carter, 1985). It was demonstrated that the plant could grow and reproduce 150 km beyond its present distribution. It seems likely that the species has not reached the limits of its potential range. A number of other examples of experimental manipulations of seeds and seedlings are reviewed by Mack (1996).

Turning to other factors that influence the introduction process, it must be stressed that introduced species may require particular soil conditions for their establishment and growth and, if those fundamental edaphic requirements are not fulfilled, the plants may fail to establish. Also, there has been much speculation as to whether introduced species are only able to invade disturbed ecosystems. Are natural ecosystems less or not invasible (Crawley, 1987; Rejmánek, 1996)? Debate about this question soon encounters the usual problems of deciding whether or not a particular ecosystem is natural. In reality, so many of the world's ecosystems are grossly influenced by human activities and alien species have been widely introduced into disturbed areas. Also, as we have seen, many species have been deliberately planted, and those represent a bridgehead of populations from which invasion of other sites can occur. Finally, Rejmánek (1996) points out the difficulties of assessing the invasibility of habitats and ecosystems, stressing that one-point-in-time studies may be particularly unreliable. Searching for generalisations, he considers that mesic ecosystems are more invasive than xeric, because the latter are particularly unfavourable to seed germination and seedling survival. Open water ecosystems are 'notoriously open' to invasions. Overall, invasion would appear to be promoted by disturbance, eutrophication and fragmentation of habitats.

Establishment of introduced species: the importance of mutualisms

As mutualistic associations between plant roots and fungi are widespread in the plant kingdom, the availability of the appropriate fungal associates in the introduced range is clearly of prime importance (Richardson *et al.*, 2000). Many introduced species appear to form mycorrhizal associations in their new range. However, the lack of appropriate symbionts would appear to be a barrier to the establishment of some species out of their native areas.

In economic crops this barrier has been overcome. For instance, the initial establishment of pines in the southern hemisphere has sometimes been retarded by the lack of suitable soil fungi. However, pine trees have been widely introduced and their eventual success as plantation crops, and in some cases as invasive plants, is apparently due to the introduction from the native range of *plants in soil*, which provided an inoculum of the appropriate fungi (Richardson *et al.*, 2000). The widespread intentional and unintended distribution of soil across the world would appear to offer a means of dispersal of symbiotic fungi. For economically important nitrogen-fixing plants, some of which have become invasive, the necessary symbionts appear to be widely distributed, but where establishment initially failed, success has been achieved by deliberately introducing the appropriate rhizobial strains, e.g. the Oxford Forestry Institute has distributed many seed stocks abroad and success in establishment has been assured by the provision of a mixture of up to six rhizobial strains (Richardson *et al.*, 2000).

Turning to other mutualisms, animal/plant associations are crucial in many species for pollination and seed/fruit dispersal. There is abundant evidence that plant species in their introduced range are often exploited, opportunistically, by the 'resident native organisms' as food sources, shelter, nesting sites etc. (Low, 2002a, b). Alien plant species are often serviced by generalist native insect or bird pollinators (Richardson *et al.*, 2000). However, in some cases introduced species have suffered severe pollen limitation until the arrival of an appropriate introduced pollinator. Thus, the failure of the introduced Red Clover (*Trifolium pratense*) to set seed in New Zealand was overcome by the introduction of bumblebees (Hopkins, 1914). Another example is provided by the invasion history of *Ficus* in Florida. Only 3 of 60 introduced species have become invasive, following the accidental introduction of their specific wasp pollinators (Nadel, Frank & Knight, 1992).

Turning to the question of seed/fruit dispersal of species in their introduced range, evidence suggests that native birds and mammals may quickly establish 'facultative mutualistic relations' with alien species (Richardson *et al.*, 2000). Some of these prove to be novel associations. For instance, several species of pine from North America and the Mediterranean basin have wind-dispersed seeds in their native range. But, in Australia, some seed of these species is also dispersed by the feeding of cockatoos (Richardson & Higgins, 1999). Sometimes the introduction of an alien disperser facilitates the dispersal of introduced species. For instance, the blackbird (*Turdus merula*) is an important disperser of introduced species in Victoria, Australia (Carr, 1993).

Development of introduced populations: the lag phase

In studies of invasive species, it has frequently been observed that there may be a lag phase both in build-up of populations and in the extensions of territory (Mack & Lonsdale, 2001). A lag phase is not limited to invasive species. It is part of normal natural population increase and range expansion (for mathematical models of lag effects see Crooks & Soulé, 1999). A number of additional factors may account for lag phases in introduced populations.

Allee effects

When populations are small, as is likely on first introduction, it is possible that plants may be able to increase by vegetative propagation, but the species may not be able to reproduce sexually because they lack the appropriate sexual partners. Thus, in the case of *Elodea* (see above) the absence of male plants in Britain prevented sexual reproduction: the spectacular increase in populations was due to vegetative propagation.

In other species, small initial colonising populations may lack the necessary genetic self-incompatibility alleles. A good example is provided by Purple Loosestrife (*Lythrum salicaria*) in North America. While the plant can grow from detached portions of root or stem, the species is widely spread by seed that is dispersed by wind and water, and in mud attached to wildlife, tyres, boats and footwear (Bossard *et al.*, 2000, 237).

Regarding sexual reproduction, it is important to note that individual plants are self-sterile: the presence of more than one genetic individual is necessary for sexual reproduction. However, the breeding system is complex as there are three visibly different genetic variants in the species with different flower types (morphs), and the presence of all three is necessary to release the full potential of sexual reproduction. It was Darwin, in one of his most famous experiments, who showed that self-pollinations and crosses between individuals of the same morph usually fail to produce seed. Investigations by Stout (1923) revealed the genetic basis for this self- and intra-morph incompatibility system, which functions to promote outcrossing and prevent inbreeding. European surveys have revealed that generally populations are uniformly trimorphic roughly in the ratio 1:1:1 (Schoch-Bodmer, 1938; Halka & Halka, 1974). It is of great interest, therefore, that a survey of introduced North American sites found that some populations did not have all three morphs (Eckert & Barrett, 1992). 'Twenty per cent of the populations they sampled in Ontario were dimorphic', while 'three per cent proved to be monomorphic'. Given that European populations are claimed to be trimorphic, the difference in the newly colonised North America may be the result of repeated founder events and periods of small population size. Either the three style morphs did not all 'arrive' at the initiation of the new population (founder effects) or, as populations may have remained small for a period, by chance one or two of the style morphs may have been lost from populations by genetic drift.

Reflecting on 1:1:1 morph ratios found in surveys of European populations, Eckert and Barrett (1992) question whether they are truly representative. There is no reason why stochastic forces should not influence morph frequencies in Europe. It seems possible, therefore, that those carrying out surveys in Europe might have avoided sampling small populations. More recent investigations in south-western France (native range) and Ontario (introduced range) have examined population structure in more detail (Eckert, Manicacci & Barrett, 1996). Summarising their findings, Barrett (2002, 124) noted:

> French populations surveyed were predominantly tristylous, whereas those in Ontario were often missing mating types. This pattern was associated with differences in ecology and metapopulation structure between the two regions. French populations of *L. salicaria* occur primarily in roadside ditches associated with the agricultural landscapes of the region. The distribution of populations

results in a high level of connectivity, providing opportunities for gene flow among populations ... In contrast, introduced Ontario populations are more isolated from each other and opportunities for missing morphs to establish in non-tristylous populations through gene flow are restricted.

Dispersal vectors and the availability of new habitat

A species may suddenly increase its range and number of populations if dispersal is suddenly unwittingly facilitated by human actions, and new exploitable habitat becomes available (Crooks & Soulé, 1999).

Wartime bomb damage

Galinsoga parviflora, a native of South America, escaped from the Royal Botanic Gardens at Kew. At first, it was found only locally around Kew. However, abundant habitat and dispersal opportunities occurred during the Second World War, as large areas of London were affected by bomb damage and subsequent redevelopment (Salisbury, 1964).

Road-building

The Cut-leaved Teasel (*Dipsacus lacinatus*), first detected in North America in the 1800s, remained localised in northern New York State around Albany for a number of years. However, a dramatic increase in this species has occurred in the upper mid-western states of the USA, associated with the building of interstate highways and other roads (Solecki, 1993).

Railways

Senecio squalidus was introduced into Oxford Botanic Garden about 300 years ago. However, it is only in the last 150 years that the species, escaping from the garden, has expanded throughout most of urban Britain principally along the railway tracks (for details see Brennan, Harris & Hiscock, 2005). Various factors might have delayed its expansion. The Botanic Garden is some distance from the railway station. University authorities were worried about undergraduates leaving the city for the delights of London, and, thus, there was some delay in connecting Oxford to the Great Western Railway system. When the connection was made, it provided *S. squalidus* with ready access to rail-side habitats and waste ground, where it now grows in abundance.

Watercourses

Purple Loosestrife (*Lythrum salicaria*), the breeding system of which we have discussed above, provides an excellent example of time lag from introduction to the 'full' ecological impact of an invasive species. The date of first arrival of this species in North America is unknown, but it was 'well established' by the 1830s (Thompson, Stuckey & Thompson, 1987). It is possible that it arrived in ship's ballast, or as seeds on imported wool. It was widely used as a medical herb and in the form of robust horticultural varieties it provides an attractive display in garden and border plantings. It is highly probable therefore that there is a diversity of genetic stocks in North America originating from different parts of Europe, and, in addition, horticultural variants escaped into the 'wild'. Barrett notes: 'out of such a diverse "hybrid soup" inevitably comes genetic combinations with novel phenotypes' (Barrett, 2000, 132). The spread of this species across North America has been investigated by Thompson *et al.* (1987). *Lythrum* was firmly established along the New England seaboard by the 1830s, but major expansion only began with the opening of a number of canal systems in eastern USA, the Erie Canal (completed 1825) and more than 4,800 km of canals in New England, New York, Pennsylvania, New Jersey and Ohio (completed by 1840). These waterways provided not only a habitat for the species but also further opportunities for dispersal. *Lythrum* has now invaded wetlands across North America, including the arid west, where it has spread through irrigation schemes. In addition, the plant occurs in damp patches besides highways, turnpikes and railroad networks, and from these colonies has spread into river systems and artificial drainage channels.

There was a lag phase too in the spread of *Myriophyllum aquaticum*, a native of South America that was introduced into the USA in the 1800s for use in water gardens and aquaria. The plant does not reproduce by seed in North America and the phenomenal spread of the plant in freshwater lakes, ponds and slow-moving rivers is entirely by vegetative propagation. The plant is spread by water currents, aided by pumping operations, recreational boating, the dumping of aquarium plants in water courses and by the movements of wild fowl (Bossard *et al.*, 2000).

Lag phases: causes unknown

In Florida, two introduced species – *Melaleuca quinquenervia* (Paperbark) and *Schinus terebinthifolius* (Brazilian Pepper) were both deliberately introduced in treeless Everglades in the early 1900s. Several decades elapsed before populations began to expand. Major increases in Brazilian Pepper populations began in the 1950s. What was the cause(s) of these lag phases? Ewel (1986) lists a number of possibilities. Perhaps the ecology conditions in the Everglades had changed in such a way that made the area more 'invasion prone'? Also, many biologists see parallels between the invasive behaviour of plants and the building up of 'infection potential' that is necessary to initiate a major fungal or bacterial epidemic (Mack *et al.*, 2000). Thus, it may have taken several decades to build up the reproductive potential of the two invasive woody species. Another possibility is that the 'new colonists were confined to restricted habitats until mutations favourable for further colonisation became available' (Crooks & Soulé, 1999, 106). The natural mutation rate is very low, and any changes that occurred might be very difficult to detect as their effect may be subtle.

However, as the next section reveals, some mutations involving fitness have been detected and studied in introduced plants.

Natural selection in naturalising populations of plants

Despite the fact that introduced species provide excellent potential opportunities for the investigation of microevolution, many studies of weeds/invasive species have focused narrowly on weed management. It is of great interest, therefore, that weed-control studies have revealed highly convincing examples of 'selection in action' in the evolution of herbicide-resistant variants in introduced plants. As we saw in Chapter 9, evidence suggests that herbicide tolerance in several weed species has evolved independently at different sites. In addition, there is evidence that herbicide resistance has arisen independently – i.e. polytopically – in populations of a species not only in its native areas *but also in its introduced range*. For example, triazine-resistant variants of *Senecio vulgaris* were discovered within a few years of the first introduction of the herbicide in Washington State, British Columbia and Western Europe (Ryan, 1970). Likewise, triazine-resistant plants of *Chenopodium album* were discovered more or less at the same time in Washington State, eastern Canada and France (Lebaron & Gressel, 1982).

Have self-fertile variants evolved in introduced species?

In a species that is self-incompatible, a single founder individual introduced into new territory is self-sterile and therefore unable to set seeds. It might be supposed therefore that self-fertile variants might evolve and be at a strong selective advantage in their introduced range. A recent study of *Senecio squalidus* has examined this possibility (Brennan *et al.*, 2005). As we saw above, this species was introduced into Oxford Botanic Garden about 300 years ago and, after a lag phase, it has now colonised most of urban Britain. Did this spectacular colonisation involve self-fertile variants? A range of population samples was collected and the breeding system closely studied. In the native range the species is strongly self-incompatible. Crosses between different plants indicated that there were only a small number of self-incompatibility alleles in British populations, almost certainly as a consequence of one or a succession of founder effects, or the effects of genetic drift in very small populations. Studying plants from its introduced range in Britain, it was discovered that self-pollination resulted in few progeny, and the species was very sensitive to inbreeding depression. Overall, therefore, the investigations revealed that the strength of the self-incompatibility system had not weakened in the remarkable range extension in Britain.

Ecotypic differentiation

Chapter 3 provides an account of the discovery of the phenomenon of ecotypic differentiation. In investigations with plants collected from diverse climatic and edaphic situations in one or a range of common gardens, together with transplant experiments (or sown material), it has been repeatedly confirmed that intraspecific habitat correlated genetic variation occurs in species in growth habit, germination behaviour, length of life history, reproductive behaviour and output etc. While the action of natural selection is not directly observed in these investigations, the patterns of variability are explicable in terms of selection. An important question now arises. When a species enters its introduced range, do the processes of ecotypic differentiation take place? Evidence has become available from a number of investigations.

As we have seen above, studies of *Capsella* suggest that a wide range of 'preadapted' races or ecotypes have been introduced into North America, and selection has acted and is acting on this pool of variability, with the result that ecotypic patterns have been established (Neuffer & Hurka, 1999; Neuffer & Linde, 1999).

A second example concerns the movement of species in the reverse direction. Two species of goldenrod (*Solidago altissima* and *S. gigantea*) were introduced into Europe from North America about 250 years ago (Weber & Schmidt, 1998). Population samples were collected from across Europe (in a north to south transect from 61 degrees to 44 degrees from Sweden, through Denmark, Germany, Switzerland to northern Italy). An experiment with cloned material in an experimental garden was carried out and it was discovered that there were highly significant differences in morphological and life-history traits. Also, northern populations flowered earlier than populations from the southern parts of the transect. The cline in flowering time discovered resembles that found in North America. Weber and Schmidt offer a number of possible conclusions regarding variability and selection. Firstly, 'the original plant material introduced from the native range may already have contained all the genetic variation that was later expressed among populations in the new range. In such a case, selection would act on an array of introduced genotypes, filtering out the ones that match the new environments. The process would not require the *in situ* formation of new genotypes.' However, they also consider another possibility, as there is no evidence for multiple introductions on a large scale. It seems possible that after initial colonisation 'newly adapted genotypes' evolved 'by mutation and/or recombination, followed by differential mortality and reproduction among the different populations'. If the second possibility is accepted, ecotypic differentiation must have occurred with great rapidity considering the clonal nature of the growth of individual plants and that the effective generation time might be of the order of years or decades. Perhaps only 10–20 sexual generations have elapsed since introduction, in which case the microevolution in *Solidago* could have been very rapid indeed. Further studies are required to test these two possibilities.

Experiments on *Echinochloa crus-galli* have revealed another arena for microevolution in introduced plants: namely the question of selection as an introduced species moves into climatically different territory. Recently this species has invaded the Quebec region of Canada from the USA. Comparative investigations involving physiological studies under a range of conditions have revealed that Canadian material has evolved enhanced catalytic efficiency of some enzymes (Hakam & Simon, 2000).

Another study of the invasion of an introduced species from the USA to Canada has been carried out by Warwick and Black (1986), who studied *Abutilon theophrasti*, a native of China, originally introduced to the USA in the mid 1700s as a potential fibre crop. It is now a major weed throughout the USA, and was also found in Canada in small populations on waste ground and garden habitats. However, since the 1950s it has spread onto agricultural land, and, although once thought to be too frost-sensitive to expand its range in Canada, the species has recently expanded northwards into eastern Canada (Ontario, Quebec, Nova Scotia). This plant represents an ideal subject for studying the microevolutionary consequences of the northwards extension of the range of a colonising introduced species. Warwick and Black (1986) grew plants from seeds collected at sites from Ohio (39 degrees) to Ontario (45 degrees) in a common garden trial. They also investigated genetic variability as revealed by electrophoretic studies of 16 enzyme systems. They discovered low levels of genetic variability in the enzyme studies that are perhaps indicative of founder effects. However, the cultivation experiment revealed a good deal of variability in germination, growth, morphology and life-history traits such as dormancy. In many cases the variability was correlated with latitude and climatic factors. This case history gives some indication of what is likely to happen in 'the first stages of differentiation in response to local environment'. Several other studies of weed species invading northwards into Canada have been carried out and these all show patterns of variability interpretable in terms of recent microevolution (Warwick, 1990; Clements *et al.*, 2004).

General-purpose genotypes

The examples so far discussed have provided evidence that suggests the importance of local adaptation in introduced species through the action of natural selection. Baker (1965), however, considered that general-purpose genotypes may be of importance in the colonisation process (see Crooks & Soulé, 1999). Thus, introduced plants may succeed in establishing a new colony through their capacity for phenotypic plasticity and physiological adaptability rather than through their precise genetic fit to the new conditions. Baker's ideas have had wide coverage in the botanical literature.

A recent study of *Verbascum thapsus* makes use of Baker's concept (Parker, Rodriguez & Loik, 2002). This monocarpic perennial invasive species has been introduced many times into North America from Europe, both as a medicinal and a garden plant. By 1880, it had become naturalised in California and is now common on roadsides and waste ground. The species is becoming invasive in the vicinity of Yosemite National Park, where it is one of the few non-native species to colonise high ground. In a study of the variability of *V. thapsus*, seeds were collected from ten populations in California from sites ranging from 75 to 2,260 m (Parker, Rodriguez & Loik, 2002). Plants were grown in a garden trial to study a number of morphological traits, and also in growth chambers to examine physiological traits such as freezing tolerance and responses to different temperature regimes. Analysis of variance of the data sets revealed significant genetic difference between populations that may have resulted from founder effects and genetic drift. However, overall there was evidence of considerable phenotypic plasticity. The authors conclude that colonisation was 'not driven by rapid adaptation', but interpreted their results in term of the presence of general-purpose genotypes at this early stage in the invasion of higher elevations. They note that a reciprocal transplant experiment would cast further light on the question of adaptation versus plasticity. Clearly, while such an experiment would be of considerable theoretical interest, there may be difficulties in persuading those that are trying to control populations in higher ground of the merits of introducing into a mountain garden a range of populations from different parts of California (Parker *et al.*, 2002).

General-purpose genotypes in apomictic plants

Van Dijk (2003) reviews recent advances in our understanding of variation in *Taraxacum* (introduced into North America) and the closely related genus *Chondrilla* (now introduced into America and Australia). He employs Baker's concept of general-purpose genotypes in interpreting the patterns of variation in these two groups, which consist of mixtures of sexual and widely dispersed apomictic variants in Europe. He notes: 'allozymes and DNA markers indicate that apomictic populations are highly polyclonal. In *Taraxacum*, clonal diversity can be generated by rare hybridisation between sexuals and apomicts, the latter acting as pollen donors . . . Some clones are geographically widespread and probably represent phenotypically plastic general-purpose genotypes.' With regard to the potential of apomictic clones he concludes that although they may be considered as 'evolutionary dead ends', because of their breeding system, 'the genes controlling apomixis can escape from degeneration and extinction via pollen in crosses between sexuals and apomicts. In this way, apomixis genes are transferred to a new genetic background, potentially adaptive and cleansed from linked deleterious mutations' that might have accumulated during successive generations of apomictic reproduction.

Evolutionary changes in animal populations linked to introduced plants

As we have seen above, there is abundant evidence that plant species in their introduced range are often exploited, opportunistically, by the resident native organisms (Low, 2002a, b). However, pioneering studies have revealed that, in some cases, genetic-based microevolutionary changes are involved.

A change in host is documented in Soapberry Bugs (*Jadera haematoloma*) in Florida (Carroll & Dingle, 1996). These insects now feed on the Asian Golden Rain tree (*Koelreuteria paniculata*), which was introduced as an ornamental, particularly since the 1940s, and bears smaller fruits than those of its native host, the Balloon Vine (*Cardiospermum corindum*). This change in behaviour has been accompanied by an inherited change in their stylets: those feeding on the fruits of introduced plants having shorter stylets (*c.* 25% smaller) relative to those of Soapberry Bugs on their native host (Carroll, Klassen & Dingle, 1998).

Another example is provided by microevolution in the native Edith's Checkerspot butterfly (*Euphydryas editha*) in Nevada (Singer, Thomas & Parmesan, 1993). The insect is switching its preference from Chinese Houses (*Collinsia parviflora*) to Ribwort Plantain (*Plantago lanceolata*), which has probably been in Nevada as an introduced weed for *c.* 100–150 years. This microevolutionary change has been studied by examining egg-laying behaviour in cage experiments, and has occurred in *c.* 100–150 of generations of the butterfly. The counter adaptation to new food plants is not universal in the species, for, in locations where ribwort plantain is not yet found, no change of preference is discovered in insects presented with a choice.

Natural selection: which species are likely to succeed as invaders?

Turning to broader questions of microevolution, there has been considerable interest in trying to discover what characteristics invasive species might have in common, with the aim of predicting which other species might become serious invaders in the future. Rejmánek (1996) makes a very strong case for moving from individual case histories to the search for generalisations. Do invasive species as a group share certain traits and behavioural characteristics? The first step is to prepare a list of invasives and a companion list of non-invasive species. While biologists might agree on the former, there may be difficulties in agreeing a definitive list of non-invasive naturalised species for a particular region. Because of the widespread occurrence of 'lag' effects, recent and even longstanding introductions may not yet have revealed their full potential in the new area.

Many invading species have traits listed in Baker's list of characteristics of an 'ideal weed' (see Table 6.3), such as continuous seed production, high seed output, lack of specialised germination requirements etc. Useful information has also come from comparing invasive/non-invasive species of single genera. In a comparison of alien *Bromus* in the USA, Hulbert (1955) discovered that some were locally dominant, infrequent or rare: only *B. tectorum* was spectacularly invasive. Other brome species lacked one or more of what appear to be key adaptations contributing to the success of *B. tectorum*; namely, autumn germination, rapid root growth, and tolerance to repeated clipping and frost. In another study, Rejmánek (1996) examined the characteristics of 24 species of *Pinus* – 12 invasive and 12 non-invasive. Three characters in particular discriminated between the two groups. Invasive species have small seed weight, short juvenile periods and short interval between large seed crops. The non-invasive *Pinus* species have larger seeds, and longer life cycles. One prediction that is still being explored is that of genome size. The amount of DNA in plant species is very variable: in some groups, invasive species have a small genome size (see Rejmánek, 1996, 173).

Examining whether it might be possible to extend these observations and characterise the traits that distinguished invasive species from those that are merely naturalising, Rejmánek concluded that there might be the beginnings of a predictive theory, but no simple picture has emerged. Crawley (1987) is sceptical about finding generalisations. This conclusion is perhaps not surprising as invasive species occur in most plant groups (fungi, ferns, algae, flowering plants), and exist in a wide range of different growth-forms (trees, perennials, annuals), they reproduce through a wide range of breeding systems (asexual, apomictic, self-fertile and self-incompatible), and exhibit wide differences in propagule dispersal, dormancy etc.

The success of invasive plants: interrelationships with other species

Many invasive species reach a level of superabundance that is truly remarkable. While some have looked for species-specific attributes, which might promote invasiveness, it is important to consider the interactions of invasives with other species in an ecosystem context.

A number of hypotheses, which are not mutually exclusive, have been proposed to account for the success of invasive species within ecosystems. These have been comprehensively reviewed by Hierro, Maron and Callaway (2005) and their account forms the basis of the following brief outline. Firstly, it must be emphasised that none of these hypotheses have yet to be properly tested: too few comparative studies of native and introduced species and ecosystems have been made.

The natural enemies release hypothesis

Species in their native range are subject to specialist pests, diseases etc. When a species establishes itself in new territory these 'enemies' are left behind. In contrast, the introduced plant is surrounded by native species all of which have their specialist enemies intact. It is supposed, therefore, that the newly arrived introduced species may obtain a competitive advantage over the native species in its new habitat and become invasive. In reviewing the literature, Hierro *et al.* (2005) consider that this hypothesis is poorly tested. This is not surprising, as a full investigation would require detailed knowledge of the pests/diseases of all the members of an ecosystem. Furthermore, it is not clear how far, or indeed if, native plants are constrained by pests and diseases in their native habitats (Crawley, 1989). Keane and Crawley (2002) consider how the enemy release hypothesis might be tested using herbicides, molluscicides, pesticides and fencing to test the growth and performance of native and introduced populations of a species.

The evolution of invasiveness

In an extension to the previous hypothesis it is supposed that, when they are introduced, species are released from 'specialist enemies'; variants that relocate 'resources from the maintenance of resistance to traits such as size or fecundity' might be at a selective advantage in their introduced range (Hierro *et al.*, 2005). This hypothesis receives some support from a number of studies. For example, in a garden trial of St John's Wort (*Hypericum perforatum*) in the UK, it was discovered that North American invasive plants were more robust than native material from Europe (Pritchard, 1960). However, while Maron *et al.* (2004) obtained a similar result after growing exotic and native plants of St John's Wort in a common garden in Washington for two years, 'this pattern disappeared after a further year, and there was no evidence that exotic genotypes were larger or more fecund than natives when grown in California, Sweden and Spain' (quoted from Hierro *et al.*, 2005, 9).

If invasive species in their introduced range are indeed more robust than their counterparts in the native range, then the origin of these larger/more fecund plants is an important question. Do they arise from new mutations 'in the wild'? Another explanation should be considered. Could larger invasive plants in their introduced range have arisen, in part, from 'showy' horticultural stocks, originating in European gardens, from which they were dispersed as ornamental plants to different parts of the world, and subsequently escaped to form feral colonies?

Definitive tests of the 'evolution of invasiveness' would require the testing of a range of genotypes – both native and invasive – in several common gardens in the introduced and native ranges. To minimise maternal effects, Hierro *et al.* (2005) consider that plants should be raised from seed through one generation in a common garden, and then the material should be tested in several common gardens in the native and introduced range. However, serious ethical issues are raised by common garden experiments of this type. The accidental release of invasive plants could have serious economic consequences. Perhaps experiments could be carried out in isolated plots (so that there is no chance of gene flow into the wild), with all the plants/seed destroyed at the end of the experiment. However, in many cases the risks might be judged to be too great and the investigations should be performed in contained controlled environments, such as glasshouses or growth chambers.

The empty niche hypothesis

The concept of niche is extremely complex. By definition a niche 'can only exist in the presence of an organism', therefore, the hypothesis of there being an 'empty niche' presents some logical difficulties (Johnstone, 1986; Hierro *et al.*, 2005). However, setting these aside, in essence this hypothesis proposes that an invasive species is able to have access to an un-used or under-used resource that is not being exploited by native species. An example is provided by the highly invasive species *Centaurea solstitialis*, which dominates huge areas of Californian grasslands. Native plants – annuals and perennials – are shallow rooted. In contrast, *C. solstitialis* has deep roots and is able to exploit unused water resources below 60 cm in the soil (Dyer & Rice, 1999 and references cited therein).

While observations and experiments offer some support to the empty niche hypothesis, Hierro *et al.* (2005, 10) consider that studies must go beyond the examination of what happens in the introduced range and investigate the notion of 'free or underused resources' by 'parallel studies of both the native and introduced range'. Moreover, such studies should include measurements of abiotic aspects of the ecosystems.

The novel weapons hypothesis

This hypothesis is based on the phenomenon of allelopathy. It is proposed that: 'exotics exude allochemicals that are relatively ineffective against well-adapted neighbours in original communities but highly inhibitory to naïve plants in recipient communities' (Hierro *et al.*, 2005). In testing these ideas, Callaway and Aschehoug (2000) compared the inhibitory effects of *Centaurea diffusa*, an invasive Eurasian herb in North America, on three grasses that co-exist with it in Eurasia with three North American grasses matched for size and morphology. They discovered that activated charcoal inactivated the allochemicals. In other tests (quoted from Hierro *et al.*, 2005, 10):

> *Centaurea diffusa* had much stronger negative effects on North American species unless activated charcoal was added to ameliorate the action of allochemicals, when Eurasian species were more affected. The overall effect of activated carbon on North American species was positive, but it reduced dramatically the biomass of all Eurasian grass species growing with *C. diffusa*. These results suggest that *C. diffusa* produced chemicals that long-term and familiar Eurasian neighbours have adapted to, but its new North American neighbours have not.

Progress in characterisation of the allochemicals secreted by plants has been made. For instance, Bais *et al.* (2003) discovered that roots of invasive *Centaurea maculosa* secreted catechin that inhibited the growth of native species.

The disturbance hypothesis

This hypothesis proposes that native plants have not 'experienced the type and intensity of disturbances to which exotics are adapted' (Hierro *et al.*, 2005). This hypothesis has received little detailed investigation. It could best be tested by carefully designed field experiments mimicking disturbance, in which seed additions of native and invasive species would be tested as 'treatments'.

The species richness hypothesis

Elton (1958) proposed that the more diverse ecosystems might be more resistant to invasion than species-poor communities. A number of observations and experiments have studied this proposition with 'mixed results' (Hierro *et al.*, 2005). Further investigations are needed. One way this hypothesis has been tested is by pioneering investigations of the behaviour of invasive and non-invasive species within artificial assemblages of organisms (see Dukes, 2002; Stachowicz *et al.*, 2002).

The propagule pressure hypothesis

It has been suggested that the degree of invasiveness observed in many species, and the number of invasives found in a particular area, could be a function of a single factor – the amount of seed that arrives in a community (Mack *et al.*, 2000 and references cited therein). This hypothesis has yet to be fully explored. It could be tested by sowing different amounts of seed in study areas, as one of a number of factors in a properly designed field experiment, with subplots subject to different levels of disturbance and resource supply etc.

Co-evolution

A number of the hypotheses outlined above stress the importance of interactions between species. This is particularly true if allelopathy is widespread in invasive/native plant interactions. Hierro *et al.* (2005, 12) draw attention to a major issue:

> Plant communities are widely thought to be 'individualistic', composed primarily of species that have similar adaptations to a particular physical environment. This traditional view downplays any persistent and powerful role of co-evolution in shaping the structure of interactions. The novel weapons [hypothesis] suggest that interactions among plant species may drive natural selection in communities and imply that natural biological communities may evolve in some way as functionally organised units.

The co-evolutionary aspect of plant invasions has been stressed particularly by Crosby (1986, 1994). In the colonisation of the New World, to found Neo-Europes, Europeans did not travel alone, but brought with them what Crosby termed 'portmanteau biota' – crop plants, domestic animals, weeds, seeds and diseases, which in many cases exploded into huge populations (Melville, 1994). These introduced organisms transformed New World ecosystems; and the imported diseases, to which Europeans had co-evolved, devastated indigenous New World societies. According to Crosby (1986), co-evolution played a major role in the evolution of portmanteau biota in their native range in Eurasia. Regarding the relationship between forage plants and domesticated animals, he notes: 'for thousands of years, Old World grazing animals and certain grasses, plus the other weeds of Eurasia and N. Africa, have been adapting to each other'. As we have seen in earlier chapters, microevolution in plants growing in grazed pastures, in hay meadows and in the weeds of arable fields has involved co-evolution. There has been 'mutual adaptation' of grasses and grazing animals, and indeed between arable crop plants and weeds. Crosby (1986, 288) points out that, in the comparatively recent colonisation of the New World, 'the co-evolution of Old World weeds and Old World grazers gave to the former a special advantage after the two spread in the Neo-Europes'. He envisages the impact of introduced plants and animals on the pre-existing ecosystems of the New World as follows:

> The Old World quadrupeds, when transported to America, Australia and New Zealand, stripped away the local grasses and forbs, and these, which in most cases had been subjected to light grazing before, were often slow to recover. In the meantime, the Old World weeds, particularly those from Europe and nearby parts of Asia and Africa, swept in and occupied the bare ground. They were tolerant of open sunlight, bare soil, and close cropping and of being constantly trod upon and they possessed a number of means of propagation and spread.

In the Neo-Europes:

> . . . the success of the portmanteau biota and its dominant member, the European human, was a *team effort by organisms that had evolved in conflict and co-operation over a long time.* [my emphasis]

More research is needed to examine the co-evolution in plants of human-dominated ecosystems. However, careful analysis is necessary as it is clear the successful aliens are widespread and also 'plant species which grow together did not necessarily evolve together' (Crawley, 1987, 448).

Ecological consequences of introductions

There is abundant evidence that natural ecosystems have been influenced, or greatly modified or radically transformed by introduced organisms. In some cases the changes are so great it could be claimed that new ecosystems are produced.

This book is about winners and losers, but it is essential in the present context to stress the interactions of all the organisms in ecosystems, and illustrate the manifold changes that can occur following the introduction of alien plants, animals, pests and diseases. Here, space prevents a comprehensive treatment of ecosystem changes wrought by introduced species. A number of key issues are identified, together with selected case histories. For more detailed information the following sources may be consulted (Elton, 1958; Drake *et al.*, 1989; Cronk & Fuller, 1995; Williamson, 1996; Sundlund, Schei & Viken, 1999).

The critical point to acknowledge is that invasive introduced species may change the collective properties of ecosystems (Vitousek, 1990). Three major groups of changes have been identified. Firstly, the abiotic environment may be altered. Secondly, invasive species might change the frequency and/or the intensity of ecosystem disturbances, e.g. fire regimes, hydrology and pests and diseases. Thirdly, invasive species may enter native ecosystems at different trophic levels. Thus, some invaders 'dominate' at the plant producer level and give 'bottom-up' effects, while the entry of 'new' carnivores produces 'top-down' changes. Other invasives appear in native ecosystems as 'new' herbivores. In some cases, invasive organisms have entered ecosystems at several trophic levels, either more or less simultaneously or in succession. In all cases, cascade effects occur through major changes to food webs.

It is clear, therefore, that as a consequence of the establishment and spread of introduced species every element of ecosystem structure and functioning may be altered. There is a very large literature on the subject: the review by Levin *et al.* (2003) considers more than 150 papers on the ecosystem impacts of invasive plant species. In presenting a few illustrative case histories, it is helpful to distinguish between direct and indirect effects of invasive species (Simberloff, 2001).

Direct effects

Introduced species may have direct effects on others, through changes to habitats, competition, predation, herbivory, parasitism and disease. Also, there may be effects through hybridisation between native and introduce species (see Chapter 13).

Introduced herbivorous animals

Domesticated grazing animals and stock introduced for sport have directly modified ecosystems. For example, rabbits, native to the Iberian Peninsula, have been widely introduced in Europe, Australia, New Zealand, South America etc., and to over 800 small islands and island groups (Flux & Fullagar, 1992, see Williamson, 1996). Imported repeatedly into Australia as game in the early nineteenth century, rabbits had a profound effect on the vegetation (Fenner & Ratcliffe, 1965).

On the voyage of the *Beagle*, Darwin (1839) noted the loss of forests in St Helena, following the introduction of goats in 1502. In the nineteenth century, there were multiple introductions of fallow deer in New Zealand. These herbivores have contributed to massive changes in the composition and regeneration of forest ecosystems by their preferential browsing of palatable species (Husheer & Frampton, 2005).

Shading and overgrowth

Invasive introduced species may shade out or overgrow the native species in aquatic and terrestrial communities, e.g. the floating aquatic fern *Salvinia molesta*, a native of South America, widely introduced across the world, from southern Africa to south-east Asia, from the Philippines to New Zealand (Lee, 2002a, b); and Chinese Tallow (*Sapium sebiferum*) woodland has replaced prairie in Texas (Cox, 1999).

Modified fire regimes

The presence of invasive species may increase the frequency and intensity with which damaging fires occur in an ecosystem, e.g. in western grasslands of the USA invaded by *Bromus tectorum*; and in those areas of the Everglades dominated by Australian Paperbark (*Melaleuca quinquenervia*), a species which has highly flammable foliage and leaf litter (Simberloff, 2001).

Modified hydrological and nutrient regimes

Changes in hydrology, evapo-transpiration and nutrient status may occur in ecosystems invaded by populations of introduced plants. For instance, there have been significant changes in topography and hydrology in areas invaded by salt cedars in the arid south-western USA (Cox, 1999). In addition, invasive nitrogen-fixing species may change the nutrient status of soils they invade. For example, species composition, nutrient cycling and plant succession have been altered in an area of Hawaii invaded by the nitrogen-fixing Firetree (*Myrica faya*) (Simberloff, 2001). Invasive introduced species may change other soil properties, for instance, marked acidification of soils occurs in grasslands in New Zealand dominated by *Hieracium pilosella* (Lee, 2002a, b).

Parasitism and diseases

In some cases introduced plant diseases have had profound effects on natural and semi-natural vegetation. In the 1890s, the American Chestnut (*Castanea dentata*), a major component of many forests in the eastern USA was attacked by the pathogen Chestnut Blight (*Cryphonectria parasitica*) introduced from Asia into North America on ornamental nursery material. In less than 50 years, chestnut blight spread throughout 91 million hectares of hard-wood forest dominated by chestnut or where chestnut represented a significant component of the forest (more than 25% of the tree species). Harper (1977) considers that this represents 'the largest single change in a natural population that has ever been recorded by man'. Some chestnuts remain, as they are capable of sprouting from roots (Williamson, 1996, 135), but developing plants are cut back by the blight before they can seed.

Dutch elm disease, which may have originated from Asia, is caused by the fungus *Ophiostoma ulmi*, and has devastated populations of elms (*Ulmus*) in North America and Europe. The disease is spread by bark beetles (*Scolytus* spp). An epidemic, caused by the virulent *O. novo-ulmi*, which probably originated from infected material of North America origin, devastated most of the elms in Britain in the 1960s. Elms still survive, as non-flowering suckers growing from stumps (Williamson, 1996).

Some fungal diseases have a very restricted host range, but the 'dieback' fungus *Phytophthora cinnamomi* has a very wide host range of perhaps 1000 different species in western Australia. This fungus, which is thought to originate in the tropics, was probably introduced to western Australia on diseased nursery material from eastern parts of the country. A major effect of the disease in the west has been the loss of more than 5,000 hectares of Jarrah (*Eucalyptus marginata*) forest. Spread of the fungus seems to be facilitated by road making, logging and mining operations. In some areas of eastern Australia more than 'half the species in plant communities have been destroyed' (Broembsen, 1989).

Indirect effects

As a consequence of invasion by introduced species, complex 'chains of reaction' occur throughout ecosystems.

Introduced animals

Profound changes to ecosystems have followed the introduction of carnivorous animals (and/or animals that are able to eat a wide range of animal/plant materials). Such is the interrelatedness of ecosystems that the introduced carnivores impacting on the herbivores bring changes in the frequency of different species, and changes in community development. Invasive organisms have brought about profound changes to island floras. For instance, populations of feral dogs (a carnivore), pigs and black rats (feeding on a wide range of organisms), together with feral populations of herbivores – cattle, goats, donkeys and horses – have had major influence on the ecosystems of the Galapagos Islands. These effects have been compounded by the introduction of 240 plant species, of which 15 are invasive (Brockie *et al.*, 1988). The presence of such a wide range of introduced organisms threatens many of the native endemic fauna and flora and native ecosystems, such as the *Scalesia*-dominated forests.

Changes in status

As a consequence of the loss of dominance of chestnut in much of the eastern USA (see above), there have been changes in the relative frequency of other tree species: for example, as an indirect consequence red oak (*Quercus rubra*) has greatly increased in areas formerly dominated by chestnut (Simberloff, 2001).

Diseases

Indirectly, human diseases have also influenced ecosystems. So many people died of the Black Death in Europe in the fourteenth century that many areas of agricultural land were abandoned, at least temporarily, and semi-natural vegetation returned to former farmland (Diamond, 1997). Also, in the European conquest of the New World, diseases brought from the Old World devastated the indigenous populations with dramatic changes in the balance of agricultural and semi-natural vegetation.

Diseases of domesticated animals

As a consequence of animal diseases, natural and semi-natural vegetation has been influenced indirectly in chain reactions. Rindepest, a viral disease of domestic and other ruminants, was introduced from India to Eritrea in north-east Africa in 1887. As it moved south to the Cape in 1900, large numbers of African cattle were killed (perhaps over 90%), and there was mass starvation in human populations. The virus also infected a number of herbivore species – antelope, giraffe, buffalo, wildebeest etc. – and many animals died, leading to the starvation and death of many native carnivores. Thus, the disease brought profound changes to the vegetation cover and composition in the ecosystems of Africa, through changes to human populations and in decreasing the numbers of wild carnivores/herbivores (Williamson, 1996).

Coda

As a consequence of human activities, there has been widespread dispersal – accidental and deliberate – of a wide range of introduced organisms and new assemblages of plants and animals occur worldwide. The invasive behaviour of a subset of the naturalised species has transformed many native ecosystems, resulting in the widespread occurrence of 'new' ecosystems. While some investigations of the microevolution of naturalised and invasive species have been carried out, there is enormous potential for further insights into all microevolutionary issues, especially if researchers combine molecular studies with the common garden trials, reciprocal transplant experiments, tolerance testing and genetic experiments (Novak, 2007; Lavergne & Molofsky, 2007). Baker's concept of general-purpose genotypes should be examined further, especially in the light of the apparent variability of many introduced species and the evidence for genetic-based microevolutionary differentiation in many species, rather than survival through phenotypic and developmental plasticity (Clements *et al.*, 2004). Given that new species are still being introduced into North America, Australia etc., there is the potential to study the genetics of the invasion process in more detail. Also, further investigations of genetic variability and breeding behaviour of introduced and weedy species could be highly revealing. Interspecific hybridisation and polyploidy are microevolutionary processes of importance in the evolution of some weeds. Many native and introduced weeds are allopolyploid in origin, typically showing wider ecological amplitude than parental taxa, and in having multiple genomes they often have many heterozygous loci and produce multiple enzyme phenotypes. As we see in Chapter 13, major advances have recently been made in our understanding of the evolution of polyploid species involving crosses between introduced and native species, and further insights are certain if populations of these new species continue to be investigated.

What will happen in the future? Firstly, it seems highly likely that the full impact of introduced and invasive species has not yet been witnessed. Taking but a single example, Parker *et al.* (2002) point out that there are huge areas of suitable territory, including high ground, for the continued expansion of the invasive *Verbascum thapsus* in western USA. They conclude that the 'struggle' to control the species has 'only just begun'. Secondly, despite the best efforts of those in charge of quarantine services in different countries, many additional species are being accidentally introduced. At present, only about 10,000 species of plants are seriously invasive species. It has been estimated that perhaps 10% of the 260,000 vascular plants, so far named, are 'colonising species' (Williamson & Brown, 1986). Thus, there is a high potential for a larger number of species to become invasive. For instance, by studying the information in foreign floras, Reed (1977) estimated that there are *c.* 1,200 weedy species capable of invading the USA, that have not yet done so. Finally, as we see in later chapters, global climate change models predict that there may be widespread opportunities for the further spread of invasive plants and animals.

Will biological invasions subside in time? The infestation of British waterways by *Elodea*, an account of which opened this chapter, suggests that this might be a possibility. Indeed, Cox (1999, 17) speculates that over hundreds or thousands of generations natural selection may 'be working towards "evolutionary control" of detrimental exotics . . . in time evolutionary adjustments may occur and counter adaptation may transform the exotic into a "peaceful" member of the community'. Evidence on this issue is difficult to obtain, as biological invasions are generally so damaging to crop production, range management and forestry etc. that pesticides and/or biological control (see Chapter 15) are used to destroy or control invasive populations. Clearly, management may change the course of the invasion and prevent us from witnessing what might be the 'natural' outcome.

Invasive species often dominate any discussion of the role of alien taxa in ecosystems. They are the clear Oscar winners on the evolutionary stage and deserve their celebrity. However, it is important not to underestimate the major effect of a very large number of introduced species that are only locally or regionally abundant. These have not yet been accorded global celebrity status, but their capacity to influence microevolution at a population level must be recognised. This is implicit in what Darwin wrote in the *Origin*.

He was clear that, to use his own words, there was a 'struggle for existence', a 'war of nature' and a 'great battle of life'. Considering the impact of introduced species, it is important to acknowledge a key Darwinian insight that is sometimes overlooked: 'as natural selection acts by competition, it adapts and improves the inhabitants of each country *only* in relation to their co-inhabitants' (Darwin, 1901, 389). Restating this in modern population biological terms and viewing the situation from a plant's-eye perspective, the appearance of *any* introduced species has the potential to alter the rules and influence the 'result' – which species are winners and which losers – in the microevolutionary game, as it is being played out locally.

In later chapters, a number of further issues relating to introduced taxa will be confronted. For instance, what will be the eventual outcome of the mass introduction of taxa across the globe? Will introduced naturalised and invasive organisms be the inevitable winners in human-dominated ecosystems? What will be the fate of endangered native species? Will they be the losers in a changing world or will conservation efforts in their protection ultimately prevail? As a prelude to these wider issues, in the next chapter we examine recent progress in our understanding of the microevolutionary pressures on endangered species.

12

Endangered species: investigating the extinction process at the population level

The evidence presented in previous chapters makes it plain that human activities create profound and highly complex changes in ecosystems through the imposition of new selection pressures. Darwin predicted that selection favours those species and populations that have the necessary genetic variability to survive and prosper under changing conditions. In contrast, species and populations that lack the necessary Darwinian fitness for the new conditions are at a selective disadvantage, and are in danger of becoming extinct.

This chapter analyses the extinction processes at the population level, providing the necessary background to a key issue to be examined below. Can the extinction of species/populations be prevented by appropriate conservation management?

As a prelude to the discussion on extinction, it is important, as always, to understand Darwin's own view of the vulnerability of rare species and appreciate his insight that extinction is often the end point of a process of decline. In the *Origin* (Darwin, 1901, 79–80), he writes:

> Natural selection acts solely through the preservation of variations in some way advantageous, which consequently endure. Owing to the high geometric rate of increase of all organic beings, each area is already fully stocked with inhabitants; and it follows from this, that as the favoured forms increase in number, so, generally, will the less favoured decrease and become rare. Rarity, as geology tells us, is the precursor to extinction. We can see that any form which is represented by few individuals will run a good chance of utter extinction, during great fluctuations in the nature of the seasons, or from a temporary increase in the number of its enemies. But we may go further than this; for, as new forms are produced, unless we admit that specific forms can go on indefinitely increasing in number, many old forms must become extinct . . . We have seen that the species which are most numerous in individuals have the best chance of producing favourable variations within any given period . . . Hence, rare species will be less quickly modified or improved within any given period; they will consequently be beaten in the race for life by the modified and improved descendants of the commoner species . . . New species in the course of time are formed through natural selection, others will become rarer and rarer, and finally extinct. The forms which stand in closest competition with those undergoing modification and improvement, will naturally suffer the most.

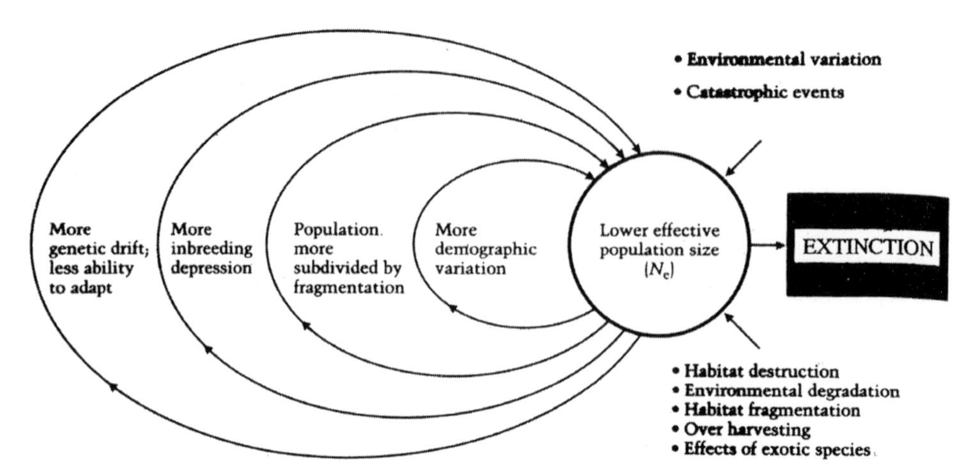

Fig. 12.1 Extinction vortices progressively lower population sizes, leading to local extinctions of species. Once a species enters a vortex, its population size becomes progressively smaller, which in turn enhances the negative effects of the vortex. (Adapted from Guerrant, 1992, and Gilpin & Soulé, 1986, from Primack, 1993. Reproduced with permission of Sinauer Associates, Inc.)

A general model of declining populations

Firstly, it is important to make it plain that there are different types of rarity. Thus, Rabinowitz, Cairns and Dillon (1986) showed that rare species have different characteristics in relation to three factors: geographic range (broad or narrow); habitat specificity (wide or specialised); and local population size (large or small). In addition, they emphasise the point that not all rare species are in decline, at least over the time frame for which records of population numbers are available. However, newly arrived introduced species may be rare, but later, as we saw in Chapter 11, some may expand their range, a few becoming invasive. Despite these exceptions, and risking a generalisation, it is abundantly clear that throughout the world many rare species are threatened with extinction, through the loss of populations and the severe decline in numbers in others. Natural causes, such as disease, storms and floods, may contribute to declines and extinctions, but many species are directly or indirectly at risk from human activities.

Gilpin and Soulé (1986) proposed a general model of extinction (Fig. 12.1) that likens the interaction of many factors to a vortex that progressively lowers population size and leads through genetic depauperation to population extinction. Turning to the details of the model, many human-influenced/determined factors are incorporated, including habitat destruction and change, the introduction of alien plants and animals, the over-exploitation of natural resources and pollution. These factors may result in population decline through one or more severe bottleneck effects. Unless these declines are temporary, demographic problems may then arise, through the disruption of the reproductive processes and/or the failure of seedling establishment etc., and population numbers may decline further. Human activities often lead to fragmentation and subdivision of plant populations to such a degree that gene flow is curtailed or prevented, and populations become genetically isolated. Reduction in population size also leads to the possibility of inbreeding depression and genetic drift. These factors too contribute to the 'vortex effect', leading to a further reduction in population size, loss of genetic variability, and the eventual extinction of the population.

Habitat loss and ecosystem changes

It is important at this stage to examine some of the assumptions behind the vortex model. In considering the nature of habitat fragmentation, Schwartz (1997, xiii) makes a very important point: 'Most of the theoretical underpinnings of conservation biology were developed under a conceptual model of landscapes *that are not yet entirely fragmented*' and '*assume an ongoing process of habitat loss*' [my emphasis].

Thus, it is often supposed that immediately prior to contemporary human disturbance the habitat was *completely natural untouched wilderness*. A case in point is the destruction of tropical forests in Amazonia and elsewhere. Conservationists have visualised such deforestation as the loss and fragmentation of 'virgin' forest. However, as we have shown in earlier chapters, in many parts of the world tropical rain forest has been very strongly influenced by human activities. Thus, as Schwartz (1997, xiii) makes clear, in many parts of the world, 'habitat conversion' is 'complete, extensive, and typically in the realm of the historical'.

The vortex model and cultural landscapes

Developing this idea, it is important to note that very large areas of the world have, for centuries, been evolving as cultural landscapes. In many regions, the loss of natural ecosystems occurred long ago, and the ecosystems we see today have been serially modified by human agency – either deliberate or accidental – through many stages of exploitation, management and sometimes abandonment.

Rare plants in cultural landscapes

Turning now to rare plants, it is important to see their decline not only in a Darwinian framework, but also within the context of cultural landscapes. Following the theory of natural selection, it is proposed that at each change in human land use, selection favours those species/populations that have the highest Darwinian fitness (winners) in that particular habitat in the cultural landscape, while other species of lesser fitness decline, and perhaps become extinct. This situation is fluid and, if further changes in the cultural landscape occur that exert different selection pressures, selection will again occur *with the possibility of new winners and losers*.

This concept receives support from the evidence reviewed in previous chapters and it is very important in considering the vortex model. It is a point of supreme importance that often a species becomes endangered when a particular habitat, long-managed as part of the cultural landscape, is decisively modified by a further change in human land use. Endangered species often show very high habitat specificity to particular elements of cultural landscapes managed by traditional practices. Risking a generalisation, throughout the world the endangerment of many plant, and indeed animal species, is associated with the neglect, change or abandonment of a whole array of traditional practices.

Thus, many weed species that were highly successful under 'traditional' agricultural practices have been driven to extinction locally or regionally with the arrival of mechanised agriculture, with its intensive use of seed cleaning, fertilisers, herbicides, and changes to the timing of sowing, harvest, treatment of stubbles etc. Another example is provided by the presence of the many rare and endangered species in traditionally managed European hay meadows. When such land use was first established, it may be supposed that those species unable to survive this type of land use were lost. However, many species did survive and, as we have seen in Chapter 8, adaptation to specific haymaking procedures and dates has occurred through species selection and intraspecific ecotypic differentiation. In the current landscapes, such species are highly endangered if haymaking is abandoned and new forms of grassland management are adopted, for instance, sowing specific cultivars of grasses, together with fertiliser treatment, and multiple cutting dates.

The nature of fragmented populations

In the light of the very widespread, indeed almost worldwide occurrence of cultural landscapes, it is unwise to assume, without investigation, that habitat fragments in regions long settled by humans are the unchanged remnants of natural ecosystems. At first sight, fragments of former natural ecosystems might be expected to survive, but, when their history is examined, it is apparent that many have been slightly or greatly modified by human activities as part of cultural landscapes. This is true not only for those lands comparatively recently exploited in the European colonisation of North America, Australia, Argentina etc., but also in areas such as China, India and Africa, where ecosystems have been subject to centuries of human exploitation.

An additional point to stress is that each area of forest, grassland etc. has its own *site-specific history*. Each is different in the history of its fragmentation, extent and duration of its exploitation, and the sequence of decisive land-use changes to which it has been subjected.

Populations declining to extinction

Conservationists have abundant opportunity to study declining populations, but rarely has the process been recorded in detail. Some case histories have been published (Harper, 1977). In his book *Britain's Rare Flowers*, Marren (1999) cites the case of a rare ecotype of *Veronica spicata* at a site in the East Anglian Breckland, where details of population size have been reported at various dates before it became extinct.

In 1910, there were several hundred specimens of *Veronica* on Garboldisham Heath, but, by 1922, after the area had been used as a military rifle range in the First World War, not a single specimen could be found. However, in 1924, several hundred plants were discovered. During the Second World War the Heath was again used for military purposes, and, at the end of the war, a huge population with 2,000 flowering spikes was found in one area. In 1953, a large part of the Heath was ploughed and the population was then restricted to the edge of the new arable area and to the chalk rubble of the earthworks of an archaeological monument – the Devil's Ditch. Thereafter, despite habitat management – auto-scything part of the dyke – the population declined: zero (1964); three (1965); one (1971); zero (1972 & 1973); and five (1975). Only a few vegetative shoots were found in 1976, but in 1977, eight spikes were found in bud in a group of twelve plants. Finally, the area became a pig farm and the population became extinct. On the road to extinction, the population of *Veronica* apparently passed through several bottleneck events of decline and recovery. However, recovery was not sustained and the population became extinct.

Concerning the cause of decline, Marren (1999) points to the loss of habitat, and the changes in vegetation composition and height that accompanied the decline in grazing not only by sheep, but also by rabbits, populations of which were greatly reduced from 1954/55 onwards, following the introduction of myxomatosis. In addition, with the introduction of arable agriculture on the Heath, the soils and vegetation of the area changed under the impact of herbicides and fertilisers. This example illustrates the point emphasised above, namely that populations often become endangered following changes in land use.

In his account and summary tabulation of the extinction of 20 British species, Marren (1999) provides a *suggested cause* in each case. These range from over-collecting, drainage, eutrophication, loss of old walls, decline in grazing etc. While these suggestions are helpful, it is important to make it clear that, where detailed examination has been possible, population decline and species extinction are often caused by several interacting factors rather than a single cause. Nonetheless, the notion of single factors is strongly held in conservation circles. This point is emphasised by Gurevitch and Padilla (2004, 447) in considering the effects of introduced species. They note: 'Ecologists, conservation biologists and managers widely believe that invasions by non-native species are a leading cause of recent species extinctions'. However, after reviewing the evidence, Gurevitch and Padilla came to the conclusion: 'existing data on causes of extinctions and threats are, in many cases, anecdotal, speculative, or based upon limited field observation. Although it is clear that obtaining quantitative and experimental data are impossible under many circumstances, the problem remains that correlation is often assumed to imply causation.' They stress that endangered plants, generally, 'face more than one threat'.

Not only is it important to examine critically the notion of 'single determining factors', it is crucial to consider how, often, human activities influence the prospects for the populations of plant species through 'distant' disturbances in food chains and webs. Thus, it has been argued that the numbers of two species of palms in central Panama (*Attalea butyracea* and *Astrocaryum standleyanum*) are strongly influenced by the level of poaching of monkeys, peccaries, tapir, deer and certain rodents (Wright *et al.*, 2000). The loss of these herbivores alters the numbers of non-poached species, and, through chains of interactions, adversely affects the palms not only through increased seed predation, but also by reducing seed dispersal and seedling recruitment.

Plant species may be endangered by many factors. In teasing out the key issues in population decline and extinction in any particular case, it is important to make the distinction between the immediate and nearest factor(s), the so-called 'proximate cause(s)', and recognise the activities of humans as 'a common ultimate factor' acting in multifarious ways in ecosystems across the globe.

Studying populations of plants

As a means of introducing key concepts and confronting their practical implications, it is helpful to imagine the position of those who have accepted responsibility for the investigation of populations of a suite of rarities in a local nature reserve. The most important point to stress is that conservation often involves immediate action in the face of a crisis. However, the success of conservation efforts may depend on a thorough knowledge of demography, ecology and history of the threatened populations that can only come through intensive long-term studies.

Counting plants

In order to establish that populations of rarities are indeed in decline through bottleneck effects, it is necessary to make a census of the plants. At first sight, it would seem easy to count plants, as they are generally rooted to the ground. But there are a number of complexities.

Given that population numbers may vary considerably year on year (Harper, 1977), there are great advantages in moving from a 'snapshot' of population numbers, taken on a single visit, to a more secure knowledge, through repeated observations of marked plots in different seasons and over several years. Given the problems of relocating study sites in areas of habitat change, the use of permanent quadrats should be considered, the 'corners' of which are marked by buried metal rods or concrete posts. By close study, it is possible to plot the positions of seedlings, juvenile and mature plants in a selected piece of vegetation. The overlaying of transparent plastic record sheets makes it simple to examine changes over time. Thus, by visiting the study areas repeatedly, it has proved possible to make careful studies of the demography of plant populations by recording the arrival of new seedlings, and the growth, performance and eventual demise of adult plants. It is also possible to assess the effects of pests and diseases. By adding or removing plants it is also possible to investigate the interrelations of species with the ecosystem (Harper, 1977).

Here, our concern is to confront the issues that must be faced in making an accurate census of plant numbers. As in demographic studies of populations of the higher animals, it has proved comparatively easy to count the number of individuals in some plant species (certain annuals, some tree species etc.), but other species present difficulties. As we have seen in Chapter 3, some plant species are clonal and it is difficult to determine what is an individual. In other species, individuals may fuse together, e.g. *Ficus* species (Thomson *et al.*, 1991). Further complexities in studying population numbers arise in particular groups. For example, members of the Orchidaceae have a subterranean post-germination phase of the life cycle involving symbiosis, saprophytism or parasitism in complex host–fungus relationships (Pritchard, 1989). As a consequence, above-ground parts may not emerge for many years. Even after 12 years' study, Hutchings (1989) found it impossible to estimate the length of the subterranean phase in Early Spider Orchid (*Ophrys sphegodes*). In addition, first flowering of the emergent orchid plants may not take place for several years. Surprisingly, his demographic studies also revealed that even when above-ground parts of plants have 'emerged' from their underground organs, they do not necessarily reappear every year. The ability to remain in a dormant state below ground for one or more years has also been discovered in demographic investigations of other species, e.g. Marsh Gentian (*Gentiana pneumonanthe*) (Oostermeijer *et al.*, 1992) and *Delphinium* species in North America (Epling & Lewis, 1952).

Turning to another issue, demographic studies of tropical species may be especially difficult, as there may be a huge 'seedling bank' of suppressed tree seedlings beneath the forest trees, and careful taxonomic work is necessary to 'match' seedlings to their seed parents.

Soil seed banks

Of special importance in assessing population numbers is the discovery that there may be a bank of dormant viable seeds in the soil. Information on soil seed banks has come from three basic types of investigation. (1) Artificial seed burial experiments. (2) Observations of seedlings emerging from cleared plots of ground. (3) Investigations of seedling emergence from field collections of soil taken into glasshouses and other artificially controlled environments. Sometimes, as pre-treatments in laboratory experiments, sub-samples of soil were exposed to cold temperatures, and in some studies attempts were made to 'concentrate' the seed by sieving or flotation treatments in salt solutions (see Thompson & Grime, 1979).

A number of classifications of seed bank behaviour have been proposed (see Thompson & Grime, 1979; Poschlod & Jackel, 1993; Thompson *et al.*, 1997). For the purposes of publishing a database of published information for species in north-west Europe, and taking account of the limitations of some of the data sets, Thompson *et al.* (1997) recognised three different types of seed behaviour: *transient* – viable seeds persisted in the soil for less than 1 year; *short-term persistent* – viable seeds survived for more than 1 year but less than 5 years; and *long-term persistent* – viable seeds survived in soil for at least 5 years. These studies have provided information on the type of seed bank founding different species, and sometimes, where the experiment was appropriately designed, information on the stratification and seed density in different horizons of the soil.

The database gives details of 1,189 members of the north-west European flora and draws not only on European information but also on the behaviour of species in regions such as North America, Australia etc., where they have been introduced. There are some remarkable differences between the species: amongst the top 100, ranked in order of their maximum recorded longevity, more than 50% are reported to have survived more than 40 years in the soil. Others have only transient soil seed banks.

Returning to the decline in small populations, it is clear that the type of a seed bank may determine whether, and how quickly, populations recover from bottleneck effects. This is illustrated by a study by Willems (1995), who examined soil seed banks in a small fragment of species-rich calcareous grassland at a site in Savelsbosch Nature Reserve, Netherlands. After the Second World War, the area of such grasslands declined with the gradual cessation of sheep grazing. For a period the site was not managed, but, in the 1950s, the area became a reserve and the grassland areas were mown. Then, in the late 1980s, the grassland was fenced and sheep introduced in conservation management. As the site is *c.* 5 km from the nearest area of calcareous grassland, it is considered that in almost all cases spontaneous seed immigration must be ruled out, apart from a few species, such as orchids, that are capable of long-range dispersal because of their minute seeds.

The species composition of the seed bank was examined by counting the number and identity of seedlings emerging from soil samples in seed trays. Detailed study of ground vegetation was also carried out, and historical records examined. Three species – *Cirsium acaule*, *Primula veris* and *Carlina vulgaris* – recorded in 1944, were not detected in a survey carried out in the 1970s. Willems was interested to discover whether these species were still represented in the soil seed bank. He failed to detect the species either in the above-ground vegetation or in the seed bank. The database accumulated by Thompson *et al.* (1997) gives information on all three species. *Cirsium acaule* and *Carlina vulgaris* both have transient soil seed banks, while *Primula veris* is reported to have either a transient or short-term persistent soil seed bank. Reflecting on the situation, Willems concludes that the three species – all of rosette habit – were lost in the period 1944–70, when sheep grazing was abandoned, and the sward became higher and denser. Also, any seed of the three species in the seed bank did not remain viable for long enough 'to take advantage' of the conservation management established after the period of abandonment. This example provides yet another example of the vulnerability of species at transitions in the evolution of cultural landscapes, and illustrates the value of long-term monitoring of plant populations.

Thompson *et al.* (1997) provide valuable information on seed banks for NW Europe. But for most of the world's ecosystems information is very limited. Leck, Parker and Simpson (1989) have brought together useful reviews of the seed banks in arctic and alpine areas, coniferous forests, tropical ecosystems, grasslands, Californian chaparral, deserts, wetlands and arable land.

A number of generalisations are emerging from studies of seed bank floras. Harper (1977) suggested: '(i) long-lived seeds are characteristic of disturbed habitats, (ii) most long-lived seeds are annuals or biennials, (iii) small seeds tend to have much greater longevity than large ones, (iv) aquatic plants may have great seed longevity, and (v) seeds of mature tropical forest trees have very short lives.' This summary is quoted from Thompson *et al.* (1997, Preface), who consider each point a hypothesis for further study.

In an evolutionary context, it has been proposed that seed banks have evolved in short-lived plant species that are faced with unpredictable highly variable environments, fitness being maximised, and the likelihood of extinction being reduced, by what is interpreted as a 'hedge-betting' strategy. Following a study of *Phacelia dubia*, a species with considerable seed banks whose populations fluctuate widely in size, Del Castillo (1994) stressed a number of general points important to the understanding of a wide range of species. The seed bank acts as a kind of 'evolutionary memory' of past events. In some years, very large seed crops may be produced and much seed added to the seed bank, the proportional contribution of different genotypes depending on their individual output. In other years, little or no seed may be added. The genetic constitution of each generation of above-ground plants will be influenced by which individuals survive and emerge from the pool of genotypes in the seed bank, as well as other factors such as gene flow between populations.

Demography studies

A number of investigations of rare species have revealed important characteristics of small and declining populations. For example, studies of the Marsh Gentian (*Gentiana pneumonanthe*) by Oostermeijer *et al.* (1992) revealed that some populations are 'dynamic', with high seedling turnover and low numbers of adult flowering plants, while other populations are 'static' or 'senile', consisting of flowering individuals, but hardly any seedlings. These differences are related to the current management of the populations. 'Dynamic' populations are associated with a high level of disturbance associated with mowing, grazing and sod-cutting (a traditional management in the Netherlands, where cut turf is used in animal husbandry etc.). Regular high annual seedling recruitment occurs in these disturbed areas. In contrast, where traditional management has been abandoned, 'senile' populations are found that hardly show any recruitment, even where there are large populations of several thousand adult plants. Even though demographic studies suggest that individual plants may live for 10–20 years, senile populations are at risk of extinction in the longer term if there is no effective renewal of the populations through seedling establishment.

Oostermeijer *et al.* (1992) report that 'rejuvenation of senile populations' may be achieved by reintroducing grazing and mowing, but these managements must be carefully timed to prevent damage to the flowering individuals. In addition, the successful establishment of seedlings is encouraged if the ground around flowering individuals is disturbed by small-scale sod-cutting just before the release of the seeds.

So far we have been considering the demography of comparatively short-lived plants. Longer-lived species may also occur in 'senile populations'. For instance, in the northern wheat belt of Western Australia, forest has been dramatically cleared and only small remnant patches of woodland remain. In a long-term study of one such patch (from 1929 to 1997), no regeneration of *Eucalyptus salmonophloia* and *E. loxophleba* occurred and the condition of the surviving trees greatly deteriorated (Saunders *et al.*, 2003, 245). The lack of regeneration in this case is due to stock grazing of the woodlands and active management, including fencing, will be necessary to 'counter the present regime of benign neglect that characterises most of Australia's management of native vegetation in agricultural landscapes'.

Demographic studies of other long-lived plants have revealed different behaviour. Some species may persist through their capacity for clonal propagation and/or ability to sprout. Bond and Midgley (2001) point out that many woody plants are capable of resprouting after severe damage, surviving droughts, flooding, herbivory, landslides and anthropogenic disturbances etc. that kill many other species. Such species are able to persist for many years by vegetative regeneration, even though they may be poor recruiters by seed establishment. Indeed, it has been proposed that sprouters might allocate more resources to storage of reserves to support re-growth, and that these allocations represent a cost traded against growth and reproduction. Field studies offer some support for this hypothesis (see Bond & Midgley (2001) for a review of the evidence). However, while some species are able to survive for long periods by sprouting, and others persist through clonal growth, both these groups of species are at risk of extinction in the face of overwhelming anthropogenic habitat destruction and exploitation.

This conclusion is supported by a number of studies. An interesting example illustrates the point and also illuminates another issue – the threat to demographic health and survival of many species collected in the wild for horticultural purposes. In South African populations there are many threatened species of cycads. Raimondo and Donaldson (2003) investigated two species. *Encephalartos cycadifolius* is a very slow growing grassland species producing abundant seed that also can re-grow, from underground parts, after fire and damage by animals. In contrast, *E. villosus* is a non-suckering forest species regenerating by seed. Demographic evidence revealed that the collection of seed had minimal impact on population growth in both species. In contrast, the largely illegal collection of material from adults for the horticultural trade has led to rapid population decline in both species, despite the fact that *E. cycadifolius* is potentially immortal through its capacity to regenerate from underground stems. This species grows so slowly that populations suffering a substantial loss of adult plants could not be expected to recover within the time span (<50 years) of a typical conservation programme.

Pollen limitation in plant populations

Turning to other elements of the vortex model, a recent review by Ghazoul (2005c) stresses that pollination failure can arise from a variety of factors in fragmented anthropogenic habitats. It may be a function of 'breeding system, life history, the pollination vector, the degree of specialisation amongst plants and their pollinators, and other indirect effects of habitat change acting on plants and pollinators'. These factors are examined in turn.

In areas transformed by human activities, rare plants are often found in isolated fragments of semi-natural vegetation. As a consequence of changes in insect faunas of such islands, it is possible that insect-pollinated plant species may not receive sufficient visits to cross-pollinate the flowers.

In a test of this hypothesis, two populations of *Dianthus deltoides* in south-west Sweden were compared to see if seed set was lower in a small island population than in a larger one, because of pollen limitation (Jennersten, 1988). The flowers of this self-compatible insect-pollinated species secrete nectar at the base of the stamens. Flowers are protandrous, anthers being ripe two days before the stigmatic surfaces are receptive. The study revealed that the diversity and abundance of insects visiting plants was much lower in the small population. Seed set was higher in the larger population. Hand-pollination increased seed set four-fold in the small population, but not in the large. It was also discovered that the two populations did not differ in the amount of nectar crop and the ovule number per flower. In addition, there was no evidence that either population suffered from resource limitation, as reproductive capacity did not increase at either site by watering or fertiliser addition. Therefore, Jennersten concluded that the differences in seed production at the two sites were the result of differences in pollinator service, the island population being pollen limited. If *Dianthus* flowers are unpollinated for several days, self-pollination may occur. Clearly, if pollen limitation occurs for a number of seasons, regular self-fertilisation will increase the level of inbreeding in small populations with implications for their long-term fitness (see below).

A recent review considers a number of cases where pollen limitation was demonstrated and others where no effects were detected (Ghazoul, 2005c). Generally, only in very small populations, usually less than 50, were effects on fecundity reported.

Pollination disruptions

The very close relationship between plants and their mutualistic pollinators has been widely studied (Proctor, Yeo & Lack, 1996). These relationships are under extreme strain in many parts of the world (Kremen & Ricketts, 2000). But, while there have been serious declines in crop-pollinating honey bees in the USA (Holden, 2006), and declines in butterflies and bumble bees in Europe (Stokstad, 2006), Ghazoul (2005a, b) questions whether there is a global crisis. However, regional problems have been identified. In Hawaii, Samoa etc., many alien species have been introduced and human activities have damaged natural ecosystems. Elmqvist (2000, 1238) paints a 'frightening picture of pollinator disruption that emerges from oceanic islands' with 'reciprocal extinction of entire pollinator sets and the plant guilds that support them . . . entire pollinator sets, including indigenous birds, bats and insects are disappearing. In some cases, introduced organisms begin to fill the role of pollinators, but all too often pollinator niches for indigenous plants (but not invasive species) remain empty.'

In other places, disruption to pollination systems is less severe but still under strain. For example, in a detailed study of populations of two bird-pollinated endemic mistletoe species in New Zealand, pollen limitation was demonstrated, both *Peraxilla colensoi* and *P. tetrapetala* receiving too few visits from their pollinating mutualists (Robertson *et al.*, 1998). Pollination failure has coincided with declines in densities of important pollinators of mistletoes, Bellbirds and Tuis, whose decline, in turn, coincided with the introduction of the Ship Rat (*Rattus rattus*) to New Zealand.

In South Australia, 80–90% of the native vegetation has been replaced by farmland. Only isolated fragments of native woodlands and heathlands remain. In investigations of natural and assisted pollination of flowers of various bird-pollinated species of *Grevillea*, *Astroloma* etc., Paton (2000) found evidence of pollen limitation related to a serious decline in nectarivorous birds (honey eaters) associated with a major increase in populations of the European honey bees introduced into Australia in the 1820s. Feral and managed colonies of bees are now widespread and are present in such numbers that their consumption of limited floral resources contributes to the reduction of honey-eater populations, as birds are food-limited in the summer and autumn. Honey bees visit a range of native plants including bird-pollinated species and their activities have a bearing on pollen limitation. For instance, *Banksia ornata*, which was formerly pollinated by native nectar-feeding birds now present in reduced numbers, is effectively pollinated by introduced honey bees (Paton, 1997).

Allee effects in plants: dioecy and gynodioecy

Returning to the fate of plants in declining populations, it is important to note that reproduction may be curtailed or prevented by a failure to find a mate – the so called Allee effect (Stephens & Sutherland, 1999). For example, in the Kentucky Coffee Tree (*Gymnocladus dioica*) – a rare forest canopy legume species in Canada – sexual reproduction does not occur in most of the populations (Ambrose, 1983). Indeed, only one population of the species has both male and female plants. While seed production has been observed in this population, seedling establishment has not been discovered, and reproduction is entirely by clonal growth from the root systems.

Some plants have a gynodioecious breeding system; populations containing female and hermaphrodite plants. A comparative study of plants from small and large populations of *Lobelia spicata* at various sites in Illinois has been carried out by Byers, Warsaw and Meagher (2005). The genetics of gynodioecy is complex: the original paper may be consulted for details. Here, we note that in small populations the 'normal' functioning of the breeding system was distorted resulting in an increase in female progeny and 'a decrease in male function'. Such changes, which may be due to losses of alleles through genetic drift in declining populations, have 'the potential to impede seed production' in the future.

Allee effects in plants: self-incompatibility

A lack of reproductive success in declining and small populations may arise from self-incompatibility. The phenomenon is best introduced by an example (Demauro, 1994). There are five different variants of the Lakeside Daisy (*Hymenoxys acaulis*) in North America. A rare variant of the plant (var. *glabra*) is endemic in the western Great Lakes Region. While Canadian populations are large, there has been a history of population loss and decline in the USA, and, in the early 1970s only one population of *c.* 30 plants remained at a site in Illinois along the DesPlaines River in Will County. Although insects visited flowers, no seed was produced in the wild or in hand pollinations of material removed from the site. More detailed investigations of the breeding system followed, involving crossing experiments with plants from Illinois and from a site in Ohio. Hand-pollination experiments revealed that the species is strongly self-incompatible.

A self-incompatibility system prevents hermaphrodite plants from producing seeds by self-fertilisation through shared genetic identity, in possessing the same S alleles in the pollen and the style. The system also functions to prevent certain crosses between different plants, namely those possessing the same mating type (mediated by S alleles). For details of the genetics of self-incompatibility mechanisms see Richards (1997). The crossing experiments with *H. acaulis* var. *glabra* detected 15 mating types. Significantly, all the plants from the single small Illinois population were of the same mating type, a finding that explains the lack of sexual reproduction despite visits by pollinating insects.

It is clear that, historically, the population in Will County was once part of a much larger system of populations, almost all of which were destroyed by intense industrial development in the DesPlains river valley. Only one population survived and persisted through clonal growth.

Populations of self-incompatible plants generally have a large number of S alleles (Richards, 1997), and, as a consequence, many compatible matings are possible between members of a population. However, in populations subject to one or more severe bottleneck events, S alleles may be lost through genetic drift, and sexual reproduction may become impossible. In the absence of comprehensive studies of populations of self-incompatible species through their decline from boom to bust, biologists have made a number of computer simulations, modelling the effects of decline in species with different types of self-incompatibility (e.g. Byers & Meagher, 1992; Vekemans, Schierup & Christiansen, 1998). All the models predict that S alleles will be lost from very small populations. For instance, Imrie, Kirkman and Ross (1972) have shown in computer simulations for *Carthamus flavescens* that, in isolated populations with no migration and no seed bank, populations of 16 or fewer plants became extinct following the loss of S alleles. Moreover, it has been discovered in several studies that, if populations are of scattered individuals and there is little or no gene flow between them, then even though the total population may be high, S alleles may be lost by drift to a significant extent in small sub-populations.

Turning to other examples, the first point to stress is that reproductive failure due to loss of S alleles in small populations might be quite common. This follows from the observation that approximately 50% of all angiosperms are genetically self-incompatible (Nettencourt, 1977). Detailed investigations have revealed a number of important insights.

There are less than 50 extant populations of *Aster furcatus* in the USA. By means of self- and cross-pollination experiments, computer simulation studies and investigations of genetic variation using allozymes, Les, Reinartz and Esselman (1991) and Reinhartz and Les (1994) established that this is a self-incompatible species, and that some populations had very little allelic variation in S alleles limiting seed set. In fact, some populations consisted of only a single clonally propagating individual. They also discovered that self-compatible individuals were present in certain populations, a finding of considerable interest to conservationists. It was concluded: '*Aster furcatus* appears to be evolving self-compatibility as a result of bottleneck-induced losses of S-alleles' (Reinhartz & Les, 1994, 446).

Populations of *Rutidosis leptorrhynchoides*, a multi-stemmed herbaceous perennial daisy found in the temperate grasslands of south-east Australia, have been studied by Young *et al.* (2000). The species has declined since the nineteenth century, as grassland ecosystems have been reduced, following European settlement and now only *c.* 0.5% of their original area of 2 million hectares survives (Kirkpatrick, McDougall & Hyde, 1995). Furthermore, human activities within the grassland areas have contributed to the endangering of many native species through deliberate and accidental fires, and the introduction of exotic weeds.

Young *et al.* (2000) investigated material from 22 populations of different size (from 5 to 100,000 plants). Controlled crossing experiments – including bagging, pollinator exclusion and manual pollinations – confirmed that the species is self-incompatible. Small populations showed a significant loss of S alleles relative to large populations. Thus, there were 16 S alleles in a population of 70,000, but only 3 in the smallest population of only 5 plants. This significant genetic erosion of the S alleles has a number of implications. There is severe mate limitation in the smaller populations, with mate availability falling to below 50% in populations smaller than *c.* 60 plants. Although seeds are wind dispersed, it is estimated that dispersal distance is very limited, usually less than 0.5 m. Thus, fragmented populations may be effectively isolated from gene flow between populations. Young and associates also detected the presence of self-compatible individuals (*c.* <15%) in some populations. As in the case of *Aster furcatus* (see above), selection may favour these self-compatible plants in small populations, with the breeding system moving from obligate outcrossing to a mixed mating system. The study also revealed that some populations were diploid, while others were tetraploid. Young *et al.* (2000) discuss the implications of this for S allele erosion in small populations.

Another species of 'daisy' from the south-east Australian grasslands has also been studied, namely, *Leucochrysum albicans* subsp. *albicans* var. *tricolor* (Costin, Morgan & Young, 2001). This species occurs in very small isolated populations persisting only on roadsides, railway verges and in cemeteries. A study of breeding system, seed production and germinability was carried out on material from 14 populations of different sizes (74–50,000). As in the case of *Rutidosis*, *L. albicans* also proved to be self-incompatible. However, experiments revealed that outcrossing rates were uniformly high and seed production was not straightforwardly related to population size. Overall there is no evidence that reproductive success is in decline in fragmented populations. The reasons for this are not clear. However, *Leucochrysum* has a short flowering period and particularly showy flowers, and this may maximise outcrossing rates. In contrast, *Rutidosis* flowers over an extended period, and may be neither showy enough to attract generalist pollinators nor have enough potential mates flowering at any one time to maximise outcrossing rates. Thus, the outcome of population fragmentation and decline in the two species is not the same even though they share the same ecosystem. A general point to stress is that plant species respond differently to the effects of fragmentation (Costin *et al.*, 2001, 283).

A number of studies have determined self-incompatibility to be the cause of reproductive failure in remnant fragmented populations and have recommended how appropriate conservation action could secure the future of the species concerned. Two examples may serve by way of illustration.

Wilcock and Jennings (1999) report that in some populations of *Linnaea borealis*, a dwarf spreading perennial found in boreal forests in Britain, Scandinavia and North America, flowers are well visited by insects, but seed set is variable and sometimes non-existent. Investigations of the underlying cause(s) of reproductive failure were carried out in populations in Deeside, Scotland. Samples of flowers were self- and cross-pollinated, and transfers of pollen were made between individuals and populations in the field. It was concluded that the species is strongly self-incompatible, and populations contained only one mating type. Many Scottish populations are therefore genetically incapable of sexual reproduction, and colonies consist of clonally spreading individuals. As a conservation measure, the partner limitation found in *Linnaea* could be remedied by translocation of plants of appropriate genotypes between populations.

Translocation has also been recommended in other situations, for instance in the case of the endangered *Ziziphus celata* – a woody clonal shrub of Lake Wales Ridge, Florida (Weekley & Race, 2001). Only six populations remain, four of which are sterile, perhaps consisting of a single clone. The species occurs in degraded fragmented populations, the surrounding area having been converted to pasture. Crossing experiments revealed that the species is self-incompatible and Weekley and Race recommend that 'compatible mating types' should be identified and translocated to the different hitherto sterile populations to create 'reproductively viable populations'. The issue of translocations, to promote the survival of endangered species, will be examined further in Chapter 16.

Climatic limitations on reproduction

In addition to the factors discussed above, evidence suggests that there may be other reasons for reproductive failure that are unconnected with human activities. Investigations reveal that, in favourable climatic periods, certain plants may have extended the limits of their latitudinal or altitudinal range and can survive vegetatively in locations where, at present, their reproduction is prevented by climatic factors etc. For example, in studies of *Tilia cordata* in northern Britain, Pigott and Huntley (1981) discovered that the key factor preventing reproduction by seed is the temperature sensitivity of pollen tube growth and the short period of receptivity of the stigma and style. Pollen tubes did not grow sufficiently rapidly to allow fertilisation before the abortion of the styles. Concerning the past history of the species, evidence suggests that this tree reached its present northern limit in the warmer climate of *c*. 5,000 BP. Woodward, (1987) gives details of other cases where climate plays a crucial role in determining the altitudinal or latitudinal limits of a species, through the sensitivity of life-cycle stages.

The extinction vortex: stochastic events

A declining population may be severely damaged by stochastic events and be driven to a lower population level or to extinction. 'The sobering message . . . is that even very rare events can doom a population to extinction, if they are severe enough . . . conservation biologists may seriously underestimate the hazards faced by small populations if they neglect these low-frequency events' (Holsinger, 2000, 68).

In the extinction process some authorities stress the importance of 'natural' stochastic events, such as flooding, fires, droughts, landslips, tornadoes and other storms (Young & Clarke, 2000). However, for Shafer (1990, 40), 'the term environmental stochasticity will be taken to include variations attributable to competition, predation, disease and natural disasters'. Essentially, random variables are involved. In essence, the next state of the environment is not fully determined by what has previously happened.

While it could be argued that some stochastic events are entirely 'natural', clearly, in other cases, human influences cannot be ruled out. For example, huge numbers of alien animals, plants, diseases and pests have been introduced and their arrival may be a catastrophic event for native fauna and flora. Landslides and floods may be more frequent in areas of deforestation. Also, the frequency and severity of storms has been linked to anthropogenic global climate change (see Chapter 18).

The frequency and severity of fires has a major impact on ecosystems. While some are 'natural' fires from lightning strikes, others are the result of human activities. Fire has a major influence on the establishment and growth of many species. For instance, the endangered western Australian shrub *Verticordia fimbrilepis* ssp. *fimbrilepis* is now restricted primarily to roadside verges, as the ecosystem in which it formerly grew has been largely lost in the development of an agricultural landscape (Yates & Ladd, 2005). *Verticordia* has a seed bank of viable seeds in the soil that declines after 30 months. It is instructive to consider the regeneration niche for this species, as it is different from the adult niche (see Grubb (1977) for more details of this important concept). It has been discovered that fire may kill adult *Verticordia* plants, yet germination of seeds of the species is stimulated by smoke, and mass recruitment of seedlings occurs in first and second winters after fire. There is currently some germination between fires, but it is not yet clear whether this is enough to sustain the populations long term at all sites, for most populations are in decline and cannot recover without periodic fires. This raises a considerable practical difficulty for conservationists, as fire suppression has been a major concern for land managers in the agricultural landscapes where *Verticordia* grows.

Concerning the effect of catastrophic events, it is helpful to consider an example, reported in studies of *Collinsia verna* (Scrophulariaceae), a winter annual found on the flood plains and damp woodlands in eastern USA (Kalisz, Horth & McPeek, 2000). The species has a soil seed bank of dormant seeds that can last for at least 4 years. Random stochastic events sometimes influence the populations of the species, which may be as large 20 million or as small as seven plants. For instance, in 1992–3 there was a major flood at a study site at Raccoon Grove Forest Preserve near Monee, Illinois. All adult plants were destroyed, and also a great deal of soil containing some of the seed bank was washed away. However, some of the seed bank remained on site and germinated in the following autumn. Thus, the soil seed bank may play a crucial role in the recovery and persistence of this species in areas subject to rare catastrophic flooding.

The vortex model: genetic effects

The concept of the idealised population and how declining populations depart from the 'ideal'

As we have noted in Chapter 3, natural populations of plants differ greatly in their structure and also in their breeding systems, which may involve obligate outbreeding mediated by dioecy (separate male and female plants) or genetic self-incompatibility; mixed mating systems of outcrossing and selfing; self-fertilisation predominantly; or facultative and obligate apomixis. Species also differ greatly in the degree to which they are able to reproduce by asexual vegetative means.

Threatened and endangered species can be found within each of these different major categories of breeding behaviour, presenting the conservationist with a bewildering array of different situations. How have geneticists approached this complex area? What follows is based on the excellent account of Frankham *et al.* (2002).

Population genetics is built on the concept of an 'idealised population', and much has been learned about the theory and practice of conservation management of small and declining populations (most particularly in animals) by comparing the behaviour of 'real' populations against this standard. The idealised population has the following properties: it is a large population drawn at random as a sample from a very large base population. This idealised population is of constant size, in which generations are distinct and non-overlapping, all individuals have the potential to breed, sex ratios are equal, mating is at random (including selfing), there is no selection at any part of the life cycle, and mutation is ignored. The idealised population is visualised as a closed system with no migration.

Starting from the notion of an 'idealised' system, a large number of models have been devised to reflect the population biology of 'real' animals and plants, by examining what happens when one or more factors are changed from the 'ideal', e.g. overlapping generations, etc. In addition, it has proved possible to test and refine many of these models in experiments with wild, experimental and captive populations. Frankham *et al.* (2002) provide the most recent, comprehensive and accessible account of conservation genetics. Acknowledging the extreme complexity of plant and animal populations and the gaps in our understanding, they predict that there are three inescapable 'genetic' properties of small and declining populations that have to be confronted if such populations are to be managed to ensure their long-term survival.

Firstly, in nature, populations differ from the ideal in a number of respects. For instance, in the idealised population all the individuals take part in reproduction, but in natural populations, however, not all the individuals may take part. Moreover, there may be unequal numbers in different generations, unbalanced sex ratios, overlapping generations etc. In addition, in a major departure from the behaviour of individuals in the idealised population, different seed parents may make different contributions to the next generation. For example, in a study of *Papaver dubium*, it was discovered that in a population of 2,316 plants, 50% of the seed was produced by only 2% of the population, with 4.6% being produced by one very fecund individual (observations of Mackay, quoted in Crawford, 1984). In recognition of the difference between the 'ideal' and 'real-life' populations, geneticists estimate the effective population size (N_e), which may be much smaller than the absolute (census) number in the population (Wright, 1931). Frankham *et al.* (2002, 189) note:

The effective size of a population is the size of the idealised population that would lose genetic diversity (or become inbred) at the same rate as the actual population. For example, if a real population loses genetic diversity at the same rate as an ideal population of 100, then we say the real population has an effective size of 100, even if it contains 1000 individuals. Thus, the N_e of a population is a measure of its genetic behaviour, relative to that of an ideal population.

For details of how to estimate N_e see Crawford (1984) and Frankham *et al.* (2002). The fundamental point to conclude from these observations is that the effective population is likely to be much smaller than the census population and this is highly significant in small and declining populations of conservation interest and concern.

A second inescapable genetic consequence of smallness and decline in population size is the greater impact of chance effects through genetic drift. In recognition of the great importance of this phenomenon, it is important to stress once again the changes in genetic composition that can occur as a consequence of irregular random fluctuations in gene frequency. Thus, as plants reproduce 'each offspring receives one allele selected at random from each parent. Just by chance, some alleles, especially rare ones, may not be passed on to the offspring and may be lost' (Frankham *et al.*, 2002, 178). Thus, allele frequencies may change from generation to generation. In addition, separate fragments of a once single united population may diverge genetically. The effects of genetic drift are predicted to be significantly greater in the small and declining populations, whose fate is such a major concern to conservationists (Peters, Lonie & Moran, 1980; Moran & Hopper, 1983).

In testing these ideas, some carefully designed experimental studies with animals have been carried out. For example, the effects of chance have been estimated in experiments with artificial populations of flour beetles, comparing replicate lines of large and small populations separately maintained through 20 generations (McCauley & Wade, 1981). In practice, separating the effects of genetic drift from those of selection in wild populations presents many difficulties. In order to advance beyond simple mathematical models and confront the complexities of wild populations, geneticists have devised computer programmes that model a 'reasonable level of reality and complexity' allowing the evaluation of 'the likely effects of drift, selection, migration and population subdivision' (Frankham *et al.*, 2002, 194).

The third, inescapable problem in small fragmented and genetically isolated populations is inbreeding (see Chapter 3). In the course of time sexual reproduction amongst relatives becomes inevitable in small populations, and, as a consequence, all the individuals in a population come to be genetically related. Inbreeding in normally outcrossed species leads to inbreeding depression, as deleterious recessive alleles are expressed, homozygosity increases and heterozygosity is reduced. Symptoms of inbreeding depression include lower survival and much reduced reproductive success, leading to an increased risk of extinction.

From theoretical considerations, computer modelling and controlled experiments, the three concepts outlined above have all been tested, as hypotheses, in investigations of declining populations of wild plants 'in nature'.

The genetics of small and fragmented populations

Remarkable insights into the genetics of wild populations have been made by employing allozyme techniques and DNA marker systems (see Chapter 7). It is beyond the scope of this book to examine details of these different techniques. Full accounts are provided by Karp *et al.* (1998) and Baker (2000).

Starting with the notion of the idealised population, new insights have followed the pioneering studies of Wright (1931), and it is now possible to estimate a range of genetic measures from allozyme and molecular marker data sets, including the number of loci that are polymorphic (P); the number of alleles per locus (A); the number of alleles per polymorphic locus (AP); the proportion of individuals or loci that would be expected to be heterozygous (H_e) and the observed heterozygosity (H_o). In characterising the variability in particular studies, species-level measures include the subscript 's', e.g. P_s, A_s etc., and population-level comparisons have subscript 'p', e.g. P_p, A_p etc. In the use of F statistics, F_{is} 'relates individuals to subpopulations'; F_{it} ' relates individuals to total population'; and F_{st} 'relates subpopulations to the total population' (Beebee & Rowe, 2004, 302). Thus, F_{st} provides some estimates of gene flow through migration of seeds/fruits and nuclear gene flow (N_m). Comparative studies of genetic markers have added greatly to our understanding of rare species. For the testing of some hypotheses, studies of allozyme and/or molecular markers have been carried out in combination with cultivation trials, crossing experiments, progeny testing etc. Hundreds of papers have been published worldwide on conservation genetics of plants and animals. Here, based on the analyses set out in Frankham (1996), the results of testing a number of key hypotheses are presented.

1 Genetic variation is greater in species with wider ranges

Hamrick and Godt (1989, 1504) tested this hypothesis and discovered that 'populations within plant species with wider geographic ranges have higher allozyme variation'. Endemic species exhibited the least variability. This finding is of considerable conservation interest as many endemic species are threatened with extinction.

2 Rare species are less variable genetically than common plants

To examine general trends, Cole (2003) analysed data from 247 plant species in 57 generic-level comparisons of rare and common species. He discovered that 'all species-level measures of variation (P_s, A_s, AP_s, H_{es}) and mean population-level measures (P_p, A_p, AP_p, H_{ep} and H_o) show reductions significant at the $p < 0.001$ level'. This provides clear evidence that, within this data set and for these marker genes, rare plants are generally less variable than common species overall and show reduced variability for a range of different measures. However, it is important not to extend the generalisation too far. For only a relatively few species of plants is there an adequate genetic database. The sample size is small relative to the projected number of endangered species – *c*. 10% of the world's flora (a minimum of, say, 25,000 species).

It is important to stress that the results of some studies do not conform to a hypothesis. For example, in an investigation studying allozymes, Ge *et al*. (1998) detected high levels of variability in both the endangered Chinese species *Adenophora lobophylla* and in its widespread congener *A. potaninii*. In another investigation, employing RAPDs techniques, higher levels of variation were detected in the endangered South African Cone bush (*Leucadendron elimense*) than in the common *L. salignum* (Tansley & Brown, 2000). At first sight that result seems counterintuitive. However, one key factor, likely to be important in the maintenance of genetic variability, is the occurrence of a large long-lived bank of seeds in the soil. This seed bank acts as a store of genetic variability in a species that is often subject to bottleneck events.

3 Small population often have low genetic variability

Some species are to be found only in very small populations. In the following case studies it was discovered that populations exhibit low or no genetic variability.

- *Allozyme studies*: Waller, O'Malley and Gawler (1987) investigating the endemic *Pedicularis furbishiae* from northern Maine, USA; Godt, Hamrick and Bratton (1994) studying the lily *Helonias bullata* found in the wetlands of south-eastern USA; and Shapcott (1998) examining variability in the rare palm *Ptychosperma bleeseri* in the Northern Territory, Australia.
- *Molecular markers*: Gustafsson and Gustafsson (1994) and Black-Samuelsson *et al*. (1997) investigating Swedish populations of the rare *Vicia pisiformis* using RFLP and RAPD techniques; Swensen *et al*. (1995) examining variability in the endangered island endemic *Malacothamnus fasciculatus* var. *nesioticus* from Santa Cruz island off southern California, USA (allozymes, RAPD and ribosomal DNA techniques employed); and Calero *et al*. (1999) investigating *Lysimachia minoricensis*, a species endemic to the island of Minorca, Spain.

4 Genetic variation within species is positively correlated with population size

Theoretical considerations suggest that small populations will be less variable genetically than large populations of the same species. This hypothesis has been tested by Frankham (1996), in an analysis of data sets from allozyme studies of 23 investigations of plants and animals (of which 16 are plant species). In the comparisons, population size varied greatly, e.g. from a single individual to 20,000 flowering plants in *Gentiana pneumonanthe* (Raijmann *et al*., 1994) and 20 individuals to 400,000 in *Halocarpus bidwilli* (Billington, 1991). Frankham discovered that correlation between allozyme genetic variation (H_e or H_o) and the logarithm of population size was positive in 22 of the 23 studies, 'representing a highly significant excess of positive correlations ($X^2 = 19.17$, df = 1, $p < 0.000.025$)'. Since Frankham's review, investigations of other species have confirmed that small populations are generally less variable than large (see Frankham *et al.*, 2002; Frankham, 2005a, b), but it is unwise to generalise (see Shapcott, 1994).

5 Patterns of genetic variability detected in space reflect what happens to declining populations in time

As we have just seen, intraspecific genetic variation has been examined in large and small populations of plants collected from different sites. Do these comparisons of different populations in space reflect what is likely to happen in any particular population as it declines through time? This hypothesis, as far as I am aware, has not been properly tested, for it would require long-term historical and genetic studies. This issue has been discussed by Matocq and Villablanca (2001) in the context of their study of kangaroo rats in California, in which they compared genetic variability in modern samples with that revealed by a collection of museum specimens of skins collected in 1918. Evidence from gene sequencing (a fragment of cytochrome b gene) suggested that the low genetic diversity detected in modern samples appeared to be a long-established 'historic' feature of the species rather than a recent event. With respect to plants, the authors suggested that, for plant species, comparison of recent collections with historic material housed in herbarium collections might shed light on the magnitude and rate of loss of variation in declining populations. However, they drew attention to the fact that generally only very small and perhaps unrepresentative collections are found in herbaria.

6 Adaptive genetic variation decreases in declining populations

Bekessy *et al.* (2003) consider that studies of neutral genetic markers (allozymes and RAPDs) have made a major contribution to our understanding of genetic variability in species and populations. However, they point out that often only small samples are examined. Moreover, they emphasise: 'estimates of genetic variability and differentiation may differ widely depending on the choice of marker (Geburek, 1997), statistics employed (Bossart & Prowell, 1998), the life-stage sampled (Alvarez-Buylla *et al.*, 1996) and even the laboratory in which experiments are conducted (Jones *et al.*, 1997, 267)'. Bekessy *et al.* (2003) also confront a major issue: do neutral markers provide a reliable indication of what will happen to adaptive significant genetic traits in declining populations? This question is posed in the context of studies of genetic variability of the Monkey Puzzle tree (*Araucaria araucana*), an iconic species endemic to southern Chile and Argentina. This culturally significant species is found from 600 m to almost 2,000 m above sea level, in a range of ecological conditions. In a search for effective conservation management of this vulnerable species, genetic variability was assessed using molecular markers (RAPDs). This data set proved useful, but Bekessy *et al.* (2003) point out that the study of neutral markers failed to detect differences in one important trait, namely population differences in drought tolerance. They stress the importance of discovering whether populations contain variability for adaptively significant morphological, physiological and behavioural traits, and recommend an integrated approach to assessing genetic variability using both molecular and classical approaches. Such investigations would involve crossing experiments, progeny trials and tolerance testing. Also, they recommend that garden trials and reciprocal transplant experiments be carried out. Such tests can be particularly revealing of the presence/absence of traits of high adaptive significance, such as the timing of flowering and fruiting, and heat, drought, frost tolerance etc. However, it is important to stress that, while studies of neutral genetic markers may be carried out relatively quickly, the analysis of adaptively significant traits is a much more daunting long-term task, as many of these traits may be polygenically controlled (Podolsky, 2001).

7 *Declining populations are subject to an increasing risk of inbreeding depression*

The effects of inbreeding are estimated by comparing the performance of progenies of self-pollinated individuals with progenies of the same stocks derived from deliberate controlled cross-fertilisation. A range of characters have been examined in studies of inbreeding depression, including the percentage of filled seed, germination rate, plant size, seed production and various measures of fitness (see Frankel, Brown & Burdon, 1995, and references cited therein).

Many investigations of plants carried out in controlled environments have revealed significant inbreeding depression (Charlesworth & Charlesworth, 1987). For example, in a glasshouse experiment, Dudash (1990) discovered 55% inbreeding depression in selfed progenies of *Sabatia angularis*. It has long been suspected that inbreeding depression may be more severe in wild populations. This hypothesis was tested by Dudash in his investigation of *S. angularis*, and it was discovered that there was 75% inbreeding depression in the companion field experiment using the same stocks. This figure is well in excess of the 55% inbreeding depression estimated in the glasshouse. It is suggested that conditions for growth are optimised in the glasshouse environment, while plants in the wild are subject to more stressful environments. Thus, inbreeding depression is not a fixed quantity, and 'is likely to be substantially greater in nature than estimates from captive populations of animals and plants would lead us to believe' (Frankham *et al.*, 2002, 288).

Few investigators have studied inbreeding depression in wild populations of different sizes. Van Treuren *et al.* (1993), investigating remnant populations of *Scabiosa columbaria* in the Netherlands, discovered no clear relationship between population size and the level of inbreeding depression, despite their finding that the species is highly susceptible to inbreeding depression. However, in a study of 17 Dutch populations of *Succisa pratensis*, Vergeer *et al.* (2003) discovered that smaller populations had less genetic variation and higher inbreeding coefficients.

Theoretical considerations, computer simulations and empirical investigations all suggest that inbreeding depression is an important phenomenon in declining populations. Further investigations are necessary to determine the speed with which inbreeding depression might seriously impact plant populations with different types of population fragmentation, breeding systems, life form, site histories etc. Moreover, our understanding of the process in wild populations is incomplete. It has been suggested that, in time, declining populations are 'purged' through inbreeding and selection of their genetic load of deleterious alleles (Lande, 1988). Frankham (2005a) has reviewed the literature on the subject and concluded that theoretical work 'indicates that purging will have only modest effects in small populations' and empirical studies have 'typically found only moderate effects of purging' (Byers & Waller, 1999; Crnokrak & Barrett, 2002; Paland & Schmidt, 2003).

Further studies of declining populations are necessary to increase our understanding of the extent and implications of inbreeding depression in wild populations, especially the purging of genetic loads. This is particularly true of plant species that reproduce predominantly by self-fertilisation. It is possible that species reproducing regularly by selfing will be less affected by population decline in small isolated habitat fragments, as deleterious recessive alleles will already have segregated out and been lost by natural selection (Charlesworth & Charlesworth, 1987). The evidence so far available suggests that 'selfers' are subject to inbreeding depression, but to a lesser degree than in those species that are regularly outcrossed (Husband & Schemske, 1996).

How large do populations have to be to ensure long-term survival?

To estimate the probability of long-term persistence of populations, a wide range of Population Viability Analyses (PVA) have been developed. Early models examined future population survival or extinction from demographic information on birth, survival and death rates, population sizes, and the stochastic effects of natural events, often ignoring risks associated with genetic factors and catastrophes (Reed *et al.*, 2001). Thus, PVAs project 'forward in time' through carrying out replicate runs with data sets with computer packages or tailor-made software. As the models incorporate stochastic factors, replicate runs do not give exactly the same result, but collectively reveal important information on extinction risk. There is no single definition of a PVA: risk is calculated under a range of different assumptions by different investigators, usually in relation to the probability of a minimum viable population (MVP) surviving for a defined time period. Thus, Meffe and Carroll (1994, 678) define the minimum viable population as: 'the smallest isolated population size that has a specified statistical chance of remaining extant for a specified period of time in the face of foreseeable demographic, genetic and environmental stochasticities, plus natural catastrophies'.

Examining the development of the concept, early estimates of MVPs for conservation were devised from population genetics theory in combination with experience in breeding domestic and laboratory animals (Frankel & Soulé, 1981). It was discovered that the maximum tolerable rate of inbreeding is *c.* 1%, and in order to offer protection against immediate loss of fitness a minimum effective population size of 50 (N_e) is necessary in zoos and other holding operations. Populations of such size would, however, offer no long-term security: an effective population of 500 (N_e) was necessary for populations in the wild and in nature reserves to prevent erosion of genetic variability for an indefinite period and to allow continuing evolution. These 50/500 'rules of thumb' (Frankel & Soulé, 1981), which were widely circulated in the conservation community, were later reconsidered (Soulé, 1987).

In later investigations, MVPs have been determined for particular species of plants. For instance, in the case of *Asarum canadense*, calculations of PVA including environmental and demographic stochasticity revealed that an MVA of 1,000 individuals would be necessary to give a 95% probability of surviving for 100 years (Damman & Cain, 1998).

In a second example, the demography of wild Ginseng (*Panax quinquefolius*) was studied at several sites in central Appalachia, USA. This species is harvested in the wild as a medicinal plant. McGraw and Furedi (2005) carried out a PVA and determined that a MVP of *c*. 800 was necessary to give a 95% probability of surviving for 100 years. It is of interest that this value is greater than that found in any one of the existing populations currently being monitored (present largest population census 406 plants). However, the analyses carried out so far have not included all the factors that might influence ginseng populations. Thus, pressures on populations of wild ginseng may increase by deforestation, disease introduced from cultivated plants into the wild, an increasing demand for the plants as a medicinal herb and by increasing deer browsing. Alternatively, pressures on the species may be eased by factors tending to decrease the deer population (the reintroduction of top carnivores in forests and the spread of ungulate diseases). In addition, conservation and other management decisions may increase the extent of forest habitat allowing for greater populations of wild ginseng. In addition, the introduction of simulated wild populations of plants (cultivated seed sown in woodlands) may decrease pressure on the wild populations. A third example is provided by the studies of the Mexican palm (Fig. 12.2).

A different approach to the question of MVPs was made by Reed (2005, 563). He examined the results of 11 studies in which plant populations of different size were studied and fitness measured. 'The evidence suggests that there is a linear relationship between log population size and population fitness over the range of population sizes examined.' He concludes that populations will have to be 'maintained to numbers >2000 to maintain fitness for the long-term' and that 'conservation efforts should ultimately aim at maintaining populations of several thousand individuals to ensure long term persistence'.

Overall, relatively few PVAs have been calculated for plants. It is extremely difficult to collect the long-term field data necessary for the PVA modelling. For example, even after 11 years monitoring of populations of the rare cactus *Pediocactus paradinei* in 14 permanent plots, there were uncertainties about the reliability of PVA assessments in this long-lived perennial species with its erratic and infrequent seedling recruitment (Frye, 1996). Moreover, PVA studies face a number of major challenges in adapting models designed primarily for animals to take account of the 'peculiarities' of plants, in particular the sessile habit of most higher plant species; the widespread occurrence of phenotypic plasticity and clonal growth; the complexities of plant population dynamics; sub-structuring

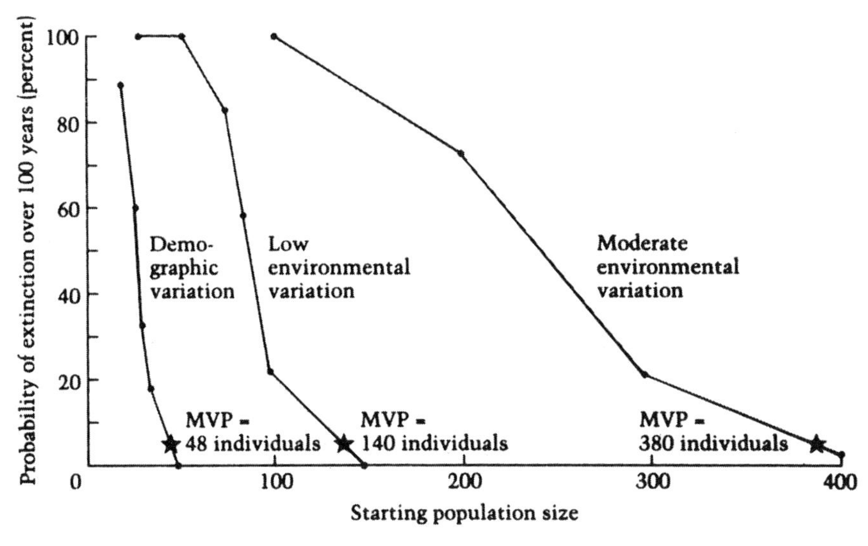

Fig. 12.2 The effects of demographic variation, low environmental variation and moderate environmental variation on the probability of extinction of a population of Mexican palm, *Astrocaryum mexicanum*. In this study, the minimum viable population size, shown as stars, was defined as the population size at which there is a less than 5% chance of the population going extinct within 100 years. (After Menges, 1991, from Primack, 1993. Reproduced by kind permission of Professor E. Menges)

and gene flow; the wide range of plant breeding systems; the occurrence of a wide range of mutualisms; and the presence and dynamics of soil seed banks in many species.

Several hundred PVAs have been published for a range of mostly animal species (Menges, 2000; Reed *et al.*, 2002). Conservationists have employed PVA in the defence of potential habitat loss, for example by predicting the minimum viable populations necessary for the survival long term of Californian Spotted Owls (*Strix occidentalis occidentalis*), in the face of proposals for changes in land use (LaHaye, Gutierrez & Akçakaya, 1994).

PVA predictions and conservation

Some authorities have questioned the accuracy and practical value of computer simulations (Possingham, 1996). Burgman and Possingham (2000) make three important points. Firstly, 'the role of population modelling in conservation is suffering a backlash from early over-enthusiasm and an unrealistic expectation that PVA would solve all single species conservation problems'. Secondly, some critics are of the opinion that PVA is largely of theoretical academic interest with only limited practical application. Thirdly, it has been claimed that the wide availability of software presents investigators with a 'loaded gun' that is 'dangerous in the hands of the untrained'. In countering these criticisms, Burgman and Possingham, (2000, 102) contend that PVA should be used with other approaches for predicting the outcome of different conservation management strategies. Essentially, PVA provides a 'decision-support tool rather than a decision-making tool', and is best used in 'comparing risks rather than measuring risks'. Reed *et al.* (2002) have also made a critical review of the subject. They point out that computer packages to examine PVA are widely available and require no mathematical or programming skills. They too question the 'appropriateness of the software' chosen by some investigators. Frankham *et al.* (2002, 520) offer an important overview. PVAs may be of value, in the same way as weather and economic forecasts, but only if they are based on accurate data sets. However, they conclude that even if the PVAs are inaccurate to a degree, investigations using this approach are important in focusing attention on areas of ignorance in the population biology of species of conservation concern.

What of developments in the longer term? Overall, Reed *et al.* (2002) consider that the devising of more sophisticated PVAs is retarded less by constraints in modelling than by the lack of availability of sufficient reliable field data on all aspects of plant biology, ecology and genetics. However, there are some major weaknesses in current PVA models; for example, many models assume random mating (Reed *et al.*, 2002). Thus, PVA models were developed for 'idealised' populations and major challenges lie ahead to develop more 'realistic' models appropriate to real-life populations, such as those of ginseng that we considered above. Responding to these challenges, recent developments in PVA take account of genetic concerns such as inbreeding, habitat fragmentation and other factors (Burgman & Possingham, 2000; Reed *et al.*, 2002). Increasingly, populations and species are subject to conservation and other management, and adapting PVA to take account of the likely effects of different management regimes presents a formidable challenge. In addition, a far more difficult series of problems is facing the conservation community; namely, the modelling of PVA in relation to projected global climate changes.

Metapopulations

So far we have discussed what happens in individual declining populations. However, groups of related populations may exist in the wild, as for example when human activities disrupt once continuous natural habitat into smaller fragmented populations. As we saw in Chapter 3, such population systems may function as metapopulations.

The metapopulation concept recognises the presence of localised patches of suitable habitat for a particular species. At any one time, some of these patches are occupied by a species in question, but not necessarily all of them. Considering an unoccupied patch, colonisation eventually occurs by seed/fruit dispersal from an occupied patch. The newly established colony expands, but later numbers again decline, and extinction follows. The metapopulation model, therefore, highlights the interaction between populations, emphasising that each patch has a finite life. The success of a species in an area depends upon there being, at any point in time, some patches with healthy colonies, for it is from these that empty patches are successfully recolonised. A range of metapopulation models has been developed. Brussard (1997) characterises three different variants: (i) 'classical (Levins type) metapopulations consist of several small and extinction prone local populations connected by a moderate amount of migration'; (ii) another model considers 'mainland-island metapopulations', which 'are made up of one or more large habitat blocks that supply colonists to small outlying patches'; and (iii) 'non-equilibrium metapopulations [that] are declining to extinction because dispersal is too infrequent for reestablishment after local extinctions occur'.

Support for the concept of metapopulations has come from studies of population in many animal groups – from kangaroo rats to monk seals and mountain sheep to butterflies (Thrall, Burdon & Murray, 2000). Some botanical studies have also employed the concept in characterising the dynamic of populations. For example, there are 28 colonies of *Pedicularis furbishiae* along the St John River in Maine, USA (Menges, 1990, 1991). Mapping these colonies over 4 years established the patterns of extinction and colonisation and confirmed that the species existed as an interconnected series of populations with metapopulation characteristics. In addition, the metapopulation concept was tested in groups of island populations of *Silene dioica* in Sweden (Giles & Goudet, 1997; Giles, Lundqvist & Goudet, 1998; Ingvarsson & Giles, 1999) and several species in the scrublands of south-central Florida, USA (Quintana-Ascencio & Menges, 1996).

Metapopulation models are still the subject of lively debate (see Harrison, 1994; Freckleton & Watkinson, 2003). Firstly, there are questions about the universality of the concept. Following a review of the literature Harrison (1994) writes: 'Are all species truly metapopulations, if a long enough time scale is considered? It is sometimes argued that this is so.' She concludes that the concept is valuable in that it sets out important hypotheses to test and 'important data to gather' but, overall, she is 'suspicious of powerful generalisations' and 'ready to use formulas' (Harrison, 1994). Freckleton and Watkinson (2003) emphasise 'a patchy population is not a metapopulation' and raise the important question: 'Do researchers try to "shoehorn" all populations into the restricted framework of the metapopulation model?' Several investigators stress the importance of studying different plant species on a case-by-case basis. A recent example is revealing. Jäkäläniemi *et al.* (2005) have studied the population dynamics of the endangered species *Silene tatarica* in Finland. They discovered that this riverside plant does not exhibit the extinction–recolonisation dynamics predicted from metapopulations at equilibrium. They found high rates of recolonisation, with establishment tracking the available habitat, which was not resolved into discrete habitat patches.

The reviews of the metapopulation concept highlight several issues with practical implications. It is important to emphasise that not all extinction is caused by human activity: some, according to the metapopulation concept, is part of the natural extinction/recolonisation cycles found in the wild. In addition, there are some special concerns about studying plant populations. It is difficult enough in animal studies to be completely certain that a species/population has become extinct. In plants, this may be even more problematic, as, for example, in long-lived populations of woodland and grassland species, many of which have extensive vegetative reproduction and some have long-lived seed banks. Given these characteristics, extinction/recolonisation events are likely to occur over a very extended time frame, well beyond the time period of the average research project.

Another problem has also been identified. It is very difficult to be sure that a habitat once occupied by a particular species is still suitable for that species. Given the many ways in which human activities can impinge on habitat, apparently 'suitable' habitat may have been made 'unsuitable' directly or indirectly by human activities. This problem is of real concern to conservationists. Should they wait to see if natural recolonisation occurs at a site of the local extinction of a species, or should they take immediate action to reintroduce the extinct species? There is also a problem of the conservation of single populations of rare species, as it could be argued from the metapopulation model that such populations are doomed to extinction, only those species with metapopulation structure being viable in the long term. Metapopulation models place great emphasis on recolonisation of 'vacant' but suitable territory. In order to encourage gene flow/migration processes, the establishment of wildlife corridors has been encouraged. Are such corridors necessary in all cases? Finally, several authors have extended the metapopulation concept and now include, within the concept, captive animals in zoos and, by implication, plants in botanic gardens (see Foose *et al.*, 1995). In effect the stocks in zoos and botanic gardens may be viewed as a 'mainland' resource in a metapopulation model from which new populations can be created in the wild to replace those that have gone extinct. Given that many animal and plant populations are under threat in the wild, 'natural' extinction and recolonisation cycles may cease to operate and artificial reintroduction of stocks of zoo and gardens into the wild may become necessary. We consider these conservation issues further in later chapters.

Concluding comments

This chapter has been largely concerned with analysing the vortex model of declining populations. The model is an important tool for the study of anthropogenic habitat fragmentation. It is important to stress that while such fragmentation may occur as humans exploit natural ecosystems, commonly, across the world, fragments of ecosystems of interest to conservationists are elements of former or contemporary cultural landscapes (Keitt, 2003).

With regard to the fragmentation process itself, Ewers and Didham (2006, 117) in a recent review, point to the many advances in our knowledge, but also highlight the limitations of our current understanding and the need for further research. They stress that fragmentation is most evident to the observer as a 'landscape phenomenon', but clearly, there are profound implications for the ecosystems the fragments contain, which may be revealed only by very close long-term study. Thus, 'species that survive in habitat remnants are confronted with a modified environment of reduced area, increased isolation and novel ecological boundaries'. Individual species react in different ways to fragmentation, depending on their life history strategies, dispersal ability, degree of habitat specialisation etc. Complex changes occur at fragment boundaries often producing clear 'edge effects' that may penetrate deep into the fragment. Some effects of fragmentation appear comparatively quickly, but many researchers have stressed that longer term there may be 'reductions in species richness'– so-called extinction debts – as populations of some species decline to extinction. Ewers and Didham (2006, 117) also emphasise: 'the synergistic interactions of fragmentation with climate change, human-altered disturbance regimes, species interactions and other drivers of population decline [that] may magnify the impacts of fragmentation', and they conclude: 'anthropogenic fragmentation is a recent phenomenon in evolutionary time' and emphasise: 'the final, long-term impacts of habitat fragmentation may not yet have shown themselves'.

Returning to the vortex model, we have seen there is now an expanding base of theoretical knowledge and an increasing body of scientific results that reveal the interaction of many factors in population decline. And there is enough evidence to conclude that declining populations of rare and endangered species are in danger of genetic erosion and extinction. Such a conclusion raises many complex issues on how far species decline might be halted and reversed by conservation management, and these concerns will be considered in Chapter 15.

Considering the direction of future research on the vortex model, there is an urgent need to discover more about the behaviour of populations in the field. For example, it has been suggested that the effects of inbreeding predicted from models employing random mating may not serve to reveal what happens in 'wild' populations of plants (Amos & Balmford, 2001). If most of the progeny come from the mating of the most outbred plants in the population, perhaps, inbreeding depression may not be such a pressing issue. Also, there is another issue worthy of further investigation. Many plant species are polyploids and, in comparison with diploids, repeated selfing may not lead quite so swiftly to inbreeding depression.

Further research would also be valuable on the interaction of different factors in the vortex model. A recent investigation of small and large populations of *Succisa pratensis* in the Netherlands points the way (Vergeer *et al.*, 2003, 18). In a study of 17 populations of different sizes, it was discovered that there were strong interactions between population size, genetic variability and habitat quality (in particular the degree of eutrophication of the soil). Larger populations were found in less eutrophic conditions. Small populations were found to be less variable, and had higher inbreeding coefficients. It is predicted from these studies that 'there will be a continuing decline in small populations due to deteriorating habitat conditions, decreased genetic variation and a reduced reproductive capacity'.

Overall, it must be stressed that the different factors involved in the vortex model of population decline have been tested on different species. For no individual plant species have all the elements of the decline and fall of populations been fully investigated. Given that plant species are so individualistic in their ecology, population biology, breeding systems, site histories etc., a great deal of research is needed to understand the patterns and processes of population decline for a species being considered for conservation.

In Chapters 8–12, many aspects of microevolution in response to human activities have been discussed. However, one key area – hybridisation within and between species – has yet to be examined, and these interactions will be reviewed in the next chapter.

13

Hybridisation and speciation in anthropogenically influenced ecosystems

This chapter considers a number of questions relating to hybridisation and speciation in ecosystems disturbed by human activities. Many species of plants are currently endangered. If there are many extinctions in the near future, will new species quickly evolve to take their place? As a consequence of human activities, a large number of species of plants have been introduced accidentally or deliberately across the world. What are the consequences of such human-induced breakdown of geographical isolation between species? Agricultural landscapes cover a large area of the globe. What have been the microevolutionary effects of the interaction of crop, wild, garden and weedy plants? Finally, what are the implications for conservation of species hybridisation in anthropogenic ecosystems?

Will new species quickly evolve to take the place of those that become extinct?

Human activities produce many patchy and mosaic habitats. In situations where a genetically polymorphic species occurs in two or more adjacent habitats, disruptive selection is likely to take place and this could provide the setting for the early stages of speciation. An example shows the possibilities. In many parts of Britain, within the extensive pasture lands there exist islands of heavy-metal contaminated debris and spoil left by the mining industry. As we saw in Chapter 10, where a species occurs both on the mine debris and in the pasture, there is evidence of gene flow and disruptive selection. Pasture plants are heavy-metal sensitive and cannot grow on the mine: mine plants perform less successfully in the pasture. Gene flow will produce a mixture of progeny, and mine × mine and pasture × pasture crosses are likely to produce the fittest offspring. Theoretical considerations suggest that selection would favour any mechanism that encouraged within-habitat crosses. Cultivation and other experiments by Antonovics (1968) and McNeilly and Antonovics (1968) of samples of *Agrostis capillaris* (*A. tenuis*) and *Anthoxanthum odoratum* drawn from a transect across the mine-pasture boundary revealed that, although there was an overlap in flowering time between pasture and mine plants, those on the mine flowered about one week earlier, a difference that persisted in cultivation. In addition, some plants on the mine and in the pasture proved to be self-fertile, even though both species are generally self-incompatible. A difference in flowering time and a degree of self-fertility could be seen as the beginnings of isolating mechanisms and, therefore, represents the first steps on the road to speciation. It is of considerable interest that patterns of differentiation have persisted for over 40 years at this site (Antonovics, 2006). Given that these two species have a high turnover of individuals and different patterns of flowering difference (shown to have a genetic basis in the original studies), the persistence of the pattern argues for strong selective forces maintaining the differences.

Thus, we can see that the initial stages of speciation – population differentiation involving all the processes of microevolution – will certainly occur over time in anthropogenically fragmented areas, including scattered nature reserves and in mosaic situations. However, what we know of gradual speciation suggests that it is a long process, perhaps involving very many generations of plants (Hendry, Nosil & Rieseberg, 2007), and it seems unlikely that new species will evolve quickly by this route, to take the place of those that have recently become extinct, or are on the verge of extinction.

One of the major advances of our understanding in the post-Darwinian period has been the recognition that polyploidy is a frequent means of plant speciation. By this route new species may be created instantaneously through somatic mutation or the fusion of unreduced gametes. Polyploidy has been a significant factor in the evolution of crop and garden plants (Simmonds, 1976; Walters, 1993). Is polyploidy a route to rapid speciation of many new species of plants, 'to replace' those that might soon become extinct?

Minority-type disadvantage

Research has revealed that unreduced gametes are often produced at low frequency by many species, yet polyploidy is comparatively rare. To achieve species-hood the new individual polyploid must establish itself and reproduce successfully within an ecosystem. Here, it faces competition from other plants, not least from its own probably more numerous, parental species. New polyploids may, therefore, suffer from 'minority-type disadvantage': their Darwinian fitness may be insufficient for them to survive. As we shall see below, there are only three case histories where new successful 'wild' polyploids have been produced, as a consequence of human activities.

Evidence reveals that the interaction between species in the three cases involved first hybridisation and then polyploidy. Two of the case histories concern a native and an introduced species. The third case history reveals hybridisation and subsequent polyploidy between three introduced species of the same genus.

The evolution of a new polyploid in *Spartina*

In the nineteenth century, the eastern North American coastal species *Spartina alterniflora* ($2n = 62$) was accidentally introduced into Western Europe. Most likely it came in ship's ballast. On tidal mud in Southampton Water, England, *S. alterniflora* hybridised with the native *S. maritima* ($2n = 60$). The first confirmed record of the sterile cross between the two species, *S.* × *townsendii* ($2n = 62$) was in 1870, and the plant still grows vegetatively in the area of origin. A fertile 'hybrid variant' (*S. anglica*) was later detected with

$2n = 120$, 122 and 124 chromosomes. The first certain record of this plant was in 1892. It was postulated that this had arisen by chromosome doubling from *S.* × *townsendii* to give a vigorous allopolyploid (Marchant, 1967, 1968; Salmon, Ainouche & Wendel, 2005). As the parental species are hexaploids, *S. anglica* is a dodecaploid allopolyploid.

A series of investigations have confirmed these hypotheses, through studies of seed proteins and isozyme markers (Raybould *et al.*, 1991a, b), and later by means of DNA markers. The inheritance of chloroplast DNA (cpDNA) in most angiosperms is strictly maternal and therefore provides a means of detecting parentage in hybrid taxa. As *S. alterniflora*, *S.* × *townsendii* and *S. anglica* all have the same chloroplast genome, *S. alterniflora* was the female parent in the initial cross (Ferris, King & Gray, 1997).

Using molecular markers (random amplified polymorphic DNA markers: RAPD and inter-simple-sequence repeats: ISSR), Ayres and Strong (2001) discovered that the DNA profiles in *S.* × *townsendii* and *S. anglica* were identical in their aggregates of diagnostic fragments characterising *S. alterniflora* and *S. maritima*, thus confirming their postulated origins. Despite early studies that revealed a lack of genetic variability in *S. anglica* (Raybould *et al.*, 1991a, b), the molecular investigations of Ayres and Strong (2001) detected 'widespread genetic variation within *S. anglica*' that could indicate: (a) the species could have been produced more than once from the same but genetically different parents; (b) alleles could be lost through recombination; or (c) as *S. anglica* has different chromosome numbers, genetic variability was due to the loss of entire chromosomes.

In 1892, another sterile *Spartina* hybrid, somewhat different in morphology from *S.* × *townsendii*, was discovered in the Bidassoa estuary of the Basque region of France and this taxon was named *S.* × *neyrautii*. It was suggested that this plant was the reciprocal hybrid between *S. alterniflora* (male) and *S. maritima* (female) (Raybould *et al.*, 1990). However, molecular investigations (RAPD, ISSR and cpDNA) have disproved this hypothesis, revealing that *S.* × *townsendii* and *S.* × *neyrautii* both share the same maternal (*S. alterniflora*) and paternal (*S. maritima*) parental species. Thus, two independent hybridisation events occurred, but in the production of the two hybrids, in England and south-west France, parents of different nuclear genotype were involved (Baumel *et al.*, 2003).

Spartina anglica is characterised by vigorous rhizomatous clonal propagation. Formed from two parental species that lacked variability in the area where they hybridised, the new polyploid was able to overcome minority-type disadvantage as it has greater morphological plasticity and a wider ecological amplitude than either of its parents in tolerating both highly anoxic and polluted sediments. This plant has been widely introduced in Northern Europe, North America, China, Australia, New Zealand etc. It has spread rapidly from sites of introduction, and has become an invasive 'weed' in some areas.

Spartina anglica provides geneticists with an opportunity to examine what happens in the early evolution of a 'new' polyploid and its interaction with parental and other species of the genus. It is of great interest, therefore, that the investigations of Salmon *et al.* (2005) have detected rapid genetic changes in *S. anglica*. Furthermore, studies have also examined the rapid decline in *S. maritima* in the northernmost parts of its range. The species reproduces almost exclusively by vegetative spread, and molecular studies have revealed that populations lack genetic variability. Only one nuclear marker out of 98 investigated showed polymorphism (Yannic, Baumel & Ainouche, 2004). Ainouche *et al.* (2004) provide a very valuable review of more than 100 years of research on hybridisation, speciation and polyploidy in *Spartina*, including the interactions between native North American *S. foliosa* and the introduced *S. anglica* (see below).

The origin of *Senecio cambrensis*

Senecio cambrensis ($2n = 60$) was first recorded on roadsides at Cefn-y-Bedd, near Ffrith, in North Wales, in 1948 (Rosser, 1955; Ingram & Noltie, 1995). Later, the species was discovered near Colwyn Bay, North Wales, and Ludlow in England. It was similar to the sterile *S.* × *baxteri* ($2n = 30$), a hybrid from the cross between the introduced *S. squalidus* ($2n = 20$) and native species *S. vulgaris* ($2n = 40$), but *S. cambrensis* produced viable seeds. The ancestry of the 'new' polyploid became clear when it was 'recreated' in the laboratory by treating the synthesised hybrid *S. vulgaris* × *S. squalidus* with colchicine, a chemical that by interfering with spindle formation produces polyploid cells (experiment performed by Harland and reported by Rosser, 1955). The resynthesis of *S. cambrensis* has also been successfully carried out in experiments by Weir and Ingram (1980).

Concerning the ancestry of *S. cambrensis*, another hypothesis has been proposed and tested (Lowe & Abbott, 1996). The species could have arisen directly from the related species *S. teneriffae* ($2n = 60$), but studies of patterns of isozymes in the two species do not support this possibility (Abbott & Lowe, 2004).

Further studies of the new polyploid have revealed the presence of *S. cambrensis* on derelict land at Leith, Edinburgh, a long way from the original site in Wales (Abbott, Ingram & Noltie, 1983; Abbott, Noltie & Ingram, 1983). Studies of cpDNA showed that *S. cambrensis* had different profiles at the two sites, providing convincing evidence for polytopic origin of the species.

Molecular studies are continuing on *S. cambrensis*, which may have undergone rapid genome evolution since its origin (Abbott & Lowe, 2004). *Senecio cambrensis* (and the taxa involved in its origin) is also being employed in studies of the effects of hybridisation and polyploidy on floral gene expression (Hegarty *et al.*, 2006).

As we have noted above, when new polyploids arise they may not survive because of minority-type disadvantage. Studies of *S. cambrensis* reveal that it is highly self-fertile, thus, at point of origin, it was reproductively isolated from the more numerous parental populations and was able to produce fertile offspring by self-pollination. Initially, studies of *S. cambrensis* in Wales revealed that populations were expanding. In contrast, the *S. cambrensis* population in Edinburgh has become extinct following redevelopment in the area (Abbott & Forbes, 2002). Recent research, however, has also shown a decline in *S. cambrensis*, both in population number and population size in Wales, despite the detection of high genetic variability in AFLP molecular markers (Abbott, Ireland & Rogers, 2007). The cause(s) of this decline are, as yet, unclear.

Further investigations have detected another new species of *Senecio* in York, UK (Lowe & Abbott, 2004). Investigations using isozyme markers, together with experimental resynthesis, have revealed that this new species – *Senecio eboracensis* ($2n = 40$) – also owes its origin to hybridisation between the native species *S. vulgaris* ($2n = 40$) and the introduced species *S. squalidus* ($2n = 20$). All newly arising species face minority type disadvantage. *Senecio eboracensis* presumably passed the initial test of survival and reproduction amongst individuals of its parental species because it is self-fertile. From a conservation viewpoint, this new species presents an interesting problem. The long-term survival of the taxon is in doubt, as it is in an urban area, and could be swept away with city development.

Evolution of new *Tragopogon* species

Three European species of *Tragopogon* (all diploids with $2n = 12$ chromosomes) have been introduced into North America, becoming weeds of roadsides and disturbed ground. *Tragopogon dubius* arrived about 1928 while *T. porrifolius* and *T. pratensis* came earlier, around 1916. Although largely allopatric in Europe, these three species can produce highly sterile hybrids. Accidental introduction into the USA brought the three species together in the Palouse region of eastern Washington and adjacent Idaho. Ownbey (1950) studying mixed populations found not only sterile interspecific hybrids ($2n = 12$), but also detected small groups of fertile plants with intermediate characters that proved on investigation to be tetraploids with $2n = 24$ chromosomes. On morphological grounds, Ownbey suggested that *T. mirus* was the hybrid derivative *T. dubius* × *T. porrifolius*, while *T. miscellus* was the cross between *T. pratensis* and *T. dubius*.

A long succession of investigations using a range of techniques have confirmed Ownbey's hypotheses – studies of karyotypes (Ownbey & McCollum, 1954; Brehm & Ownbey, 1965), biochemical characteristics (Belzer & Ownbey, 1971), isozymes (Roose & Gottlieb, 1976), and DNA markers (a large number of contributions by different investigators are reviewed by Soltis *et al.* (2004)).

On morphological grounds, Ownbey postulated that the two polyploids had both arisen polytopically. Each had arisen more than once at separate sites: three in the case of *T. mirus* and two in the case of *T. miscellus*. Further investigations of origins have been carried out using DNA and isozyme markers and these have estimated that, in the Palouse area, *T. miscellus* has between 2 and 21 independent origins, while *T. mirus* has 5 to 9 (Soltis *et al.*, 1995).

Molecular studies have also explored the direction of crossing that gave rise to the new polyploids (Soltis & Soltis, 1989). Studies of cpDNA revealed that *T. miscellus* in the Pullman area had *T. dubius* as the maternal parent, but in other places *T. pratensis* was the female parent. All the populations of *T. mirus* tested in this study proved to have *T. porrifolius* as the female parent.

There have been dramatic increases in the distribution of both the parental and new polyploid species since the pioneering studies of Ownbey. Studies at some sites have revealed increases in the populations of the polyploids, but at others there has been a decline in numbers and some populations have become extinct, e.g. the building of a new house and garden in Moscow, Idaho, destroyed one population of *T. miscellus*. Further hybridisations are possible 'in the wild' as 25 non-native species obtained by Ownbey have formed hybrids in the garden where they were grown, and three species of *Tragopogon* have escaped 'into the wild'.

Further studies of the *Tragopogons* have been carried out (Cook *et al.*, 1998; Pires *et al.*, 2004; Soltis *et al.*, 2004; Kovarik *et al.*, 2005). In particular, the 'new' polyploids appear to have overcome minority type disadvantage, as they are more successful ecologically than their parental taxa. In addition, advanced molecular and cytological techniques are being employed to examine the microevolutionary changes that have occurred since the polyploids were formed. All the chromosomes of the diploid progenitors have been characterised and it appears that the diploid complements are additive in both polyploids with no major chromosome rearrangements. However, changes in gene expression have been detected in *T. miscellus* and its diploid parent as well as between populations of the polyploid from different origins. Other studies have examined gene silencing and novel gene expression in the polyploid relative to the parental diploid stocks (Soltis *et al.*, 2004) and rapid evolution of nuclear ribosomal DNA (Kovarik *et al.*, 2005; Matyasek *et al.*, 2007).

Evolution of new homoploid species through hybridisation

Evidence from molecular and other studies has shown that in a few cases new species have arisen sympatrically by hybridisation between species with no evidence of polyploidy (see Hegarty & Hiscock, 2005; Chapman & Burke, 2007; Rieseberg & Willis, 2007). In most situations, it is not clear whether human activities, either direct or indirect, have been involved in the origin of these new taxa. However, in the case of *Senecio squalidus* the evidence for human influence on the evolution of a neospecies is very strong.

Investigation of diploid European *Senecio* suggests that the original plants of '*S. squalidus*', collected on Mt Etna and brought to Oxford Botanic Garden in the eighteenth century, may have been hybrids between *S. chrysanthemifolius* and *S. aethnensis* (Abbott, Curnow & Irwin, 1995; James & Abbott, 2005). Recent molecular studies employing random amplified polymorphic DNA/intersimple sequence repeats (RAPD/ISSR) have confirmed this hypothesis. The role of human intervention is clear in this case. Hybrid plants were transported to Oxford, providing long-distance separation and reproductive isolation from parental stocks on Etna. Thus, S. *squalidus* is a diploid hybrid of mixed ancestry. In considering the eventual spectacular extension of the range of the species, it is possible that *S. squalidus* was 'preadapted', or, perhaps, significant microevolutionary changes may have occurred in the Oxford garden, in the city or elsewhere, as the plant began to colonise habitats across the city and beyond through the railway network.

The rarity of recent polyploidy in wild plants

Concerning the incidence of 'new' polyploidy, it is important to reiterate the point that there are few confirmed cases of very recent origin of 'wild' polyploids. However, the floras of many areas have not yet been properly examined and polyploids may be more common than present information indicates. The probability of finding more 'new' polyploids is high, as a very large number of species have been introduced across the world by human activities. Taking the longer perspective, following the break-up of Pangaea to form the current continental land masses, the enormous diversity of plant species developed in geographical isolation. In addition, natural rare long-range dispersal populated isolated volcanic islands, where speciation in isolation contributed to further diversification. The ancient patterns of plant distribution are being dramatically altered, both accidentally and deliberately, by human activities. There would seem to be increasing opportunities for both hybridisation and polyploidy to take place. However, current evidence suggests that neither gradual speciation nor the abrupt speciation associated with hybridisation and polyploidy will lead to rapid evolution of 'sufficient' new plant species to 'take the place' of those currently threatened with extinction.

What happens when human activities cause a breakdown of reproductive isolation?

We have considered three cases where new species have developed by polyploidy following hybridisation, as a consequence of the breakdown of geographical isolation. Turning now to an allied phenomenon, what happens when there is breakdown of ecological isolation between evolving taxa? As we have seen earlier, such a breakdown may occur naturally, but it now occurs with great frequency as a consequence of human activities.

High levels of interspecific hybridisation (Stace, 1975), and intergeneric hybridisation are reported. Knobloch (1971) listed 23,675 putative interspecific and intergeneric hybrids. Following a study of regional floras, Ellstrand *et al.* (1996) estimated that *c.* 11 per cent of species were involved in hybridisation. Stace (1989), extrapolating from the incidence of hybrids in the British Flora, estimated that there might be 78,000 naturally occurring interspecific hybrids in the Angiosperms. Some groups have a particularly high incidence of hybrids, willowherbs (Onagraceae), orchids (Orchidaceae), pines (Pinaceae), the rose family (Rosaceae), willows (Salicaceae). Ellstrand, Whitkus and Rieseberg (1996) added the figworts (Scrophulariaceae), grasses (Poaceae), composites (Asteraceae) and sedges (Cyperaceae) to the list of over-represented groups.

Reflecting on these figures, a number of important caveats must be entered (Maunder *et al.*, 2004).

1. The figures are heavily biased: only information from well-investigated temperate floras is available.
2. Concerning the incidence of hybridisation in ecosystems disturbed by human activities, it is not always clear whether the figures relate to 'natural' events or 'human disturbance' (Maunder *et al.*, 2004, 325).
3. There is a much more fundamental issue to face. What is a hybrid? Maunder *et al.* (2004) draw attention to the different usages of the term hybrid. Geneticists refer to the offspring obtained by crossing two genetically distinct lines as hybrids, while taxonomists generally restrict the term to offspring of crossing between taxonomically distinct species. As we have seen in Chapter 6, taxonomists and others have different definitions and working concepts of 'what is a species'.

Thus, there are very considerable difficulties in interpreting the number of hybrids reported in the literature.

4. There is much debate about the extent and significance of intra- and interspecific gene flow.

Hybridisation: the extent of gene flow

'According to the Neo-Darwinist view of evolutionary biology in the mid-1900s, natural selection held centre stage, but gene flow and its interspecific counterpart, hybridisation, played major supporting parts . . . Both gene flow and hybridisation were thought to be common and important mechanisms for Darwinian change' (Ellstrand, 2003, 1163). However, a major reappraisal of gene flow and hybridisation occurred in the 1970s and 1980s, when experiments on plants revealed that the dispersal of both pollen and seed from individual plants was often restricted, being highly skewed with most pollen/seeds dispersing close to the pollen/seed parent, with a rapidly diminishing amount 'travelling' some distance from the source. Technically, the distribution of pollen/seed was not normally distributed, but is of leptokurtic form. Thus, it was concluded that gene flow could be very restricted. As a consequence, the importance of hybridisation and introgression were probably underestimated. However, in the last 20 years or so, a great deal of research has been carried out on gene flow and hybridisation using molecular markers. A variety of different studies have contributed to our current understanding. In the following pages the recent insights into interspecific hybridisation, including introgression, and intraspecific hybridisation will be reviewed. In essence, the 1980s consensus on the extent and rate of gene flow and hybridisation has been overturned (Ellstrand, 2003, 1163):

> Twenty-five years ago, both were considered rare and largely inconsequential. Now gene flow and hybridisation are known to be idiosyncratic, varying with the specific populations involved. Gene flow typically occurs at evolutionary significant rates, and at significant distances. Spontaneous hybridization occasionally has important applied consequences, such as stimulating the evolution of more aggressive invasives and increasing the extinction risk for rare species.

The potential for the hybridisation to produced new weeds and invasives is also examined.

Hybridisation and introgression: the use of molecular markers to test hypotheses

In the past, hypotheses concerning hybridisation/introgression/gene flow in wild plants were examined by means of morphological studies and investigations of pollen and seed dispersal. Such studies do not provide critical tests of hypotheses (Briggs & Walters, 1997). As the following case histories reveal, molecular techniques have provided crucial insights into the role of hybridisation in microevolution patterns and processes (Rieseberg & Wendel, 1993). Various interactions are examined: (a) between wild species; (b) between native wild and introduced species; (c) between wild and crop plants; and (d) between wild, weedy and crop species. A total of 165 proposed well-documented cases of introgression have been examined. (A longer list could have been assembled, if all the 'proposed' cases of introgression had been included.) In 37 cases, the evidence comes from studies with

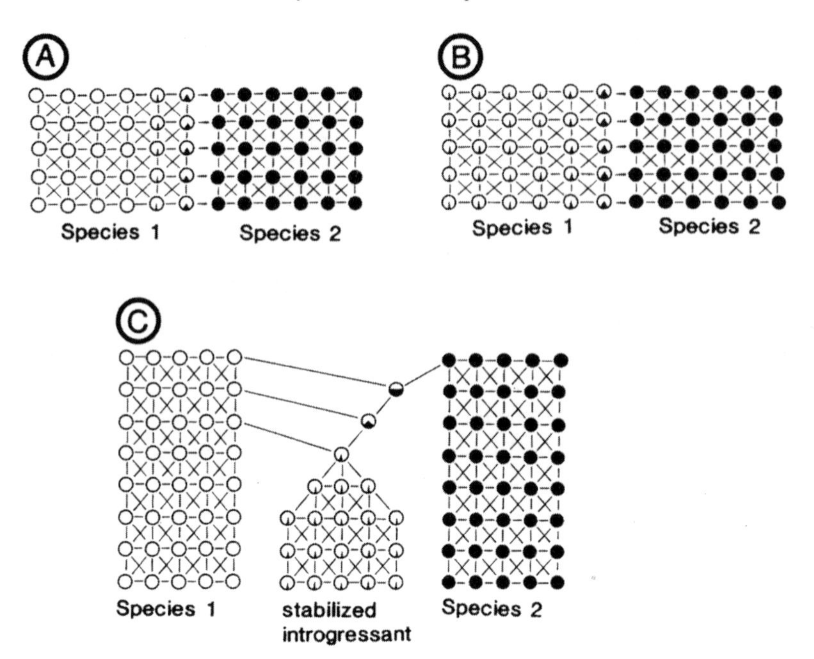

Fig. 13.1 (A) Localised introgression, (B) dispersed introgression and (C) the origin of a stabilised introgressant between two species, populations of which are indicated by black and white circles, with lines indicating the direction of crosses. (From Rieseberg & Wendel, 1993. Reproduced with permission of Oxford University Press)

molecular tools and is regarded as 'robust'. Since this review, further investigations have been published (see below). The role of humans is not always clearly indicated in studies of plant introgression. Here, the concern is to examine models and case histories where introgression has been studied in relation to anthropogenic factors (Fig. 13.1).

Breakdown of ecological isolation in *Iris*: species interactions

In Southern Louisiana, a number of perennial clonally spreading *Iris* species occur. Early investigation by Riley and others concerned two species: *I. fulva* with brick-red flowers found in semi-shaded understorey areas on wet clay soils along the elevated levee banks of the 'bayou' water channels of the Mississippi river; and the violet-flowered *I. hexagona* (called var. *giganti-caerulea* in the early literature) found in full-sun on open interconnected freshwater marshes and swamps near the artificial water channels. In areas of human disturbance there are apparent hybrids between the two species with a range of flower colour. In areas where no disturbance was apparent, flower colour was more uniform (Fig. 13.2).

Morphological studies of these *Iris* populations supported the hypothesis of hybridisation (Riley, 1938), and this case history was used as a major example by Anderson (1949), who argued that introgressive hybridisation was important in microevolution. In the case of *Iris*

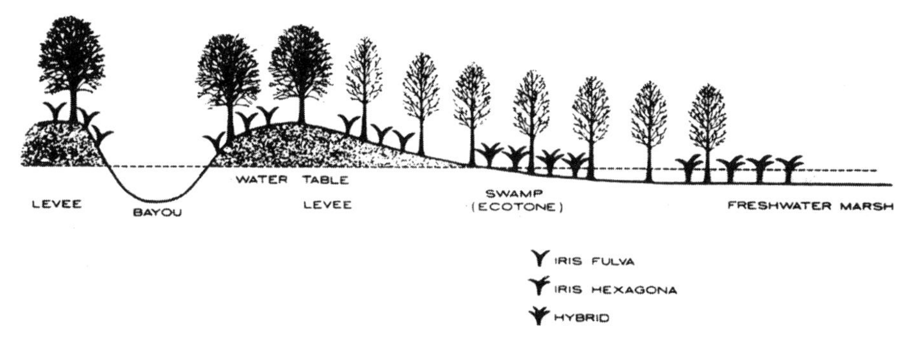

Fig. 13.2 Habitat associations of *I. fulva*, *I. hexagona* and natural hybrids. This illustration advances the hypothesis that natural hybrids occur in the ecotone between the parental habitats. (After Viosca, 1935, from Arnold & Bennett, 1993. Reproduced with permission of the American Iris Society)

the evidence for his conclusion came from morphological investigations. A sample of each of the supposed parental taxa was collected at sites where they did not come into contact, and at two sites where apparent hybridisation was occurring. Seven floral characteristics in which *I. fulva* and *I. hexagona* differed were examined, and displayed as a hybrid index in which the presumed parental species gain the highest (in this case *I. hexagona*) and lowest (*I. fulva*) scores, with plants of intermediate morphology having intermediate values.

Anderson (1949) considers the ecological conditions in the area with the hybrid populations. He points out that the delta region was settled mainly by the French, whose small farms lined the rivers and bayous. Property boundaries were at right angles to the rivers in long narrow holdings, set out as fields, pasture and woodland managed for cordwood and timber. No two farms were treated in the same fashion. In undisturbed areas the two *Iris* species grew in ecologically different areas. However, in areas of human disturbance, putative hybrid swarms of the two *Iris* species were growing – in a mosaic of second-growth woodland, grazing areas with good grass cover and heavily overgrazed pastures with bare soil and cattle wallows etc.

Anderson (1949, 15, 62–63) considers the site to be an example of a general phenomenon, the breakdown of natural ecological isolation of species by human activities in the landscape. He notes: 'when he [man] digs ditches, lumbers woodland, builds roads, creates pastures, etc., man unconsciously brings about new combinations of light and moisture and soil conditions'. In a memorable phrase, Anderson considered that at many sites, and in many ways, humans may be said to 'hybridize the habitat'. He speculates on the possibility that introgression might have regional as well as local effects. With regard to the significance of introgression in evolution, he notes: 'nearly all the published data on introgression demonstrates its importance in areas where man has upset natural forces'. But he goes on to emphasise:

We might logically expect that introgression would be equally effective when nature herself does the upsetting. Floods, fires, tornadoes and hurricanes must certainly have operated upon natural vegetation long before the advent of man. Like man himself all these phenomena alter conditions catastrophically, break down barriers between species, and provide unusual new habitats in which hybrid derivatives may for a time find a foothold, thus serving as a bridge by which groups of genes from one species can invade the germplasm of another.

Anderson's interpretation of the variation in *Iris* was questioned by Randolph, Nelson and Plaisted (1967), who concluded that, while there was evidence for hybridisation at some sites where there were narrow zones of contact between the species, there was no evidence that introgression had blurred the boundaries between species generally. Moreover, it also emerged that the two species had different chromosome numbers: *I. fulva*, $2n = 42$: *I. hexagona*, $2n = 44$.

Early investigations of introgression in 'wild'species, such as those on which Anderson constructed his model, were based on morphological studies alone. A critical test of the introgression hypothesis requires evidence of incorporation of genes from one species into another. The study of wild populations has been transformed by the use of molecular markers. The classic studies of the Louisana Irises has been re-examined by Arnold and associates using a variety of nuclear (rDNA, allozyme and RAPD) and cytoplasmic (cpDNA) genetic markers (Arnold & Bennett, 1993). Studies have been made of sites where the 'parental' species occur in isolation. Species-specific markers have been identified.

In an area where putative hybrid swarms have been described in earlier investigations – Bayou L'ourse – the proportion of *I. fulva*- and *I. hexagona*-specific molecular markers was determined for each individual plant (Fig. 13.3). Multiple genotypes were discovered and patterns of genetic variation indicated 'ongoing hybridisation and introgression' with a 'preponderance of advanced generation *I. hexagona* backcross individuals' (Arnold & Bennett, 1993, 122). These investigations also confirm Anderson's speculation about the wider significance of introgression. Not only was introgression detected where the two species occurred together; there was also evidence for dispersed introgression, as Arnold and co-workers detected 'introgression of markers well outside present-day zones of sympatry' (Arnold, Bouck & Cornman, 2004).

In most flowering plants, cpDNA markers are maternally inherited, being carried through the egg cells from the seed parent to its seeds. This mode of inheritance provides a means of detecting whether longer-range introgression was by seed dispersal or by pollen. Evidence suggests that introgression was by pollen gene flow, most probably facilitated by long-range foraging of bees (Fig. 13.4). In contrast, 'seed-mediated gene flow' was not detected, 'even over very short distances' (Arnold *et al.*, 2004, 145).

Studies by Arnold and associates have investigated other aspects of introgression.

1. In its natural habitat *I. hexagona* is subject to brackish waters, while *I. fulva* grows in freshwater areas. Experiments on the effect of different salt concentrations revealed a fitness hierarchy: *I. hexagona* > hybrids > *I. fulva*.
2. Shade tolerance has been examined for the two parental and two hybrid classes. *Iris fulva* was most shade tolerant and therefore had the highest inferred fitness in shade, with *I. hexagona* having the lowest fitness. Hybrids were of intermediate tolerance. Those morphologically similar to

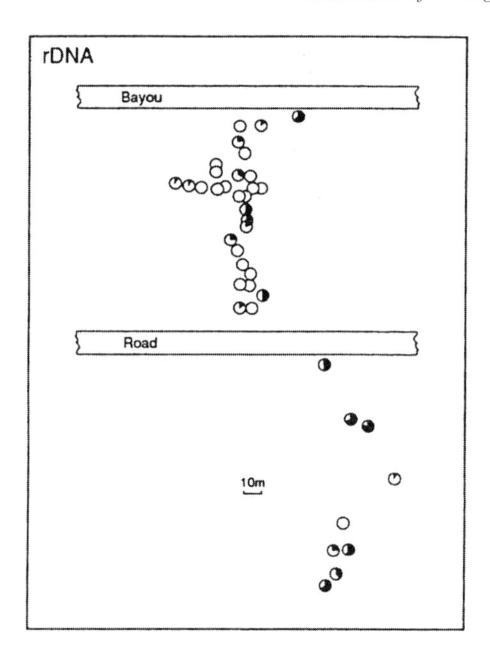

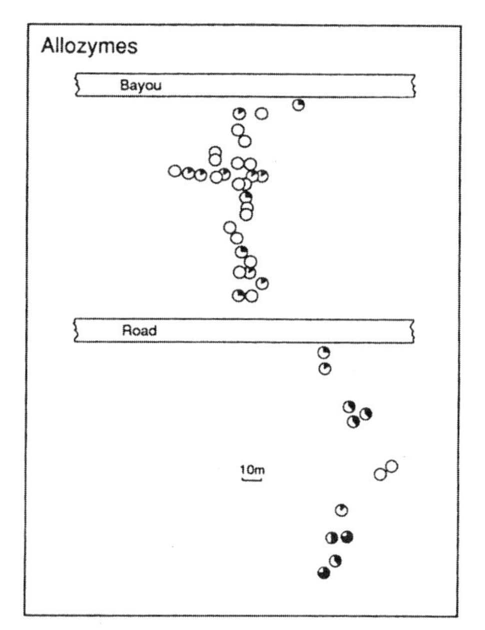

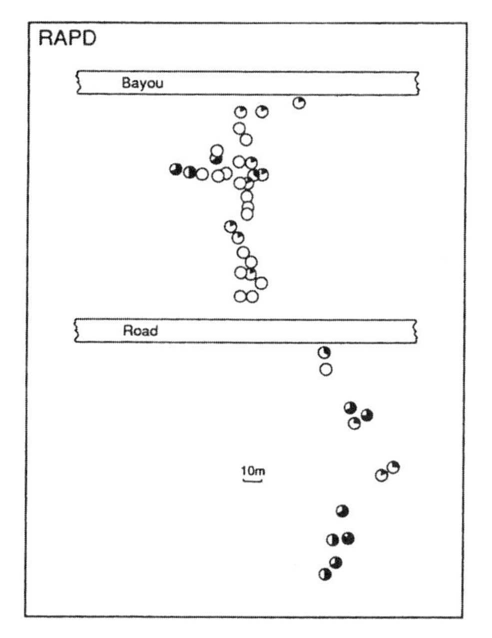

Fig. 13.3 Genetic variation in a sample of 42 individuals from the Bayou L'ourse population. Each circle represents an individual plant. The filled and open portions of the circle represent the proportion of *I. fulva* and *I. hexagona* markers, respectively. The top-left panel, the top-right panel, and the bottom panel illustrate the rDNA, allozyme and RAPD variations, respectively. There are three missing data points (plants) for the rDNA analysis. (From Arnold & Bennett, 1993. Reproduced with permission of Oxford University Press)

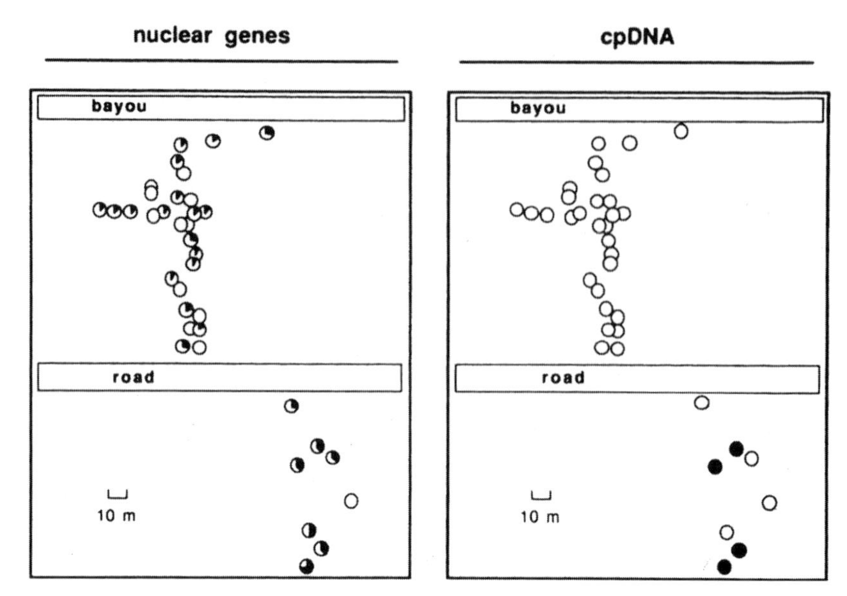

Fig. 13.4 Asymmetrical introgression between *Iris fulva* and *I. hexagona* likely to result from pollen flow. Each circle represents a single plant. Left: relative proportion of *I. fulva* (shaded) and *I. hexagona* (unshaded) nuclear markers. Right: similar representation for maternally transmitted chloroplast DNA markers. Note in particular the population between the road and the bayou, where multilocus nuclear genotypes suggest the presence of advanced generation hybrids or backcrosses, despite the apparent absence of seed dispersal that would be registered by chloroplast DNA from *I. fulva* in this area. (After Arnold, 1992, from Avise, 1994)

I. fulva had, in all but one comparison, a higher fitness (greater shade tolerance) than those closer to *I. hexagona*.

3. Competition experiments revealed that *I. hexagona* and hybrids close to this taxon were equivalent in their competitive ability. Both were superior to *I. fulva*.
4. Anderson (1949) speculated that in disturbed areas (either natural or human-disturbed) hybrid plants may have higher Darwinian fitness than either of the parents, and, therefore, be at a selective advantage. This hypothesis has been tested by Emms and Arnold (1997) using material of the two parental species – *I. fulva* and *I. hexagona* – and artificial F_1 and F_2 hybrid plants, assayed for species-specific nuclear and cytoplasmic markers and divided into two classes: *I. fulva*-like plants with low frequencies of *I. hexagona* markers and *I. hexagona*-like plants with low frequencies of *I. fulva* markers. Plants were set out in a reciprocal transplant experiment, and examined for survivorship, leaf production, rhizome growth and flowering. Overall their findings suggested 'F_1 and some F_2 and back cross progeny demonstrate higher fitness than parental taxa' and, therefore, in microevolution, hybrid individuals through extensive introgression, could replace their parental species.
5. Investigations of *Iris* hybrids have revealed that their sexual reproductive capacity is low relative to the parental taxa. In contrast, hybrids are long-lived and capable of considerable vegetative spread. Thus, over time, genotypically different clones fragment and separate individual colonies establish up to 10 m apart (Burke *et al.*, 2000). Morphologically, these clones are often distinctive, and taxonomically may be treated as groups of microspecies (*sensu* Grant, 1981).

6. Investigations have revealed a further dimension to the patterns of introgression in the Louisiana *Iris*. In some places, on drier habitats – unimproved pastures, hardwood forests – a third species *I. brevicaulis* occurs with *I. fulva* and *I. hexagona*. Molecular studies of a population at Bayou Teche revealed that some individuals had markers characteristic for all three species, confirming that introgression had occurred between all three (Arnold, Hamrick & Bennett, 1990; Arnold, 1993).
7. Further complexities were discovered in the *Iris* populations, when a new homoploid hybrid species was discovered – *I. nelsonii* (Arnold, 1993). Genetic molecular markers diagnostic of *I. fulva*, *I. hexagona* and *I. brevicaulis* were present in this stabilised introgressant, revealing, astonishingly, that three *Iris* species were involved in the origin of *I. nelsonii*. While human activities have clearly influenced the patterns of introgression in the Louisiana Irises, it is not clear what direct role, if any, humans had in the evolution of *I. nelsonii*.

Introgression in *Rorippa* in Germany

Investigations of isozymes and cpDNA have detected introgressive hybridisation between three species of Yellowcress (*Rorippa* species – *R. amphibia*, *R. palustris* and *R. sylvestris*) in northern Germany (Bleeker & Hurka, 2001). Studies of populations of *R. amphibia* and *R. sylvestris* detected hybridisation and bidirectional introgression at sites along the banks of the River Elbe, the last river in Central Europe to have natural erosion and sedimentation patterns. The two species have different ecological niches: *R. amphibia* is found in and by waterways, while *R. sylvestris* grows on wet ground and areas where water stands in the winter. Periodic flooding of the Elbe occurs, leading to the breakdown of natural ecological barriers, but there are no natural permanent riverside sites where hybrids can persist. However, along man-made drainage channels there is a permanent contact zone between *R. amphiba* (found in waterside areas) and *R. palustris* (found on damp areas and waste ground) and unidirectional introgression of *R. palustris* markers (cpDNA and isozymes) into *R. amphibia* has been detected.

Introgression between introduced species: is hybridisation a stimulus to invasiveness?

In Chapter 11, a number of factors were considered that might account for the invasiveness of introduced species. An additional factor has been proposed, namely, that hybridisation may be a stimulus to invasiveness (Abbott, 1992; Ellstrand & Schierenbeck, 2000, 2006). It has often been noted that some introduced species do not achieve invasive status on their arrival overseas. They remain for a time as ‘sleeper weeds’, only later may some become invasive (Groves, 2006). Ellstrand and Schierenbeck (2006) consider that hybridisation may be the key to the origin of subsequent invasiveness. The stimulus of hybridisation may involve native and non-native species with the production of new polyploids or stabilised introgressants may be produced (Abbott, 1992). In addition, it is proposed by Ellstrand and Schierenbeck that the stimulus to invasiveness could come from intraspecific crossing between multiple independent introductions of a species. They carried out a review of the botanical literature in search of convincing cases of hybridisation as a stimulus to invasiveness. Only those examples were accepted where thorough investigations had been carried out including genetical studies (allozymes DNA markers etc.), crossing experiments, and evidence of spontaneous generation of the hybrids and their persistence in the wild. Only 28 case histories were found. Most were herbaceous, perennial species of disturbed ground. In half the cases, a non-native species was one of the parents of the cross.

How might hybridisation stimulate invasiveness?

Ellstrand and Schierenbeck (2006) consider that the hybridisation event could generate 'evolutionary novelty' by increasing variability and through the subsequent genetic recombination in hybrids. Hybridity is stabilised by various means, e.g. by allopolyploidy, apomixis or clonal spread. In addition, hybridity may be the means by which plants from small isolated populations are brought together and thereby 'escape' the effects of accumulated detrimental mutations. Here, a number of examples illustrate the way hybridisation might stimulate invasiveness.

Rhododendron ponticum is an ornamental species, first introduced into Britain in 1763. It is one of 500 species of rhododendron grown in British gardens. The species has become extensively naturalised and is now spreading invasively in North Wales, parts of Scotland, Ireland etc. A survey of British nature conservation and forestry authorities, wildlife trusts and private landowners revealed that more than 52,000 ha was affected by the species, of which 30,000 ha was in nature reserves. In 2001, more than £670,000 was spent on controlling the plant (Dehnen-Schmutz, Perrings & Williamson, 2004).

The source(s) of the material first introduced is a matter of interest. Stace (1991) considers that British material originated from south-east Europe and south-west Asia. But there are also populations of *R. ponticum* in the Iberian peninsula. Using more than 260 accessions from across the British Isles together with material of a range of Rhododendrons sampled from sites in Europe, Turkey and elsewhere, Milne and Abbott (2000) have carried out investigations using molecular markers (cpDNA, rDNA and RFLPs) to determine the origin(s) and variability of *R. ponticum*.

Two cpDNA variants were detected indicating multiple introductions. Of the accessions, 89% had a cpDNA profile indicating an origin in Spain, while 10% were of a variant unique to Portugal. As judged by these samples, introductions from south-east Europe have not influenced the cpDNA genetics of British populations. Evidence from DNA markers also indicated that introgressive hybridisation had occurred between the *R. ponticum* and another species – the North American *R. catawbiense*. Introgressants from this cross were significantly more abundant in material from eastern Scotland. It is possible that such a pattern could owe its origin to one or more of the following factors: the aesthetic choices made by gardeners; the selection of 'hardy hybrids' by breeders; or natural selection, operating through the rigours of the Scottish climate, may have favoured introgressants with genes from the more cold-tolerant North American species. More investigations are required to resolve these issues. There was also evidence for introgressive hybridisation between *R. ponticum* and two other species – *R. maximum* and another as yet unidentified taxon. *R. ponticum* in the British Isles has crimson blotches on the flowers. Milne and

Abbott (2000) suggest that these floral traits could have come through introgression with *R. arboreum*, a species in a different sub-grouping of the genus *Rhododendron*. However, having considered all the available evidence, Dr James Cullen, an expert on *Rhododendron*, has come to a different conclusion (personal communication, 2008). He considers that *R. ponticum* in Britain is an 'entity distinct from wild *R. ponticum*', being 'an old, mostly man-made hybrid swarm or neospecies', involving four closely related species introduced into Britain in the late eighteenth century (*R. ponticum*, two eastern North American species *R. catawbiense* and *R. maximum*, and *R. macrophyllum* from western USA). Further investigations, employing molecular markers, could be helpful in testing these ideas.

Invasiveness following interspecific hybridisation between introduced fungi

Although fungi are now classified as a separate group from plants, it is helpful to consider the effects of hybridisation and invasiveness. Interspecific hybrids appear to be rare in fungi, but with increasing world trade in plants and plant products, many pathogens are being introduced to new geographic areas, thereby offering the opportunities for hybridisation. It is of great interest that hybrids have been found in certain fungal groups, for example, between native (*Ophiostoma ulmi*) and introduced (*O. novo-ulmi*) species of Dutch Elm disease in Europe (Brasier *et al.*, 1998). In addition, a new *Melampsora* rust species in New Zealand proved to be a hybrid between two introduced species (Spiers & Hopcroft, 1994) and hybridisation appears to have played a significant role in the evolution of modern strains of *Fusarium* and *Epichloe* taxa (see O'Donnell & Cigelnik (1997) and references cited therein).

Another disease has received special attention. In 1990, a previously unknown fungal disease killed many Alder (*Alnus*) trees in Western Europe and this new disease has been closely investigated (Brasier, Cooke & Duncan, 1999). Taxonomic and molecular studies have revealed that the fungus is probably a hybrid between *Phytophthora cambivora* and an unknown taxon close to *P. fragariae*. Other studies have revealed that neither *P. cambivora* nor *P. fragariae* are pathogenic to alder: the interspecific hybrids, however, are likely to survive if they have 'a fitness advantage over the parent species, such as increased aggressiveness or the ability to exploit a new host' (Brasier, Rose & Gibbs, 1995). Both species would appear to be introduced into Europe, most probably brought in on infected plants, possibly *Rubus* on which both species occur. Quarantine measures are in place to try to prevent the spread of fungal pathogens, but Brasier *et al.* (1999) conclude: 'many hybrids or introgressants are unlikely to be detected by conventional, mainly morphologically based diagnostic methods used in international quarantine'.

Invasiveness following intraspecific hybridisation

In Chapter 11, evidence was presented that introduced species are sometimes genetically depauperate because of founder effects. Invasiveness may be stimulated in cases where well-differentiated populations hybridise in new territories (Abbott *et al.*, 2003). In such cases, the within-population variability may be dramatically increased. Ellstrand and Schierenbeck (2006) cite two cases where invasiveness may have been stimulated by intraspecific hybridisation. *Echium plantagineum* is a noxious weed in Australia. The average population there was found to be more diverse than were those genetically analysed in its native range in Europe (Burdon & Brown, 1986). This species has been introduced more than once to Australia, both intentionally and unintentionally (Briggs, 1985). Similarly, North American populations of the introduced weed Cheatgrass (*Bromus tectorum*) were found to have increased within-population genetic variation as compared with populations from its source range in Europe and northern Africa (Novak & Mack, 1993). Again, there is ample evidence of multiple introductions (Novak, Mack & Soltis, 1993).

Crop–wild–weed interactions

Crop plants have been taken across the globe. Hybrids may be produced if crops and their wild relatives become sympatric, have overlapping flowering times, share pollinating mechanisms, are reproductively compatible, their hybrids are viable and have a degree of cross-fertility. Moreover, if F_1 hybrids have some fertility, then further generations of hybrids may be produced, and backcrossing to one or both parental taxa may result in introgression. In early studies of crop–wild species interactions, investigators used morphological evidence and sometimes crossing experiments. But, to provide a more critical test of apparent hybridisation and introgression, increasingly the interactions of crop–wild (and indeed weedy relatives) are being investigated by the use of appropriate molecular genetic markers.

Ellstrand, Prentice and Hancock (1999) discovered evidence for hybridisation with wild relatives in 12 of the 13 most important crop plants: wheat, rice, maize, soybean, barley, cotton, sorghum, millet, beans, rapeseed, sunflower and sugarcane. Only in the case of groundnuts was there no evidence of spontaneous hybridisation between crop and wild relatives, but experimental crosses produced fertile hybrids. Investigation of crop–wild relative hybridisation has recently been extended to additional crops, and, having reviewed the literature, Ellstrand (2003, 1166) concludes that, across the globe, at least 48 cultivated species cross with one or more of their wild relatives.

The rate of hybridisation between crop plants and compatible wild relatives has been intensively studied in a number of cases. In comprehensive studies of hybridisation between Oilseed Rape (*Brassica napus*) and one of its wild relatives Wild Turnip (*B. rapa*) in the UK, it was estimated that as many as 32,000 hybrids could occur in waterside habitats and 17,000 at agricultural sites (Hails & Morley, 2005).

While the incidence of hybridisation is clearly important, the rate at which genes from crop plants might enter populations of wild species depends upon the relative fitness of F_1, F_2 and subsequent generation hybrids and backcrosses. On the face of it, it would appear that crop genetic traits, such as lack of seed dormancy, might be selectively disadvantageous in the wild. Reviewing the evidence, Hails and Morley (2005, 245) conclude: 'although hybrids often do have reduced fitness relative to both parents, this is not always the case . . . In a few cases, hybrids had enhanced fecundity compared with their wild relatives.' However, 'some degree of outbreeding depression appears to be more common than is hybrid vigor'.

Microevolution in action through hybridisation in weeds

There are many examples of introgression of crop genes into wild plants. For example, in India a red-pigmented cultivar of rice was developed to allow the crop seedlings to be distinguished from the green seedlings of weeds including those of weedy rice. But this strategy for weed control failed for, after a few seasons, through crop–weed hybridisation and introgression, the red pigment appeared in high frequency in weedy rice plants (Oka & Chang, 1959).

There is evidence from allozyme and RAPD molecular markers that spontaneous hybridisation occurs at a substantial rate and over distances within <1,000 m between the cultivated sunflower (*Helianthus annuus*) and wild variants of the same species (Arias & Rieseberg, 1994). Studies have also revealed that crop-specific RAPD markers (presumably neutral in effect) persist in the wild sunflower for at least five generations following their creation, when crop and weed last grew together for one season (Whitton *et al.*, 1997). In situations where the crop and weed had a history of long-term contact (up to 40 years), evidence of substantial introgression of crop genes into the wild variants was found, with every individual tested having at least one crop-specific allele marker (Linder *et al.*, 1998). The fitness of crop–weed hybrids in sunflowers has been examined by Mercer *et al.* (2007). They discovered in experiments that hybrids may have high fitness in stressful agricultural environments, through the possession of 'domestication traits', such as rapid growth and early flowering.

New weedy species arise through crop–weed–wild interactions

There is evidence that crop–weed interactions may result in other microevolutionary changes leading to the evolution of new weed species. For example, RFLP molecular studies have also revealed that species-specific alleles of crop sorghum (*Sorghum bicolor*) are present in wild *S. bicolor* that occurs with the crop in Africa (Aldrich & Doebley, 1992; Aldrich *et al.*, 1992). In addition, the weedy *Sorghum almum* appears to be a hybrid product between crop sorghum and *S. propinquum*. Also, 'the evolution of enhanced weediness' in *S. halepense* – one of the world's worst weeds – may have involved introgressive hybridisation with crop sorghum (Holm *et al.*, 1977a, b). Looking at other cases, Ellstrand *et al.* (1999) report that crop-to-weed gene flow has resulted in increased weediness in the wild relatives of 7 of the 13 most important crops.

Hybridisation increases the extinction risk in endangered species

The endemic wild rice of Taiwan (*Oryza rufipogon* ssp. *formosana*) hybridises with cultivated rice (Kiang, Antonovics & Wu, 1979). Over the past century, pollen and seed fertility have declined in the wild rice, and there has been a pronounced shift towards characters of cultivated rice. In many parts of Asia, other subspecies of *O. rufipogon* are similarly threatened, and populations of another wild species – *O. nivara* – are being modified through extensive hybridisation with the crop rice (Chang, 1995). Small (1984) gives other examples where the wild relatives are threatened with extinction through extensive hybridisation with their domesticated relatives.

Transgenic crops and their interactions with wild and weedy relatives

Genetically modified (GM) crops are of increasing importance in the modern agriculture of some but not all countries. In 1996, 1.7 m hectares of GM crops were grown (www.afan.com.au); by 2007, this figure had increased to 114 m ha (www.gmo-compass.org). In traditional crop breeding, one way of introducing new genes into crop plants is by crossing existing cultivars with their wild relatives and rejecting those derivatives that do not contain the desired trait. In contrast, GM methodologies present a speedier route to crop improvement as genetic material specifying the desired trait is introduced, and this DNA can be from an entirely unrelated species. Thus, the bacterium *Bacillus thuringiensis* (Bt) naturally produces toxins that act as pesticides against insects. Transgenic Bt constructs have been introduced into several crops, as has herbicide resistance. In the future, suitably designed GM plants may provide cost-effective vehicles for the production of vaccines, antibiotics, industrial proteins etc. (Giddings *et al.*, 2000).

Interest in hybridisation between crop–wild–weedy related plants has been greatly increased with the development of transgenic crops, as there is a concern that such hybridisation might present an avenue for the 'escape of transgenes' from transformed crops into other species (Ellstrand *et al.*, 1999). This possibility has been carefully examined in sunflowers and oilseed rape.

As we have seen above, crossing between non-GM crop and weedy/wild sunflowers *(Helianthus)* is well established. Two situations involving transgenic sunflowers have been examined. Bt crop × wild crosses have been backcrossed to wild plants. In 'uncaged natural' conditions insect damage to hybrids was reduced by the presence of the Bt transgene, resulting in an increase in seed production of 14% in Colorado and 55% in Nebraska (Snow *et al.*, 2003). A glasshouse experiment showed that the Bt transgene did not affect fecundity in the hybrid plants; apparently they paid no cost in this component of fitness (Snow *et al.*, 2003). In a second study, the effect of a disease resistance transgene (OxOx) was examined in hybrids. This transgene confers resistance to fungal pathogen white mould (*Sclerotina sclerotiorum*). There was no detectable effect on fitness even in the face of severe pathogen attack. In an experiment involving growth of the transgenic hybrid derivatives at several sites, it was concluded that 'should the transgene escape, it would do little more than diffuse neutrally throughout the recipient population' (Burke & Rieseberg, 2003).

The potential for transgenes to escape has been examined in Oilseed Rape/Canola (*Brassica napus*), a polyploid species ($2n = 38$: genomic constitution AACC) that can hybridise with both its parental species *Brassica rapa* (AA $2n = 20$) and *Brassica oleracea*

(CC $2n = 18$). It can also hybridise with Wild Radish (*Raphanus raphanistrum*) and other weedy relatives (Warwick *et al.*, 2003; Chèvre *et al.*, 2004).

Considering non-GM stocks, hybridisation between *B. napus* × *B. rapa* has been observed in the wild in Denmark (Jørgensen & Anderson, 1994) and in the UK (Scott & Wilkinson, 1998). The hybrid plants in the F_1 and later generations are on average less fit than parental stocks, but some individual hybrids were as fit as the parents. In studies of GM material, Mikkelsen, Anderson and Jørgensen (1996) found that transgenic oilseed rape × *B. rapa* hybrids were similar to *B. rapa* in chromosome number and morphology and had relatively high fertility. Thus, interspecific hybridisation provides a possible pathway for transgenes in oilseed rape to pass to wild *B. rapa*. Confirmation that this transgenic hybridisation route occurs under field conditions has been established by Warwick *et al.* (2003). Studying polyploidy levels, herbicide resistance, male fertility and species-specific ALFP markers, they established that glyphosate-resistant transgene in the rape crop ($2n = 38$) had passed to the weed (*B. rapa*, $2n = 20$) through hybridisation. Furthermore, not only were F_1 triploid ($2n = 29$) hybrids produced; in monitoring the persistence of the transgene at two sites over a 6-year period, Warwick and associates discovered evidence for hybrid × *B. rapa* backcross generations – effectively the transgenic herbicide resistance has been transferred from rape through introgression hybridisation to its weedy relative and has persisted in the absence of recent herbicide treatment. Long-term studies will be necessary to determine whether *stable* incorporation of the transgene into the weed occurs. This will depend upon many factors, including whether glyphosate is used in cropping the land, and the fitness relationships between the transgene containing hybrid derivatives and weed genotypes without this trait.

In another study of the interaction of transgenic oilseed rape and the weed *B. rapa*, investigations of BtGM oilseed rape × *B. rapa* first generation hybrids revealed that, in the presence of herbivores, the Bt construct conferred some advantage, but in the absence of herbivores, there was a detectable fitness cost (Vacher *et al.*, 2004).

Reviewing all the evidence so far available, Chapman and Burke (2006) consider that wide generalisations concerning the fitness of hybrids between GM crop and wild/weedy relatives should be avoided. Each situation should be considered on a case-by-case basis.

In considering the likelihood of new weeds arising through crossing between transgenic crops and their wild relatives, the possibility that the crop itself might produce weedy variants should not be overlooked (Warwick & Small, 1999). A case in point is feral oilseed rape, which is a common weed of agricultural land where it appears as a volunteer. It is becoming clear that farm machinery may distribute the plant. Moreover, volunteers, so distributed, may not grow immediately, but, by secondary dormancy, may take their place in the soil seed bank, with germination at a later date either within a crop of rape or some other crop. Controlling such plants by herbicide treatment is compromised by the fact that some volunteers are herbicide resistant. Recently, it has been discovered that some volunteers are resistant to two herbicides, having been formed as a consequence of hybridisation following gene flow between two oilseed rape crops transgenically transformed to different herbicide resistance. Gruber, Pekrun and Claupein (2004) predict that volunteers will persist in future crops, and, as a consequence, contamination of seed lots is highly likely and weed control becomes even more complex. The microevolutionary impact of the potential long-term presence of volunteers in the non-agricultural landscape has not yet been fully explored (Warwick, Beckie & Small, 1999).

The microevolution of weed beets

So far, our account has focused on two North American case histories of crop–weed interaction in sunflowers and oilseed rape. Investigation of weed beets in European sugar beet crops have provided a fascinating glimpse of microevolution in action.

Beta vulgaris ssp. *maritima* occurs on the coasts of Europe from the Mediterranean through western Europe to the Baltic, and there are also inland populations. This taxon is ancestral to the many variants of cultivated beet, sugar beet, fodder beet, red table beet and leaf beet. All are classified within *B. vulgaris* ssp. *vulgaris*. Wild beet is a self-incompatible diploid ($2n = 18$) that is generally perennial, needing vernalisation for flowering. However, populations of the wild species contain a dominant allele of gene B that suppresses completely any requirement for vernalisation. Northern populations of wild beet are of bb genotype, with the frequency of B increasing in southern coastal and inland populations (Fig. 13.5).

Sugar beet is the most widely grown cultivated beet, the major cultivars of which are triploid biennials that mostly remain vegetative throughout their cultivation until the swollen underground beets are harvested. However, annual 'bolters' (flowering individuals) may be found in the fields. Recently, the incidence of bolters has greatly increased and a serious weed problem has developed. The origin and microevolutionary significance of these weedy bolters have been investigated by garden trials, genetic experiments and molecular studies, and involve hybridisation between wild and crop plants.

Wild and cultivated beets are sympatric in many places, especially where seed production of beet seed is carried out (Fig. 13.5 and Fig. 13.6). In the production of such seed, diploid male sterile seed-bearing plants are crossed with pollen-bearing tetraploid plants giving triploid plants that are fully or partially male sterile. Investigations have revealed that wild plants carrying the B allele may grow close enough to the crop for some hybridisation to occur. Thus, F_1 bolters are produced and their seed is present for many years in the soil seed bank, often surviving until the next crop of beet is grown. In 1978–81, 18–27% of the UK beet fields had annual forms of weed beet (Maughen, 1984). At first, the eradication of weedy beets was often neglected, and weedy lineages have increased, selection favouring the weed plants that were able to reproduce before harvesting of the crop, i.e. those variants that have no vernalisation requirement. In some areas of France, the effect of this unwitting selection has been an increase in the frequency of B allele in weed populations from the initial 50% in F_1 hybrids to 60–80% (van Dijk, 2004).

It has proved extremely difficult to control the weedy beets in sugar beet fields, as there are no weed-specific herbicides: any herbicide that kills the weed will also damage the crop, as both belong to the same species (Soukup & Holec, 2004). Reflecting on the situation in

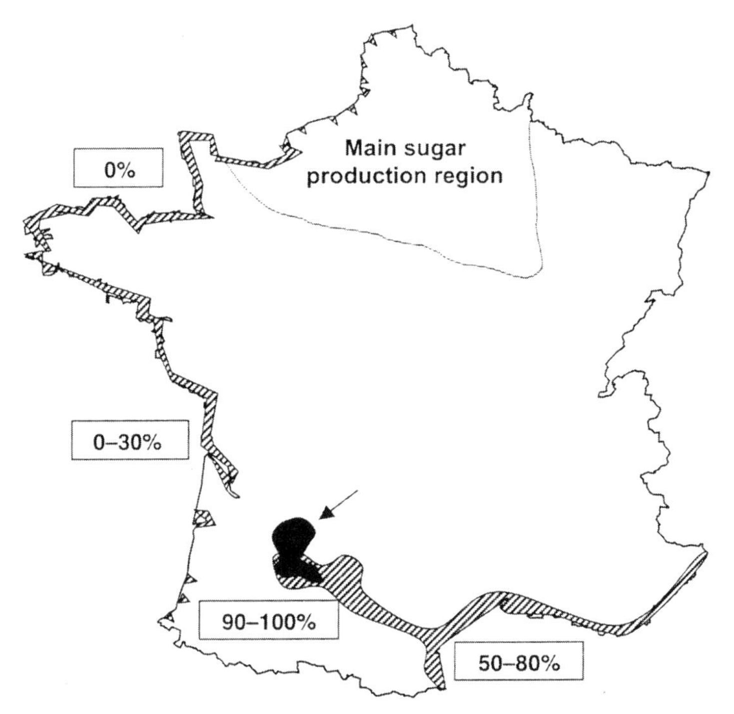

Fig. 13.5 The geographical distribution of wild coastal and inland variants of *Beta vulgaris* (hatched area); the main sugar beet seed production area (arrowed and indicated in black) and the main sugar beet growing region of France. Wild beet have a variable percentage of B- (which means that all plants are without a requirement for vernalisation, whether BB or Bb); while cultivated beets do have a requirement for vernalisation and are bb (diploid) or bbb (triploid in genetic constitution). (From van Dijk, 2004. Reproduced with permission of CAB International, Wallingford, UK)

the Czech Republic, they report that, since the 1980s, weedy beet has become an increasing problem, with *c.* 50% of fields infested by the 1990s. It is interesting to consider why weed beets are important. When harvested with the crop, the fibrous underground parts of weed beets are also included, and these present mechanical problems in sugar processing. In addition, beet fields are sometimes sown with other crops to disrupt the life cycles of important pathogens e.g. *Rhizomania*. However, the strategy of disruption may fail, as weedy beets are hosts to the same pathogens, and these may grow from the seed bank to infest other crops sown in rotation with beet.

In trying to control the weed, labour-intensive methods were first employed, but, because hoeing is time consuming and expensive, other methods have been devised. Herbicide applications may help to reduce weeds occurring between the crop rows. In addition, cutting machines and herbicide 'wiping systems' have been developed to decapitate or poison the weedy beets that have extended their flowering stalks above the vegetative sugar beet crop. But, investigations have revealed an intriguing complication. Not all weedy beets have an erect growth habit – low-growing prostrate variants escape decapitation

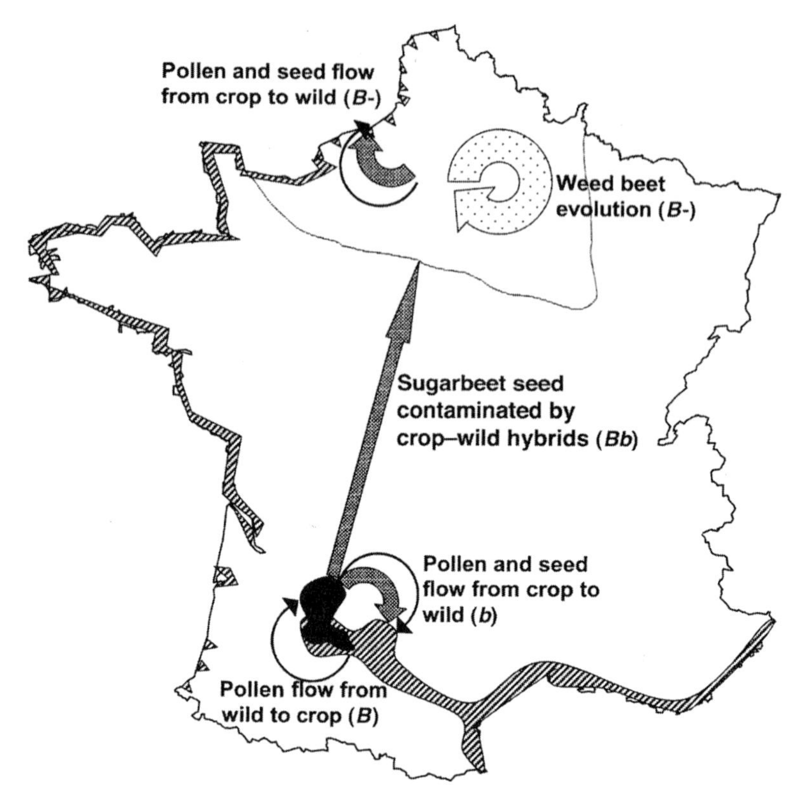

Fig. 13.6 Evolution of weedy beet. Gene flow by seed (thick arrows) and pollen (thin arrows) between crop and wild. Weed beets occur in the sugar beet fields and are the progeny of crop–wild hybrids formed in the seed production area. (From van Dijk, 2004. Reproduced with permission of CAB International, Wallingford, UK)

and/or poisoning from contact with herbicide pads. It seems possible that, if these types of weed control are practised frequently, low-growing variants will be at a selection advantage and increase in frequency in response to directional selection for prostrate growth habit.

Transgenic beet

In some areas, infestations of weed beet have made it impossible to grow sugar beet. In an attempt to overcome the problem, transgenic herbicide-resistant sugar beet cultivars have been produced. Transgenically modified sugar beet with 'built-in' resistance to glyphosate is expected to be in widespread use in the USA in 2008. Then, it should be possible to control the weedy beets, which are herbicide sensitive. However, there is a concern that herbicide resistance could be transferred from crop to weed. Such concerns have been shown to be justified, as molecular studies are revealing the extent of the two-way processes of gene flow between wild populations and crop. Using DNA genetic markers, Cuguen *et al.* (2004) report the results of a study of populations from the French coast near Boulogne (Fig. 13.7).

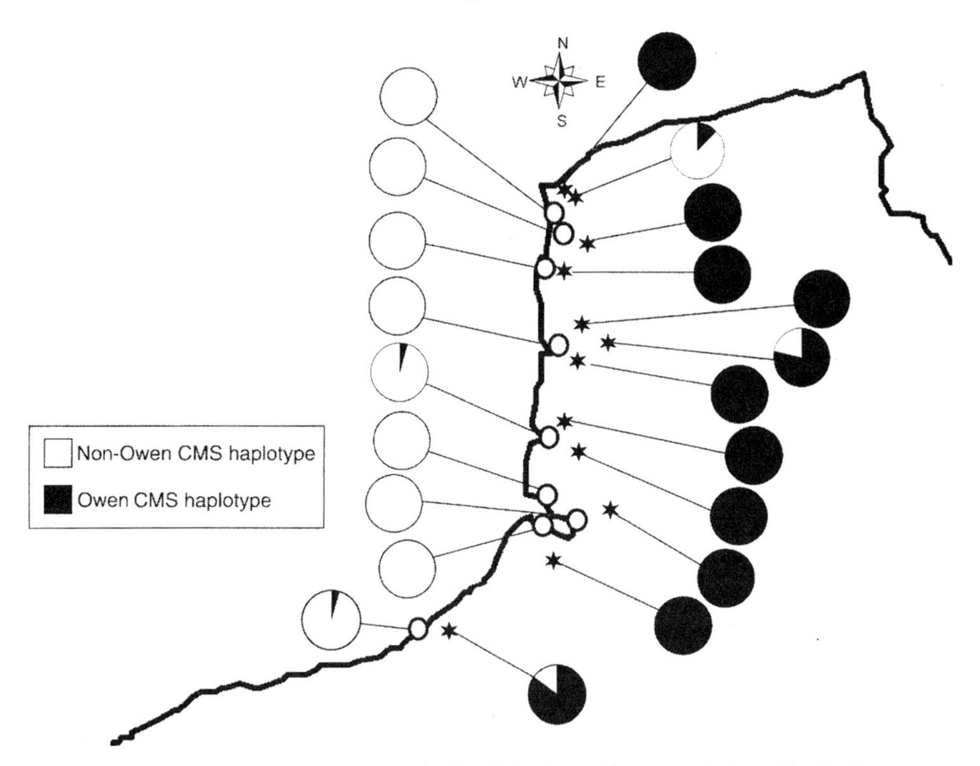

Fig. 13.7 Distribution of chloroplast diversity in wild and weed beet populations. The black segments of the pie diagrams indicate the percentage of Owen CMS haplotypes detected in samples taken in the study area in northern France. CMS is specific for cultivated lines of beet. White indicates the percentage of non-Owen CMS haplotypes. (Cuguen *et al.*, 2004. Reproduced with permission of CAB International, Wallingford, UK)

There were 9 sea beet and 12 weed beet populations, the latter being chosen as those nearest to the French coast. Most of the weedy populations had a high incidence of the Owen CMS type cytoplasm that is typical of cultivated beets (89.5% overall). This reflects the origin of the weedy beets from wild pollen fertilising eggs of cultivated plants (carrying Owen CMS cytoplasm). In addition, three of the weedy populations were found to be polymorphic: some of the populations having non-Owen CMS variants. The results also revealed traces of gene flow from crop to wild populations, as some of the coastal populations had a low frequency of Owen CMS types (0.9 overall). As the cytoplasm type (Owen or non-Owen) is maternally inherited, it is clear that gene flow by seed has occurred in both directions – from sea beet into weedy populations and vice versa.

Cuguen *et al.* (2004) also report the investigation of patterns of variability of the same populations using nuclear DNA markers. There was very clear genetic differentiation between the coastal and weedy beets. In general, the presence in weedy beet of markers known to occur in cultivated material was confirmed. In addition, there was also further evidence of limited gene flow between wild and weedy plants, in markers that could be transmitted either by seed or pollen. Thus, there is strong evidence that gene exchanges link weed and wild populations through pollen and seed flow. This finding has implications for the growing of transgenic crops of beet in regions where there are populations of wild beet, albeit some distance away. It has often been asserted that gene flow between transgenic crops and their wild relatives could be prevented, if there was sufficient distance between them. In this study, of a wind-pollinated plant, it has been shown that gene flow by pollen from crop to crop and from crop to wild relatives may occur over considerable distances, estimated as exceeding 1,000 m (see Cugen *et al.*, 2004).

Given this evidence of gene flow between the different variants of beet, it seems very likely that any transgenic elements introduced into cultivated beet will escape through hybridisation. To minimise this possibility it has been suggested that transgenic herbicide tolerance should be built into the male pollen-bearing tetraploid lines. This tolerance could then be introduced into the triploid sugar beet cultivars. Introduction into weedy beet would theoretically be difficult, as weedy beets crossing with tetraploids would generate sterile 3x hybrids, and the diploid male sterile lines crossing with wild plants would not produce herbicide-resistant progeny. Many transgenic cultivars have been produced, and some are in crop production. It will be interesting to follow the continuing microevolution of weedy beets to see if transgenes are introgressed from crop plants to weed, perhaps by crossing between any crop bolters and weedy variants.

Keeping the transgene within the crop

A number of approaches are being explored to prevent transgenes from escaping from the crop into related species. Some researchers have suggested that wide spacing of GM crops from each other and from conventional crops of the same species could reduce gene flow by pollen, but studies have shown that pollinating pollen may travel large distances and wide spacing would be necessary.

Others are exploring the possibility of creating apomictic variants of normal sexually reproducing crops. In apomixis, normal sexual reproduction is by-passed and seeds are produced automatically. Yet another route is possible. Some flowering plant species produce cleistogamous flowers, e.g. *Viola* spp., in which self-pollination occurs within the closed bud. No pollen escapes from the flower. Both these approaches offer the possibility of containing transgenes within the crop, but neither is yet available.

Other means of confining transgenes within the crop have been proposed. If transgenic elements of importance were genetically linked to cultivated traits, such traits would have high fitness within the agricultural sphere but give plants of very low fitness elsewhere. A model system has been tested in *Arabidopsis*, in which the herbicide-resistant transgene was coupled with a dwarf gene. Dwarfing genes are advantageous in many agricultural situations but could be at a selective disadvantage elsewhere. Such a system has yet to be tested out-of-doors and in a crop species. Molecular approaches to containing transgenes are also possible; for example, by creating and planting male sterile transgenic crops interspersed with non-transgenic pollen donors (Chapman & Burke, 2006).

Crop–weed–wild species hybrids: estimates of their fitness

Reflecting on the evidence so far available, Hails and Morley (2005) conclude that issues relating to fitness in non-GM and GM crop interactions with related cultivars, species etc. are far from settled, as the tests have often been carried out under short-term experimental conditions that minimised plant competition. To estimate the longer-term effects of hybridisation between crops and wild weedy plants, more experiments are needed to estimate fitness of hybrids in conditions approaching those found in the man-made landscapes and semi-natural vegetation (Hails & Morley, 2005; Chapman & Burke, 2006). Ideally, such comparative experiments of parental and hybrid derivatives should explore fitness in different sites and, as there is much variability seasonally and annually, these too should be explored. Technically, the requirement is to investigate in detail genotype × environmental interactions.

Hybridisation and conservation

Ex situ *conservation*

Many botanic gardens were founded before conservation was an issue, and therefore they were not designed to maximise their effectiveness in *ex situ* conservation. The planting arrangements employed present an immediate problem.

Botanic gardens and arboreta contain very large collections of species from across the world, set out in geographic displays, order beds and themed areas. Such plantings bring related, formerly geographically isolated, allopatric species into close proximity, where there is a risk of spontaneous hybridisation. In the past, botanic gardens were also involved in the deliberate hybridisation programmes to develop new cultivars of species of horticultural interest, e.g. orchids, magnolias, roses etc. In most countries, however, such horticultural breeding is now the province of commercial breeders. But, accidental interspecific hybridisation in botanic gardens is still a live issue for conservationists.

Many gardens contain material of high conservation interest, but accidental hybridisation may occur in garden collections. Therefore, the use of open-pollinated seed (or progenies from such seed) carries the risk that hybrid plants could be introduced into the 'wild' in reintroduction and restoration projects. In addition, hybrids may be unwittingly formed, propagated and displayed in gardens. The scale of the problems with botanic garden stocks is hard to judge, until the status of botanic garden collections is checked more thoroughly using molecular markers. A number of preliminary investigations point the way to further studies (see Maunder *et al*., 2004).

Echium pininana is endemic to the Canary Islands where it is endangered in the wild. Molecular studies have revealed that stocks of this species in gardens have hybridised with other *Echium* species. There are also potential problems with many other groups, e.g. palms, willows, poppies, primulas, saintpaulias etc. Thus, interspecific hybridisation in botanic garden stocks may render them useless in reintroduction and restoration projects.

Long-term conservation of rare and endangered species can only succeed if the genetic integrity of collections is protected.

Gene flow from gardens into the wild

The vogue for naturalistic plantings and wild gardens can cause difficulties if these are set out in 'the wild', as for example in California, South Africa, South America, Australia etc. For instance, the presence of European cultivars of the epiphytic cactus *Schlumbergera* in Brazilian gardens threatens, through hybridisation, a number of rare wild species of the genus growing nearby (N. Taylor, personal communication to Maunder *et al.*, 2004, 349). Also, introgressive hybridisation would seem a possibility between horticultural selections of *Echinacea* grown in the Chicago Botanic Garden and wild populations of two threatened species growing in adjacent wild areas (Floate & Whitham, 1993).

In addition, garden escapes can also influence wild populations. For instance, the Californian endemic *Oenothera wolfii* is at risk through hybridisation with 'escaping' plants of *O. glazioviana* (Imper, 1997). Crossing between garden and native plants is also a particular threat to endemic species on islands. For instance, on Juan Fernandez Islands, hybridisation has been detected between the introduced *Acaena argentea* and the endemic *Margyricarpus digynus* (Crawford *et al.*, 1993).

Considering our present knowledge of the effect of gardens in wild places, Maunder *et al.* (2004) consider that the risk to the integrity of rare and endangered species might be small as spontaneous hybrids are often highly infertile and may not spread beyond the confines of the botanic garden. Moreover, gardens are often on small sites relative to wild areas and propagule pressures from garden to the wild may be insignificant. As we have seen earlier, botanic garden stocks have contributed to the evolution of highly invasive species. Much vigilance is necessary as 'it would be a terrible irony if the long-term legacy of *ex situ* conservation included new hybrids and invasives rather than recovered species' (Maunder *et al.*, 2004, 354).

Extinction of endangered species through hybridisation and introgression

Rhymer and Simberloff (1996) point out that many rare and endangered plants and animals could be threatened by hybridisation, with or without introgression with more abundant, often introduced taxa. Studies with molecular markers have explored the genetic interactions between native/introduced species of fish, birds, mammals etc. and have demonstrated adverse effects of hybridisation. Endangerment through hybridisation and introgression also seems a possibility in rare plant species. Rhymer and Simberloff (1996) cite a number of examples (based on morphological evidence) suggested to them by conservationists in different parts of the USA. For instance, native *Lupinus littoralis* is hybridising with aggressive introduced *L. arboreus* in a dune reserve in California. In Texas, the endangered White Flywheel (*Gaillardia aestivalis*) is hybridising with Indian Blanket (*G. pulchella*), a species widely planted on roadsides. The endangered Bakersfield Saltbush (*Atriplex tularensis*) is apparently being lost in introgressive swarms with widespread *A. serenana*. Californian Sycamore (*Platanus racemosa*) appears to be threatened by hybridisation with the introduced London Plane (*P.* × *acerifolia*). The Californian Black Walnut is hybridising with species imported for commercial purposes. In addition, the endangered Western Bog Lily (*Lilium occidentale*), Nelson's Sidalcea (*Sidalcea nelsoniana*) and Peacock Larkspur (*Delphinium pavonaceum*) are all suspected of hybridisation with commoner species of their respective genera.

Morphological evidence alone provides neither a test of the presence of hybrids nor a sound basis for predicting the long-term consequences of any hybridisation. Recently, a number of studies employing molecular markers and studies of fitness of parents and hybrids have greatly enlarged our understanding.

Introgression in Spartina

Twenty-five years ago, Smooth Cordgrass (*S. alterniflora*), a species of the east coast of North America, was introduced into San Francisco Bay, California, where it now hybridises with its native sister species California Cordgrass (*S. foliosa*). Investigations of cpDNA have confirmed that hybridisation occurs in both directions to give not only F_1 plants but also backcrosses. The two *Spartina* species have different flowering times and F_1 plants are rare. However, some introgressive hybrids of high fitness, with intermediate flowering behaviour have been produced. It was postulated that these derivatives had a higher potential to spread than *S. alterniflora* and represented a threat to the native *S. foliosa* and other species (Anttila *et al.*, 1998).

These predictions have been further tested (see review by Ayres, Zaremba & Strong, 2004). In molecular studies, nuclear DNA markers diagnostic of the two species have been identified and nine categories of hybrids genetically recognised. There have been several generations of hybrids in San Francisco Bay and bidirectional hybrids were identified near the sites where plants of *S. alterniflora* were set out. Reviewing a number of experimental investigations, Ayres *et al.* (2004) conclude: 'a subset of hybrid genotypes out-competes the native species [*foliosa*], overgrows its niche space, produces much seed, and sires the majority of seed on native flowers'. While S. *foliosa* alone is present on other shorelines along the west coast, the local extinction of native *foliosa* is possible where *S. alterniflora* has been established.

Endangerment of a rare endemic through hybridisation

The Catalina Island Mountain Mahogany (*Cercocarpus traskiae*) is native to one site on the Californian coast, where its population has recently fallen to seven adult plants and *c.* 70 seedlings. Investigations using allozyme markers have confirmed that hybridisation with the common *C. betuloides* is a major threat to the species, as four out of the seven plants are interspecific hybrids (Rieseberg *et al.*, 1989). These findings were confirmed in studies using RAPD markers (Rieseberg & Gerber, 1995).

Conservation of hybridising species

Interspecific hybridisation occurs in many groups of organisms. Sometimes hybrids are sterile, but in other cases introgressive hybridisation occurs. More or less stable long-standing hybrid zones have been discovered in some groups of organisms, while in others, transitory or recurrent mosaic zones are found (Harrison, 1993). Traditionally, the theory and practice of conservation management has been focused on taxa at the species level and on securing their survival. If an endangered species is apparently being hybridised out of existence by crossing with an introduced or commoner native species, should management attempt to remove the commoner species and hybrids? Some take a different view, arguing that hybridisation complexes should be managed to allow evolutionary processes to continue. A number of case histories indicate the range of options open to conservation managers.

A. As we have shown earlier in this chapter, a number of new species have arisen as a result of hybridisation and polyploidy. It could be argued that sites occupied by new polyploids should be preserved to allow us to study their continued evolution. However, they have arisen in human-disturbed areas outside nature reserves and it is not possible to 'conserve' all the sites in urban disturbed areas, e.g. *Senecio cambrensis* in Edinburgh and *S. eboracensis* in York. In the case of *Spartina anglica*, it could be argued that sites where it has been introduced should be maintained in a condition that would allow their continuing evolution. However, *S. anglica* has become seriously invasive in many areas and control measures have been employed. For example, in Washington State a wide range of control measures have been tested, including weeding, mowing and herbicide applications (Hedge, Kriwoken & Patten, 2003).

   Instead of allowing microevolutionary processes to take their course in animal populations, culling is being used as a means of trying to prevent the extinction of endangered species as a result of their crossing with introduced taxa. For example, feral populations of North American Ruddy duck (*Oxyura jamaicensis*) have been established in the UK. They have invaded the continent and are hybridising with the extremely endangered White-headed duck (*O. leucocephala*) in its last European outpost in southern Spain. Culling the Ruddy duck is in progress (Smith *et al.*, 2005). This example prompts the question whether action to remove invasive species and hybrids might be a strategy worth employing in plant conservation. However, it is important to note the difficulty, perhaps impossibility, of removing all hybrids, as some are cryptic and only detectable by molecular techniques.

B. Molecular studies of *Orchis mascula* and *O. pauciflora* populations in the Mediterranean indicate that F_1 hybrids occur, but later generations and backcrossing are rare or absent. F_1 hybrids are less fit than the parental species (Cozzolino *et al.*, 2006). Regarding conservation management they conclude: 'hybridisation is a natural phenomenon in food-deceptive orchids and it does not pose a threat to their survival. Furthermore, sympatric zones provide the stage for the evolutionary processes in orchids, and this peculiarity should be taken into account when devising orchid conservation strategies.'

C. Where geographical and ecological barriers are removed, introgressive hybridisation may take place. How should hybrid swarms be managed in conservation? One approach is to manage the habitat to allow microevolutionary processes to continue without trying to manipulate the genotypic composition of the hybridising populations themselves. For example, *Primula elatior* and *P. vulgaris* come into contact in coppiced woodland in Britain, and morphological studies support the hypothesis of introgressive hybridisation. Confirmation of this hypothesis has come through the use of allozymes in studying populations in Britain (Gurney, 2000). In Buff Wood in Cambridgeshire, and elsewhere, the continued existence of hybrid swarms has been secured by re-imposing coppicing cycles in the management of these ancient woodlands (Rackham, 1975).

Concluding remarks

The increasing use of molecular tools to study plant microevolution has provided important evidence on a number of issues.

As a consequence of the human-influenced breakdown of ecological isolation, interspecific hybridisation is quite common in plants, and, in some cases, introgression has been confirmed. In terms of the concept of winners and losers, in areas where the habitat has been hybridised, the fittest genotypes may be hybrid derivatives not the parental species.

The breakdown of geographical barriers by human agency has also contributed not only to the incidence of hybridisation, but also to polyploidy. A few new wild polyploid species have been detected that have arisen recently as a direct consequence of plant introductions. It is possible that many more cases of polyploidy have occurred following hybridisation, but we do not detect them because the plants fail to survive the immediate selection pressures imposed by minority-type disadvantage. Perhaps wider surveys might reveal more 'new' polyploids, but so far the evidence suggests that they are rare.

On the basis of present knowledge, it is highly unlikely that these speciation processes will produce new species at a sufficient rate to replace those that might soon become extinct as a result of human activities. This conclusion follows from the fact that only a very small number of new wild polyploid species, and a handful of new homoploid species, have been discovered. Moreover, while in habitats fragmented by human activities there is evidence that genetic differentiation occurs, this represents only the first stage of gradual speciation, which is thought to involve many generations.

It is important to stress that, almost always, crops are not genetically isolated from their wild and weedy relatives, and many complex interactions have been documented. Hybrid derivatives may be of limited fitness, but situations have been described where hybrids have high fitness.

The arena of possible gene flow within populations and 'interactions' between crop–weed–wild variants may be large, as molecular studies have detected widespread pollen dispersal. For example, there are clouds of pollen of high genetic diversity emanating from rape fields, indicating long-distance dispersal and possible gene flow (Devaux *et al.*, 2005). As we have seen above, gene flow via seed dispersal on farm machinery has the potential to encourage the wide dispersal of 'volunteers'. As our present knowledge of gene flow comes mainly from situations involving large monocultures of various crops, it may be an unreliable guide to gene flow in plant populations generally. More research is needed on gene flow in species-rich grassland and forest communities: generalisations should be avoided and each situation of interest should be explored individually. The possibility that new weeds may evolve from transgenic crops or through the hybridisation of transgenic crops and related weedy species has been widely debated. Looking for an 'entirely new weed' arising by this route would be inappropriate however: weeds evolve from pre-existing plants often through complex hybrid interactions. It is important to stress the common ancestry and interlinked futures of plants within wild–weedy–crop complexes.

Across the world, those managing the land – farmer, conservationist, landscape designer etc. – are facing the problems of invasive species. Evidence is emerging that hybridisation – both intraspecific and interspecific – may contribute to evolution of increased invasiveness in certain taxa. Also, introgression is generating hybrids that extend species' ranges and facilitate the invasion of new habitats. For example, hybridisation between wild and crop sunflowers is generating a wide range of crop–wild sunflower hybrids. Invasive behaviour of some of the extreme hybrid derivative variants is resulting in range extensions and the colonisation of sand dunes, desert floor and salt marsh (Rieseberg *et al.*, 2007).

The presence of sites where there are hybrid populations, hybrid swarms, new polyploid species etc. present critical challenges for the conservationist. The approach of some would be to try to 'preserve' site and plants. Others insist that management should allow the possibility of future evolution, but it should be immediately acknowledged that the management regime chosen *will* determine the course this takes. This point has been emphasised in relation to the management of endemic *Sorbus* species found on the Isle of Arran, Scotland (Robertson, Newton & Ennos, 2004). Another critical issue in management concerns the presumed stability or movement of hybrid zones (Buggs, 2007). He has called for more long-term investigations (reciprocal transplants, crossing experiments, surveys of molecular markers and population studies) to explore such zones of species contact and hybridisation, which are 'natural laboratories for studying the origin, maintenance and demise of species'.

Growing new cultivars produced by transgenic technology has changed the agriculture of many regions, with important consequences for conservation, especially if transgenic plants/pollen/seeds enter the wider landscape. For example, some farmers employ organic methods, in which no herbicides, pesticides or GM crops are used. The 'status' of food produced organically could be put at risk if there is gene flow from transgenic to organic crops. It has been claimed that buffer zones will prevent gene flow, but investigations have revealed that pollen flow and movement of seeds on farm machinery may cause significant and widespread gene flow in some crops.

If transgenic crops genetically engineered to be resistant to a particular herbicide are planted, it is claimed that weeds sensitive to the same herbicide will be more efficiently controlled using lower doses than in conventional agriculture. Moreover, crops can be bioengineered against insect damage using the Bt gene. Insecticidal transgenic plants based on proteinase inhibitors and lectins have also been developed (Velkov *et al.*, 2005). While farmers are in favour of making weed and insect control as effective as possible, it is clear that certain GM crops may have dramatic effects on non-target animals, plants and soil organisms; for instance, populations of beneficial insects including pollinators could be affected. Thus, conservationists are concerned that biodiversity (birds, insects and plant life) could further decline under the impact of a farming landscape dominated by transgenic technology. Reliable evidence of the effects of GM crops can only emerge from properly designed experiments. For example, farm-scale trials carried out in the UK to investigate the separation distances necessary to prevent crossing between GM and conventional crop varieties. The experiments were also designed to test whether biodiversity declined under GM cropping regimes. A decline was detected in some cases, but in others no decline was detected (see Weekes *et al.* (2005) for estimates of crop-to-crop gene flow and access to the literature on biodiversity under GM cropping). Although such trials are revealing, generalising from experiments to the wider landscape is highly problematical, especially as the range of transgenic plants is increasing with the prospect of the planting of crops engineered to produce pharmaceutical and other chemicals.

Transgenic grass cultivars are being developed for sports turf (see Wang, Hopkins & Mian, 2001; Ge *et al.*, 2007) and for wider use. For example, transgenic glyphosate-resistant (GRCB) Creeping Bent Grass (*Agrostis stolonifera*) has been developed and was planted out in 2002 in a carefully regulated planting of 162 ha in Oregon, USA (Zapiola *et al.*, 2008, 486). After the field had been 'taken out of production', subsequent testing around the site established that gene flow had occurred to the species growing in fields outside the site. Also, high frequencies of GRCB plants had survived in the 'production' area. Overall, for this outbreeding wind-pollinated grass, it was concluded that 'it was unrealistic to think that containment or eradication of GRCB could be accomplished'. These findings 'highlight the potential for transgenic escape and gene flow at a landscape level'.

Transgenes are being introduced into a number of fruit trees/shrubs (Laimer *et al.*, 2005) and a range of trees (Kuparinen & Schurr, 2007). In many cases, the intention of transgenic modification is to increase yield, but there are other motives, for example the restoration of populations of forest trees – American Chestnut and American Elm – both devastated by fungal diseases. Thus, Merkle *et al.* (2007) report that gene transfer systems have been devised to produce transgenic chestnut and elm, and the resulting plants await full evaluation. However, they acknowledge that such restoration, employing transgenic trees, raises a number of 'technical, environmental, economic and ethical questions'. Looking to the future, there could be many unwitting effects of the widespread introduction of transgenic trees and grasses. The distinction between wild and cultivated plants is not clear-cut. Many populations of grasses and trees could be considered as 'wild' although managed. Others are definitely crops, with grasses grown for silage, and trees of selected provenance set out in plantations. Between these two extremes are a whole array of human-exploited grassland and forested ecosystems. Gene flow from cropped areas stocked with grass species or trees could easily occur, and transgenes could be widely introduced into populations whose status might lie anywhere along the continuum between fully cultivated and 'wild'. Species hybridisation could also occur. For example, there has been a great deal of research on transgenic herbicide-resistant poplars (Meilan *et al.*,

2002) and *Eucalyptus*, as these are important plantation crops in many parts of the world. However, it could be seen as irresponsible to plant transgenic *Eucalyptus* in Australia – a centre of diversity for the genus. To overcome these difficulties it has been suggested that, in designing transgenic constructs to introduce into crops, genes for domesticated traits should be combined with the desired trait(s). Then, any hybrids with wild relatives growing on non-cultivated land would have such reduced fitness in the wild that they would be quickly eliminated. Whether such a strategy would be successful remains to be investigated.

14
Ex situ conservation

In their introduction to the *Plant Red Data Book* of endangered species, Lucas and Synge (1978, 31) concluded: 'Botanic gardens . . . are poised to play a major role not only in cultivating the [endangered] plants concerned, but also in their re-introduction, in habitat management and even in owning and maintaining small reserves for particular species.' Moreover, they predict: 'once the individual facts on threats, habitats sites and populations are known, successful conservation of most plant species is likely to prove far less difficult and costly than that of animals'. After 30 years, it is timely to review the role of botanic and other gardens in the conservation of endangered species, and to examine current views on these predictions.

The use of the phrase 'poised to play a major role' suggests that the use of botanic gardens for the conservation of wild endangered species represented something of a new direction for gardens in the 1970s and 1980s (Lucas & Synge, 1978). For what purposes were botanic gardens founded in the first place, and how have they developed historically? What have theoretical investigations and practical studies revealed of the 'strengths and weaknesses' of gardens in their proposed new role? Given the 'costs' associated with *ex situ* conservation, endangered species are generally grown in botanic gardens and arboreta. What are the likely consequences of such cultivation?

Botanic gardens

Botanic gardens have been defined as: 'Institutions holding documented collections of living plants for the purposes of scientific research, conservation, display and education' (Cheney, Navarro & Wyse Jackson, 2000, 7). However, this simple definition is not entirely satisfactory, as gardens differ widely in the degree to which they take part in these activities. There are also differences in size and number of personnel, and in the extent to which their collections are properly documented and correctly labelled (Heywood, 1987). Some gardens are managed by state or local authorities, while others are part of universities, or are privately funded. In some cases, gardens bear the prestigious name 'botanic' for purely historic reasons, and may now be managed as public parks.

History of botanic gardens

For present purposes we note that the first gardens that could justifiably claim to be 'botanic' were probably those developed in the Arab world, pre-Spanish Mexico, and China; the first botanic gardens in the 'western tradition' being founded as 'medicinal gardens whose primary role was to provide material and instruction for students of medicine' (Heywood, 1987). These were established in Italy (Pisa, 1543; Padua, 1545; Florence, 1545; Bologna, 1547), France (Paris, 1597; Montpellier, 1598), Switzerland (Zurich, 1560), the Netherlands (Leiden, 1577) and other western countries (Oxford, 1621; Uppsala, 1655). Later, collections were expanded and some botanic gardens became centres for the study of taxonomy and systematics, using not only the living collections, but also preserved material in newly created museums and herbaria.

In the nineteenth century, European and specially created satellite colonial gardens assumed another role, namely the transfer of economically important plant material to colonies overseas. For example, 'through its research, its dissemination of scientific information, and its practical activities, which included plant smuggling, Kew Gardens played a major part in the development of several highly profitable and strategically important plant-based industries in the tropical colonies. These new plantation crops complemented Britain's home industries to form a comprehensive system of energy extraction and commodity exchange' (Brockway, 1979, 6). Other European powers – the Dutch, French, Germans, Belgians etc. – also used botanic gardens as conduits for 'plant transfers and development'. In the case of cinchona (a treatment for malaria), rubber and sisal, which were all protected species indigenous to Latin America, plants were transferred 'by Europeans to Asia or Africa for development as a plantation crop in their colonial possessions'. The Latin American countries lost native industries in these transfers. Other territories 'acquired them only in a geographical sense, the real benefits going to Europe' (Brockway, 1979, 8). The role of the European and colonial tropical botanic gardens in these transfers is well documented and a number of reviews provide detailed accounts of the transfers (often illegal) of cloves, chocolate, tea, coffee, oil palm, sugar, spices, cotton, indigo, tobacco, breadfruit etc. (Purseglove, 1959; Holttum, 1970; Brockway, 1979; Heywood, 1983, 1987). The aim was to discover appropriate and profitable crops for different colonies. For instance, the St Vincent Botanic Garden was established in 1765 and, in an attempt to find suitable exploitable crops for this Caribbean island, a number of economic plants were introduced from overseas. To this end, nutmeg and black pepper were introduced from French Guiana in 1791, and breadfruit was brought from Tahiti, Polynesia, by Captain Bligh in 1793 (Anon., 2000). As we noted in an earlier chapter, bringing plants from overseas to determine their potential as economic crops, ornamental plants etc. was widely practised, by establishing state-sponsored or private acclimatisation gardens, for example in France, Britain and Australia. Amongst the most famous of the surviving gardens of this type is the acclimatisation garden of La Orotava on the Canary Island of Tenerife, set up by Spain in 1788. Later, as agricultural institutes and commercial companies took over the role of plant breeding, forestry research etc. in colonial territories, many botanic gardens overseas became neglected (Heywood, 1987). However, some colonial tropical botanic gardens expanded, and included in their activities the scientific studies of the local flora.

In the nineteenth and twentieth centuries many new gardens were developed in Europe and the USA. Some were centres of horticultural research, while others, such as the Missouri Botanic Garden (founded in 1859), were important scientific institutions from their inception. Others, bearing the name botanic, 'had no real claim to the title at all' (Heywood, 1987, 3).

While many botanic gardens have flourished and new gardens have been opened, others, particularly in tropical counties, have declined. At the present time there are of the order of 1,800 botanic gardens and arboreta in 148 countries worldwide (Wyse Jackson & Sutherland, 2000), but estimates depend upon how botanic gardens are defined. However, it is important to stress that there is a conspicuous mismatch between the floristic richness of different areas and the number of botanic gardens they contain. Relative to the size of the flora, there are few gardens in Central and South America and Africa.

Traditional botanic gardens: what do they contain?

As a prelude to examining the theory and practice of *ex situ* conservation, it is important to make the important point that, while a number of new botanic gardens have been developed in the last 20–30 years with conservation as a major objective (e.g. the Chicago and Welsh Botanic Gardens), *ex situ* conservation represents a 'new' activity in established gardens, whose design and plant collections were devised for other purposes.

Gardens are many and various, but some general comments may be made. Botanic gardens are usually designed to display trees, shrubs, perennial and annual plants in a variety of settings. Some are in formal plantings, but also more naturalistic gardens have been devised, e.g. wetland areas, and rock, herb and scented gardens etc. Often, where climate permits, these displays are set in lawns. Plants are usually named, labelled and maintained as spaced plants in weed-free plots. Sometimes there are special displays, for example, glasshouse collections, and there may be order beds displaying the different families of flowering plants.

In a very useful historical review, Stearn (1984) has discussed the ever-widening number of plants available for cultivation in botanic and private gardens (Table 14.1). The choice of plants displayed in any particular garden reflects its date of origin, the acquisition/disposal policies of its directors and governing bodies, the research activities and travels of its staff, and sometimes the eccentricities of its staff and benefactors. Botanic gardens receive their plants by exchange, gift, purchase and, for the larger institutions, some were collected on expeditions. Many are the result of exchanges between gardens either informally or by selecting seeds offered free-of-charge in their official seed list (Heywood, 1976). Botanic gardens have traditionally been centres of horticultural research, and many gardens include displays of cultivars developed from wild collections by artificial selection.

Major botanic gardens have always maintained large collections of plants, and it has been estimated that more than 85,000 species of plants are in cultivation (Raven, 2004, xiii).

Table 14.1 *The date of the introduction and origin of different species in botanic gardens (See Stearn, 1984 for details of researches by Kraus (1841–1915), who detected six main periods of plant introductions in the collections in the Halle Botanic Garden in Germany. To bring this research up-to-date, Stearn added the three final phases)*

1. Introductions from within Europe to 1560.
2. Following the expansion of the Ottoman Empire, there was an influx of plants from the Balkans and western Asia 1560–1620.
3. Plants from Canada and Virginia arrived in the period 1620–1686.
4. From 1687 to 1772, there was a large influx of plants from the Cape Region of South Africa.
5. As exploration continued in North America (1687–1772), a large number of trees and shrubs found their way into botanic gardens.
6. In the period 1772–1820, there were many introductions from Australia.
7. Then followed a period (1820–1900) in which heated glasshouses were developed for growing frost-sensitive tropical and other plants. Also, plants from Japan and North America were brought into cultivation.
8. Expeditions to western China resulted in the introduction of many species from this region in the period 1900–1930.
9. From the 1930s onwards, gardeners no longer expected 'new plants from the wild' (Stearn, 1984), and more and more hybrids were created between the plants already in cultivation.

Heywood (1987) considers:'the expansion of the collections themselves was stimulated by generalised pretensions of building up comprehensive samples of plant diversity for general scientific study... rather than to any clearly articulated policy'. Some garden displays appear to be based on the proposition that as many specimens as possible should be grown, especially in glasshouses. In the late nineteenth and early twentieth centuries, gardeners became aware that some of the species they grew were likely to be extinct in the wild, e.g. *Ginkgo biloba* from China (Wilson, 1919); *Amherstia nobilis* from India (Blatter & Millard, 1993). This did not, however, begin an immediate campaign for conservation. A change of attitude came later.

Botanic gardens: Victorian relics or twenty-first century challenge?

In 1985, Professor J. N. Eloff gave an inaugural lecture with this title at the University of Cape Town, on his appointment as the new director of the world famous Kirstenbosch Botanic Gardens in South Africa (Eloff, 1985). In the 1970s and 1980s the subject of the lecture was highly topical, as botanic gardens in many countries were questioning their traditional role, and were arriving at the conclusion that conservation of endangered species should become one of their key activities. Heywood (1987, 15–16), examining the roots of what he calls a *crise d'identité* in botanic gardens in the 1970s, concludes: 'the scientific reputation of a botanic garden is judged, not by the living collections' but by the publications of its staff. Thus, in large botanic gardens with associated herbaria, there was often a move 'away from the formerly dominant living collections to the herbarium collections', staff being employed to work on a series of Floras and other publications. Sometimes, the living collections were 'neglected or even ignored by the scientific staff'. Also, financial problems affected even the most prestigious institutions. For instance, in the 1980s, as the political climate changed, Kew had to justify its government funding. It was against this background that directors of botanic gardens began 'to look to conservation as one of the major goals and indeed justifications' for maintaining the living collections (Heywood, 1987).

As an example, it is instructive to examine the changes at the Royal Botanic Gardens at Kew. According to Desmond (1995), Kew first became involved in an international conservation project when Melville, in his retirement, and then others, began researches that led to the production of the *Red Data Book of Threatened Flowering Plants and Gymnosperms* (1970), which we have examined in some detail above. To promote wider debate about the role of botanic gardens, Kew sponsored an international conference on 'The Practical Rôle of Botanic Gardens in the Conservation of Rare and Threatened Plants' (1978). Kew is now a famous centre for conservation of plants. The extent of the transformation at Kew is evident from a comparison of two histories of Kew. Turrill (1959) does not mention conservation as a part of the work of Kew. In contrast, Desmond (1995) reports on its many conservation initiatives and activities.

Reflecting on the general change of outlook, Cheney *et al.* (2000, 10) conclude: 'in the last 20–30 years there has been a renaissance in botanic gardens world-wide, largely as a result of the developing concern for the loss of biodiversity and the need for many more institutions to become active in plant resource conservation'.

Gardeners and conservation

At this point it is worth stressing that, while it might be assumed that gardeners are the natural allies of those promoting conservation, in reality they use enormous quantities of peat, rocks, stones, pebbles etc., the collection of which may result in 'damage' to ecosystems. In addition, the collection of wild plants, seeds, bulbs, corms etc. for the horticultural trade may lead to the endangering of species, e.g. orchids, palms, cycads, cacti etc. Sometimes gardeners unwittingly contribute to such endangerment. For example, Read (1989, 1993) and Read and Thomas (2001) drew attention to huge numbers of wild-collected bulbs transported to Europe, e.g. from Turkey. Gardeners were being offered wild-collected material, but it was fraudulently labelled as coming from 'horticultural' sources. It is argued by Read and Thomas that, as wild collecting is threatening populations of many bulb-forming species, a more sustainable approach would be for exporting countries to set up their own nurseries to grow and export bulbs.

Ex situ conservation in botanic gardens

The urgent need for conservation measures

In the 1970s, a number of influential books on environmental issues and the detailed information in the *Red Data Books* provided new and important insights about endangered species, and a strong stimulus to action. Botanists were shocked at the revelation that *c.* 10% of the world's flora could be endangered. As *ex situ* conservation began to assume a higher profile in the activities of botanic gardens, existing collections in gardens, drawn as we have seen from so many worldwide sources, were examined to see which endangered species they contained. The first *IUCN Plant Red Data Book* by Lucas and Synge (1978) provides details of a representative selection of 250 endangered species, and the case histories make reference as to whether species are to be found in botanic gardens.

In investigating endangered species, with small surviving populations in the wild, Lucas and Synge (1978) confirmed that sometimes plants were in cultivation in a botanic garden. For example, only four widely separated ancient specimens of the tree *Punica protopunica* survived in the wild. The species is endemic to Socotra, an island to the east of the Horn of Africa. The flora evolved in the absence of large mammals, but because of excessive grazing by introduced animals (goats and cattle), 132 of the 216 species of endemic flowering plants were rare or threatened, with 85 in immediate danger of extinction. *Punica protopunica* is now being protected at the Royal Botanic Gardens, Kew (Lucas & Synge, 1978).

In another group of cases, field studies confirmed the precarious position of the surviving populations of some endangered species, but the move to bring the plant into cultivation, although urgent, had yet to be made. For example, the island of Réunion, in the Indian Ocean, has many endangered endemic species, threatened by agricultural and forestry developments, grazing by feral deer, and numerous vigorous introduced plants etc. On the island, the small tree *Badula crassa* (Myrsinaceae), was found to be critically endangered with only three scattered individuals (Lucas & Synge, 1978).

Investigations also revealed that some species were already extinct in the wild. The only remaining plants were in botanic and sometimes other gardens. The Hawaiian palm 'Loulu' (*Pritchardia macrocarpa*) provides an interesting example. Demand from gardeners for specimens of this species had been so great that this plant had become extinct in the wild. However, plants survive in private gardens, and a single individual was found in the collections at the Honolulu Botanical Garden.

Aims of ex situ *plant conservation*

Since the 1970s a number of major objectives have been defined for *ex situ* conservation by botanic gardens (Wyse Jackson & Sutherland, 2000).

- To prevent the immediate extinction of endangered species.
- Where it is not possible to return species to the wild because of permanent loss of habitat, to conserve species in the longer term, whether as living plants or, as stored seed, in gene banks.

- As we shall see in a later chapter, with the reimposition of appropriate management in nature reserves, conservationists hope to reinvigorate depauperate populations of endangered species. In devising appropriate management, botanic gardens, through their experience in cultivating rare and endangered species, could provide very valuable information on the autecology, breeding system etc.
- It has been proposed that *ex situ* conservation should be combined with *in situ* activities to produce an integrated approach to secure the future of rare and endangered species (Falk, Millar & Olwell, 1996; Maxted, 2001). Thus, botanic gardens can provide a temporary refuge for species that are threatened by immediate danger, and then, by multiplication of stocks, provide material for the reintroduction of the species to restored ecosystems at their original sites or elsewhere.
- To provide material for educating the public about conservation.

With reference to these aims, a number of key issues will be considered in turn.

- By comparison with animals in zoos, the *ex situ* conservation of plants in gardens would appear to be simple and effective.
- There are a number of strategic and practical issues to face concerning the strengths and limitations of growing any collection in botanic gardens. To what extent has it proved possible to conserve collections of endangered species long term in such gardens?
- Can endangered species be conserved in gene banks? How should populations of plants/seeds be sampled and maintained in *ex situ* conservation?
- Some of the early literature on conservation states that endangered species can be 'preserved' in botanic gardens. Such terminology hints at the possibility of maintaining species unchanged in botanic gardens. Will species change genetically in botanic gardens and, if so, what are the evolutionary implications of such changes?

Botanic gardens and zoos

The *ex situ* conservation of animals and plants together in the same site is not generally attempted. For a variety of reasons, it is not possible to grow plants with their specific pollinators, fruit dispersers etc. Concerning the conservation of animals, zoologists are faced with a mass of problems (logistical, legal and practical) and high costs in the collection of animals in the wild, their shipment and housing/ maintenance in zoos (see Hancocks, 2001). The zoo environment brings with it behavioural changes in animals, and there may be problems in supplying appropriate food for each species, and it may not be possible to persuade the inmates to breed in captivity, and, even if charismatic megafauna do mate successfully, they produce few progeny. Moreover, there is the ever-present problem of attending to the welfare of the animals. In contrast, *ex situ* conservation of plants in botanic gardens would appear to be a cheap and effective means of conservation (Hancocks, 1994). Botanic gardens are able to maintain a wide range of environments for displaying their collections, including: wet through to dry habitats; sunny and shady areas; and indoor and outdoor facilities. By the addition of peat, sand, gravel or lime etc., the range of soil conditions in the garden may be extended beyond those found naturally on the site. The ability of many plants to regenerate from vegetative fragments allows stocks to be built up quickly and, given the fecundity of individual plants relative to endangered vertebrates, potentially gardeners can raise large numbers of progeny from seed. Moreover, many (but not all) plants are self-fertile and viable seed can be produced in quantity from a single individual. It has proved possible to grow most species in gardens, only a few have proved as yet impossible to propagate, e.g. *Hibiscadelphus woodii* from Hawaii (Prance, 2004, xxiv).

Some inherent limitations of botanic gardens

To set against these advantages, there are a number of general difficulties in maintaining any collections of plants in gardens. Indeed, amongst the mainly positive and optimistic views of early reviews of *ex situ* conservation, as we shall see, there have been many forthright comments exposing the limitations of gardens.

Space and resources

Typically, botanic gardens are often full to capacity, and most species are represented by only a very few individuals (Hurka, 1994). Where a new use of botanic garden space and resources is proposed in an established institution, there is likely to be a conflict of interest between those who wish to maintain older established displays and landscape features, and those who propose that space and resources be used for other purposes, such as the *ex situ* conservation of wild species.

Ex situ conservation require the assurance of long-term finance. While it is true that conserving plants may be cheaper than that for animals in zoos, the cost may still be considerable. For example, it has been calculated (at 1987 rates) that it cost $64 per tree per year at the Arnold Arboretum, Boston (Ashton, 1987, 126). This is not an insignificant sum considering that it might take *c.* 50 years for a tree to come into flower. Finding the resources to maintain living collections continues to be a concern, indeed, some are warning of a 'collections crisis' brought about by 'dwindling support' (Dosmann, 2006).

Problems of continuity

In considering the practical possibilities of *ex situ* conservation of living plants in botanic gardens, it is important to note problems of ensuring continuity. If the past is a guide to the future, changes of policy, natural disasters, wilful damage and wars may all determine the fate of collections.

Effective long-term conservation of living plants, especially those that are extinct in the wild, requires that such plants have an assured future. The lifespan of individuals in gardens varies very widely. Maunder (1997) has analysed the survivorship of accessions of three groups of plants at Kew, some of which have been in the garden for many years. 'No *Caralluma* (Asclepiadaceae) accessions survived for as long as 25 years, and all had annual mortality rates greater than 16% . . . In contrast, some accessions of both *Nepenthes*

(Nepenthaceae) and *Primula* (Primulaceae) were maintained in an *ex situ* context for up to 35 years with a mortality rate of <1 per cent.' Other groups of plants, in particular trees and other woody plants may be longer lived, and a few notable plants have survived in botanic gardens for hundreds of years. For example, at the Hortus Botanicus at Padua, the oldest plant, until 1984, was a specimen of *Vitex agnus-castus* known in the collections since 1550 (Masson, 1966). At present, the oldest specimen – 'Goethe's Palm' – dates from 1585. This palm – *Chamaerops humilis* var. *arborescens* – was carefully studied by the German scholar, while he was preparing his famous essay *Metamorphosis of Plants* (1786). At Kew, the oldest plant under glass is a cycad (*Encephalartos longifolius*) collected in Natal in 1775 (Hepper, 1989). It is difficult to determine the oldest tree at Kew, but a possible candidate may be an eighteenth-century Sweet Chestnut (*Castanea sativa*) (Desmond, 1995).

Despite the survival of certain venerable specimens in botanic gardens, evidence suggests that for a number of reasons the long-term maintenance of collections cannot be taken for granted. For instance, in the nineteenth century Professor Babington assembled a large number of *Rubus* species in the Cambridge Botanic Garden: this collection has been lost. Reflecting on this issue, from the perspective of one who has an unrivalled knowledge of botanic gardens worldwide, Raven (1981) writes:

> Unfortunately, such [special] collections are often dismantled or simply deteriorate after the specialists who built them up are no longer active at the respective institutions. Although they are often of very great value internationally, they may, if they are not actively utilized, come to be viewed as a drain upon limited resources of the institution where they are housed. Even when financial considerations are not limiting, it is difficult to provide for such collections the meticulous and sustained care that is essential for their survival without the attention of the specialist who is deeply concerned for them.

What will be the long-term fate of collections of endangered species assembled by enthusiasts in botanic gardens?

Disastrous events

All stands of trees are vulnerable to exceptional storms. For instance, on the night of 16 October 1987, hurricane force winds in southern Britain uprooted or badly damaged about 10% of the 9,000 trees in the Royal Botanic Gardens, Kew. At Kew's satellite garden at Wakehurst Place, Sussex, 17–20,000 trees were blown down or irreparably damaged, and in some exposed places more than 95% of the trees over 10 metres were destroyed (Langmead, 1995). The same storm destroyed 25% of the collection at the National Pinetum, Bedgebury, Kent (Morgan, 2001). Thus, as a consequence of this storm, many trees of conservation interest were lost or damaged at several sites in southern England. Trees in botanic gardens in hurricane zones are especially vulnerable to damage. For instance, in 1992, the Fairchild tropical garden in Florida lost 8,000 accessions when it was hit by Hurricane Andrew. Given that global climate change may result in increased number of storms (see below) the long-term survival of endangered species of trees in botanic gardens obliges institutions to face the costs and space implications of maintaining reserve stocks of saplings etc.

Botanic gardens and their collections have been subject to damage from other sources. For instance, in February 1913, the Orchid House at Kew was raided by militant suffragettes, who broke windows and destroyed some of the plants. In a second raid they made a more effective protest – they burned down the tea pavilion (Turrill, 1959). Botanic gardens often fare badly in wartime. For example, at its peak, the famous garden of the Komarov Botanical Institute of the Russian Academy of Sciences, the forerunner of which was founded in 1713 by decree of Peter the First, had 28,000 taxa in its collections. The garden, now restored, suffered huge losses in the Second World War and especially in the Siege of Leningrad (Smirnov & Tkachenko, 2001). More recently, during the war in Bosnia, the collections of the botanic garden in Sarajevo were virtually all destroyed by direct damage and fire, as more than 400 shells fell onto the site (S. M. Walters, personal communication).

Documentation and misidentification

Even the most celebrated gardens have problems with documentation and labelling of the plants in the collections. Many plants grown in botanic gardens have been obtained through seed lists, a tradition that goes back to the nineteenth century. In a review of the origin and development of the seed list (*Index Seminum*), Heywood (1976, 226) considers in detail the questions of documentation and misidentification. He concludes:

> as botanic gardens grew in number, size and importance, and as the range of taxa cultivated increased . . . the lists tended in many cases to become overlarge, unselective, inaccurate and repetitious in the sense that they often contained the same material, often received shortly before from some other garden . . . The same often inaccurately identified seeds were exchanged, cultivated and exchanged and cultivated year after year, often ending up in the garden which first issued them. The often considerable percentage of taxonomic misidentifications of plants grown in botanic gardens until recently is a direct consequence of the receipt of misidentified seed received through the *Index Seminum* system . . . Over many years I have worked on *Digitalis* and calculated the percentage of incorrectly identified seed received [from seed lists] as 70–80% despite the small number of taxa in the genus and the ease of their identification.

Heywood makes the strong case that botanic gardens should not exchange seeds 'whose identity has not been verified'.

In a more recent review, documentation and labelling problems are again highlighted. Hurka (1994, 377), director of a university botanic garden in Germany, concluded: 'the original locality of the accessions and their subsequent handling is often not known', and, furthermore, he came to the depressing conclusion: 'a high percentage of plants is mislabelled, either from erroneous determination or inadvertent misplacement of the correct label'. Clearly, while botanic garden staff are keenly aware of the problem, the labelling and documentation of material remains an issue.

Investigations using molecular methods are providing new information on the question of misidentifications. Using RAPD and SSR analysis to study *ex situ* material of *Sophora toromiro* (endemic to Easter Island), Maunder *et al.* (1999) detected a number of misidentified trees in botanic garden collections. In another molecular study, Goodall-Copestake *et al.* (2005) report the finding of some misidentified specimens of elms (*Ulmus*) in *ex situ* collections.

Plant diseases in botanic gardens

Plant diseases may damage or kill plants wherever they grow. Some diseases affect particular species or genera. For instance, 'Dutch elm disease has had a devastating effect on elms in western Europe and North America through much of the twentieth century' and a fresh outbreak was reported in the late 1960s that affected most of 23 million elms in southern England (Brasier & Gibbs, 1973; Burdekin, 1979). The elm collection in Cambridge University Botanic Garden was infected, and trees were felled on safety grounds. Two very damaging diseases – lethal yellowing disease (a mycoplasma) and cadang cadang (a viroid-induced disorder) have seriously damaged palm collections in gardens (Maunder *et al.*, 2001b).

Other diseases are not so host specific and can damage or kill many different groups of plants. For example, an infestation of the honey fungus (*Armillaria*) affected many adjacent species growing together in the Cambridge University Botanic Garden and part of the arboretum had to be relocated with new stock. Amongst the other plant diseases that can affect plant collections is the bacterial disease fireblight (*Erwinia amylovora*), which damages Rosaceous plants of different genera. This disease has been introduced into western Europe and Australia (McCracken, 2001). There is a large number of other exotic introduced plant pathogens (Randles, 1985), and many of these have the potential to damage botanic garden collections.

Ex situ conservation: seed banks of wild species

As we have seen, the very long-term maintenance of living plants in collections is problematic. However, a major advance in conservation came with the recognition that in crop plants the viability of dried seeds could be greatly extended by storage at low temperature in a gene bank (see Linington & Pritchard, 2001, on which this account is based). Thus, it has been proved possible to conserve a wide range of wild and semi-wild species at major centres in many parts of the world, e.g. Australia (Department of Conservation and Land Management of Western Australia); Belgium (National Botanic Garden); Spain (Cordoba); Tanzania; UK (Royal Botanic Gardens Kew, Millennium Seed Project, Wakehurst Place: Linington, 2001; Van Slageren, 2003) and USA (Berry Botanic Garden, Oregon). There are smaller facilities in many other places.

Using the methodologies devised for crop plants, about 10% (150) of botanic gardens now have seed banking facilities for wild species (Hong, Linington & Ellis, 1998). However, few have adopted FAO/IPGRI standards (1994) for long-term base collections storage, which 'requires fresh seed to be dried within the range 3 to 7% moisture content, packaged in a moisture-proof way, and placed at subzero (and preferably −18 °C) temperatures' (Linington & Pritchard, 2001). Some storage facilities are based on the use of cold rooms and/or domestic freezers. Other institutions store at ultra-low temperatures (−160 °C) achieved by means of liquid nitrogen.

Research has revealed that many, but not all, species (see below) produce seeds that can remain viable when dry at low temperature. These so-called orthodox seeds 'can be dried to low moisture contents (where less than 5% of the seeds' fresh weight is water) without loss of viability' (Linington & Pritchard, 2001, 171). Moreover, a '1% reduction in moisture content roughly doubles the seed longevity'. It has been discovered that this applies to storage in the range 15–25% moisture content. With regard to temperature it has been discovered that 'a 5 °C reduction in temperature doubles the seed longevity'. Whether this relationship is strictly correct is debatable (Walters, 2004). A predictive model relating temperature and moisture was also developed by Ellis and Roberts (1980), based on the accelerated ageing of seed lots by subjecting them to increases in moisture content and increases in temperature, and extrapolating the graphs to colder and drier conditions. The results for rice revealed that, at 5% moisture and −20 °C, there was a theoretical potential for the seed to survive for at least 1,900 years (Linington & Pritchard, 2001, 171).

While the seed of many species exhibits 'orthodox behaviour', the seed of other species (so-called recalcitrants) is desiccation-sensitive – if the moisture content of seeds falls below *c.* 40%, the seed of such species dies. Interestingly, some genera, such as *Acer*, contain both orthodox and recalcitrant species (Linington & Pritchard, 2001). Further research has revealed that seed behaviour does not fall neatly into two categories – orthodox versus recalcitrant. Intermediate behaviour has been discovered, for some species have seeds that can withstand limited drying, but the seed dies, if moisture content falls below about 10% (Linington & Pritchard, 2001, 172).

Given the variation in the response of different species to seed desiccation, there has been speculation on the evolution of seed behaviour. It is not possible to determine the desiccation tolerance of fossil seeds, but it seems likely that the immediate ancestors of land plants were desiccation sensitive and that desiccation tolerance emerged early and possibly independently in different evolutionary lines, as such tolerance would allow the invasion of unoccupied terrestrial niches (Pammenter & Berjak, 2000). Thus, it seems possible that desiccation sensitivity could be a primitive condition or one derived secondarily from desiccation-tolerant ancestors. Recalcitrant behaviour would appear to be adaptively significant in habitats continuously conducive to germination and seedling growth, such as occur in tropical rain forests etc. Desiccation-tolerant species often have a persistent seed bank within the soil. In contrast, desiccation-sensitive species may have a persistent seedling bank. As desiccation tolerance seems to vary, it has been suggested that there is a continuum of desiccation tolerance amongst non-orthodox species (Pammenter & Berjak, 2000), while others are content to classify behaviour into three categories: orthodox, intermediate and recalcitrant (Walters, 2004). Studies of seed size in relation to desiccation tolerance suggest that recalcitrant species often have larger seeds; however, the overall relationship between seed category (orthodox, recalcitrant, intermediate) and seed size is not strong (Hong & Ellis, 1996). In the absence of 'morphological markers' of seed behaviour, the desiccation tolerance of endangered species has to be determined by appropriate tests

(Linington & Pritchard, 2001). It follows from these observations that not all endangered species can be conserved *ex situ* as dried seeds in cold storage.

At this point it is interesting to comment on the history of seed bank facilities in gardens. We may take as an example the developments at Kew. Seed banks were not, initially, explicitly designed for *ex situ* conservation. According to Brennan (1977, 6), 'the Kew Seed Bank at Wakehurst Place... grew from a need to avoid the often unpredictable annual harvesting of seed' that was being collected to provide material for the official Kew Seed List, which, as we have seen above, provided a means of exchanging seeds with other botanic gardens. At present, the specially designed Millennium Seed Bank at Kew (Wakehurst Place) is aiming to collect and store more than 24,000 species of economic plants of arid and semi-arid lands, some of which are endangered (Linington, 2001). This represents about 10% of the higher plants of the world. Also, as part of English Nature's Species Recovery Programme, they are engaged in a project to collect and store samples of all the British native species. This material will be available for *in situ* conservation of endangered species through reintroduction and restoration projects.

Seed banks with cold storage must have assured supplies of electricity and, should the public supplies fail, the large-scale seed storage installations at Wakehurst and elsewhere in the developed world make use of designed emergency facilities. However, there are important seed banks of crop plants in the developing world, where the electricity supply is uncertain. The Director of the International Plant Genetic Resources Institute of the World Bank has reported that many seed banks 'are in a perilous state'. For example, many samples were lost recently at the Koronovia seed bank on Fiji, when ancient refrigerators broke down (Anon., 2002a). Important seed banks have been destroyed in war zones in Afghanistan (Anon., 2002b). Some of this material of old cultivars and landraces is irreplaceable. However, it is possible that some of the stocks are represented in seed banks elsewhere. A new gene bank, available to store seed of crop plants from all the nations of the world, opened in February 2008. The facility, under the auspices of the Global Crop Diversity Trust and financed by the Norwegian government, has been constructed deep underground in the permafrost of the mountains at Svalbard, Norway (www.norwaypost.no).

Micropropagation

Very many species of plants can easily be propagated in gardens by exploiting the plant's natural capacity for vegetative reproduction (e.g. dividing plants; rooting cuttings from leaf, stem or root; planting offsets and runners etc.). To increase the efficiency of the process, and to maximise the chance of success with endangered plants, rare hybrids etc., another series of technical advances were made in the 1970s when micropropagation techniques were beginning to be used by botanic gardens. In essence, isolated surface-sterilised tissues are grown in liquid or agar culture in flasks or jars containing sterile media prepared from salts, sugars, vitamins and growth hormones (Sugii & Lamoureux, 2004). To an extent, the appropriate medium for each species is determined by trial and error. After a period of initial growth in vitro, material is then transplanted into soil (see Sarasan *et al.*, 2006, for a review of recent progress in this field). Micropropagation techniques have been used in the vegetative propagation of endangered species; for example, at the Andromeda Botanic Garden, Barbados, for producing plantlets of the rare Farleyense Fern (*Adiantum tenerum* var. *farleyense*), a variant that does not produce viable spores and has been reduced to four individuals in the wild (Anon., 2000, 28).

Micropropagation protocols have also been developed to allow the efficient germination of orchid seeds. This has involved the isolation and growth in pure culture of fungi with which the orchids have symbiotic partnerships. In experiments at Kew, it has proved possible to grow vigorous seedlings of such British rarities as *Orchis laxiflora* and *Cypripedium calceolus*, by sowing the minute orchid seeds in specialised media containing the appropriate fungus. Plants raised in this way have been successfully used in a reintroduction programme (Ramsey & Stewart, 1998).

Other types of gene bank

Advances in tissue culture and molecular research have made it possible to store non-seed material in gene banks under controlled conditions (Linington & Pritchard, 2001). Thus, in research on a range of species, it has been shown that plant tissues and structures may be successfully stored at low moisture content at subzero temperatures, e.g. pollen, plant embryos, spores (of pteridophytes and bryophytes), fungal hyphae, micro-organisms (bacteria, viruses and fungal spores), and vegetative tissues (of bryophytes, ferns and flowering plants). Furthermore, DNA banks have been established in Japan, USA (Missouri Botanical Gardens) and the UK (Kew) to store samples from plant sources (Linington & Pritchard, 2001).

These advances in micropropagation and gene banking provide new opportunities for *ex situ* conservation of crop plants. They are beginning to be employed in the conservation of wild plants. A number of key issues are being explored in species that do not reproduce by seed, species whose seeds are recalcitrant and therefore not bankable, and species that have such long life cycles that seed collection is impossible (Linington & Pritchard, 2001). How far will cryopreservation techniques make it possible to establish gene banks of vegetative material? Further issues of importance to conservation must also be considered. Can 'mass micropropagation' of rare and endangered plants be achieved as part of major reintroduction projects? While some progress has been made in growing orchids with their symbiotic fungi, will it be possible to extend these studies to include the symbiotic relationships of other groups of plants? Other organisations have huge and varied collections of plants, but lack the facilities for modern gene banking, e.g. the National Trust in Britain (Simpson, 2001). The devising of collaborative projects could be a way forward. A good example of this is provided by concerns about endangered bryophytes in Britain, where at least 50 mosses are on the verge of extinction. In 1999, Kew and English Nature signed a contract to grow mosses and liverworts under sterile conditions and store them in liquid nitrogen (www.rbgkew.org.uk).

Genetic changes in cultivation and in seed banks

In contemplating the possibility of genetic changes it is helpful to trace the steps involved in the introduction of plants to gardens.

Typological approaches

Traditionally, botanic gardens have often grown a single accession of as many species as possible, and common species are not represented. To some extent this strategy flows from the problem of space, but also it may reflect a typological approach to diversity amongst gardeners (see Chapter 7). In devising *ex situ* strategies, some conservationists may also exhibit a typological attitude. Their aim, if it is made explicit, is to make sure that some material of each species is conserved, but with no special attention being paid to conserving the variability within species. It may also be assumed that all the populations of a species have a common germination pattern and that one sample is truly representative. Recent studies by Martin *et al.* (2001) have revealed intraspecific variability in germination characteristics, undermining the assumption that one population may be representative of all.

A typological approach does not satisfy those who have studied the intraspecific geographical and ecological variability of rare and endangered species. They argue that the unit of conservation must be the population. Thus, a decision must be made at the outset on what will be the unit of conservation in a particular case, as this will determine the sampling strategy employed. While it is clear that there is a case for conserving population samples, rather than species samples, it is important to stress that there are practical and cost implications. Collecting populations rather than bulk species samples is more time consuming and expensive, and requires more storage spaces in gene banks. It is also clear that it will be impossible to collect seeds from all threatened species. In considering how priorities might be established, Farnsworth *et al.* (2006) present a wide-ranging review based on a case study of rare New England plants.

Collecting samples

Unless conservationists are dealing with a situation where all the surviving plants of an endangered species are brought into cultivation, it is necessary to collect representative samples of plants/seeds in the wild. It is also important to recognise at the outset that when samples are taken from wild populations, and brought to gardens and seed banks, that founder effects are possible. A new colony in a garden and a species sample in a seed bank established from one or a few individuals are likely to contain only a fraction of the genetic variation present in the source population. It is very important, therefore, to devise a strategy for collecting a sample that 'captures' the variability of the wild population from which it came. Thus, while some unplanned collecting undoubtedly occurs as conservationists come upon a rare and endangered species, ideally, sampling for long-term conservation of wild plants must be designed in advance (Anon., 1991a; Guarino, Rao & Reid, 1995; Guerrant *et al.*, 2004a) taking into account theoretical models of genetic variability of plants and the practical experience of sampling populations of cultivated plants – landraces, ancient cultivars etc. – and the wild ancestors of crop plants. The basis of planned sampling is to collect such samples that will include a certain proportion of the genetic variability of the population. A number of different approaches have been considered (see Guerrant *et al.*, 2004a). For instance, 'Marshall & Brown (1975) suggested that the objective of a conservation collection for a genetically represented sample should be to include at least one copy of 95% of all alleles that occur in a (large) population at frequencies greater than 0.05 (5 percent) (quoted by Guerrant *et al.*, 2004a, 423).

Concerning the collecting process, a number of practical issues have to be confronted. Firstly, it is essential to obtain permission to sample populations and to make sure that the collection of plants and/or seeds does not damage small populations of rare and endangered species. Secondly, if population samples are being collected it is important to face the challenge and accept the limitations of our understanding of what constitutes a population of interbreeding plants. Some notion of population structure for a particular species may be possible, if the genetic variation, breeding system and patterns of gene flow have been determined by experiment. However, the collection of material for conservation must often be done in circumstances where we do not have enough information to define populations or estimate their genetic variation with any precision. Thirdly, there are some practical and biological issues to face in sampling. It may be necessary to make some initial assumptions, based on the behaviour of related species, about whether a species has orthodox, intermediate or recalcitrant seed behaviour. Only later can these assumptions be tested. Concerning the collecting process itself, it is essential to make representative collections, not just collect from the most accessible places. Thus, while it may be easy to collect a species near a road, other populations of the same species should be considered, even if they are more difficult to collect. Plants produce different amounts of seed. Unless care is taken, the seed of larger and more seed-rich individuals may be over-represented in samples. Often conservationists are presented with a single opportunity to collect from a site. However, not all plants in a population seed at the same time. Thus, early flowering individuals might have shed their seed before the day of sampling, and others may not yet have seeded. It is important that ripe seed is collected, for it has been discovered that unripe seed of orthodox species behaves in a recalcitrant fashion (Walters, 2004). In some cases there may be a significant proportion of non-flowering individuals. For instance, some populations may be biennials. In a particular year some plants flower, while others are at an immature stage and will flower the following year. In effect, there are two distinct gene pools. A single collecting expedition will not sample all the variability in the population. There may also be a long-term buried bank of viable seed in the soil. Often, this potential variability is neither sampled nor explored.

A number of sampling schemes have been proposed. However, Hawkes (1987) considers that overcomplicated rules will lead to no collecting. He proposes: 'Within a population walk backwards and forwards as if ploughing a field taking seed heads at random.' Heads should be collected randomly or at least non-selectively so many paces apart, with the aim of collecting 50–100 heads containing *c*. 2,500–5,000 seeds for a bulk sample. In sampling plants, he recognises the distorting effects that might flow from multiple sampling of individual widespread clones and suggests that spaced sampling is necessary. He also recognises the importance of sampling a species across its ecogeographical range and in different microhabitats. If seeds cannot be found, he proposes that vegetative plants, bulbs, corms etc. should be collected and grown on in botanic gardens. These can be sib-mated with bulked pollen in an insect-proof glasshouse and composite seed sample should be collected for banking. Finally, Hawkes stresses the vital importance of taking full details of the site's location, ecological, topographical and status details and provides proforma documents, as examples of good practice.

A sampling strategy, based on the experience of earlier collectors, has also been developed by the Center for Plant Conservation (Wieland, 1993). They recommend that a population sample of 1,500–3,000 seeds be collected from at least 10–50 individuals per population, with pooling of seeds. Collecting activities may be spread over a number of years, providing that this does not interfere with the demography of the plant populations. They too recommend that species are represented by collections from several populations, especially where species are restricted to three or fewer sites, where intraspecific ecotypes have been described or suspected, and where there are indicators of danger to plants, e.g. at sites where there are 'for sale signs or bulldozers on site'. As a general recommendation, it is suggested that up to five populations be collected per species. They stress that sampling strategies must be carefully designed, especially if the collections are to be used in species reintroduction and habitat restoration.

In a number of cases, molecular methods have been employed to check whether previous sampling for *ex situ* conservation has been 'adequate' to capture the variation in the wild, and some 'deficiencies' have been detected (see, for example, the studies of the critically endangered Iberian *Borderea chouardii*: Segarra-Moragues, Iriondo & Catalán (2005); investigations of 675 accessions of the black poplar (*Populus nigra)* in nine gene banks in Europe: Storme *et al*. (2004); and *ex situ* collections of Texas wild rice (*Zizania texana*): Richards *et al.* (2007)).

Growing plants from wild-collected samples in gardens

Many plants have well-developed means of vegetative reproduction, runners, stolons, rhizomes, detachable bulbils etc. Some water plants 'fall apart' and the separate pieces regenerate. Gardeners exploit this property of plants by artificially dividing plants and rooting the detached fragments. Species differ in the capacity of their different structures to root. Gardeners have wide experience in 'rooting' cuttings, taking the roots of some species, the stems of others and the leaves of a third group. Potentially, each fragment has the capacity to root and produce an independent plant, which, barring mutation, is of the genotype of the mother plant from which it came. This facility provides a cheap and successful means of propagating plants that can be exploited in *ex situ* conservation. In this regard, the behaviour of many plants provides a vivid contrast to that of higher animals, whose lifespan is more clearly defined by the progressive decline associated with the ageing process (Turker, 2002).

However, it should not be assumed that an everlasting cycle of asexual reproduction is possible in plants. Very little is known about the decline with age in the viability of plants, but recent studies have shown that ageing processes occur, e.g. in studies of *Cistus* there was an increase in oxidative stress in chloroplasts and a reduction in photosynthesis as plants became older (Munne-Bosch & Alegre, 2002). Furthermore, it has also been suggested that the propagation of plant stocks, through very long cycles of vegetative propagation, may not be possible because of the build-up of virus infections and other diseases. For example, in 1949, the sterile hybrid *Primula* × *scapeosa* was produced by crossing two Himalayan species. This garden plant was widely propagated by vegetative means, but by 1982 it had ceased to be grown as it was debilitated by cucumber mosaic virus (Richards, 1986).

Gene banks: sampling and the regeneration process

There has recently been a critical review of the sampling methods used in conservation in relation to the regeneration of seed stocks when viability declines in storage (Lawrence, 2002). For details of the rigorous and detailed analysis the original paper should be consulted. Here the main conclusions are presented. In collecting seed from the wild, Lawrence (2002) makes it clear that attention must also be paid to the loss of viability that will occur, with time in seed banks. Studies on dried seeds have revealed that biochemical, physiological and chromosomal changes occur in time (see Abdul-Baki & Anderson, 1972; Fenner & Thompson, 2005). The ageing process in dried seeds kept under very low temperatures is under active review in seed banks.

As seed ages naturally at ambient temperatures, genetic mutations and chromosomal changes are known to occur. Mutations may perhaps occur as a result of the quality checks performed on the seed. Thus, in seed banks, newly arrived samples are checked using X-rays to make sure that the seed has properly formed embryos and contains no insects. As seed is regularly sent around the world and insects might survive as eggs at low temperature, checking the seed is a necessary safeguard against the introduction of potential insect pests to new territories. Could X-rays induce mutations in the seeds? Research is in progress on the genetic effects of long-term storage, but as yet little is known of the changes that might occur in the very dry and intensely cold conditions of the seed and other types of gene bank.

In setting up seed bank collections, the original sample is usually divided into the main collection and a working collection, which is used for research, viability testing etc. Also, there are withdrawals for research, for *in situ* conservation activities etc. Taking subsamples from a genetically diverse main sample will produce batches that differ genetically. Moreover, stored seed represents a diminishing deposit. The name bank is not entirely appropriate, as interest is generally added to bank accounts. The storage of seed in a seed bank does not create any additional seed. When viability falls, say to 85% or lower, new stocks are grown to produce new seed. The original seed samples were collected in their natural habitat, and Hondelmann (1976, 219) makes the very important point that, at the rejuvenation stage in a garden, considerable new selection 'pressures' will be exerted on the stocks if plants are grown in climatic, edaphic or biotic conditions different from those of their wild habitat.

Given the evidence that genetic changes occur as seeds age, when, in time, testing reveals that loss in viability has reached a critical point (usually set at 85%, but sometimes at a lower viability, perhaps 50%), seeds are taken from the store and plants are grown to provide a new batch of seed for storage. Lawrence makes the important point that the design of the initial sampling should take account of how genetic variability might be protected in the subsequent regeneration process.

Lawrence calculates that if the unit of conservation is the species then, at a minimum, a bulk sample of seeds from '172 plants drawn at random from a population of a target species, is of sufficient size to capture all or very nearly all of the polymorphic genes that are segregating in a population, provided that their frequency is not less than 0.05'. However, if the population is the unit, then Lawrence (2002) recommends that seed destined for seed banking be collected separately from a range of seed parents and kept in separate packets. Having seeds from separate seed parents within populations, rather than bulk samples, provides a more efficient means of preserving the genetic variability at the regeneration process. He recommends that in the initial sampling seed be collected from a minimum of 12 individuals from 15 populations ($15 \times 12 = 180$ in total), the samples from different seed parents being kept separate in storage.

When it becomes necessary to regenerate new seed from the samples in storage, it is essential to do so in a manner that conserves the gene frequencies against the possibility of genetic drift. As we saw in Chapter 3, genetic drift over generations can cause gene frequencies to change, such that some alleles become fixed and other(s) lost. Also, with loss of variability, drift can cause 'populations to become progressively more and more inbred even though individuals in each generation mate at random' (Lawrence, 2002, 200). To retain the genetic integrity of the material when it becomes necessary to regenerate samples of seed, Lawrence recommends that for each population one plant is raised from the seed of each of the 12 seed parents, and pairs of plants from each population are mated at random in controlled pollinations. This involves artificial transfer of pollen, following emasculation and protection of the female flowers. Seed produced by this controlled bi-parental mating procedure will minimise the effect of drift, as every one of the original seed parents is represented by descendants (barring accidents, differential responses to selection pressures etc.).

Reassessing the size of samples in gene banks

The major concern of collecting samples has been to collect a genetically representative sample from populations and to maintain this variability once it has been 'captured'. However, recently, as we see in Chapter 16, there have now been many reintroductions of species into the wild, and it has been realised that large numbers of seedlings may be required for success. Indeed, several attempts have to be made before successful establishment is achieved. As reintroduction may only be accomplished at high demographic cost, Guerrant *et al.* (2004a) consider that larger samples should be collected and stored, perhaps an order of magnitude greater than those currently conserved.

Larger samples would prevent another problem. To determine the viability of cold-stored seed samples, testing is necessary. In trying to be economical with the seeds in storage, only a few seeds may be available to determine viability. As a consequence, it is possible, by chance, to draw false conclusions. Thus, in statistical testing it may be concluded that a decline in viability has occurred when it has not (a Type I, false change error) or there may be failure to detect a decline when one has occurred (a Type II, missed change error). If large initial samples are in cold storage, then viability could be tested using larger samples, thereby reducing the possibility of error. While it may be possible to collect large 'species' samples from the wild, the separate collection and storage of maternal lines, as recommended by Lawrence (see above), will provide far fewer seeds for viability testing.

Wild plants in botanic gardens: selective forces acting on living plants

While there are few experimental studies of the selective forces experienced by wild plants as they are brought into gardens, Jones (1999) makes it clear: 'there will never be a rest from natural selection. Genes are checked at all times for their ability to cope.' Plants in gardens may also be subject to artificial selection involving both conscious and unconscious elements. Even if those tending the plants are aware of the desirability of 'keeping wild plants' wild, it is certain that selective forces will be at work on any material brought into gardens; likewise, selection acts on animals living and reproducing in zoos.

Developing this key point, natural selection takes place on material collected in one climatic zone and subsequently cultivated *ex situ* in another (e.g. alpine to lowland; Mediterranean to northern Europe etc.). Also, the diverse activities of gardening will present newly arrived plants and seedlings grown from seed samples with artificial selection, both conscious and unconscious, as introduced plants encounter new soils, changed water regimes, altered microclimates in pots, glasshouses or shade houses, and new pests and diseases etc. In germinating seeds, gardeners may make no allowance for the fact that some of them could be dormant, and those seedlings that grow immediately may be 'selected', the ungerminated could be discarded with the soil. The tallest seedlings may be chosen for potting on. Unwittingly, those plants with the largest, most horticulturally desirable flower colour may be selected. Whilst those carrying out *ex situ* conservation might protest that they are not consciously making choices from a sample of plants that would constitute artificial selection, Darwin stressed the power of unconscious selection, which by definition is impossible to avoid. Many plants produce huge quantities of seed and seedlings and, therefore, inevitably gardeners select, and exercise choice to select the one or few from the many, an exercise that will change the genetics and evolutionary potential of those stocks.

Moving from these theoretical concerns, it is helpful to consider the experience of those responsible for botanic garden collections. Townsend (1977, 1979) makes some observations from his experience at Kew, indicating that the selective forces imposed on plants in transit and in gardens are many and various. In justifying the building of a micropropagation unit at Kew, he reviews the cultivation techniques then practised (Townsend, 1979, 189). 'Despite a constant stream of often haphazard plant accessions', the total taxa in cultivation rose only marginally since (a) the condition on arrival was poor (due to insufficient knowledge, poor storage and poor transport) and (b) appreciation of the plants' needs by the propagator was often lacking since field notes were usually unavailable. Thus, plants and seeds were usually given a 'standard' treatment, dependent on an alpine, temperate or tropical origin; 'so that plants were usually tolerant of considerable abuse'. Those that survived formed the basis of a 'botanic garden flora'. Townsend concludes that, under such circumstances, 'many rare and unique plants failed to grow'. Reflecting on the success of the micropropagation approaches and more recent garden practice, Townsend (1979) makes it clear that at Kew: 'nowadays collectors in the field collect on the basis of specific requests from botanists and horticulturists'; and 'improved storage techniques and more efficient transport allied to meticulous field notes means that the majority of plants on arrival at Kew have the potential for growth'.

While it is clear that practices at Kew have evolved, many botanic gardens throughout the world have limited facilities and the particular garden practices employed at each garden will act as powerful selective forces on newly arriving plants. In particular, it has been stressed that there are perils in using commercial prepared composts for all plants arriving from the wild. Varley (1979) provides an excellent example of the importance of growing plants in the correct soil. The shrub *Trochetia erythroxylon* (St Helena Redwood) is a very rare endangered species, and difficulties had been experienced in growing it in gardens. Varley discovered, in an experiment, that the plant is calcifuge, thriving only on acid soil, watered with distilled water. This finding explains why the species became endangered. Since the arrival of the Europeans in 1502, deforestation on the island of St Helena occurred not only by clear felling trees but also by the action of introduced goats eating tree seedlings. Extensive erosion of the acid forest topsoil accompanied deforestation and exposed the underlying basic rocks on the island, producing a soil of alkaline reaction, inimitable to the growth of the redwood, which, as we have seen, requires an acid soil.

Ex situ conservation of whole ecosystems

Generally, *ex situ* conservation involves individual plants maintained as spaced plants. However, a pioneering attempt to conserve 'collectively' a threatened ecosystem *ex situ* has been made. Cranston and Valentine (1983) give details. In 1964, there was a proposal to build a reservoir in Upper Teesdale, UK. Although vigorously opposed by conservationists, this proposal was finally approved. In 1970, from the site of the dam, cores of turf were collected containing the Teesdale rarities (*Gentiana verna*, *Primula farinosa* etc.). The material was taken to gardens at Durham University, UK, and Jodrell Bank Garden of Manchester

University, UK. Concerning the Manchester site, the turf was first planted into pots, but later the materials were transferred to large concrete boxes containing soil resembling that of the native habitat. Some individual plants, of a variety of species, were removed from some of the cores and grown as individual plants in a variety of conditions, but, as many of the species were delicate, many cores were left intact. Initial establishment in the boxes was successful, and an attempt was made to maintain the communities by weeding and the density of the rarities was maintained by judicious in-planting. Two major conclusions were drawn from these experiments. Firstly, while there was initial success in maintaining the rarities *ex situ*, eventually all the material was lost, and nothing remains of these experiments today. A second major finding concerns the attempt to maintain the rarities in the concrete boxes. Two types of weed proved problematic. Garden weeds, such as *Poa annua* and *Stellaria media* invaded the containers and were removed. However, in addition, several of the Teesdale species themselves became troublesome weeds, e.g. *Equisetum variegatum*, *Viola riviniana*, *Thymus drucei*, *Campanula rotundifolia* and *Helianthemum chamaecistus*. Reflecting on the fate of the experiment, it is clear that in effect the concrete boxes contained a 'selection experiment' and over time, despite management, there were winners and losers.

Selection in gardens can also change the breeding behaviour of a species

For instance, *Phlox drummondii*, a common spring-flowering species from central and south Texas (Levin, 1976), was introduced to Europe in 1835. Seeds were distributed across Europe and nurserymen selected true-breeding strains with 'different structures, branching patterns, corolla colours and eye markings'. By 1915, there were more than 200 cultivars. In the process of domestication, the breeding behaviour changed. In wild material, plants exhibit genetic self-incompatibility, and seed is only produced following the crossing of genetically distinct variants. In growing plants in gardens, the natural self-incompatibility of *Phlox* was breached by unconscious selection for self-fertility, a process that could occur in other species.

Hybridisation

Another class of genetic changes may occur in botanic gardens, as they often contain plants of related species and genera offering the opportunity for interspecific hybridisation. This is especially true of those related plants grown in close proximity, as for example in order or display beds. Thus, there may serious difficulties in maintaining pure stocks of a rare or endangered species if other related species are within gene flow. As an illustration, the researches of Snogerup (1979) provide a specific example of this general problem. He reports the difficulty in outside plantings of preventing crossing between related Crucifers, even plants of different genera and different basic chromosome numbers. To maintain pure stocks, the only solution is to grow material in insect-proof glasshouses.

The probability of interspecific hybridisation in trees maintained in arboreta is also very high. To overcome this difficulty at the National Pinetum, Bedgebury, Kent, UK, it has been decided to concentrate on the conservation of 12 endangered species, which will not produce interspecific hybrids, as each is the sole member if its genus (i.e. each species is from a monotypic genus, for example *Araucaria araucana*) (Morgan, 2001). However, this cannot be a universal remedy, for many of the larger genera of conifers contain endangered species, and interspecific hybridisation may occur if they are grown close together. Maunder *et al.* (2004) provide a very useful review of the incidence and implications of hybridisation in *ex situ* plant collections.

How far has *ex situ* conservation of plants in botanic gardens succeeded?

Success in the short term

Living plants of many endangered species threatened with extinction or already extinct in the wild are now being grown in botanic gardens, and this must be regarded as a major achievement (Havens *et al.*, 2006). The presence of so many rare and endangered species in cultivation offers some insurance against total loss in the wild (see Botanic Gardens Conservation International – BGCI – on www.bgci.org.). However, there are too few gardens in the many developing countries and endangered species, especially in tropical and sub-tropical areas, are not yet being conserved as living plants. Moreover, it is suspected that many of these same species have recalcitrant or intermediate seed behaviour, which means that they cannot be stored in dry/cold conditions. In contrast, many endangered species with 'orthodox' seed behaviour are now being stored as seed in gene banks. It remains to be seen how far the recalcitrant and intermediate material might be amenable to storage using cryotechniques, of spores, pollen, shoot tips, embryos etc.

Assessments of the effectiveness of ex situ *conservation in particular groups of threatened species*

So far we have been concerned with a number of general, practical and theoretical issues. It is helpful now to turn to three very informative surveys that have examined the conservation value of certain *ex situ* collections.

1 Botanical expeditions: how far have gardens succeeded in conserving the field collections of germ plasm?

In planning the conservation of species *ex situ*, expeditions are often mounted and living material is then propagated and distributed to botanic gardens. In many cases excellent and detailed expedition reports are published. Dosmann and Del Tredici (2003) cite a number of examples, including expeditions to Asia, Yugoslavia, Japan etc. In the case of economic plants, the fate of germplasm collection has sometimes been examined. For example, the location of germplasm of 4,451 collections of soya bean from eastern Asia has been investigated by Hymowitz (1984). However, there have been few published accounts of the fate of expedition collections of wild plants. How far have botanic gardens succeeded in their efforts to collect, transport, grow and maintain material collected in the wild?

Research has revealed what happened to plants collected on the 1980 Sino-American Botanical Expedition (SABE) to Hubei Province in China (Dosmann & Del Tredici, 2003). In this area of high species diversity in temperate flora, herbarium material was collected together with living plants and seeds. This 'germplasm' was distributed to many botanic gardens and arboreta in North America and Europe. Enquiries, of more than 30 gardens in North America and others overseas, have revealed what had happened to the living collections 22 years after they were brought into cultivation. The survey established that only 258 of the original 621 collections had survived. Investigating why 'these survival percentages seem rather low' Dosmann and Del Tredici discovered that some of the material collected had failed to survive the 4–5 months it took to arrange transit to America. Moreover, the authors were 'astonished' to discover that in the various botanic gardens 115 of these survivors were represented by only a single accession, and often by only a single plant. In addition, there were sometimes difficulties with the documentation. Some botanic gardens had failed to record the original SABE collection numbers, and, therefore, as some species had been collected more than once on the expedition, it was now impossible to determine where the material had been originally collected. Furthermore, some gardens had retained only the accession number in their records: there were no habitat or field details. In addition, it was impossible to determine a full list of gardens growing the material, because in 1982 more than 3,300 plants from the collections were offered to attendees at a national meeting (the American Association of Botanical Gardens and Arboreta (AABGA) meeting at Millbrook, New York), who were allowed to take freely what they wanted. Unfortunately, no records were kept of what was taken, and by whom.

2 *How far has* ex situ *conservation of palms been successful?*

A major study of the effectiveness of conservation of palms in botanic gardens was carried out by Maunder *et al.* (2001b). 'Palms are a prominent and popular component of botanic garden collections', and a survey was carried out to determine the holdings in 35 collections in 20 countries. It was discovered that 902 of the 2,700 palm species are in *ex situ* collections. Enquiries revealed that 130 of the 222 threatened palm species are available from commercial trade, and 77 are present in gardens. Two species extinct in the wild are conserved *ex situ*. There are specimens of *Corypha taliera* from Bengal (which has not been seen in the wild since the 1980s) at Kew, the Montgomery Botanic Garden in Florida and Singapore Botanic Garden. The last known specimens of Mauritian endemic *Hyophorbe amaricaulis* are in the Curepipe Botanic Gardens, Mauritius.

Despite the presence of threatened palms in *ex situ* collections, Maunder *et al.* (2001b) concluded that gardens in temperate areas were not, in general, acting as sources of palm material for conservation, but rather were in effect 'sinks', dependent upon the importation of replacement stocks from tropical or commercial nurseries. In addition, there were cases of interspecific hybridisation in gardens (Noblick, 1992), some of which was deliberately generated by enthusiasts (Glassman, 1971). Molecular studies have also detected hybridisation between wild populations of the threatened Australian palm *Ptychosperma bleeseri*, and nearby exotic congeneric taxa growing in urban gardens (Shapcott, 1998). The survey also revealed that botanic gardens have the potential to introduce invasive species of palm into new areas. For instance, the Chinese Fan Palm (*Livistona chinensis*), introduced to Mauritius through the Pamplemousse Botanic Garden, is proving a threat to wild populations of the endangered endemic *Acanthophoenix rubra*. Chinese Fan Palm is also invasive in Hawaii, Bermuda and Florida.

Reflecting on the results of their survey, Maunder *et al.* (2001b) draw some important conclusions that point to limitations in the *ex situ* conservation of palms.

1. 'Botanic gardens at present can only maintain a small proportion of threatened palm species, often at low levels of genetic and numerical representation.'
2. 'Effective *ex situ* conservation of threatened plants will only be feasible in the source countries where genetical diverse *ex situ* populations can be managed in tandem with wild populations.'
3. 'The extensive temperate botanic garden collections, mostly housed in glasshouses with severe space constraints and heavy recurrent maintenance and labour costs, maintain taxonomically diverse but numerically restricted stocks of limited potential value for reintroduction. To contribute effectively to plant conservation it is suggested that the conservation focus for northern garden activity should shift away from the collection as a strategic conservation objective. Thus the northern display collections become primarily resources for public education, scientific research and fund raising to support the imperatives of in-country habitat conservation.'

This important analysis deals with plants, many of which have a tropical or sub-tropical distribution. How far have botanic gardens been successful in *ex situ* conservation of plants from more temperate conditions?

3 Has ex situ *conservation of threatened European species of plants been a success?*

A questionnaire was sent to 624 botanic gardens/institutions in 40 European countries (Maunder *et al.*, 2001a), enquiring about *ex situ* conservation of 573 European threatened species (so called Bern Convention species). Only 119 (19%) gardens in 29 (73%) countries responded. The survey revealed that a majority of accessions of these species were not of wild origin (61%). The rest were presumably of unknown origin or derived from other botanic garden stocks. For example, while 12 gardens had specimens of *Dracaena draco*, the famous 'dragon-tree' tree from the island of Tenerife, Canary Islands, only two had stocks of documented wild origin. Looking at the records of plants held at the Royal Botanic Garden, Kew, there were 226 accessions of 119 species of the Bern Convention list. However, only 90 out of 226 accessions (40%) were of wild origin, while 65 (29%)

were of garden origin and 71 (31%) were of unknown origin. For only 2% of the accessions were full records available, including latitude and longitude. For those species grown from seed, pedigree history was not available. With regard to botanic garden seed lists, Maunder *et al.* (2001a) conclude that these are 'heavily skewed towards taxa that reproduce by seed in sufficient quantities to enable collection and distribution'. The survey established that some 'Bern Convention' species that have been lost in the wild are present in botanic gardens, e.g. Greuter (1994) discovered that 37 plant species had become extinct in the wild in the Mediterranean region, but four of these species were in cultivation in botanic gardens. A good example is *Tulipa sprengeri* from Turkey.

Maunder *et al.* (2001a) draw some important conclusions from their study.

1. The majority of taxa are held in a 'small number of collections, dominated by non-wild accessions and are not adequately documented'.
2. Some species are present in many collections, but this does not indicate genetic diversity. For example, *Lysimachia minoricensis*, extinct in the wild, is available in 20 botanic gardens' lists, but all are thought to be from a single founder cultivated at the Jardín Botánico de Barcelona, as studies have revealed isozyme uniformity in the material held in botanic garden stocks (Ibáñez *et al.*, 1999). This lack of variation could reflect the true state of the original population, but, as concluded in the earlier studies, uniformity might owe its origin to the sampling procedures used in collecting the material, before it became extinct.
3. Some of the stocks of the Bern Convention species have been affected by hybridisation in botanic gardens. For instance, in a number of gardens, *Chionodoxa luciliae* has hybridised with *Scilla* species. In addition, it seems likely that selection in gardens for amenable stocks has altered some species, e.g. it is suspected that *ex situ* material of *Aldrovanda vesiculosa* is 'dominated by horticultural amenable tropical provenances rather than the threatened wild European stocks'.
4. Reflecting on the broader picture, they conclude: 'the conservation utility of existing botanic garden collections should be questioned; most accessions are held out of the range country in mixed collections with insufficient accession data'. Therefore, 'the majority of stocks held in botanic gardens are of limited conservation value, accordingly, the most practical contribution to conservation for many botanic gardens, at least in the short term, will be through effective interpretation of conservation issues to the visiting public'.

Botanic gardens and the conservation message

Some gardens have developed excellent education programmes, special displays and labelling to promote the conservation of rare and endangered plants. In my personal experience, visiting gardens in Europe, North and Central America, Australia, the Far East and South Africa, many have yet to reach the standard of promotion of the best.

Many botanic gardens have chosen to convey the conservation message by labelling individuals or groups of plants. For instance, there is a rare plants trail at Conservatoire National de Brest, France (Cheney *et al.*, 2000, 47). Gardens were established within areas of natural or semi-natural vegetation, and, to this day, include such areas within their boundaries, e.g. the famous gardens at Kirstenbosch, Cape Province, South Africa, and the Botanic Gardens at Singapore. In parts of these gardens it is possible for the visitor to see rare plants in a semi-natural setting. In contrast, other gardens were established on what had been agricultural land, and many have created 'natural-looking wild gardens' by naturalistic plantings of wild plants in specially landscaped sites, to illustrate ecological communities and succession and to display rare and endangered species in ecological context. For example, a raised bog at the University of Salzburg; dry grasslands at the University of Vienna Botanic Garden; a wetland at Cambridge University Botanic Garden; and a 'Fernarium' has been developed at the University of Kebangsaan, Bangi Forest Reserve, Malaysia (Bidin, 1991). A new and very interesting departure from the traditional garden has been the development of Queen Elizabeth Botanic Park within a nature reserve on Grand Cayman. Also, a new botanic garden has been established within a reserve on Belize. The newly established Chicago Botanic Garden has been developed within a forest reserve. A point of great interest, to consider in a later chapter, is the potential blurring of the distinction between the managed nature reserve and the wild-looking areas in gardens. To what extent are these equivalent in microevolutionary terms?

Future prospects for *ex situ* conservation in botanic gardens

Experience over the last 25 years or so has shown that many endangered plants have been successfully rescued by being brought into gardens and conserved *ex situ* as living plants. Investigations have also revealed that for those species of plants with orthodox seed behaviour it is possible to conserve seeds in seed banks, although little is known about the very long-term effects on their genetic variability of such storage, and of cultivation in the future to regenerate seed stocks. Research is in progress to try to devise storage conditions for intermediate and recalcitrant species.

Some of the criticisms of garden policy and practice made over the years by interested parties have been addressed and improvements made. Thus, it is widely appreciated that the conservation value of *ex situ* collections could be increased by improving the quality of information held on species, ensuring accurate labelling of plants and the collecting and storage of information on cultivation practices etc. Botanic Gardens Conservation International (BGCS), established to promote co-operation between gardens, has encouraged the continuing development of databases, e.g. PlantNet. Generally, lack of space and resources limit the numbers of plants that can be grown, but where circumstances permit, botanic gardens have grown many more individuals of some species. Thus, genetically structured populations have been set up in satellite arboreta, each species being represented in a network of sites. This conservation strategy is being applied to rare temperate rainforest conifer tree species in arboreta in Britain and Ireland, with a major contribution being made by the arboreta associated with the Royal Botanic Garden, Edinburgh (Page & Gardner, 1994).

Despite these efforts by botanic garden staff to improve the efficiency of *ex situ* conservation, serious doubts have been raised about the efficiency in sampling and long-term conserving of the genetic variability of wild plants brought into cultivation, or whose seeds are to be stored in seed banks. For example, Hamilton (1994) poses the very important question: how far is the sampling seed for seed banks effective in capturing and conserving genetic diversity found in wild populations? In the past such questions have been approached by applying theoretical models, and extrapolation from crop plants, Recently, there have been major advances in our understanding of the genetic variability of crops and seed bank stocks through studies using molecular markers (Hammer, 2003).

Molecular methods are now beginning to be employed to study genetic variability in populations of 'wild' plants, to examine whether samples taken from them for *ex situ* conservation accurately represent this original variability. For instance, in studying the endangered species *Vatica guangxiensis* in China using RAPD markers, Li, Xu and He (2002) compared the genetic variability in *ex situ* stocks with that found in the three remaining natural populations. They discovered that plants maintained *ex situ* contained representative sufficient genetic variation to maintain long-term survival and evolutionary process in *Vatica*, but that further sampling would be necessary to conserve more exclusive alleles represented in the wild but not yet present in the *ex situ* collection.

A study of *Naufraga balearica* by Fridlender and Boisselier-Doubayle (2000) checked the variability in *ex situ* collections in relation to that detected in the wild, using RAPD markers. They discovered that material from Corsica, cultivated in botanic gardens at Lyon, Brest and Porquerolles, contained only a few patterns of variability – most individuals showed a dominant pattern indicating clonal origin. In contrast, five wild populations from Majorca had distinctive genetic patterns. The Corsican material conserved *ex situ* appeared to be genetically related to individual samples at Cala San Vicente, Majorca, prompting the researchers to suggest that the plant was not native to Corsica, but had been introduced from Majorca some time in the past.

These studies reveal the power of modern molecular techniques. Further research is urgently needed to study genetic variability – in wild populations of endangered species; in samples collected for *ex situ* conservation; in material propagated in gardens and/or storage in seed banks; in stocks regenerated from seed banks; and in material taken from gardens for *in situ* reintroduction programmes. Also, it would be revealing to study possible changes in adaptive characteristic in *ex situ* collections. Flowering time, seed set, disease resistance etc. should be examined in comparing samples of 'wild' and *ex situ* stocks in common garden and other experiments.

Will *ex situ* conservation lead to domestication?

In bringing plants into gardens – in effect into 'captivity' – geneticists point out that selection pressures do not disappear, they change. Plants are likely to be subject to less and different interspecific competition if grown as spaced individuals or in separate pots, and they may be isolated not only from the herbivores, pests and diseases they encountered in the wild, but also their mutualists, including mycorrhizae, pollinators, and seed/fruit dispersers

(Hancocks, 1994; Havens *et al.* 2004, 464). Overall selection pressures in gardens may be less severe than in the wild, as plants are subjected to fewer and different stresses. For example, to combat fungal infections at the seedling stage, gardeners often use fungicides, and insect pests may be controlled. Such treatments may increase the survivorship of seedlings, and 'less fit' individuals might be brought to maturity. Furthermore, over a number of generations, this could include genetic mutants segregating out in progenies. Considering further the question of the selection pressures in gardens, Darwin's insights into evolution mean that there is no escape from natural selection. In addition, he made it clear that plants will also be subjected to artificial selection – both conscious and unconscious. Such is the impact of cultivation that Prance (2004, xxiii) concludes that *ex situ* conservation will 'halt or distort the natural process of evolution', and in the longer term Ashton (1987, 125) concludes: 'for all practical intents *ex situ* species conservation leads irreversibly to domestication'.

Ex situ conservation of animals in zoos and aquaria provides important insights into what is likely to happen to plants maintained in gardens. For example, Frankel and Soulé (1981, 150) write:

> It can be convincingly argued (Spurway, 1952, 1955) that domestication is an insidious and corrupting force in every CP [captive propagation] programme and that it is impossible to avoid selective breeding by the humans who manage such programmes.

In the same vein, Jones (1999, 37) writes:

> A zoological garden bears the unwelcome message that, because of man's inadvertent selection, any animal taken from the wild becomes domestic, a travesty of its natural self. Evolution is as hard at work on caged animals as on those born free. In time they will emerge as beings quite different from what they were. Those who conserve animals in the hope of returning their descendants to Nature may be disappointed by what they let loose. Their failure shows how descent with modification is impossible to avoid.

Evidence that animals change rapidly in captivity has been obtained from studies of fish and insects. For instance, after two or three generations in hatcheries, genetic changes have been detected in endangered species of fish being bred for reintroduction to the wild (Meffe, 1995). Probably the captive animal about which most is known is *Drosophila*. Stocks of flies managed to minimise changes, which is the aim of *ex situ* conservation, have been compared with the large populations of wild stocks from which they came. After 11 generations, populations of captive stocks had lower genetic variability and much lower reproductive fitness (Frankham & Loebel 1992; Ralls & Meadows, 1993). Further researches have revealed the many changes brought about by keeping *Drosophila* under 'zoo' conditions (see Frankham, 2005a, b, for a review of the evidence).

Because of the many difficulties encountered in *ex situ* conservation, should endangered species be allowed to become extinct? Many would agree with the sentiments of Frankel and Soulé (1981, 167) when they write: 'it is recognised that the introduction into cultivation is likely to narrow the genetic diversity of a species and to change its variation pattern in response to drastically altered selection pressures. Yet it is obviously preferable to keep alive a seriously endangered species as a semi-domesticate, rather than allow it to disappear altogether.' While there is much still to be learned about what happens to wild plants introduced into gardens, it is clear that in the longer term plants passing through several generations in gardens will enter an uncertain domain between wild and domesticated.

Conclusions

Since the 1970s, a very large number of theoretical studies of different aspects of *ex situ* conservation have been published. Although many key issues were identified in the early years, exhaustive theoretical studies in the last 40 years have 'raised the stakes'. If conservation is likened to a horse race, then the requirement for effective *ex situ* conservation – the jumps and fences of sampling, cultivation, documentation etc. – have all been raised, over the years, as geneticists, ecologists etc. have considered the issues involved. Looking at the practical side of *ex situ* conservation, it is possible to report successes – some taxa extinct in the wild are now in botanic gardens. In addition, many gardens are active in conservation and display of rare and endangered species. However, as we have seen above, there are practical and logistical difficulties in maintaining any stocks long term. These relate to space, costs, resources, labelling, record-keeping, plant diseases etc. Furthermore, there are too few botanic gardens in tropical and sub-tropical regions, where the risks of biodiversity loss are greatest.

In earlier chapters, we saw how an increasing number of plant species are being threatened with extinction. Given that *ex situ* facilities are limited, how do we decide which species to conserve? If the habitat of a species is effectively lost, should an increasing number of species be maintained in botanic gardens indefinitely? If so, the possibility of the domestication of wild stocks through their long cultivation in gardens increases. However, if suitable habitat for restoration and reintroduction projects still exists, then 'rescued' plant species could be propagated in gardens before being returned to the wild. But, the use of stocks propagated through many generations *ex situ* may be problematic, as a degree of domestication may have occurred, and plants may not be able to survive, in what was once their native habitat. A possible example is provided by the case of *Lotus berthelotii*, a legume native to the Canary Islands, that has long been grown in botanic gardens. Conservationists attempted to reintroduce the plant to its native habitat in Tenerife, but plants brought back to the islands from Europe died in the nursery on Gran Canaria (D. Bramwell, personal communication, 2002, in Havens *et al.*, 2004, 464). It is possible that the *Lotus* stocks had become adapted to European gardens, and could not survive the higher temperatures in the Canaries.

Finally, the availability of *ex situ* conservation facilities in many botanic gardens raises other issues. If an endangered species has reached very low numbers in the wild and its habitat is under threat, conservationists may take the decision to bring the material into botanic gardens. In effect, the natural population could be lost as a consequence of the steps taken to save it (Guerrant *et al.*, 2004a). Also, if it is assumed by the general public that *ex situ* conservation (in garden and/or seed banks) is an adequate and successful means of conservation in itself, then developers' plans may not be resisted, and the last vestiges of some particular wild habitat may be destroyed, denying all possibility of habitat restoration and species reintroduction (Rolston, 2004, 33).

In the next three chapters, we examine conservation of species '*in situ*' – in reserves and other protected areas – where the endangered species concerned are part of their own natural ecosystem. Also, the theoretical and practical implications of the call for co-ordination between *in situ* and *ex situ* approaches will be considered.

15

In situ conservation: within and outside reserves

A call for parks to preserve forests

Wallace, the co-founder, with Darwin, of the theory of evolution by natural selection, understood the value – and the vulnerability – of the natural world. Writing in 1910, he made an impassioned plea (quoted in Berry, 2002, 147):

> It is really deplorable that in so many of our tropical dependencies no attempt has been made to preserve for posterity any *adequate* portions of the native vegetation, especially of the virgin forests . . . Surely before it is too late . . . a suitable provision shall be made of forest or mountain 'reserves', not for the purpose of forestry and timber cutting alone, but in order to preserve adequate and even abundant examples of those most glorious and entrancing features of our earth, its native forests, woods, mountain slopes, and alpine pastures.

The concept of nature reserves has a long history. As we shall see, important advances were made in the nineteenth century, with the development of national parks in North America. Currently, it has been estimated that *c*. 7.9% of the Earth's land surface and *c*. 0.5% sea area are protected in reserves of various kinds (Balmford *et al*., 2002, 952). Ten different types of parks are listed by Given (1994, 96; Table 15.1), ranging over closed reserves, national parks, extractive reserves etc. The establishment of reserves for nature conservation has been one of the most important and 'enduring strategies for the conservation of global diversity' (Hopper, 1996, 253). According to Given (1994), while reserves of different kinds play a major role in conservation, they are only one element of *in situ* conservation, for many endangered species and ecosystems are found outside reserves.

This chapter considers the origin, development and management of national parks and reserves, with particular reference to Yellowstone National Park, USA, and Wicken Fen Nature Reserve, Cambridgeshire, UK. Taking a wider view, the strengths and weaknesses of reserved areas in conserving biodiversity are considered, and the microevolutionary implications of *in situ* management are reviewed.

Early parks and 'reservations'

Concerning the history of *in situ* conservation, it is sometimes considered that reserves and national parks were first established in the USA from the 1830s onwards (see below) and,

Table 15.1 *Different types of protected areas*

Primary conservation objective[a]	I	II	III	IV	V	VI	VII	VIII	IX	X
Maintain sample ecosystems in natural state	1	1	1	1	2	3	1	2	1	1
Conserve genetic resources	1	1	1	1	2	3	1	3	1	1
Maintain ecological diversity and environmental regulation	3	1	1	2	2	2	1	2	1	1
Provide education, research and environmental monitoring	1	2	1	1	2	3	2	2	1	1
Produce timber, forage on a sustained basis	–	–	–	3	2	–	3	1	3	–
Conserve watershed condition	2	1	2	2	2	2	2	2	2	2
Protect sites and objects of cultural heritage	–	1	3	–	1	3	1	3	2	1
Stimulate rational, sustainable use of marginal areas and rural development	2	1	2	2	1	3	2	1	2	2

[a] Types of protected areas: I, strict nature reserve; II, national park; III, monument/landmark; IV, managed reserve; V, protected landscape; VI, resource reserve; VII, anthropological reserve; VIII, multiple use area; IX, biosphere reserve; X, world heritage site.

Ratings: 1, primary objective for management of area and resources; 2, not necessarily primary but always included as an important objective; 3, included as an objective where applicable and whenever resources and other management objectives permit.

From Given (1994) after MacKinnon *et al.* (1986). Reproduced with permission of IUCN, Gland, Switzerland.

In an attempt to increase the area under conservation management, additional categories of protection have been designated in different countries. For instance, in England, in addition to National Parks, National Nature Reserves (NNR) and Local Nature Reserves (LNR), a number of forms of 'conservation' have been devised; www.naturalengland.org.uk gives the following details. 4,000 sites of Special Scientific Interest (SSSIs) on public and private land contain significant wildlife, landscape and geological features (7% of the land area). Areas of Outstanding Natural Beauty (AONB) (36 areas covering about 15%). Special Areas of Conservation (SAC) in special protection under the EU Habitats Directive. Heritage Coasts – 33% (1,057 km) of the English coastline. Marine Protection Areas.

inspired by these developments, similar reserves and parks were established worldwide. While there is some truth in this assertion, in many parts of the world, land use involving elements of conservation predate the founding of the American parks.

Investigations have revealed that for centuries there have, in effect, been protected areas in different parts of the world that, to some extent, have conserved wildlife. Such sites were designated and maintained for different purposes, as holy sites, hunting preserves of various types, game parks and forest 'reserves' of various kinds.

Holy places

From prehistoric times, in Asia, Africa and other parts of the world, certain forest groves and, indeed, whole landscapes have been regarded as sacred. For example, Ramakrishnan (1996) gives an account of sacred groves in India, which date back to before the Vedas era (1500 to 500 BC). Of varying sizes, some are less than a hectare. But others are a few square kilometres in extent, for instance, in western Kerala in southern India, in the Chhotanagpur region of Bihar in northern India, and elsewhere. In other cases, whole landscapes are regarded as sacred, e.g. the Sikkim region. Access to sacred groves and landscapes was often restricted, and they were subject to careful traditional management. Some of these areas are now degraded, owing to a decline in value systems.

Hunting and forest reserves

As long ago as 700 BC, Assyrian noblemen hunted in designed training reserves (Runte, 1997, 2). In France, there is a long history of protected areas for forestry and hunting back to the Middle Ages, e.g. the forests at Fontainebleau have been protected for centuries. From 1853, areas within the forest were designated as reserves, and by 1904 this had risen to about 10% (Rackham, 1980). It is significant that the word park comes from the Old French and Middle English 'parc', meaning 'an enclosed piece of ground stocked with beasts of the chase, held by prescription by king's grant' (Runte, 1997, 2).

In Britain, one of the oldest extant protected areas, the New Forest – recently designated as a national park – was declared a royal hunting preserve in 1079 (Anon., 1991b). In Russia, records dating back to the twelfth century reveal the existence of wildlife reserves, where native and exotic wild animals were kept for hunting by the tsar, princes and nobility. For instance, 'Kubanskaya Okhota' has a long history as a hunting preserve, and the area is now within the Caucasian Biosphere Reserve.

Protection of forests and preservation of sites of scenic beauty

As a reaction to the ruthless exploitation by European powers of their colonial territories overseas, concerns were raised about environmental change and the fate of game species. For example, in many overseas colonies, such as St Helena, the Canary Islands, the Caribbean etc., the marked microclimatic changes, particularly in rainfall, accompanying deforestation led to the development of regulations on forest conservation. In 1764, mountain ridges of the island of Tobago were 'reserved in woods for rains' (Grove, 1995, 272). The forest on the hilltops of Singapore was protected from the 1840s (Anon., 1991b). In the eighteenth century, the French legislated to protect some of the forested areas of Mauritius (Grove, 1995), and the first directive for the protection of Indian forests was enacted as early as 1865. There is abundant evidence of the ruthless hunting of wild animals in many colonial territories. And, from the seventeenth century onwards, legislation was enacted to protect forests, game and scenic areas, and many game and forest reserves were established. From the late 1600s, colonial administrators in South Africa attempted to control hunting. And, in 1697, the first game reserve was set up (Reid & Steyn, 1990). There were also moves to protect areas of outstanding landscape, e.g. in 1866 a reserve was established at Jenolan Caves, in the Southern Blue Mountains, New South Wales, Australia.

The establishment of American National Parks

The American National Parks were founded to provide 'a mighty system of national museums of the primitive American wilderness' (Runte, 1997, 110). The first park, a huge area of 2 million acres (810,000 hectares) on the Wyoming–Montana frontier, containing the majority of the world's geysers and hot springs, was established at Yellowstone in 1872 (Anon., 2004). While Yellowstone is regarded by many as the very first national park, historians point to other earlier reserves. In the USA, the oldest national reservation is Hot Springs, Arkansas, established in 1832. Furthermore, a State Park, to be administered by California for public use and recreation, was established in 1864 to secure the future of the Yosemite Valley (Anon., 1990). A reserve was also established to protect the Mariposa Grove of Big Trees (*Sequoiadendron giganteum*).

Regarding the aims of national parks, Runte (1997, xxii) considers that the 'idea evolved to fulfil cultural rather than environmental needs' – part of a search for 'cultural identity'. He argues: 'when national parks were first established, protection of the "environment" as now defined was the least of preservationists' aims' (p. 11). Unlike the Old World, the USA lacked 'an established past', particularly in art, architecture, and literature (p. 7). He continues: the landscapes in the eastern part of the country were 'nothing extraordinary', with the exception of Niagara Falls, but private ownership and the 'onslaught of commercialism robbed the cataract of credibility as a cultural legacy' (p. 8). However, with the discovery of the spectacular scenery in the west, Americans decided that Yellowstone must be preserved from abuse and the establishment of the first park, and others in due course, could then 'be truly convincing proof of the New World's cultural promise' (p. 9). However, Sellars (1997, 4, 9) makes another important point. While 'scenery has provided the primary inspiration for national parks . . . tourism has been their primary justification'. Indeed: 'from the first . . . the national parks served corporate profit motives, the Northern Pacific [Railway Company] having imposed continuous influence on the Yellowstone Park proposal'.

Regarding the question of which areas might be included within national parks, Runte (1997, xi) concludes that Yellowstone and later parks were so circumscribed as to include 'nothing of proven commercial value'. In essence, they were 'worthless lands' (p. xv) containing spectacular scenery and offering unique visual experiences. The insistence that parks be on 'useless ground' ensured that there were no grassland national parks in the USA until 1985. Concerning the setting of the boundaries of the park, Runte (1997) emphasises that Yellowstone, despite its very large size, was not designed for wildlife. This is clear from the fact that, in winter, large mammals were driven by the cold from Yellowstone Park into adjacent lower valleys outside park boundaries (p. 139).

The aims of national parks

The American National Park movement has provided not one but a series of powerful models, influencing conservation through protected areas throughout the world. The wording of the Act that brought the park into existence is very revealing. The Yellowstone National Park Act of 1872 reads: 'The headwaters of the Yellowstone River . . . is hereby reserved and withdrawn from settlement, occupancy, or sale . . . and dedicated and set apart as a public park or pleasuring ground for the benefit and enjoyment of people' (Anon., 2004, 7).

In 1916, the National Parks Service was created with a re-evaluated mission that included reference to the conservation of wildlife. The aim was to:

> promote and regulate the use of the Federal areas known as national parks, monuments and reservations . . . by such means and measures as conform to the fundamental purpose to conserve the scenery and the natural historic objects and the wildlife therein and to provide for the enjoyment of the same in such manner and by such means as will leave them unimpaired for enjoyment of future generations. (Anon., 2004, 7)

Commenting on this new statement of aims, Runte (1997, 196) saw this as part of a process of exchanging romanticism for environmentalism.

The concept of human exclusion

One of the key concepts enshrined in the national parks and wilderness movements in the USA is the principle that humans should be excluded from living in the protected areas (Spence, 1999). Those establishing the parks failed to recognise, or wished, for their own purposes, to deny the fact that they were often dealing with cultural landscapes not pristine wilderness. Thus, soon after the founding of Yellowstone National Park in 1872, the indigenous Indian tribes – the Sheepeaters – were moved out to reservations in Wyoming and Idaho (Schullery, 1997). Likewise, in 1957, Papago Indian farming was prohibited at the Organ Pipe National Monument in Arizona, and, in 1962, all non-historic Papago structures were cleared to make way for the Organ Pipe Wilderness Area (Nabhan, 1987, 89).

Elsewhere, people have been removed and forbidden to use wilderness areas, for example in the creation of a network of reserves in India to protect the tiger (see Callicott & Nelson, 1998). In the development of Kruger National Park in South Africa, the Wankie Reserve in Southern Rhodesia (Zimbabwe) and the extensive parks of Kenya and Tanganyika (Tanzania), indigenous graziers were excluded (Christopher, 1984, 186). Land was also expropriated in the establishment of national parks in Canada (MacEachern, 2001, 19).

The notion that humans should be excluded from living in protected areas is a very complex and contentious issue. As we saw in an earlier chapter, Darwin concluded that humans were the product of evolution, a species amongst species, and could therefore, logically, be construed as part of 'nature'. Extending this line of reasoning, it could be claimed that as humans are part of the natural world, logically anything humans do is natural, including the development of agricultural land, the management of grasslands and forests, the building of towns and cities. Despite the notion that humans and their activities be excluded from parks and reserves, such areas receive increasing numbers of visitors and, far from being 'preserved' from human interference, there is abundant evidence that such areas have often been subject to considerable management.

Changing aims

As we have seen, the parks were not an untouched pristine wilderness area at the time of their designation (see Pyne, 1982, 71–83). 'Most national parks came into existence already altered by intensive human activity, Yellowstone being the least affected. All had experienced some impact from use by Native Americans' and, therefore, were 'probably not in a truly pristine condition' (Sellars, 1997, 256). At first establishment the avowed aim of the park was 'preservation'. How did park staff act on this categorical imperative? Sellars (1997, 22) considers that in the early years 'management of the parks under mandate to preserve natural conditions took two basic approaches: to ignore, or to manipulate'. 'Inconspicuous species' were of 'little concern'. Furthermore, 'managers sought to enhance the parks' appeal by manipulating the more conspicuous resources that contributed to public enjoyment, such as large mammals, entire forests, and fish populations'.

Concerning the origin and execution of these early management plans for national parks, Wright (1999) concludes that park staff drew on the models of English and American game parks. The aim was 'to protect species considered to be desirable, primarily large ungulates, and to eliminate species or processes which threatened the desired species', and 'the management methods used to achieve these goals included artificial feeding, control of predators, fire suppression and the elimination of disease and pathogens'. While many of these activities were directed at controlling the animals of the park, given the emphasis here placed on plants, it is important to note that this management is certain to have had profound effects on the structure, composition and dynamics of vegetation.

Predator control

Beginning in the 1880s, when the army looked after Yellowstone, wolves and coyotes were controlled, because they sometimes preyed on livestock, and attacked what were considered to be the more desirable wildlife species such as deer and elk (Sellars, 1997, 25). The army also acted to control the poaching of elk and illegal grazing that occurred in the early years of the park.

Feeding the animals

For a long period 'the Park Service conducted a kind of ranching and farming operation to maintain productivity and presence of favored species' such as elk (Sellars, 1997, 70–81). Such operations required hay, and this was grown, until 1956, on *c*. 600 acres of ploughed, seeded and irrigated parkland. The introduced timothy grass now recognised as an 'aggressively invading species' was a component of this hay. (Other species of non-native trees, shrubs and grasses were also introduced as part of landscaping schemes.) Yellowstone has very severe winters and, to survive, elk must move to lower ground outside the boundary for the winter. Ironically, animals cared for inside the park 'as 'wildlife', suddenly became 'game', once they left the park (Sellars, 1997, 71). To protect the species from over-enthusiastic hunters, many animals overwinter in the National Elk Refuge near Jackson Hole, established 1912, where they are protected and provided with feed and shelter (Boyce, 1989). Other species of animals were also fed in the park (deer, bighorn sheep and other ungulates). In the early decades of the park, visitors also enjoyed viewing the feeding of bears with garbage. Bears are now differently managed, the aim is to wean the bears from human food and garbage, and encourage them to eat a more natural diet.

Stocking Yellowstone with fish

Today, Yellowstone contains 'one of the most significant, near-pristine aquatic ecosystems found in the United States' (Anon., 2004, 117) but, to provide sport for fishermen (currently about 75,000 per annum), from 1881 a number of fish species were introduced into certain areas of the park (Sellars, 1997, 23). Such measures were in keeping with the concept of the park enshrined in the founding Act, in providing 'pleasuring grounds'. Subsequently, using hatcheries to increase stocks, more than 310 million fish had been introduced to park waters by the mid twentieth century. These interventions have greatly influenced the freshwater ecosystems of the park. Recently, without permission, lake trout have been introduced into the Yellowstone Lake itself, an aquatic ecosystem hitherto composed entirely of native species (Anon., 2004, 118).

Protection of forests against fire

For the first 100 years of its existence, Yellowstone had an active fire suppression policy. This had a profound effect on the vegetation of the area, as the vegetation is fire adapted, as a consequence of the long occurrence of natural fires through lightning strikes. Indeed, some species are fire-dependent. For example, lodge pole pine produces some cones that only open when heated in fires to at least 113 °C.

Protection of forest against insects

From the mid 1920s, the US Congress was 'appropriating funds for the control of insects in the Parks. Principal targets at the time included the lodge pole sawfly and spruce budworm in Yellowstone, the needle miner and bark beetle in Yosemite, and the pine beetle in Crater Lake' (Sellars, 1997, 84). In addition: 'To combat such attacks, the [Park] Service used chemical sprays and also felled infected trees, peeled off their bark, and burned them.' Also, from the 1930s, there were disease eradication programmes in national parks, with white pine blister rust, a non-native fungus, as a major target (Sellars, 1997, 83). As part of insect control, DDT was later employed in Yellowstone and Yosemite parks and elsewhere (Sellars, 1997, 162).

It is instructive to consider the control of insects in the national parks in Atlantic Canada (MacEachern, 2001). In the 1950s, an infestation of spruce budworm occurred attacking spruce and balsam fir. This is a natural disease infestation that peaks every 27–70 years, leaving areas of dead and dying trees. MacEachern (2001, 213) quotes the views of Deputy Minister Robertson on how to react to the spruce bud infestation:

> The question is how best to conserve. It could be argued that maintenance of the parks 'so as to leave them unimpaired for the enjoyment of future generations' means that the forests must be left in an utter state of nature, and with nature left to do with them just what she will – destroy them by disease, by fire, or anything else. We do *not* take that position as to destruction by fire, and rightly so. Equally, I think we cannot take that position where a risk of destruction by disease occurs. In other words, we do not hold to the view that the right course is 'hands off'. What we should aim at, it seems to me, in the case of fire, disease and all other aspects of forest concern, is an intelligent and scientific policy of management and protection, the object of which is to see, insofar as we can, that the areas now covered by healthy forest remain covered by healthy forest, year in year out, in perpetuity.

Instead of allowing the infestation to take its course and subside naturally, the solution to the problem was spraying infected areas with insecticides, including DDT. Although management of the parks is now less interventionist, in the past the control of insect infestations with insecticides was the accepted policy of Canadian National Park management for decades.

Policy of natural regulation

The policies followed at Yellowstone had by the 1960s led to a superabundance of elk and other ungulates. The effect of 'overgrazing' of aspen groves was demonstrated by monitoring fenced exclosures that excluded the elk (Budiansky, 1995). Some culling was carried out at Yellowstone, together with relocating some elk to other sites (Wright, 1999). The total fire-suppression policy led to the build-up of enormous quantities of dead wood, litter and fallen trees. Thus, huge quantities of combustible material accumulated in the ecosystem. Also, increasing resources were being expended on the control of insects and diseases. The use of pesticides was called into question in 1962 with the publication of Rachel Carson's famous book *Silent Spring*.

In the light of the effects of past management, a reassessment of the aims of the national parks was carried out, leading to the publication of the Leopold Report of 1963. This report recommended more ecosystems-based approaches to management, with 'natural regulation' the guiding principle. Also, there was to be a major new emphasis, namely re-creation of lost landscapes. They recommended: 'that the biotic associations within each park be maintained, or where necessary recreated, as nearly as possible in the condition that prevailed when the area was first visited by the white man', in short the Leopold Report proposed: 'a national park should represent a vignette of primitive America' (Runte, 1997,

198). The notion of re-creation of lost landscapes is an important theme that will be examined in more detail below.

Natural management

Concerning the recommendations of the Leopold Report, the Parks Service was initially reluctant to abandon totally fire and pest controls. In time, however, these management practices were modified. Firstly, in 1972, it was decided to allow some natural fires to run their course, and to begin 'controlled' burning. However, in 1988, after droughts, the park authorities faced an exceptional situation. Many fires burned in the park, some were natural, others were apparently started by humans (Anon., 2004, 68). The situation became so dangerous that steps had to be taken to control the fires. In all, $120 m was spent on putting out this fire, in which 793,888 acres (36%) of the park was burnt. Now, millions of lodge pole pine seedlings, in more or less even-age stands, have become established in areas damaged by the fire.

With regard to pest control, the widespread use of chemical sprays was 'replaced by the restrictive Integrated Pest Management Program. Intended to avoid use of chemicals except where absolutely necessary, the Program would emphasise natural controls with minimal environmental effects, including the use of naturally occurring predators and disease agents' (Sellars, 1997, 254). Concerning the management of animals, the decision was taken to allow elk populations to be regulated naturally by the amount of food available, and eventually to reintroduce the wolf. Animals, including grizzly bears, would not be allowed to feed at garbage dumps. Park staff would make renewed efforts to control invasive alien plants such as *Linaria dalmatica*, *Centaurea maculosa*, *Cirsium arvense*, *Leucanthemum vulgare*, *Cynoglossum officinale* and *Euphorbia esula* (Sellars, 1997, 258). In the 1990s the wolf, driven to extinction by former management, was successfully reintroduced in the park (Anon., 2004, 164).

The establishment of parks and reserves in Europe

At this point, it is instructive to examine the development, aims and management of nature reserves and national parks in Europe. In Britain, in the 1890s, there was widespread concern about the possible loss of open spaces, especially in and near cities (Hunter, 1890). Moreover, there was a call for the creation of bird sanctuaries in London parks (Chipperfield, 1895), and on private estates (Whitburn, 1898).

Few nature reserves were established in Britain before 1900. For example, the Breydon Society in East Anglia bought Breydon Water to establish a bird reserve (Evans, 1997, 45). In addition, nature reserves were set up at Wicken Fen, Cambridgeshire (1899) and at Blakeney Point in Norfolk (1912) by the National Trust. This organisation was influenced by the celebrated conservation organisation – the Trustees of Reservations – founded in Massachusetts, New England (in 1891) by the landscape architect Charles Eliot (www.thetrustees.org).

In 1913, at the winter meeting of the newly formed British Ecological Society, Professor F. W. Oliver gave an address on 'nature reserves' (Oliver, 1913, 55). He reported that in England 'there had not been much progress', bemoaning the fact that in England: 'with the leisured classes of to-day the Natural History instinct takes the form of the chase, the garden and golf, while the masses of people seem to be preoccupied with other matters'. In contrast, progress had been made in other parts of Europe. With regard to the setting up of reserves, he clearly had the model of the American National Parks in his mind when he noted:

> Extensive closing of areas which it is desired to reserve is resented by many people as a kind of game preserving. Any area which has not been under long-continued private ownership carries with it public rights, either ill- or well-defined. The land at Blakeney Point belongs to the National Trust but it would be a blunder to enclose it, for from time immemorial people have had access for collecting samphire, shooting and picnicking: it is impossible to cancel these rights summarily and live at peace with our neighbours.

Overall, his approach was to encourage private owners to take the initiative in establishing reserves.

A useful account of the establishment of reserves in other parts of Europe is provided by an address given to the Berne International Conference on the Protection of Nature, November 1913, by Professor Conwentz (reprinted 1914, 112). In 1898, a report was drawn up concerning the 'endangering of primitive nature' in Prussia and setting out 'proposals for protecting it' through the establishment of reserves. In 1906, this led to the establishment of an Institute for the Care of National Monuments by the Prussian Ministry of Education to preserve 'formations of nature', including plant and animal associations, that are 'still in their primitive location and remain completely or almost completely untouched by civilisation'. Conwentz gives details of reserves already set up following the 1898 initiative. These included: fenland (Plage Fen, Brandenberg); beech groves (Sadlowo, East Prussia); beech and oak woodland (in Reinhard Forest, Sababurg, chosen as model forest tracts for landscape painters); lime groves (Colbitz, Magdeburg); moorland (Neulinum, on the River Vistula); and marshland (Zehlau, East Prussia) etc. Towns and cities had also joined the initiative, for instance Dresden and Cottbus purchased forests from private owners to create reserves. Reserves were also created in other parts of Germany; for example, heathland at Sempter Heide was purchased in 1877 by the Botanical Society of Landshut, and reserves were also established in Saxony and Baden.

The drive to set up reserves was part of a general concern about threats to wildlife, which led to a number of international meetings (e.g. Paris, 1883; agreements on the protection of seals in the Behring Sea: Paris, 1895; the Protection of Birds: London, 1900; Preservation of African wildlife: Paris, 1909). Also, books were written calling for the setting up of reserves; for example, Massart (1912) reviewed conservation efforts in America, England, Switzerland, Germany, Denmark, Java etc. and suggested sites for reserves in Belgium.

Despite the many calls for the establishment of reserves in Britain, many landowners were antagonistic. In May 1912, the naturalist Charles Rothschild hoped to change attitudes with the establishment of the Society for the Promotion of Nature Reserves (the SPNR). This organisation drew up a list of potential sites that was presented to the Board of Agriculture (by 1915, 284 sites). Unfortunately, the government was not persuaded by his arguments in a time of war, and the movement lost momentum when Rothschild died in 1923 (Rothschild & Marren, 1997). It is of considerable interest that Rothschild's sites were so well selected that many were eventually conserved in the National Parks and National Nature Reserves that were declared after the Second World War. Unfortunately, however, some of his sites were destroyed, damaged or greatly modified in the interwar period (Rothschild & Marren, 1997). For details of the long drawn-out battle to conserve Britain's habitats and wildlife, see Sheail (1976, 1981,1987, 1998), Rackham (1986), Evans (1992, 1997) and Rothschild and Marren (1997).

Managing reserves in cultural landscapes: Wicken Fen as a case study

This reserve is often said to be 'a remnant of the undrained fenland' once abundant in East Anglia (Rowell, 1997, 194). This description hints at 'wilderness', but, as Rowell makes clear, undrained 'is a relative concept; as part of the Fenland, the reserve cannot escape the effects of the massive drainage scheme that controls water-levels and allows agriculture to flourish in these low-lying lands. The whole reserve area has been under the influence of drainage schemes for at least three and a half centuries.'

Wicken, where Charles Darwin collected specimens in the 1820s, is famous for its wide range of wetland wildlife, but it was to secure the future of insects that influential entomologists purchased parcels of land and presented them to the fledgling National Trust (Friday, 1997). The original idea for the management of the reserve was made clear by Farren (1926) who writes: 'When parts of Wicken Fen first came under control of the National Trust, there was a general idea – perhaps a natural one – that it should be allowed to run wild' and, thereby to 'return to its original state' (quoted in Lock, Friday & Bennett, 1997, 216).

As we have seen, the notion that reserves should be left alone was a commonly held view in the later nineteenth and early twentieth centuries. Such a view was advanced by the London *Times* newspaper, which published a leading article on 18 December 1912, drafted in part by Charles Rothschild, describing the aims of the Society for the Promotion of Nature Reserves (the SPNR) that had been founded in the same year. The article emphasised that there is an 'urgent need to preserve the last relics of unspoiled nature . . . with their old native flora and fauna' by setting up reserves. What was feared was 'a sort of universal suburbanism'. It was considered 'futile to try to move threatened species elsewhere', the 'right way to protect species was to protect their homes'. In addition, 'the only effective method of protecting nature is to interfere with it as little as possible' (Rothschild & Marren, 1997, 18).

As late as the 1950s, Diver, the first Director of the British Nature Conservancy was also influenced by the same ideas, being in favour of 'letting nature take its course' on reserves. However, experience in practical management on reserves was proving otherwise: management was essential. At Wicken, leaving the reserve to nature was an inappropriate way to conserve those species for which the fen was famous. Traditionally, Wicken Fen had been intensively managed. In essence, it was an important part of the cultural landscape of East Anglia, supplying specialised crops of sedge and reed for thatching, and sedge and peat for fuel, etc. At the end of the nineteenth century these activities were in transition or decline. In the early decade as a reserve 'left to nature', the almost treeless fen (as revealed in contemporary photographs) was massively invaded by impenetrable scrub. As Farren points out, letting the Fen run wild ignored the question: 'what is the state most suitable to the economy of the species of insects of all orders for which the sedge fen is famous' (quoted in Lock *et al.*, 1997, 216).

It was not until the 1930s that experience in trying to manage reserves in different parts of Britain indicated that, if particular species associated with an area of cultural landscape were to be conserved, it was essential to manage them by continuing or reintroducing the appropriate traditional management. This very important principle is detailed in the following quotation from a report prepared for the British Ecological Society. This report considers British habitats, but enunciates crucial general guidelines that have proved important in devising appropriate management for conserving cultural landscapes worldwide (Anon., 1944, 83–115):

> In considering the actual administration of Nature Reserves it is essential to recognize, first of all, that informed and careful management is in fact necessary for nearly all reserves as well as for National Parks. A great deal of the vegetation which it is desired to preserve is partly the result of human activity through the centuries. In other words it is 'semi-natural' and not wholly 'natural' in the sense of being the product of 'Nature' alone. This is true for example of downland, of the hillside grasslands and many of the moors of the north and west, of much heathland, of undrained and partially drained fenland, of most of our deciduous woodland, whether the trees have been planted or not. These types of country together make up by far the greater portion of the regions of which relatively large samples should be preserved. In order to maintain any area of this semi-natural vegetation more or less in its present condition the existing human activities (grazing, burning, mowing, coppicing or selective felling) must generally be continued. This applies to the strictest nature reserve or sanctuary as much as it does to any area to be established as a National Park. If the human activities cease the vegetation immediately begins to change. The formerly grazed downland, the periodically burned heath, the fen which was regularly mown, becomes colonized by shrubs and often by trees. When the human activities cease, the character of the area is wholly altered, and the plant communities with their characteristic animals, which it is desired to preserve, disappear. If coppicing is discontinued in a wood of 'coppice-with-standards' type, the shrubs become 'overgrown' and the ground vegetation is seriously impoverished owing to the dense shade created. Thus if you 'let Nature alone' in such areas you frustrate the very aim you have in view. With the changes in the plant communities which she immediately initiates the conditions of life are altered and the interesting or rare species which depended for their existence on the former (man-made) environment may disappear altogether.

Conservation management: a Darwinian perspective

The final point made in this quotation is of crucial importance for the effective conservation of threatened ecosystems and endangered species. Essentially, it confirms the central

Darwinian message of this book. Human activities have resulted in modified ecosystems, in which new selection pressures have been and are at work. In such circumstances selection has favoured some species, but not others, within each of the elements of the cultural landscapes. However, if traditional practices decline or are abandoned, through human-mediated management changes, or neglect, new selection pressures will be brought into play. Unless species have the appropriate genetic variability to respond to the new habitat conditions, the Darwinian fitness of populations will be reduced and populations will decline in numbers, with the possibility of genetic depauperation through the processes of selection, genetic drift, inbreeding depression etc. (see Chapter 12). Thus, the appropriate conservation management is to counter the newly emerging selection pressures associated with change by re-imposing those former conditions, in which the now endangered species were 'winners'. The appropriate management/restoration route, therefore, is to resort to *precedent* by maintaining appropriate long-term traditional practices where they are still employed, reinforcing the practices where they are faltering, and re-imposing them where they have been abandoned (see Rackham, 1998).

This approach can be very effective if some of the individual plants associated with the former practices still survive in a somewhat changed ecosystem, e.g. trees, shrubs and long-lived ground flora species that have vigorous vegetative reproduction, through stolons, rhizomes, bulbs, corms etc. In these cases, genetic continuity is maintained between the former traditional practices, through a transition of different land use and a return to the use of traditional management. However, in other cases habitats have been so altered by human activities that the species associated with traditional land use have all disappeared.

Modern farming practices have endangered particular species through the loss of heathlands, downland, hedges, ponds, wetlands and coastal communities etc., resulting in damage to many SSSIs in the UK. Also, most lowland species-rich grasslands have been improved by the use of fertilisers (with only 4% remaining unimproved in 1994). As we saw in Chapter 8 in discussing the Park Grass Experiment, increasing the nutrient status of the soil encourages some agriculturally important species at the expense of others. For example, experiments with *Orchis morio* have revealed that the species is very adversely affected by application of inorganic and organic fertilisers (Silvertown *et al.*, 1994).

Countering these losses, the traditional management appropriate to a particular site has often been continued or successfully reinstated, not only in national parks and reserves but also at many other sites, sometimes through management agreements with private owners supported by legislation to protect endangered species (Sutherland & Hill, 1995). Thus, in reserves on coastal, upland grasslands and heathlands, and on certain arable lands by agreement, traditional methods of land preparation, crop cultivation, stock rearing, game production etc. have been maintained or reintroduced, including appropriate fencing and hedges. Also, forests and woodland have been brought back into traditional management, using methods of coppicing with standards. In addition, forest-grazing systems – wood pasture – have been reintroduced with pollarded trees and grasslands beneath. Each of these traditional practices – coppicing, burning, grazing – requires carefully defined periodic intervention into the future, rather than a single act of management.

Examples of 'resort to precedent' in the management of individual species

A. The successful restoration of populations of the Starfruit (*Damasonium alisma*), one of the rarest plants in the UK, was accomplished by cleaning out overgrown neglected village ponds and allowing seed long-buried in the mud to germinate. Management in the future will require periodic action.

B. Restoration of populations of the very rare Strapwort (*Corrigiola litoralis*) at Slapston Ley, Devon, provides another interesting example of management to take advantage of buried seed banks (Marren, 1999). Until 1968, the species grew in footprints of cattle coming to the waterside to drink. However, conservation managers decided that the cattle should be withdrawn from the area 'in the belief that their presence was unnatural'. As a consequence, the muddy habitat was replaced by tall sedges and reeds and the population of strapwort declined. However, with the removal of the fences and the return of cattle to the site, the population recovered.

C. Restoration also rescued the colony of Adders Tongue Spearwort *(Ranunculus ophioglossifolius)* at Badgeworth, the only colony in Britain (Marren, 1999). The pond in which it grew was fenced as a protected nature reserve. However, the plant declined. Fencing out cattle was an inappropriate strategy, for the species requires open muddy areas for germination from the seed bank. Thus, the species is well adapted to shallow pools that flood in autumn and winter, and partially dry out in summer. Trampling by cattle brings the seeds to the surface and submergence in the winter offers protection from frost. Many other rare species require disturbance, formerly provided by the activities of domestic animals, to allow plants to grow from buried seed banks. In the absence of livestock, rotavating has been employed to increase populations of *Filago pyramidata* and *Corynephorus canescens*.

D. The very rare Fen Violet (*Viola persicifolia*), long thought to be extinct at Wicken Fen, has been encouraged to return by germination from dormant seed in the seed bank provoked by habitat management mimicking the abandoned practice of peat digging. The species requires seasonally flooded land, where the underlying peat is exposed (Marren, 1999). In other contexts, see below, a number of other examples of resort to precedent are presented.

Reintroducing traditional management

The reintroduction of coppicing, pollarding, burning and other traditional practices results in dramatic changes in reserves, which may be misunderstood by the general public, who see the actions as wilful damage. In recognition of this reaction, Marren (1999, 282–3) concludes that it 'needs confidence to make a mess on a nature reserve', but this is exactly what is required for the successful conservation of some endangered species, where recent human-induced habitat changes are acting against the long-term survival of these species. It has to be recognised that many British rare plants are 'attuned to a world of shepherds, horses, farmers, labourers and low-tech machines'. Regarding the means by which 'resort to precedent' is achieved, Marren makes the important point that rural life has completely changed: we live in a world of industrial agriculture. Thus, to achieve the necessary traditional management, 'we must now find substitutes from our own world of conservation volunteers, contractors, power tools and management agreements'. However, while some conservationists are content to use these modern 'tools', there is a distaste amongst some ecologists for the tools technology provides. 'Bulldozers, herbicides, pesticides, chainsaws and high explosives are, for many conservation-minded ecologists, the instruments of the Devil. It is using precisely these means that the damage they wish to put right was created. This is an attitude, which, while perhaps understandable, is none the less a barrier to progress. No tool in itself is bad or good; what matters is how it is used' (Edwards *et al.*, 1997, 385).

Although the practice is controversial, selective herbicides have been used in some management projects, particularly in the control of invasive plant species (see below). Also they have been employed in the conservation of rare cornfield weeds, which is the largest group of declining species in the UK (Marren, 1999). Some species are effectively lost (Thorow-wax: *Bupleurum rotundifolium*, Corn Cockle: *Agrostemma githago*), while others are scarce (Corn Buttercup: *Ranunculus arvensis*, Red Hemp-nettle: *Galeopsis angustifolia*). Many of these species have declined in the face of herbicide use in agriculture. However, some rare arable weeds have been successfully conserved by Somerset Wildlife Trust on unsprayed ploughed headlands around arable fields. But there is a serious problem with perennial weed grasses, and these are controlled by the use of Roundup (containing, as active ingredient, the broad spectrum herbicide glyphosate) and other selective weed killers. Pesticides have also been used elsewhere; for instance, in the effective management of *Lilium martagon* populations in southern England. Scrub has been cleared, and areas fenced to exclude deer. But it also proved necessary to protect populations from the recently arrived beetle *Lilioceris lilii*, by dusting the flowers with malathion (Mackworth-Praed, 1991).

Ecosystem management in the past

The concept of resorting to precedent requires a thorough knowledge of the landscape management in the past. In devising such management schemes it is worth attending to the insights offered by Lowenthal (1985): 'Memory, history and relicts of earlier times shed light on the past. But the past they reveal is not simply what happened; it is in large measure a past of our own creation, moulded by selected erosion, oblivion, and invention.' Reflecting on other issues important in conservation management, Rackham (1998) concludes: 'for some reason historical ecology is particularly productive of canards and factoids'. A factoid is 'a statement which looks like a fact, is believed as a fact, and has all the properties of a fact except that it is not true'. For instance, 'the public believes, in defiance of everyday experience, that trees are necessarily killed by cutting them down'. And, taking another example, 'in Scotland, the "restoration" of the Great Caledonian Wood has become a political issue, regardless of the protests of historical ecologists who point out that the Great Caledonian Wood is mythical and never existed in historic times' (Dickson, 1993).

In order to re-impose so-called traditional practices it is important to examine the evidence very carefully. For example, under conservation management for most of the twentieth century, sedge has been cut outside the growing season at Wicken Fen, as early managers thought this was traditional practice. This is unlikely to be the case, however, for in Norfolk, where commercial cropping has long been continued, sedge is cut in the growing season (Lock *et al.*, 1997, 234). Thus, to manage for conservation it is essential to carry out appropriate historical and ecological research to establish not only the appropriate management, but also to determine precise details of timing and frequency. In order to re-establish former land use it is necessary to confront the very wide diversity of traditional practices.

Restoration of traditional practices

The re-imposition of rural traditional practices is not a simple issue, but a prescription fraught with complexity. Conservation organisations, with limited funds and sometimes relying on volunteer labour, must consider the practicality and recurrent costs of reintroducing coppicing, peat digging, ditching, fencing, scrub clearance, grazing regimes etc. For instance, there may not be an entirely secure market for the traditionally harvested plant materials, such as firewood, sticks, reeds, sedge etc. Charcoal, traditionally produced from coppice may be more expensive than imported charcoal made from tropical timber.

There are many ways in which animals have been subject to 'traditional practices'. These include hunting for food (including fishing, trapping, snaring, wild fowling, egg collecting etc.), sport, trophy hunting and pest control. Here, we are primarily concerned with management as it relates to plants. But it is important to consider these practices, as they profoundly influence plant components of ecosystems, e.g. through altering the balance of carnivores/herbivores etc. These traditional practices could simply be re-imposed. However, circumstances change. For instance, hare coursing, fox hunting, seal culling etc. are now thought to be cruel by some, but not all, of the general public. Local and international pressure has resulted in the banning of certain traditional practices, such as whaling.

Animal welfare is also a major concern. For example, in order to secure the future of the Bush Mallow (*Malacothamnus clementinus*) on San Clemente Island, California, 6,000 introduced goats were transported to the mainland and about 15,000 were shot. The removal of the goats was successfully accomplished, but only in the face of petitions and lawsuits from animal rights advocates who protested that the action was 'cruel and that the welfare of the goats ought to take priority over that of a few endemic plants' (Rolston, 2004, 30). Animal rights activists have come into conflict with conservationists in other situations; for example, in their opposition to the removal of feral pigs from some of the Hawaiian Islands (Simberloff, 2001).

Welfare concerns are also raised on other issues. In reserves and parks many would consider it unacceptable for animals to starve to death. Over time, such an event is certain in the natural evolutionary drama, but, in a well-visited park or reserve, there would be public pressure 'to save the animals' by providing food and water. There are special problems in dealing with domestic animals, especially in reserves in heavily populated regions. It may be uneconomic to graze sheep for limited periods in pocket-handkerchief-sized reserves in chalk grassland, but of greater concern is the vulnerability of unattended sheep to savaging by domestic dogs, and, in the extreme case, to rustling. In addition, there are also concerns about the health and safety of the visiting public, whose support is vital for conservation efforts. Thus, it may be impossible in certain well-visited sites to retain diseased, damaged and fallen trees that are necessary parts of a naturally functioning ecosystem.

The use of fire has a very long history in many parts of the world, e.g. the aboriginal peoples of Australia have been burning the landscape for thousands of years (Bowman, 1998). Conservationists are continuing or reintroducing such traditional burning in reserves in Australia (Morrison *et al.*, 1996). In cultural landscapes, management by fire can be hazardous not only to humans in the area, but also to their property, domestic and farm livestock and crops etc. It is stressed that prescribed burning for conservation is never risk-free. For instance, in 1980 a spring burn in Michigan, carried out to improve the habitat for the Kirkland warblers, 'raged over nearly 50,000 acres, took one life, and destroyed numerous houses' (Pyne, 1982, 122). There are many other issues. As Pyne acknowledges, prescribed fires may produce a great deal of smoke, in quantities incompatible with legislation enacted to control atmospheric pollution. Also, deliberate burning of vegetation on poorly consolidated slopes may later lead to soil erosion and flash flooding during periods of heavy rain. In re-imposing traditional practices, there is another complicating factor. Many reserves and other areas of conservation interest are not owned outright, but are in multiple-ownership, and obtaining prior permission for a prescribed burn (or other management activity) from all owners may be difficult.

The devising of non-traditional practices by conservation managers

While many traditional practices have been reinstated by conservation managers, they do not hesitate to establish new landscape features or practices that have little or no historical precedent. For instance, in the Norfolk Broads, lagoons have been created near Barton Broad to 'store' nutrient-rich silt extracted from the Broad, and thereby improve its water quality. Also, areas of marshland are regularly flooded by the RSPB for the benefit of wading birds: such flooding would have been strenuously avoided in former agricultural practice (Williamson, 1997, 164).

The role of experiments in conservation management

How is conservation management to be devised and carried out? Some activities have clear historical precedents and to an extent predictable outcomes. In many cases the reintroduced practice has to be modified in the interests of economy or safety (volunteers working on ladders etc.). Thus, burning or mowing is substituted for grazing. Cattle grazing may take the place of sheep grazing etc. Some consider that 'common sense' is a good guide to the likely effects of such 'substitute treatments': others are concerned that scientific experimentation should play a major role in testing the effects of proposed management changes.

It is undeniable that conservation often involves rapid action with inadequate resources in the face of crisis. Is scientific experimentation too time consuming and expensive?

Sutherland (1998) makes a very strong case for the use of properly designed experiments in conservation of national parks and nature reserves. Firstly, he stresses the need for management plans with clear statements of aims and objectives. In the formulation of such plans it is important that those who will carry out the management should be involved, otherwise the work on the ground will not necessarily be satisfactory. In Britain, there have been a number of 'conservation accidents' in cases where those undertaking the work were ill-informed. For example, in Wales, there are perhaps six plants of the endangered *Cotoneaster cambricus* (plus material introduced by conservationists). By accident the best specimen was cut and burnt in a clearance of shrubs. Another series of incidents have threatened the British endemic species *Sorbus wilmottiana*. This species occurs at only one site with a population of 'only a few dozen trees' (Marren, 1999, 9). Rich and Houston (2004) reveal that one tree was stolen, another was damaged in an attempted theft. In the present context, it is important to note that several trees have been cut down during conservation management; some, but not all, have re-grown from the cut stumps.

In the drawing up of plans for reserves, Sutherland (1998) notes that a common mistake is to manage for maximum diversification of ecosystems within reserves. His view challenges the widespread notion that every reserve deserves a lake! Regarding the management, Sutherland (1998, 206) makes a powerful case for long-term controlled experiments based on hypothesis testing, with proper recording and analysis of long-term trends.

> Land managers should regularly carry out randomised, replicated, controlled and monitored experiments but in practice very rarely do . . . At the absolute minimum there should be a control (for example the old management technique), the consequences of the change in management should be monitored and the results made available for others to consult. In practice, even this bare minimum is rarely achieved. Instead, the norm is to alter the management (often making a number of changes simultaneously), not have a control area, subjectively decide if the changes have been successful and pass on the opinions verbally.

An example of a situation where proper experimentation was not carried out involves the management of grasslands at a site in the UK containing the rare orchid *Ophrys sphegodes* (Hutchings, 1989). Traditionally the site was sheep grazed, but, for economic reasons, traditional practices were abandoned. A change to winter cattle grazing caused much trampling of the vegetation and the orchid population declined, probably because of severe mechanical damage to the underground parts of the plant. From 1980, sheep grazing was reintroduced, but, as the animals were allowed to graze continuously, all the orchid flower stalks were eaten by sheep. Only when periodic and carefully timed sheep grazing was introduced did the populations recover (Hutchings, 1989).

In contrast, the management of the Hayley Wood nature reserve, Cambridgeshire, provides an example of the elegant use of an experiment to inform management. This famous ancient woodland has been subject to traditional woodland management for centuries, through the practice of coppicing with standard trees (Fig. 15.1). There was a period immediately before the wood became a reserve when traditional practices were abandoned, but coppicing has been reintroduced, in particular to encourage the special ground flora including the Oxlip (*Primula elatior*). However, a new factor had to be taken into consideration in the management of the wood. Over a long period, but especially in the period

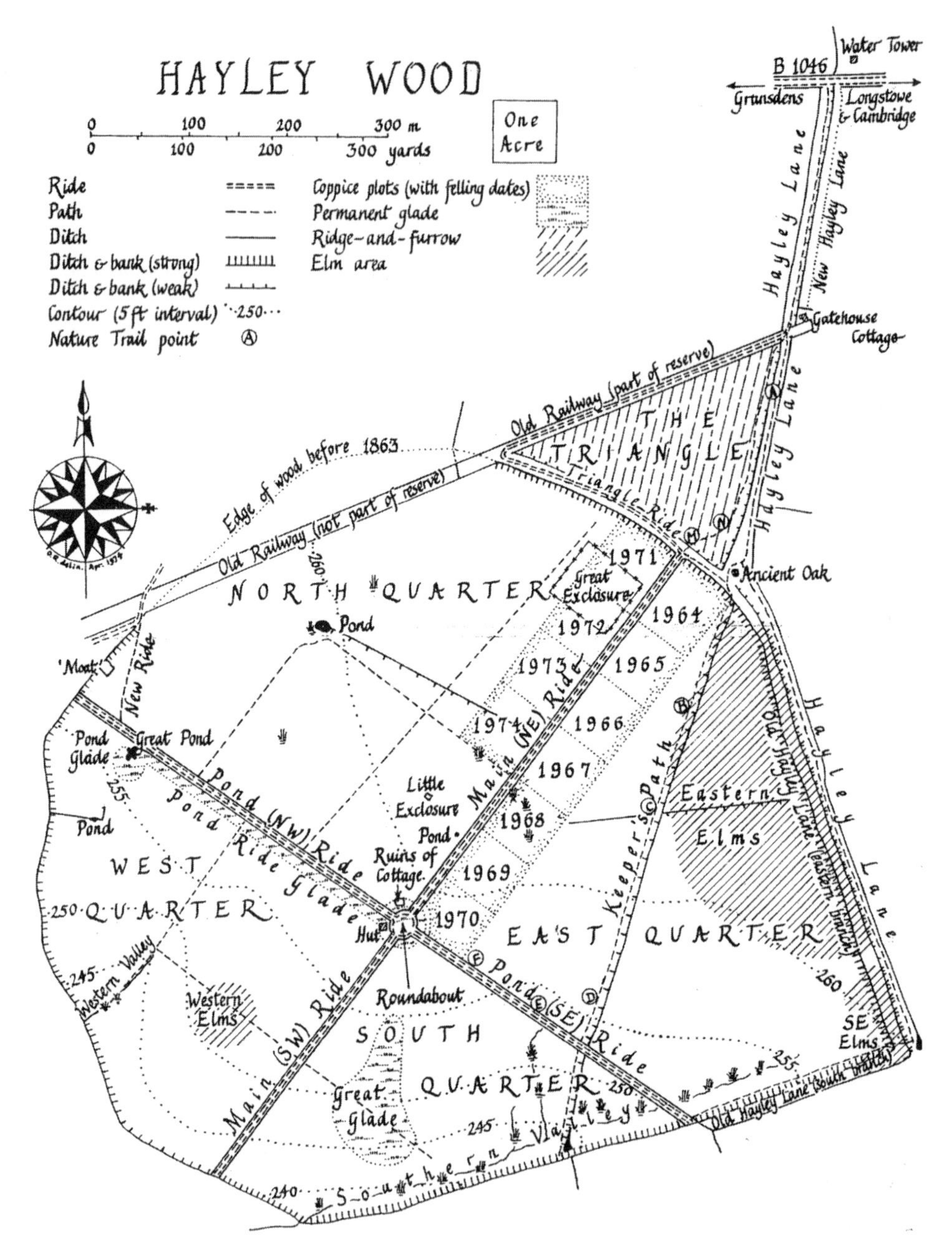

Fig. 15.1 Map of Hayley Wood, a reserve belonging the local Wildlife Trust, Cambridgeshire, UK. Rackham (1975) gives full details of this famous ancient woodland, which has a long history of management by the traditional coppice with standards system, and supports a unique rich fauna and flora. Note that with the building of the Cambridge–Oxford railway a portion of this famous ancient wood was lost and a portion gained – the Triangle – which is traversed by ridge and furrow indicating its former agricultural use as plough land. Woodland species have colonised the triangle

1939–45, fallow deer escaped from the nearby Waresley Park, and now range over considerable territory including Hayley Wood (Rackham, 1975). Changes in the woodland ground flora, especially losses of the inflorescences of the famous oxlips, were attributed to deer browsing. To test this hypothesis, two small deer-proof exclosures were erected and the flowering of oxlip inside and outside these structures was compared. Evidence confirmed that deer browsed the oxlip flowers to an extent that threatened the populations long term. As a consequence of this experiment, a larger area of the wood was enclosed by deer-proof fencing.

A call for evidence-based conservation

Returning to the role of experimentation, Sutherland *et al.* (2004) conclude: 'current conservation practice faces the same problems as did old-fashioned medical practice', where 'most decisions were based not upon evidence but upon anecdotal sources'. However, modern medicine owes much of its success to long-term properly designed drug and treatment trials. In contemporary conservation management, very little evidence is collected on the consequences of current practice so that future decisions cannot be based on what does and what does not work (Sutherland *et al.*, 2004). They suggest the setting up of a central database of information on conservation practice and believe 'a greater shift to evidence-based conservation would not only be highly effective, but is also likely to result in enhanced funding by actively demonstrating this effectiveness to funders and policy formers'.

Developing these ideas further, two influential concepts have been developed in relation to longer-term conservation ecosystem management. The first is adaptive management in which there is a 'continuing process of action-based planning, monitoring, researching and adjusting with the objective of improving the implementation and achieving the desired goals and outcomes' (Szaro, 1996, 750; Fig. 15.2). This type of planning can be very helpful if it is combined with properly designed experiments. Peterson, Cumming & Carpenter (2003) have developed a second concept used by conservationists, namely scenario planning. This allows a number of possible plans for the future to be sketched out and evaluated.

Fig. 15.1 (*cont.*) from the ancient woodland, but such migration, for example by Oxlip (*Primula elatior*), is slow. After experiments with a number of small exclosures indicated the dramatic effect on the ground flora and coppice of grazing by introduced Fallow deer (*Dama dama*), a section of the wood was fenced (to the east of the Main (NE) Ride), and coppicing of about one acre plots has been carried out (at the dates indicated) and continues at the same intervals to the present day. To the west of this Ride, there are a number of coppice plots unprotected by fencing. The Great Exclosure is another section of fencing and straddles two adjacent coppice plots. Here it is possible to compare the effect of open access and exclusion of deer. The design and layout of the coppicing plots and fencing allows experimental investigation of the effects of grazing, which have been increased by the arrival of another introduced deer species, the Muntjac (*Muntiacus reevesi*). The Fallow deer appear to have originated from Waresley Park, from which animals escaped, particularly in the 1939–45 period when the fences around the Park were neglected (Rackham, 1975, 171. Reproduced with permission of the Wildlife Trust, www.wildlifebcnp.org).

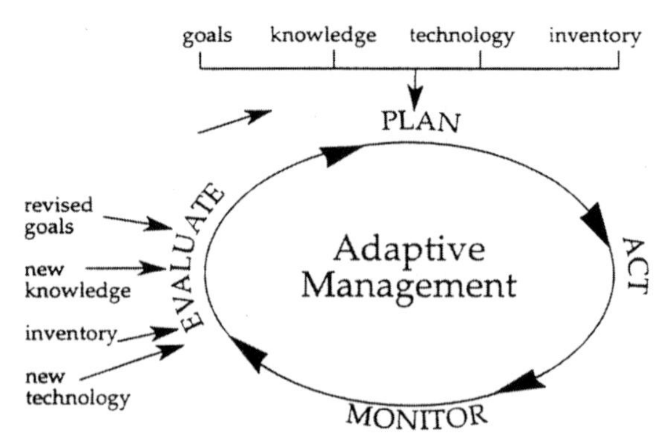

Fig. 15.2 The principles of adaptive management (From Szaro, 1996. Reproduced with permission of Oxford University Press)

In predicting the effect of different management regimes on rare and endangered species, the RAMAS GIS package of population models and Population Viability Analysis (PVA) (a concept introduced in Chapter 12) has proved helpful. A few examples illustrate their use in different management situations.

A. The Monkey Puzzle tree (*Araucaria araucana*) in Southern Chile. Populations of this long-lived, slow-growing iconic species are now protected from logging, and trade in plants is covered by CITES regulations. PVA modelling has used information on population distributions, possible metapopulation structures, demographic data, mating system (the tree has separate male and female individuals), seed dispersal and predation, and the catastrophic effects of fires and volcanic activity (Bekessy *et al.*, 2004). Families of risk curves have been used in management planning. They reveal that populations are likely to be relatively stable over the next 100 years, but, if populations are to be maintained and recovery to larger populations be achieved, human-set fires must be controlled (a most difficult requirement to fulfil) and seed harvest (for human consumption) must be lower than at present.
B. *Grevillea caleyi* in New South Wales, Australia. In this endangered understorey shrub, germination from a soil seed bank is triggered by heat and smoke from fires. Seeds fall from the canopy, and were perhaps once dispersed by emus. Under human influence these birds have now gone from the ecosystem. Seeds of *Grevillea* are sometimes heavily predated by native mammals. Modelling has examined the effects of prescribed burning at different intervals, in combination with accidental fires and arson. Also, PVAs have examined the combined effect of fire, and current and reduced seed predation (Regan & Auld, 2004).
C. The influence of introduced diseases, in combination with other factors, has been modelled using PVA. For instance, in the conservation of the Tasmanian shrub *Epacris barbata*, viability assessments have been made of the effects of management for fire frequency, and limiting human access to prevent the spread of fungal infection by *Phytophthora cinnamomi* (Keith, 2004). The Whitebark Pine (*Pinus albicaulis*) in Mt Rainier National Park, Washington, USA, is under threat from an introduced fungus – white pine blister rust (*Cronartium ribicola*), which is native to the Far East and was introduced in a batch of seedlings of white pine from Europe in 1910. Modelling, using PVA, has revealed that populations of the pine are likely to increase in the absence of the fungus, but, where the rust occurs, dramatic decline is likely, despite the removal of *Ribes* species on which this fungus completes a part of its complex life cycle (Ettl & Cottone, 2004).

Conflicts in conservation management

Experiments and modelling are helpful in estimating the likely effect of certain actions. But conservationists rarely have a free hand in deciding how to manage a reserve: there are conflicts of interest to resolve. They are supported in their work by the general public, but must take into account the funds available and the views of their supporters and opponents in the appreciation and resolution of management dilemmas and conflicts. For example, in the case of Hayley Wood, the problem of deer browsing, mentioned above, could have been solved by culling the animals. But the general public see introduced deer as an acceptable, indeed welcome, addition to the wildlife of Hayley. It is significant that introduced fallow deer figure more prominently on the reserve notice boards than the famed oxlip. Thus, the most acceptable solution to the problem was the erection of deer-proof fencing around the coppice plots. At first sight, fencing would appear to be an intrusive modern artefact within the wood, but traditionally coppice plots were protected by fencing, ditches etc. against browsing domestic and 'wild' animals (Rackham, 1975).

Another example of 'conflict of interest' is provided *Delphinium montanum*, an endemic species in the Pyrenees, of which only about 4,000 individuals survive (Simon *et al*., 2001). Botanists have attempted to conserve this species, but have discovered that the plant is at risk through grazing by Pyrenean chamois (*Rupicapra rupicapra pyrenaica*) – a threatened species currently increasing in numbers though population reinforcement. As some chamois die through larkspur poisoning and a cull of chamois is out of the question, the obvious solution is to provide and maintain chamois-proof fencing around the larkspur populations. However, Simon and associates report that it is very difficult to resolve this conflict of interests, as the erection of fences has proved ineffective – chamois pull them down.

Another interesting conflict of interests involves different groups of plants in fragmentary nutrient-poor calcareous grasslands in Holland, traditionally grazed by sheep (Oostermeijer *et al*., 2002). In order to encourage and satisfy the requirements of a number of flagship groups of rare orchids, butterflies and invertebrates, present management involves either sheep grazing for a period in the late summer or, where the areas are small, mowing the reserves at the end of the season. This conservation management, in its timing and intensity, is different from traditional grazing, which involved large areas of continuous habitat. One group of species is particularly disadvantaged by this conservation management – late-flowering *Gentianella germanica* and *Gentianopsis ciliata*. Present conservation grazing and mowing practices, which aim to prevent dominance by the grass *Brachypodium pinnatum*, occur 'too early in the season' for these late-flowering gentians, and their seed crop is lost. Moreover, the practice of non-rotational mowing of entire reserves does not accurately mimic traditional grazing for it leads to habitat uniformity. Also, there is 'no escape for canopy-inhabiting invertebrates into undisturbed parts with higher vegetation' (p. 347). This case study points to another issue of general concern: management could be 'too tidy' in its execution. Many species of conservation interest might survive and flourish in the unkempt seemingly neglected corners of reserves.

Turning to wider issues, it is important to recognise the many areas of potential conflict between conservation organisations and the farming communities. Concerning plants, conserved areas may harbour weeds that affect agricultural land. In terms of animals of interest to conservationists, there are many situations of conflict – e.g. elephants and other animals moving out of reserves to forage in adjacent croplands. In the UK, a number of areas of concern and conflict have been identified, especially the use of farm chemicals, and the 'control' of mammals, seen by conservationists as wildlife, but viewed as pests, involving high economic costs, by the farming communities (Tattersall & Manley, 2003). These include the impact of agri-chemicals, especially pesticides; crop damage by rabbits; the role of badgers in the spread of TB in cattle; the effects on wildlife of rodenticides used on farms; the damage caused by deer to crops and conservation areas; and the control of fox populations.

Is resort to precedent sufficient to ensure the survival of endangered species and ecosystems in cultural landscapes?

While it is possible to restore *some* traditional practices to national parks, nature reserves and other sites of conservation interest, will such management return the area to its original condition? There are many factors that make this unlikely, as the cultural landscapes change under pressure of human influences. Increasingly, reserves and parks are islands near and within developed land and they may be adversely influenced by human activities near or within the reserve boundaries. It is not possible within the scope of this book to provide a comprehensive account of these influences. Here, by way of illustration of general principles, we consider the effects of the use of water resources, pollution, invasive species, and legal and illegal activities.

Water resources

Many reserves are small and most do not contain complete watersheds within their boundaries. Human use and control of the water resources, including the extraction of water from underground sources by bore holes etc., may have a profound effect on the hydrology of the reserved area. Thus, the Lopham Fen Nature Reserve in Norfolk, UK, was profoundly and adversely affected by the sinking of boreholes outside the reserve (Harding, 1993). Steps have now been taken to ensure that the reserve receives adequate water supplies, but in regions of severe water shortage, wetland reserves may be given low priority and may dry out. Even in situations where national parks are very large, their ecosystems may be threatened by hydrological alterations outside their boundaries (e.g. many US National Parks and other Public Lands of high conservation interest (Pringle, 2000)). Such difficulties may be part of a serious wider problem. Recently, after studying a wide range of data sets in climate modelling, Barnett *et al.* (2008) predicted that human-induced changes would result in a future 'crisis in water supply for the western United States'.

Pollution

Officially designated parks and reserves and other areas with important ecosystems and endangered species are impacted by human pollution from various sources. For instance, atmospheric pollution, both gaseous and in the form of acid rain, drifts from urban industrial areas to affect terrestrial, freshwater and marine ecosystems hundreds of miles away. In Scandinavia, lime has been added to try to restore severely affected lakes in areas where acid groundwaters are the product of local geology (Wallstedt & Borg, 2003). In an attempt to counter the effects of acid rain on wetlands, various treatments, largely unsuccessful, have been tested in the Netherlands, including flooding areas with base-rich groundwater, topsoil removal and liming (Bootsma *et al.*, 2002). As we saw in Chapter 4, urban industrial complexes, modern transport systems and agricultural methods (with their reliance on fertilisers and intensive stock rearing) produce widespread environmental pollution, leading to the eutrophication of terrestrial, marine and freshwater ecosystems, both inside and outside reserves. Locally, reserves surrounded by agricultural land may be impacted with spray drift of fertilisers and pesticides, including herbicides. While there has been considerable progress in controlling pollution in some regions of the world, areas of conservation interest are not immune from the effects of pollution. In areas of outstanding natural beauty in the UK, including conserved lands, chemical methods have been developed to strip nutrients (particularly phosphates) from water entering lakes and rivers. Also, pumps have been developed to remove eutrophic mud from wetlands. Such operations have been carried out at Barton Broad in the Norfolk wetlands, UK (Madgwick, 1999).

Invasive plants

The presence of invasive introduced organisms in conservation sensitive areas complicates the 'resort to precedent' model of conservation management. Looking first at invasive plants, it is interesting to discover that some problems owe their origin to former park officials. The second national park in Australia, Belair, near Adelaide, was founded in 1891. To shade roads and picnic grounds, the park commissioners planted foreign poplars, willows, oaks, redwoods and horse chestnuts. 'Not until 1923 did the park adopt a natives-only policy. Belair today is perhaps the most weed-infested national park in Australia' and 'according to the park management plan nearly 400 exotic species grow in the park . . . Belair's Tree Planting Committee never imagined they were sowing heartbreaking problems for the future' (Low, 2002a, 298).

In an attempt to control invasive plants, fire, weeding and mechanical removal have been employed. However, hand weeding is extremely time consuming and often herbicides are used. For example, since its introduction 125 years ago, *Ammophila arenaria* has spread along the entire western coast of North America, creating wide foredunes dominated exclusively by this species. There is no evidence that foredunes existed prior to the introduction of this species: many areas had transverse dunes built up in winter. Thus, the presence of major foredunes has changed the coastal ecosystems in Oregon and elsewhere, as *Ammophila* out-competes native species and blocks movement of sand inland to feed the original dune systems (Wiedemann & Pickart, 1996). In experimental trials, manual removal of *Ammophila* proved to be extremely expensive at *c.* US$20K per ha. Applications of the herbicide Roundup gave only limited control of *Ammophila*. A soil fumigant – Vapam^{TM} – was more effective, but its use raises many issues. Along the same coastline four introduced *Spartina* species have invaded several major estuaries, transforming the ecosystems on mudflats and salt marshes (Daehler & Strong, 1996). Some estuaries have yet to be invaded. Effective control of these grasses has been achieved by herbicide applications, but only if 'voluminous' quantities are used (Aberle, 1990).

Hand weeding, herbicide treatment and removal of the plants with heavy machinery have all been employed to remove invasive introduced trees and shrubs from parks, reserves and other areas of conservation interest. *Rhododendron ponticum* is a common woody species in woods and copses, having been planted widely throughout Britain as cover for game birds and for decorative purposes during the Victorian era. Usher (1986) and Tyler, Pullin and Stewart (2006) provide details of the problem and the attempts at a solution. It is a serious threat in acidic oak woodland in Killarney, where the dense shade cast by its foliage threatens the ground flora, especially bryophytes. Cutting, hand pulling of re-growth and herbicide treatment may contain the problem. A more radical solution is to use machinery to tear the plant out by the roots. However, it is difficult to achieve total eradication: indeed, yearly treatments must be carried out, as the removal of the adult plant disturbs the soil and prepares the ground for the germination of the *Rhododendron* seed, which is produced in abundance and is widely dispersed by the removal of the adult plants.

Sometimes there are special difficulties in controlling *R. ponticum*, for example on the island of Lundy in the Bristol Channel, UK. Here it grows on sea cliffs threatening populations of the endemic species Lundy Cabbage (*Coincya wrightii*) found only on this island. Controlling the *R. ponticum* – by herbicide and cutting etc. – has proved especially hazardous, requiring the professional skills of contractors working on ropes on the cliff face (www.english-nature.org.uk).

Invasive disease organisms

In many cases 'resort to precedent' may not restore ecosystems to their former condition because of the past and present effects of introduced disease organisms (e.g. Chestnut Blight in eastern USA, Dutch Elm disease in Europe and elsewhere, *Phytophthora cinnamomi* in Australia etc.).

Introduced animal species

Animals, especially those that are highly invasive, have had profound effects on ecosystems in many parts of the world, influencing the plant components through changes in the balance of carnivores and herbivores, changes to food-webs, predator/prey relationships etc. For instance, in the USA, Gypsy moth defoliated 59 million acres of forest between 1924 and 1990. Individual outbreaks have had devastating effect, for instance over a million oak trees were killed in New Jersey (Winston, 1997, 29). In combating the effects of introduced animals, a wide range of tactics has been employed including exclusion by fencing, trapping, shooting and poisoning. Which strategy to use has to be very carefully considered, especially if there are native species to conserve in the same area (Simberloff, 2001). In New Zealand, dramatic ecosystem changes have taken place following the introduction of goats, deer, brush tail possums and a wide range of other mammals. The impact of predatory mammals, especially rats, mustelids and cats, on the vertebrate fauna has been 'catastrophic', a situation that has been described as 'an ecological collapse' (Towns & Atkinson, 1991). In areas of high conservation interest, possums and introduced species of rodents (including house mice) are being successfully controlled by a network of bait stations. At forest margins, traps are set to catch feral cats and mustelids, and feral goats, deer and cattle are being controlled by culling. Introduced wasp populations have been controlled to low levels by pesticides. With these measures it has proved possible to control the numbers of many introduced species, but total eradication has proved impossible in many cases. However, the eradication of larger introduced mammals has been achieved on some off-shore islands of New Zealand. Also, an intensive poisoning programme has successfully eradicated Norway rats from one of the Queen Charlotte Islands, British Columbia. Mass culling has also proved an effective tool in the eradication of invasive mammal species on Sarigan Island, North Pacific (goats and pigs), Monito Island (rats), and islands in the Seychelles (rats and rabbits) (Simberloff, 2001). It is proposed to eradicate the populations of North American Beaver (*Castor canadensis*) that were introduced into Tierra del Fuego in the 1940s to start a fur trade (Choi, 2008). From 50 introduced animals, the population has grown to more than 100,000 and they have made profound changes to 16 million hectares of indigenous forest. Choi points out that just as the extremely expensive eradication project is being assessed in South America, European beaver have now been reintroduced to parts of Scotland, having been extinct there for the last 400 years.

An alternative strategy to combat introduced organisms in areas of high conservation interest is to use biological control, i.e. the use of living organisms as pest control agents. In many cases such control mechanisms have proved effective (Vincent, Goettel & Lazarovitis, 2007), but there is an element of risk as the agent may damage non-target populations of other organisms (Thomas & Willis, 1998). A case in point is provided by the use of the moth *Cactoblastis cactorum*, which was so successfully used to control invasive *Opuntia* species in Australia. The same insect has been used to control infestation of *Opuntia* elsewhere, for instance, in the Caribbean. However, the moth has moved to Florida, where it is now attacking five native species of cacti, including the 12 surviving plants of the rare Semaphore cactus (*O. spinosissima*). Conservationists acted just in time: 'cages were lowered over the plants and cuttings of all twelve were taken to Fairchild Tropical Garden' (Low, 2002a, 270). In the wild, this species is highly threatened by development projects, but its 'death knell may be sounded by *Cactoblastis*.

There is considerable disagreement as to whether widespread invasive species can be eradicated at reasonable cost. Simberloff (2001) cites four investigations then in progress. In 50 attempted eradications in California, none succeeded, if the infected areas exceeded 100 ha. Persistence may be the key, however, for a 10-year eradication programme had almost eliminated the annual grass *Cenchrus echinatus* on the 411 ha Laysan island, Hawaii. It is reported that the introduced vine (*Pueraria phaseoloides*) has been successfully removed on one of the Galapagos Islands.

However, the successful removal of introduced biota does not necessarily permit ecosystems to return to the status quo, as invaders may cause decisive change to the habitat (Brockie *et al.*, 1988). For instance, invasion by *Tamarix* species resulted in soils of high salinity, unsuitable for native species to re-establish (Zavaleta, Hobbs & Mooney, 2001). There may be other surprises too. *Hypericum perforatum* is an invasive introduced weed in many parts of North America. Biological control has been successful with the introduced insect *Chrysolina quadrigemina*. But studies of permanent plots in Shasta County, California, revealed that the control of *Hypericum* resulted in the simultaneous emergence of *Centaurea solstitialis* as a future regional pest and the greater incidence of the invasive introduced grass *Bromus mollis* (Mack, 2000, 160). There have been other unexpected consequences of the control of invasive organisms. Removal of the feral water buffalo in Kakadu National Park, Australia, encouraged regeneration of wetlands as predicted, but alien grass species have also flourished, and the introduced Para grass (*Brachiaria mutica*) now covers about 10% of the flood plain of the park (Cowie & Werner, 1993; Petty *et al.*, 2007). The removal of feral cattle from San Cristobel Island in the Galapagos allowed the previously suppressed exotic introduced guava (*Psidium guajava*) to grow into dense thickets (Hamann, 1984).

National parks and reserves: threats from illegal activities

In well-regulated reserves and parks, the aim is to secure legal protection for ecosystems, and to control the numbers and activities of visitors. The impact of visitors within the parks is minimised by the provision of footpaths, boardwalks etc. Some areas may be fenced and be seasonally out-of-bounds. Regulations control camping, hunting, fishing, harvesting of natural resources etc. To educate the public, many sources of information are provided, e.g. notice boards, booklets, nature trails and websites. In many countries there is legal protection for certain wildlife. In the UK, the Wildlife and Countryside Act of 1981 makes it illegal to uproot any plant without the owner's permission, and consent is also required to introduce any organisms at any Site of Special Scientific Interest (SSSI). While legal protection has been enacted to protect species/habitats in many parts of the world, this legislation has proved partial at best, totally ineffective at worst, because of insufficient funding. Bruner, Gullison and Balmford (2004) estimate that, across the globe, the shortfall is of the order of US$1–1.7 billion per year. The costs of expanding and managing a wider protected area system would be of the order of $4 billion per year for the next decade.

Political corruption, theft, poaching, illegal logging, mining, firewood collecting, illicit grazing etc. are also serious problems in some reserves. Terborgh (1999) has reviewed the effects of these activities and concluded that they result in major damage to many ecosystems, especially in the tropics. Indeed, the situation is so grave that many national parks in the tropics are, in effect, paper parks – parks in name only. Given that tropical ecosystems contain a huge range of biodiversity not known elsewhere on Earth, and acknowledging the major role played by the National Parks Service in the protection and management of the national parks of the USA, Terborgh proposed that a similar service be established to protect national parks overseas. The service could be paid for as foreign aid, and to prevent any hint of colonialism, the scheme could be operated through national governments by the United Nations. The proposal is an example of a top-down solution to a major issue.

Others have approached the problem from a different perspective, by investigating the history of park development, trying to understand the root causes of illegal behaviour and reach solutions involving compromise. The management of the Arusha National Park in Tanzania, established in the 1960s, illustrates some of the key issues. What follows is based on the work of Neumann (1998). Adopting the national park model developed in the USA, Meru peasants and livestock herders were excluded. The area designated was viewed by colonists as a remnant of vacant African landscape where Nature was "unspoiled" until recent times (p. 2). Over the years, park officials and conservationists have had to contend with many illegal activities such as 'livestock trespass, illegal hunting, wood theft' etc. From the perspective of the Meru, Arusha is typical of parks across Africa, in being 'created out of lands with long histories of occupancy and use' with 'forced relocations and curtailment of resource access' (p. 4). Thus, the Meru have 'reduced access to ancestoral lands, restriction on customary resource uses, and the predation of wildlife on [their] cultivated lands [outside the Park boundaries]'. Examination of the historical context of the development of Arusha Park reveals that conservationists lack an understanding of (or are in denial about) the history of the area. It could be said that they are a party to the elimination of the record of 'indigenous history and culture'. 'In essence, the establishment of national parks and associated protected areas criminalised many customary land and natural resources uses for communities across Africa. As a result, most protected areas have become arenas for struggles over resources between state conservation agencies and the local peasants and pastoralists' (p. 5). It is clear from these comments that issues of crime, resistance and protest are very complex. 'Violations of resource laws in Arusha National Park that conservationists and park authorities attribute to "poverty", "ignorance" and "population pressure" contain a political dimension. While acts such as illegal firewood collection, grazing trespass, and forest encroachment may have multiple meanings and intensions, in the context of the park's criminalization of local customary rights of access, they are political insofar as they represent a rejection [by the Meru] of the state's claims of ownership and management' (p. 49).

Over the last few years, thinking on national parks has begun to change (Magome & Murombedzi, 2003, 100). For example, in South Africa, apartheid officially ended on 27 April 1994 and land reform, often involving the courts, has begun under democratic rule, including lands reserved for conservation (Magome & Murombedzi, 2003, 110). For instance, the Richtersveld National Park was established in 1991 to conserve the succulent Karoo vegetation of the Namib desert. The Nama peoples would have lost grazing rights, firewood, access to medicinal plants and honey. In the dying days of apartheid, they negotiated with the park authorities and obtained significant concessions. The Nama were recognised as the rightful owners of the land. A reduction in the size of the proposed park was agreed with a shorter lease and payment of a lease fee. Also, grazing was to be allowed in the park with a specified number of animals (6,600, mainly goats and sheep) with guaranteed job opportunities and majority representation on the planning committee. In this instance, the park authorities and the Nama forged an agreement on the park as it was being established.

Across Africa, much greater challenges must be faced as the management of long-standing parks is being reassessed. In South Africa, changes have been made in the Kruger National Park, originally set up as a game reserve in 1898. By the 1970s, Kruger had reached its current size (20,000 km^2 of land – the size of the state of Massachusetts). There were forced removals of 1,500 local Makuleke people, and in 1995 they lodged a land claim for 250 km^2 in the northern part of the park. The land restitution commission upheld the claim of the Makuleke on condition that the area continued in its current land use of protecting biodiversity, with a prohibition on mining, agricultural activities and building of houses. Moreover, any commercial development they initiated must be linked to conservation. The Makuleke were allowed limited harvesting of wildlife species, and were granted equal representation on the board of management.

The rights of indigenous peoples are being recognised in other countries. For instance, 14% of Australia – more than 1 million square miles – has been restored to aboriginal peoples, who now own the parks at Kakadu and Uluru. These territories have been leased back to the park agencies.

The major focus of this book is to examine the patterns and processes of microevolution in human-managed areas. It is clear that the nature of this management, both in its legal and illegal forms, will determine how ecosystems change and which species will be winners or losers. What must be stressed is that conservation of biodiversity, through parks, reserves or elsewhere, is a highly political arena, not only in the developing world, but also across the globe. The long-term aims of conservationists often conflict with the short-term interests and needs of individuals, political groups and commercial interests. Extreme difficulties, often involving military or paramilitary conflict, occur where the status of reserved land and its resources are being contested by displaced indigenous peoples, squatters, landless peasants, refugees etc., who are striving to survive in conditions of abject poverty and disease. Compromises are necessary in conservation, but they 'can easily reach the point where they undermine conservation goals and even strengthen anti-conservation forces' (Adams & Mulligan, 2003). We return to some of these issues in later chapters.

Conclusions

As we have seen, many national parks, reserves and other conservation areas have been successfully established across the world. While the original aim in establishing reserved areas was to 'preserve' scenery, landscapes and later ecosystems, in reality much active management has been carried out. There has been both legal and illegal management. Evidence from a range of sources reviewed in earlier chapters supports the view that human activities, in their management of ecosystems, exert new powerful selection pressures to which species respond. Risking a generalisation, across the world, traditional rural practices in many areas are now in decline and large areas of cultural landscape have been greatly modified by modern industrial agriculture and forestry. Concerning conservation of endangered species and ecosystems in cultural landscapes, vestiges of habitat and small populations sometimes remain. In some cases these are long-lived species and it is assumed that at least some of the original genotypes have persisted. Whether such assumptions are justified has rarely been examined. Where investigations have been carried out, as we saw in Chapter 12, there is evidence that declining populations are very vulnerable to genetic erosion. In these remnant areas, the aim has been to re-impose the traditional practices of haymaking, grazing, forest and wetland management etc. to which these species were adapted. In effect, resort to precedent aims to nullify the recent anthropogenic selection pressures that adversely affect wildlife following the decline of traditional management. Research has revealed evidence of past practices, but there are many challenges in trying to unravel the details. Where it has not proved possible to return to the exact traditional management, conservationists have tried to employ a close match or compromised to find an acceptable alternative, e.g. mowing instead of grazing etc. Sometimes new 'seemingly traditional practices' have been adopted.

Looking critically at the model of resort to precedent, it is clear that an exact match to previous conditions is not possible in practice, as newly arising human influences – such as pollution, introduced species, and legal and illegal resource use originating outside and inside the reserves – cannot be completely nullified or controlled. Further discussion of this point appears below.

Since management of reserves and parks became accepted practice, another debate has raged. Are there limits to what is acceptable in management and what types of management interventions were appropriate? For instance, in the early days of conservation, the reintroduction of species was not an acceptable official practice in Britain and elsewhere. However, as we shall see in the next chapter, management in reserves, and restoration of damaged or destroyed ecosystems, has recently evolved to the point where 'creative conservation' involves major interventions to remodel the landscape, and restore abiotic and biotic components (Sheail, Treweek & Mountford, 1997). Interventions also include the reinforcement of declining populations of endangered species, and the reintroduction of extinct populations (Bowles & Whelan, 1994).

16

Creative conservation through restoration and reintroduction

This chapter considers the microevolutionary implications of creative conservation, which involves habitat restoration and species reintroduction. It considers underlying concepts and practice, including which stocks to use in projects.

Bradshaw (1987) has devised a general model of ecosystem restoration (Fig. 16.1) to take account not only of the aims of conservationists in parks and reserves, but also the schemes of a wide range of professionals, including landscape architects and landscape gardeners, who have undertaken projects for government departments, local authorities, industrial corporations and private landowners etc. Not all these restorations aim to promote wildlife conservation, although they may do so indirectly. Some are/were concerned with restoring damaged areas to productive farmland, removing pollutants from rivers and lakes, restoring watersheds to prevent erosion etc. (Bradshaw & Chadwick, 1980; Jordan, Gilpin & Aber, 1987; Falk, Palmer & Zedler, 2006). Restoration for nature conservation is, therefore, only one strand in the complex relationship of humans to their landscape.

Restoration presents the ecologists with powerful challenges, providing the 'acid test' of the ecologist's understanding of vegetation (Bradshaw, 1987). Harper (1987) likens ecological restoration to the repair of a watch: taking it to pieces provides an understanding of how it works and how it might be rebuilt.

Turning to the details of the model, the diagram indicates how human activities have changed natural habitats. Sometimes ecosystems have been severely degraded, and such areas may be rehabilitated towards their original state by natural plant succession. In the past, natural processes have been largely responsible for any habitat recovery. For instance, in the Severn Valley, England, woodlands have recovered as a period of industrialisation was followed by the transfer of most of the operations elsewhere. Heslop-Harrison and Lucas (1978, 298) give details. The earliest coke-fired iron-smelting blast furnace was opened in 1630 and prints from the 'late 18th century and the early 19th century show that much of the woodland, even on the steepest valley sides was devastated, with heavy fall-out of pollution from kilns and furnaces, all of which were coal or coke fired. The recovery here has been almost entirely a natural process, with the seeding in of native species from neighbouring woodlands which were not affected by industrial operations in the river valley.'

However, such processes are slow and the outcome uncertain, and political pressures in many countries of the industrialised West require accelerated rehabilitation of degraded

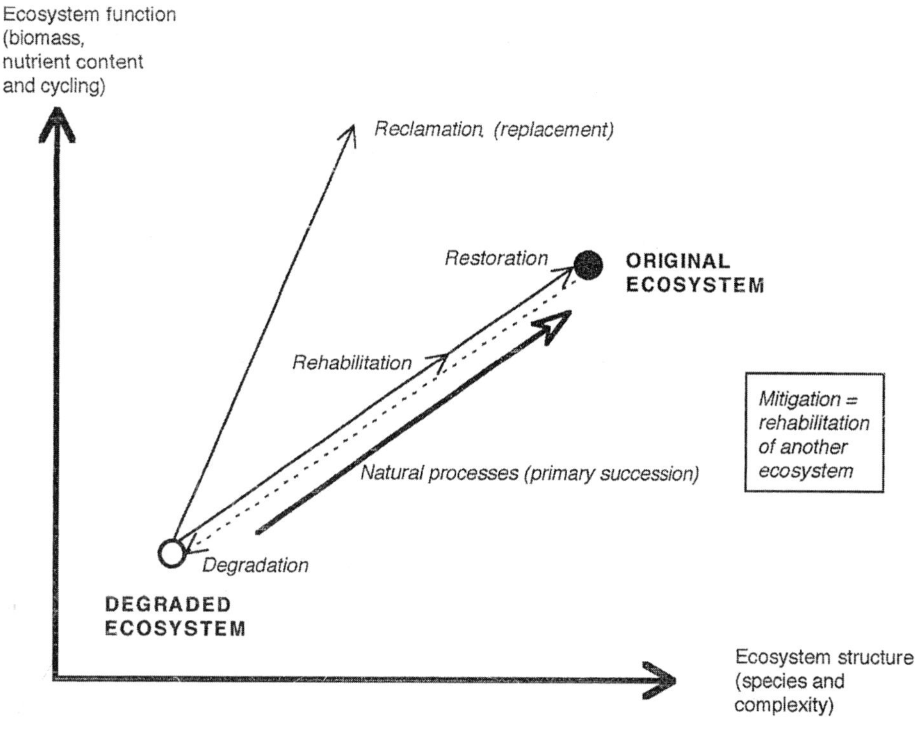

Fig. 16.1 The different options for the improvement of a degraded ecosystem can be expressed in terms of the two major characteristics of structure and function. When degradation occurs, both characteristics are usually reduced, although not necessarily equally. Used in its narrow sense, *restoration* implies bringing back the ecosystem to its original or previous state in terms of both structure and function. There are a number of other alternatives, including *rehabilitation*, in which this is not totally achieved, and *replacement* of the original by something different. For some, all these possibilities are covered by the general term *reclamation*. (From Bradshaw, 1987. Reproduced with permission of Cambridge University Press)

lands and rivers through ecological restoration towards its original state. Alternatively, restoration may aim to achieve other end points. For example, the objective may be the reclamation of derelict industrial land to create urban parks, sports fields, amenity woodland or agricultural land etc. However, a much more elaborate diagram would be necessary to accommodate the many different types of restorations that conservationists are attempting to achieve in their increasing management of damaged elements of cultural landscapes (Maunder, 1992; Anderson, 1995: Gilbert & Anderson, 1998; Falk *et al.*, 2006). In these situations, the aim is to restore ecosystems created by humans (often involving domesticated animals and plants), and not to reproduce the original vegetation of the region (Bradshaw & Chadwick, 1980).

The choice of accelerated restoration, instead of relying on natural processes, has interesting consequences. Sometimes restoration may sweep away significant biodiversity that has colonised derelict sites. For example, a thorough botanical survey of Glasgow revealed a wide range of interesting plant species in the wastelands of abandoned mines and industrial sites (Dickson, Macpherson & Watson, 2000). Some of these species have been lost in the later remodelling of these areas.

Creative conservation through restoration projects: some examples

Table 16.1 provides examples of general approaches to restoration. It is not intended to be comprehensive. Given that the major focus of this book is plant conservation, more space is devoted to the botanical elements of projects. In preparing this section, the *Handbook of Ecological Restoration* (eds. Perrow & Davy, 2002) has been a major source of information, and this excellent work should be consulted for further details.

Which stocks of plants should be used in creative conservation?

Creative management and restoration projects involve the sowing or planting of a range of plant species, and the choice of material for such schemes has microevolutionary implications. Two different basic approaches may be recognised – 'mix' or 'match' (Gray, 2002). Either, a wide 'mix' selection of genotypes of the appropriate species may be used to provide genetic variability on which 'selection can act'. Or, alternatively, native stocks may be chosen that match, as closely as possible, those that were formerly present, or still exist in small numbers at the site in question. In essence, the 'match' approach emphasises the restoration of what was present in former times: the 'mix' looks to the future in providing variability to allow continued microevolution.

Native stocks

Most practitioners of restoration emphasise the importance of using native stocks. At first sight this looks like a simple prescription. However, the word 'native' is open to different interpretations (McKay *et al.*, 2005). Thus, stocks of native British species could be imported from other parts of Europe for use in projects in the UK. Alternatively, ecologists could insist that the plants must be from British seed sources, or make a further restriction, namely, that the plants originate from specified regions or similar habitats close to the restoration site.

Many conservationists insist that only local native stocks be used in restorations – the 'match' approach. This view is supported by a very considerable body of genecological research. Cultivation trials, reciprocal transplant experiments and cytogenetical studies on many species have revealed that each contains a range of habitat-specific local genetically based ecotypic races. Such races result from the actions of natural selection in genetically variable species. Some ecotypes are morphologically distinct (e.g. dwarf alpine races), others are physiologically different, being adapted to different soils and to live in different latitudinal and altitudinal climatic zones. Also, there is abundant evidence that natural selection, operating in human-modified habitats, also generates distinct 'biotic' ecotypes, as, for example, the herbicide-resistant strains of weeds, metal-tolerant variants on mines,

Table 16.1 *Restoration of ecosystems through 'Creative Conservation'. A range of examples are given, based on the* Handbook of Ecological Restoration *(Perrow & Davy, 2002) and* Restoration Ecology *(Van Andel & Aronson, 2006), where full references to the literature are provided.*

This table indicates something of the variety of approaches: it is not intended to be comprehensive. Some additional management tools are also indicated – with appropriate reference to source(s).

Applying these approaches in a wide range of restoration projects has resulted in many successful landscape transformations, but many problems and intractable issues have come to light. There are many challenges ahead for the developing science of restoration ecology.

Abiotic factors

- In coastal areas throughout the developed world, beaches, dunes and shingle areas have been stabilised and reformed, sometimes as a conservation measure but more often as coastal defences, by transferring sand and shingle, and constructing stone, clay and concrete sea defences.
- By restoring tidal flows and the excavation of flood-borne sediments, salt marshes have been restored in e.g. Connecticut and Oregon.
- As an adaptation to sea level rise associated with climate change, some sea walls and defences protecting agricultural land have been removed, allowing new coastal habitats of conservation significance to develop.
- Mangrove forests have been restored in Australia and elsewhere.
- Artificial reefs and islands have been constructed using discarded tyres, derelict cars, bamboo, piping, concrete etc. in the Pacific, Puerto Rico etc.
- At many sites in the developed world, rivers and streams have been restored by removing hard structures canalising rivers and waterways, and allowing seasonal inundation of adjacent flood plains.
- Areas of former wetland have been restored to functioning by manipulation of the water levels, and the removal of excess nutrients by dredging out phosphate-rich sediments. Hydrosere successions have been reinstated by excavating ponds and clearing waterways. Artificial manipulation of the hydrology of wetlands near the sea has been carried out. For instance, at Minsmere RSPB reserve, scapes have been excavated to produce water bodies (with artificial islands) of different controlled salinities from mixing seawater and freshwater (see Fig. 16.2).
- Restoration of freshwater lakes has involved the reduction of nutrient loads from feeder streams and rivers, removal of phosphate-rich sediments, and liming treatments to restore acidified water bodies.
- A multitude of restoration projects have rehabilitated former industrial sites, mine spoil, domestic and industrial waste dumps (Bradshaw & Chadwick, 1980).

Animals

As part of restoration, reintroduction and introduction projects, many different management techniques are use to manipulate animal numbers.

- *Restoration of populations.* By preventing poaching and controlling hunting, conservation managers and others are increasing the numbers of endangered wild animals in many areas.
- *Reintroduction/introduction etc.* In the conservation of animals, many restoration projects have involved the reintroduction or restocking of populations by controlled releases from wild stocks (e.g. wolves have been introduced into Yellowstone National Park), *ex situ* sources or from parks

(*cont.*)

Table 16.1 (*cont.*)

and reserves where populations have become too large. As part of the restoration of lakes, fish have been introduced. Many bird species have been reintroduced, e.g. the red kite in the UK. There is a growing trade in wild animals to stock newly formed game parks and reserves on former agricultural land in many parts of the world, e.g. in South Africa. Experiments have also revealed the possibility of transplanting corals to damaged reefs (Okubo, Taniguchi & Motokawa, 2005; Piniak & Brown, 2008).

- *Provision of habitat and food.* Conservationists have taken action to provide nest sites and food for wildlife, e.g. the inoculation of decay fungi into trees to produce sites for cavity-nesting birds, providing artificial nesting sites for birds, bats, solitary bees and wasps, encouraging food plants for butterflies and other insects, and the artificial feeding and provision of water for birds and mammals.
- *Encouraging habitat connectivity.*To allow wildlife to access between habitat fragments isolated by human activities, underpasses beneath roads and pipelines have been devised, e.g. badger pipes, amphibial and reptile tunnels; and to allow fish access around dams, weirs etc. fish ladders have been constructed.
- *Culling of animals.* This has proved necessary in the management and restoration of wildlife and control of invasive species both within and outside reserves and parks. For example, in the 1960s, wells were dug to provide artificial water supplies in Tsavo National Park, Kenya. With success in controlling poaching and the provision of artificial water supplies, elephant populations conserved under 'natural management' increased in number to such an extent that about 6,000 starved to death and to avoid such situations elsewhere culling has proved necessary (Budiansky, 1995).
- *Protection against disease.* As *ex situ* populations of rare and endangered species of mammals are routinely inoculated to prevent disease, steps have now been taken to protect animals 'in the wild', e.g. cheetah and black rhino have been successfully 'darted' with vaccine to protect against anthrax (Turnbull *et al.*, 2004).
- *Contraception.* To control numbers of feral and wild carnivores that have become 'pests', or reservoirs of disease, birth control methods – currently employed on domestic and captive exotic carnivores – are being examined as possible means of controlling populations of feral animals (Jewgenow *et al.*, 2006; Bradford & Hobbs, 2008).
- *Restriction of access.* The use of fencing to control grazing/predation by wild, feral and domestic animals in reserves and parks (and in the wider landscape) is informed by exclosure and other experiments. For instance, electric fences are used in the Minsmere RSPB reserve in Suffolk, UK, to prevent predation of birds by foxes (Axell, 1977). Fencing has been erected to prevent trampling of plants of endangered *Astragulus cremnophylax* (Maschinski, Frye & Rutman, 1997); and to examine the effects of deer populations on forest understorey in North America and different areas of Europe. To protect populations of endangered *Echium* in the Canary Islands it has been proposed that fencing be erected around the plants (Marrero-Gomez *et al.*, 2000). In an attempt to find a way of controlling feral pigs in Texas, experiments have been carried out with different types of electric fencing, but no fence was found to be entirely pig-proof (Reidy, Campbell & Hewitt, 2008).

Plants

- Restoration projects often involve re-enforcement, re-establishment or creation of new populations of both common and rare and endangered species (see text for case histories).

Table 16.1 (*cont.*)

In addition to the species selected to grow in the long term, the degradation of many urban/industrial and mining areas has proved so extreme that it is necessary to sow nurse crops to ensure initial stabilisation of the soil, prevent the growth of weeds and, in the case of leguminous nurse plants, to provide additional nutrients to the infertile soils through nitrogen fixation. Such nurse plants may or may not be designed to be part of the final ecosystem.

- In the restoration of grasslands and heathlands, a number of approaches have been used singly or in combination, including scrub removal, control of frequency and severity of fires, the reintroduction of grazing by domestic stock and rabbits. In areas where stock cannot be employed, mowing has been used as a substitute for grazing.
- In the restoration of grasslands and heathlands on former arable areas, the nutrient status of the soil may have to be reduced (see Gilbert & Anderson, 1998). In brief, this may be achieved by burning; grazing, mowing or cropping without the addition of any fertilisers, and taking the 'harvest' offsite; or by removing topsoil. Also, it is recommended that, to keep nutrient status low, plant material (cut vegetation and woody material from scrub clearance) and animal waste (dung from grazing animals and domesticated pets) should be removed from the site.
- Restoration of appropriate vegetation cover may be achieved by relying on the soil seed bank in the soil, sowing seeds broadcast or in slots cut in the turf, or if suitable material is available by the translocation of turves of vegetation cut from an existing grassland or heathland. Such restoration often involves the use of nurse crops.
- Restoration of prairie has been achieved at many sites in the USA by sowing or planting of appropriate species. To control weeds, especially of introduced invasive species, various treatments have been used – mowing, fire and the use of selective herbicides. To reduce the soil nutrient status in restorations of prairie on former arable land, a number of treatments have been applied – stripping topsoil, cutting, leaching and carefully controlled grazing with removal of animals at night. In the continuing management of restored prairie, further plant species have been added and, to simulate natural processes, grazing, burning and occasional tilling have been carried out.
- Restoration of dry land vegetation has been attempted in many parts of the world by control of grazing and by planting, e.g. nitrogen-fixing trees etc. In the semi-arid lands in Australia, overgrazed by feral rabbits, goats and sheep, areas under restoration have been fenced to control grazing, appropriate plants have been sown, and attempts have been made to make the best use of rainfall by controlling run-off water with carefully constructed systems of earth banks.
- In the restoration of temperate woodlands, traditional management practices have been re-established; for instance, coppicing with standards or wood pasture (woodland with grazing areas beneath managed pollarded trees). Also, restorations have been established that rely on natural regeneration with minimum intervention. Management aims to mimic natural disturbance regimes including processes of gap formation and subsequent colonisation. Some restorations have involved the expansion of existing woodlands into adjacent territory by new plantings or natural regeneration. Many restoration projects in the UK and elsewhere involve the removal of introduced conifers planted within deciduous woodland and also the replacement or remodelling of plantations of introduced conifer species. (Rackham (2006) and Newton (2007) provide comprehensive up-to-date reviews of techniques in forest ecology and conservation.)

(*cont.*)

Table 16.1 (*cont.*)

- Restoration projects have been carried out in tropical moist and dry forests, involving the planting of trees, with non-native species as 'nurse trees'. To encourage the dispersal of fruits and seeds by birds, restoration sometimes involves the provision of perches, feeding and nesting sites such as logs and woodpiles.
- The building of oil pipelines, roads etc. has caused damage in many ecosystems across the globe. For instance, in Alaskan tundra, rehabilitation has involved controlling human access to allow natural regeneration, and in some cases the planting of native species.
- Along over-used trails, recreation areas and footpaths, and around eroded ski slopes (Urbanska, 1994; 1997), damaged swards have been stabilised by the use of open-weave biodegradable matting, followed by sowing and planting. Restorations of alpine ecosystems have involved additional techniques, including the mulching of prepared ground with hay (containing seed), the laying of grassy turf transplants, the insertion into swards of individual plants raised elsewhere, and the laying of ready-made swards grown off-site.
- Restorations have also been undertaken in aquatic ecosystems, e.g. sea grass planting in marine areas damaged by boats; and transplanting macrophytes in lakes and rivers.
- Many projects involve the extensive planting of appropriate water plants in freshwater ecosystems. For instance, in order to attract rare species of birds, the RSPB has planted many reed beds in their wetland reserves.

under pylons etc. (see Chapters 8–10). Research has also demonstrated that in many cases site-specific ecotypes occur, e.g. in hay meadows at different altitudes.

If this picture of ecotypic differentiation is correct then it should be possible to demonstrate 'home site advantage' in situations in which the growth and reproduction of native plants is compared with plants from other areas. An early test was devised by McMillan (1969), who collected material of four prairie grasses (*Andropogon scoparius*, *A. gerardii*, *Panicum virgatum* and *Sorghastrum nutans*) from many parts of North America. The growth and survival of 682 clones was examined in a garden trial at the Plant Ecology Research Laboratory, University of Austin, Texas. For all four species, McMillan discovered that population samples collected from the central Texas grasslands showed the greatest survival, with plants originating from distant northern and eastern sites of the USA being eliminated.

The home advantage hypothesis was also investigated in experiments carried out by Montalvo and Ellstrand (2000). They investigated Californian Coastal Sage (*Lotus scoparius*) from 12 sources. Allozyme variation was studied and two common garden experiments were set out with seedling populations. Again, there was convincing evidence that local plants performed better than those from distant localities. Employing reciprocal transplant techniques, studies of the possibility of 'home advantage' were carried out on samples of Wiregrass (*Aristida beyrichiana*) from different sites in Florida (Gordon & Rice, 1998). Taking into account the results of this and earlier studies, Gordon and Rice concluded that there were local adaptations in the grass particularly

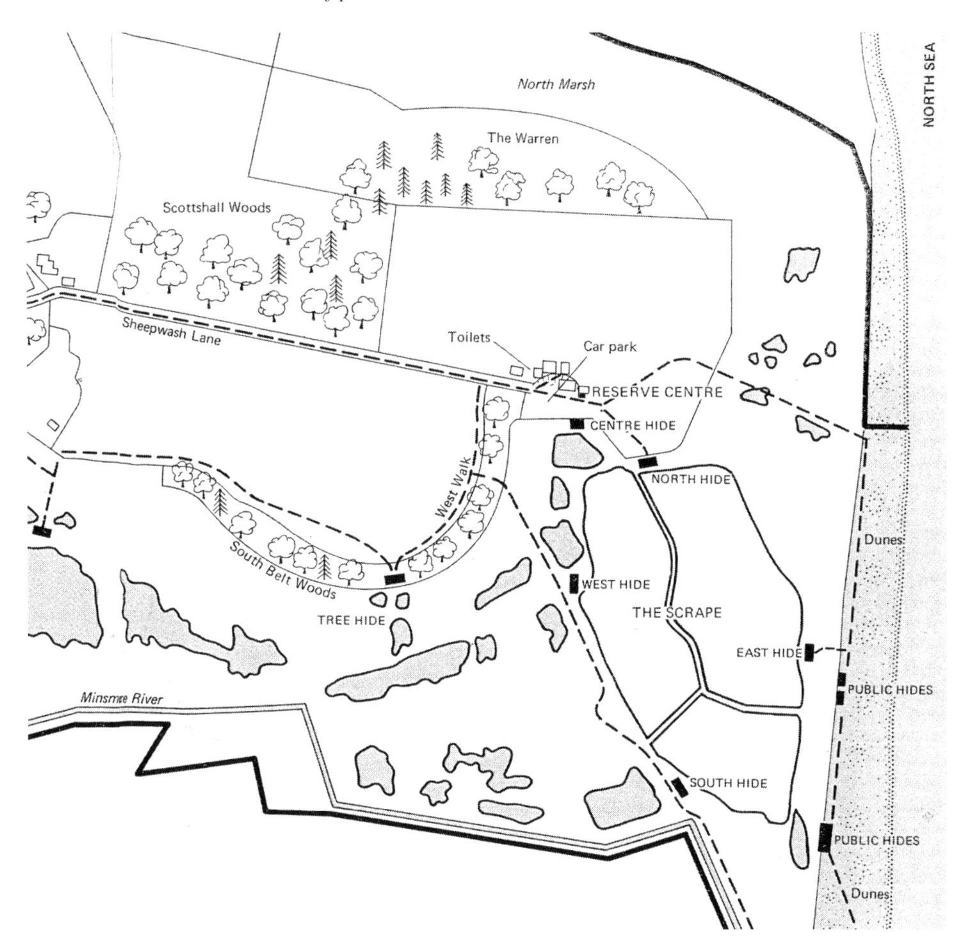

Fig. 16.2 Creative conservation in action: the Minsmere Reserve of the Royal Society for the Protection of Birds, Suffolk, UK. In the eastern part of the reserve 'The Scape' was excavated to provide shallow pools with artificial islands and muddy areas. Water levels are regulated, as is the 'salinity' of the water, through the mixing of gravity-fed piped supplies coming from a freshwater reservoir to the west of the reserve and seawater from the North Sea. (From Axell, 1977. Reproduced with permission of Random House Group Ltd.)

in later life stages and that restoration projects involving this species 'should use local seed sources'. However, some other experiments have yielded different results. For instance, no consistent home-site advantage of local genotypes was detected in studies of *Lotus corniculatus* employed in the restoration of limestone quarries in the UK (Smith *et al.*, 2005).

One route to effective restoration would be to sow or transplant the appropriate local ecotypic variant(s) for all the plant species (common and rare) appropriate to the ecosystem it is aimed to recreate. A number of seed sources are available and it is interesting to examine how they conform to the match model.

1. In the restoration of shingle habitats disturbed by the building of Sizewell Nuclear Power Station, Suffolk, UK, stocks of seeds were collected before the work began and, after testing, they were satisfactorily maintained in storage until the construction project was complete and the restoration of the shingle communities could begin (Walmsley & Davy, 1997a, b, c).
2. Seeds may be collected from fully functioning traditionally managed sites, such as hay meadows, and used either directly or, after a period of *ex situ* conservation, in the restoration of other areas. There are concerns about this practice, as the collection of seed/fruits from threatened or declining populations may put them at increased risk. In addition, as we saw above, the seed mixture from any individual hay field does not provide a 'universal hay' type appropriate for restoration of all sites. A further point of interest is whether all the species represented in the hay community in the field will be present in any particular sample of seeds.
3. Seed may be obtained from wildflower farms, where stocks of many species are grown and mixtures prepared for sale for restoration projects. Likewise, some native trees and shrubs from known local sources are available in specialist nurseries and offered for sale. Considering the worldwide interest in restoration, it is important to stress that only a relatively few easily grown species are available. Thus, the seeds of many tropical and sub-tropical and slow-growing species from arid or alpine regions are not yet available in commercial quantities. This makes it difficult, for example, to restore montane areas damaged by skiing etc. (Urbanska, 1997). In addition, there is a concern about the limited genetic variability of wildflower seed stocks, especially if material came originally from only one or few sources. Moreover, growing 'wild' flowers as a crop through many generations on wildflower farms has important microevolutionary implications (see below).
4. Plants and seed may be raised *ex situ*, from seeds collected in remnant ecosystems and relict populations for use in restorations and species recovery programmes. For instance, stocks of many prairie species were collected from tiny refuges of 'original prairie' grassland discovered in cemeteries, along railway tracks etc.
5. Seeds or plants may be available from botanic and other specialist gardens. But, as we saw in Chapter 14, such stocks may contain a restricted range of genetic variability, related to initial sampling or propagation *ex situ*.
6. There are sometimes difficulties in the restoration of areas highly contaminated with heavy metals. As a result of experimental investigations, heavy-metal tolerant variants have been identified and some of these have been developed as commercial varieties, e.g. red fescue variety 'Merlin' (Bradshaw & Chadwick, 1980).

It is clear from these comments that the concept of using 'matching' stocks for creative conservation has firm scientific support, as it recognises the importance of ecotypic differentiation. However, in practice, there are complications. For instance, wildflower seed mixtures may contain unexpected and inappropriate components.

An excellent case history is provided by the creation, on set-aside arable plough land, of a restored chalk grassland on the Gog Magog Hills near Cambridge, UK (Akeroyd, 1994). In order to reduce food surpluses, the concept of permanent or temporary set-aside agricultural land was pioneered in the USA in the 1930s. The European Union introduced 'set-aside' schemes in the 1980s (Clarke, 1992). In the Gog Magog Project, wildflower seed mixtures were purchased from reputable suppliers. However, detailed taxonomic investigation of the species in the resulting sward suggested that some of the plants that grew were probably of garden/horticultural origin (*Achillea millefolium*, *Centaurea cyanus*, *Chrysanthemum segetum* and *Leucanthemum* × *superbum*), while others were fodder/agricultural variants (*Medicago lupulina*, *Sanguisorba minor* ssp. *minor*, *Trifolium pratense*). There were also weeds of southern European origin (*Picris hieracioides* ssp. *grandiflora*, *Ranunculus marginatus* var. *marginatus*). These observations raise two important general issues. Many of the variants sown were of the correct species for the purpose, but wholly inappropriate infraspecific variants were included in the seed mixtures. Also, as we have seen in Chapter 11, it is very common for commercial seed stocks, even of 'wild' flowers, to contain contaminant weeds.

Restoration using stocks from diverse sources: the 'mix' strategy

A second approach to restoration is to use seed stocks that come from a variety of sites. This approach is recommended where the breeding system of endangered species needs to be 'restored'. For instance, as we saw in Chapter 3, some species are dioecious and both male and female plants must be present for sexual reproduction to occur. In many other species breeding behaviour is controlled by genetic self-incompatibility mechanisms and, for these to function properly, a mixture of S allele genotypes must be present. Thus, the only way to restore sexual reproduction in a population of a self-incompatible species, where only one or a few S allele(s) are represented, is to introduce other S alleles from different populations. Effective creation of new populations by involving mixed plantings is very well illustrated in the case of the rare endemic *Hymenoxys acaulis* var. *glabra* in the Great Lakes area in the USA (Demauro, 1994). These populations were created using plants of different mating type from different populations, together with their F_1 and open-pollinated F_1 progenies. In addition, plants were grouped to maximise functioning of the self-incompatibly system.

Mix or match?

There are powerful arguments for the use of matching stocks in restorations. Local populations have survived selective forces and to a greater or lesser degree have demonstrated their fitness. For some naturalists, our local ecotypic variants are part of our national heritage and these should be conserved in the same way as historic buildings, treasured cultural landscapes etc. (Akeroyd, 1994). In planting the wrong variants, local populations may be harmed and, potentially, 'introduced' taxa could become weeds by out-competing the local native plants or swamping them in hybridisation. Furthermore, the widespread addition of mixed stocks in the countryside could cause confusion in determining the natural plant distributions. These concerns have been seen by some as protecting the genetic purity of local variants, and preventing 'genetic pollution' and contamination.

However, the long-term goal of conservation, of which restoration is a part, is to secure the future of species, and in this longer perspective it is important to consider the capacity of the population to evolve in the future. Mixed stocks have some advantages in that they can bring novel variation into potentially depauperate populations. In this light, hybridisation and introgression may be seen as creative forces. Without the infusion of new genetic variability, local populations may prove to be unable to respond to the selection pressures imposed by future habitat changes. In later chapters, it is argued that the effects of predicted global climate change will greatly increase selection pressures on plant populations. Will small populations of rare and endangered species be able to respond? However, there is a further issue: the potential advantage of greater variability in mixed plantings is widely acknowledged, but the introduction of non-local stocks into a population risks the occurrence of outbreeding depression, in which maladapted variants are produced though hybridisation. This possibility is real as has been shown by a number of studies (Falk *et al.*, 2006, and references cited therein). An investigation highly relevant to the present concerns was carried out on Swiss plants by Keller, Kollmann and Edwards (2000). Local variants of three arable weed species (*Agrostemma githago*, *Papaver rhoeas* and *Silene latifolia* ssp. *alba*) were crossed with stocks obtained from wildflower mixtures. F_1 plants exhibited heterosis (hybrid vigour) but, as judged by biomass production, the F_2 plants exhibited reduced fitness relative to local parents. While there are potential negative effects of outbreeding depression, it is difficult to predict what their impact might be in restoration schemes. Such results could be used as an argument for rejecting mixed plantings. However, the possible harm of outbreeding depression in mixed populations must be weighed against the better-known dangerous consequences of inbreeding depression that can occur in small local genetically depauperate populations. In addition, while progenies resulting from outbreeding in mixed plants may be maladapted, some variants may be at a selective advantage in changing circumstances. Gray considers that both approaches – mix and match – have validity. Local matched plantings may be the correct approach for restoring certain key elements in cultural landscapes. In contrast, in employing tree and grass species in wide-scale regional restoration projects, mixed stocks have clear advantages, especially if they are carefully selected. For instance, in the effective use of *Pinus ponderosa* in regional plantings, conservationists recognise that Californian stocks of the tree are killed by frost when planted in Colorado. Therefore, appropriate mixed seed batches are best collected from climatic zones similar to that of the restored site. Sowing with mixed stocks is also recommended for the restoration of highly disturbed sites, such as quarries or roadsides, in the creation of new ecosystems without exact counterparts in nature (Lesica & Allendorf, 1999).

These examples point to important issues that must be faced with any restoration project. If the use of 'wild seed' best fits the aim of the project, then where possible such collections should be made. In collecting such seed, attention must be paid to the sampling techniques employed, including equalising the contribution of different seed parents, and the methods of bulking-up seed etc. In essence, from initial field collection, through to seed/plantings on site, the aim should be to avoid deleterious genetic changes. It may be possible to follow this advice if molecular studies are carried out on the original populations and any seed/plants derived from these. But for most species the genetic variability in original and derived population remains unknown. With regard to commercial stocks, Millar and Libby (1989) recommend that seeds and plants should not be purchased unless absolutely necessary, but if the use of wildflower seeds and nursery-grown shrubs and trees is unavoidable, then only those of certified known wild origin should be employed.

Creative conservation of endangered species

A number of different types of intervention have been devised, which are often referred to as 'reintroduction', but many conservationists insist on the use of separate terms (indicated in brackets).

A. Increasing the numbers of plants in declining populations (augmentation, reinforcement).
B. Re-establishing populations at sites where they have become extinct (reintroduction, restocking).
C. Creating populations at new sites within or outside their historic distribution (introduction).

In creative conservation, theory meets practice in interesting ways. As we saw in Chapter 12, there have been many important advances in our understanding of declining populations. Ideally, a carefully worked out plan confronting the following issues should be devised before restoration is attempted (Anon., 1991a; Menninger & Palmer, 2006). How have these understandings been translated into workable practice? To illustrate important general issues, brief details of a number of plants species restorations are presented (Bowles and Whelan (1994) may be consulted for detailed information on a number of case histories of animal and plant restorations).

How many populations should be established and in what spatial configuration?

Decisions on these important issues should be part of the design of the project. For example, in the case of *Cirsium pitcheri*, a threatened short-lived perennial on the Great Lakes dunes, the recovery programme was designed to take account of the behaviour of interlinked populations within the dynamic landscape. The plant colonises the early- to mid-successional stages, but populations can become extinct following catastrophic erosion, or later successional development. Generalising from this example, it is clear that the appropriate restoration model for many plant species might involve reinstating a functioning metapopulation system (see Chapter 12).

Site preparation and selection

It is also important to select appropriate ecological 'safe sites' within habitats, where reintroductions can not only establish themselves, but also produce self-sustaining populations (Urbanska, 1997). Such decisions require accurate knowledge of the autecology of the species. This point is clearly evident in the reintroduction of populations of *Rutidosis leptorrhynchoides* in Victoria, Australia. This is a gap-sensitive species, recruitment and survival being unlikely within canopies (Morgan, 1997).

Also, in order to achieve success in creative conservation of endangered species, adverse selection pressures precipitating loss or decline in the original populations should be properly identified and steps taken to restore the habitat or choose another site free from adverse influences. The case of *Lambertia orbifolia* (Proteaceae), a threatened shrub in Western Australia, is revealing (Cochrane, 2004). Molecular studies revealed that a population at Narrikup had high genetic diversity, despite being reduced to 169 plants. However, overall the population was in poor condition, being affected by weed invasion, and infections of aerial canker and *Phytophthora cinnamomi* (a devastating soil-borne fungus spread on the feet of animals and humans). Therefore, a new population was established in a nearby conservation reserve away from *Phytophthora* infection. Seed was collected and, in carefully prepared fenced ground, 216 seedlings were planted, mulched and shaded, with extra protection from wind and predators provided by plastic cones. Frequent monitoring revealed that, over the first 12 months, 98% of the plantings survived, but Cochrane recommends that access to the site should be carefully controlled to ensure that *Phytophthora* does not arrive and kill the new planting.

Decisions on population size

Many naturalists and others pay special attention to rarities. They know where rare plants grow, having enjoyed hunting for them. An important question is whether those designing restoration see such projects as a way of maintaining *rare species as rarities* – present in small numbers in few locations – or is the aim to make rarities more common and with larger populations sizes. This issue is rarely confronted, but has been discussed by Harper (1981, 201). Neo-Darwinian insights make it plain that small declining populations are under threat of extinction, and, therefore, management in such a way as to keep rare species rare but self-sustaining is probably impossible in the longer term.

Franklin (1980) and Soulé (1980) introduced the 50/500 rule, which proposed that an effective population size (N_e) of 50 was necessary in the short term to protect against inbreeding depression, while an effective population of 500 was needed to counter genetic drift and provide the population with a genetic capacity to evolve in the longer term. These 'rules of thumb' were later replaced by individual assessments of Minimum Viable Populations (MVP). Estimates for MVP sizes in plants are extremely difficult to quantify, but many geneticists consider that N_e should be of the order of 1,000 as a minimum, with higher numbers desirable.

Many attempts to restock the populations of endangered species do not address the question of MVPs. For instance, in the Species Recovery Programme in the UK, initiated in 1991 and carried out by English Nature, the objective was to secure the future of a number of rare plants and animals. For example, for Strapwort (*Corrigiola litoralis*) the aim was to 'establish four self-sustaining populations of at least 50 plants each around the shores of Slapton Ley [Devon]' (Deadman,1993). In the case of the Lady's Slipper Orchid, *ex situ* propagation was 'to allow restocking of circa 30 plants at the native site' and the introduction of native plants at up to 'five former sites' or suitable alternatives (Deadman, 1993; see also Ramsey & Stewart, 1998).

Demographic costs of reintroduction

In restoration projects the seed sown may not germinate and there may be significant losses in the seedling stage. In estimating such demographic costs, modelling has been carried out by Guerrant and Fiedler (2004, 375). For example, they estimate: 'in the most extreme case, an outplanting of 1000 *Panax* seedlings would on average be expected to drop by more than 98% to just 15 individuals within 3 years before the populations began to rise'. Thus, '67 times as many plants as are needed to reach sexual maturity will need to be planted. At this level of attrition, if the reintroduction goal is to have 1000 founder *Panax* plants reach sexual maturity, it would be necessary to plant almost 67,000 seedlings.'

The demographic costs of establishment may be reduced if bulbs, corms, clonally propagated individuals or seedlings are transplanted. Thus, Bell, Bowles and McEachern (2003) estimate that restoration of a viable population of *Cirsium pitcheri*, a threatened short-lived perennial on the Great Lakes dunes, would entail the sowing of about 250,000 seeds, but could be accomplished by planting 1,600 seedlings. In studies of the success of reintroduction of four species in Massachusetts, Drayton and Primack (2000) discovered that plants established better than seedlings, and seedlings were more successful than sown seeds. In the reintroduction of the endangered *Echinacea laevigata* in south-eastern USA, it was discovered that planting adult plants was more successful than setting out seedlings (Alley & Affolter, 2004).

The demographic cost may be lowered if steps are taken to nullify 'adverse' ecological factors. Sometimes it is not possible to remove such influences and steps are taken to try to minimise their effects. For instance, McEachern, Bowles and Pavlovi (1994) discovered that the use of insecticide reduced the level of insect herbivory in *Cirsium pitcheri* and greater numbers of seeds were produced from transplants. In other reintroduction projects, herbicide treatments have been applied to suppress weeds. For instance, *Amsinckia grandiflora*, a narrowly distributed Californian endemic species, two of the three existing populations were declining by 1991. A reintroduction project was initiated, by the US Fish and Wildlife Service, to establish three self-sustaining populations. At a new site, introduction was carried out as an experiment (Pavlik, 1994) with several different treatments: burning, grass-specific herbicide and hand clipping. 3,460 nutlets were sown and the new population was monitored regularly. Demographic studies revealed that over a thousand plants survived to reproduction and that competition from introduced grasses was most effectively reduced by herbicide treatment.

The restoration of endangered species has involved other procedures to protect the plants. *Stephanomeria malheurensis*, known from only a single locality in Oregon, USA, became extinct less than 20 years after its discovery. The area where it grew was damaged by fire in 1972, followed by an invasion of the aggressive introduced *Bromus tectorum*. Using plants generated from stocks maintained *ex situ* at the University of Davis, California, 1,000 seedlings, protected by rodent-proof exclosures, were set out in an artificially watered area cleared of *B. tectorum*. In the first year, 1,000 plants survived and produced 40,000 seeds: numbers fluctuated thereafter (Guerrant, 1992).

Turning to another case, Demauro (1994) reports that in the founding of a new population of *Hymenoxys acaulis* var. *glabra* in the Great Lakes area, 1,000 transplants were set out, a figure chosen with the MVP concept in mind. However, this proved to be inadequate, as 95% of the first transplants were lost through drought, and further plantings were necessary because of severe competition from introduced grasses (*Avena fatua* and three *Bromus* species).

The Torrey Pine (*Pinus torreyana*), the rarest pine in the world, is only found in California. Populations are conserved both in the Channel Islands National Park and in the Torrey Pines State Reserve – an area being 'engulfed by the spread of San Diego and its suburbs' (Ledwig, 1996, 265). Many factors are implicated in a serious decline in the species, including the Californian ips (*Ips paraconfusus*). This beetle can kill trees, especially those stressed by drought, disease or wounding. Lines of pheromone traps have been used to halt the advance of ips invasions. Given the difficulties facing the remaining populations of Torrey Pine, a reintroduction programme has been initiated using native stocks of seed. Both sowing and seedling transplants were employed. In total, 513 seedlings were planted out, each protected with a plastic sleeve to prevent animal damage. Seedling survival was excellent and Ledwig reports that 'restoration seems assured'.

The outcome of the reintroductions just outlined raises several important issues. Using all the plant material raised *ex situ* in a first attempt at reintroduction may be a mistake. Adequate reserve stocks must be maintained (Hunt, 1974; Falk *et al.*, 2006). Moreover, if various adverse habitat factors cannot be eliminated, it may be necessary to 'protect' the transplants from pests, herbivores or competitors. Should such protection be seen as part of the establishment stage only, or should it be a continuing management tool? This issue is highly relevant to the conservation and restoration of populations of the very rare Florida semaphore cactus (*Opuntia corallicola*). This species is threatened by the arrival of the cactus-feeding moth *Cactoblastis cactorum*, and cages have been erected to contain and protect the plant (Stiling, Rossi & Gordon, 2000). Will it be possible to grow the cactus in unprotected outplantings? Experiments have shown that, so far, no outplantings using fallen cactus pads were destroyed by the *Cactoblastis*, but many died of browning, probably by a plant pathogen, and some were damaged by trampling, probably by deer.

Co-ordinated efforts involving ex situ *and* in situ *conservation*

Often conservationists have no facilities of their own for *ex situ* propagation of plants, and botanic gardens are seen as providing an important role in creative conservation. For example, *Saxifraga cespitosa* is a very rare plant occurring on Cwm Idwal, North Wales. In 1975 the population of this montane species was reduced to four plants. From seeds collected on site, plants were raised in Liverpool Botanic Garden and in 1978 the population was restocked with 130 mature plants, 195 seedlings and 1,300 seeds. In 1980, there were 48 mature plants in the population (Parker, 1982). While the number of plants + seeds approached the size recommended for a MVP, there was only a modest recovery in population size. It is interesting that the stated intention was to produce a small population equal to the estimated size in 1796, when the species was first discovered.

Avoiding bottleneck effects in creative conservation

Geneticists emphasise the importance of maintaining genetic variability in managing populations of endangered species (Chapter 12). In the past there was no simple means of estimating genetic variability. Now, by using molecular tools there is the potential to monitor the genetic variability at all stages of the restoration process, including: any 'original' populations that have survived; samples taken from these and other populations for *ex situ* conservation; stocks in seed banks or cultivation; material planted out in the wild; and the fate of reintroduced populations (Falk *et al.*, 2006). Retrospective analysis of restoration projects reveals valuable insights into possible deficiencies in the sampling and propagation procedures used in restoration project. For instance, Williams and Davis (1996) discovered that restored beds of eelgrass (*Zostera marina*) were less variable than those that were undisturbed. Analysis of the restoration process revealed that the methods of collecting the material for restoration had produced an unintentional genetic bottleneck effect limiting variability, with possible consequences to long-term population fitness.

A second example also draws attention to some other issues of sampling to be faced in carrying out restoration. The endangered Mauna Kea Silversword (*Argyroxiphium sandwicense* ssp. *sandwicense*) has suffered severe population decline following the introduction of grazing animals on the island of Hawaii. Since the 1980s, conservationists have been carrying out a restocking programme to reinforce the existing population, and, by 1997, 450 plants had been propagated *ex situ* and transplanted onto Mount Kea volcano. However, molecular investigations of 90 RAPD loci detected 11 polymorphic loci in the original population, but only 3 in the out-planted population. Unintentionally, a population bottleneck had been created by the restocking process: all the out-planted population appear to have originated as F_1 or F_2 offspring of only two maternal founders. To redress this genetic imbalance, the authors recommended the addition of new genotypes to the out-planted population (Robichaux, Friar & Mount, 1997).

Problems were also discovered in conservation of the Corrigin grevillea (*Grevillea scapigera*) of south-western Australia, one of the rarest plants in the world with only 47 individuals in the wild. Using micropropagation techniques many plants were generated and, by January 1999, more than 200 were successfully transplanted to a secure site 'in the wild'. While at first sight the translocation process appeared to be a success, molecular studies revealed a number of troubling issues. Using AFLP molecular techniques, Krauss, Dixon and Dixon (2002) examined the genetic variability of the resulting populations. They discovered that 8 clones, not 10 as first thought, were represented in the new population, but worryingly 54% of the plants were of a single clone. Moreover, studies of the variability of seeds produced by these plants revealed that there were problems with inbreeding as 'the F_1s were on average 22% more inbred and 20% less heterozygous than their parents, largely because 85% of all seeds were the product of only 4 clones'. The results 'highlight the difficulty of maintaining genetic fidelity through a large translocation program'. The authors recommended that the genetic base of the translocated population should be widened and that population structure should be altered to encourage greater levels of outcrossing between different genotypes.

Does restoration require ongoing management?

Restoration is often designed and executed as a commercial enterprise of comparatively short duration. Site works are carried out to restore abiotic functioning, and seeds are sown, trees planted etc. The project completed, the contractors move on. However, as more conservation projects are being carried out, it is becoming clear that monitoring and continuing management may be required to ensure the future of the 'new population'.

In this regard, the fate of restoration projects in Hawaii is illuminating. Widespread deforestation has occurred, as land was cleared for cattle grazing and other agricultural activities. Nearly 10% (101) of species were made extinct and 358 species are threatened with extinction. Mehrhoff (1996, 104) notes:

> Introduced avian malaria and pox have decimated endemic Hawaiian honeycreepers and left most of the forests below 600 metres (2000 feet) in elevation virtually without native birds. The loss of bird pollinators and seed dispersers can only be guessed at, but it is likely to be a significant factor in the decline of some, at least, of the more than fifty extinct or endangered species found in this habitat.

Since the early 1900s, attempts have been made to counter the threats to endangered endemic species by transplanting experiments. Thus, between 1910 and 1960, foresters planted not only 78 endangered species back into the wild, but also stocks of 948 alien species (Skolmen, 1979). Some of the attempts at 'reintroduction' involved small numbers of plants, but in the restoration of uhiuhi (*Caesalpinia kavaiensis*) and kauila (*Colubrina oppositifolia*) larger numbers were transferred. Despite these interventions, Mehrhoff (1996) reports that none of these early transplantings survive to the present day. More recently, new transplantation programmes have been undertaken. In the case of *Chamaesyce skottsbergii* var. *skottsbergii*, endangered as a result of habitat destruction in the construction of a deep-water harbour facility, 218 nursery-grown specimens were out-planted to a new site, but all died. Subsequently, 748 plants were established at another site, but these also died as management was not maintained and weeds overtopped the transplants.

It is clear from early efforts that some restorations are comparatively inexpensive. However, reintroductions that require fencing and long-term weed and pest control may be very costly. For instance, in the case of reintroduction of uhiuhi (*Caesalpinia kavaiensis*) in Hawaii, 2.5 kilometres of fencing would be required to exclude pigs and goats, at an estimated cost of $30,000. This is not the only management needed. To clear fountain grass from the 'restoration' area would cost $9,000 per acre (total $900,000). This sum does not include the expenses of continuing management of invasive weeds, or the control of introduced rats that eat the seeds of uhiuhi.

How might success in reintroduction be assessed?

There has been considerable debate about how to determine whether a project has been successful (see Higgs, 1995). Perhaps some will judge a project to be a success, if an endangered species of concern has been successfully returned to the restoration site. However, others propose a sterner test, namely whether the restored populations have become self-sustaining populations, with both effectively functioning breeding systems and mutualisms, which do not require ongoing intervention (White, 1996, 81). Applying this standard, it is not clear whether any of the examples cited above have been completely successful. Thus, for some long-lived species it will be many years before they reach reproductive maturity. For instance, reintroduction populations of the Sargent's Cherry Palm (*Pseudophoenix sargentii*) have been planted in the Florida Keys in the period 1991–4 (Maschinski & Duquesnel, 2006). These reintroduced plants are still at the vegetative stage, as it takes more than 30 years for them to be sexually mature. Evaluating the outcome of restorations is a complex undertaking involving the analysis of the breeding behaviour of the new population with respect to pollinator visitation, inbreeding etc. (see Lofflin & Kephart, 2005).

We are clearly in the early days of programmes of species reintroductions/restocking etc.: time will tell how successful this element of creative conservation might be. With regard to future management of 'restored' species – both common and endangered – many more of the advances in our theoretical understanding must filter through to conservation managers designing, organising or carrying out projects. In many cases, well-designed long-term experiments are needed (Sutherland, 1998), especially to understand the possible effects of global climate change.

Complex ecosytems: understanding succession

Restoration ecology presents many challenges. In a few cases, whole ecosystems under threat have been transplanted to new sites. For instance, a 0.4 ha grassland in the Harz Mountains, Lower Saxony, Germany, was moved to a nearby area using a bulldozer with a special excavating shovel (Bruelheide & Flintrop, 2000). Over a 5-year period, the project was successful in conserving some rare species, but the maintenance of the original spatial relationships between the plants failed completely.

Generally, restoration and reintroduction involves the sowing of seeds or transplanting into new ground or established vegetation. The important issue here is whether it is possible to restore whole ecosystems? Experience shows that woodland, of sorts, can be established by planting trees (Ferris-Kaan, 1995), but it is a live issue whether a complete forest ecosystem can be recreated in all its complexity? For example, could such an ecosystem be reconstructed quickly on arable land by planting late colonist trees, shrubs etc. at the outset, or would it be necessary to plant/sow in sequence, particularly to establish the ground flora? Given the pressures to restore highly degraded areas quickly, could the very long processes of natural succession through which forest is created be speeded up by intervention?

The effective restoration of complex communities requires a clear understanding of the succession processes. Management and restoration of areas of conservation interest involves intervention in natural processes, to speed or direct successions. Is there an over-arching model of succession that can guide restoration?

In such interventions, the concept of there being a 'balance of nature' that should be maintained or restored is a view widely held by many conservationists (Budiansky, 1995). The 'balance of nature' metaphor is part of a classical view of the nature of communities, which considers that succession processes lead to a stable end-point climax vegetation appropriate to any particular area under study (Clements, 1916). Such systems were considered to be functionally complete in themselves and, when disturbed, they quickly return to the same equilibrium. Humans and their activities are not regarded as part of the system and, therefore, for many conservation activities the aim is to protect and maintain areas of natural climax vegetation, with humans excluded. The model also predicts that, even if the system is disturbed, if left alone, vegetation will return to the path of succession to the appropriate climax vegetation.

Detailed studies of the dynamics of communities have not confirmed the climax model, and the notion of the balance of nature is now dismissed by many as a 'myth' (Budiansky, 1995). Contemporary ecologists are more likely to employ another metaphor, namely the 'flux of nature' (Pickett, Parker & Fiedler, 1992), for evidence suggests that many succession pathways may occur with multiple persistent states of vegetation, rather than a single climax. Advances in our understanding comes from the work of the celebrated ecologists Gleason and Ramensky, who emphasised the individualistic behaviour of species and considered that the vegetation at any particular site was the result of the interactions of those species that arrived by chance and succeeded in establishing themselves (Shugart, 2001). Moreover, in their view, interactions between species produced continuously varying vegetation cover related to environmental conditions rather than distinct communities. The results of a great deal of subsequent ecological research emphasise the 'openness' of vegetation systems, pointing to the widespread occurrence of shifting mosaics and patchiness. Thus, ecologists are now more fully aware that the early notions of succession were simplistic. Instead of succession always following a predicted track, ecologists point to the hugely disruptive natural forces at work in the wild, such as fires, windstorms, floods, plagues of herbivorous insects etc.

In addition, detailed studies of succession have revealed the complexity of the process. The 'climax theory' of succession postulates the widespread occurrence of 'facilitation'. Thus, the primary stages of succession provide the necessary conditions – through the stabilisation of the terrain, the increase in fertility of the soil etc. – for colonisation of other species representing the later stages of succession. While there is clear evidence of such facilitation, from a Darwinian perspective it is difficult to understand why early successional species have evolved to allow or help other species to take over the sites they occupy. A considerable body of research now shows that facilitation is only part of the picture. Some species, once established, rather than providing the opportunities for the colonisation of others persist and flourish *excluding and suppressing* their competitors. 'Succession' in such circumstances only proceeds if the early colonists are damaged and replaced by other species with greater Darwinian fitness. Clearly, plant successions are complex and there is much still to be learned about succession processes in restored areas (Falk *et al.*, 2006).

While professional ecologists have rejected the classical climax paradigm, the notions of 'balance of nature' and climax vegetation are still influential in some conservation circles. The eminent conservationists Pickett *et al.* (1992) insist that these outmoded ideas should be abandoned. However, notions of the flux of nature do not provide a simple model for restoration projects, because a certain outcome is not predicted. Some conservationists are reluctant to abandon the climax theory, for it provides a clearer goal and simpler end point for their activities in managing and restoring vegetation.

Aims and objectives in restorations for conservation: differing views

It is important to stress that many possible models may be devised for the restoration of any particular tract of land, especially if the area is designed as a multiple purpose landscape, incorporating agricultural, conservation and leisure activities in an aesthetically pleasing setting (Fig. 16.3). However, if conservation management is to be a significant element (or indeed the major objective), two different approaches are influential – re-wilding and management of cultural landscapes. As we have seen in earlier chapters, each has strong historic roots, and each seeks a different outcome.

Re-wilding

In the first model, the aim of restoration is to restore human-disturbed ecosystems to their natural 'wilderness' state through appropriate management. This approach – sometimes referred to as 're-wilding' – is championed by some North American and European conservationists (Forman, 2004; Donlan, 2005, 2006). Re-wilding proposals generally focus on animals, but clearly they have major implications for plants and vegetation. From the outset, there are difficulties with the re-wilding approach. Evidence suggests that in the current post-glacial period there have been many different 'natural' ecosystems at any particular site, as vegetation expanded northwards to colonise land left by the retreating ice. Which of these 'natural' ecosystems is to be preferred as the model for restoration? There are other difficulties. Some of the species of these past ecosystems have become extinct, preventing the possibility of a fully authentic restoration. In addition, multitudes of alien species of animals, plants and micro-organisms have been introduced that were not part of the former natural ecosystems of the region. While, for some, there are insurmountable difficulties in recreating the original pristine wilderness, others are convinced of the possibility of restoring damaged landscapes to 'wild lands' that have the aesthetic properties of wilderness. This could be achieved by reintroductions, locally or regionally, of certain animals and plants, together with a withdrawal of almost all management. Thereafter, ecosystems would be allowed to change naturally. Turning to specific proposals and schemes, Donlan *et al.* (2005) point out that North America lost much of its megafauna in the post-glacial Pleistocene period (see Chapter 5) and note the grave dangers faced by

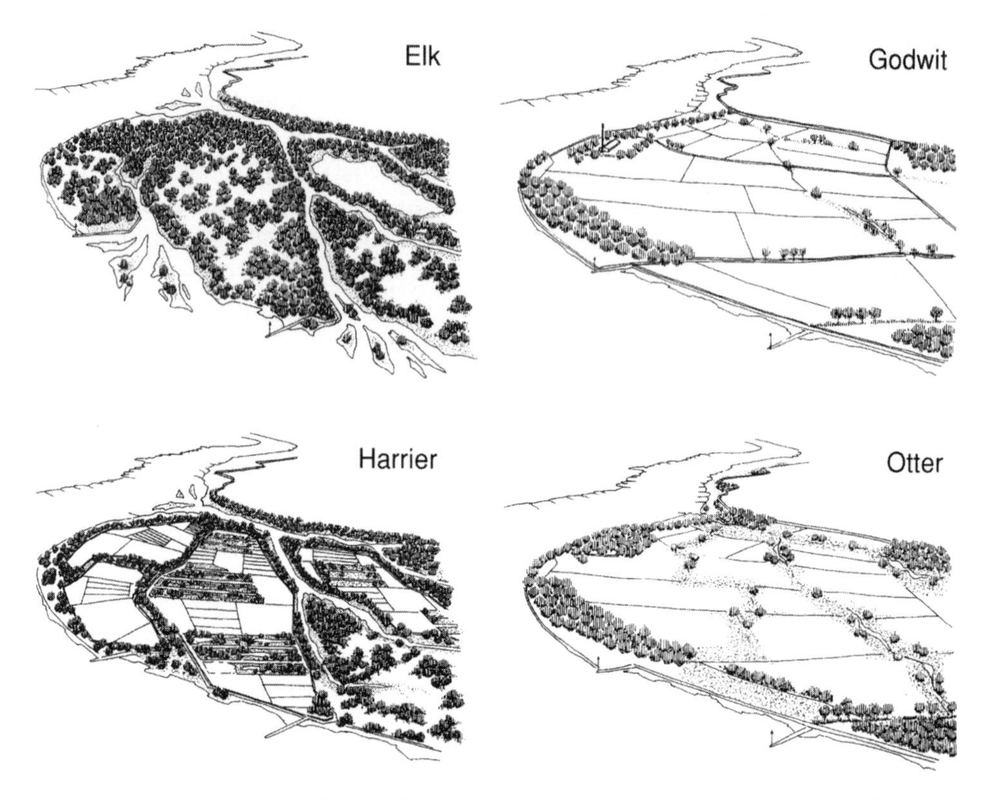

Fig. 16.3 Alternative scenarios for restoring a region in central Netherlands. Elk: restoring wilderness; Harrier: restoring a cultural landscape; Godwit: functional landscape to take account of the needs of multiple users; and Otter: taking elements from these three models, but with the objective of increasing the connectivity of elements of the landscape. (From Harms, Knaapen & Rademakers, 1993. Reproduced with kind permission of Springer Science and Business Media)

many extant populations of large animals in Asia and Africa. They also draw attention to the declining human populations in certain areas of North America (for instance the Great Plains); the great public interest in wildlife parks; and the successful restoration of wolf populations in Yellowstone. The re-wilding proposal seeks to recreate the Pleistocene heritage by introducing large animals – proxies of those that formerly lived in North America. Thus, they propose to introduce wild horses and Bactrian camels, and, more controversially, African cheetahs, Asian and African elephants and lions. The aim would be to create free-roaming populations of these megafauna, in 'ecological history parks' suitably fenced to reduce animal–human conflicts. Prototypes of such areas already exist in the USA. For instance, some '77,000 large mammals (most of them Asian and African ungulates, but also cheetahs, camels and kangaroos) roam free on Texas ranches' (Foreman, 2004; Donlan *et al.*, 2005, 914). Further details of the proposals are set out in Donlan (2007).

Restoration through re-wilding also has its supporters in Europe and elsewhere (Taylor, 2005). Such moves have been championed by the Dutch ecologist Vera (2000). He challenges the view that prehistoric European landscapes were totally dominated by forest. In his view the grazing pressures exerted by large herds of animals – bison, deer, wild cattle and wild horses – produced a dynamic savannah-like landscape of grasslands and woodlands. While his ideas have been challenged (see for example Birks, 2005), they have provided a model for re-wilding. For example, 20 miles from Amsterdam is the famous Oostvaardersplassen, which since 1968 has been a pseudo-wilderness. Management is dictated by the grazing activities of primitive breeds (Heck cattle and Konik ponies), creating a dynamic landscape of grasslands, marsh and woodland. Beaver have also been introduced, with plans to bring bison to the reserve (Marren, 2005). Rewilding is in progress in Ennerdale, Lake District, UK through felling exotic conifers, planting native juniper and broad leaves, and changes in grazing (more cattle/fewer sheep). At Wicken Fen Nature Reserve, the National Trust has ambitious plans to increase the size of the reserve. Konik ponies have been introduced, the eventual aim being 'to preside over a wilderness of re-created wetland managed not by conservation volunteers but by wild animals' (Marren, 2005). A database of re-wilding projects in the British Isles has been organised by V. Ward of Leeds University (www.wildland-network.org.uk). In other parks and reserves the notion of re-wilding is gaining momentum, with non-intervention being one of the possible management strategies, as, for example at Lady Park Wood, Gloucestershire (Peterken & Mountford, 1998).

Restoration of cultural landscapes

The second major model of restoration is very different, as is made plain by the following quotation from Hall (2000), considering Italian ecosystems. 'Over the last 100 years in the United States the goal of restoration has often been to *re-wild* the land excluding detrimental human elements, whereas the usual goal in Italy has been to *re-garden* the land by including beneficial human elements.' Such a view would be endorsed by ecologists living in other countries dominated by cultural landscapes. Thus, in countries where reserves and national parks were created from elements of cultural landscape, restoration has often had different aims, namely to recreate habitat conditions and promote the species associated with particular, usually traditional, land use.

Changes in management style: implications for microevolution

It is clear from these observations that there is no single concept driving conservation management. Management strategies have changed dramatically over time in longstanding reserves and parks, such as Yellowstone. This is not surprising as the earliest parks were established while the science of ecology was in its infancy, and managers were searching for the most appropriate way to manage the areas. Rackham (1998) has drawn attention to changing 'fashions' in management, e.g. plant trees, 'let nature take its course', intervene to promote particular species etc. Thus, there have been dramatic abrupt changes in many old-established reserves. For example, in the 1950s at the Beinn Eighe National Nature Reserve, Wester Ross, Scotland, a policy of natural regeneration was adopted in the management of the Atlantic Pine Forest. Later, management was more interventionist, but in the 1980s, the policy was non-interventionist. However, given the many changes in fashion in conservation management, it is important to stress that each change in style may drastically change the selection pressures to which species are subjected. Our knowledge of the genetics of endangered species is in its infancy, but it is clear that genetically depauperate populations may lack the genetic variation to respond to a succession of dramatic changes in their environment imposed unwittingly by humans, or deliberately applied by repeatedly changing the management plans of parks and reserves.

Restoration and management: gardening the wild

Management to recreate cultural landscapes has often been likened to gardening, and the management of endangered species within these areas to the raising of crops. Many conservationists have dismissed these ideas, but they are worth a closer look. We are familiar with the notion of growing crops in arable fields, grasslands and vegetable plots. Furthermore, many plants that some would consider 'wild' are part of the crop. Thus, while grasses predominate in hay, other species contribute to the nutritional value of the crop. Various species – sedge, rushes, litter – are also harvested as crops from wetlands. Likewise, 'bags' of game birds, fish etc. could be viewed as crops. The products of woodland management – fence poles, timber, charcoal etc. – could also be seen as crops. While some species are 'the crop' or 'part of the crop', there are also many fellow travellers in the different elements of cultural landscapes. Other species are not sufficiently threatening to be regarded as weeds or pests and they may be ignored. Also, some fellow travellers might have a minor role in cultural life as medicinal or food plants.

As we have seen above, 'resort to precedent' involves management and restoration to retain and encourage threatened and endangered species that were co-adapted or tolerant of former cultural practices by re-imposing the appropriate conditions. These interventions use precisely the same methods used in growing and harvesting the diverse crops noted above. If the argument just formulated is accepted, then wildlife may be viewed as a specialised crop. The end product is a desire, not only for so many tons of hay, but the continued existence of rare and endangered species in the different facies of cultural landscapes.

The notion of wildlife as a crop is clearly presented in the classic book by naturalist Richard Mabey (1980), *The Common Ground*, written at the instigation of the Nature Conservancy Council of the UK to 'widen public debate about nature conservation'. While conservation within parks and reserves is a key strategy of conservationists, the encouragement of wildlife outside reserves is crucial to the long-term survival of species. As much wildlife is found in traditionally managed farmlands, it is threatened by the imposition of highly mechanised farming supported often by state subsidy with its reliance on herbicides, fertilisers etc. Subsidies are often paid to farmers, compensating them for retaining wildlife on their land by not maximising their yields everywhere. Mabey (1980, 237) is keen to change this perspective. He writes: 'I think that we should present the financial incentives to conservation as *rewards*' and he goes on to make what to some will be a radical statement '*and promote the idea that wildlife is itself a vital crop*' [my emphasis]. Interestingly, steps in this direction have recently been taken in the European Union with changes to farm support. While some conservationists might resist these radical ideas, from the microevolutionary perspective, there is merit in the notion that endangered plant species are being managed as a specialist crops by methods akin to gardening.

The idea that wildlife could (or should) be viewed as a crop is not new. In the 1920s, the Earl of Onslow was a major figure in the early conservation movement in Africa. As Adams (2004, 217) notes, the world Onslow knew was 'of pheasants, grouse and deer, of society house parties and well-managed rural estates' and he saw national parks 'as analogous to shooting estates'. Thus, wildlife cropping has been part of the conservation management of several areas including national parks (Adams, 2004, 218–20). For instance, from 1959, hippopotamus were culled in Queen Elizabeth National Park in Uganda, in an attempt to control soil erosion. The meat from these animals was then sold.

In the 1960s, the British Nature Conservancy was responsible for advising the government on colonial territories in East and Central Africa (Adams, 2004, 219–20). Following field studies in the region, it was recommended that 'wildlife should be regarded and managed as a resource' through 'game cropping and controlled hunting'. These ideas have been very influential in post-colonial times. For instance, in Zambia in the 1980s, the fees from controlled hunting were used to support local development as well as 'game guards'. In the 1990s these ideas were incorporated into a wider concept of 'community-based natural resource management'. Adams notes that 'the approach has been widely used, although with very different degrees of community involvement'. But he considers that such operations may be 'over centralised' and confer little benefit and make only a 'minor contribution to the economy' of local villages. As the 'cropped meat is expensive for poorer households, illegal hunting continues to be extremely attractive and economically important'.

The notion that wildlife could be treated as a specialist crop raises important issues. In his studies of evolution, Darwin considered domestication of plants and animals, and contemplating this he concluded: 'man . . . may be said to have been trying an experiment on a gigantic scale'. It is the contention of this book that Darwin's observation should be recast to include the present tense and to extend the scale of the experiment to include entire cultural landscapes. More and more areas of the world's surface are being influenced decisively by human activities: and in addition to natural selection processes, ecosystems are being impacted by human-induced selection pressures. Taking the argument a step forward, Darwin defined a role not only for conscious deliberate selection in domestication, but also stressed the effects of unconscious selection (now often called unwitting selection). As we have seen above, there is increasing convincing evidence that human activities present plants with 'new' selection pressures in agricultural (as well as industrial) landscapes. While the selective effects of conservation management in plants have yet to be examined, there can be little doubt that intensive management and rapid changes in management regimes exert powerful selection pressures on populations. Support for this notion comes from a study of the unwitting evolutionary consequences in trophy hunting of bighorn sheep (*Ovis canadensis*) (Coltman *et al*., 2003). Over a 30-year period, body weight and horn size have declined in this species as a consequence of trophy hunters targeting animals with the largest rapidly growing horns. Rams with large horns have been removed from the population before they could achieve 'high reproductive success', with the consequence that rams are now lighter, smaller-horned and fewer trophy rams are found in the wild.

Turning to another example, studies have been made of populations of African elephants in South Luangwa National Park and adjacent game management areas in Zambia. Jachmann, Berry and Imae (1995) consider that the increasing incidence of the heritable condition 'tusklessness' in female elephants, that rose from 10% in 1969 to 38% in 1989, was apparently the result of selective illegal ivory hunting. More recently, tusklessness in females has declined to 28% in 1993, as a result of migration from game areas into the park. However, in the Addo National Park, South Africa, historic records suggests that the high incidence of tusklessness is not attributable to hunting, but is more likely to be the result of non-selective influences – genetic isolation, drift and bottleneck effects (Whitehouse, 2002).

As a working hypothesis for further study, populations of plants and animals subjected to intensive human management will be subject to unwitting microevolutionary change. Given our present state of knowledge, it would seem impossible through management to keep plants and animals in 'a state of nature'. As a consequence, managed species, including those that are endangered, are likely to occupy or will come to occupy the ambiguous territory between the truly wild and the fully domesticated. Plant species will be closer to the domesticated pole, if they have been subject to longstanding intensive management through several generations in reserves. As we have noted in Chapter 14, incipient domestication is also a risk in *ex situ* conservation, not only for stock raised on wildflower farms, but also for reserve stocks of material and plants raised for restoration projects that have been propagated through several generations from seed in botanic gardens.

Concluding remarks

In closing this chapter, it is important to make a number of key points from the perspective of microevolution. Firstly, two dominant models have been devised to manage reserves and parks – re-wilding and restoring cultural landscapes. Reflecting on these models in the context of microevolution, it is important to note that *both* approaches involve very considerable long-term human intervention in management, and the notion that national parks are untouched wilderness areas separate from cultural landscapes must surely be abandoned. Indeed, there is a case for regarding reserves and national parks as *part* of cultural landscapes, deserving of special attention to conserve endangered species and ecosystems.

The above account has focused on officially managed reserves. However, many reserves are essentially parks in name only – so-called paper parks. Terborgh (1999) in a survey of the state-of-play of parks in tropical countries reported that while some were successfully managed, others were under-funded, under-staffed or subject to civil strife, with illegal logging, hunting, mining and land clearance for agriculture etc. From the perspective of the conservationist, paper parks are often grossly '*mismanaged*', putting their valuable biodiversity at risk. However, looking at the situation from a different 'microevolutionary' perspective, paper parks are being subjected to '*different management*' with profound consequences for the survival of endangered ecosystems and their biota.

From a microevolutionary perspective, management and restoration activities in parks and reserves consist of identifying those human activities that are endangering species through the imposition of new selection pressures. Management then seeks to counter these adverse selection pressures. Sometimes single factors may be of supreme importance and, as we have seen in Chapter 15, strategies have sometimes successfully counteracted adverse selection pressures. However, in other cases, multiple interacting factors might be involved, presenting much more complex issues in management with many conflicts of interest. In addition, areas of conservation interest are greatly influenced by human-induced changes originating from outside the reserves – in the form of altered hydrology, pollution and invasive species etc. These present much more complex selection pressures, and counteracting these might not be possible, or may only be partially successful through regional or international agreements and actions. With regard to the reintroduction or reinforcement of populations of endangered plants, can ecologists be certain that the appropriate conditions have been fully re-established and can they be maintained?

Turning to the conservation of individual species, action plans have been drawn up to 'rescue' endangered plant and animal species; for example, in Britain, Species Recovery plants are being funded by English Nature. There are difficult questions to face in the restoration of plant populations of rare and endangered species. Is the aim to maintain rare species as rarities? Given our current understanding of the genetics of very small populations, it is clear that such populations may easily decline further, leading to genetic depauperation and loss of evolutionary potential.

Considering the prospects for enacting specific plans to secure the future of the huge number of endangered plant species, it is important to face the fact that much of the conservation funding is directed towards charismatic animals, mammals, birds and butterflies. The notion that conservation of animals will somehow *automatically* secure the future of threatened plants etc. must be strongly questioned, especially if ecologists focus too closely on the plight of individual animal species. Given the widespread occurrence of mutualisms and other interactions within ecosystems, the prospects for all endangered species will be increased if the management of animal populations for conservation is more holistic (see Chapter 17).

In the last two chapters, the concept of 'resort to precedent' in the management and restoration of reserves in cultural landscapes has been discussed. By such means endangered species and ecosystems may be restored through the re-imposition of such treatments as burning, grazing, coppicing, haymaking etc. Also, more complex management may be imposed through creative conservation. A microevolutionary question of great significance is whether conservationists can find the funds and public support to manage, repair and maintain these different human-made communities *in perpetuity*, through affluent times, as well as periods of national crisis. Will the management of cultural landscapes command public support indefinitely, especially if some endangered species are lost?

This chapter has considered the theory and practice of habitat and species restorations. A major issue to be faced is the fact that many conservation projects appear to be concerned to recreate something that existed in the past. Given that we have an imperfect knowledge of past ecosystems, and many new selection pressures are at work in contemporary communities, it is surely impossible to recreate the past. Moreover, the insights obtained from studies of microevolution force us to consider the future, and consider the 'fitness' of populations of species, and the 'adaptability' of communities in a changing world. Thus, while there is a case to be made for using 'matching' stocks of plants in restorations, it may be that 'mixed' stocks offer greater microevolutionary potential.

Concerning restoration, many searching questions must be asked. Is it possible to plant an area with most of the necessary plant species? Will the full complement of associated species of animals, plants and soil micro-organisms become re-established naturally or should steps be taken to manage these processes too? If so, is it possible to manage soil development, plant successions and animal migrations to produce an end product indistinguishable from the natural? In the present state of knowledge, it would seem impossible to recreate precisely the complex ecosystems we think existed in the past. What can be achieved is likely to be through 'naturalistic plantings' (Cairns, 1998, 218). If such areas are provided, perhaps much wildlife will arrive through 'natural' dispersal. This concept is widely held in conservation circles, especially by ornithologists, and is expressed by the view that 'if you build it, they will come' (Stockwell, Kinnison & Hendry, 2006, 130). Judiciously, other animals and plants can be introduced.

Overall, despite these reservations, restoration has huge potential to change ugly damaged unproductive wastelands, polluted wetlands and rivers into aesthetically pleasing landscapes that provide various services for mankind, such as clean water, forest products etc. Some sites are better candidates for restoration than others. Miller (2006, 356) has recently pointed out that restoring large reserves may be possible in rural areas, for instance to prairie, but he considers that the restoration of small reserves may be 'an unrealistic goal, however, in the midst of urban development and rapidly escalating prices'.

Of particular significance in restoration ecology is the question of mitigation. In order to secure permission to undertake development that will damage or destroy areas of conservation importance, developers sometimes offer something in mitigation. For instance, a proposed development might damage or destroy a wetland of conservation significance, in which case, developers might promise to recreate a new wetland elsewhere on a different piece of land. Given the complexity of ecosystems, the general public, politicians and those concerned professionally with landscape planning should be wary of claims that it is possible to recreate a new wetland, and perhaps try to protect the original threatened wetland. While much progress has been made in restoring ecosystems, some are concerned that fake communities are being created, especially in circumstances involving mitigation (Elliot, 1997).

Another point of microevolutionary significance is that restoration will often create new habitats without exact counterparts in nature, thus, 'wildlife friendly' habitats have been created in the 'restoration' of quarries, landfill sites, opencast mines etc. In many areas of Britain, former gravel workings are often restored to provide wetland and aquatic habitats, together with recreational activities such as fishing and sailing. Other 'wildlife friendly' habitats are also being produced in the development of more naturalistic enclosures in zoos and large-scale ecological areas in botanic gardens and arboreta. Such developments raise interesting microevolutionary questions. We may choose to make a distinction between the highly managed nature reserve and the large 'naturalistic' enclosures in wildlife parks and gardens being devised to exhibit plants and animals in naturalistic ecological settings, but from the plant's-eye view there may be little difference (Cooper, 2000, 1135).

Having discussed the origin, management and restoration of national parks and reserves, the next chapter considers the microevolutionary significance of the fact that reserved lands are set in a matrix of lands used for other purposes.

17

Reserves in the landscape

As we saw in Chapter 15, the first national parks were developed in the nineteenth century. They were the forerunners of many nature reserves and wildlife sanctuaries. As the conservation movement has gathered momentum, *in situ* management of reserved areas has become the key strategy employed by conservationists in their efforts to secure the future of biodiversity. There are many types of reserves of different sizes: all are fragments within a matrix of surrounding territories that might have different land use. Considering the size, geographical location and 'connectedness' of lands set aside for conservation, how far has it proved possible to secure the future of endangered species and ecosystems? Also, what 'strengths and weaknesses' have been identified after more than a century of experience in managing reserves?

Reserve design

The theory of island biogeography proposed by MacArthur and Wilson (1967) drew on important earlier research by Preston and others (see Mann & Plummer, 1996). The theory explored the implications of the observation that species richness tended to be low on small isolated islands far from land. In contrast, biodiversity was often greater on large islands close to the mainland. On the basis of these ideas conservationists proposed a number of model systems for reserves and national parks.

MacArthur and Wilson (1967) examined the species/area relationships by considering the extinction of species on islands in relation to immigration from the mainland and elsewhere. They concluded that, as the dispersal distance is smaller, there would be a higher rate of immigration on islands closer to the mainland (or close to other islands) than to more isolated islands. In addition, a larger island would support a larger number of species and have a higher probability of intercepting immigrants than a smaller island. Therefore, for any given island, with the passage of time, an equilibrium would be established between immigration and extinction. Mannion (1998, 378) stresses: 'the resulting equilibrium is dynamic in so far as species are continually arriving and continually becoming extinct, though as time progresses the rates of arrival and extinction diminish'. Thus, in comparative terms, other factors being equal, large islands near the mainland will have the highest biodiversity. MacArthur and Wilson also introduced the concept of a 'rescue effect'. Any

Fig. 17.1 Farming landscape in Wisconsin, USA: fields and 'island' woodlots. (From Richard C. Davis (ed.) *Encyclopedia of American Forest and Conservation History* (2006) © 1983, Gale, a part of Cengage Learning Inc. Reproduced with permission, www.cengage.com)

species becoming extinct on an island could be rescued by the arrival of individuals of the same taxon arriving from the mainland or another island source.

The theory of island biogeography was developed to investigate species/area relationships on oceanic islands. Conservationists were quick to see a parallel with terrestrial nature reserves and other fragmentary habitats (Figs. 17.1 and 17.2), which are, in effect, island ecosystems within a sea of developed lands (Diamond, 1975, 1976; Diamond & May, 1976; Shafer, 1990). The results of their investigations have provided a series of predictions about reserve design, size, location etc. (Fig. 17.3). These 'better' and 'worse' situations relate to the number of species in a particular area and do not address the question of what would be judged to be best/worse for any particular species. Firstly, a number of issues about individual reserves will be considered and then the relationships between areas will be examined.

The application of the theory of island biogeography to conservation

A reduction in the size of any ecosystem, such as the Amazonian rain forest, will lead to extinctions (Mann & Plummer, 1996). In considering the implications for biodiversity in reserved areas such as national parks, it is important to recognise that such areas are

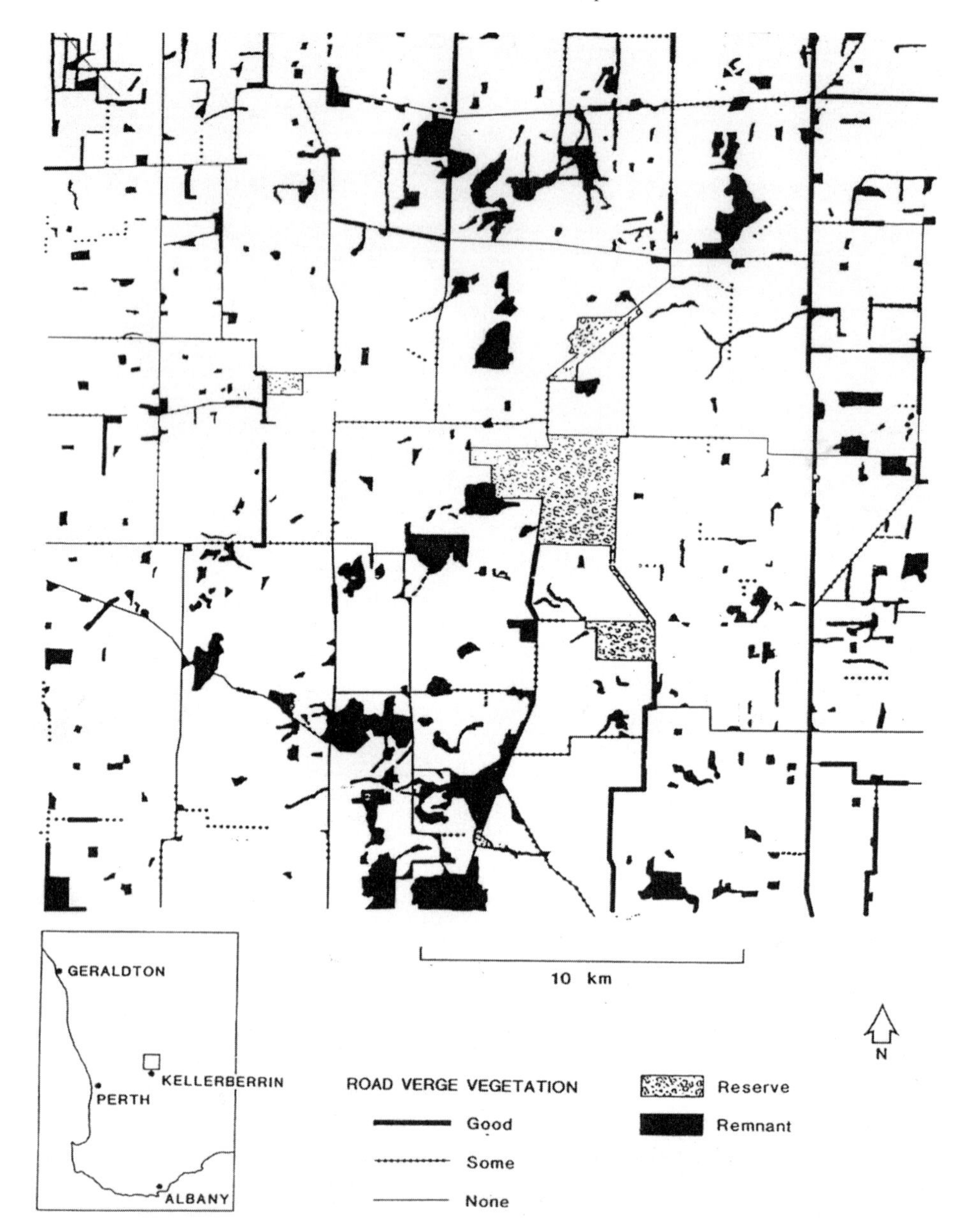

Fig. 17.2 A section of the landscape of the Western Australia wheat belt, north of Kellerberrin, showing the locations of reserves, and fragments of the 'original' vegetation that survive in fields and along roadsides. (From Hobbs, Saunders & Arnold, 1993. Reproduced by permission of Elsevier Ltd.)

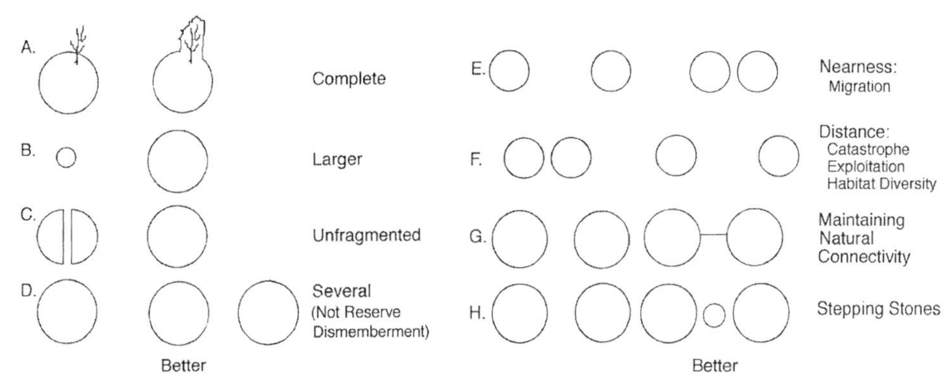

Fig. 17.3 Reserve design. It is proposed that the option on the right is better than the one on the left. A: It is preferable that complete watersheds, migratory routes and feeding grounds are within the reserve. B: Larger is better than smaller. C: Unfragmented is better than fragmented. D: It is preferable to have more than one reserve, as this guards against catastrophe and human exploitation, and may provide habitat for patchily distributed species including endemics. E: Nearness is better than being farther apart, because it allows greater opportunity for migration, providing that the landscape is traversable by a species of interest. F: However, greater distance may sometimes be better as this may reduce the effects of catastrophes, disease etc. G: Maintaining natural connectivity through effective usable corridors is far better than no connection. H: Small stepping stone reserves, if exploited by a species of concern, are better than none at all. (From Shafer, 1997, with kind permission of Springer Science and Business Media)

fragments of what was once a larger ecosystem. In all situations, it follows that the reduction in area will lead to extinction of some species, a process known as 'relaxation'. This hypothesis applies to very large parks, e.g. Yellowstone, as well as to very small reserves, and it has been tested in a number of investigations (see Collinge, 1996). Here we examine a number of examples.

During the construction of the Panama Canal, which was opened in 1914, a dam was constructed on the Chagres river and rising water levels created the Barrro Colorado Island, now administered as a reserve by the Smithsonian Tropical Research Institute. Willis (1974) documented the losses of bird species from this new island, which is a fragment of a once much wider ecosystem. Newmark (1995) has examined patterns of mammalian extinction in western North American national parks. He discovered that the 'number of extinctions has exceeded the number of colonisations since park establishment' and 'the rate of extinction is inversely related to park area'.

Species loss from reserves has also been investigated by botanists. By comparing nineteenth century records with those of the present day, Turner *et al.* (1994) discovered that, from a fragment of once-widespread tropical rain forest conserved as part of the Singapore Botanic Gardens, significant losses in plant species had occurred. Evidence suggested that species with shorter life cycles were more extinction prone than longer-lived species. Turner (1996, 200) reviewed investigations on the fate of other forest fragments and found, in nearly all cases, 'isolated fragments suffer reductions in species richness with time after excision from continuous forest'. However, the relaxation times for tropical forest fragments could be extremely long, as radiocarbon dating and investigations of tree rings have established that many tropical tree species may be more than a thousand years old (Chambers *et al.*, 2001).

Comparison of species lists also supports the concept of relaxation in temperate ecosystems. Using this method, Drayton and Primack (1996) examined changes in the western 4,000-acre part of the Middlesex Fells Woodland Park in Boston, USA, in the period 1894 to 1993. Of the original 422 species, 155 were no longer present – a loss of *c.* 0.36% per year over the last century. Sixty-four new species arrived in the area, the majority being 'exotics'.

While these examples are consistent with the concept of 'relaxation', a note of caution is necessary. The reserved lands in question have been influenced not only by their location and area at the time of their designation, but also by many anthropogenic factors. For example, over the study period there has been a marked 'increase in human activity' at Middlesex Fells. Easier access is provided by a greater number of trails and roads, leading to more fires. There has also been 'deliberate thinning' of the forest canopy to make the area more beautiful, and, with increased use, there has been greater trampling of the ground vegetation etc. An additional point must be stressed concerning processes. At conservation sites around the world, immigration/extinction processes are not given free rein: there has been a great deal of deliberate management to prevent the extinction of rare and endangered species and to control immigrant species, especially introduced taxa.

Nonetheless, on the basis of species/area considerations, it seems clear that some species in reserves are destined to become extinct. They have been characterised as the 'living dead' (Janzen, 2001). Likely candidates are isolated single individuals of self-incompatible or dioecious species marooned in a fragment beyond the range of natural gene flow from their nearest con-specific populations and, therefore, unable to reproduce sexually. However, investigations have shown that it is unwise to make generalisations: assumptions about lack of gene flow may not be correct. For instance, a molecular study of the tropical tree *Swietenia humilis*, using micro-satellite markers, discovered that remnant isolated trees were successfully pollinated. In addition, as we saw earlier, many species may persist for long periods through vigorous asexual reproduction. A good example is provided by the rare multi-stemmed *Eucalyptus argutifolia*, which is found in only 15 widely disjunct populations in Western Australia (Kennington *et al.*, 1996). Allozyme studies and molecular investigations, using the M13 DNA fingerprinting techniques, revealed that the plant produces large clonal patches extending over wide areas, the largest detected being 160 m^2.

Fragmentation

Whether any particular ecosystem or species will survive depends upon a number of factors, including the size of the fragments, the variety and area of different habitats it contains and the precise detail of the fragmentation process, which may have been extreme in many parts of the world. For instance, in the state of Illinois, USA, there were 21,000,000 acres of tall grass prairie at the time of European settlement (Robertson, Anderson & Schwartz, 1997,

63). The exploitation of the prairie lands with their deep soils full of densely tangled roots depended upon the development of the self-scouring steel-bladed plough by John Deere in 1837, and the opening of the railroads in the 1850s and 1860s. In the following decades, the prairies were rapidly converted to croplands and now there are only about 250 high-grade relict tall grass sites. Significantly, 83% of these are less than 10 acres with 30% less than 1 acre.

Returning to questions of reserve design, some species may be lost immediately because the settling of park boundaries decreases ecosystem heterogeneity and their habitat may disappear entirely (Gascon, Laurance & Lovejoy, 2003). In part this is because the setting of park boundaries is a highly political issue, as potentially exploitable resources are involved. As we saw earlier, reserves have generally been established on what were considered 'useless' lands at the time of the founding of the park (Runte, 1997). Thus, parks have often been delimited in such a way as to exclude the areas with accessible forests, recoverable mineral deposits and soils of high fertility etc. In some cases the politicians have hedged their bets. For instance, when the Wollemi National Park was established in Australia in 1979, the lands were 'protected' to a depth of 400 m to allow the possibility of coal mining from pits situated outside the reserve (Mercer, 1995).

The effects of fragmentation through deforestation have been critically studied in the Biological Dynamics of Forest Fragments Project (BDFFP) carried out in Amazonia, 80 km north of Manaus, Brazil. The project investigated a series of isolated replicated forest reserves of 1, 10 and 100 ha created by clearing the forest (Bierregaard *et al.*, 2001). Commenting on patterns of deforestation in the Amazon basin, they make the very important point that deforestation is not spatially random, but depends upon soil fertility, topography and ease of access by river or road.

Turning to another issue, it seems likely that many species have complex metapopulation breeding structures, and these may be damaged or lost if only a portion of an ecosystem remains as a 'reserve', and all populations beyond the boundary are destroyed. In situations where change is less dramatic, populations of the same species may survive in adjacent lands outside the park. However, it would be unwise to assume that metapopulation structure will be maintained, as there is evidence that elements within the anthropogenic transformed matrix – roads, farmland, urban developments – may represent effective barriers to gene flow between populations (see below).

Impact of fragmentation

Much more research is necessary to understand the manifold effects of fragmentation and its aftermath, for a major conclusion of the BDFFP was that the ecosystems of fragmented areas were 'substantially altered', especially if top predators – such as the harpy eagle and jaguar – were no longer present. 'Cascade effects' follow, impacting species down the food chain, indeed throughout the entire ecosystem. In addition, there is evidence that crowding effects may occur as species of birds displaced from the deforested area move into the remaining forest fragment(s) (Bierregaard *et al.*, 2001).

Whether a particular species has a secure future within and outside reserve boundaries depends, crucially, on whether all the conditions necessary for establishment, growth and reproduction are satisfied in the reduced area, including the presence of other species on which these processes depend. Thus, the survival of many plant species will depend on the maintenance of the plant–animal mutualisms necessary for pollination, seed dispersal etc., and the fragmented forest areas may not contain all the ecosystem elements necessary for these mutualists. Thus, some animals require different ecosystems for feeding and roosting, and make daily movements between them. Some species migrate to different areas seasonally to reach feeding or breeding grounds in non-reserve territory. These requirements must be met for the pollinators/seed dispersers to survive. Summarising the results of the BDFFP the authors concluded that the Amazon forests had many 'fragmentation sensitive' species and pointed to the 'reluctance of many animal species to leave the forest cover'. Many forest-dependent animals 'avoid even narrow (less than 100 m) clearings'. Indeed, some would not cross the 'narrowest (less than 50 m) forest clearings'. Thus, the cleared ground associated with roadways and power lines presents 'significant barriers' to some animal movements. Recovery might be slow or uncertain in plants and animals that are poor dispersers, have low fecundity, or very exacting habitat requirements. The BDFFP also revealed that the nature of the matrix around forest fragments was of the utmost importance in considering their fate and ecosystem dynamics. Fragments surrounded by re-growth of forest were likely to be closer to the original vegetation cover than those set in an 'inhospitable' matrix of agricultural lands.

The matrix surrounding reserves

It is important to stress that the matrix varies greatly from place to place. Fragments and matrix should be viewed as part of the evolving dynamic cultural landscape of the people who live there. In the mid-western USA, the 'semi-natural matrix has undergone a profound conversion from grazed open woodlands to closed woodlots bounded by intensive agriculture'. Thus, secondary woodland has developed 'as a result of farm abandonment' leading to 'large changes in species composition and large declines in species diversity. At the same time remnant prairies have been almost eliminated, leading to large declines in native grassland species. Orchards, hedgerows, and brushy, tree-filed fencerows, once a habitat of major importance in supporting wildlife have been greatly reduced because they interfered with the efficient operation and mechanization of farms as row crops were increasingly planted' (Brown, Curtin & Braithwaite, 2003, 331).

Fragmentation may yield mosaics of different land use, e.g. grasslands managed for pasture/hay; hayfields given different fertiliser treatments; different types of grasslands on sports fields and golf courses; contaminated/non-contaminated land (Vos & Opdam, 1993; Hansson, Fahrig & Merriam, 1995). As we saw earlier, population differentiation has often been detected in mosaic areas. Sometimes 'crisp' patterns of differentiation are detected. But in other cases, more complex patterns of ecotypic genetic variation have been discovered (Bradshaw, 1959b).

Molecular studies have revealed that different populations existing as patterns of fragments may each have a different spectrum of genetic variation. For instance, rRNA gene copy number has been examined in eight populations of *Pinus rigida* sampled from the Pine Barrens of New Jersey (Govindaraju & Cullis, 1992, 133). rRNA plays a central role in protein synthesis. It was therefore of great interest that 'numbers of rDNA copies were found to be lower among populations subjected to environmental stress' such as local fire regimes. It seems possible that variants with different rates of growth and development may be at a selective advantage in different areas.

As we saw in Chapter 12, the relationship between population size and genetic variability has been examined. How do these variable relate to fragment size? White Box (*Eucalyptus albens*) occurs in south-east Australia, where forested areas have been fragmented by agricultural development (Prober & Brown, 1994). Population sizes are closely related to fragment sizes. Genetic investigations revealed: 'the number of alleles, polymorphism and heterozygosity at 18 allozyme loci were all positively correlated with the number of reproductively mature individuals within populations which ranged from 14 to >10,000' (from Young (1995), who has provided a valuable review of other investigations of the landscape and genetic variability in plants). Thus, the small populations of White Box typical of the smaller fragments were less variable. This lowered genetic variability in small populations could put them at risk of genetic erosion and eventual extinction, especially if fragments were so far apart that gene flow did not occur.

Edge effects

In Chapter 15, the effects of external influences on reserves were considered. Here, our concern is with factors impacting on the 'raw' edge of a fragment exposed to whatever changes occur in the matrix. Edge effects have been investigated in many regions of the world, including the BDFFP (Bierregaard *et al.*, 2001). As part of this project, the depth of penetration of such effects into forest remnants was examined (Fig. 17.4).

A first point to examine is the assumption behind model systems. A newly isolated fragment of natural vegetation has a perimeter exposed to anthropogenic effects, and therefore larger reserves are to be preferred (Fig. 17.3), as they have a larger heart of unimpacted ecosystem than small reserves and support a large number of species. In order to protect this central unimpacted zone the ideal 'biosphere reserve' should have a buffer zone, with human dwellings, agriculture etc. activities restricted to a further zone beyond the protected interior. However, as we have seen in earlier chapters, in reality many apparently pristine ecosystems have been or are currently impacted by human activities. Considering the history of the sites, the notion of a pristine unimpacted wholly natural preserved ecosystem at the heart of the reserve may not be sustainable. In many cases the whole of the reserve is or has been subject to historic anthropogenic disturbance.

Returning to the BDFFP investigations, partial destruction of tropical rainforest gives rise to forest fragments, the edges of which are newly exposed to changed microclimatological conditions – increased light and temperature, decreased humidity etc. (Fig. 17.4). Trees

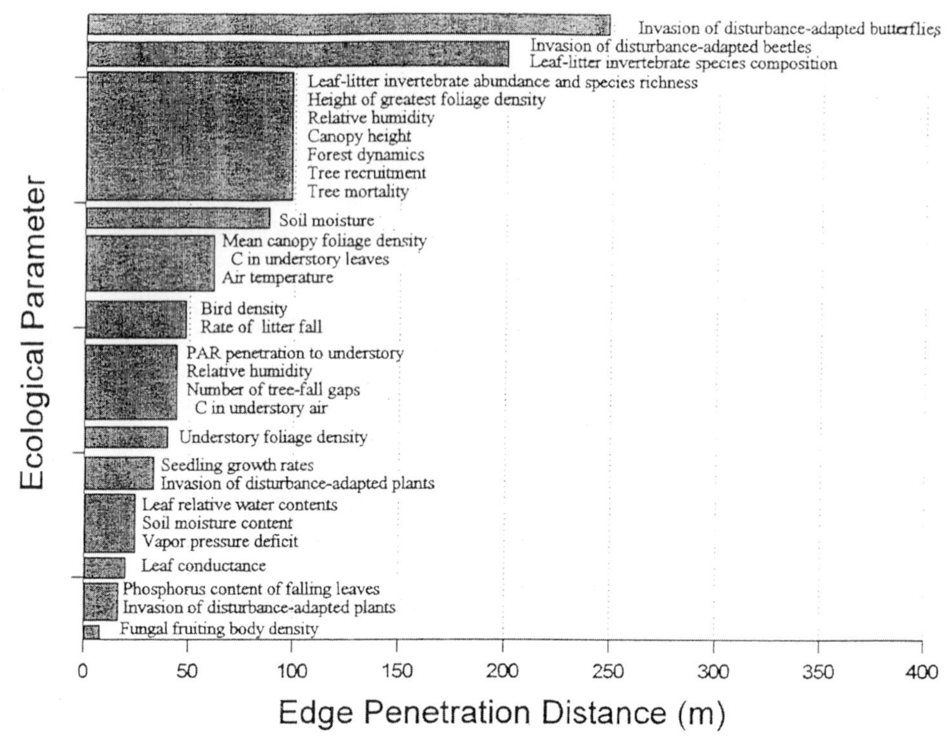

Fig. 17.4 Penetration distances of various edge effects into forest remnants recorded in the BDFFP studies. (From Bierregaard *et al.*, 2001. From *Lessons From Amazonia: The Ecology and Conservation of a Fragmented Forest*, Fig. 29.1 © Yale University Press)

that were once within a continuous canopy are exposed to storm damage at the raw edge. Also, there are many edge effects involving plants and animals, including the invasion of disturbance-adapted biota, changes to seedling recruitment, fungal fruiting bodies and changes to the soils.

Investigations of tropical forest at other sites have produced interesting results. Studies of tropical forest in north-east Queensland, Australia, have revealed that changes to the canopy can be detected up to 500 m into the fragments with very obvious changes up to 200 m (Laurance, 1991). They also discovered that edges of fragments were characterised by 'exceptional abundance of heavy lianas, climbing rattens (*Calamus* spp.) and other disturbance-adapted species (*Dendrocnide* and *Solanum* spp.)'.

Edge effects have also been described in temperate ecosystem fragments (Forman, 1995; Collinge, 1996). Strong microclimatic changes have been detected at the edges of forest fragments in Wisconsin, Indiana, Pennsylvania and the Pacific North-west. In addition, there is evidence that edge effects vary according to the direction of prevailing winds and particularly with regard to aspect and orientation. In the northern hemisphere, south-facing edges are generally drier, warmer and wider than north-facing edges. Near to suburbia, edge effects in forested areas arise from damage to trees in the collecting of firewood, and the fly-tipping of grass clippings, garden plants, Christmas trees, builders' rubble etc.

Turning to other examples, many reserves are embedded in agricultural land and may be subject to pesticide spray drift and eutrophication by fertilisers, e.g. the small chalk grassland reserves such as Devil's Dyke, Cambridgeshire. This site is also a favourite place for local people to exercise dogs, a factor leading to further nutrient enhancement of the soils. In another investigation it was discovered that roadsides through heathland in the New Forest, UK, are characterised by an enhanced growth of vascular plants, a predominance of grasses and adverse effects on lichens, probably as a consequence of eutrophication through the deposition of nitrogen compounds from vehicle exhaust systems (Angold, 1997). Studies of the 'road effect' zone have also been carried out at the edges of a four-lane highway west of Boston, Massachusetts (Forman & Deblinger, 2000). They discovered that very obvious effects could be seen more than 100 m from the road, but roadside avoidance by grassland birds and the effect of salt was apparent outwards to a distance of over 1 km. Overall, the road zone averaged *c.* 600 m with convoluted boundaries and a 'few long fingers'. They concluded that 'busy roads and nature reserves should be well separated'.

The conclusion from these and other studies is that all habitat fragments are subject to edge effects. Moreover, as fragment size increases, the proportion of the area impacted by edge is reduced (Collinge, 1996). In addition, shape is a critical issue. Thus, long thin fragments have a greater proportion of their area impacted by edge effects than square fragments. Indeed, as in the case of Devil's Dyke, the whole of the nature reserve is subject to such effects.

Recognising the problems of edge effects, including human impacts at the reserve–matrix boundary, it has been suggested that, ideally, all parks and reserves should be surrounded by a buffer zone. For example, Shafer (1999a) considers the biological, social and legal aspects of proposals to create buffer zones round national parks in the USA. There was a strong call, in the 1930s, for buffer zones to be declared around parks as 'human impacts such as poaching' were a problem. While, in some cases, there were changes in the use of adjacent Federal Lands, in others, the obstacle to the creation of buffers was that land adjacent to the park was in private hands, and the 'social climate' opposed the intrusion on the 'rights of private landowner'.

The proximity of reserves to migration corridors

A great deal of debate has centred on the question whether it is better to have one large or several small reserves (the so-called SLOSS controversy) (see Shafer, 1990). Taking the precautionary principle, many conservationists think that it is wise to 'not put all the eggs in one basket', but clearly there is more edge effect in two smaller fragments than one large area.

What would be considered the 'ideal' number and geographical location of two or more reserves? If they are close, there is a greater chance that species will migrate between them (Fig. 17.3). As we saw in Chapter 12, occasional gene flow between populations may be highly significant in maintaining and enhancing the genetic variability of species 'confined' within reserves. Extending the argument, in the extreme case, a species may become extinct following a localised catastrophic event and recolonisation could occur from another adjacent unaffected reserve. While acknowledging these advantages, Shafer (1990) stresses that they are not a sufficient reason to 'dismember' a large reserve. He also points out some of the perils of having *in situ* conservation areas too close together (Shafer, 2001): 'Since the mid-1970s, reserve planners have been advised to locate reserves in close proximity to facilitate biotic migration.' However, natural catastrophes (such as hurricanes, storms, flooding, fires, disease) and anthropogenic events (such as the introduction of exotic invasive species and pest organisms) can affect whole regions, and endangered species in reserves in close proximity might be subject to the same dramatic event.

Corridors in the landscape

Inspired by the theory of island biogeography, models of the 'ideal' reserve suggest that gene flow and migration may be more reliably assured if natural connectivity between sister reserves is maintained through the provision of functional migration corridors. This term refers to 'a linear landscape element composed of native vegetation which links patches of similar native vegetation' (Collinge, 1996). However, it is important to note that the term 'corridor' is used to mean different things by ecologists, landscape architects and geographers.

Seen from the air there are many linear vegetated elements that might act as corridors across matrix linking reserves or areas of conservation interest. Some 'vegetated' patches and strips may appear to be entirely 'natural', but, on close inspection, there is clear evidence that they have been modified by human activities. Other 'corridor' systems are clearly of human origin, such as fire breaks, wind breaks (orientated according to the direction of the prevailing wind and snowstorm direction), power transmission lines, oil and gas pipelines, hedges, woodland patches and strips maintained for game, tree-covered and grass areas as part of strip cultivation patterns, tracks, trails, greenways, metalled roads, motorways, railways, etc.

Vegetated strips may be simple, being essentially uni-functional, long and continuous. But in other cases, they are short, interrupted, multi-functional elements that intersect at 'nodes' with other linear vegetated strips and patches. Particularly in the regions of urban industrial development, such elements are highly individualistic, being intermixed and interconnecting with complex patchworks of private and public housing, gardens, roads, canals, railways, derelict and waste land, dumps, allotments, illegal housing etc.

In the English countryside, hedgerows are very conspicuous linear elements, serving a number of functions. For instance, they mark legal boundaries, provide a means of confining and protecting stock and crops, and regulating access etc. (Rackham, 1986). While hedges might appear simple in structure, on inspection they provide, potentially, several different types of habitat corridors. Typically, the central element of the hedge consists of woody plants, but often there are adjacent parallel strips of grassland, ditches, access tracks and uncultivated headlands around cultivated fields etc. (Pollard, Hooper & Moore, 1974).

Aerial photographs and satellite images reveal that river systems across the world provide the most obvious green linear elements of waterways, fringing and riverside vegetation, flood plains etc. Again, such systems are often far from natural. In most settled areas, they are highly modified, and there is fierce competition between users. Thus, river systems provide many goods and services. Along some of their length, they are generally subject to canalisation and embankment etc., to control flow and flooding, in the provision of drinking and irrigation water, fish, power generation, navigation, waste disposal, cooling liquid for power stations etc.

Evidence for functioning corridors

While linear or more complex vegetated strips may run between island ecosystems of conservation interest, the question to confront is whether they are functioning as wildlife corridors or, indeed, as barriers to dispersal for some species (Forman, 1995).

Collinge (1996) has reviewed the evidence of the use of corridors by various animal species. Beier and Noss (1998) have also made a critical assessment of whether habitat corridors provide connectivity, in which 32 case studies of animals were critically examined. Fewer than half supported the utility of corridors; but some well-designed experiments suggest that corridors could be important. Recently, further evidence has been collected (Haddad *et al.*, 2003), but there is still a high degree of uncertainty about animal use of corridors.

A direct test as to whether corridors facilitate the movement of plants has been devised by Tewksbury *et al.* (2002). Within a landscape of 40–50-year old forest dominated by pine species (*Pinus taeda* and *P. palustris*), eight 50-hectare early-successional areas were created by felling and burning at the Savannah River site of the National Environmental Research Park in South Carolina (Fig. 17.5). Within each landscape, a 1 ha central patch was surrounded by four peripheral patches, only one of which was connected to the central patch by a 25 m corridor. Of the three isolated patches, one had wings of dead-end corridors on either side. This made it possible to test the effect of patch area and investigate whether 'wings' acted as corridors, even though they were not connected to patches. A number of investigations were carried out. Those concerned with plants studied the pollen movement between individuals of dioecious species of holly (*Ilex verticillata*). Male plants were planted in the central patch with females in the peripheral patches. Plants were visited by a wide variety of insects. It was discovered that where peripheral plots were provided with a corridor connection, there was greater fruit set.

Seed movements were also investigated from fruit-bearing cuttings and transplanted live trees of holly (*Ilex vomitoria*) located on the central patch: all other fruiting material of the species having been cleared from the area. Seed movement from central to peripheral plots was also investigated in wax myrtle (*Myrica cerifera*). Transplanted trees (and fruiting branches wired in the canopy) were provided on the central plot. As it was impractical to

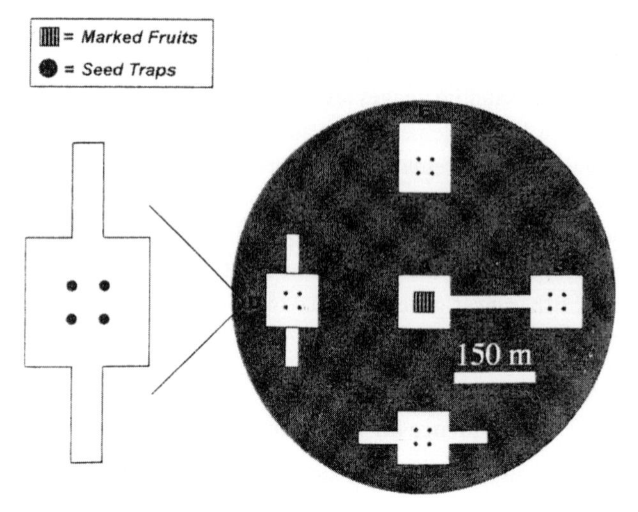

Fig. 17.5 One of eight experimental landscapes at the Savannah River Site National Experimental Research Park, showing the four patch types. Each experimental landscape had a source patch where marked fruits were placed in the centre and three types of receiver patches, where seed traps were placed. One receiver patch in each landscape was attached to the source patch by a corridor. In four landscapes two of the remaining receiver patches were winged and one was rectangular as pictured here. In the other four landscapes two receiving patches were rectangular and one was winged. (From Levey *et al.*, 2005. Reproduced by permission from *Science*, © American Association for the Advancement of Science)

remove all fruiting myrtles from the experimental area, fruits in the central plot were sprayed to give them a coating of fluorescent powder. Fruiting plants and the cut branches were visited by a range of bird species. The dispersal of fruits was examined by studying faecal samples collected in seed traps from beneath artificial perches in each peripheral patch. It was discovered that holly seed consumed in the central patch was more likely to be found in the corridor-connected peripheral patches. Considering the dispersal of marked myrtle fruits, a greater proportion of faecal samples in connected patches contained fluorescent powder. The study provides a large-scale demonstration that habitat corridors can facilitate movement between patches through plant–animal interactions. However, the authors make it plain that the 'extension of these results to even larger scales and landscapes must proceed with caution'.

Further investigations in this experimental area have studied the behaviour of seed-eating eastern bluebirds (*Sialia sialis*) (Levey *et al.*, 2005). Instead of flying down the middle of patches, bluebirds tended to use the edges of the forested edges of patches. Such behaviour might offer more perches and protection against birds of prey. Perhaps, for this species, the edges of patches are more important than the corridor width. However, again, the authors make it plain that it would be unwise to generalise from this experiment: other species may behave differently. There is also another key point to stress. The experimental landscape investigated is unusual in having open patches surrounded with woodland. More commonly, patch/corridor systems involve islands of woodland, fen or grassland within a matrix of agricultural land or urban development (Stokstad, 2005b).

There are a number of other studies relevant to movement of plants within corridors. For instance, Gurnell *et al.* (2005) highlight the active transport of 'downed trees and sprouting drift wood' by rivers running through corridors of riparian woodland. In another study, Bullock and Samways (2005) investigated the narrow remnant grassland corridors remaining within pine plantations at a site in South Africa. They concluded that arthropod assemblages associated with indigenous grassland species were qualitatively similar throughout the network. These associations could be threatened, however, if cattle grazing were to increase.

Strengths and weaknesses of corridors as linkages between reserves

Just as reserves are impacted by edge effects, so too are corridors. Reflecting on the results of the BDFFP, Bierregaard *et al.* (2001) concluded that corridors should be a kilometre wide or more. Moreover, as edge effects penetrated at least 100 m into fragments, the minimum effective corridor would be at least 300 m wide. This bare minimum 'would protect only 100 m of unaffected forest'. To maximise the 'benefit' they proposed that such corridors should run along major rivers and streams.

Turning to another issue, there may be problems with reserves situated in close proximity or linked by corridors. Close proximity and the presence of corridors could facilitate the migration of disease organisms, invasive introduced species etc. between sister reserves. For example, in South Africa, France and north-west USA, river systems are a major route for range expansion of invasive introduced species (Hood & Naiman, 2000). Ditches have also proved to be a conduit for the migration of invasive species, for instance, *Lythrum salicaria* in the USA (Wilcox & Murphy 1985, 602). Trails too may contribute to the spread of invasive species. Studies by Campbell and Gibson (2001) revealed that the seeds of 23 exotic species were found in horse dung collected from trails used for recreational riding in southern Illinois. In another investigation, Ghersa *et al.* (2002) examined the invasion by introduced species of the agricultural landscape of the rolling pampas grasslands of Argentina. They conclude that landscape corridors – highways, intersecting dirt roads and streams – are conduits for the invasion by more than 40 woody species, including *Gleditsia triacanthos*, *Morus alba* and *Melia azedarach*.

Reserves and the conservation of particular species

In Chapters 15 and 16 the evolution of management in reserved lands was discussed, including devising of recovery plans for individual species. It was concluded that, while such plans have often been successful in the short term, as a general approach, they are open to criticism. Budiansky (1995, 170) notes: 'trying to tackle the problems of wildlife conservation one species at a time is cumbersome, time consuming'. Both for management efficiency and a focus for fund-raising, efforts have often focused not on ecosystems as a whole, but on the conservation of charismatic species of animals. Three different categories have been recognised by Simberloff (1998):

- A flagship species, usually a large vertebrate, may be chosen to 'anchor a conservation campaign because it arouses public interest and sympathy'.
- An umbrella species, again a large animal species, needs very large tracts of land to survive and, in satisfying these needs, conservationists hope to secure the future of other species in the ecosystem.
- Conservation efforts could be focused on a keystone species, i.e. a species that underpins the functioning of entire ecosystems, e.g. the beaver (*Castor canadensis*) in its building of dams.

Many conservationists believe that the successful management of these key players in their ecosystems will secure the future of all the species in the habitat. To an extent, therefore, conservation of plants is often dependent upon the management plans devised for charismatic animal species, and it is appropriate, therefore, to examine briefly the number and size of reserves necessary to maintain viable populations of emblematic mammals and birds.

Here, as an example, it is helpful to consider the area of nature reserves necessary to secure the future of the Florida panther (*Puma concolor coryi*) (Shrader-Frechette & McCoy, 1993). Field work has established that the animal has a home range of perhaps 30,000 to 50,000 acres. Population biologists estimate that the minimum viable population would be about 50 individuals. Astonishingly, this gives a minimum viable area in the region of 2,500,000 acres (Shrader-Frechette & McCoy, 1993). Recently, however, it has been pointed out that setting a MVP of less than 100 might allow the animal to survive for 100 years, but this number is 'quite low' and the panther could be subject to 'continued genetic problems' (Kautz *et al.*, 2006). Within the context of strong economic development in Florida, the debate continues about how much territory could or should be set aside to enlarge present reserves, and provide corridors between reserves that might or might not be used by the animals. Reflecting on questions relating to fragment size in the BDFFP, Bierregaard *et al.* (2001) conclude that to provide habitat for top predators such as harpy eagles in their study zone, 'a million or more hectares would be required to support a MVP of say 500 pairs'.

It is clear from these examples that even very large reserves may not be of sufficient size to conserve wide-ranging top-predator animals. However, if the focus is on the conservation of particular plants and many smaller animal species, then many conservationists consider that a few well-chosen smaller reserves could have an important role to play (Soulé & Simberloff, 1986), but stress the uncertainties associated with MVPs and PVA estimates of plant populations. Thus, it is difficult to determine how small a reserve might be and still secure the future of the plant populations growing there (Young & Clarke, 2000).

Returning to questions concerning the size of reserves, there may be considerable problems with small areas. Long-term *in situ* conservation of many plant species requires natural habitat disturbance. In large pristine landscapes, such disturbance may affect particular areas, but not all, leading to mosaics of different ecosystems, each of which is of a significant size. However, many reserves are too small to 'capture the full scope of natural disturbance processes. Fire, wind-storms, landslides, floods, disease outbreaks – all are necessary to maintain the mosaic of habitats' (Budiansky, 1995, 134). If major natural disturbance events occur across all or a large part of their territory, then managers of small reserves may be facing a 'catastrophe'. Another related point should be stressed, if there is only a single reserve, an endangered species may be at extreme risk from catastrophic events (see Shafer, 2001). For example, following a drought, the extremely rare *Stephanomeria malheurensis* was lost from the single locality where it grew in eastern Oregon, USA (Guerrant, 1992).

The location of present-day reserves in relation to biodiversity 'hotspots'

As we saw in Chapter 15, many national parks were developed to safeguard scenery and landscapes, but as the conservation movement gathered momentum, there have been major efforts to establish reserves in areas with significant biodiversity. Has a 'sufficient' number of reserves been established in all the major areas of high biodiversity under threat?

In the identification of such areas, Myers (1997, 125, and references cited therein) made an important contribution by developing the concept of hotspots. Reflecting on the issues, he noted: 'much biodiversity is located in small areas of the planet. As much as 20% of plant species and still higher proportion of animal species are confined to 0.5% of Earth's land surface.' Given that funding for conservation is limited, the concept of hotspots provides a potential means of identifying key areas for conservation through the setting up of reserves and other measures (Myers, 1997).

This concept has attracted wide attention and a great number of research papers have been published. In brief, early research identified 18 territories containing nearly 50,000 endemic plant species. These included 14 areas in tropical forests in Amazonia, Philippines, Madagascar, Borneo, Peninsular Malaysia etc. In addition, four hotspots were recognised with Mediterranean ecosystems – the Cape region of South Africa, central Chile, California and south-west Australia. Later research extended the number of areas of high biodiversity (see Ceballos & Ehrlich, 2006).

The concept of hotspots has been critically examined by Ceballos and Ehrlich (2006) who investigated the detailed distribution of bird species and all the 4,818 non-marine mammal species. They studied how much overlap was evident amongst hotspots defined in terms of (a) species richness, (b) species at risk and (c) measures of threat. They concluded: 'hotspots of species-richness, "endemism", and extinction threat were noncongruent'. Overall, they considered that, 'in assigning conservation priority', hotspots were 'at best a limited strategy'.

Grenyer *et al.* (2006) have also made a detailed study of hotspots. They examined the hypothesis that 'different taxonomic groups show congruent geographical patterns of distribution' and that the 'distribution of extinction-prone species in one group can therefore act as a surrogate for vulnerable species in other groups when conservation decisions are being made'. Information on the distribution and rarity of every known species of mammal, bird and amphibian was examined on the databases held as the World Conservation Union's Red Data Lists. In total, information on 19,349 species was investigated. Maps were prepared of total species richness, rare species richness and threatened species richness. Although distributions of total species richness were similar in the three groups of animals, 'congruence in the distribution of rare and threatened species is markedly lower. Congruence is especially low amongst the very rarest species. Cross-taxon congruence is also highly scale dependent being particularly low at the finer spatial resolutions relevant to real protected areas . . . Hotspots of rarity and threat are therefore largely non-overlapping across groups.'

The lack of coincidence between the different elements of diversity and threat is a complex issue, but there is another concern of importance. Different groups are subject to different types of threat. For instance, many bird species are threatened by habitat loss; tigers are at serious risk from poaching, and amphibians are endangered by disease (see Stuart *et al.*, 2004). Undoubtedly, more reserves are needed to conserve the world's biodiversity effectively, but it has to be concluded that the concept of hotspots does not yield a simple list of target areas.

Aquatic freshwater ecosystems in parks, reserves and matrix

Can the principles of reserve design for terrestrial areas be used for designing freshwater aquatic reserves? This issue has been examined in detail by Dudgeon *et al.* (2006) – a multi-author paper written by experts from China, Australia, Switzerland, Japan, Canada, France, USA, Chile and the UK that synthesises worldwide experience. They point out that freshwater systems are subject to multiple threats from 'overexploitation, water pollution, flow modification, destruction or degradation of habitat, and invasion by exotic species . . . The particular vulnerability of freshwater biodiversity also reflects the fact that fresh water is a resource for humans that may be extracted, diverted, contained or contaminated in ways that compromise its value as a habitat for organisms.' In many parts of the world 'fresh water is subject to severe competition among multiple human stakeholders, to the point that armed conflicts can arise'. Moreover, 'in the vast majority of disagreements over multiple uses of water, whether they are international or on a local scale, allocation of water to maintain aquatic biodiversity is largely disregarded'. Turning to strategies for the protection of biodiversity, the authors conclude:

> Terrestrial conservation strategies tend to emphasize areas of high habitat quality that can be bounded and protected. This 'fortress conservation' is likely to fail for fresh waters and may even be counterproductive, for river segments or lakes subject to conservation protection are often embedded in unprotected drainage basins unless the boundaries are drawn at a catchment scale, which is virtually never the case. This problem of boundary definition impedes sensible local management of

freshwater diversity because protection of a particular component of river biota (and often habitat) requires control over the upstream drainage network, the surrounding land, the riparian zone, and – in the case of migrating aquatic fauna – downstream reaches . . . The catchment scale is generally appropriate for all types of freshwater management of freshwater habitats . . . However, this approach can be problematic in practice, as relatively large areas of land need to be managed in order to protect relatively small water bodies.

In conservation practice at the ecosystem level there is a need to focus not only on rivers but also on their dependent flood plains. Thus:

Large animals such as bear, swamp deer, rhinos, and elephants make seasonal use of riparian areas and floodplains for feeding and breeding. Maintaining proximity to water is essential for large animals during the tropical dry season, which can be a period of severe ecological stress for herbivores. Effective preservation of biodiversity associated with freshwater habitats must therefore take account of the year-round habitat use . . . as well as the needs of the strictly aquatic biota. Maintenance of some semblance of the natural flow variability and the flood/drought cycle of rivers and their flood plains, vernal pools, and water-level fluctuations in wetlands and along lakeshores, also will be essential.

Marine reserves

In acknowledging the 'ecological linkages' between land and sea, sometimes combined terrestrial and marine areas are included in a single park around islands or off-shore of a mainland reserve (Shafer, 1999b). Also, marine reserves have been established or are under consideration (Allison, Lubchenco & Carr, 1998). However, Shafer (1999b) emphasises that the problems of marine conservation will not be solved by parks alone.

Concluding comments

There are hundreds of reserves and parks worldwide, of many different types: strictly managed wilderness areas; national parks managed to protect ecosystems and provide recreation; national parks containing 'working' cultural landscapes based on agriculture, rural industries etc.; national monuments; habitat and species management areas; protected landscapes; and areas for resource use, sometimes under sustainable management (Holdgate & Philips, 1999). Those in charge of reserved areas in different parts of the world are in the best position to judge their strengths and weaknesses. An honest appraisal is essential. While many reserves throughout the world are offering a degree of protection to the ecosystems they contain, as we saw above, many reserves are essentially paper parks. With regard to size, it is undeniable that many reserves are very small, which raises the questions about the security of the biodiversity they contain, especially as small and declining populations are at risk of loss of genetic variability through genetic drift etc. (Boecklen, 1986). Furthermore, small reserves seem destined to lose some of their biodiversity, not only because of increasing human activities but also through the processes of 'relaxation'.

Concerning the effective size of reserved areas, in some cases co-ordinated management of adjacent lands may be possible. For instance, as we saw earlier, the Yellowstone National

Park lacks areas where major herbivores can over-winter etc. The formation of the Elk Refuge and, later, the Greater Yellowstone areas (by including National Forest areas) has greatly increased the effectiveness of the national park itself. On a much smaller scale, many reserves have been enlarged, e.g. Buff Wood, Cambridgeshire (Fig. 5.1). However, in highly settled regions of the world it may be effectively impossible to increase the size of reserves.

While on biological grounds it could be argued that small reserves should be avoided because of their limitations, another issue must also be considered. This is admirably expressed by Schwartz and van Mantgem (1997) in their consideration of the role of the many small reserves in Illinois, USA. The point they emphasise applies across the globe:

> To build a conservation constituency we need to provide opportunities for participating in the management of native species and natural lands. A large network of small sites afford us the best opportunity to develop this connection to natural areas among the broadest spectrum of people. Further, it is the caring and nurturing of local biotic resources that often drives our concern for more global conservation issues. To accomplish global conservation goals, we first need to get many people personally involved in their local environment.

Great public loyalty to local nature reserves is found across the developed world, with particular focus on endangered species. However, some hard truths must be faced: the cherished rarity may decline and survival may depend on intensive management. If the species are lost, reintroductions are often carried out. But such is the vulnerability of small populations that the survival of a particular species might involve repeated reintroductions from *ex situ* stocks. As we have seen earlier in the book, intense and continued management *in situ* and the maintenance and propagation of *ex situ* stocks carries with it the risk of domestication.

Evidence suggests that reserves are significantly influenced by major edge effects. However, as we have seen in earlier chapters, the notion of a pristine untouched heart at the centre of the reserve may be erroneous in the light of advances in our understanding of the history of human activities. Thus, any consideration of edge effects should include boundary issues. Conflicts arise between conservationists and local people in situations where territories with a long history of human occupation and resource use were 'cleared' and managed as if they were pristine 'wilderness'. Thus, there can be very uneasy relations between conservationists, park managers and those who live at the boundary of a reserve and surrounding matrix. The notion that parks are 'protected paradises' has consequences. A 'purely preservationist approach, where Parks are considered "fortresses" under siege, invincible or soon eradicated, carries great political risks. It requires an essentially militaristic defense strategy and will almost always heighten conflict' (McNeely, 1989, 155).

The aim of reserved areas is to give protection to wildlife, especially those charismatic animals that have iconic status for ecotourism. Around some parks, boundary fencing has been erected. For instance, fences have been erected at the Haleakala National Park to protect native species from damage by feral goats (Shafer, 1999b). But park boundaries are generally unfenced and animals are free to move outside the park where they may destroy crops, damage property and even kill or injure villagers, In many parts of the world, the failure to control animals at the boundaries of national parks is seen by local people as another injustice arising from their colonial past, resulting in antipathy or hostility to the park and its aims (Anderson & Grove, 1987; Adams & Mulligan, 2003). There is increasing recognition that 'conservation projects cannot be successful when implemented in isolation from neighbouring human populations. Maintaining nature reserves has undergone a major strategy shift from trying to keep people out of reserves to allowing some economic development by local communities as an integral part of conservation strategies' (Vogt *et al.*, 2001, 872).

Another issue to confront is that many reserves are geographically isolated, and, as we have seen, a number of conservationists have advocated the creation of wildlife corridors between reserves, but, as yet, there is only limited evidence that corridors are effective as habitats for wildlife or routes of dispersal. Rare plant species have been studied by generations of naturalists. As yet there is very little evidence to support the notion that the apparent corridors provide habitat in which rare species can grow. Furthermore, the necessary conditions for transit, via gene flow/migration – by wind, animal–plant mutualisms – may not be satisfied. In addition, other conditions for effective extension of range may not be fulfilled, for many plants have very exacting requirements for seedling establishment (see Grubb, 1977). From the 'plant's-eye view' the matrix of human-modified ecosystems, even with vegetated corridors of various kinds, may represent an insuperable barrier to dispersal between reserves.

Risking a generalisation, the fate of many plant species is inexorably linked with that of charismatic species of birds or mammals. In defining priority areas, the concept of hotspots has recently come to prominence. It is important to acknowledge, therefore, that recent major investigations have revealed that the concept has limitations. Hotspots may be defined using different criteria – overall species richness, endemism and threat levels to the species of different taxonomic groups. Unfortunately, plotting hotspots on different criteria does not yield 'coincident all-encompassing' areas on which conservation efforts can be concentrated. This is particularly the case when hotspots are plotted at a scale appropriate to that of most reserves. Thus, securing the future of populations of charismatic animal species and their habitat will not necessarily provide security for the full range of endemic, rare and endangered plant species. However, if provision for top predators in the ecosystem is excluded from the design, many authorities consider that effective reserves for plants could be quite small. But perhaps such areas would be regarded as second-best, lacking the full interlocking complexity of a more natural ecosystem.

Studies of hotspots have focused attention on a major issue, namely that many key ecosystems and their rich biodiversity are found outside the protection offered by reserves; e.g. a very large number of rare Australian plant species are found outside reserves (Kirkpatrick, 1989). In addition, many conservationists now conclude that rather than focus exclusively on 'spectacular wild areas and emotionally appealing endangered species' increased attention must be paid 'to the rather unglamorous semi-natural matrix' (Brown *et al.*, 2003, 338).

Considering the characteristics of wild species that thrive in human-modified ecosystems, Western (1989, 162) has listed a number of categories. (1) Species may be useful or profitable, e.g. certain plant products are harvested as food, medicine etc.; and African wildlife is being transferred from 'the wild' to naturalistic vegetated areas on private ranches as part of ecotourism enterprises. (2) Some species are considered to present 'zero costs' in habitats exploited by humans, e.g. species that are not seen to compete with grazing domestic animals etc. (3) Some species are 'fugitive rarities' that do not register as worthy of pest control. (4) Many species survive in areas of 'unused, little used or abandoned' territories. For instance, areas beyond water sources that are marginal for stock grazing, territories that are unfit for humans and or domestic stock because they are 'disease ridden', or troubled by warfare. (5) Some species are tolerated because they are ecologically complementary to human use. According to Western, in some areas the presence of elephants on grazing lands may be tolerated as they improve an area for stock grazing by controlling shrubs and trees. (6) Some species survive because they are given favoured status, as, for example, in providing 'sport', or are subjects for 'aesthetic appreciation' etc. (7) Many species may survive through the indifference of landowners, or their ignorance of the wildlife on their property. (8) Across the globe, there is an increased chance of species survival if they are protected by law, and such protection is recognised and enforced (Sheail, 1998). To this list may be added weedy and invasive species. Such species may be controlled on agricultural lands, but are rarely exterminated.

Throughout the world, considerable areas have already been designated as national parks and reserved areas. In many parts of the world the scale and intensity of human development is so great that the creation of very large new reserves and extensive corridors is highly unlikely. However, in some territories relatively undamaged ecosystems still persist, although they may be highly fragmented. New conservation areas have been established and research is being carried out on the best way to select sites representative of ecosystems not yet within reserves and to increase the effectiveness of conserved lands by designing and creating reserve networks (Coates, 1988; Hobbs *et al.*, 1993; Kirkpatrick & Gilfedder, 1995; Pressey, Possingham & Margules, 1996; Cabeza & Moilanen, 2001). Thus, many new reserves have been developed in Australia to provide more comprehensive coverage. In addition, private parks and reserves are being developed in many parts of the world. To an extent these initiatives owe something to the ideas generated by the Theory of Island Biogeography. But many authorities consider that, on balance, the theory has 'not lived up to its promise of serving practical conservation biology' (Gascon *et al.*, 2003, 43). Having written many of the key papers on reserve design, Shafer (1997) concludes: 'the theory served as a foundation for thinking about nature reserve design in the 1970s and later'. But, 'the empirical basis supporting such use is slim, and respect for the theory's conservation usefulness has declined'. Moreover, there is support for Shafer's view that, for many species, total reserve area is less important than the provision of sufficient and appropriate habitat.

In closing this account of the size, shape and distribution of reserves and parks, a final point of great significance must be stressed. Parks and reserves aim to conserve biodiversity in perpetuity, but already, some major weaknesses, as well as undoubted strengths, have been identified. What of the future? Predictions concerning global climate change indicate that the worldwide system of reserved areas is likely to be subjected to the severest of tests. This issue will be considered in the next chapter.

18

Climate change

There is increasing evidence that climate change is occurring, and the anthropogenic contribution to such change is being clarified (IPCC, 2007a, b, c, d). While most experts in the field accept this proposition, a diminishing number of vocal sceptics are to be found in academia, in political circles and in the media. This chapter considers the evidence for climate change. Chapters 19 and 20 examine the microevolutionary consequences of such changes, and the implications for conservation.

The greenhouse effect and climate change

Solar energy is received on Earth from the Sun. Some of this energy is reflected back from clouds and the Earth's surface, but some is trapped by the so-called greenhouse gases (water vapour, carbon dioxide, carbon monoxide, methane and nitrous oxide) resulting in the warming of the planet. As we shall see, there is mounting evidence that anthropogenic activities are increasing the levels of these gases in the atmosphere resulting in global warming.

This account draws on the recent *Fourth Intergovernmental Panel on Climate Change Assessment Report* (IPCC, 2007a) on the physical sciences basis for accessing climate change. It has been prepared by over 1,200 expert authors, and reviewers from 40 countries (Giles, 2007). At the time of writing only the *Summaries for Policy Makers* are available.

The aim is to provide a definitive assessment, in which the role of anthropogenic and natural factors in climate change are critically examined. Evidence from many different disciplines is considered, including computer modelling and the results of a wide range of observational and experimental studies. The result is a report set out in measured prose to present the current state of knowledge. Bearing in mind the differences of opinion concerning the causes of climate change, it is important here to be aware of precisely what the Report concluded. Hence, a number of quotations have been taken from the Summary, in which the following terms are used 'to indicate assessed likelihood, using expert judgement, of an outcome or a result: *Virtually certain* >99% probability of occurrence, *Extremely likely* >95%, *Very likely* >90%, *Likely* >66%, *More likely than not* 50%, *Unlikely* <33%, *Very unlikely* <10%, *Extremely unlikely* <5%'. For those who wish to consult recent key peer-reviewed researches, a number of references are also provided that are not specifically mentioned in the Report.

The IPCC (2007a, 2) conclude:

> Changes in the atmospheric abundance of greenhouse gases and aerosols, in solar radiation and in land surface properties alter the energy balance of the climate system. These changes are expressed in terms of 'radiative forcing' which is used to compare how a range of human and natural factors drive warming or cooling influences on global climate. The term radiative forcing [expressed in watts per square metre ($W\ m^{-2}$)] is a measure of the influence that a factor has in altering the balance of incoming and outgoing energy in the Earth-atmosphere system.

In studying anthropogenic effects, historical and contemporary studies are available and, to provide a longer-term perspective over the last 650,000 years, the changing concentrations of greenhouse gases in air bubbles trapped in ice have been quantified.

Carbon dioxide is a potent greenhouse gas. Since the beginning of the industrial revolution, atmospheric carbon dioxide has increased from about 280 parts per million by volume (ppm) to 379 ppm (IPCC, 2007a, 2). This figure exceeds by 'far the natural range over the last 650,000 years (180 to 300 ppm) as determined from ice cores'. Moreover, 'the annual carbon dioxide concentration growth-rate was larger during the last 10 years (1995–2005: 1.9 ppm per year) than it has been since the beginning of continuous direct atmospheric measurements (1960–2005 average: 1.4 ppm per year)'. The IPCC Report concludes that the 'primary source of the increased atmospheric concentration of carbon dioxide since the pre-industrial period results from fossil fuel use, with land use change providing another significant but smaller contribution'.

Another important greenhouse gas, methane, 'increased from a pre-industrial value of about 715 parts per billion (ppb) to 1732 ppb in the early 1990s' and reached '1774 ppb in 2005.' 'The atmospheric concentration of methane in 2005 exceeds by far the natural range of the last 650,000 years (320 to 790 ppb) as determined from ice cores . . . It is *very likely* that the observed increase in methane concentration is due to anthropogenic activities, predominantly agriculture and fossil fuel use, but relative contributions from different source types are not well determined' (IPCC, 2007a, 3).

'The global atmospheric nitrous oxide concentration increased from a pre-industrial value of about 270 ppb to 319 ppb in 2005.' The Report concluded: 'More than a third of all nitrous oxide emissions are anthropogenic and are primarily due to agriculture.'

In addition, nitrogen oxides, carbon monoxide and hydrocarbons have contributed significantly to global warming effects. The deposition of black carbon on snow also made a contribution through the changes in the albedo of land surfaces. The IPCC Report (IPCC, 2007a, 3) also notes: 'Anthropogenic contributions to aerosols (primarily sulphate, organic carbon, black carbon, nitrate and dust) together produce a cooling effect.' Thus, the optical properties of clouds are changed by atmospheric pollutant particles, with the result that more solar radiation is being reflected back into space. This factor – global cooling – is now 'better understood' through ground observations, satellite measurements and modelling, but remains 'the dominant uncertainty in radiative forcing'.

'The understanding of anthropogenic warming and cooling influences on climate have improved since the IPCC Third Assessment Report (2001) leading to very high confidence [later in the Report estimated at 9 out of 10 chance of being correct] that globally average net effect of human activities since 1750 has been one of warming, with a radiative forcing of +1.6 [+0.6 to +2.4] W m^{-2}.' The values in square brackets indicate statistical confidence intervals. 'There is an estimated 5% likelihood that the value could be above the range given in square brackets and 5% likelihood that the value could be below that range.' For further information on the calculation of best estimates and confidence intervals, the Report should be consulted.

Could the increase in temperature be the result of solar activity? The IPCC Report (2007a, 3) considers that possibility and concludes: 'Changes in solar radiation since 1750 have contributed to global warming estimated at +0.12 [+0.06 to +0.30] W m^{-2}.'

Direct observations of climate change

The Report (2007a, 4) concludes: 'Warming of the climate system is unequivocal, as is now evident from observations of increases in global average air and ocean temperatures, widespread melting of snow and ice, and rising global mean sea level.' Each of these elements is considered in turn.

Air temperatures

'Eleven of the last 12 years (1995–2006) have been among the 12 warmest' since records of air temperature began in 1850. The trend in warming 'over the last 50 years is nearly twice that for the last 100 years'.

The upper atmosphere

Warming rates are 'similar to those of surface temperature record'. Moreover, the average water vapour content of the atmosphere has increased, and is 'broadly consistent with the extra water vapour that warmer air can hold'. In quantifying atmospheric changes, the Report emphasises that uncertainties remain.

Ocean temperatures

There is evidence that the average temperatures have increased. Moreover, 'such warming causes seawater to expand, contributing to sea level rise'.

Sea level rise

'There is *high confidence* that the rate of observed sea level rise increased from the 19th to the 20th century'. In total it is estimated that sea levels rose 0.17 [0.12 to 0.22] m in the twentieth century with contributions from thermal expansion: a decline on average in glaciers and snow-cover, and losses from the ice sheets in the Arctic and Antarctic.

Long-term changes in climate

A number of trends are noted in the IPCC Report. In summary, there have been large increases in average temperatures in the Arctic; shrinkage of Arctic sea ice; decrease in the area of frozen ground; higher precipitation in many areas of the world (eastern parts of North and South America, northern Europe, and northern and central Asia); drying in the Sahel, Mediterranean, southern Africa and parts of southern Asia; changes in precipitation and evaporation over the oceans; strengthening of winds in some regions; more intense and longer droughts over wider areas than in the 1970s; frequent heavy precipitation events; widespread changes in extreme temperatures; and an increase in tropical cyclone activity in the North Atlantic.

The likelihood that humans are contributing to climate change

The Report concludes (2007a, 8): 'Most of the observed increase in globally averaged temperatures since the mid-20th century is *very likely* due to the observed increase in anthropogenic greenhouse gas concentrations . . . Discernable human influences now extend to other aspects of climate, including ocean warming, continental-average temperatures, temperature extremes and wind patterns.'

Projections of climate change in the future

Comprehensive and intensive modelling studies in different countries, by a wide range of experts, have yielded many different projections of likely climate change in the future. The assumptions and time frames of these models vary. Drawing on the most up-to-date information available, the IPCC Report provides projections to the year 2100 for six different so-called 'storylines'. These are detailed in Box 18.1 with information about the assumptions of each model. Predictions are graphically displayed in Fig. 18.1.

It is informative to examine the two scenarios that represent the most divergent forecasts, both of which assume a population growth peak in mid century and then a decline. At one extreme, A1F1 assumes the continued intensive use of fossil fuels to promote rapid economic growth. It is estimated that by 2100 this scenario will lead to a global average surface temperature increase of 4.0 °C (likely range 2.4 to 6.4 °C). In contrast, the B1 scenario assumes rapid reductions in the use of fossil fuels, widespread use of 'clean and resource efficient technology', and reductions in the intensity of the use of raw materials. 'The emphasis is on global solutions to economic, social and environmental sustainability, including improved equity, but without additional climate initiatives.' The best estimate for global mean surface temperature rise is 1.8 °C (likely range 1.1 to 2.9 °C).

Assessment of climate change in the future

Warming is expected to be greatest over land surfaces and at high northern latitudes. Permafrost melting will increase in depth and snow cover will decrease. Under all

Box 18.1 The Emission Scenarios of the IPCC Special Report on Emission Scenarios (SRES)

A1. The Al storyline and scenario family describes a future world of very rapid economic growth, global population that peaks in mid-century and declines thereafter, and the rapid introduction of new and more efficient technologies. Major underlying themes are convergence among regions, capacity building and increased cultural and social interactions, with a substantial reduction in regional differences in per capita income. The Al scenario family develops into three groups that describe alternative directions of technological change in the energy system. The three Al groups are distinguished by their technological emphasis: fossil intensive (AlFI), non-fossil energy sources (A1T) or a balance across all sources (A1B) (where balanced is defined as not relying too heavily on one particular energy source, on the assumption that similar improvement rates apply to all energy supply and end use technologies).

A2. The A2 storyline and scenario family describes a very heterogeneous world. The underlying theme is self-reliance and preservation of local identities. Fertility patterns across regions converge very slowly, which results in continuously increasing population. Economic development is primarily regionally oriented and per capita economic growth and technological change more fragmented and slower than other storylines.

B1. The B1 storyline and scenario family describes a convergent world with the same global population, that peaks in mid-century and declines thereafter, as in the A1 storyline, but with rapid change in economic structures toward a service and information economy, with reductions in material intensity and the introduction of clean and resource efficient technologies. The emphasis is on global solutions to economic, social and environmental sustainability, including improved equity, but without additional climate initiatives.

B2. The B2 storyline and scenario family describes a world in which the emphasis is on local solutions to economic, social and environmental sustainability. It is a world with continuously increasing global population, at a rate lower than A2, intermediate levels of economic development, and less rapid and more diverse technological change than in the B1 and Al storylines. While the scenario is also oriented towards environmental protection and social equity, it focuses on local and regional levels.

An illustrative scenario was chosen for each of the six scenario groups A1B, A1FI, A1T, A2, B1 and B2. All should be considered equally sound.

The SRES scenarios do not include additional climate initiatives, which means that no scenarios are included that explicitly assume implementation of the United Nations Framework Convention on Climate Change or the emissions targets of the Kyoto Protocol. Reproduced with permission from the *IPCC Fourth Assessment Report*, February 2007.

'storylines' ice cover will contract in the Arctic and Antarctic. 'It is *very likely* that hot extremes, heat waves, and heavy precipitation events will continue to become more frequent.' Tropical typhoons and hurricanes are likely to become more intense, but there is uncertainty about their likely frequency. 'Extra-tropical storm tracks are projected to move

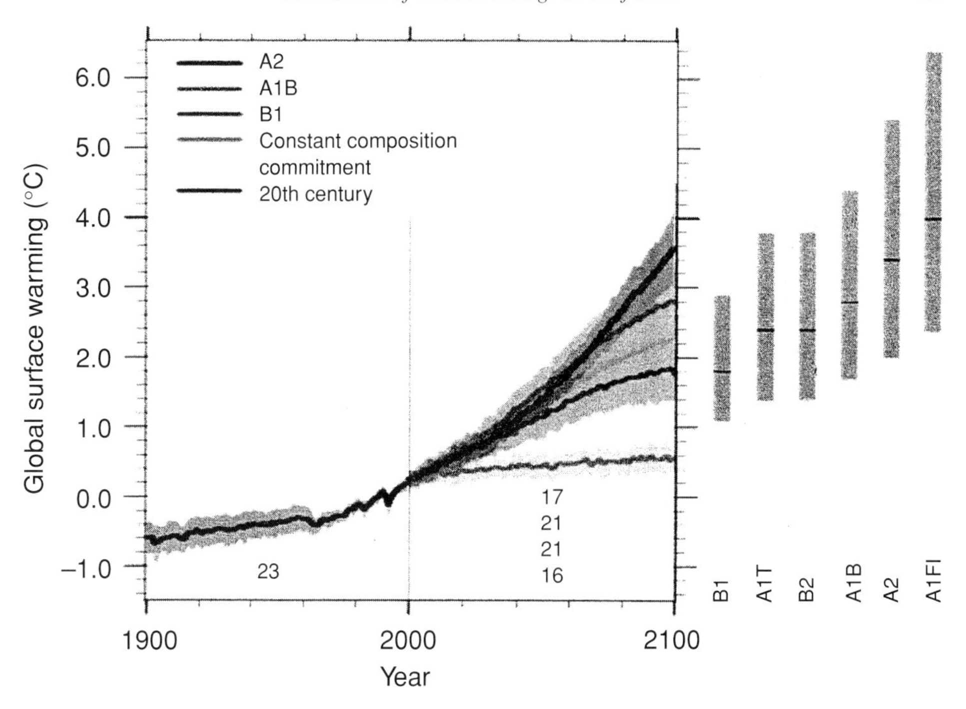

Fig. 18.1 Solid lines are multi-model global averages of surface warming (relative to 1980–99) for the scenarios A2, A1B and B1, shown as continuations of the twentieth century simulations. Shading denotes the plus/minus one standard deviation range of individual model annual means. The number of climate models (AOGCMs) run for a given time period and scenario is indicated by the numbers in the lower part of the panel. The lowest line to the right of the graph is for the experiment where concentrations were held constant at year 2000 values. The bars to the right of the graph indicate the best estimate (solid line within each bar) and the likely range assessed over the six SRES marker scenarios. The assessment of the best estimate and likely ranges include the AOGCMs in the left of the figure, as well as the results from a hierarchy of independent models and observational constraints. (Reproduced with permission from the *IPCC Fourth Assessment Report*, February 2007)

polewards.' Also, 'increases in the amount of precipitation are *very likely* in high-latitudes, while decreases are likely in most sub-tropical land regions'.

Northern Europe is at the same latitude as Labrador, but it is not subjected to the harsh Canadian winters because the warm waters of the Gulf Stream flowing from the Gulf of Mexico keep the temperatures at least 5 °C warmer. This current has another component. As the waters cool in the North Atlantic, they sink and a counter 'conveyor belt' deep current of seawater flows south. Recently, there has been great concern as evidence suggested that warm currents were weakening (Alley *et al.*, 2005). It is postulated that extra freshwater is being released by global warming from melting ice into the North Atlantic. Freshwater is of lower density of than seawater, and as a consequence, melt waters were reducing the capacity of cold water to sink and sustain the Gulf Stream ocean circulation system. If this circulation system is interrupted, an abrupt change in climate could occur. There is evidence from the examination of ice cores that such abrupt changes have occurred in the past, e.g. precipitous North Atlantic climate changes occurred during the last ice age. Some authorities have concluded: 'the threat of an abrupt circulation switch in the North Atlantic seems to be receding but researchers are still worried' (Kerr, 2005, 432). The IPCC Report offers the most up-to-date assessment, concluding that the Gulf Stream, technically the meridional overturning circulation system (MOC), might slow or halt, leading to a fall in temperatures.

On balance, the present evidence (IPCC, 2007a, 12) suggests: 'it is very likely that the meridional overturning circulation of the Atlantic Ocean will slow during the 21st century'. However, 'temperatures in the Atlantic region are projected to increase despite such changes due to the much larger warming associated with projected increases of greenhouse gases'.

The Report also stresses several other very uncomfortable findings.

A. Even if concentrations of greenhouse gases were 'frozen' at the levels of the year 2000, a further warming of about 0.1 °C per decade will be expected from the burden of gases already present in the atmosphere.
B. Global warming will lead to increasing acidification of the oceans. A major report by Orr *et al.* (2005, 681) stresses that there might be important consequences for ecosystems and food chains in the sea, as corals and some planktonic species may have difficulty in maintaining their external calcium carbonate skeletons.
C. There is likely to be considerable sea level rise through thermal expansion and through melting of the Greenland ice sheet, but as yet there is 'no consensus' on the magnitude of sea level change.
D. The Report makes gloomy reading about the distant future. 'Both past and future anthropogenic carbon dioxide emissions will continue to contribute to warming and sea level rise for more than a millennium, due to the timescale required for the removal of this gas from the atmosphere' (IPCC, 2007a, 13).

Concluding remarks

Predictions, certainty and proof

Computer modelling is a very important tool for studying climate change and many different models have been devised based on different assumptions (Goudriaan *et al.*, 1999). What is the status of predictions about climate and ecosystems derived from such techniques?

Given the impossibility of trying to study the full complexity of the Earth climate and ecosystems in their entirety, models allow the investigation of what is hoped are the essential components of systems. However, it has been stressed by Oreskes *et al.* (1994, 1511) that it is impossible to incorporate 'all influencing factors': models are essentially 'abstractions of reality and necessarily ignore many components of reality'. They continue: 'philosophers have argued that models provide useful tools for formulating hypotheses and exploring "what if" questions'. It is important to note, however, in the words of the famous statistician George Box, 'models are never true'. It is 'only necessary that they be useful. For this it is usually needful only that they are not grossly wrong' (Box, 1979).

Concerning the relationship of model to reality, it is also important to note: 'strict validation of models is never possible – models represent hypotheses about reality, and hypotheses can only be falsified' (Goudriaan *et al.*, 1999, 132). This is a key issue emphasised in Chapter 2. Moreover, as they point out: 'The situation is even more peculiar for models of global phenomena because there is not a single independent set of observations to test the models against, since this would require a second planet with comparable features.'

Araújo *et al.* (2005) note that models are generally tested by resubstitution: 'the data used to calibrate (or train) models are also used to validate (test) them'. However, a growing number of studies employ partitioned data, some of which is used to devise and calibrate the model, while another portion is used for validation. Many models generate predictions concerning the future, but they can also be used to 'predict' backwards in time and such predictions can be checked against historical data sets. However, regarding the response of species to climate change, there is 'no assurance that models that show good predictive ability for past range shifts will give reliable predictions of future shifts' (Araújo *et al.*, 2005, 1510). Validation, therefore, remains a difficult issue. However, in some cases, independent data sets, collected at different dates, have been used to validate models of range shifts (e.g. Araújo *et al.* (2005) have used two data sets from studies of birds in the UK in the 1960s and 1990s).

Reflecting on the IPCC Report's assessments of future climate change and its effect on species and ecosystems, it is important to stress that projections are from model simulations. They are couched in the language of probabilities: they estimate *likelihoods not certainties*.

Climate change sceptics

In the prestigious American journal *Science*, Oreskes (2004, 1686) considers the attitudes of climate change sceptics: 'Policy-makers and the media, particularly in the United States, frequently assert that climate science is highly uncertain. Some have used this as an argument against adopting strong measures to reduce greenhouse gas emissions.' In addition, 'some corporations whose revenues might be adversely affected by controls on carbon dioxide emissions have also alleged major uncertainties in the science. Such statements suggest that there might be substantive disagreement in the scientific community about the reality of anthropogenic climate change. This is not the case.' In the past few years, further scientific evidence has accumulated, culminating in the major IPCC reports.

An editorial in *Nature* (Anon., 2007a, 567) celebrates the release of the first section of this four-part report as 'an important milestone'. It notes: 'there has been twenty years of frustrating trench-warfare' surrounding the topic of anthropogenic climate change. The editorial concludes that the IPCC Report 'has served a useful purpose in removing the last ground from under the climate-change sceptics' feet, leaving them looking marooned and ridiculous'.

In the light of these predictions, we must take action, but Oreskes (2004) faces an important issue: 'The scientific consensus might, of course, be wrong. If the history of science teaches anything, it is humility, and no one can be faulted for failing to act on what is not known. But our grandchildren will surely blame us if they find that we understood the reality of anthropogenic climate change and failed to do anything about it.'

Tackling the problems of climate change

In a reaction to the IPCC Report, an editorial in *Nature* (Anon., 2007a) looks ahead at the major challenges confronting mankind.

> The world now broadly accepts that we have a problem, if not a crisis. So what can be done? . . . at present the political response to the situation is, in large part, incongruous. We need to restrict emissions in the developed world, and some steps are being undertaken to do just that, chiefly through the much-maligned Kyoto Protocol. We need to develop clean energy sources, although each one – nuclear power, biofuels, wind power and hydropower, for example – creates its own environmental battlefield . . . Even the most progressive governments continue to put the issue of climate change on the back seat behind their fundamental commitment to strong economic growth . . . the fundamental difficulty here is that it has been politically impossible to accept that fighting global warming may involve some economic sacrifice, at least while the sceptics were in the picture. As these are vanquished, it becomes possible – and indeed necessary – to start the discussion.

While some may consider that the sceptics have been defeated, it is important to look deeper into the issues. Science is a 'truth-seeking' endeavour and, therefore, it must be stressed that present knowledge is always provisional. 'Professional' scepticism for political or commercial ends should be seen for what it is. However, society should value 'true' sceptics, who challenge the received orthodoxy and through their ideas, observations and theories advance our scientific understanding. As we saw in Chapter 2, scientific research should always involve 'a search for error' (Landau quoted in Caldwell, 1996, 402).

19

Microevolution and climate change

As we celebrate the 200th anniversary of his birth and the 150th anniversary of the publication of *On the Origin of Species* in 2009, it is interesting to realise that Darwin made predictions about the outcome of a change in climate, a natural event in his speculation, rather than the anthropogenically induced changes we now face (Darwin, 1901, 59):

> We shall best understand the probable course of natural selection by taking the case of a country undergoing some slight physical change, for instance, of climate. The proportional numbers of its inhabitants will almost immediately undergo a change, and some species will probably become extinct. We may conclude, from what we have seen of the intimate and complex manner in which the inhabitants of each country are bound together, that any change in the numerical proportions of the inhabitants, independently of the change in climate itself, would seriously affect the others. If the country were open on its borders, new forms would certainly immigrate, and this would likewise seriously disturb the relations of some of the former inhabitants. Let it be remembered how powerful the influence of a single introduced tree or mammal has been shown to be. But in the case of an island, or of a country partly surrounded by barriers, into which new and better adapted forms could not freely enter, we should then have places in the economy of nature which would assuredly be better filled up, if some of the original inhabitants were in some manner modified; for, had the area been open to immigration, these same places would have been seized on by intruders. In such cases, slight modifications, which in any way favoured the individuals of any species, by better adapting them to their altered conditions, would tend to be preserved; and natural selection would have free scope for the work of improvement.

How far have Darwin's predictions concerning microevolution been realised? Each of a number of propositions will be examined in turn, including responses to elevated carbon dioxide; changes in life cycles and range shifts; and microevolutionary changes in response to climate.

Responses to increased carbon dioxide in the air, rising temperatures and drought

Carbon dioxide

Photosynthesis in plants involves one of three metabolic pathways. In a recent review, Ziska and Bunce (2006) consider the implication of the fact that plants evolved when atmospheric CO_2 was greater than at present (four or five times present levels). They conclude: 'because CO_2 remains the sole source of carbon for plant photosynthesis, and because at present CO_2 is less than optimal, as atmospheric CO_2 increases, photosynthesis at the biochemical level will be stimulated accordingly'. This prediction is confirmed using a variety of experimental approaches, including studies of leaves in enclosed systems, the behaviour of plants exposed to elevated CO_2 in open-top chambers compared with control chambers with ambient CO_2 etc. (see Bowes, 1996).

However, about 4% of all known plant species have a different photosynthetic pathway (C_4 rather than C_3) that may have evolved 'in response to declining CO_2 and warmer climates' (Ziska & Bunce, 2006). On biochemical grounds, they report that an increase in atmospheric CO_2 'should have little effect on net photosynthesis in C_4 plants'. However, some researchers have detected enhanced photosynthesis in such plants in response to elevated CO_2. On present evidence, Ziska and Bunce (2006) conclude: 'many details regarding how the C_4 biochemical and cellular mechanisms responds to elevated CO_2 remain unclear'. Evidence suggests that increases in atmospheric CO_2 should stimulate photosynthetic rate in the third group of plants that have Crassulacean acid metabolism (*c*. 1% of all plant species).

From the microevolutionary standpoint the interactions between these different groups is of particular interest. Ziska and Bunce (2006) point out that competition can occur when species respond differently to 'resource enhancement'. In managed agricultural situations competition has been examined in studies of the interactions of C_3 and C_4 crops growing with C_3 and C_4 weeds. In some situations, elevated CO_2 favours the weeds, but in other cases the crop is at a competitive advantage (Table 19.1). It is important to note that some of these tests were carried out under glasshouse conditions. Further field-based trials are required to extend these studies to take account of other interacting factors; namely, the wide regional differences in soils, precipitation, temperature, land management regimes etc.

Less is known about the effects of elevated CO_2 on unmanaged ecosystems, but some important pioneering investigations have been carried out. For instance, in a marsh environment under elevated CO_2, the biomass production of the C_3 sedge (*Scirpus olneyi*) exceeded that of the C_4 grass *Spartina patens* (Curtis *et al*., 1989).

Only a relatively small number of studies of the effects of elevated CO_2 on interspecific competition have been carried out, and, therefore, generalisations about the behaviour of C_3 and C_4 plants should be avoided (Ziska & Bunce, 2006), especially as increasing CO_2 is only one facet of the increasing anthropogenic impacts on ecosystems (e.g. land use, water resources, nitrogen deposition etc.). They conclude: 'any experimental approach that focuses on ecosystem dynamics ... should take not only CO_2 into account, but also other rapidly changing variables'.

In experiments in which invasive and weedy species were grown in elevated CO_2 (together with ambient controls), a number of important findings were obtained. Increased CO_2 in the atmosphere may result in mounting difficulties in controlling invasive and weedy plant species.

Table 19.1 *Summary of studies on mixtures of C_3 and C_4 plants examining whether weed or crops were 'favoured' under conditions of elevated carbon dioxide. Favoured indicates that under such conditions 'winners' produced more crop or weed biomass. Pasture contained a mixture of C_3 grass species.*

Crop	Weed	Increasing [CO_2] favours?	Environment
A. C^4 crops/C^4 weeds			
Sorghum	*Amaranthus retroflexus*	Weed	Field
B. C^4 crops/C^3 weeds			
Sorghum	*Xanthium strumarium*	Weed	Glasshouse
Sorghum	*Abutilon theophrasti*	Weed	Field
C. C^3 crops/C^3 weeds			
Soybean	*Chenopodium album*	Weed	Field
Lucerne	*Taraxacum officinale*	Weed	Field
Pasture	*Taraxacum and Plantago*	Weed	Field
Pasture	*Plantago lanceolata*	Weed	Chamber
D. C^3 crops/C^4 weeds			
Fescue	*Sorghum halepense*	Crop	Glasshouse
Soybean	*Sorghum halepense*	Crop	Chamber
Rice	*Echinochloa glabrescens*	Crop	Glasshouse
Pasture	*Paspalum dilatatum*	Crop	Chamber
Lucerne	Various grasses	Crop	Field
Soybean	*Amaranthus retroflexus*	Crop	Field

Adapted from Ziska and Bunce (2006). Reproduced with permission of Blackwell Publishing Ltd.

- CO_2-induced increases in productivity of Cheatgrass (*Bromus tectorum*) were detected, and with climate change this may be leading to increased fire frequency and intensity as this invasive species is stimulated to produced higher above-ground biomass giving greater fuel loads (Ziska, Reeves & Blank, 2005).
- Enhanced CO_2 results in increased growth and toxicity of Poison ivy (*Toxicodendron radicans*), a species famous as an important cause of contact dermatitis in the USA (Mohan *et al.*, 2006).
- Increase in atmospheric CO_2, in line with projections for 2100, results in substantially increased root biomass – relative to above-ground parts – in Canada thistle (*Cirsium arvense*) (Ziska, 2002). This response may make it more difficult to control this noxious weed by glyphosate herbicide (Ziska, Faulkner & Lydon, 2004).

Rising temperatures

Studying the effect of rising temperature on plants and ecosystems presents many problems. Three pioneer studies point the way. The effects of higher temperatures have been studied in the Karoo region near Vanrhynsdorp, South Africa (Musil, Schmiedel & Midgley, 2005).

This area is a biodiversity hotspot for succulents. By means of open-top acrylic chambers placed over vegetated plots, passive daytime maximum temperatures were elevated by an average of 5.5 °C. Survival and growth in these plots was compared with that in similar-sized uncovered control plots subject to ambient air temperatures. Comparing the effect of treatment and control, there was a 2.1 to 4.9-fold increase in plant and canopy mortality of the foliage of dwarf succulents under the 'canopies'. This is not surprising, as the plants, on occasion, were subjected to temperatures beyond 55 °C (the upper temperature limit for most vascular plants). The authors conclude that it is 'likely that current thermal regimes are closely proximate to tolerable extremes for many of the 1563 almost all endemic succulent species . . . which diversified rapidly in the region during the cool Pleistocene'. Looking to the future of these species, the authors take the view that 'anthropogenic warming could therefore significantly exceed their thermal thresholds resulting in localised extinctions of particularly those specialised species range-restricted to specific habitats'.

Experiments have also been carried out in other ecosystems, e.g. in 32 plots representing forest, grasslands, high and low latitude/altitude tundra ecosystems in North America. The authors report an average 19% productivity increase in response to an average 2.4 °C of experimental warming in the 20 plots distributed in the cooler regions above 35° N (Rustad *et al.*, 2001). There have been other studies on the effects of elevated temperatures (see Stenstrom & Jonsdottir, 2006; Lambrecht *et al.*, 2007, and references cited therein). Further investigations are needed to understand this complex area, but there is reason to suppose that species groups will not all react in the same way.

The combined effects of increased temperature and elevated CO_2 have been examined in an imaginative experiment carried out in the Baltimore area, USA, involving study plots in the city centre (A), an outer suburb (B) and a rural location (C). In the city centre, atmospheric CO_2 is, on average, elevated by 66 ppm relative to rural sites. Also, because of the heat island effect of the city, the temperature was consistently higher at A (14.8 °C), than B (13.6 °C) and C (12.7 °C). Differences in nitrogen deposition at the three sites were also studied but this did not appear to affect soil nitrogen content (George *et al.*, 2007). Soil from the rural site was collected and set out in uniform beds in the three sites. Over five growing seasons, the plant populations that grew from the soil seed bank were monitored. In brief, two major observations were made. Spectacularly greater weed swards grew in the city plot. In addition, in relation to plots B and C, there was accelerated succession to shrubs and trees in the city, especially invasive *Ailanthus*, Norway maples and mulberries (Christopher, 2008). The spectacular response of weedy and invasive species is perhaps an indication of the way urban ecosystems might respond under conditions of climate change.

Drought

There is great concern about the fate of the tropical rain forests under global warming. It is predicted that droughts will increase in the Amazon region under climate change. This could lead to profound changes in the forest ecosystems, for example: altered hydrology; reductions in the capacity of the forests to act as a carbon sinks; and a greater risk of fires (Laurance, 1998; Stokstad, 2005a).

These hypotheses have been tested in an experiment set up in the Tapajós National Forest in Brazil, in an area with seasonal drought (Nepstad *et al.*, 2002; Stokstad, 2005a,b). A 1-hectare plot of the forest ecosystem was shielded from rain by building a canopy of 5,600 plastic panels. It is estimated that about 80% of the rainfall was intercepted. Towers were also erected and trenches dug to enable the ecosystem to be studied from the 'basement to the attic'. The responses of plants in the forced-drought plot were compared with those of an uncovered control plot.

While the forest proved resilient at first, trees slowed or stopped their growth under the canopy. Four years into the experiment, however, more dramatic changes were evident. There had been many deaths (up to 9% per year) amongst the high canopy trees, allowing more light to reach and dry out the floor of the forest. Fire risk was assessed and it was discovered that the control plot was highly inflammable for about 10 days per year. In contrast, the 'covered' plot was highly vulnerable for 8–10 weeks each year. Also, the covered plot was only able to store 2 tons of carbon as wood, whereas 7 tons were stored in the control plot. Overall the experiment revealed that the forest ecosystem under study has a capacity to survive regular dry seasons, presumably though evolutionary adaptation, but extended droughts can be very damaging.

It is not clear how far the results of this experiment can be used to predict the responses of forests elsewhere. The Tapajós Forest naturally experiences a dry season, but large areas of Brazilian rain forest do not have regular dry seasons and species in these ecosystems may have no strategies for coping with drought. In addition, the imposed drought treatment in the experiment was abrupt and very severe, perhaps representing 'a worst-case scenario'. Climate change might be gradual, giving the forest ecosystems an opportunity to adapt. Whether species stressed by drought and other factors are capable of such adaptation will be considered below.

Interacting factors

Long-term investigations at a site near Toolik Lake, in the Brooks Range of arctic Alaska, have attempted more complex experiments to examine the interaction of several factors likely to change under global warming. A replicated experiment with appropriate controls was set up to examine, among a range of treatments, the effect of additional nutrients (predicted to rise as a result of soil warming), and increases in temperatures (achieved by placing a plastic greenhouse over 'appropriate plots' during the growing season). It is beyond the scope of this account to examine the responses to treatments in detail. Overall the authors conclude that species did not react uniformly to the various treatments and predict that individualistic responses in growth and mortality would be expected to accompany a '3–8 °C increase in air temperature, increased cloudiness and increased nutrient availability' (Chaplin & Shaver, 1996).

A call for the wider use of reciprocal transplant experiments

While acknowledging the importance of use of multi-factorial experiments, Midgley and Thuiller (2005) have called for the wider use of reciprocal transplant techniques to investigate the responses of plants in a wide range of geographical sites with different climates. These simpler experiments would be invaluable for studying the effect of the totality of changing environmental factors, and would be very revealing in determining the range limits of species/regional ecotypes etc.

Recent changes in the timing of various life cycle events

From 1936 to 1947, the celebrated conservationist Aldo Leopold kept records at his farm in Fairfield, Wisconsin, USA, of the date of first flowering of certain species in the spring and the date of arrival of certain migratory birds. Recording was resumed by Nina Leopold Bradley from 1976 to 1998 (Bradley *et al.*, 1999). Analysis of these records reveals that, for many species, spring events have become earlier, such as the flowering of *Phlox divaricata* and *Baptisia leucantha*. For others there was no such shift to early flowering (e.g. *Penstemon gracilis* and *Hypericum perforatum*).

Several long-running dated records of spring events have been discovered in Europe (Menzel *et al.*, 2005). For example, in 1882, a Phenological Network was founded by Hoffman and Ihne at Giessen in Germany, yielding yearly records from 1883 to 1941. Their findings provide a base line with which to judge current changes. For instance, they reveal that spring does not simply progress from south to north across Europe, but 'for early, mid and late spring we found a marked progress of the seasonal onset from SW to NE throughout Europe. More precisely from WSW to ENE in early spring, then from SW to NE and finally from SSW to NNE in late spring.'

Walther *et al.* (2002, 389) have considered a range of studies of phenology in Europe and North America, and conclude that there are many examples of changes in the timing of spring activities including 'earlier breeding and first singing of birds, earlier arrival of migrant birds, earlier appearance of butterflies, earlier choruses and spawning of amphibian and earlier shooting and flowering of plants'. Later onset of autumn phenological events is also well documented, but 'these shifts are less pronounced' and reveal more 'heterogenous behaviour' than those of spring events. Thus, 'in Europe, for example, leaf colour changes show a progressive delay of 0.3–1.6 days per decade, whereas the length of the growing season has increased in some areas by up to 3.6 days per decade over the last 50 years.'

Evidence of recent vegetation changes ascribed to climate change has been detected by studying satellite images (Pearce, 2001). Over the past 20 years, land between 40 and 70° N is markedly greener (by 12% in Russia and central Asia, while Canada and northern USA are 8% greener). Greening is also occurring at an earlier date. These phenomena are linked to the length of the growing season, which has increased by 18 days in Europe and Siberia and 12 days in North America.

In northern latitudes, growth is limited by temperature: the situation is completely different in the tropics, where plants already have optimum temperature conditions. Here, climate change has resulted in droughts (through higher temperature and lowered or altered patterns of precipitation) resulting in major changes in vegetation in some places.

Providing yet another review of the expanding data sets, Schwartz, Ahas and Aasa (2006) conclude that phenological changes are occurring across most temperate land in the northern hemisphere and produce some figures of the extent and rate of change. There is 'nearly universal quicker onset of early spring warmth (spring indices (SI) first leaf date, -1.2 days decade^{-1}), late spring warmth (SI first bloom date, -1.0 day decade^{-1}), last spring day below 5 °C (-1.4 days decade^{-1}), and last spring freeze date (-1.5 days decade^{-1}) across most temperate NH [northern hemisphere] land regions over the 1955–2002 period'. However, it is important to stress two important points, (a) the heterogenous nature of species' responses; and (b) that spring events have not advanced in all areas. For instance, a later start to the growing season has been reported in the Balkans (Walther *et al.*, 2002).

Phenology: different environmental cues provide the triggers for key processes

In some 'responsive' plant and animal species the timing of spring events is critically influenced by temperature. However, not all species respond in the same way. As evidence has accumulated, it is clear that while some species have phenological adaptability in relation to temperature, others do not. Thus, in species 'non-responsive' to temperature, phenological behaviour is regulated by a variety of different genetically controlled mechanisms. For instance, photoperiod is the key factor in many species of plants and animals, controlling such processes as dormancy, growth rates and 'flowering in plants; diapause in insects; reproductive activity in vertebrates; and migration in birds' (Bradley *et al.*, 1999). The microevolutionary implications of these differences will be explored below, but clearly, without phenological adaptibilty, some species may face increasing stress under global warming.

Range shifts: vegetation zones and individual species

Climate change is predicted to produce movement of vegetation zones and species, towards the poles latitudinally and up mountains altitudinally (Figs. 19.1 and 19.2). Evidence of recent latitudinal and altitudinal range shifts comes from studies of a variety of plants and animals (Table 19.2 from Walther *et al.*, 2002). Grabherr, Gottfried and Pauli (1994) surveyed montane plants on 26 mountain summits in Switzerland and compared current species distribution with historical records. A recent pronounced shift to greater species richness at higher elevations has been detected. However, for nine species there were more detailed records, revealing a less than 1–4 m increase in altitude per decade. Altitudinal changes in ecosystems have also been found in Costa Rica, with losses of cloud forest species and invasion of species from lower elevations (Pounds, Fogden & Campbell, 1999).

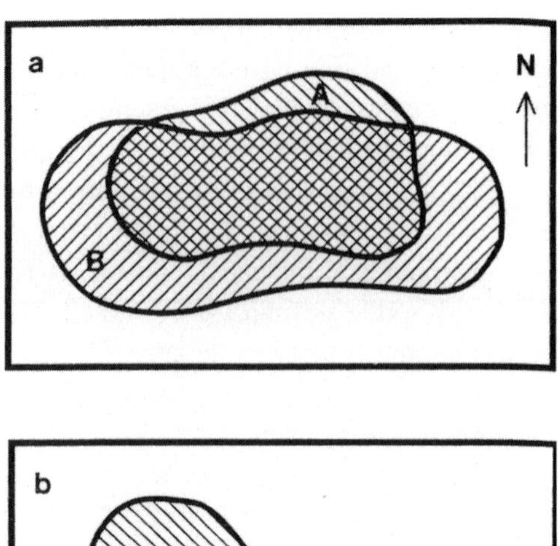

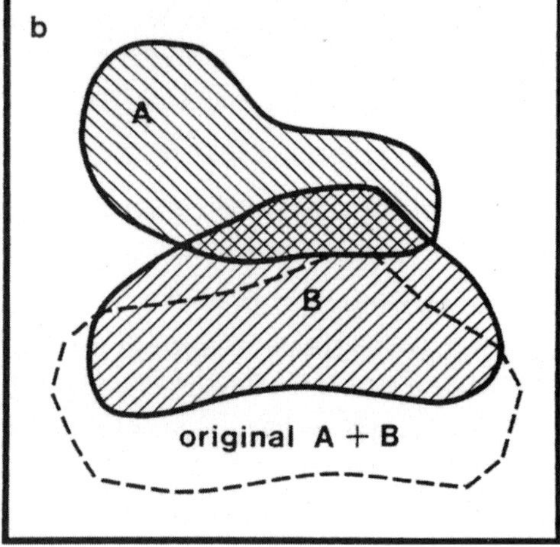

Fig. 19.1 Changes in species distributions with climate change. (a) Initial distribution of two species A and B whose ranges largely overlap. (b) In response to climate change, latitudinal shifting occurs at species-specific rates and the ranges dissociate. (From Peters, 1992. Reproduced with permission from Peters & Lovejoy (eds.) *Global Warming and Biological Diversity* © Yale University Press)

Studies of forests in western USA have revealed that establishment has increased in the sub-alpine areas, at the tree line itself, and, significantly, young trees have become established above the present tree line (Peterson, 1994). In the maritime Antarctic, there is evidence of the expansion in populations of *Colobanthus quitensis* and *Deschampsia antarctica* in response to warming trends (Smith, 1994).

While this book focuses mainly on plants, it is important to recognise that vegetation and its component plant species may be decisively influenced by expansion polewards of many animal groups. The UK fauna and flora have been intensively studied and there is a great deal of evidence of past and present distribution of a wide range of species. Using data sets for a wide variety of vertebrate and invertebrate species – from spiders to freshwater fish, from soldier beetles to woodlice – fine-scale changes in distribution have been examined by Hickling *et al.* (2006). They conclude that a variety of species in a wide diversity of groups

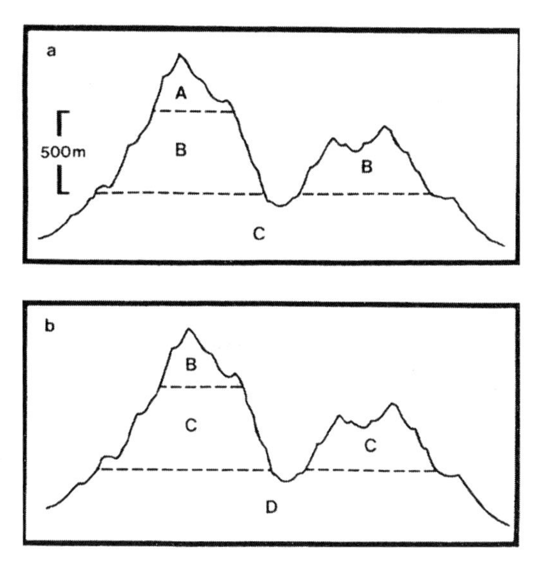

Fig. 19.2 (a) Present altitudinal distribution of three species A, B and C. (b) Species distributions after a 500-meter shift in altitude following a 3 °C climate change. Species A becomes locally extinct, B colonises higher ground and comes to occupy less total area, while another species D comes to occupy the lower slopes. (Peters, 1992. Reproduced with permission from R. L. Peters & T. E. Lovejoy (eds.) *Global Warming and Biological Diversity* © Yale University Press)

have 'moved northwards and uphill in Britain over approximately 25 years, mirroring, and in some cases exceeding, the response of better-known groups'.

Are apparent range changes the result of changing climate?

The changing distribution of plants and animals can be influenced by many factors. It is important to consider these in any study of the apparent effects of climate change. A study of butterflies in the UK provides important insights. The investigation examined the distribution of 46 butterfly species that approach their northern climatic range margins in the UK. Warren *et al.* (2001) examined the question: has recent climate change led to recent range extension? They discovered that half the species – those that are habitat generalists and mobile – have increased their distributions in line with expectations. However, other generalists and 89% of habitat specialist species show reductions in distribution. The authors consider that these reductions in species area are the result of habitat loss, rather than the direct result of climate change.

The effects of climate change are more concentrated with elevation than with latitude. Thus, 'a 3 °C change in mean annual temperature corresponds to a shift in isotherms of approximately 300–400 km in latitude (in the temperate zone) or 500 m in elevation' (Hughes, 2000). Appreciating this point, a number of researchers have studied and detected upward colonisation of trees and shrubs at/near/above the tree line in alpine areas. It is important to stress, however, that the elevation of tree lines on mountains should not be

Table 19.2 *Ecological responses to recent climate change: latitudinal and altitudinal shifts*

Species[a]	Location	Observed changes	Climate link
Treeline	Europe, New Zealand	Advancement towards higher altitudes	General warming
Arctic shrub vegetation	Alaska	Expansion of shrubs in previously shrub-free areas	Environmental warming
Alpine plants	European Alps	Elevational shift of 1–4 m per decade	General warming
Antarctic plants and invertebrates	Antarctica	Distribution changes	Liquid water availability and increased temperature
Zooplankton, intertidal invertebrate and fish communities	Californian coast, North Atlantic	Increasing abundance of warm-water species	Warmer shoreline ocean temperature
39 butterfly species	North America and Europe	Northward range shifts up to 200 km over 27 years	Increased temperatures
Edith's Checkerspot butterfly (*Euphydryas editha*)	Western United States	124 m upward and 92 km northward shift since the beginning of the twentieth century	
Lowland birds	Costa Rica	Extension of distribution from lower mountain slopes to higher areas	Dry season mist frequency
12 bird species	Britain	18.9 km average range movement northwards over a 20-year period	Winter temperatures
Red fox (*Vulpes vulpes*), Arctic fox (*Alopex lagopus*)	Canada	Northward expansion of red fox range and simultaneous retreat of Arctic fox range	General warming

[a] Where possible, numbers of species that showed a response to climate change are given.

From Walther *et al.* (2002).

automatically interpreted as a response to climate change. In many parts of the world, human activities have had a major influence on the tree line. For example, in Europe, grasslands above abrupt tree lines are often the result of traditional summer grazing by domestic stock. The care needed to analyse causes of changes at the tree line is illustrated by studies of high-mountain vegetation in the Spanish Central Range (Sanz-Elorza *et al.*, 2003). Comparison of aerial photographs, from 1957 and 1991, reveals that high-mountain grasslands are being invaded by scrub patches of species characteristic of lower wooded slopes. The authors conclude that these changes are *probably* due to climate change. However, they stress that all possible contributing factors must be evaluated, such as fire and grazing regimes, contamination of the soil by mining, air pollution, including the possibility of changes in acid rain etc. In this case, there is no evidence of fires for two centuries, and no detectable pollution. At the end of the nineteenth century, sheep grazing (by small numbers of animals) was replaced by cattle grazing at equivalent stocking rates. But, overall, the authors conclude that changes in grazing practices may have possible synergistic effects with climate change.

Human-induced climate change induces species changes

In order to test whether changes in phenology and range shifts can be attributed to an anthropogenic component of climate change, modelling and statistical analyses were carried out by Root *et al.* (2005). Data from 29 published studies were selected for close examination. Only species that exhibit statistically significant relationships to the warming signal were selected for study. Models were devised to examine three scenarios based on estimates of: (i) underlying natural climate forcing (NF); (ii) forcing attributable to anthropogenic greenhouse gases and aerosols (AF); and (iii) coupling of both natural and anthropogenic forcings (CF). Reflecting on their findings, the authors make it plain that: 'using modeled climatic variables and observed species data . . . we demonstrate statistically significant "joint attribution", a two step linkage: human activities contribute significantly to temperature changes and human-changed temperatures are associated with discernable changes in plant and animal traits'. Exploring the emerging patterns very closely, Menzel *et al.* (2006) have confirmed that the 'European phenological response to climate change matches the warming pattern.'

Phenological and range-shift changes are microevolutionary responses

There is evidence that both the degree of 'adaptability' and the extent to which the phenotype of the plant can change are responses under genetic control (Sultan, 1987). It follows, therefore, that species/ecotypes/genetic individuals in populations will be subjected to natural selection by the growing signals of climate change, and that they are highly likely to respond individualistically.

Under conditions of continuing climate change, it is highly likely that populations of species will reach the limits of their development adaptability and phenotypic plasticity, and organisms will be subjected to directional selection. A simple model of evolutionary responses to climate change has been proposed by Bradshaw and McNeilly (1991, 9).

Populations contain a 'pool of genetic variation', and, if it contains appropriate genetic variation, plants and animals 'may be able to respond to selection, change genetically and thereby survive'. If no genetic variability is available 'by which the species can cope with the new selection pressure', the population will become extinct (Bradshaw & McNeilly, 1991). They continue: 'exactly what happens may be complicated: the species may have sufficient variability to respond to a certain extent, and then be able to respond no further ... limits to selection must always be reached'. Species may acquire new variation appropriate to the changing situation through hybridisation (see Chapter 13) or 'by the slow process of mutation – or become extinct'. They also point out the complexity of the selection processes under climate change, which presents multi-stress situations involving not only changing temperature regimes, but also changes in precipitation, storm frequency, sea level etc.

Plant responses to previous climate changes

With increasing climate change, alternative outcomes are possible. Populations may adapt '*in situ*' or survive through migration. Bradshaw and McNeilly (1991) stress this point: 'If a species is evolving in relation to climate at a time when major changes are occurring, then, in the simplest hypothesis, it might be expected to remain in the same geographical area without migrating.' Alternatively, while populations of species might become extinct in their original sites, range shifts may occur though migration.

Evidence of the responses of species in the past is provided by detailed investigations of geological periods of dramatic climate changes, as occurred in the Quaternary period (Fig. 19.3). The post-glacial responses of species over the last 12,000 years have been reconstructed from the records of pollen, seeds and other plant remains preserved in dated anaerobic deposits of various kinds – peat, lake and marine sediments etc.

Considering the responses of species to climate change, Bradshaw and McNeilly note: 'what is remarkable is that almost all species have behaved as though they were evolutionary fixed: to cope with climate change they moved over hundreds or thousands of miles'. Patterns of behaviour reveal 'simultaneous retreat of their range limits along one front and extension of their range limits on another front'. They also consider the evidence from the changing distributions of arctic-alpine species such as *Dryas octopetala* and *Betula nana*. 'In the late-glacial period these were abundant over much of Britain, yet their distributions are now very restricted. They have obviously failed to evolve in relation to the demands of a warming climate.' In contrast, some species, such as *Betula pubescens*, *Ranunculus repens* and *Succisa pratensis*, 'have been able to persist in the same lowland locations despite the warming climate. Whether these have done so by evolutionary change has never been tested'. Bradshaw and McNeilly (1991) conclude:

> Most species, but perhaps not all, are unable to evolve, or evolve sufficiently, to cope with all aspects of the climate change. Although species may be able to evolve to some extent, they are certainly not able to evolve enough, to all the different aspects, to be able to remain in their original habitats as climate changes; they will be forced to migrate. Then, if geographical features prevent migration, they will become extinct, for example as *Tsuga* and *Pterocarya* are in Europe.

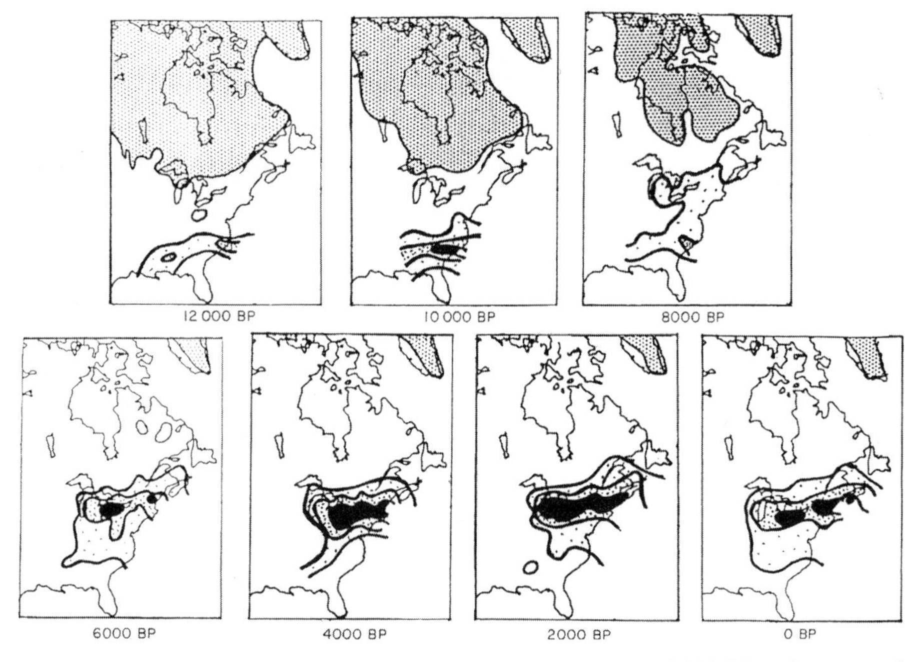

Fig. 19.3 Isopoll maps of *Fagus grandifolia* in the eastern USA from 12,000 BP to the present day (isopolls at 1, 5 and 10%). (From Davis *et al.*, 1986. Reproduced with permission of Springer Science and Business Media)

Adaptation and migration of species: is there a significant role for selection?

In a highly illuminating review, Davis and Shaw (2001) provide some important insights into migration and adaptation, by questioning assumptions that are sometimes made. Their comments are also helpful in considering the concept that species have a 'climatic envelope', which is considered below. It is sometimes supposed that: (i) migration is essentially a passive affair mediated by seed and pollen dispersal; (ii) the responses of the species will remain stable; (iii) intraspecific genetic variability can be ignored; (iv) species are genetically undifferentiated 'comprising individuals with broad tolerances'. Davis and Shaw point out that this assumption 'resembles the concept of the "general purpose genotype" proposed 40 years ago for weedy plant species'. However, as we saw in Chapters 8 and 9, many weedy species are genetically differentiated. And (v) finally, it could also be assumed that 'evolutionary change only occurs on long time scales'.

Davis and Shaw (2001) insist that migration in relation to climate change involves 'a great deal more than dispersal of seed ahead of the advancing species front'. By means of common garden experiments, provenance trials and reciprocal transplant experiments etc., it has been demonstrated that present-day populations of a wide range of species are ecotypically differentiated. In essence, populations have been (and indeed, are being) subjected to the action of natural selection in local habitats and those with the greatest Darwinian fitness survive. There is very strong evidence that ecotypic differentiation has occurred in many forest species such as white spruce, lodgepole pine, red alder, and Scots pine in Sweden.

If present-day populations are adaptively differentiated, then it can be assumed that this also occurred in the past. As all populations at all times are subjected to natural selection, Davis and Shaw deduce that plant and animal migration during the many climate changes of the Quaternary was accompanied by adaptation throughout the species' range. Moreover, as climate change alters local conditions across the range of the species, in the course of time, poorly adapted genotypes will occur throughout a species' range, and be subject to natural selection.

Davis and Shaw consider that gene flow would have been important throughout the range of the species. Regarding the advancing front, founder effects are highly likely.

> Dispersal is likely to be random with respect to a seed's adaptation to conditions where it lands, differential survival during seedling establishment selectively 'sieves' out genotypes that do not tolerate local conditions. Differential growth and reproduction further promote adaptation of physiological characteristics. The arrival of seeds that are somewhat 'preadapted' to the novel climate (e.g. seeds from more southerly populations during periods of climate warming) may contribute to adaptation, yet selection would also promote new genetic combinations, for example of photoperiod and temperature responses suited to the novel growing season.

Migration in the post-glacial

Advances in our understanding of the routes of migration in the post-glacial have come from molecular studies (Avise, 2000; Hewitt, 2000; Hewitt & Nichols, 2005). At the present day, species of plants and animals often have regional variants, and studies of DNA markers have revealed that many of these are genomically distinct. By examining the intraspecific similarities and variability of the DNA, it has been possible to reconstruct patterns of post-glacial migrations revealing 'from which refugia particular genomes emerged' to reach their present distributions. For example, in summary, the results of such analyses reveal that 'Britain received oaks, shrews, hedgehogs and bears from Spain, and grasshoppers, alder, beech and newts from the Balkans' (Fig. 19.4). However, while different species came from the same broad region, the pathways in the different groups are highly individualistic.

These findings, together with a mass of information from Quaternary studies, reveal that the members of present ecosystems in northern Europe and North America did *not all migrate together in concert* in their post-glacial expansion from southern areas and refugia. The behaviour of each species was individualistic. Some of this individualism comes from the wide diversity of breeding systems in plants. 'An organism's means of dispersal, which produce gene flow and allows it to reach new suitable places, are genetically determined and the products of considerable evolution from ancient to modern times' (Hewitt & Nichols, 2005). Molecular studies have also made it possible to study genetic variability across territories. Some very important trends in genetic diversity have been identified with implications for conservation (Dolan, 1994). Thus: 'northern Europe has less genetic

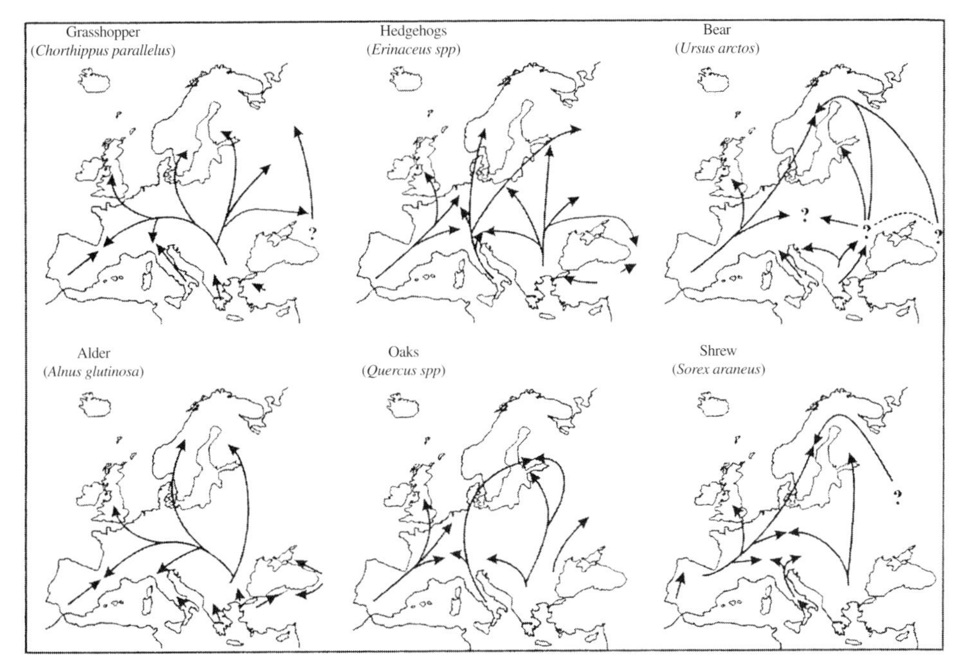

Fig. 19.4 Migration after the Ice Age: the post-glacial colonisation patterns of species from southern European refugia. The grasshopper, the hedgehog and the bear are three paradigm patterns emerging from different combinations of Iberia, Italy and the Balkans; likewise, the alder, oak and shrew. (From Hewitt & Nichols, 2005. Reproduced with permission of Yale University Press ©)

variability than southern Europe in terms of numbers of species, subspecies divisions, and allelic diversity' (Hewitt & Nichols, 2005).

Major questions arise from consideration of migration in the past. Given that adaptation *in situ* might occur under climate change, is there any direct evidence for the action of natural selection? Regarding the migration of plant and animal species, are past events a guide to future events? What are the implications for conservation of climate changes? Each of these issues will be considered in turn.

Microevolutionary responses to climate change

Evidence comes from a variety of experiments.

Comparison of ancestral and descendant populations

Franks, Sim and Weis (2007) studied the potential for an evolutionary response to multi-season drought by examining the speed of development in the life cycle of field mustard (*Brassica rapa*) in southern California. In a set of common garden environments, the behaviour of 'ancestral' plants (raised from stored seed collected in 1997) was compared with 'descendant' populations collected (in 2004 from the same sites), after extreme droughts from 2000 to 2004 produced abbreviated growing seasons. Descendant populations proved to be earlier flowering 'by 1.9 days in one study and 8.6 days in another'. The intermediate behaviour of the cross between ancestral and descendent plants supports the view that differences in behaviour are heritable. As a means of escaping the effects of the drought, the quicker maturity of the descendant populations provides evidence of a rapid adaptive response.

Artifical selection experiments

In an experiment by Potvin and Tousignant (1996), on the short-lived species *Brassica juncea*, plants were raised under controlled conditions. Artificial selection was imposed on one set of plants by raising them through seven generations of conditions simulating climate change. A companion set was raised under ambient conditions. Two sets of plants, representing the selected and unselected material, were grown in a common garden experiment. In comparing the growth and behaviour of the two sets of material the climate-change selection lines did not show any response for five reproductive traits. It was concluded that this self-fertilising species was unable to respond to artificial selection, a limitation most probably resulting from inbreeding depression. It would be revealing to extend these experimental approaches to species with other types of breeding system.

Use of reciprocal transplant experiments

With climate change, northern populations of a widespread species are likely to experience a climate similar to that found in the southern parts of their range. This model was tested with *Chamaecrista fasciculata*, which has genetically different regional variants. Three populations were reciprocally transplanted in the Great Plains area, USA. Growth and reproduction of the plants were examined. The investigations revealed that northern populations are likely to be severely challenged by climate change, as their Darwinian fitness was significantly reduced 'in hotter drier climates as a result of lower seed production'.

Molecular approaches

Jump *et al.* (2006) have investigated European populations of *Fagus sylvatica*, a tree found in different climatic zones. In association with dendrochronological investigations, molecular studies revealed a DNA marker that is associated with 'temperature-related adaptive differentiation'. Gene frequency at this locus varied predictably with temperature. Jump and associates concluded that while *Fagus* 'may show some capacity for an *in situ* adaptive response to rising temperatures', this is not 'enough to allow all populations to persist in all of their current locations'.

Microevolutionary changes in animals

There have been a number of long-term studies of animal populations that were not initially concerned with climate change. From these detailed 'base-line' studies, insights into the effects of climate change have been obtained. Such experiments provide encouragement to botanists to initiate more long-term studies on the effects of climate change.

For 30 years, researchers have studied the pitcher plant mosquito (*Wyeomyia smithii*), which completes its development within the pitchers of the insectivorous purple pitcher plant (*Sarracenia purpurea*). Bradshaw and Holzapfel (2001) detected genetic differences between southern and northern populations of the mosquito. Adaptively significant differences were detected reflecting the shorter season in the north. However, recent climate change is producing a longer growing season in the northern part of the range of the mosquito, and there is evidence of adaptive genetic responses in life history of northern populations over a time interval as short as 5 years. In *Drosophila melanogaster* regional differences in alcohol dehydrogenase polymorphism have been studied for many years in eastern coastal Australia. There have been genetic shifts over the past 20 years. Reflecting recent climate change, southern populations now have the genetic constitution of more northerly populations (Umina *et al.*, 2005). Genetic shift to earlier births has been discovered in red squirrels in the Yukon in response to increasing arctic temperatures (Réale *et al.*, 2003).

Interactions amongst species and the effects of climate change

Through their complex interrelationships, the plants and animals of an ecosystem are constantly being tested by natural selection. A number of pieces of evidence point to altered biotic relationships under climate change, with considerable implications for microevolutionary change. For example, warmer springs in the Netherlands have resulted in a mismatch between the peak availability of insects and the peak food demands of nestlings of the Great Tit (*Parus major*) (Visser *et al.*, 1998). In the UK, winter warmth has 'disrupted the synchrony between the winter moth (*Operophtera brumata*) hatching and oak bud burst' (Walther *et al.*, 2002). Shifts in beak morphology have been detected in Darwin's finches after their food supplies were affected by drought (Grant & Grant, 2002).

Edith's Checkerspot butterfly (*Euphydryas editha*) is found on the west coast of North America from Mexico to Canada. The food plant for larval stages is *Collinsia torreyi*. The butterfly has been intensively studied for more than 40 years by many researchers. Parmesan (2005) provides a detailed review of the large literature on the species. Here, it is important to note that the timing of the life-history stages and the success or failure of populations is linked to the seasonal and yearly behaviour of the food plant. With recent climate change towards warmer/drier climate conditions, there have been significant changes in the distribution of the butterfly. In comparison with northern populations, extinctions in southern populations have been four times greater. Extinction is related to the 'shortening of the window of time in which the host is edible'. Overall, the species distribution has shifted 94 km towards the north and 124 m upwards in elevation.

Further research on plant/animal interactions is certain to increase our understanding of ecosystem reactions to climate change (Parmesan, 2006).

Migration in the face of climate change: microevolutionary speculations

Plant species have different capacities for migration

A good deal of information on breeding systems and dispersal has appeared in earlier chapters. Here, we note that quick responses may be possible in annual or short-lived species with a capacity for wide seed dispersal. Other species may be at a selective disadvantage in terms of migration, as they rarely produce seed.

Ecotypes and species with special habitat requirements

Many species/ecotypes have very particular edaphic and other habitat requirements. For example, some only grow on serpentine soils, in woods on boulder clay etc. Successful migration to another site would depend upon the availability of the precise habitat conditions required by the species/variant in question, e.g. the very rare North American species *Aconitum noveboracense* only grows in a few geographically isolated sites on sandstone in cool, shaded ravines on seeps or talus slopes, by rock shelters or on vertical cliff faces in close proximity to running water (Dixon & May, 1990).

Invading occupied territory

For a species to migrate from A to B and successfully found a new self-sustaining population, it is necessary that plants establish themselves, and, for longer-term success, develop to reproductive maturity. However, potential new territories may already be occupied by vegetation, and the appropriate regeneration niche may not available (Grubb, 1977).

Invasive species

In invading disturbed areas, a migrating species might encounter new selection pressures associated with the presence of weedy or invasive introduced species. Such species are also likely to migrate to new territories. For instance, Kudzu, the Japanese vine (*Pueraria montana*), introduced into the USA from Japan in 1876 as an ornamental plant, has become a serious invasive species across the southern states. Recent studies of this leguminous plant have revealed that the northern limit of the species is determined by winter frosts, the plants being restricted to a zone with less than 80 days of frost each winter. It is predicted that under climate change the plant could spread much further north (Reilly, 2007).

Co-evolved mutualisms

If species have exacting requirements for pollination and seed/fruit dispersal, these may only be met if the appropriate animals are present at the new site. There may be a 'delay' in the arrival of these mutualists, and pollination and dispersal of fruits and seeds may be compromised in their absence.

The speed and rate of migration

By means of detailed Quaternary studies, it has proved possible to estimate the rate of migration of some species of plants. Thus, in northern USA, *Tsuga canadensis* migrated 20–25 km per century, while oak and pine covered a distance of 30–40 km per century (Davis, 1990). Considering the rate of present climate change, it has been estimated that migration will have to be much more rapid than in the post-glacial period, as the rate of warming is forecast to be between '10 and 100 times faster than the rate of deglacial warming' (Huntley, 1991, 19). Such a rate of unaided migration may be beyond the capacity of many species. The possibility that assisted migration could be achieved by judicious sowing and plantings will be considered in the next chapter.

Migration of individual species and communities

Evidence from studies of the post-glacial migration reveals that the behaviour of different species is individualistic. There is no evidence that whole communities migrate together either as an advancing wave or by long-distance dispersal and back filling. It is clear, then, that migrant species will form 'new' plant communities, based on the responses of individual species. Thus, present-day communities may dissociate, with their constituent species migrating at different rates or not at all, and because at migrant sites 'new' communities are created, former assemblages may not be reconstituted. On a long time scale, former, existing and future community types are seen by many ecologists as ephemeral associations. This notion receives support from the studies of modern forest communities, many of which were recently developed. As a consequence of migration of species into existing vegetation, new associations will be created that have 'no past or present counterparts' (Lovejoy & Hannah, 2005).

Changes to present-day communities that have 'lost' species through climate change

If species migrate successfully elsewhere and become extinct in their former sites, the residual communities 'left behind' may change very radically, not only through the effect of climate change but also through other anthropogenic disturbance. Such sites might be vulnerable to invasions by weeds and invasive species. For example, the widespread prolific species *Melaleuca quinquenervia* that has colonised disturbed areas in Florida is predicted to spread under climate change (Peters, 1992).

Migration in man-disturbed environments

In the post-glacial period, extensive migrations of species occurred as the climate warmed. In the present day, many semi-natural and natural ecosystems occur as isolated fragments within anthropogenically developed areas. For successful dispersal, species must successfully migrate across hostile territory such as roads, urban industrial areas, plantations, grazing lands etc. Also, in many regions of the world, they are 'confronted' by encircling farmland, where migrants could encounter pesticides and highly fertilised soils in intensively managed cropping regimes.

Migration and corridors

If areas of semi-natural vegetation are already linked by corridors of appropriate size and community composition, then species migration might be facilitated. Many species of plants could be dispersed to appropriate new territory by animals. Even if the corridor territory itself is uncongenial to plant migrants, animals that used the corridors as dispersal routes could disperse seeds they had ingested. To counter warming trends, corridors running north/south would be needed to facilitate range extension. Under climate change it is predicted that vegetation zones will move upwards in mountainous areas. To facilitate the migration of rare and endangered alpine species, effective corridors would need to be orientated up the mountains, to enable species to migrate to greater altitudes. Ultimately, under greater global warming, vegetation zones will move higher, and summit vegetation may be lost from all but the highest mountain tops. Natural migration of high-altitude species between high peaks is likely to be highly problematical.

The likely long-term effect of migrations

Many factors are significant in plant distribution. The climatic envelope to which species are adapted is of paramount importance. It is predicted that as climate changes in the longer term, species will be climatically stressed in their present locations. And, if they are incapable of genetic adaptation *in situ*, they will become extinct, unless successful migration occurs to a site with an appropriate climatic envelope (Fig. 19.5).

It is helpful to examine an example of modelling the effects of climate change. Thuiller *et al.* (2005) examined climate envelopes – studied, as usual as a grid of pixels – of 1,350 European plants under seven different climate scenarios. For modelling they made the simplifying assumptions: (a) 'current envelopes reflect species' environmental preferences which will be retained under climate change'; (b) instantaneous change in distribution would occur; and (c) the responses of species to increased carbon dioxide were ignored. As the modelling does 'not capture details of population dynamics or biotic interactions nor the lags in spatial range shifts associated with processes of dispersal, establishment and local extinction', the study examined the effect of two contrasting assumptions: either species were assumed to be unable to disperse in the time frame under consideration or there were no constraints on dispersal. The authors noted, at the outset, that in reality most species would fall between these two extremes in their behaviour.

On the basis of these studies the authors concluded that more than half the species studied would be 'vulnerable or threatened by 2080. Expected species loss and turnover per pixel proved to be highly variable across scenarios (27–42% and 45–63% respectively, averaged over Europe) and across regions (2.5–86% and 17–86% averages over scenarios).'

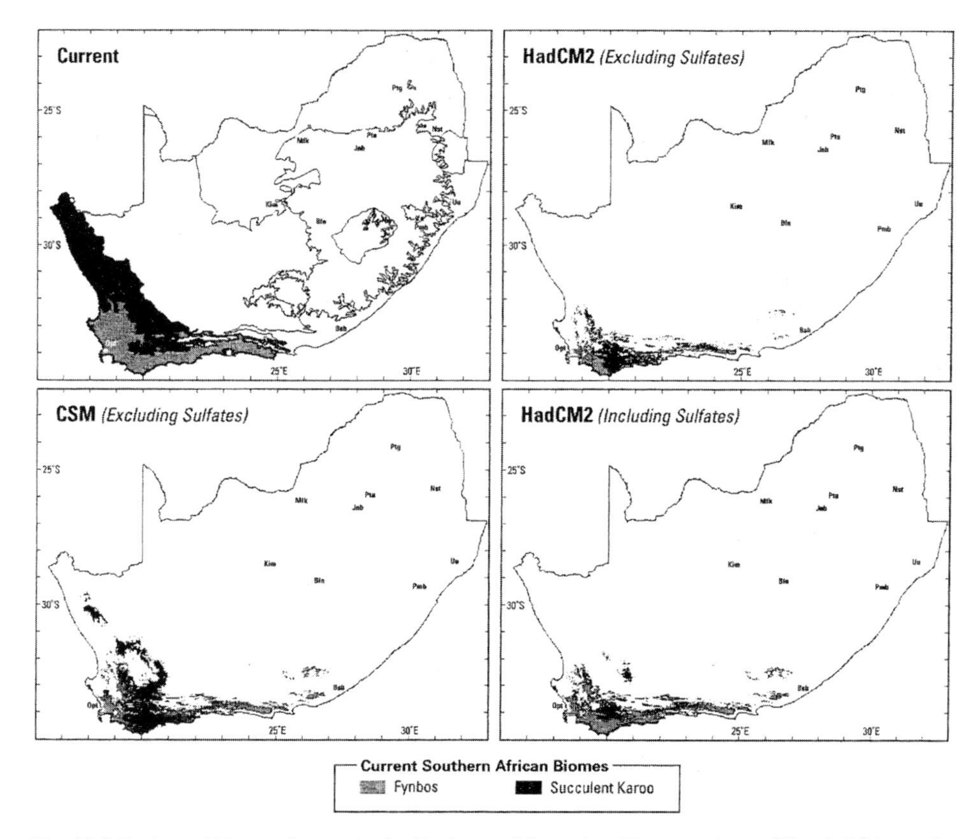

Fig. 19.5 Projected biome changes in the Fynbos and Succulent Karoo regions of South Africa under various climate change scenarios. The vegetation that is likely to replace displaced ecosystems of the two biomes is likely to have no current counterparts in the South African ecosystems. (From Hannah, Lovejoy & Schneider, 2005. Reproduced with permission of Yale University Press ©)

Looking at some elements of the results in more detail, it is clear that montane species are 'disproportionately sensitive to climate change (*c.* 60% species loss). The boreal regions were projected to lose fewer species although gaining many others from immigration'; great changes are likely to occur at the junction of Mediterranean and other zones. Overall they concluded that the 'risks of extinction for European plants may be large, even in moderate scenarios of climate change'.

The scale of the effect of changing climate envelopes has been examined in the flora of the USA (Kutner & Morse, 1996). For the purposes of modelling present and future distributions in this case, the assumption was made that mean annual temperature is adequate for defining the climatic envelope; that systems are at equilibrium now; and that climate determines the range of a plant species. Modelling revealed that, on these assumptions and with a +3 °C temperature change in the future, many species would be subject to climate shifts, with *c.* 10%+ of plant species having a potential distribution completely outside their current climatic envelope. Thus, 'with a mean global warming of 3 °C about 7 to

11 percent (1060–1670) of over 15,000 plant species in North America were entirely out of their climatic envelopes and thus considered vulnerable to extinction' (Kutner & Morse, 1996, 33).

In an investigation of 117 native Florida tree species, Crumpacker, Box and Hardin (2002) discovered that even small climate warming resulted in adverse effects. At a wider scale, Thomas *et al.* (2004) carried out climate envelope modelling to examine the extinction risk for a set of species in a number of regions representing 20% of Earth's terrestrial surface. For each species, they calculated the current and future climatic envelopes, and considered the alternative possibilities that species have either no limits to dispersal or are incapable of dispersal. Three methods were employed to estimate extinction risk based on species/area relationships under a range of future climate scenarios. They discovered: '15–37% of species in our sample of regions and taxa will be "committed to extinction" . . . Minimal climate-warming scenarios produce lower projections of species committed to extinction (*c.* 18%), than mid-range (*c.* 24%) and maximum change (*c.* 35%) scenarios.'

Concluding remarks

Evidence suggests that climate change is already having major impact on species and ecosystems, through adaptation and changing distributions. Looking to the future, the most important conclusion to draw from this chapter is that, for species of interest or concern, each of the different outcomes of the impact of climate change – phenotypic plasticity, developmental flexibility, adaptation *in situ*, migration to new territory, extinction – should be seen as the working out of microevolutionary processes acting at the population level (Davis, Shaw & Etterson, 2005). Thus, it is important once again to stress that evolution is not something that took place in the past. Darwin's insight is worth repeating. As he saw it: 'Natural selection is daily and hourly scrutinising, throughout the world, the slightest variations; rejecting those that are bad, preserving and adding up all that are good; silently and insensibly working, whenever and wherever opportunity offers' (Darwin, 1901, 60). It follows therefore that, potentially, all the microevolutionary processes identified in previous chapters – mutation, recombination, breeding systems, natural selection, chance effects, gene flow or lack of it etc. – are involved in the individualistic responses of plant species to climate change.

It may be concluded that, if the predictions concerning climate change are realised, some species will adapt *in situ*, others will migrate successfully to more congenial territory, and a final group will become extinct. Overall, it seems possible that 'vulnerable, less dispersible rare species' will be replaced by 'resilient, dispersible, widespread species' (Kutner & Morse, 1996, 29).

While the study of climate changes in the past, particularly in the Holocene, has highlighted the likely direction of future changes in plant and animal species, there are some important differences. Potentially, the speed of future climate change could greatly exceed that in the post-glacial. Furthermore, in many parts of the world, the potential for species migration has been greatly reduced by major anthropogenic changes to the environment.

Predictions concerning the range changes under climate change are derived from models. It is important to emphasise once again the simplification of real life situations is necessary to allow modelling. For example, Ohlemüller *et al.* (2006, 1789) point out that many models are based on the assumption of either complete or zero dispersal. Furthermore, it is assumed 'species will die out within regions that are predicted to become climatically unsuitable for them'. This second assumption may be invalid, as once established, many species might persist vegetatively for long periods. Ohlemüller and associates also recommend that other refinements should be introduced into models, namely, considering 'the distance to the nearest and to all projected climatically suitable areas', and the degree of climatic hostility a species would experience in migrating from current to future territory. Notwithstanding these caveats, modelling has provided important insights that could not be obtained by other means.

This chapter has considered microevolutionary responses of plants to anthropogenic climate change. Further detailed investigations are needed to expand the limits of our present rather meagre knowledge (see Jump & Peñuelas, 2005). Also, the difficulty of attributing observed change entirely to climatic factors must be appreciated. A very recent investigation confirms this point. In 1851–1858, the philosopher and conservationist Thoreau recorded the plant species and their flowering times in the woods around Concord, Massachusetts. Willis *et al.* (2008) have studied the accumulated records for the area and conclude, 'climate change has led to major changes in some species and not others'. However, McDonald *et al.* (2009) take the view that the change in abundance of these species has another cause, namely, the intensified grazing pressures imposed by increasing numbers of White-tailed deer (*Odocoileus virginianus*).

The next chapter examines the implications of climate change for conservation.

20

The implications of climate change for the theory and practice of conservation

Conservation efforts aim to prevent the extinction of endangered species and to provide them with a long-term evolutionary future. In 2000, a colloquium of the National Academy of Sciences of the USA, meeting in Irvine, California, considered 'The biotic crisis and the future of evolution' (papers given at the meeting were published in *Proceedings of the National Academy of Sciences, USA*, **98**, 2001). Myers and Knoll (2001, 5389), speaking at the meeting, concluded that 'human activities have brought the Earth to the brink of biotic crisis' and 'in decades to come a large number of species will be lost'. This pessimistic assessment was echoed by many others at the colloquium. It is significant, however, that, while several speakers mentioned climate change, they did not highlight it as a major issue. In less than a decade, there have been major advances in our understanding. As we shall see in this chapter, anthropogenic climate change presents a dramatic and potentially catastrophic threat to the world's biodiversity with profound implications for the theory and practice of conservation.

Doubts about modelling

Some naturalists, accustomed to a life outdoors, are antagonistic to the study of ecological/conservation issues through computer modelling, and doubt the value of such studies in the prediction of future trends. However, it is important to accept that modelling provides important helpful insights that *cannot be obtained by any other means*. While acknowledging the limitations of models with their simplifying assumptions, it would be foolish not to take their conclusions seriously.

The Fourth IPCC Assessment Report, *April 2007*

A further instalment of the *Fourth IPCC Assessment Report* of climate change (IPCC, 2007b) provides the most up-to-date and definitive succinct review of the likely impacts of climate change on the biosphere. As in earlier reports, assessments are presented in terms of probabilities rather than certainties. Tables 20.1 and 20.2 quote some of their major findings pertinent to the themes of this book. Very precise measured statements have been

Table 20.1 *Projected impacts of climate change in selected regions of the world. Selected findings of the Working Group on Climate Change Impacts, Adaptation and Vulnerability.* Intergovernmental Panel on Climate Change 4th Assessment Report, *6 April 2007 (IPCC, 2007b). Quotations, by permission of IPCC, are from the* Summary for Policymakers, *in which the following terms have been used to indicate the assessed likelihood of an outcome or a result:* Virtually certain *>99% probability of occurrence,* Extremely likely *>95%,* Likely *>66%,* More likely than not *>50%,* Very unlikely *<10%,* Extremely unlikely *<5%.*
The following terms have been used to express confidence in a statement: Very high confidence *At least a 9 out of 10 chance of being correct,* High confidence *About an 8 out of 10 chance*, Medium confidence *About a 5 out of 10 chance*. Low confidence *About a 2 out of 10 chance*, Very low confidence *Less than a 1 in 10 chance. Level of confidence is indicated as follows: *** Very high confidence, ** High confidence, * Medium confidence.*

Freshwater resources

By mid-century, annual average river run-off and water availability are projected to increase by 10–40% at high latitudes and in some wet tropical areas, and decrease by 10–30% over some dry regions at mid-latitudes and in the dry tropics [**].

Drought-affected areas will likely increase in extent. Heavy precipitation events, which are very likely to increase in frequency, will augment flood risk [**].

Water supplies stored in glaciers and snow cover are projected to decline, reducing water availability in regions supplied by meltwater from major mountain ranges, where more than one-sixth of the world population currently lives [**].

Ecosystems

The resilience of many ecosystems is likely to be exceeded this century by an unprecedented combination of climate change, associated disturbances (e.g. flooding, drought, wildfire, insects, ocean acidification) and other global change drivers (e.g. land-use change, pollution, over-exploitation of resources) [**].

Approximately 20–30% of plant and animal species assessed so far are likely to be at increased risk of extinction if increases of global average temperature exceed 1.5–2.5 °C [*].

For increases in global average temperature exceeding 1.5–2.5 °C and in concomitant atmospheric carbon dioxide concentrations, there are projected to be major changes in ecosystem structure and function, species' ecological interactions and species' geographic ranges, with predominantly negative consequences for biodiversity, and ecosystem goods and services, e.g. water and food supply [**].

Food production

Globally, the potential for food production is projected to increase with increases in local average temperature over the range 1–3 °C, but above this it is projected to decrease [*].

Crop productivity is projected to increase slightly at mid to high latitudes for local mean temperature increases of 1–3 °C . . . at lower latitudes, especially seasonally dry and tropical regions, crop productivity is projected to decrease for even small local temperature increases (1–2 °C) which would increase the risk of hunger [*].

(cont.)

Table 20.1 (*cont.*)

Coastal ecosystems and low-lying areas

Coasts are projected to be exposed to increasing risks, including coastal erosion [***].

Coastal wetlands including salt marshes and mangroves are projected to be negatively affected by sea level rise especially where they are constrained on their landward side, or starved of sediment [***].

Many millions more people are projected to be flooded every year due to sea level rise by the 2080s. Those densely populated and low-lying areas where adaptive capacity is relatively low, and which already face other challenges such as tropical storms or local coastal subsidence, are especially at risk. The numbers affected will be largest in the mega-deltas of Asia and Africa while small islands are especially vulnerable [***].

Industry, settlement and society

In the aggregate net effects will tend to be more negative the larger the change in climate [**].

Projected climate change-related exposures are likely to affect the health status of millions of people, particularly those with low adaptive capacity [**].

prepared and agreed by a large group of international experts, and, for clarity, these are quoted verbatim.

National parks and nature reserves as the major focus of conservation efforts

For decades, through the establishment of national parks and nature reserves, *in situ* conservation has been the major strategy to safeguard endangered ecosystems and species. In earlier chapters the strengths of this worldwide system of reserved lands were outlined, but also some serious weaknesses of reserve design and protection were highlighted. Thus, many reserves and parks are very small and isolated, and are managed as defended 'fortresses' against poaching, illegal logging, squatters etc. They are also impacted by many factors originating outside the reserve boundaries that are beyond the control of park managers, such as pollution, altered hydrology etc. The emergence of climate change as a dramatic, and perhaps catastrophic, additional 'force' interacting with other known problems, raises the question of whether reserves and national parks, in many cases already under siege, will be able to secure the future of the world's biodiversity.

Climate change in areas of high conservation significance

As we have seen in the last two chapters, ecosystems across the world, and the reserved areas they contain, are likely to be destroyed and others greatly modified under the influence of climate change. Here, by way of example, a number of situations are highlighted.

In many parts of the world, coastal area reserves and other non-reserved ecosystems of conservation interest, which have no appropriate hinterlands to which ecosystems can

Table 20.2 *Implications of climate change for conservation in particular regions of the world (Terms as in Table 20.1).*

Africa

By 2020, between 75 and 250 million peoples are projected to be exposed to an increase of water stress due to climate change. If coupled with increased demand, this will adversely affect livelihoods and exacerbate water-related problems [**].

Asia

Freshwater availability in central, south, east and south-east Asia particularly in large river basins is projected to decrease due to climate change which, along with population growth and increasing demand arising from higher standards of living, could adversely affect more than a billion people by the 2050s [**].

Australia and New Zealand

As a result of reduced precipitation and increased evaporation, water security problems are projected to intensify by 2030 in southern and eastern Australia and, in New Zealand, in North Island and some eastern regions [**].

Significant loss of biodiversity is projected to occur by 2020 in some ecologically rich sites including the Great Barrier Reef and Queensland Wet Tropics. Other areas at risk include Kakadu wetlands, south-west Australia, sub-Antarctic islands and alpine areas of both countries [**].

Europe

Negative impacts will include increased risk of inland flash floods, and more frequent coastal flooding and increased erosion (due to storminess and sea level rise). The great majority of organisms and ecosystems will have difficulties adapting to climate change. Mountainous areas will face glacier retreat, reduced snow cover and winter tourism, and extensive species losses (in some areas up to 60% under high emission scenarios by 2080 [***].

In Southern Europe, climate change is projected to worsen conditions (high temperatures and drought) in a region already vulnerable to climate variability [**].

In Central and Eastern Europe, summer precipitation is projected to decrease, causing higher water stress . . . Forest productivity is expected to decline and the frequency of peat land fires to increase [**].

In Northern Europe, climate change is initially projected to bring mixed effects, including some benefits such as reduced demand for heating, increased crop yields and increased forest growth. However, as climate change continues, its negative impacts (including more frequent winter floods, endangered ecosystems and increasing ground instability) are likely to outweigh its benefits [**].

Latin America

By mid-century, increases in temperature and associated decreases in soil water are projected to lead to gradual replacement of tropical forest by savannah in eastern Amazonia . . . There is a risk of significant biodiversity loss though species extinction in many areas of tropical Latin America [**].

North America

Warming in western mountains is projected to cause decreased snowpack, more winter flooding, and reduced summer flows, exacerbating competition for over-allocated water resources [***].

(*cont.*)

Table 20.2 (*cont.*)

Disturbances from pests, diseases, and fire are projected to have increasing impacts on forests, with an extended period of high fire risk and large increase in area burned [***].
Coastal communities and habitats will be increasingly stressed by climate change impacts interacting with development and pollution [***].

Polar regions

In the polar regions, the main projected biophysical effects are reductions in thickness and extent of glaciers and ice sheets, and changes in natural ecosystems with detrimental effects on many organisms including migratory birds, mammals and higher predators [**].
In both polar regions, specific ecosystems and habitats are projected to be vulnerable, as climatic barriers to species invasions are lowered [**].
Very large sea level rises that would result from widespread deglaciation of Greenland and West Antarctic ice sheets imply major changes in coastlines and ecosystems, and inundation of low-lying areas, with greatest effect in river deltas. . . . There is medium confidence that at least partial deglaciation of the Greenland ice sheet, and possibly the West Antarctic ice sheet would occur over a period ranging from centuries to millennia for a global average temperature increase of 1–4 °C (relative to 1990–2000), causing a contribution to sea level rise of 4–6 m or more [*].

Small islands

Small islands, whether located in the tropics or higher latitudes, have characteristics which make them especially vulnerable to the effects of climate change, sea level rise and extreme events [***].
With higher temperatures, increased invasion by non-native species is expected to occur, particularly on middle and high latitude islands [**].

retreat, are likely to be lost through sea level rise. In other coastal areas, incursions of salt water are already damaging freshwater habitats, some of them in reserved areas. Long episodes of drought (in many cases inevitably followed by fires – either natural or set by people) are likely to damage, alter or destroy many ecosystems, including forests and grasslands, both inside and outside reserves. Warming is predicted to lead to major changes in the higher latitudes and higher altitudes, and it is highly likely that certain ecosystems will be lost or modified. Species losses at particular sites seem inevitable (Fig. 20.1), and the interrelationships within communities are set to change, as native and introduced taxa are likely to invade. Of course, the most extreme effects of predicted climate change may not be felt if action to reduce emissions of greenhouse gases is effective. However, as we have seen earlier, evidence suggests that some change is inevitable given the amount of anthropogenically generated greenhouse gases already emitted.

Given the stark warnings that are now being made about climate change, how should conservationists react? It would be helpful if simple and straightforward prescriptions could be formulated to guide new conservation strategies. However, no such simple remedies are immediately apparent. Here, the aim is to identify and comment on important issues.

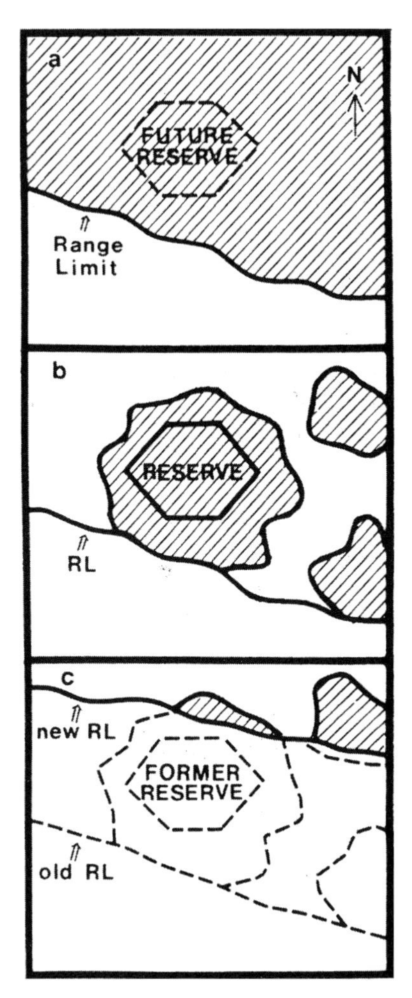

Fig. 20.1 Climatic warming may cause species within biological reserves to disappear. Hatching indicates (a) species distribution before either human habitation or climate change, (b) fragmented species distribution after human activities but before projected climate change, (c) species distribution in the cultural landscape after climate change. (Peters, 1992. Reproduced with permission from R. L. Peters & T. E. Lovejoy (eds.) *Global Warming and Biological Diversity* © Yale University Press)

The mandates of national parks

National parks in different parts of the world are managed under a great variety of mandates. In the early days the aim was to 'preserve' the ecosystems within the park and many conservationists still cherish these notions. However, the cumulative evidence from more than a decade of intensive study of climate change forces us to an uncomfortable conclusion. The hope that somehow the ecosystems of a reserve can be 'preserved' intact must be abandoned. This is not a new realisation. Detailed ecological and palaeobotanical evidence confirms (and past experience in management emphasises) that the communities of parks/reserves, indeed, of any ecosystem, cannot be 'preserved' in a particular state for any length of time. Ecosystems are by their very nature dynamic. In many countries, the management of national parks is informed by a wide knowledge of the ecology of natural systems. However, conservation under climate change will need to be reassessed. For instance, Scott (2005) considers the case of the Canadian National Parks, which are being managed under amendments formalised in 1988. These require the maintenance of' 'ecological integrity' in the parks. In setting this aim it is recognised that 'ecosystems are inherently dynamic and change does not necessarily mean a loss of integrity'. Thus, 'a system with integrity may exist in several states, but the change occurs within acceptable limits' (Scott, 2005, 342). Climate change is predicted to be particularly strong in the northern latitudes of Canada, and here ecosystems may undergo profound changes well beyond the 'acceptable limits' recognised by present management protocols. How should management respond, given the uncertainties of the speed and direction of change?

Management and restoration in reserves and the wider environment

As we saw in Chapter 15, national parks and nature reserves have been managed with different aims. Put simply, the North American model seeks to maintain or recreate wilderness leaving it unimpaired for future generations. In contrast, in areas where longstanding cultural landscapes are dominant (with their associated rare and endangered species), the aim is to manage and restore former cultural practices such as coppicing, haymaking etc.

In earlier chapters, examples were given of the wide range of conservation management aimed at recreating or restoring the past ecosystems. Not only have restoration projects been carried out in reserves but there has also been a great deal of 'restoration' in the surrounding matrix of human-dominated landscapes.

Under both models of *in situ* conservation, management is designed to recreate lost or damaged ecosystems that *formerly* existed in particular sites. Such schemes are often costly, and therefore the aim is to produce self-sustaining communities that will persist into the future. The underlying assumptions on which projects are planned are rarely spelled out in detail. Is it assumed that a degree of climatic and environmental stability is likely in the future? However plausible such assumptions might have seemed in the early days of conservation management and restoration, there has been a revolution in our understanding. It is forecast that unprecedented climate changes could occur and that managed and restored ecosystems may be severely impacted. All ecosystems, including those of conservation interest, will be subject to a multitude of new selection pressures. There will be winners and losers. Conservationists must acknowledge the high likelihood of *dramatic changes in the future* to the ecosystems they are striving to sustain, or recreate.

Choice of stocks to use in restoration and management

In planning for and responding to climate change, the selection of appropriate species and stocks of plants becomes the key issue in management and restoration. As we saw in Chapter 16, plants and animals from local sources have been employed in restorations, as there is evidence that they are best-fitted to the local conditions now prevailing. But will this be the best strategy in the future? Will locally sourced stocks have enough genetic variability to respond to the new selection pressures associated with changing climate? The use of genetically diverse stocks would seem more logical in the face of future uncertainties. There is a probability that mixed stocks from different areas would have greater genetic variability on which selection can act. Whether this proposition is correct could be tested, before the restoration is attempted, by subjecting material to artificial selection by mimicking critical climatic factors predicted to change in the future. Perhaps such experiments will reveal that some but not all of the available stocks contain appropriate variability.

To allow for the long-term survival of species and ecosystems, future management and restoration projects must be designed with climate change in mind; otherwise newly completed restorations might be short lived. A programme of continuous restoration management may be necessary, rather than the single time-limited project. Yet another serious question has to be faced. Should plans for some managed restorations be abandoned? For example, at certain sites the re-creation of freshwater marshland near the sea may be a high-risk project, as the local topography may predispose the area to flooding in future storm surges leading to salt water inundation.

The withdrawal of management and restoration

As we saw in previous chapters, some conservationists have in the past favoured policies of 'non-intervention' in nature reserves. Given the problems of predicting the rate and direction of changes locally, it seems possible that there could be a widespread return to this strategy. Such an approach could be attractive to the risk averse, and could be promoted as a more cost-effective approach. Taking *in situ* conservation in this direction would, however, put at risk many endangered species, especially in cultural landscapes. These require the traditional management to be continued, theoretically, in perpetuity.

The conservation of endangered species

Conservationists have concentrated a great deal of their attention and resources to the management of rare and endangered plants and animals. In many cases, research has revealed the particular management required to secure their short-term survival. This may take the form of re-imposition of former cultural practices, or more dramatic interventions, such as restocking and reintroduction. In some cases, these approaches have achieved a degree of success, while others have failed. However, looking to the future, there are many problems with the intensive management approach if populations are subjected to changing climate.

Endangered species, hitherto successfully managed, could be at great risk under climate change. Dramatic events such as storms, serious droughts, flooding etc. are predicted (Table 20.1). Change is likely to be cumulative. It is possible, therefore, that individual episodes are not recognised by reserve managers as part of developing long-term trends. Thus, a period of, say, drought may be *taken* to be an isolated one-off event, rather than *recognised* as the forerunner of a climate shift to frequent droughts.

The key issue facing conservationists is how to react to change. The natural instinct is to protect endangered species and 'nurse' them through difficult times. For instance, plants could be watered in a period of drought. Likewise, animals could be provided with water and food etc. In such cases, conservationists could find themselves managing year-by-year events and crisis-by-crisis episodes against what might be a trend of relentless and inevitable change. Alternatively, managers could conclude that potential climate-induced change could be so overwhelming that management interventions are futile. However, abandoning wildlife would not be acceptable in many societies, which in varying degrees have embraced environmentalism.

Relocation of reserves

In the light of these difficulties, various approaches have been explored by conservationists. Peters (1992) has suggested that, where possible, reserves should be greatly enlarged to accommodate species range shifts. Alternatively, reserves could be relocated to take account of new distributions. Thus, in some cases, reserve territories could be de-gazetted and used in whole or in part for other purposes. More suitable areas elsewhere could be brought under conservation control, for example, by private or public purchase or covenanting of privately owned land; through management agreements with landowners; or by land-trading between parks/ reserves, and state forests etc. Such solutions might be possible in countries blessed with abundant territory and willing and generous patronage, but, politically, expansion of reserves and trading of lands could be impossible to achieve in densely populated areas of the world. Moreover, the notion that reserves/parks could be relocated on the grounds of biodiversity changes does not take into account popular sentiment. National parks and reserves protect not only biodiversity but also cherished landscapes that will continue to provide enjoyment, inspiration and aesthetic experiences to locals and visitors as leisure or holiday destinations, even if some biodiversity is lost.

Wildlife corridors

In the post-glacial period, species migration took place in natural ecosystems. In contrast, in the face of current climate change, organisms 'on the move' are faced with all the complexity of human agricultural and urban industrial ecosystems (Walther *et al.*, 2002). Agricultural, urban, transport and industrial landscape elements cover much of the anthropogenic landscape and present formidable challenges to migration between scattered elements of more natural vegetation. Many conservation authorities consider that, to allow range shifts under climate change, corridors should be designated or created between reserved lands. Some of the theoretical considerations concerning corridors have been examined earlier in the book: here we are concerned with more practical issues. Three questions are considered. What types of corridor might effectively link different fragmentary ecosystems? Would such systems be certain to facilitate the migration of plants? And, is there the political will to bear the cost of constructing such corridors?

Plants differ widely in the efficiency and methods of dispersal of seeds/fruits/ vegetative structures such as bulbs etc. For some species, corridors with a variety of vegetation types might act as conduits for transit of plants distributed by animals or water. However, some species lack wide dispersal capability, and to enable them to take up 'residence' in corridors, as they migrate from A to B, the specific niche requirements for particular species would have to be met, and vegetation within each corridor is then more critical. Also, 'residence' corridors would have to be wide enough to minimise edge effects, and thereby provide a central strip of congenial appropriate vegetation for the transit via residence.

Facing the complexity of climate change, it is clear that it would be necessary to provide more than one corridor system within a region (Kutner & Morse, 1996). By way of example, they consider in outline the situation in the USA. As temperature changes with latitude, corridors of N/S orientation would be necessary to facilitate migration. However, such arrangements would not provide for the predicted changes in precipitation/moisture gradients, which might require corridors of E/W orientation. Other corridors will be needed to encourage migration from coastal sites that might be lost by sea level rise to inland areas on higher ground. Also, corridors should run from low to high ground to allow range expansions in montane regions. Considering that temperature changes are more gradual latitudinally than altitudinally, longer ones will be necessary in lowland areas: in contrast, shorter corridors might be effective in the mountains.

Given the cost and political complexities that would have to be overcome in creating corridors, a key question is how effective might wildlife corridors be in facilitating migration? Most of the investigations of corridors have been carried out to examine their use by animals. Such studies are relevant to some plant migrations, as many species have animal-dispersed seeds and fruits. In their critical review of the extent to which corridors were effective conduits for animal movement, Beier and Noss (1998) examined 32 case histories. Fewer than half the studies supported the utility of corridors; but some well-designed experiments suggest that corridors could be important. Recently, there has been a resurgence of interest in connectivity between fragmented animal populations, of which corridors are an important element (Crooks & Sanjayan, 2006). In this volume, Noss and Daly (2006) debate the practical questions to be faced: should corridors take the 'shortest route', or if necessary the 'only remaining route', however long? Should corridors link similar ecosystems, say wetlands, or should they connect dissimilar ecosystems of conservation interest? Should corridors be designed to fit the needs of charismatic animals, with the hope that other species will also use them? How should the routes of corridors be planned? Should they be designed by experts on the particular species who are guided in their design of an appropriate route by records of 'animal presence, movements or signs' (Noss & Daly, 2006, 597)? Notwithstanding the biological requirements of conservation corridors, should they be designed on the 'least cost basis'? Others have posed a different question: should corridors be designed primarily with wildlife in mind, or should they be an element in the design of landscape features serving many purposes, such as hiking trails, cycle tracks and greenways (Jongman & Pungetti, 2004)?

Turning to another practical issue, effective corridors are required *now*, as range shifts are occurring in response to climate change. Given the urgency of the situation, it might be possible to complete a corridor system if significant elements already exist. However, if long corridors have to be created *de novo*, it might be many years before the land can be acquired, and, if necessary, appropriate vegetation could be created. Finally, a complete 'network' of corridors would be needed to provide a theoretical (if not realised) opportunity for the many different endangered species to migrate.

The design of corridors for animal movement has been the prime concern of conservationists. While the provision of corridors might encourage the successful migration of some plant species, they might fail to facilitate the successful movement of species or local races with highly specialised edaphic requirements from their present sites to new areas with appropriate climatic envelopes, e.g. species/ecotypes restricted to serpentine, chalk or limestone outcrops.

In territories with sparse populations, it may be possible to move from theory to practice in the provision of corridors. But in areas with high human populations, it seems highly unlikely that new corridors would ever be constructed for wildlife. In terms of cost, it would be cheaper to arrange multiple reintroductions of species from A to B, rather than make an expensive corridor that might not be used. However, another approach is being considered.

Stepping stone areas to encourage migration

Given the impossibility of securing land for continuous corridors in many areas, perhaps, in some cases, incomplete systems might be effective. Could migration be facilitated between two reserves by the provision of 'wildlife friendly areas' to act as 'stepping stones' across hostile territory? This idea receives support from many naturalists, especially ornithologists.

Few detailed investigations of such systems have been carried out. Therefore, it would be unwise to assume that 'stepping stones' will be an effective means of encouraging all species. Indeed, it has already been reported that some animals are shy of crossing even short stretches of uncongenial territory, and bigger gaps might deter others (Beier & Noss, 1998).

A second point to stress is that the advent of industrial farming and intensive forestry has severely reduced such 'stepping stone' systems in the wider countryside. To illustrate this point, in Britain, there have been major losses of ponds, small woodlands, marl pits, grazed roadsides, hedgerows, wetlands, isolated landscape trees, orchards, and, as we reported above, large-scale movements of stock on the hoof have been abandoned, thereby limiting a traditional method of dispersal of seeds. In addition, many fields are often cultivated to their boundary lines. Also, transitory elements of agricultural ecosystems have been lost in some areas, e.g. crop stubbles, which provided food for birds in the winter, are now often recultivated immediately after cropping.

There have also been losses of 'stepping stones' in modern forestry management. In many sites, plantations of non-native trees have replaced native species, yielding equal-aged stands with abrupt rather than graded margins and lacking natural clearings, wet areas, relatively undisturbed streams sides etc, Treating forests as a crop has led to reduction of important stepping stone habitats on a micro-scale with an overall reduction in the numbers of ancient decrepit and veteran trees (important to insects, birds and epiphytic plants such as lichens and mosses) and reductions in the amount of decaying timber on the forest floor (crucial for insect life, fungi etc.).

However, there are welcome signs of change (Firbank, 2005). At some sites in Britain, regimented plantations of non-native tree species are being replaced by more wildlife-friendly planting schemes using native species. There are also important changes in the agricultural landscapes. In many countries, farming land has been taken out of cultivation in the short or longer term, and some has been used for conservation purposes (see, for example, the use of set-aside land in Germany; Schroder, 2004). In some areas, across the EU, the opportunity has been taken to increase the size of reserves and areas of conservation interest by incorporating permanent set-aside lands. However, agricultural policies change. In 2007/8, in response to food shortages, rising cereal prices and the drive to produce increasing amounts of biofuels, an EU directive resulted in a drastic reduction in set-aside land.

More far-reaching changes are also being introduced in EU countries. From 2005, instead of receiving production premiums, farmers have been offered payments for managing their land to higher standards of food safety, animal health and welfare, and for making production 'more environmentally friendly' (Schmid & Sinabell, 2007). This approach is resulting in the reinstating of some of the lost 'stepping stones' and providing others. In some cases, such practices are being combined with organic farming: in others cases, more conventional methods are being continued, but with modified management.

Farmers are encouraged by the EU:

- To reinstate and undertake better management of hedgerows.
- To establish uncultivated headlands and field margins on their land (Fig. 20.2), and use equipment that confines sprays, fertilisers etc. to the cropped areas, thereby minimising spray drift into hedges and adjacent vegetation nearby.
- To encourage wildlife including beneficial insects, unploughed beetle banks are being created across cropped fields (Fig. 20.3).
- To plant more trees and management of existing woodland.
- To use buffer zones in the more efficient disposal of farm wastes and to undertake more targeted use of fertilisers. By these measures it is hoped to control eutrophication that has such a devastating effect on rivers, lakes, estuaries etc. For example, the EU has promulgated the Nitrates Directive, and declared Nitrate Vulnerable Zones as part of initiating the more sustainable use of fertilisers. The impact of this legislation on a nitrate vulnerable area in Scotland is considered by Macgregor and Warren (2006).

Other organisations are implementing management to make the farm landscape more 'wildlife friendly'. Following research by the Royal Society for the Protection of Birds in Britain, nesting and feeding sites for birds are being provided, by leaving unsprayed areas within cropped fields. In addition, by agreement, certain farming activities have been rescheduled to prevent loss of chicks, with stubbles being left to provide weed seeds.

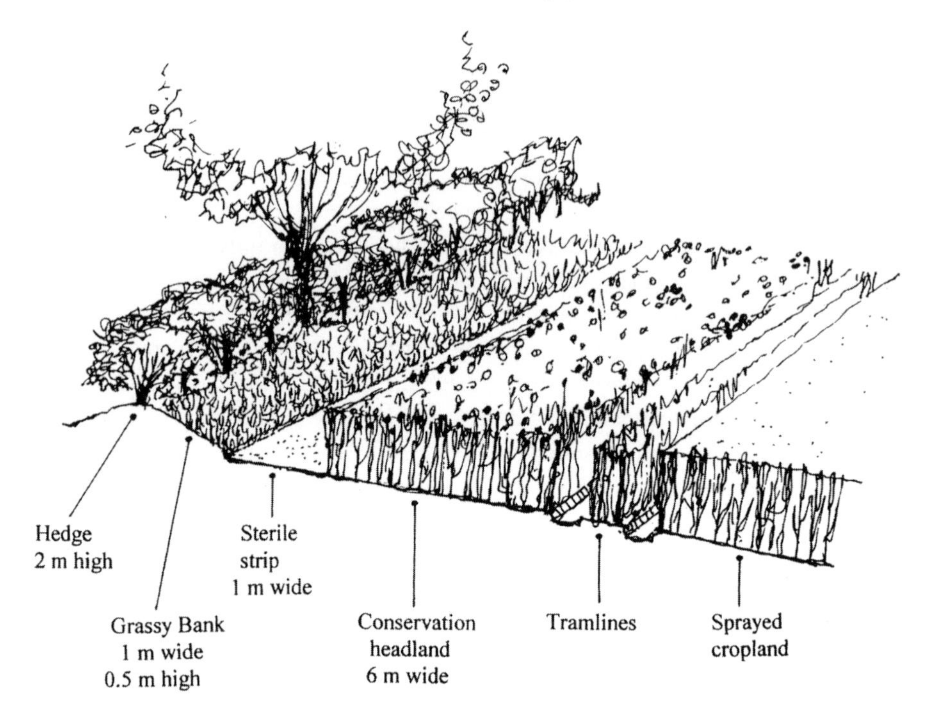

Fig. 20.2 A field margin designed to encourage the nesting and feeding of wild birds (including game), the conservation of arable weeds and the prevention of the spread of troublesome grasses into the crop. (From Gilbert & Anderson, 1998. Reproduced with permission of the Game and Wildlife Conservation Trust, Fordingbridge, Hampshire, UK & Oxford University Press)

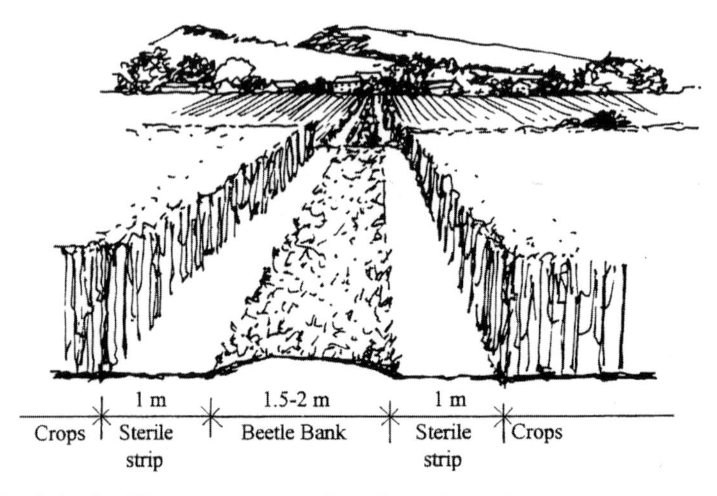

Fig. 20.3 Beetle banks. These are narrow strips of uncultivated grassland crossing arable fields. These support insect predators of crop pests and favour such game birds as the grey partridge (*Perdix perdix*). (From Gilbert & Anderson, 1998. Reproduced with permission of the Game and Wildlife Conservation Trust, Fordingbridge, Hampshire, UK & Oxford University Press)

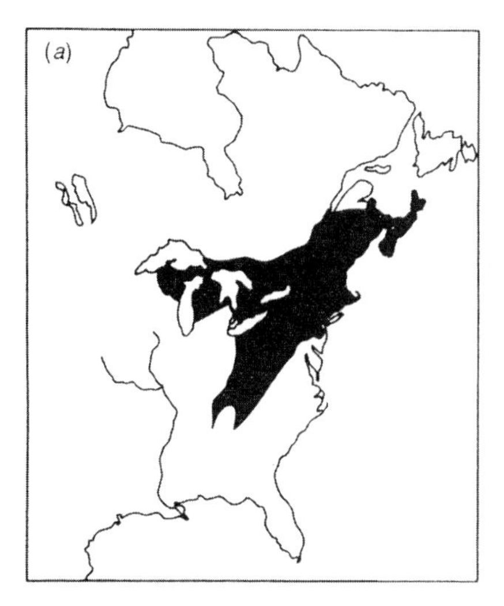

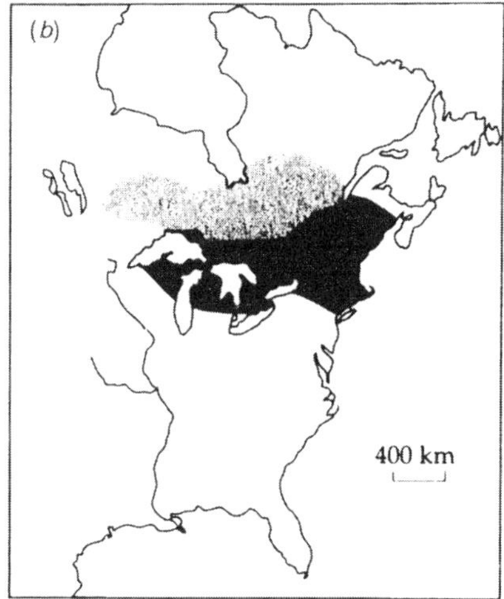

Fig. 20.4 Present and future geographic range for Hemlock (*Tsuga canadensis*). (a) Present range. (b) Range in AD 2090 under the Goddard Institute of Space Studies model with two-fold CO_2 scenario. The black area is the projected occupied range considering the rate of migration. The grey area is the potential range with climate change. (From Zabinski & Davis, 1989. Reproduced with permission of Taylor and Francis Group)

In Britain, the Game Conservancy is encouraging the planting of marginal strips in arable fields to provide cover and food for game birds. Different mixtures are supplied for use in the UK, containing the following species: buckwheat, canary grass, kale, mustards, quinoa, millet, clover, sunflower etc. (Suppliers of such mixtures include Spratt's Game Foods.) In addition to providing food for game birds, these crops also feed wild birds, and encourage butterflies and other insect visitors.

While newly emerging policies may make some agricultural areas more wildlife friendly, in the longer term, 'stepping stones' have been created or recreated in only relatively few areas. In many part of the world, farmland is still intensively managed as 'industrial agriculture', stepping stones are few and far between, and migration of wildlife through such terrain is highly problematical. Moreover, given the food shortages being experienced across the world, the era of significant areas being 'set-aside' in the EU and elsewhere is probably drawing to a close.

Assisted migration

Under climate change, it is predicted that endangered species will be stressed in some or all their populations. Marginal populations are particularly at risk. Under changing circumstance, and given the difficulties of providing land for corridors, should humans take over the role of species dispersal? Some authorities have considered the possibility (Hulme, 2005). Interventions could involve the sowing of seeds or planting of material in new locations in their potential but unoccupied range under climate change (Fig. 20.4).

Falk, Millar and Olwell (1996) have provided the definitive handbook for those involved with plant restorations. In this volume, Kutner and Morse (1996) consider that introducing species to sites outside their historic range may be necessary as a response to 'potential climate change'. However, a number of issues of concern have been raised.

1. Generally, restorations have only reintroduced species to sites within their historic range. There are different interpretations of the term historic. Usually, it refers to the distribution as revealed by published records and herbarium specimens. But, logically, on a longer time frame, the 'historic' distribution could be deduced from palaeological records, and may reveal that the historic range was different from that of today.
2. Traditionally, field botanists have been antagonistic to the notion of widespread human transfers 'in the wild', as this will blur what they take to be the 'natural' distribution of species. However, as we have seen in previous chapters, human activities have made it difficult, if not impossible, to determine the natural distribution of many species in long-settled regions.
3. Falk *et al.* (1996, 477) see climate change as a major issue, but, essentially, they regard it as a problem for the future. Thus, in the 'Guidelines for Reintroductions' they write: 'Over the next 50 years changing climates *may just begin to affect many plant populations*, primarily populations at the ecological margins of species distributions' (my emphasis). However, a decade after the publication of this volume, there is increasing evidence that species and ecosystems are being impacted by climate change, and how to react has become a much more urgent issue. Should transfers only be contemplated as a last resort prompted by a crisis? Or should new populations be established outside the historic range now or in the very near future, as an insurance policy against future climate change?
4. The selection of sites outside the historic limits of a species presents some challenges. Firstly, it is necessary to make projections about the changing climate envelopes of species (Kutner & Morse, 1996). Next, should assisted migration be restricted to transfers between nature reserves, or could they also involve private land, where their future may be less assured? In some cases legislation protects endangered species in their last strongholds. Should such legislation be redrafted to provide opportunities for more dramatic management, and who should be charged with the responsibility of assisting plants and animals in their dispersal? Should transfers be made only by professional ecologists employed by local or state authorities?
5. In addition, there is a great deal more to effective establishment of a new population than mere transfer of seeds. Our knowledge of the autecology of rare plants is sometimes rudimentary. Many plant species require highly specific niches to germinate, survive and reproduce, and a transfer site must satisfy these, sometimes exacting, requirements. The success of the project at the new site can only be judged when plants have shown that they are capable of reproducing unaided through several generations.
6. There is a concern that the transfer of species to other sites outside their current range could generate 'new' local or national invasive species. As we have seen in earlier chapters, there is the probability that some highly dispersible native plants could, on transfer, produce large populations and thereby threaten species at the new site, especially if they were transferred without their pests, diseases and current competitors. This raises the question of multiple species transfers. Should the focus be on securing the future of individual species or should suites of species be transferred, for example to maintain co-evolved mutualisms or indeed other significant species interactions such as parasitism etc.? Given that climate change could be an on-going phenomenon,

conservationists are very likely to be faced with the need for successive transfers to different locations.

7. Any transfer will require extensive plant materials. For common species, whose dispersal capacity is insufficient to fill the newly changing climate envelope, seeds, say of tree species, could be collected and sown, perhaps from the air. In the case of rare and endangered species, the collection of extensive material from struggling wild populations might result in their extinction. Thus, the availability of *ex situ* facilities for plant propagation will be critical (Falk *et al.*, 1996). Seed bank collections in cold storage are a crucial insurance policy against total loss of a species. Generally, these collections are not of sufficient size to allow them to be used as a direct source of material for major restorations. For cold-stored seed stocks, and others collected directly from the wild, *ex situ* propagation is required. Not only will it be necessary to grow material for the transfer project, but it would also be prudent to retain reserves of material *ex situ* to make sure that transfers 'in the wild' have been successful. As we saw in Chapter 14, there are a number of difficulties in the use of *ex situ* stocks raised through several generations under garden conditions. Here 'wild' plants will be subjected to major new selection pressures in gardens and genetic changes may occur, including loss of variation if only few plants are grown. Does it matter if plants, changed somewhat in cultivation, are planted in the wild? From an aesthetic point of view, perhaps not, but from a microevolutionary perspective there are major risks. Plants that have been subject to one set of selective processes and chance events in gardens, leading to reduced variability, will be suddenly faced with other selective forces in the wild. Having perhaps adapted to the garden environment, it seems possible that such plants may retain insufficient genetic variation to respond to their new situation. More experimental investigations are necessary to study these crucial issues.

Some climate change scenarios predict very widespread extinction of animals and plants in some parts of the world. If the policy of transfers is adopted, decisions will have to be made on which species to move. Even if a small subset is selected, questions arise concerning the availability of sufficient *ex situ* growing facilities. In addition, future climate change will have profound effects on botanic gardens and other sites that are used to maintain *ex situ* collections. Such changes will also impact domestic gardens and city parks, having major effects on what, in many predominantly agricultural areas, are important reservoirs of wildlife. Thus, increasing temperature and problems with availability of water supplies are likely to stress the long-lived specimens in displays, and climatic change will make it necessary to review and restrict the selection of species chosen for cultivation to those best adapted to the developing conditions.

Returning to issues concerning *ex situ* conservation. Cold seed storage facilities could be increased to cope with a greater number of threatened species, but it must be remembered that the seeds of many species, especially tropical taxa, cannot be conserved under cold-dry conditions of seed banks (see Chapter 14).

Climate change: our response to the warning signs

Thanks to intensive studies by a range of experts, the threats posed by climate change, both to humankind and biodiversity, are becoming clearer. If emissions are not curbed, further warming is predicted and a number of irreversible catastrophic 'tipping points' could lie ahead (Lynas, 2007, 276). Above a 2 °C degree rise in temperature, it is predicted that droughts could cause the collapse of the Amazon tropical rainforests with release of enormous quantities of carbon dioxide. Further warming could take us to a 4 °C world, resulting in the thawing of hitherto permanently frozen arctic soils. Oxidation of these drying soils will release very large amounts of CO_2, and in the same regions, where thawed areas remain water-logged, anaerobic bacteria will produce extremely large amounts of methane – a potent greenhouse gas. The scale of the potential releases can be judged from the fact that the Siberian permafrost region alone contains an estimated 500 billion tons of carbon. These huge releases of greenhouse gases could provoke another tipping point. A rise in ocean temperatures could result in the decomposition of methane hydrates found at the bottom of the oceans. These ice-like combinations of methane and water are only stable at high pressure and low temperature. Releases of vast amounts of methane from this source would add greatly to further warming.

The findings of climate scientists present a bleak and depressing picture, but are the worst effects inevitable? Many authorities think not. If emissions are severely curbed, there may still be time to prevent the worst scenarios. Lynas (2007, 276) notes that, to achieve the target of a rise of no more than 2 °C, 'global emissions of all greenhouse gases must peak by 2015, and drop steadily thereafter with an ultimate CO_2 stability target of 400 ppm (or 450 ppm for CO_2-equivalent)'.

The use of CO_2-equivalents requires some explanation. Some of the targets for emissions control are formulated in terms of CO_2 in the atmosphere; other authorities, recognising the role of other greenhouse gases, convert the combined effects of all these gases to produce 'carbon dioxide equivalents' (Lynas, 2007). The figures for 'equivalents' are of course greater than those for CO_2 alone and provide a more realistic estimation of 'real' risks posed by accumulating greenhouse gases. With regard to public debate and political policy, serious confusion is possible if it is not made clear whether emissions targets are calculated in terms of 'CO_2' or the 'CO_2-equivalents'.

How can cuts in greenhouse gases be achieved? Some put their faith in the devising of new technologies, but Pacala and Socolow (2004) have calculated that the wider adoption of a number of well-understood technologies, and the scaling up of others at present in development, offers a way forward, in meeting very challenging targets. In brief, they point out that the combined effect of many initiatives would reduce greenhouse gases. These include: making vehicle engines, buildings and electricity generation more fuel efficient; substituting the use of natural gas for coal in power plants; building more nuclear power stations; providing power stations and other facilities with carbon capture capacity to enable the CO_2 produced to be liquefied and pumped into geologically secure underground storage reservoirs. In addition, there should be massive increases in renewable technology, including wind turbines, and arrays of solar photovoltaic panels set out in desert environments. The electricity from such arrays could be used to hydrolyse water, providing hydrogen as a clean fuel for vehicles. It is also claimed that carbon emissions reductions could also be achieved by protecting existing forests in the Amazon, and elsewhere, by extensive reforestation schemes, and by cultivation of agricultural soils without tillage to prevent loss of organic matter (Pacala & Socolow, 2004).

In May 2007, the IPCC published their *Fourth Assessment Report* focusing on how mitigation measures might reduce greenhouse gas emissions in the short and medium term (until 2030) and in the longer term. It is beyond the scope of this chapter to provide full details of the proposals. The Report also outlines the policies, measures and financial instruments that can be applied to mitigate climate change. These include the formulation of appropriate emissions regulations and standards, and the devising of frameworks of taxes, charges, tradable carbon permits, financial incentives and voluntary agreements. It is acknowledged that a different mix of approaches would be appropriate in different countries depending on their circumstances. Some of these proposals, if implemented, could have important implications for conservation. Thus, a global warming treaty could reward those nations with biodiverse forests for policies that avoided deforestation (Tollefson, 2008). Considering the economics of adjusting to climate change, the *Stern Review* (2006) stressed the urgency of the situation and put the cost of immediate action at *c.* 1% of the world's gross national product. Doing nothing now could result in 5 to 20 times greater expenditure in the future.

Human adaptation: the threats this poses to areas of conservation interest

All the adaptive measures, proposed by Pacala and Socolow (2004) and the IPCC Report, have major implications for conservation. Space permits only a few selected examples.

1. In order to feed the growing mass of humanity, pressure to bring more land into cultivation will increase and this will threaten many semi-natural ecosystems and their biota. In addition, if steps are not taken to limit greenhouse gases, climate changes will accelerate and, increasingly, populations escaping from rising sea levels, droughts etc. will become refugees. Their struggles for survival will put at risk the wildlife of many regions of the world.
2. In areas subject to increasing drought through climate change, freshwater resources are likely to be directed towards human needs rather than to areas of conservation interest. Thus, the first call on supplies will be to provide drinking water, service irrigation schemes, and water for industrial and commercial use. Also, a very great deal of water is used to maintain parks, gardens and sports facilities such as golf courses etc. In the future, if, as predicted, droughts become more frequent in certain areas, the use of water will have to be critically examined. For instance, the famous parks and European-style gardens of Perth in Western Australia are maintained by drawing large amounts of water from aquifers below the city. Over the past decades, diminishing rainfall has not been sufficiently recharging these sources. Recharging aquifers with recycled treated wastewater is being examined, but there are concerns about whether pollutants in used-water could enter the aquifers.
3. As farmers adapt to climate change, there are likely to be major changes in agricultural lands, areas that form the matrix between reserves. Responses are certain to involve new crops, including GM varieties and new farming methods. Climate change is projected to have major adverse impact on the wine-growing regions of California (White *et al.*, 2006) and may force changes in crops grown. In addition, some initiatives are aimed at reducing the air miles involved in importing food. In many parts of Britain there has been a huge increase in the use of poly-tunnels for growing soft fruit. This allows an extension of the growing season, but radically changes the landscape qualities of the matrix of lands between areas of conservation concern. It is predicted that, unless

greenhouse gas emissions are brought under control, food production in many parts of the world will falter. For instance, in spring 2007, many agricultural areas of Australia suffered the worse droughts for 100 years.

4. Our use of fossil fuels has, in large part, generated the climate change crisis. Adaptation to climate change, through reductions in greenhouse gases, involves profound changes to energy generation through the use of renewable energy resources based on the Sun, wind and water. The widespread adoption of such schemes will have many effects on ecosystems of conservation importance. For instance, barrage schemes across estuaries have the capacity to generate electricity, but they will greatly modify present-day estuarine ecosystems. Many of the major rivers of the world already have dams to provide hydroelectricity and more are likely to be built. Wind farms are being constructed in many countries. In order to play their part in the reduction of greenhouse gases, many installations would be required. For example, Pacala and Socolow (2004) calculate that wind farms covering 30 million hectares would be required in the USA, some of which would be on land and others offshore (*c*. 3% of national territory). Such developments face opposition from those who wish to conserve existing landscapes and ecosystems.
5. To contribute to the reduction of greenhouse gases and provide enhanced energy security, the sustainable use of biofuels is seen as an important priority. Currently ethanol is being made from sugar cane, oil seed rape, palm oil etc. These fuels are generally mixed with others from fossil sources. Also, to provide fuel for power stations, willow coppice (*Salix* spp.) and elephant grass (*Miscanthus* spp.) are being grown. It is proposed to produce second-generation biofuels from plant biomass. As these sources of 'energy' become more important there will be competition for land to grow conventional versus biofuel crops. This competition effect is very clear from calculations of Pacala and Socolow (2004), who point out that, for the use of ethanol to make a serious contribution to greenhouse gas reductions, an area one-sixth of the world's croplands would be required to raise the necessary biomass. If biofuels were grown to the extent proposed, further acreages of land hitherto not intensively farmed may be drawn into cultivation, with the likely loss of areas of conservation interest. Whether biofuels always lead to a reduction in the use of fossil fuels is open to question. For example, very large quantities of carbon dioxide are released in the clearing of forests in south-east Asia to plant palms. Also, conventional fuels, from fossil sources, are used to refine and transport the palm oil across the world. It seems unlikely, therefore, in this case, that there is an overall net reduction in carbon dioxide arising from the switch from fossil to biofuels. As I write, very serious food shortages are being reported in many parts of the world. Various factors are implicated, including droughts in Australian wheat lands, increasing demand for meat and milk in the emerging economies of such countries as China and India, and rising fuel costs etc. But many commentators see the diversion of food crops to biofuel production as a major contributor, leading to acute shortages and unaffordable food prices in many developing countries.
6. Climate change is having a major impact on alpine areas. As snow cover on lower slopes becomes more unpredictable, there is more intensive use of higher alpine regions for outdoor winter sports, leading to even greater pressures on montane ecosystems.

Conclusions

Bringing anthropogenic climate change under control presents humanity with a major challenge. The degree of impact on biological resources, national parks, reserves and areas of matrix containing significant biodiversity will depend crucially upon which storyline in the IPCC Report is followed. Will it be 'business as usual', or will low carbon economies be successfully developed? Clearly, successful adjustment in reducing the use of fossil fuels is the preferred outcome, but, as we have seen, many adaptive responses are likely to involve dramatic changes in land use in agriculture, forestry, flood control etc. Adaptations will have to be wide ranging, involving responses to sea level rise, the cultivation of new crops such as biofuels, the devising of barrage schemes to generate electricity, the wider establishment of wind farms on land and sea etc. All these likely developments have implications for those concerned with conserving landscape and ecosystems. Extending the ideas set out in Pontin (1993), humankind has successfully navigated two major transitions – the adoption of agriculture and industrialisation. Can the Third Major Transition to a low carbon economy be accomplished successfully without putting civilisation at risk?

The predictions concerning climate change introduce a great deal of uncertainty in the management and future planning for ecosystems and endangered species (Malhi & Phillips, 2005; Bhatti *et al.*, 2006). Should managers be optimistic and 'hope for the best' or plan for worst-case scenarios? How can probable climate change effects be factored in to restoration and reintroduction projects? There are no easy answers to these questions.

Reflecting on the history of the rise of environmentalism and conservation in the nineteenth century reveals that European settlers in the New World, Australia etc. were able to take significant conservation action through national governments to create national parks and reserves. In long-established areas of cultural landscape, individual initiatives often preceded government action. Thus, in 1912, Rothschild saw his role in trying, unsuccessfully, to persuade landowners to conserve rare and endangered species on their large estates in England. Only later did the UK designate national parks and nature reserves etc. Now, the situation is different: success in conservation will increasingly depend upon whether *global agreement* can be achieved and *effective international action taken*. To bring climate change under control will require changes in behaviour in very large numbers of people worldwide. Conservationists, experienced in acting locally and nationally, must join the fight to achieve effective reduction in greenhouse gases, otherwise much that has been achieved in conservation will be overturned. To minimise the impact of climate and its damaging effects on the environment, adaptation must be made without delay. However, hard choices will have to be made, and, while conservationists may whole-heartedly favour the overall aim of adaptation, with regard to specific projects, they may find themselves opposing change on the grounds that cherished landscapes, threatened ecosystems and significant habitats for wildlife (including endangered species) may be damaged in the process of constructing wind farms, estuarine barrages etc.

Two recent major reports consider aspects of climate change and conservation. Ehrlich and Pringle (2008) review a range of strategies for the future safeguarding of biodiversity. They identify population growth, 'habitat conversion, environmental toxification, climate change, and direct exploitation of wildlife' as major factors. They conclude, 'these threats are formidable', But they take the view:

> Seven strategies . . . if implemented soundly and scaled up dramatically, would preserve a substantial portion of global diversity. These are actions to stabilize the human population and reduce its material consumption, the employment of endowment funds and other strategies to ensure efficiency and permanence of conservation areas, steps to make human-dominated landscape hospitable to biodiversity, measures to account for the costs of habitat degradation, the ecological reclamation of degraded lands and repatriation of extirpated species, the education and empowerment of people in the rural tropics, and the fundamental transformation of human attitudes to nature.

A second report by Heller and Zavaleta (2009) examines the range of individual recommendations made by conservationists for management under climate change of endangered species and ecosystems. These include restorations, reintroductions, and increasing landscape connectivity and resilience etc. However, both reports highlight the need to widen the horizons of conservation biology, which, I would argue, has hitherto had too narrow a focus on biodiversity.

Taking a more holistic view of environmental issues in their social, economic and political context will be crucial if humanity is to rise to the formidable array of major challenges, including bringing anthropogenic climate change under control.

21
Overview

Microevolution and conservation: Darwin's insights

Darwin provided not only the major element of our present understandings of microevolution through his theory of natural selection, but also key insights into particular issues relevant to the themes of this book – the nature of species, the co-evolutionary relationships between species, the evolution of humankind, the insights into domestication processes, the vulnerability of rare species, the impact of invasive species, and the evolutionary responses to climate change etc. Also, in his approach to testing hypotheses, Darwin was a major figure in the development of experimentation. For all these contributions, and many others not mentioned here, there is cause to celebrate the 200th anniversary of Darwin's birth and the 150th anniversary of the publication of *On the Origin of Species* in 2009.

Cultural landscapes

Historically, the initial influence of humans on the biosphere was likely to have been very small, but, as populations migrated across the world and increased in numbers, humankind has become a major, indeed decisive, force, influencing all the world's ecosystems. In the future, human impact is certain to increase, as population growth is predicted to increase dramatically.

There has been a complex transition from a 'natural world', in which humans played an insignificant role, to a world of cultural landscapes, where human activities dominate. Archaeologists are discovering that areas of apparent wilderness have been subjected to major human impacts in the past. In addition, such areas are increasingly being influenced by human activities. For instance, anthropogenically produced greenhouse gases are responsible for climate change. These gases are, by and large, the product of areas of high human population, but the effects are not confined to these regions, for evidence suggests that such pollution is changing the climate in remote areas of apparent wilderness, such as the polar regions.

These conclusions are emphasised by Walton and Bridgewater (1996, 15) who note: 'human activity has touched almost every conceivable part of the biosphere'. Indeed, they take the view: 'all landscapes (and seascapes) are to some degree anthropogenic'. They emphasise: 'most of the global land surface comprises cultural landscapes', and, in effect, plants and animals live in a 'cultural potpourri'. They consider: 'a new perspective is justified, that of the biosphere as the global garden and people as the gardeners'. It could be argued that lands reserved for nature conservation, and widely regarded as natural, often had a long history of human management and exploitation before they became nature reserves. Furthermore, in many cases, their subsequent development has often involved considerable human intervention, involving predator, fire, pest and disease control. Management has also been strongly influenced, in many cases, by principles of game management, the design of parks and gardens, and the requirements of tourism, leisure pursuits and sports.

Conservation strategies

Historically, the conservation movement has made valiant efforts to 'protect' and conserve endangered species and ecosystems through the development of national parks, and nature reserves. But, as Walton and Bridgewater (1996, 16) have pointed out: 'protected areas' have emerged 'rather opportunistically' without 'any grand national plan to capture a representative set of extant landscapes'.

In the early days in the management of nature reserves and parks, there was an 'apparent assumption' that 'the quality of protected areas can persist over time as if they are static displays of nature' (Walton & Bridgewater, 1996,16). With the subsequent development of the science of ecology, research has revealed that ecosystems in protected areas, as elsewhere, are dynamic. Also, active 'conservation management' has replaced 'protection'. Simple management was a forerunner to more 'creative conservation', which has seen the development of large and small restoration projects, not only to recreate former ecosystems and repair damage, but also to create new ecosystems important to human well-being. It is important to stress that much ecosystem restoration has been driven not by concerns about rare and endangered species, but by a desire to counter the adverse effects on human welfare of unfettered *laissez faire* anthropogenic niche construction. Following major campaigns by environmentalists, there has been effective legislation followed by action, in the developed world, to achieve cleaner air, cleaner water, counter soil erosion, and rehabilitate lands damaged by mining, industrial dereliction etc. In addition, restoration and management contribute to the conservation of cherished landscapes that have very important aesthetic and spiritual significance. While there is growing expertise in habitat restoration and rehabilitation, there are major challenges ahead. Can very complex ecosystems, such as woodland, be recreated? It is certainly possible to plant trees and other plants, but how far will it prove possible to recreate whole ecosystems. One of the major figures in the restoration movement has expressed an important view on this question. Cairns (1998, 218) writes:'for both scientific and economic reasons, human society must rest satisfied with a *naturalistic assemblage* of plants and animals rather than a precise replication of the species that once inhabited the area' [my emphasis]. If this assessment is accepted then it becomes clear that we must strive to conserve significant elements of cultural landscape (such as ancient woodlands and old-growth forests, etc.), as these ecosystems have their own intrinsic value as part of our cultural heritage, and if lost cannot be recreated.

In situ conservation: protected fortresses

The management of reserves as protected fortresses has proved an unworkable model in many regions of the world. Walton and Bridgewater (1996, 16) take the view that people are 'part of biodiversity', and 'cannot be segregated from biodiversity by fences. Protected areas, while dedicated to particular conservation purposes, must not become intensive care units for a few charismatic species surrounded by an ever-increasingly degraded area occupied by people.' Therefore, Walton and Bridgewater emphasise, as do many other authorities, that reserved areas should have 'a structural and functional interaction with the surrounding landscape', rather than being viewed as 'biodiversity sanatoria'. At the reserve–surrounding matrix interface, particularly in tropical areas, humanity is faced with two categorical imperatives: to conserve biodiversity and to raise the living standards of those in poverty and disease. Could multiple land use, here and elsewhere in the world, allow both imperatives to be fulfilled, through some form of sustainable development?

Sustainable development

Firstly, as we have seen in earlier chapters, it is important to stress that humans have always collected and used other species. Indeed, certain plants and animals are such important resources that defending them from 'exploitation by other groups of people has been one of the underlying motivations for tribalism and nationalism, and associated conflicts throughout history' (Milner-Gulland & Mace, 1998, 5). 'Where people can profitably make use of wild species, they will' (Milner-Gulland & Mace 1998, 349). Increasingly, technological advances have 'improved' the efficiency of harvesting, and there are many examples of the over-exploitation of fisheries, forests etc. Not only are populations of species of commercial interest being damaged, there are also unwitting effects of harvesting, such as the by-catch taken in commercial fishing, and the injury and destruction of non-timber species in commercial forestry.

Secondly, conservationists are particularly concerned to safeguard wild species in functioning ecosystems. However, Milner-Gulland and Mace (1998, 349) make a very important point:

> We live in a world where the biological status of wild species is not a motivating force for most human action, either at an individual, community or government level. Any conservationists that ignore this fact are either acting in privileged corners of the earth where the conflicting demands of wildlife and humans are low, or they are doomed to meet forces that could totally undermine their efforts.

Acknowledging these two major issues, has the concept of sustainable development been employed successfully in practical conservation? A number of case histories are reviewed by Meffe and Carroll (1997) (African game parks; and the Greater Yellowstone ecosystem). Further studies are provided by Milner-Gulland and Mace (1998) (forest use in south-east Asia; mahogany; management of wildlife in Nicaragua; the Falkland Islands squid fisheries; recreational use of coral reefs in the Maldives and Caribbean; wildlife conservation in Kenya; gorilla tourism; caribou and musk ox harvesting in the Northwest Territories; and game mammals in the Soviet Union). Here, our concerns are primarily about sustainable development in plants, but a number of important general issues are raised.

Ecotourism is widely viewed as a 'wildlife friendly' sustainable activity that will reduce habitat destruction, and diminish the negative impacts on local people in areas of high biodiversity by providing employment, improving healthcare and facilities etc. However, the financial benefits of coral reef, safari and wildlife tourism have, in many areas, failed to reach many local people (Milner-Gulland & Mace, 1998). In addition, across the world from the Galapagos Islands to the Caribbean, from Yellowstone and even to Antarctica, there are concerns about the long-term ecological impact of growing numbers of tourists, and their increasing demand for greater facilities. Each eco-tourist destination is unique, and issues and concerns must be considered on a site-by-site basis. For example, the case of gorilla tourism in the Volcanoes National Park, in the western rift valley Rwanda, Central Africa, provides a disturbing example. In addition to the threats from fires, habitat loss, political instability and civil wars, contact with 'tourists, guides, rangers, porters and researchers' has resulted in changes in gorilla behaviour and an increased risk of disease transmission. Thus, in the Volcanoes Park, it has proved necessary to vaccinate the animals against measles, and to give antibiotics to control an outbreak of broncho-pneumonia (for details see Butynski & Kalina, 1998).

Reflecting on the history of game hunting in the Soviet Union, Baskin (1998, 331) considers whether this sport has ever been carried out in a sustainable manner. He concludes: 'by the end of the Russian Empire in 1917, most game mammals were almost extinct due to unchecked exploitation'. Under Stalinist rule, game increased, but this was the result of truly draconian measures. Large areas were depopulated, leading to widespread starvation and genocide following collectivisation of the nomads. Now, with the collapse of the Soviet Union, some species (tiger, saiga, musk deer) are again under severe threat, to some extent because their body parts are used in traditional Far Eastern medicine.

Logging tropical rainforests often has disastrous effects on biodiversity and local communities. MacKinnen (1998) considers the dilemma faced by conservationists in New Guinea and other areas. Should logging be opposed, or should 'responsible' extraction of timber be allowed or even encouraged by companies that promise to manage in a sustainable fashion, with more equitable distribution of royalties, and trade in fully labelled timbers.

The degree to which individual tree species might be sustainably managed has also been considered. The case of Bigleaf Mahogany (*Swietenia* species) is particularly revealing. Gullison (1998) points out that, in the past, indigenous peoples probably increased the abundance of useful trees like mahogany through their management. Currently, the monetary value of trees is very high and, therefore, there is a strong incentive to harvest the adult trees wherever they are found. The life history of the species makes it very difficult to manage populations sustainably. Bigleaf Mahoganies mature in closed forests, yet seedling establishment only occurs in well-lit disturbance areas that are naturally produced by hurricanes and fire. Trees take a very long time to come to sexual maturity (*c.* 150 years). In the 'wild' they may fail to reach this age, as trees are commercially valuable even before they reproduce. Considering the longer-term effects, current harvesting of large and immature trees is very likely to promote genetic erosion and inbreeding in Bigleaf Mahogany populations. What is the experience of sustainable management through the establishment of plantations? Gullison reports that there are severe problems with Shoot Borer (*Hypsipyla grandella*), which damages the apical meristem resulting in trees with multiple trunks.

After an exhaustive review of theory and practice of sustainable development in a conservation context, Milner-Gulland and Mace (1998, 350–357) consider: 'the case for use being a conservation tool is not proven. Use of other species is an inevitable part of human existence. Thus, it is essential that we consider means by which use can be managed sustainably, but the overwhelming conclusion is that this is hard to achieve.' Reflecting on sustainable development in a wider context of forests, grasslands etc., their final comments offer some reassurance:

> The situation may not be as bleak as it appears. Changing values in society do influence the political process . . . the profile of environmental issues has been raised, and time and effort that governments give to environmental issues is increasing rapidly . . . Environmental degradation is the cumulative process of millions of local actions, some with local and some with global effects. An accumulation of millions of local actions will ultimately be the only way that global sustainability can be achieved . . . Where the political will exists, we are in a position to take real steps towards the sustainable use of our environment.

Extending this theme, Palmer *et al.* (2004, 1251–2) consider that while study of the 'few and rapidly shrinking undisturbed ecosystems is important . . . now is the time to focus on an ecology for the future'. They take the view:

> Our future environment will largely consist of human-influenced ecosystems, managed in varying degrees, in which the natural services that humans depend on will be harder and harder to maintain. The role of science in a more sustainable future must involve an improved understanding of how to design ecological solutions, not only through conservation and restoration, but also by purposeful intervention of ecological systems to provide vital services. Shifting from a focus primarily on historical, undisturbed ecosystems to a perspective that acknowledges humans as components of ecosystems, together with new research on ecosystem services and ecological design, will lay the groundwork for sustaining the quality and diversity of life on earth.

Human activities impose selective forces

As human populations have grown, and cultural landscapes have expanded, there is increasing evidence that mankind has become a significant influence on all aspects of microevolution. Ecosystems continue to be impacted by natural selection acting through climatic, edaphic and biotic effects, but, as we have seen in previous chapters, human actions also impose selective forces. In some cases, human activities are subtle and act 'at the margins', but often human activities in niche construction can produce abrupt, powerful and catastrophic changes. Until comparatively recently, human activities were known to alter the microclimate of certain ecosystems, but there was no suggestion that humans could alter the global climate. Now, as we have seen above, there is convincing evidence for global climate change as a result of the release of anthropogenic greenhouse gases.

Thus, the arena of current microevolution is ruled by extremely complex and endlessly changing interactions of natural and man-made selective forces. The evidence reviewed above has shed much light on microevolution in plants, in relation to agricultural activities, pollution, and through contact between introduced and native species etc.

The case histories investigating microevolution have often concentrated on situations where plants were subjected to extreme individual defined biotic factors. However, there is no reason to believe that microevolution happens only in such extreme situations.

As we have seen above, human activities deliberately manipulate ecosystems. Often, unwitting effects are produced. There is abundant support for the views expressed by Walton and Bridgewater (1996, 16): 'Biocultural landscapes, or bioregions, unquestionably evolve and people are a significant selection force.' In addition, there is increasing evidence that humans have an impact not only on selective processes but also on other elements of microevolution by deliberately and unwittingly breaking down natural geographical and ecological distributions, altering species composition and population sizes, and thereby altering the Darwinian fitness of component species. By these means, human activities may influence 'chance events', such as founder effects, genetic drift, species hybridisations and gene flow. In addition, Meyer (2006, 42) makes the very important point: 'Our most common tools for preserving biodiversity – prohibitory laws and regulations, bioreserves, and sustainable-development programmes – are themselves powerful engines of human selection.'

Considering the many studies of microevolution outlined in this book, it is clear that a fundamental truth emerges in relation to conservation. While stabilising selection can occur for particular traits, in a rapidly changing world, plants and animals are subject to ongoing microevolution through directional and disruptive selection, in a highly complex forward-moving relentless process. This results in shifting balances of advantage and disadvantage amongst the plant and animal actors on the evolutionary stage, resulting in winners and losers. On present evidence, although invasive species may share some characteristics, it is not possible to predict which species will be super-invasive in the future. Another insight from the many studies presented above is that the status of winner in one set of conditions is not permanent. With change, winners become the new losers, and other taxa, perhaps even of hybrid origin, might rise to predominance. Thus, it is important to realise that evolution does not 'work backwards' to re-evolve the past, to restore what was present earlier. In the complex arena of human niche building, evolution 'moves' forward, and managing present-day ecosystems by 'resort to precedent' cannot return ecosystems *precisely* to a past state, and then prevent further change. Likewise, the exact restoration of ecosystems that existed in the past is impossible. We can put together combinations of plants and animals, or add to damaged ecosystems those element that appear to be missing according to our vision of what was present in the past. But essentially, restoration and management create *new not old ecosystems* and these creations are not static entities, but arenas for microevolution.

Human activities and domestication

Reflecting on the many impacts of human activities, Western (2001, 5461) concludes:

> Human domestication of ecosystems greatly reduces species diversity. [Moreover,] of equal or greater importance, asymmetrical selective pressure on larger species downsizes communities. [Also] the expansion and intensification of domesticated landscapes [reduce the habitat] for non-domesticated species, reduce population sizes, and fragment their range by imposing physical and biological barriers to dispersal. The resulting population declines and barriers select against poor dispersers, including big species. Small, easily dispersed species able to tap into the production cycles of domesticated landscapes and heavily harvested natural resources are selectively favored.

Turning now to the relationship of humans to plants brings us to the question of domestication. Harlan (1975) states: 'To domesticate means to bring into the household. A domestic is one (servant) who lives in the same house. In the case of domesticated plants and animals, we mean that they have been altered genetically from their wild state and have come to be at home with man.' Domestication is now seen as an evolutionary process rather than an abrupt event. 'A fully domesticated plant or animal is completely dependent upon man for survival. Therefore, domestication implies a change in ecological adaptation, and this is usually associated with morphological differentiation.' There are 'inevitably many intermediate states' between the wild and the domesticated, depending on the degree of cultivation and the type of management. 'To cultivate means to conduct activities involved in caring for a plant, such as tilling the soil, preparing a seedbed, weeding, pruning, protecting, watering, and manuring. Cultivation is concerned with human activities, while domestication deals with the genetic response of the plants or animals being tended or cultivated. It is therefore possible to cultivate wild plants, and cultivated plants are not necessarily domesticated. Harvested plant material may be classified as wild, tolerated, encouraged, and domesticated' (Harlan, 1975). This classification of harvested plants has recently been reconsidered in studies of the use humans make of wild plants, particularly in the tropics, where the distinction between wild and cultivated is far from simple. For Amazonia, Clement (1990, following Harlan, 1975) proposed the following scheme in recognition of the many intermediates between the wild state and domesticates.

A *domesticated plant* is a genetically modified species completely dependent on humans for survival.

A *semi-domesticated plant* has been significantly modified but is still not completely dependent on humans for survival.

A *cultivated plant* has been introduced into human agro-ecosystems and is nurtured in a prepared seedbed.

A *managed species* is protected from human actions that might harm it, is liberated from competition with other species, or is planted in areas other than prepared seedbeds.

A *wild plant* may be used but is neither managed nor cultivated.

This classification has been devised for studying the multitude of ways that humans use plant products, as food resources, fibres, medicines etc. It could be construed as a series of steps along the path from the wild condition to full domestication and also steps in the opposite direction towards the feral and the wild. This classification is also very significant in connection with the major theme of this book – the conservation of endangered plants in human dominated or influenced ecosystems.

Maintaining species in a 'wild' state in managed environments

Of particular interest is the distinction drawn between the wild and the 'managed' state. Nature reserves and many other areas are now being managed to control abiotic conditions and interspecific competitive interactions, in an attempt to favour those species that are endangered. In recent years the repertoire of management interventions has increased. It must be recognised that, at the plant's-eye level, such management produces 'new' selection pressures. The key to long-term survival, in changing circumstances, is the presence in populations of appropriate genetic variability. If populations lack the appropriate genetic variability to respond to new situations, they will decline and face extinction. As we have seen in earlier chapters, winners under one set of conditions may become losers as circumstances change. Only those species that have the genetic capacity to respond to change, be it abrupt or gradual, will survive. Changes in human activities through management, exploitation, abandonment, neglect etc. have the potential, therefore, to produce genetic changes either in the direction of domestication or towards the wild or feral state.

Consideration of these issues brings into question whether species may be maintained in a wild state under human management. First, as we have seen, there are semantic difficulties with such concepts as 'wild'. As with other concepts, such as species, natural etc., it is not possible to arrive at a single universally agreed definition. What is important is to consider the usage of different words. For their own convenience, different groups have used the word in different ways. Given the evidence cited in this book, it seems impossible, from a genetic standpoint, to maintain wild plants in their original state long term through many generations, by propagation from seed under cultivation or any other management regime, as such changes present new selective arenas in which microevolutionary change will occur.

Despite the inevitability of genetic change, cultivated plants are often sold as 'wild'. For example, following a vogue begun in the nineteenth century (Robinson, 1894), 'wild' gardens have been created by sowing and planting 'wild' stocks. Also, nature reserves and many other areas are now being managed to control abiotic conditions and interspecific competitive interactions in an attempt to favour those species that are endangered. 'Wild' flower seeds, first marketed for sowing in 'wild' areas in gardens, are now widely available. They are produced as a specialist commercial field crop (or 'wild collected' from man-made habitats such as hay meadows). Wild flower seed is increasingly being used along the margins of new roads, in the rehabilitation of derelict land, the repair and restoration of damaged habitats and the creation of new wildlife areas etc. In addition, wild seeds and plants are being taken into botanic and other gardens for propagation, and these plants are being used to restore, reinforce or reintroduce populations of wild endangered species. They are also being used as reserve stocks in case of extinctions elsewhere. While it may be possible to maintain the genetic integrity of 'wild' material through several, perhaps many, cycles of vegetative propagation, as we saw in Chapter 14, long-term *ex situ* conservation of material in botanic gardens, through several generations involving propagation from seed, is likely to lead to genetic changes in the stocks in the direction of domestication.

In the face of difficulties in the long-term maintenance of stocks of rare and endangered species in botanic gardens, a variety of technologies are being explored in attempts to extend the viability of spores, seeds etc. in storage (see Chapter 14). Here, it is important to consider the implications of climate change for *ex situ* conservation. Conserved living biological material, including that available in seed banks, could become increasingly important in restoration and reintroduction projects if species become extinct in the 'wild'. However, there is now increasing evidence of range shifts in a variety of species and some cases of genetic adaptation to changing climates. It may be that, as time passes, stocks in cold storage in seed banks will no longer be appropriate to restore the sites from which they came, unless they contain a very wide spectrum of genetic variability on which selection can act. Furthermore, the outcome of the use of such stocks in the restoration of more distant areas could be highly unpredictable.

From the domesticated to the feral

As we have seen, a major line of enquiry in plant microevolution focuses on the process of domestication (Zeder, 2006; Ross-Ibarra, Morrell & Gaut, 2007). Another area of considerable interest is the evolution of feral organisms from the domesticated. This phenomenon is well known in animals – domesticated stocks having given rise to feral populations of pigs, goats, cattle, horses, cats, dogs etc. Less attention has been paid to ferality in plants. The advent of GM crops has highlighted this element of microevolution. Research has revealed that volunteer GM variants may continue to appear in landscapes after the crop has been harvested. Attention focuses on the accidental dispersal of seeds as crops are harvested and transported. Whether volunteers appear in succeeding years depends upon a number of factors, such as crop rotations, herbicide use, weed control and whether viable seed remains in the soil for a period (Pivard *et al.*, 2008). The critical question for farmers is whether such volunteers, some of which could be GM-herbicide resistant, could become feral in the area and pose problems for future weed control.

The microevolution of feral taxa is of considerable concern to the conservationist, not least because of the damage caused by feral animals, especially to endangered species and ecosystems on islands. Feral plants are clearly important too, especially those that have become or have the potential to become invasive.

Close study of feral plants is underway using molecular tools (Gressel, 2005a, b). A number of lines of enquiry are being pursued. It is proposed that the domestication process might be relatively fast and involves a limited number of recessive traits. De-domestication to the feral state is also visualised as a relatively speedy process with dominant traits being selected (Gressel, 2005a, b). The de-domestication hypothesis is being examined in situations where wild, domesticated and feral populations of the same species are all available. Progress is being made in studies of feral rye, wheat, radish, sunflowers and beet etc. in agricultural landscapes (Gressel, 2005a, b). Here, ferality in 'wild' landscapes is examined in three case studies. So far the microevolutionary context has been established: the de-domestication hypothesis has yet to be fully examined.

The olive, first domesticated from the wild oleaster in Mediterranean areas by 5,800 BP, exists as a number of cultivars propagated by cuttings or grafting. Investigating the process of domestication may be possible, as individual trees may be more than 2,000 years old. Probably, a few traits are involved in domestication, such as oil content, size of fruits and fruit abscission properties (Brevillé, 2005). Feral olives occur in the Mediterranean region and other areas with a similar climate – California, Australia etc. – to which the olive was introduced as a crop. The three groups of olive – wild, cultivated and feral – cannot be reliably separated morphologically, but, by employing molecular techniques, it has proved possible to begin to characterise the different groups, in what has hitherto been a confusing mixture of wild oleasters, different cultivars of the olive, abandoned plantations and feral trees (Brevillé, 2005). Molecular studies have also investigated gene flow between the cultivated olive and wild oleaster. In some areas, around the Mediterranean, olive trees may be pollinated by oleasters, but this does not affect the fruit, as its development is maternally determined. Hybrid fruit may give rise to feral plants, which may also arise from the fruit of abandoned plantations of cultivars (Breton *et al.*, 2005). The presence of wild olives in forest fragments around the Mediterranean has been confirmed by Lumaret and Ouazzani (2001). Extending these studies, the genetic relationship between wild and cultivated olives in the central Mediterranean basin has been examined in some detail by Baldoni *et al.* (2006), and the origin of feral trees from cultivated olives has been investigated in the Mediterranean (Breton *et al.*, 2005), and in Australia and some Pacific islands (Besnard, 2007). In the Adelaide Hills of South Australia, feral olives are in danger of becoming seriously invasive. Abandoned olive groves on marginal land provide a bridgehead for the invasion into areas where sheep grazing has been abandoned. These abundantly fruiting olives are rarely harvested, and the seeds are widely distributed by foxes and emu. Olive seeds survive for long periods in the soil.

Lythrum salicaria is another species that might be further investigated, using molecular methods, as invasive feral populations in North America may have evolved from crossing between introduced wild and horticultural variants. Of special interest would be the extent of any genetic changes involved, first in domestication and then in the subsequent evolution of ferality (see Kowarik, 2005).

Another candidate for close study, which is of particular concern to conservationists, is *Rhododendron ponticum* – a species with horticultural and invasive feral populations in the British Isles (Kowarik, 2005). Further molecular studies are necessary to test hypotheses concerning the role of species hybridisations in the origin of the cultivated plant. An observer, lacking historical background, could assume that all the material out in the British countryside is feral in origin – escapes from suburbia. However, it is important to note that the species was widely planted on large rural estates in the UK and Ireland, not only to beautify lakesides but also as a 'windbreak', 'a cheap fast-growing informal hedge' and for 'game covert' (Bean, 1976). In addition, other *Rhododendron* species were sometimes planted as grafts on '*ponticum*' stocks. In many cases, sucker-growth from such plants has 'overwhelmed the grafted material' (Bean, 1976). These direct plantings have provided a bridgehead for what have become major invasions. The degree to which genetic change has contributed to the invasiveness of *R. ponticum* remains to be investigated.

The results of investigations of domestication and ferality in plants and animals are not only of interest to geneticists; they are of immediate concern to conservationists, who are concerned with maintaining genetic integrity in stocks taken from the wild, propagated *ex situ* in gardens and zoos, and then released back into the wild. They are hoping to prevent or minimise genetic change through this process, by preventing bottleneck effects and other changes. But, as we have seen in earlier chapters, some element of domestication is likely to occur *ex situ* in stocks raised through several generations, as plants for restorations or back-up material. Should this happen, selection for de-domestication is then likely, as animals and plants are moved back 'into the wild' to be conserved *in situ*. In reality, the selection pressures involved, after the release processes, are likely to be closely akin to selection for the feral state. Will de-domestication of any traits acquired *ex situ* occur, and will this yield successful *feral* populations? Will endangered species, which might lack variability, be sufficiently genetically variable to be able to respond to two successive different selection arenas? Research is needed to track the genetic changes that occur as wild plants and animals are brought into 'captivity' in *ex situ* conservation. Here, it is emphasised that such studies should be extended to follow the genetic consequences of subsequent release back into the wild.

Co-evolution of humans, domesticated cattle and plants

Another area of interest concerning selection is yielding significant results. As we saw in Chapter 3, there is a widespread reluctance amongst many in society to accept Darwin's theory of human origins. It is hardly surprising, therefore, that the question of whether humans are continuing to evolve is a controversial and sometimes neglected research area.

Our concern here is with plant evolution, but a recent study of human genetics is revealing, as it links human evolution with domestication of cattle – through the drinking of unfermented milk – and, by extension, the evolution of grassland management for hay and pasture.

Burger *et al.* (2007, 3736) report: 'most mammals lose the ability to digest the milk sugar lactose after weaning because of an irreversible reduction in expression of the intestinal enzyme lactase'. The ability to continue to digest unfermented milk in human adults is conferred by the presence of a dominant 'lactose persistence' allele (LP). This genetic trait is common 'in populations of northern and central European descent'. Elsewhere, LP is frequent in pastoralists, but usually less common in non-pastoralist cultures. It has been proposed that the European pattern of LP distribution is a relatively recent phenomenon resulting from strong selection pressures. No selective advantage of LP would be expected in human communities where there was no assured milk supply. In contrast, 'the continuous availability of an energy- and calcium-rich drink' could influence Darwinian fitness in human populations, where it enabled farmers to survive poor and unpredictable harvests. Burger and associates have examined two hypotheses. Either, LP was originally rare in early farming communities, but increased with dairying, or, alternatively, some prehistoric populations were preadapted in having high LP and this played a role in their adoption of cattle husbandry. By examining DNA from archaeological human remains of various ages, Burger and colleagues discovered that LP was rare in early European farmers. These findings support the hypothesis that the LP allele has recently increased. There are grounds for accepting, therefore, that some human populations have co-evolved in the early farming period with the domestication of cattle. Moreover, the co-evolutionary linkages extend to grassland ecosystems managed to provide food for cattle. As we have seen in an earlier chapter, there is very strong evidence for the co-evolution of grasses and forbs, as humans adopted land management to provide not only grazing lands but also hay crops. Hay and pasture represent two very contrasting land uses and there is clear evidence for the co-evolution of genetic intraspecific races of plants adapted to each of the contrasting land uses. Furthermore, there is site-specific adaptation in hay ecosystems to different cutting dates and aftermath grazing.

Kulturfolger

Is there any over-arching concept that encapsulates the effects of human activities on the plant world? Rosenzweig (2001, 5409) highlights such a concept that originates from German-speaking Europe. Human activities have imposed severe selection pressures on plants and animals. Some have low Darwinian fitness and decline or become extinct. Others thrive in the niches constructed by humans. There are winners and losers. The winners are the 'Kulturfolger' – 'culture followers'– the organisms that thrive in human cultural landscapes. Meyer (2006, 4–9) makes this point very forcibly when he concludes: 'the land and the oceans will continue to teem with life, but it will be a peculiarly homogenized assemblage of organisms unnaturally selected for their compatibility with one fundamental force: us'. He stresses: 'we are making the planet especially hospitable for the weedy species: plants, animals and other organisms that thrive in continually disturbed, human-dominated environments'. Mooney and Hobbs (2000) also emphasise that a 'massive biotic homogenisation of the earth's surfaces' is occurring. Also, 'as a result of the breakdown of the major biogeographic barriers that have historically kept floras and faunas of the various continents quite distinct, mixing of biotas has resulted in many aggressive species having extraordinarily wide distributions across the globe, or locally very high population densities, and the consequent devastation of the endemic biota of specific regions'.

However, while it may be true that 'homogeneity' is increasing, paradoxically, at the present time, many geographic regions now have historically high levels of biodiversity.

This follows from the fact that, while many species are endangered, and could become extinct in the future, as yet, relatively few species have actually become extinct. Thus, in addition to the native flora of a given region, there are large numbers of introduced plants, and as a consequence, at the present day, the total plant biodiversity may be very high. Figures for the flora of Germany illustrate this point. The total number of species of vascular plants is approximately 3,300. Of these, 2,850 are indigenous native species, of which 850 are endangered and 47 extinct. However, in addition to these native species, a large number (*c.* 12,000) non-native species have been introduced to Central Europe, and many of these have been found in Germany (Scherer-Lorenzen *et al.*, 2000).

Turning to questions of losers or potential losers, Meyer (2006) considers: 'The broad path of biological evolution is now set for the next million years. And in this sense the extinction crisis – the race to save the composition, structure, and organization of biodiversity as it exists today – is over, and we have lost.' He predicts: 'over the next 100 years or so as many as half of the Earth's species . . . will functionally if not completely effectively disappear' (Meyer, 2006, 4 & 5). If they survive they will be restricted to neglected areas as 'relic ghost' species. 'While they may seem plentiful today and may in fact persist for decades, their extinction is certain, apart from a few specimens in zoos.'

Looking to the future, Meyer continuing his bleak assessment concludes: 'nothing – not national or international laws, global bioreserves, local sustainability schemes, or even "wildlands" fantasies – can change the current course. The broad path for biological evolution is now set for the next several million years.' Evolution in the biosphere is increasingly being driven by human activities. We decide which species to protect, which habitats and ecosystems to conserve, and how they will be managed. Humans determine the recovery goals of species and ecosystems, decide which areas are 'critical and which expendable'. Indeed, the magnitude of human impacts and management has resulted in the 'end of the wild'. This sentiment is also evident in the writings of McKibben (1990). Reviewing the evidence, Myers and Knoll (2001, 5389) also conclude that the current potential loss of species could rival the 'extinction spasms' recorded in geological history. Moreover, they also stress that current microevolutionary trends will shape future evolution and determine winners and losers. These include:

> homogenization of biotas, a proliferation of opportunistic species, a pest- and weed-ecology, an outburst of speciation among taxa that prosper in human-dominated ecosystems, a decline in biodisparity [morphological and physiological variety], an end to the speciation of large vertebrates, the depletion of 'evolutionary powerhouses' in the tropics, and unpredictable emergent novelties.

In an earlier review, Myers (1988, 31 & 33) concedes:

> We may learn how to manipulate habitats to enhance survival prospects. We may learn how to propagate threatened species in captivity. We may be able to apply other emergent conservation techniques, all of which could help to relieve the adverse repercussions. [But,] in sum, the evolutionary impoverishment of the impending extinction spasm, plus the numbers of species involved and the telescoped time scale of the phenomenon, may result in the greatest single setback to life's abundance and diversity since the first flickerings of life almost 4 billion years ago.

Considering the long-term prospects in the future evolution of biodiversity, it seems highly unlikely that new species will quickly evolve to replace those that become extinct (Chapter 13). However, over long periods, as Myers and Knoll (2001, 5390) point out, past extinctions were followed by 'rediversification'. They speculate that on the basis of past events: 'should we anticipate a minimum period of several million years (perhaps as much as 10 million) before evolution can re-establish anywhere near the biological configurations and ecological circuitry existing before the current crisis?' But they also conclude:

[given the] present biotic crisis, it is hard to envision a scenario under which the factors that are driving the biosphere towards grand scale biodiversity loss will be mitigated in the wake of such loss. On the contrary, on any time scale we can envisage (and any scenario that does not involves early mass mortality for humankind), the situation becomes bad and then stays bad for some time to come. Thus, on the time scale of the human species, environmental disruption (or at least some aspects of it) is permanent.

Will humankind make the necessary adaptation to bring climate change under control?

Turning to another major theme of this book, it is important to face the manifold effects on ecosystems that are predicted to occur though increasing climate change. Firstly, it should again be stressed that while predictions concerning climate change are probabilistic in nature, rather than certainties, the clear message they convey is that humankind and, indeed, the whole biosphere is facing a very uncertain future. How should conservationists react? Should they continue with 'business as usual', in conserving the species and ecosystems of reserves, parks and the wider matrix? What are the implications for ecosystem restoration and species reintroduction in the light of the increasing evidence for the reality of climate change? Should conservationists act as if the worst-case scenarios can be avoided, or should they factor into their management plans some degree of climate change impact? There are no simple answers to these urgent questions. One major problem, which is receiving the attention of modellers, is the lack of clear predictions for climate change at the local level. Without this information it is very difficult to develop appropriate conservation strategies.

Another important issue in the adaptation to climate change is the relationship between those concerned with the theoretical issues underpinning conservation and others who are responsible for the day-to-day management of reserves etc. In this book, as a shorthand device, the term 'conservationist' has been employed; however, it would be wrong to imagine those involved with conservation are a homogeneous group. This situation was discussed in a recent issue in *Nature* (Anon., 2007b). The Editorial points out:

Most of the time, conservation biologists describe problems, float solutions, prioritize areas and actions, and run computer models of natural ecosystems. They are cartographers of crises, producing demoralizing maps of threat and extinction. [However,] it generally falls to a separate and amorphous group, known as 'practitioners', to buy land, put up fences, set fires, put out fires, lobby politicians, negotiate with farmers, spray invasive weeds, poison rats and guard against poachers . . . The distance between these two groups creates a sometimes-yawning 'implementation gap' between theory and practice. Conservation biologists write and publish papers, which the practitioners seldom read. The practitioners, in turn, rarely document their actions or collate their data in forms useful to conservation biologists. Typically, practitioners make decisions based on personal experience and intuition. Their knowledge stays untapped by others – and can be impervious to fresh scientific findings.

The Editorial calls for efforts to bridge this gap with more contact and interaction between the different groups, and more meetings and workshops to encourage the co-operation of all those involved in conservation. 'What is needed is a concerted effort by both academic scientists and practitioners to get out of their respective ruts, open up paths of communication, share information and seek ever more efficient means to a common end.' At a practical level, the setting up of a new website – ConservationEvidence.com – by Professor W. Sutherland, University of Cambridge, UK, should prove a valuable innovation. Here, 'practitioners are encouraged to deposit reports on the outcomes of their interventions – successful or otherwise'.

The fate of contemporary ecosystems, including those in conservation management, depends upon whether the necessary steps are taken to reduce the emissions of greenhouse gases and limit global average temperature rise (Lynas, 2007). Have political leaders, conservationists and the general public recognised the urgency of the threat and understood that a significant (2 °C) rise could occur in perhaps as little as 8 years? Lynas (2007) has considered at some length the likely human responses to the challenges of adapting to climate changes and what follows is drawn from his very helpful account. Firstly, a general point should be stressed: climate change involves the pollution of the atmosphere – a resource held in common by all humanity. These changes represent another instance of the 'Tragedy of the Commons', a concept introduced in Chapter 4.

Hardin (1968), who first explored the idea, took, as an example, the behaviour of farmers using a shared pasture held in common (quoted from Lynas, 2007, 285):

Each herder stands to gain individually by adding another cow to the common – he gets more milk and beef. But if all the herders act in the same way, the result is overgrazing and the destruction of the shared resource. Psychological denial is integral to the process. Hardin writes: 'The individual benefits as an individual from his ability to deny the truth, even though society as a whole, of which he is a part, suffers.'

Lynas points out that denial is a common human response to warnings of an impending crisis, for it is the way to 'resolve the dissonance caused by new information which may challenge deeply held views or cherished patterns of behaviour'. In considering further the 'denial' response, Lynas (2007, 283) writes:

As Al Gore reminds his audience during the slideshow for his film *An Inconvenient Truth*, there is nothing so difficult as trying to get a man to understand something when his salary depends on him not understanding it. This is classic denial: no one wants to hold a mental image of themselves as bad or evil, so immoral acts are necessarily dressed up in a cloak of intellectual self-justification.

Response to new challenges also takes other forms. While many remain in ignorance of the problems of climate change, others, presented with the facts, suspect the information and/or those holding particular views. Thus, for some, climate change is the concern of the green activists and others of suspect political affiliations and, therefore, their concerns can be safely dismissed. Displacement is another response. Thus, climate change is caused by the profligacy of others, not ourselves. Lynas also emphasises that, despite the growing realisation of the emerging problems of climate change, there is a general unwillingness in developed countries to abandon the comforts and consumption patterns of modern societies. Indeed, the difficulties ahead are likely to increase, as elements of 'western life-styles', represented through advertising and the media, provide potent aspirational models for many who live in developing countries. Peoples of these nations have the natural desire of all humanity to provide a safe and secure future for their families with the comforts and benefits of a better standard of living. Looking to the future, there are complex ethical questions to face. Broome (2008) emphasises that we should be considering, very urgently, how our current behaviour will impact on the well-being of future generations.

An optimistic vision of the 'road ahead' is pictured by those who see the possibility of adaptive changes in human behaviour, led by developed nations, resulting in a complete reappraisal of the life-styles, increased use of renewable sources of energy and diminished patterns of consumption, to ensure that we all have a small carbon footprint and manage resources more sustainably. In effect, extending the idea proposed by Pontin (1993), humankind is faced with adaptation through a Third Great Transition (the first transition being the adoption of agriculture and the second the industrial revolution). It is hoped that such changes can be brought about through persuasion, by example of those in authority, by education, and by legislation. Personally, I would like to share their optimism and sometimes do so. As I complete this book there have been several encouraging developments. Proposals to develop marine reserves have been introduced by the US and UK governments. With regard to climate change, Barack Obama, newly inaugurated as President of the USA, has declared: 'Year after year, decade after decade, we've chosen delay over decisive action, rigid ideology has over-ruled sound science, special interests have overshadowed common sense. For the sake of our security, our economy and our planet, we must have the courage and commitment to change . . . the days of Washington dragging its heels are over. My administration will not deny the facts. We will be guided by them.' (Reported by T. Baldwin, *The Times*, 27 January 2009). However, it is important to recognise that we are currently in the midst of the most dramatic global economic recession for 60 years, which may provide a stimulus to change, but, at the same time, could limit our capacity to adapt. Indeed, recession is threatening many current conservation projects. For example, the Millennium Seed Bank at Kew faces a shortfall of more than £100m (*The Times*, 3 October 2008).

Many are optimistic that technologies may be developed to control carbon dioxide emissions and manipulate the Earth's climate. 'Excess' carbon dioxide could be locked away in suitable rock formations for underground storage. However, this proposal has yet to be made a practical reality. Further ideas are explored in a special edition of the *Philosophical Transactions of the Royal Society: Special Edition on Geoengineering*, Volume 366, 13 September 2008. Thus, the report examines the proposal to place a 'sunshade' in space to control the Earth's climate. Also, it reviews experiments designed to test the hypothesis that excess carbon dioxide could be removed through stimulating the photosynthetic activities of phytoplankton in the oceans. Organic matter formed by algal bloom would sink to the seabed and be sequestered there for centuries. Preliminary experiments have not proved encouraging, as zooplankton populations, feeding on the algae, multiply as quickly as the phytoplankton. Could zooplankton be excluded from the sytem by growing the algae in giant tubes in the sea?

Turning to recent advances in our understanding of climate change it is likely that the challenges ahead may have been underestimated by the IPCC Reports of 2007. At an international scientific meeting on climate change in Copenhagen (March 2009), it was concluded: 'Because of high rates of observed emissions, the worst-case IPCC scenario trajectories (or even worse) are being realised for parameters such as global mean surface temperature, sea level rise, ocean and ice-sheet dynamics, ocean acidification and extreme climatic effects'. Looking in more detail at one of these issues, the IPCC predicted that sea level rise might be of the order of 59 cm by the end of this century. After studying ocean temperatures and ice sheets in Greenland and elsewhere, scientists now predict that sea level rise might be of the order of 1 m by 2100, and, if sustained temperature rises reach 5–6 degrees, loss of ice sheets could occur in the middle of the next century and sea levels could rise by up to 13 m. Adaptation to such a rise would be impossible for the inhabitants of at least 20 major coastal cities, and the one-third of the world's population living in low-lying areas near the sea.

The IPCC report outlined some of the health effects of climate change. A major study has been carried out by the Lancet Commissions and University College London Institute (www.thelancet.com/climate-change). This report, published in May 2009, concludes that urgent steps must be taken to control and mitigate climate change, as such change is 'potentially the biggest global health threat of the 21st century', through altered patterns of disease and mortality, through changes in food and water supplies, sanitation, shelter and the effects of extreme events. Moreover, major threats to population in many parts of the world are likely to lead to massive increases in human migration.

While some are optimistic about our capacity to adapt to climate change, others take a bleaker view of future prospects and, in some moods, I fear they are correct. Charles Dickens, in Chapter 1 of *A Tale of Two Cities*, reflecting on another age, perfectly catches the range of emotions.

> It was the best of times, it was the worst of times, it was the age of wisdom, it was the age of foolishness, it was the epoch of belief, it was the epoch of incredulity, it was the season of Light, it was the season of Darkness, it was the spring of hope, it was the winter of despair, we had everything before us, we had nothing before us, we were all going direct to Heaven, we were all going direct the other way . . .

The eminent biologist Lovelock (2006, 13), writing about climate change, takes a pessimistic approach when he warns:

The prospects are grim, and even if we act successfully in amelioration, there *will still be hard times, as in any war, that will stretch us to the limit*. We are tough and it would take more than the predicted climate catastrophe to eliminate all breeding pairs of humans; *what is at risk is civilisation*. [my emphasis]

Lovelock stresses that humankind faces not only the likely effects of increasing climate change, but also many other serious issues, with implications for biodiversity (e.g. over-exploitation of natural resources, soil degradation and pollution problems, with an overriding concern about the growth of human populations). These factors will act synergistically with climate change. As we have seen, the struggle to survive in a changing world is likely to increase management (and mismanagement) of resources worldwide, as humanity tries to maintain the goods and services provided by ecosystems – for example, clean water (Baron *et al.*, 2002), pollination (Kremen *et al.*, 2007), flood prevention (Acreman *et al.*, 2007), timber production (Andersson, Salinas & Carlsson, 2006), natural products, soil fertility etc.

In the twentieth century there have been many successful efforts to conserve plant and animal diversity, natural ecosystems and significant elements of cultural landscapes. In many areas, the ravaged landscapes, poisoned soils and waterways have been successfully restored. Also, the importance of using ecosystems sustainably has been widely accepted. Efforts to introduce sustainable development are continuing in the twenty-first century in an attempt 'to balance and integrate the three pillars of social well-being, economic prosperity and environmental protection for the benefit of present and future generations' (Lovelock, 2006, 3, quoting Glaser, Senior Adviser to the International Council of Science). As ecosystems are continually changing, sustainable development should never be seen as aiming for a fixed target. Under climate change, the target is likely to move ever faster. In the future, under straightened circumstances, it seems highly likely that many ecosystems and species of conservation concern will be at severe risk, especially in regions where the struggle for survival and competition for resources is predicted to intensify. As we have seen, many conservation battles have already been fought, with some won and others lost. These might well prove to have been minor skirmishes, given the formidable challenges ahead.

References

Abbo, S., Gopher, A., Rubin, B. & Lev-Yadun, S. (2005). On the origin of Near Eastern founder crops and the 'dump-heap hypothesis'. *Genetic Resources and Crop Evolution*, **52**, 491–495.

Abbott, R. J. (1992). Plant invasions, interspecific hybridization and the evolution of new plant taxa. *Trends in Ecology & Evolution*, **7**, 401–405.

Abbott, R. J. & Forbes, D. G. (2002). Extinction of the Edinburgh lineage of the allopolyploid neospecies, *Senecio cambrensis* Rosser (Asteraceae). *Heredity*, **88**, 267–269.

Abbott, R. J. & Lowe, A. J. (2004). Origins, establishment and evolution of new polyploid species: *Senecio cambrensis* and *S. eboracensis* in the British Isles. *Biological Journal of the Linnean Society*, **82**, 467–474.

Abbott, R. J., Ingram, R. & Noltie, H. J. (1983). Discovery of *Senecio cambrensis* Rosser in Edinburgh. *Watsonia*, **14**, 407–408.

Abbott, R. J., Noltie, H. J. & Ingram, R. (1983). The origin and distribution of *Senecio cambrensis* in Edinburgh. *Transactions of the Botanical Society of Edinburgh*, **44**, 103–106.

Abbott, R. J., Curnow, D. J. & Irwin, J. A. (1995). Molecular systematics of *Senecio squalidus* L. and its close diploid relatives. In *Advances in Compositae Systematics*, eds. D. J. N. Hind, C. Jeffrey & G. V. Pope, pp. 223–237. Kew: Royal Botanic Garden.

Abbott, R. J., James, J. K., Milne, R. I. & Gillies, A. C. M. (2003). Plant introductions hybridization and gene flow. *Philosophical Transactions of the Royal Society of London, B*, **358**, 1123–1132.

Abbott, R. J., Ireland, H. E. & Rogers, H. J. (2007). Population decline despite high genetic diversity in the new allopolyploid species *Senecio cambrensis* (Asteraceae). *Molecular Ecology*, **16**, 1023–1033.

Abdul-Baki, A. A. & Anderson, J. D. (1972). Physiological and biochemical deterioration of seeds. In *Seed Biology*, ed. T. T. Kozlowski, Vol. II, pp. 283–315. New York & London: Academic Press.

Abel, A. L. (1954). The rotation of weed killers. *Proceedings of the British Weed Control Conference*, **1**, 249.

Aberle, B. (1990). *The Biology, Control and Eradication of Introduced Spartina (Cordgrass) Worldwide and Recommendations for Its Control in Washington.* Olympia: Washington Department of Natural Resources.

Abramovitz, J. N. (1996). *Imperiled Waters, Inpoverished Future: The Decline of Freshwater Ecosystems.* Washington, DC: Worldwatch.

Acreman, M. C. *et al.* (2007). Hydrological science and wetland restoration: some case studies from Europe. *Hydrology & Earth System Sciences*, **11**, 158–169.

Adams, C. D. (1972). *Flowering Plants of Jamaica*. Mona: University of the West Indies.

Adams, W. M. (2004). *Against Extinction: The Story of Conservation*. London & Sterling, VA: Earthspan.

Adams, W. M. & Mulligan, M. (2003). *Decolonizing Nature: Strategies for Conservation in a Post-colonial Era*. London & Sterling, VA: Earthspan.

Ainouche, M. L., Baumel, A., Salmon, A. & Yannic, G. (2004). Hybridization, polyploidy and speciation in Spartina (Poaceae). *New Phytologist*, **161**, 165–172.

Akeroyd, J. R. (1994). Some problems with introduced plants in the wild. In *The Common Ground of Wild and Cultivated Plants*, eds. A. R. Perry & R. G. Ellis, pp. 31–40. Cardiff: National Museum of Wales.

Akeroyd, J. R. & Briggs, D. (1983a). Genecological studies of *Rumex crispus* L. I. Garden experiments using transplanted material. *New Phytologist*, **94**, 309–323.

Akeroyd, J. R. & Briggs, D. (1983b). Genecological studies of *Rumex crispus* L. II. Variation in plants grown from wild-collected seed. *New Phytologist*, **94**, 325–343.

Akimoto, H. (2003). Global air quality and pollution. *Science*, **302**, 1716–1719.

Aldrich, P. R. & Doebley, J. (1992). Restriction fragment variation in the nuclear and chloroplast genomes of cultivated and wild *Sorghum bicolor*. *Theoretical and Applied Genetics*, **85**, 293–302.

Aldrich, P. R., Doebley, J., Schertz, K. F. & Stec, A. (1992). Patterns of allozyme variation in cultivated and wild *Sorghum bicolor*. *Theoretical and Applied Genetics*, **85**, 451–460.

Al-Hiyaly, S. A., McNeilly, T. & Bradshaw, A. D. (1988). The effects of zinc contamination from electricity pylons – evolution in a replicated situation. *New Phytologist*, **110**, 571–580.

Al-Hiyaly, S. A., McNeilly, T. & Bradshaw, A. D. (1990). The effect of zinc contamination from electricity pylons. Contrasting patterns of evolution in five grass species. *New Phytologist*, **114**, 183–190.

Al-Hiyaly, S. A., McNeilly, T., Bradshaw, A. D. & Mortimer, A. M. (1993). The effects of zinc contamination from electricity pylons. Genetic constraints on selection for zinc tolerance. *Heredity*, **70**, 22–32.

Allan, N. J. R., Knapp, G. W. & Stadel, C. (1988). *Human Impact on Mountains*. Lanham, MD: Rowman & Littlefield.

Allen, D. E. (1980). The early history of plant conservation in Britain. *Transaction of the Leicester Literary and Philosophical Society*, **4**, 277–283.

Allen, D. E. (1987). Changing attitudes to nature conservation: the botanical perspective. *Biological Journal of the Linnean Society*, **32**, 203–212.

Alley, R. B. *et al.* (2005). Abrupt climate change. *Science*, **299**, 2005–2010.

Alley, H. & Affolter, J. M. (2004). Experimental comparison of reintroduction methods for the endangered *Echinacea laevigata* (Boyton & Beadle) Blake. *Natural Areas Journal*, **24**, 345–350.

Allison, G. W., Lubchenco, J. & Carr, M. H. (1998). Marine reserves are necessary but not sufficient for marine conservation. *Ecological Applications*, **8**, Suppl. S79–S92.

Alvarez, L. W., Alvarez, W., Asaro, F. & Michel, H. V. (1980). Extraterrestrial cause for the Cretaceous-Tertiary Extinction. *Science*, **208**, 1095–1108.

Alvarez-Buylla, E. R., Chaos, A. C., Piñero, D. & Garay, A. A. (1996). Demographic genetics of a pioneer tropical tree species: patch dynamics seed dispersal and seed banks. *Evolution*, **50**, 1155–1166.

Alvarez-Valin, F. (2002). Neutral theory. In *Encyclopedia of Evolution*. ed. M. Patel, pp. 815–821, Oxford & New York: Oxford University Press.

Ambrose, J. D. (1983). *Status report on the Kentucky Coffee Tree Gymnocladus dioicus*. Ottawa: Committee for the Status of Endangered Wildlife in Canada.

Ammerman, A. J. (2002). Spread of agriculture. In *Encyclopedia of Evolution*, ed. M. Pagel, Vol. 1, pp. 19–22. Oxford & New York: Oxford University Press.

Amos, W. & Balmford, A. (2001). When does conservation genetics matter? *Heredity*, **87**, 257–265.

Amsellen, L., Noyer, J. L, Le Bourgeois, T. & Hossaert-McKey, M. (2000). Comparison of genetic diversity of the invasive weed *Rubus alceifolius* Poir. (Rosaceae) in its native range and areas of introduction, using amplified fragment length polymorphism (AFLP) markers. *Molecular Ecology*, **9**, 443–455.

Anable, M. E., McClaran, M. P. & Ruyle, G. B. (1992). Spread of introduced Lehmann Lovegrass *Eragrostis lehmanniana* Nees in southern Arizona, USA. *Biological Conservation*, **61**, 181–188.

Anderson, E. (1949). *Introgressive Hybridisation*. London: Chapman & Hall.

Anderson, P. (1995). Ecological restoration and creation: a review. *Biological Journal of the Linnean Society*, **56** (Suppl.), 187–211.

Anderson, R. C. (2006). Evolution and origin of the Central Grasslands of North America: climate, fire, and mammalian grazers. *Journal of the Torrey Botanical Society*, **133**, 626–647.

Anderson, R. N. & Gronwald, J. W. (1987). Non-cytoplasmic inheritance of atrazine tolerance in velvetleaf (*Abutilon theophrasti*). *Weed Science*, **35**, 496–498.

Anderson, D. & Grove, R. (1987). *Conservation in Africa: People, Policies and Practice*. Cambridge: Cambridge University Press.

Andersson, M., Salinas, O. & Carlsson, M. (2006). A landscape perspective on differentiated management for production of timber and nature conservation values. *Forest Policy & Economics*, **9**, 153–161.

Angold, P. G. (1997). The impact of a road upon adjacent heathland vegetation: effects of species composition. *Journal of Applied Ecology*, **34**, 409–417.

Anon. (1944). Report of Committee on Nature Conservation and Nature Reserves. *Journal of Ecology*, **32**, 83–115.

Anon. (1990). *A Guide to Yosemite National Park, California*. Washington, DC: National Parks Service.

Anon. (1991a). Genetic sampling guidelines for conservation collections of endangered plants, developed by the Centre for Plant Conservation. In *Genetics and Conservation of Rare Plants*, eds. D. A. Falk & K. E. Holsinger, pp. 225–238. New York & Oxford: Oxford University Press.

Anon. (1991b). *Protected Areas of the World: A Review of National Systems*. Gland & Cambridge: IUCN.

Anon. (1994). *IUCN Red List Categories*. Gland, Switzerland: IUCN.

Anon. (1995). *Mapping the distribution of water hyacinth using satellite imagery. Pilot study in Uganda*. RCSSMRS/French Technical Assistance, Nairobi.

Anon. (1997). *The RHS Plant Finder*. London: Dorling Kindersley.

Anon. (2000). *Conservation Action Plan for Botanic Gardens of the Caribbean Islands*. Kew: Botanic Gardens Conservation International.

Anon. (2002a). Seed banks receive vital cash boost. *New Scientist*, 30 August 2002.

Anon. (2002b). Looters wreck Afghan seed-bank. *New Scientist*, 10 September 2002.

Anon. (2004). *Yellowstone Resources & Issues 2004. An Annual Compendium of Information about Yellowstone National Park*. Yellowstone National Park, Hot Springs, WY.

Anon. (2007a). Editorial: Light at the end of the tunnel. *Nature*, **445**, 567.

Anon. (2007b). Editorial: The great divide. *Nature*, **450**, 135–136.

Antonovics, J. (1968). Evolution in closely adjacent plant populations. V. Evolution of self-fertility. *Heredity*, **23**, 219–238.

Antonovics, J. (2006). Evolution in closely adjacent plant populations. X. Long-term persistence of prereproductive isolation at a mine boundary. *Heredity*, **97**, 33–37.

Antonovics, J., Bradshaw, A. D. & Turner, R. G. (1971). Heavy metal tolerance in plants. *Advances in Ecological Research*, **7**, 1–85.

Anttila, C. K., Daehler, C. C., Rank, N. E. & Strong, D. R. (1998). Greater male fitness of a rare invader (*Spartina alterniflora*, Poaceae) threatens a common native (*Spartina foliosa*) with hybridization. *American Journal of Botany*, **85**, 1597–1601.

Araújo, M. B., Pearson, R. G., Thuillers, W. & Erhard, M. (2005). Validation of species-climate impact models under climate change. *Global Change Biology*, **11**, 1505–1513.

Archibald, J. D. (1997). Extinction. In *Encyclopedia of Dinosaurs*, eds. P. J Currie & K. Padian, pp. 221–230. San Diego & London: Academic Press.

Arias D. M. & Rieseberg, L. H. (1994). Gene flow between cultivated and wild sunflowers. *Theoretical and Applied Genetics*, **89**, 655–660.

Arnold, M. L. (1993). *Iris nelsonii*: origin and genetic composition of a homoploid hybrid species. *American Journal of Botany*, **80**, 577–583.

Arnold, M. L. & Bennett, B. D. (1993). Natural hybridization in Louisiana Irises: genetic variation and ecological determinants. In *Hybrid Zones and the Evolutionary Process*, ed. R. G. Harrison, pp. 115–139. New York: Oxford University Press.

Arnold, M. L., Hamrick, J. L. & Bennett, B. D. (1990). Allozyme variation in Louisiana Irises: a test for introgression and hybrid speciation. *Heredity*, **65**, 297–306.

Arnold, M. L., Bouck, A. C. & Cornman, R. S. (2004). Verne Grant and Louisiana Irises: Is there anything new under the sun? *New Phyologist*, **161**, 143–149.

Arutyunyan, R. M., Pogosyan, V. S., Simonyan, E. H., Atoyants, A. L. & Djigardjian, E. M. (1999). In situ monitoring of the ambient air around the chloroprene rubber industrial plant using the Tradescantia-stamen–hair mutation assay. *Mutation Research*, **426**, 117–120.

Ashby, E. & Anderson, M. (1981). *The Politics of Clean Air*. Oxford: Clarendon Press.

Ashmore, M. R. (2002). Air quality guidelines and their role in pollution control policy. In *Air Pollution and Plant Life*, eds. J. N. B. Bell & M. Treshow, 2nd edn, pp. 417–429. London: Wiley.

Ashton, F. M. & Crafts, A. S. (1981). *Mode of Action of Herbicides*. 2nd edn. New York: Wiley.

Ashton, P. S. (1987). Biological considerations in *in situ* vs *ex situ* plant conservation. In *Botanic Gardens and the World Conservation Strategy*, eds. D. Bramwell, O. Hamann, V. Heywood & H. Synge, pp. 117–130. London: Academic Press for IUCN.

Atkins, P. J., Simmons, I. G. & Roberts, B. (1998). *People, Land and Time: An Historical Introduction to the Relations Between Landscape, Culture and Environment*. London: Wiley.

Avise, J. C. (1994). *Molecular Markers, Natural History & Evolution*. New York: Chapman & Hall.

Avise, J. C. (2000). *Phytogeography: The History and Formation of Species*. Cambridge, MA: Harvard University Press.
Avise, J. C. (2001). *Captivating Life*. Washington, DC & London: Smithsonian Institution Press.
Axell, H. (1977). *Minsmere. A Portrait of a Bird Reserve*. London: Hutchinson.
Ayazloo, M. & Bell, J. N. B. (1981). Studies on the tolerance to SO_2 of grass populations in polluted areas. I, Identification of tolerant populations. *New Phytologist*, **88**, 203–222.
Ayres, D. R. & Strong, D. R. (2001). Origin and genetic diversity of *Spartina anglica* (Poaceae) using nuclear DNA markers. *American Journal of Botany*, **88**, 1863–1867.
Ayres, D. R., Zaremba, K. & Strong, D. R. (2004). Extinction of a common native species by hybridization with an invasive congener. *Weed Technology*, **18**, 1288–1291.
Babington, C. C. (1848). On *Anacharis Alsinastrum*, a supposed new British plant. *The Annals and Magazine of the Botanical Society, 2nd series*, **1**, 81–85.
Bachthaler, G. (1967). Changes in arable weed infestation with modern crop husbandry techniques. *Abstract 6th International Congress of Plant Protection*, pp. 167–168. Vienna.
Bais, H. P., Vepachedu, R., Gilroy, S., Callaway, R. M. & Vivanco, J. M. (2003). Allelopaphy and exotic plants: from genes to invasion. *Science*, **301**, 1377–1380.
Baker, H. G. (1937). Alluvial meadows: a comparative study of grazed and mown meadows. *Journal of Ecology*, **25**, 408–420.
Baker, H. G. (1954). Report of a paper presented at a meeting of the British Ecological Society. *Journal of Ecology*, **42**, 571.
Baker, H. G. (1955). Self-incompatibility and establishment after 'long-distance' dispersal. *Evolution*, **9**, 347–349.
Baker, H. G. (1965). Characteristics and modes of origin of weeds. In *The Genetics of Colonizing Species*, eds. H. G. Baker & G. L. Stebbins, pp. 141–172. London: Academic Press.
Baker, H. G. (1974). The evolution of weeds. *Annual Review of Ecology & Systematics*, **5**, 1–24.
Baker, A. J. M. (1987). Metal tolerance. *New Phytologist*, **106**, 93–111.
Baker, A. J. (ed.) (2000). *Molecular Methods in Ecology*. Oxford: Blackwell Science.
Baker, A. J. M. & Proctor, J. (1990). The influence of cadmium, copper, lead and zinc on the distribution and evolution of metallophytes in the British Isles. *Plant Systematics & Evolution*, **173**, 91–108.
Baker, A. J. M., Grant, C. J., Martin, M. H., Shaw, S. C. & Whitbrook, J. (1986). Induction and loss of cadmium tolerance in *Holcus lanatus* L. and other grasses. *New Phytologist*, **102**, 575–587.
Bakker, J. P & Londo, G. (1998). Grazing for conservation management in historical perspective. In *Grazing and Conservation Management*, eds. M. F. Wallis de Vries, J. P. Bakker & S. E. Van Wieren, pp. 23–54. Dordrecht: Kluwer Academic Press.
Baldoni, L. *et al.* (2006). Genetic structure of wild and cultivated olives in the central Mediterranean basin. *Annals of Botany*, **98**, 935–942.
Balée, W. (1987). Cultural forests of the Amazon. *Garden*, **11**, 12–14.
Balée, W. (1989). The culture of Amazonian forests. *Advances in Economic Botany*, **7**, 1–21.
Balmford, A. *et al.* (2002). Ecology-economic reasons for conserving wild nature. *Science*, **297**, 950–953.
Balter, M. (2007). Seeking agriculture's ancient roots. *Science*, **316**, 1830–1835.

Barnett, T. P. *et al.* (2008). Human-induced changes in the hydrology of the western United States. *Science*, **319**, 1080–1083.

Baron, J. S. *et al.* (2002). Meeting ecological and societal needs for freshwater. *Ecological Applications*, **12**, 1247–1260.

Barrett, S. C. H. (1983). Crop mimicry in weeds. *Economic Botany*, **37**, 255–282.

Barrett, S. C. H. (2000). Microevolutionary influences of global changes on plant invasions. In *Invasive Species in a Changing World*, eds. H. A Mooney & R. J. Hobbs, pp. 115–139. Washington, DC: Island Press.

Barrow, C. J. (1991). *Land Degradation*. Cambridge: Cambridge University Press.

Baskin, L. M. (1998). Hunting of game mammals in the Soviet Union. In *Conservation of Biological Resources*, eds. E. J. Milner-Gulland & R. Mace, pp. 331–345. Oxford: Blackwell.

Bates, J. W. (2002). Effects on bryophytes and lichens. In *Air Pollution and Plant life*, 2nd edn., eds. J. N. B. Bell & M. Treshow, pp. 309–342. London: Wiley.

Bateson, W. (1913). *Problems of Genetics*. London: Oxford University Press and New Haven, CN: Yale University Press.

Baucom, R. S. & Mauricio, R. (2004). Fitness costs and benefits of novel herbicide tolerance in a noxious weed. *Proccedings of the National Academy of Sciences, USA*, **101**, 13386–13390.

Baucom, R. S., Estill, J. C. & Cruzan, M. B. (2005). The effect of deforestation on the genetic diversity and structure in *Acer saccharum* (Marsh): evidence for the loss and restructuring of genetic variation in a natural system. *Conservation Genetics*, **6**, 39–50.

Baumel, A. *et al.* (2003). Genetic evidence for hybridization between the native *Spartina maritima* and the introduced *Spartina alterniflora* (Poaceae) in South-West France: *Spartina* × *neyrautii* re-examined. *Plant Systematics & Evolution*, **237**, 87–97.

Bawa, K. S. *et al.* (2004). Tropical ecosystems into the 21st century. *Science*, **306**, 227–228.

Bean, W. J (1970). *Trees and Shrubs Hardy in the British Isles*. 8th edn., Vol. IV, pp. 360–363. London: John Murray.

Bean, W. J. (1976). *Trees and Shrubs Hardy in the British Isles*. 8th revised edn., N–Rh, Vol. III. London: John Murray.

Beebee, T. & Rowe, G. (2004). *An Introduction to Molecular Ecology*. Oxford: Oxford University Press.

Beerli, P. & Felsenstein, J. (1999). Maximum-likelihood estimation of migration rates and effective population numbers in two populations using a coalescent approach. *Genetics*, **152**, 763–773.

Begon, M. (1984). Density and individual fitness: asymmetric competition. In *Evolutionary Ecology*, 23rd Symposium of the British Ecological Society, ed. B. Shorrocks, pp. 175–194. Oxford: Blackwell Scientific.

Behling, H., Pillar, V. D., Müller, S. C. & Overbeck, G. E. (2007). Late-Holocene fire history in a forest-grassland mosaic in southern Brazil: implications for conservation. *Applied Vegetation Science*, **10**, 81–90.

Beier, P. & Noss, R. F. (1998). Do habitat corridors provide connectivity? *Conservation Biology*, **12**, 1241–1252.

Bekessy, S. A., Ennos, R. A., Burgman, M. A., Newton, A. C. & Ades, P. K. (2003). Neutral DNA markers fail to detect genetic divergence in an ecologically important trait. *Biological Conservation*, **110**, 267–275.

Bekessy, S. A. *et al.* (2004). Monkey Puzzle Tree (*Araucaria araucana*) in Southern Chile: effects of timber and seed harvest, volcanic activity and fire. In *Species Conservation and Management*, ed. H. R. Akçakaya *et al.*, pp. 48–63. New York: Oxford University Press.

Bell, J. N. B. (1985). SO_2 effects on the productivity of grass species. In *Sulfur Dioxide and Vegetation*, eds. W. E. Winner, H. A. Mooney & R. A. Goldstein, pp. 209–226. Stanford: Stanford University Press.

Bell, R. (1987). Conservation with a human face. In *Conservation in Africa, People, Policies and Practice*. ed. D. Anderson & R. Grove, pp. 63–101. Cambridge: Cambridge University Press.

Bell, G. (1997). *Selection. The Mechanism of Evolution*. Oxford & New York: Oxford University Press.

Bell, J. N. B. & Clough, W. S. (1973). Depression of yield in ryegrass exposed to sulphur dioxide. *Nature*, **241**, 47–49.

Bell, J. N. B. & Mudd, C. H. (1976). Sulphur dioxide resistance in plants: a case study of *Lolium perenne*. In *Effects of Air Pollutants on Plants*, ed. T. A. Mansfield, pp. 87–103. Cambridge: Cambridge University Press.

Bell, J. N. B. & Treshow, M. (2002). *Air Pollution and Plant Life*, 2nd edn. Chichester & New York: Wiley.

Bell, J. N. B., Ashmore, M. R. & Wilson, G. B. (1991). Ecological genetics and chemical modification of the atmosphere. In *Ecological Genetics and Air Pollution*, eds. G. E. Taylor, L. F. Pitelka & M. T. Clegg, pp. 33–59. New York: Springer Verlag.

Bell, T. J., Bowles, M. L. & McEachern, K. A. (2003). Projecting the success of plant population restoration with population viability analysis. In *Population Viability in Plants: Conservation Management and Modelling Rare Plants*, eds. C. A. Brigham & M. W. Schwartz, pp. 313–348. Berlin: Springer Verlag.

Belzer N. F. & Ownbey, M. (1971). Chromatographic comparison of Tragopogon species and hybrids. *American Journal of Botany*, **58**, 791–802.

Ben-Kalio, V. D. & Clarke, D. D. (1979). Studies on tolerance in wild plants: effects of *Erysiphe fischeri* on the growth and development of *Senecio vulgaris*. *Physiological Plant Pathology*, **14**, 203–211.

Bennett, K. D. (1995). Post-glacial dynamics of pine (*Pinus sylvestris*) and pinewoods in Scotland. In *Our Pinewood Heritage*, ed. J. R. Aldhous. pp. 22–39. Forestry Commission/RSPB/SNH.

Bennett, K. D. (1997). *Evolution and Ecology: The Pace of Life*. Cambridge: Cambridge University Press.

Bennington, C. C. & McGraw, J. B. (1995). Natural selection and ecotypic variation in *Impatiens pallida*. *Ecological Monographs*, **65**, 303–323.

Berrang, P., Karnosky, D. F. & Bennett, J. P. (1988). Natural selection for ozone tolerance in *Populus tremuloides*, field verification. *Canadian Journal of Forest Research*, **19**, 519–522.

Berrang, P., Karnosky, D. F. & Bennett, J. P. (1991). Natural selection for ozone tolerance in *Populus tremuloides*, an evaluation of nationwide trends. *Canadian Journal of Forest Research*, **21**, 1091–1097.

Berry, R. J. (1977). *Inheritance and Natural History*. London: Collins.

Berry, A. (2002). *Infinite Tropics: An Alfred Russel Wallace Anthology*. London & New York: Verso.

Besnard, G. *et al.* (2007). On the origin of the invasive olives (*Olea europaea* L., Oleaceae). *Heredity*, **99**, 608–619.

Bhatti, J. S. *et al.* (2006). *Climate Change and Managed Ecosystems*. Boca Raton, FL: Taylor & Francis.

Bidin, A. (1991). The role of the Fernarium as a sanctuary for the conservation of threatened and rare ferns, with particular reference to Malaysia. In *Tropical Botanic Gardens: Their Role in Conservation and Development*, eds. V. H. Heywood & P. S. Wyse Jackson, pp. 223–239. London: Academic Press.

Bierregaard, R. O. Jr., Gascon, C., Lovejoy, T. E. & Mesquita, R. C. G. (2001). *Lessons from Amazonia*. New Haven, CT & London: Yale University Press.

Biever, C. (2005). Will Google help save the planet. *New Scientist*, 13 August 2005.

Bijlsma, R., Ouborg, N. J. & van Treuren, R. (1994). On genetic erosion and population extinction in plants: a case-study with *Scabiosa columbaria* and *Salvia pratensis*. In *Conservation Genetics*, eds. V. Loeschcke, J. Tomiuk & S. K. Jain, pp. 255–271. Basel: Birkhäuser.

Billington, H. L. (1991). Effects of population size on genetic variation in a dioecious conifer. *Conservation Biology*, **5**, 115–119.

Binggeli, P. (1996). A taxonomic, biogeographical and ecological overview of invasive woody plants. *Journal of Vegetation Science*, **7**, 121–124.

Birks, H. J. B. (2005). Mind the gap: how open were European primeval forests. *Trends in Ecology & Evolution*, **20**, 154–156.

Birney, E. *et al.* (2007). Identification and analysis of functional elements in 1% of the human genome by the ENCODE pilot project. *Nature*, **447**, 799–816.

Birt, T. P. & Baker, A. J. (2000). Polymerase chain reaction. In *Molecular Methods in Ecology*, ed. E. A. J. Baker, pp. 50–64. Oxford: Blackwell Science.

Bishop, J. A. & Cook, L. M. (1981). *Genetic Consequences of Man Made Change*. London: Academic Press.

Black-Samuelsson, S., Eriksson, G., Gustafsson, L. & Gustafsson, P. (1997). RAPD and morphological analysis of the rare plant species *Vicia pisiformis* (Fabaceae). *Biological Journal of the Linnean Society*, **61**, 325–343.

Bland, M. (1987). *An Introduction to Medical Statistics*. Oxford: Oxford University Press.

Blatter, E. & Millard, W. S. (1993). *Some Beautiful Indian Trees*. Bombay: Bombay Natural History Society and Oxford University Press.

Blaxter, K. & Robertson, N. (1995). *From Dearth to Plenty:The Modern Revolution in Food Production*. Cambridge: Cambridge University Press.

Bleeker, W. & Hurka, H. (2001). Introgressive hybridization in *Rorippa* (Brassicaceae): gene flow and its consequences in natural and anthropogenic habitats. *Molecular Ecology*, **10**, 2013–2022.

Boecklen, W. J. (1986). Optimal-design of nature reserves: consequences of genetic drift. *Biological Conservation*, **38**, 323–338.

Bohrer, V. L. (1972). On the relations of harvest methods for recognizing animal domestication in the Near East. *Economic Botany*, **26**, 145–155.

Bollinger, M. (1989). *Odontites lanceolata* (Gaudin) Reichenbach: ein formenreicher Endemit der Westalpen. *Botanische Jahrbücher für Systematik*, **111**, 1–28.

Bond, W. J. & Midgley, J. J. (2001). Ecology of sprouting in woody plants: the persistence niche. *Trends in Ecology and Evolution*, **16**, 45–51.

Bootsma, M. C., Van Den Broek, T., Barendregt, A. & Beltman, B. (2002). Rehabilitation of acidified floating fens by addition of buffered surface water. *Restoration Ecology*, **10**, 112–121.

Boserup, E. (1965). *The Conditions of Agricultural Growth*. London: Allen & Unwin.

Bossard, C. C., Randall, J. M. & Hoshovsky, M. C. (2000). *Invasive Plants of California's Wildlands*. Berkeley, CA: University of California Press.

Bossart, J. L. & Prowell, D. P. (1998). Genetic estimates of population structure and gene flow: limitations, lessons and new directions. *Trends in Ecology & Evolution*, **13**, 202–206.
Bossdorf, O., Auge, H., Lafuma, L., Rogers, W. E., Siemann, E. & Prati, D. (2005). Phenotypic and genetic differentiation in native versus introduced plant populations. *Oecologia*, **144**, 1–11.
Bossuyt, B. & Hermy, M. (2003). The potential of soil seedbanks in the ecological restoration of grassland and heathland communities. *Belgian Journal of Botany*, **136**, 23–34.
Boulos, L. & el-Hadidi, M. Nabil (1984). *The Weed Flora of Egypt*. American University of Cairo Press.
Bowers, J. E., Chapman, B. A., Rong, J. K. & Paterson, A. H. (2003). Unravelling angiosperm genome evolution by phylogeneic analysis of chromosomal duplication events. *Nature*, **422**, 433–438.
Bowes, G. (1996). Photosynthetic responses to changing atmospheric carbon dioxide concentration. In *Photosynthesis and the Environment*, ed. N.R. Baker, pp. 387–407. Dordrecht: Kluwer.
Bowles, M. L. & Whelan, C. J. (1994). *Restoration of Endangered Species: Conceptional Issues, Planning and Implementation*. Cambridge: Cambridge University Press.
Bowman, D. M. J. S. (1998). The impact of Aboriginal landscape burning on the Australian biota: Aboriginal fires in Australia. *New Phytologist*, **140**, 385–410.
Box, J. F. (1978). *R. A.Fisher: The Life of a Scientist*. New York & Chichester: Wiley.
Box, G. E. P. (1979). Some problems of statistics and everyday life. *Journal of the American Statistical Association*, **74**, 1–4.
Boyce, M. S. (1989). *The Jackson Elk Herd*. Cambridge: Cambridge University Press.
Boyle, C. M. (1960). Case of apparent resistance of *Rattus norvegicus* Berkenhout to anticoagulant poisons. *Nature*, **188**, 517.
Brack, W., Klamer, H. J. C., de Ada, M. L. & Barcelo, D. (2007). Effect-directed analysis of key toxicants in European river basins: a review. *Environmental Science and Pollution Research*, **14**, 30–38.
Bradford, J. B. & Hobbs, N. T. (2008). Regulating overabundant ungulate populations: an example for elk in Rocky Mountain National Park. *Journal of Environmental Management*, **86**, 520–528.
Bradley, N. L., Leopold, A. C., Ross, J. & Huffaker, W. (1999). Phenological changes reflect climate change in Wisconsin. *Proceedings of the National Academy of Science, USA*, **96**, 9701–9704.
Bradshaw, A. D. (1952). Populations of *Agrostis tenuis* resistant to lead and zinc poisoning. *Nature*, **169**, 1089.
Bradshaw, A. D. (1959a). Population differentiation in *Agrostis tenuis* Sibth.1. Morphological differentiation. *New Phytologist*, **58**, 208–227.
Bradshaw, A. D. (1959b). Studies of variation in Bent Grass species. II. Variation within *Agrostis tenuis. Journal of the Sports Turf Research Institute*, **35**, 6–12.
Bradshaw, A. D. (1960). Population differentiation in *Agrostis tenuis* Sibth. III. Populations in varied environments. *New Phytologist*, **59**, 92–103.
Bradshaw, M. E. (1963a). Studies on *Alchemilla filicaulis* Bus., *sensu lato* and *A. minima* Walters. I. Introduction and morphological variation in *Alchemilla filicaulis, sensu lato. Watsonia*, **5**, 304–320.
Bradshaw, M. E. (1963b). Studies on *Alchemilla filicaulis* Bus., *sensu lato* and *A. minima* Walters. II. Cytology of *Alchemilla filicaulis* Bus., *sensu lato. Watsonia*, **5**, 321–326.

Bradshaw, M. E. (1964). Studies on *Alchemilla filicaulis* Bus., *sensu lato* and *A. minima* Walters. III. *Alchemilla minima. Watsonia*, **6**, 76–81.
Bradshaw, A. D. (1965). Evolutionary significance of phenotypic plasticity in plants. *Advances in Genetics*, **13**, 115–155.
Bradshaw, A. D. (1971). Plant evolution in extreme environments. In *Ecological Genetics and Evolution*, ed. E. R. Creed, pp. 20–50. Oxford: Blackwell.
Bradshaw, A. D. (1976). Pollution and evolution. In *Effects of Air Pollutants on Plants*, ed. T. A. Mansfield., pp. 135–159. Cambridge: Cambridge University Press.
Bradshaw, A. D. (1984a). The importance of evolutionary ideas in ecology and vice versa. In *Evolutionary Ecology*, ed. B. Sharrocks, pp. 1–25. Oxford: Blackwell.
Bradshaw, A. D. (1984b). Adaptations of plants to soils containing heavy metals: a test for conceit. In *Origins and Developments of Adaptation*, eds. D. Evered & G. M. Collins, pp. 4–19. London: CIBA.
Bradshaw, A. D. (1987). The reclamation of derelict land and the ecology of ecosystems. In *Restoration Ecology*, eds. W. R. Jordan III, M. E. Gilpin & J. D. Aber, pp. 53–74. Cambridge: Cambridge University Press.
Bradshaw, A. D. & Chadwick, M. J. (1980). *The Restoration of Land*. Oxford: Blackwell.
Bradshaw, W. E. & Holzapfel, C. M. (2001). Genetic shift in photoperiodic response correlated with global warming. *Proceeding of the National Academy of Sciences, USA*, **98**, 14509–14511.
Bradshaw, A. D. & McNeilly, T. (1981). *Evolution and Pollution*. London: Edward Arnold.
Bradshaw, A. D. & McNeilly, T. (1991). Evolutionary response to global climate change. *Annals of Botany (London)*, **67**, 5–14.
Bradshaw, A. D., McNeilly, T. S. & Gregory, R. P. G. (1965). Industrialization, evolution and development of heavy metal tolerance in plants. In *Ecology and the Industrial Society*, eds. G. T. Goodman, R. W. Edwards & J. M. Lambert, pp. 327–343. Oxford: Blackwell.
Bradshaw, A. D., McNeilly, T. & Putwain, P. D. (1990). The essential qualities. In *Heavy Metal Tolerance in Plants: Evolutionary Aspects*, ed. A. J. Shaw, pp. 323–334. Boca Raton, FL: CRC Press.
Brasier, C. M. & Gibbs, J. N. (1973). Origin of the Dutch elm disease epidemic in Britain. *Nature*, **242**, 607–609.
Brasier, C. M., Rose, J. & Gibbs, J. N. (1995). An unusual Phytophthora associated with widespread alder mortality in Britain. *Plant Pathology*, **44**, 999–1007.
Brasier, C. M., Kirk, S. A., Pipe, N. D. & Buck, K. W. (1998). Rare interspecific hybrids in natural populations of the Dutch elm disease pathogens *Ophiostoma ulmi* and *O. novo-ulmi. Mycological Research*, **102**, 45–57.
Brasier, C. M., Cooke, D. E. L. & Duncan, J. M. (1999). Origin of a new Phytophthora pathogen through interspecific hybridization. *Proceedings of the National Academy of Sciences, USA*, **96**, 5878–5883.
Breese, E. L. & Tyler, B. F. (1986). Patterns of variation and the underlying genetic and cytological architecture in grasses with particular reference to Lolium. In *Infraspecific Classification of Wild and Cultivated Plants*, ed. B. T. Styles, pp. 53–69. Oxford: Clarendon Press.
Brehm, B. G. & Ownbey, M. (1965). Variation in chromatographic patterns in the *Tragopogon dubius-pratensis-porrifolius* complex. *American Journal of Botany*, **52**, 811–818.

Brennan, J. P. M. (1977). Recent technical advances in modern botanic gardens. In *Technical Advances in Modern Botanic Gardens*, Abstracts of papers for meeting 23 June 1977 at Royal Botanic Garden, Kew, pp. 5–6. London: Ministry of Agriculture, Fisheries & Food.

Brennan, A. C., Harris, S. A. & Hiscock, S. J. (2005). Modes and rates of selfing and associated inbreeding depression in the self-incompatible plant *Senecio squalidus* (Asteraceae): successful colonizing species in the British Isles. *New Phytologist*, **168**, 475–486.

Breton, C. *et al.* (2005). Olives. In *Crop Ferality and Volunteerism*, ed. J. Gressel, pp. 233–234. Boca Raton, FL: Taylor & Francis.

Brevillé, A. *et al.* (2005). Issues of ferality or potential for ferality in oats, olives, the Vigna group, ryegrass species, sunflowers, and sugarcane. In *Crop Ferality and Volunteerism*, ed. J. Gressel., pp. 231–255. Boca Raton, FL: Taylor & Francis.

Bridges, C. B. (1914). Direct proof through non-disjunction that the sex-linked genes of Drosophila are borne by the X-chromosome. *Science, NY*, **40**, 107–109.

Bridges, C. B. (1916). Non-disjunction as proof of the chromosome theory of heredity. *Genetics, Princeton*, **1**, 1–52, 107–163.

Briggs, D. (1978). Genecological studies of salt tolerance in groundsel (*Senecio vulgaris* L.) with particular reference to roadside habitats. *New Phytologist*, **81**, 381–389.

Briggs, L. (1985). *Echium*: curse or salvation. In *Pests and Parasites as Migrants*, eds. A. Gibbs & R. Meischke, pp. 152–159. Cambridge: Cambridge University Press.

Briggs, D. & Block, M. (1992). Genecological studies of groundsel (*Senecio vulgaris* L.). I. The maintenance of population variation in the Cambridge University Botanic Garden. *New Phytologist*, **121**, 257–266.

Briggs, D. & Walters, S. M. (1997). *Plant Variation and Evolution*. 3rd edn. Cambridge: Cambridge University Press.

Briggs, D., Hodkinson, H. & Block, M. (1991). Precociously developing individuals in populations of chickweed (*Stellaria media* (L.) Vill.) from different habitat types, with special reference to the effects of weed control measures. *New Phytologist*, **117**, 153–164.

Briggs, D., Block, M., Fulton, E. & Vinson, S. (1992). Genecological studies of groundsel (*Senecio vulgaris* L.). II. Historical evidence for weed control and gene flow in the Cambridge University Botanic Garden. *New Phytologist*, **121**, 267–279.

Broadley, M. R., White, P. J., Hammond, J. P., Zelko, I. & Lux, A. (2007). Zinc in plants. *New Phytologist*, **173**, 677–702.

Brockie, R. E., Loope, L. L., Usher, M. B. & Hamann, O. (1988). Biological invasions of island nature reserves. *Biological Conservation*, **44**, 9–36.

Brockway, L. H. (1979). *Science and Colonial Expansion: The Role of the British Royal Botanic Gardens*. London: Academic Press.

Broembsen, S. L.von (1989). Invasions of natural ecosystems by plant pathogens. In *Biological Invasions: A Global Perspective*, eds. J. A. Drake, H. A. Mooney, F. diCastri *et al.*, pp. 77–83. Chichester: Wiley.

Brooks, R. R. (ed.) (1998). *Plants that Hyperaccumulate Heavy Metals*. Wallingford, UK & New York: CAB International.

Brooks, R. R. & Malaisse, F. (1985). *The Heavy Metal Tolerant Flora of Southcentral Africa*. Rotterdam: Balkema.

Broome, J. (2008). The ethics of climate change: pay now or pay later? *Scientific American*, **298**, 68–73.

Brown, A. H. D. & Burdon, J. J. (1983). Multilocus diversity in an outbreeding weed, *Echium plantagineum* L. *Australian Journal of Biological Sciences*, **36**, 503–509.
Brown, E. & Crosby, M. L. (1906). *Imported low-grade clover and alfalfa seed.* Washington, DC: United States Department of Agriculture, Bureau of Plant Industry, Report 311, pp. 17–30.
Brown, A. H. D. & Marshall, D. R. (1981). Evolutionary changes accompanying colonization in plants. In *Evolution Today*, eds. G. G. Scudder & J. I. Reveal, pp. 351–363. Pittsburgh, PA: Hunt Institute for Botanical Documentation.
Brown, J. H., Curtin, C. G. & Braithwaite, R. W. (2003). Management of the semi-natural matrix. In *How Landscapes Change: Human Disturbance and Ecosystem Fragmentation in the Americas*, eds. G. A. Bradshaw & P. A. Marquet, pp. 327–342. Berlin: Springer-Verlag.
Bruelheide, H. & Flintrop, T. (2000). Evaluating the transplantation of a meadow in the Harz Mountains, Germany. *Biological Conservation*, **92**, 109–120.
Bruford, M. W. & Saccheri, I. J. (1998). DNA fingerprinting with VNTR sequences. In *Molecular Tools for Screening Biodiversity*, eds. A. Karp, P. G. Isaac & D. S. Ingram, pp. 99–108. London: Chapman & Hall.
Bruner, A. G., Gullison, R. E. & Balmford, A. (2004). Financial costs and shortfalls of managing and expanding protected-area systems in developing countries. *Bioscience*, **54**, 119–1126.
Brunn, H. H. & Fitzbøger, B. (2002). The past impact of livestock husbandry on dispersal of plant seeds in the landscape of Denmark. *Ambio*, **31**, 425–431.
Brussard, P. F. (1997). A paradigm in conservation biology. *Science*, **277**, 527–528.
Bryant, G. (ed.) (1998). *Botanica.* North Shore City, New Zealand: Bateman.
Budiansky, S. (1995). *Nature's Keepers. The New Science of Nature Management.* London: Weidenfeld & Nicolson.
Buggs, R. J. A. (2007). Empirical studies of hybrid zone movement. *Heredity*, **99**, 301–312.
Bullock, W. L. & Samways, M. J. (2005). Conservation of flower-arthropod associations in remnant African grassland corridors in an afforested pine mosaic. *Biodiversity and Conservation*, **14**, 3093–3103.
Burdekin, D. (1979). Beetle and fungus: the unholy alliance. In *After the Elm . . .* , eds. B. Clouston & K. Stansfield, pp. 65–79. London: Heinemann.
Burdon, J. J. & Brown, A. H. D. (1986). Population genetics of *Echium plantagineum* L.: target weed for biological control. *Australian Journal of Biological Sciences*, **39**, 369–378.
Burdon, J. J., Marshall, D. R. & Groves, R. H. (1980). Isozyme variation in *Chondrilla juncea* L. in Australia. *Australian Journal of Botany*, **28**, 193–198.
Burger, J., Kirchner, M., Bramanti, B., Haak, W. & Thomas, M. G. (2007). Absence of the lactase-persistence-associated allele in early Neolithic Europeans. *Proceedings of the National Academy of Sciences, USA*, **104**, 3736–3741.
Burgman, M. & Possingham, H. (2000). Population viability analysis for conservation: the good, the bad and the undescribed. In *Genetics, Demography and Viability of Fragmented Populations*, eds. A. G.Young & G. M. Clarke, pp. 97–112. Cambridge: Cambridge University Press.
Burke, J. M. & Rieseberg, L. H. (2003). Fitness effects of transgenic disease in sunflowers. *Science*, **300**, 1250.

Burke, J. M., Bulger, M. R., Wesselingh, R. A. & Arnold, M. L. (2000). Frequency and spatial patterning of clonal reproduction in Louisiana Iris hybrid populations. *Evolution*, **54**, 137–144.

Bush, E. J. & Barrett, S. C. H. (1993). Genetics of mine invasions by *Deschampsia cespitosa* (Poaceae). *Canadian Journal of Botany – Revue Canadienne de Botanique*, **71**, 1336–1348.

Butynski, T. M. & Kalina, J. (1998). Gorilla tourism: a critical look. In *Conservation of Biological Resources*, eds. E. J. Milner-Gulland & R. Mace, pp. 294–313. Oxford: Blackwell.

Butzer, K. W. (1992). The Americas before and after 1492: an introduction to current geographical research. *Annals of the Association of American Geographers*, **82**, 345–368.

Byers, D. L. & Meagher, T. R. (1992). Mate availability in small populations of plant species with homomorphic sporophytic self-incompatibility. *Heredity*, **68**, 353–359.

Byers, D. L. & Waller, D. M. (1999). Do plant populations purge their genetic load? Effects of population size and mating history on inbreeding depression. *Annual Review of Ecology and Systematics*, **30**, 479–513.

Byers, D. L., Warsaw, A. & Meagher, T. R. (2005). Consequences of prairie fragmentation on the progeny sex ratio of the gynodioecious species, *Lobelia spicata* (Campanulaeae). *Heredity*, **95**, 69–75.

Cabeza, M. & Moilanen, A. (2001). Design of reserve networks and the persistence of biodiversity. *Trends in Ecology & Evolution*, **16**, 242–248.

Cairns, J. (1998). Ecological restoration. In *Encyclopedia of Ecology & Environmental Management*, ed. P. Calow, pp. 217–19. Oxford: Blackwell.

Caldwell, L. K. (1996). Scientific assumptions and misplaced certainty in natural resources and environmental problem solving. In *Scientific Uncertainty and Environmental Problem Solving*, ed. J. Lemons, pp. 394–421. Cambridge, MA: Blackwell.

Calero, C., Ibáñez, O., Mayol, M. & Rosselló, J. A. (1999). Random amplified polymorphic DNA (RAPD) markers detect a single phenotype in *Lysimachia minoricensis* J. J. Rodr. (Primulaceae), a wild extinct plant. *Molecular Ecology*, **8**, 2133–2136.

Callaway, R. M. & Aschehoug, E. T. (2000). Invasive plants versus their new and old neighbours: a mechanism for exotic invasion. *Science*, **290**, 521–523.

Callaway, R. M., Kikvidze, Z. & Kikodze, D. (2000). Facilitation by unpalatable weeds may conserve plant diversity in overgrazed meadows in the Caucasus Mountains. *Oikos*, **89**, 275–282.

Callicott, J. B. (1991). The wilderness idea revisited: the sustainable development alternative. Paper reprinted in *Reflecting on Nature*, eds. L. Gruen & D. Jamieson (1994), pp. 252–265. New York: Oxford University Press.

Callicott, J. B. & Nelson, M. P. (1998). *The Great New Wilderness Debate*. Athens, GA: University of Georgia Press.

Campbell, A. H. (1965). Elementary food production by Australian aborigines. *Mankind*, **6**, 206–211.

Campbell, J. E. & Gibson, D. J. (2001). The effect of seeds of exotic species transported via horse dung on vegetation along trail corridors. *Plant Ecology*, **157**, 23–35.

Carpenter, R. A. (1996). Uncertainty in managing ecosystems sustainably. In *Scientific Uncertainty and Environmental Problem Solving*, ed. J. Lemons, pp. 118–159. Cambridge, MA: Blackwell.

Carr, G. W. (1993). Exotic flora of Victoria and its impact on the indigenous biota. In *Flora of Victoria*, Vol.1, eds. D. B. Foreman & N. G. Walsh, pp. 256–297. Melbourne, Australia: Inkata Press.

Carroll, S. P. & Dingle, H. (1996). The biology of post-invasion events. *Biological Conservation*, **78**, 207–214.

Carroll, S. P., Klassen, S. P. & Dingle, H. (1998). Rapidly evolving adaptations to host ecology and nutrition in the soapberry bug. *Evolutionary Ecology*, **12**, 955–968.

Carson, R. (1962). *Silent Spring*. London: Hamish Hamilton.

Cavers, P. B. & Harper, J. L. (1964). Biological Flora of the British Isles. No. 98. *Rumex obtusifolius* L. and *Rumex crispus* L. *Journal of Ecology*, **52**, 737–766.

Cavers, P. B. & Harper, J. L. (1966). Germination polymorphism in *Rumex crispus* and *Rumex obtusifolius*. *Journal of Ecology*, **54**, 367–382.

Cavers, P. B. & Harper, J. L. (1967a). Studies in the dynamics of plant populations. I. The fate of seeds and transplants introduced into various habitats. *Journal of Ecology*, **55**, 59–71.

Cavers, P. B. & Harper, J. L. (1967b). The comparative biology of closely related species living in the same area. IX. *Rumex*: the nature of the adaptation to a sea-shore habitat. *Journal of Ecology*, **55**, 73–82.

Ceballos, G. & Ehrlich, P. R. (2006). Global mammal distributions, biodiversity hotspots, and conservation. *Proceedings of the National Academy of Sciences, USA*, **103**, 19374–19379.

Chamberlin, T. C. (1897). The method of multiple working hypotheses. *Journal of Geology*, **5**, 837–848.

Chambers, J. Q., van Eldik, T., Southon, J. & Higichi, N. (2001). Tree age structure in tropical forests of central Amazonia. In *Lessons from Amazonia: The Ecology and Conservation of a Fragmented Forest*, eds. R. O. Bierregaard Jr., C. Gascon, T. E. Lovejoy & R. C. G. Mesquita, pp. 68–78. New Haven, CT & London: Yale University Press.

Chancellor, R. J. (1985). Changes in the weed flora of an arable field cultivated for 20 years. *Journal of Applied Ecology*, **22**, 491–501.

Chang, T. T. (1995). Rice. In *Evolution of Crop Plants*, eds. J. Smartt & N. W. Simmonds, 2nd edn., pp. 147–155. Harlow, UK: Longman.

Changnon, S. A. & Changnon, J. M. (1996). History of the Chicago Diversion and future implications. *Journal of the Great Lakes Research*, **22**, 100–118.

Chaplin, F. S. & Shaver, G. R. (1996). Physiological and growth responses of arctic plants to a field experiment simulating climate change. *Ecology*, **77**, 822–840.

Chapman, M. A. & Burke, J. M. (2006). Letting the gene out of the bottle: the population genetics of genetically modified crops. *New Phytologist*, **170**, 429–443.

Chapman, M. A. & Burke, J. M. (2007). Genetic divergence and hybrid speciation. *Evolution*, **61**, 1773–1780.

Chappelka, A., Renfro, J., Somers, G. & Nash, B. (1997). Evaluation of ozone injury on foliage of black cherry (*Prunus serotina*) and tall milkweed (*Asclepias exaltata*) in Great Smoky Mountains National Park. *Environmental Pollution*, **95**, 13–18.

Charadattan, R. & DeLoach, C. J. Jr. (1988). Management of pathogens and insects for weed control in agroecosystems. In *Weed Management in Agroecosystems*, eds. M. A. Altieri & M. Liebman, pp. 245–264. Boca Raton, FL: CRC Press.

Charles, A. H. (1966). Variation in grass and clover populations in response to agronomic selection pressures. *Proceedings of the 10th International Grassland Congress Helsinki*, pp. 625–629.

Charlesworth, D. & Charlesworth, B. (1987). Inbreeding depression and its evolutionary consequences. *Annual Review of Ecology and Systematics*, **18**, 237–268.
Chase, M. W. & Hillis, H. H. (1991). Silica gel: an ideal material for field preservation of leaf samples for DNA studies. *Taxon*, 215–220.
Chauvel, B. (1991). Polymorphism et selection pour la résistance aux urées substituées chez *Alopecurus myosuroides* Huds. Ph.D. thesis, University of Paris XI, Orsay, France.
Cheney, J., Navarro, J. N. & Wyse Jackson, P. (2000). *Action Plan for Botanic Gardens in the European Union*. Published by the National Botanic Garden of Belgium for Botanic Gardens Conservation International.
Cheptou, P. O., Carrue, O., Rovifed, S. & Cantarel, A. (2008). Rapid evolution of seed dispersal in an urban environment in the weed *Crepis sancta. Proceedings of the National Academy of Sciences, USA*, **105**, 3796–3799.
Chèvre, A.-M. *et al.* (2004). A review of interspecific gene flow from oilseed rape to wild relatives. In *Introgression from Genetically Modified Plants into Wild Relatives*, eds. H. C. M. den Nijs, D. Bartsch & J. Sweet, pp. 235–251. Wallingford: CABI Publishing.
Childe, V. G. (1925). *The Dawn of European Civilisation*. New York: Alfred Knopf.
Childe, V. G. (1952). *New Light on the Most Ancient East*. London: Routledge & Paul.
Chipperfield, H. (1895). Bird sanctuaries in the London Parks. *Nature Notes*, **6**, 23–25.
Chippindale, H. G. & Milton, W. E. J. (1948). On the viable seeds present in the soil beneath pastures. *Journal of Ecology*, **22**, 508–531.
Choi, C. (2008). Tierra del Fuego: the beavers must die. *Nature*, **453**, 968.
Christiansen, F. B. (1984). The definition and measurement of fitness. In *Evolutionary Ecology*, 23rd Symposium of the British Ecological Society, ed. B. Shorrocks, pp. 65–79. Oxford: Blackwell Scientific.
Christopher, A. J. (1984). *Colonial Africa*. London: Croom Helm.
Christopher, T. (2008). Can weeds help solve the climate crisis? *The New York Times*, 29 June 2008.
Clarke, J. (ed.) (1992). *Set-Aside*. Farnham, UK: Monograph No. 50 of the British Crop Protection Council.
Clausen, J. (1951). *Stages in the Evolution of Plant Species*. New York: Cornell University Press.
Clegg, M. T. & Allard, R. W. (1972). Patterns of genetic differentiation in the slender wild oat species *Avena barbata. Proceedings of the National Academy of Sciences, USA*, **69**, 1820–1824.
Clegg, M. T. & Brown, A. H. D. (1983). The founding of plant populations. In *Genetics and Conservation*, eds. C. M. Schonewald-Cox, S. M. Chambers, B. MacBryde & L. Thomas, pp. 216–228. Menlo Park, CA: Benjamin Cummings.
Cleland, C. E. (2001). Historical science, experimental science, and the scientific method. *Geology*, **29**, 987–990.
Clement, C. R. (1990). Fruit trees and the origin of agriculture in the neotropics. Paper presented at the Second International Congress of Ethnobiology, Yunnan, People's Republic of China.
Clements, F. E. (1916). *Plant Succession: An Analysis of the Development of Vegetation*. Washington, DC: Carnegie Institution of Washington.
Clements, D. R. *et al.* (2004). Adaptability of plants invading North American cropland. *Agriculture Ecosystems & Environment*, **104**, 379–398.

Cleuren, H. (2001). *Paving the Road for Forest Destruction: Key Actors and Driving Forces of Tropical Deforestation in Brazil, Ecuador and Cameroon*. Leiden Development Studies, New series, Vol. 1. Leiden: Leiden University.

Cloutier, D., Kanashiro, M., Ciampi, A. Y. & Schoen, D. J. (2007). Impact of selective logging on inbreeding and gene dispersal in an Amazonian population of *Carapa guianensis* Aubl. *Molecular Ecology*, **16**, 797–809.

Coates, D. J. (1988). Genetic diversity and population genetic structure in the rare Chittering grass wattle *Acacia anomala* Court. *Australian Journal of Botany*, **36**, 273–286.

Cobb, A. H. & Kirkwood, R. C. (2000). *Herbicides and their Mechanisms of Action*. Sheffield: Sheffield Academic Press and Boca Raton, FL: CRC Press.

Cochrane, A. (2004). Western Australia's ex situ program for threatened species: a model integrated strategy for conservation. In *Ex situ Plant Conservation*, eds. E. O. Guerrant Jr., K. Havens & M. Maunder, pp. 40–66. Washington, Covelo & London: Island Press.

Cohen, J. E. (1995). *How Many People Can the Earth Support?* New York: Norton.

Cohen, J. E. (2003). Human population: the next half century. *Science*, **302**, 1172–1175.

Colbach, N. & Sache, I. (2001). Blackgrass (*Alopecurus myosuroides* Huds.) seed dispersal from a single plant and its consequences on weed infestation. *Ecological Modelling*, **139**, 201–219.

Cole, C. T. (2003). Genetic variation in rare and common plants. *Annual Review of Ecology and Evolution*, **34**, 213–237.

Collinge, S. K. (1996). Ecological consequences of habitat fragmentation: implications for landscape architecture and planning. *Landscape and Urban Planning*, **36**, 59–77.

Coltman, D. W. *et al.* (2003). Undesirable evolutionary consequences of trophy hunting. *Nature*, **426**, 655–658.

Colwell, R. K. (1994). Human aspects of biodiversity: an evolutionary perspective. In *Biodiversity and Global Change*, eds. O. T. Solbrig, H. M. van Emden & P. G. W. J. van Oordt, pp. 211–224. Wallingford: CAB International.

Comes, H. P. & Kadereit, J. W. (1996). Genetic basis of speed of development in *Senecio vulgaris* var. *vulgaris, S. vulgaris* ssp. *denticulatus* (O. F. Muell). P. D. Sell, and *Senecio vernalis* Waldst & Kit. *Heredity*, **77**, 544–554.

Conrad, S. G. & Radosevich, S. R. (1979). Ecological fitness of *Senecio vulgaris* and *Amaranthus retroflexus* biotypes susceptible or resistant to atrazine. *Journal of Applied Ecology*, **16**, 171–177.

Conwentz, H. (1914). On national and international protection of nature. *Journal of Ecology*, **2**, 109–123.

Cook, C. D. K. & Urmi-König, K. (1985). A revision of the genus *Elodea* (Hydrocharitaceae). *Aquatic Botany*, **21**, 111–156.

Cook, S., Lefèbvre, C. & McNeilly, T. (1972). Competition between metal tolerant and normal plant populations on normal soil. *Evolution*, **26**, 366–372.

Cook, L. M., Soltis, P. S., Brunsfeld, S. J. & Soltis, D. E. (1998). Multiple independent formations of Tragopogon tetraploids (Asteraceae): evidence from RAPD markers. *Molecular Ecology*, **7**, 1293–1302.

Cooke, H. (1994). Chromosome structure: molecular aspects. In *The Encyclopedia of Molecular Biology*. ed. J. Kendrew, pp. 202–205. Oxford: Blackwell Science.

Coombe, D. E. (1956). Notes on some British plants seen in Austria. *Veröff. Geobot. Inst. Zurich*, **35**, 128–137.

Cooper, N. (2000). How natural is a nature reserve?: an ideological study of British nature conservation landscapes. *Biodiversity and Conservation*, **9**, 1131–1152.
Corley, M. F. V. & Perry, A. R. (1985). *Scopelophila cataractae* (Mitt.) Broth. In South Wales, new to Europe. *Journal of Bryology*, **13**, 323.
Correll, D. S. (1982). *Flora of the Bahamas Archipelago*. Vaduz: Cramer.
Costa, J. T. (2003). Teaching Darwin to Darwin. *Bioscience*, **53**, 1030–1031.
Costin, B. J., Morgan, K. W. & Young, A. G. (2001). Reproductive success does not decline in fragmented populations of *Leucochrysum albicans* subsp. albicans var. tricolor (Asteraceae). *Biological Conservation*, **98**, 273–284.
Cousens, R. & Mortimer, M. (1995). *Dynamics of Weed Populations*. Cambridge: Cambridge University Press.
Cowie, I. D. & Werner, P. A. (1993). Alien plant species invasive in Kakadu National Park, Tropical Northern Australia. *Biological Conservation*, **63**, 127–135.
Cox, G. W. (1999). *Alien Species in North America and Hawaii*. Washington, DC: Island Press.
Cozzolino, S., Nardella, A. M., Impagliazzo, S., Widmer, A. & Lexer, C. (2006). Hybridization and conservation of Mediterranean orchids: should we protect the orchid hybrids or the orchid hybrid zones? *Biological Conservation*, **129**, 14–23.
Cranston, D. M. & Valentine, D. H. (1983). Transplant experiments on rare plant species from Upper Teesdale. *Biological Conservation*, **26**, 175–191.
Crawford, T. J. (1984). What is a population? In *Evolutionary Ecology*, 23rd Symposium of the British Ecological Society, pp. 135–173. Oxford: Blackwell.
Crawford, D. J., Brauner, S., Cosner, M. B. & Stuessy, T. F. (1993). Use of RAPD markers to document the origin of the intergeneric hybrid *Margyracaena skottsbergii* (Rosaceae) on the Juan Fernandez Islands. *American Journal of Botany*, **80**, 89–92.
Crawley, M. J. (1987). What makes a community invasible? *Symposium of the British Ecological Society*, **26**, 429–453.
Crawley, M. J. (1989). Insect herbivores and plant population dynamics. *Annual Review of Entomology*, **34**, 531–564.
Crnokrak, P. & Barrett, S. C. H. (2002). Purging the genetic load: a review of the experimental evidence. *Evolution*, **56**, 2347–2358.
Cronk, Q. C. B. & Fuller, J. L. (1995). *Plant Invaders: The Threat to Natural Ecosystems*. London: Chapman & Hall. Reissued (2001), London & Sterling, VA: Earthspan.
Crooks, K. R. & Sanjayan, M. (eds.) (2006). *Connectivity Conservation*. Cambridge: Cambridge University Press.
Crooks, J. A. & Soulé, M. E. (1999). Lag times in population explosions of invasive species: causes and implications. In *Invasive Species and Biodiversity Management*, eds. O. T. Sundlund, P. J. Schei. & Å. Viken, pp. 103–125. Dordrecht, Boston & London: Kluwer.
Crosby, A. W. (1986). *Ecological Imperialism: The Biological Expansion of Europe, 900–1900*. Cambridge: Cambridge University Press.
Crosby, A. W. (1994). *Germs, Seeds and Animals: Studies in Ecological History*. New York & London: Sharpe, Armonk.
Crumpacker, D. W., Box, E. O. & Hardin, E. D. (2002). Implications of climatic warming for conservation of native trees and shrubs in Florida. *Conservation Biology*, **15**, 1008–1020.
Cuguen, J., Arnaud, J.-F., Delescluse, M. & Viard, F. (2004). Crop-wild interaction within *Beta vulgaris* complex: a comparative analysis of genetic diversity between sea beet and weed beet populations within the French sugar beet production

area. In *Introgression from Genetically Modified Plants into Wild Relatives*, eds. H. C. M. den Nijs, D. Bartsch & J. Sweet, pp. 183–201. Wallingford: CAB International.

Curtis, P. S. *et al.* (1989). Growth and senescence in plant communities exposed to elevated CO_2 concentration on an esturine marsh. *Oecologia*, **78**, 20–26.

Daehler. C. C. (1998). The taxonomic distribution of invasive angiosperm plants: ecological insights and comparison to agricultural weeds. *Biological Conservation*, **84**, 167–180.

Daehler, C. C. & Strong, D. R. (1996). Status, prediction and prevention of introduced cordgrass *Spartina* spp.: invasions in Pacific estuaries, USA. *Biological Conservation*, **78**, 51–58.

Dafni, A. & Heller, D. (1982). Adventive flora of Israel: phytological, ecological and agricultural aspects. *Plant Systematics & Evolution*, **140**, 1–18.

Daily, G. & Dasgupta, S. (2001). Ecosystem services, concept of. In *Encyclopedia of Biodiversity*, Vol. 2, ed. S. A. Levin, pp. 353–362. San Diego, CA & London: Academic Press.

Damman, H. & Cain, M. L. (1998). Population growth and viability analyses of the clonal woodland herb *Asarum canadense*. *Journal of Ecology*, **86**, 13–26.

Darmency, H. (1994). Genetics of herbicide resistance in weeds and crops. In *Herbicide Resistance in Plants: Biology and Biochemistry*, eds. S. B. Powles & J. A. M. Holtum, pp. 263–298. Boca Raton: Lewis.

Darmency, H. & Gasquez, J. (1981). Inheritance of triazine resistance in *Poa annua*: consequences for population dynamics. *New Phytologist*, **89**, 487–493.

Darwin, C. (1839). *Journal of Researches into the Geology and Natural History of the Various Countries Visited by H.M.S. Beagle from 1832 to 1836*. London: Henry Colburn.

Darwin, C. (1868). *The Variation of Animals and Plants under Domestication*. London: Murray. Popular Edition 1905.

Darwin, C. (1901). *The Origin of Species by Means of Natural Selection or the Preservation of Favoured Races in the Struggle for Life*. Popular Impression, 6th edn. London: John Murray.

Darwin, C. (1985). *The Correspondence of Charles Darwin*, Vol. 1, eds. F. Burkhart, S. Smith *et al.* Cambridge: Cambridge University Press.

Darwin, C. (1986). *The Works of Charles Darwin. Journal of Researches, Part 2*, eds. P. H. Barrett & R. B. Freeman. London: William Pickering.

Darwin, C. (1989). *The Descent of Man and Selection in Relation to Sex, Part 2*. London: William Pickering.

Darwin, F. & Seward, A. C. (1903). *More Letters of Charles Darwin*, 2 vols. London: Murray.

Darwin, C. & Wallace, A. (1859). On the tendency of species to form varieties: and on the perpetuation of varieties and species by natural means of selection. *Proceedings of the Linnean Society of London*, **3**, 45–62.

Davies, M. S. (1975). Physiological differences among populations of *Anthoxanthum odoratum* L. collected from the Park Grass experiment, Rothamsted. IV. Response to potassium and magnesium. *Journal of Applied Ecology*, **12**, 953–963.

Davies, M. S. & Snaydon, R. W. (1973a). Physiological differences amongst populations of *Anthoxanthum odoratum* collected from the Park Grass Experiment. I. Responses to calcium. *Journal of Applied Ecology*, **10**, 33–45.

Davies, M. S. & Snaydon, R. W. (1973b). Physiological differences amongst populations of *Anthoxanthum odoratum* collected from the Park Grass Experiment. II. Responses to aluminium. *Journal of Applied Ecology*, **10**, 47–55.
Davies, M. S. & Snaydon, R. W. (1974). Physiological differences amongst populations of *Anthoxanthum odoratum* collected from the Park Grass Experiment. III. Responses to phosphate. *Journal of Applied Ecology*, **11**, 699–707.
Davies, M. S. & Snaydon, R. W. (1976). Rapid population differentiation in a mosaic environment. III. Measures of selection pressures. *Heredity*, **36**, 59–66.
Davis, M. B. (1990). Biology and paleobiology of global climate change: introduction. *Trends in Ecology & Evolution*, **5**, 269–270.
Davis, R. C. (1983). *Encyclopedia of American Forest and Conservation History*. London & New York: Macmillan.
Davis, P. H. & Heywood, V. H. (1963). *Principles of Angiosperm Taxonomy*. Edinburgh: Oliver & Boyd.
Davis, M. B. & Shaw, R. G. (2001). Range shifts and adaptive responses to quaternary climate change. *Science*, **292**, 673–679.
Davis, M. B., Woods, K., Webb, S. L. & Futyama, R. P. (1986). Dispersal versus climate: expansion of Fagus and Tsuga into the Great Lakes Region. *Vegetatio*, **67**, 93–103.
Davis, M. B., Shaw, R. G. & Etterson, J. R. (2005). Evolutionary responses to changing climate. *Ecology*, **86**, 1704–1714.
Davison, A. (1996). *Deserted Villages in Norfolk*. North Walsham, Norfolk, UK: Poppyland Publishing.
Davison, A. W. & Barnes, J. D. (1998). Effects of ozone on wild plants. *New Phytologist*, **139**, 135–151.
Davison, A. W. & Reiling, K. (1995). A rapid change in ozone resistance of *Plantago major* after summers with high ozone concentrations. *New Phytologist*, **131**, 337–344.
De Vries, H. (1905). *Species and Varieties: Their Origin by Mutation*. Chicago: Open Court Publishing Co.
Deadman, A. (1993). *Species Recovery Programme: Aims and Objectives*. Peterborough, UK: English Nature.
Degan, B. *et al.* (2006). Impact of selective logging on genetic composition and demographic structure of four tropical tree species. *Biological Conservation*, **131**, 386–401.
Dehnen-Schmutz, K., Perrings, C. & Williamson, M. (2004). Controlling *Rhododendron ponticum* in the British Isles: an economic analysis. *Journal of Environmental Management*, **70**, 323–332.
Dehnen-Schmutz, K., Touza, J., Perrings, C. & Williamson, M. (2007). A century of the ornamental plant trade and its impact on invasion success. *Diversity and Distributions*, **13**, 527–534.
Del Castillo, R. F. (1994). Factors influencing the genetic structure of *Phacelia dubia*, a species with a seed bank and large fluctuations in population size. *Heredity*, **72**, 446–458.
DeMauro, M. M. (1994). Development and implementation of a recovery program for the federal threatened Lakeside daisy (*Hymenoxys acaulis* var. glabra). In *Restoration of Endangered Species*, eds. M. L. Bowles & C. J. Whelan, pp. 298–321. Cambridge: Cambridge University Press.
Denevan, W. M. (1992). The pristine myth: the landscape of the Americas in 1492. *Annals of the Association of American Geographers*, **82**, 369–385.

Desmond, R. (1995). *Kew. The History of the Royal Botanic Gardens*. Kew: Harvill Press, Royal Botanic Gardens.

Detwyler, T. R. (1971). *Man's Impact on Environment*. New York: McGraw-Hill.

Devaux, C. *et al.* (2005). High diversity of oilseed rape pollen clouds over an agroecosystem indicates long-distance dispersal. *Molecular Ecology*, **14**, 2269–2280.

Devine, M. D., Duke, S. O. & Fedtke, C. (1993). *Physiology of Herbicide Action*. Englewood Cliffs, NJ: Prentice-Hall.

Dewey, E. H. (1894). The Russian thistle. *United States Department of Agriculture Division of Botany Bulletin No. 15*.

di Castri, F. (1989). History of biological invasions with special emphasis on the Old World. In *Biological Invasions: A Global Perspective*, eds. J. A. Drake *et al.*, pp. 1–30. Chichester & New York: Wiley.

Diamond, J. M. (1975). The island dilemma: lessons of modern biogeographic studies for the design of natural reserves. *Biological Conservation*, **7**, 129–146.

Diamond, J. M. (1976). Island biogeography and conservation: strategy and limitations. *Science*, **193**, 1027–1029.

Diamond, J. (1989). Overview of recent extinctions. In *Conservation for the Twenty-first Century*, eds. D. Western & M. C. Pearl, pp. 37–41. New York & Oxford: Oxford University Press.

Diamond, J. (1992). *The Rise and Fall of the Third Chimpanzee: How our Animal Heritage Affects the Way We Live*. London: Vantage.

Diamond, J. (1997). *Guns, Germs and Steel*. London: Vintage.

Diamond, J. (2005). *Collapse: How Societies Choose to Fail or Survive*. London: Allen Lane, Penguin Books.

Diamond, J. M. & May, R. M. (1976). Island biogeography and the design of nature reserves. In *Theoretical Ecology: Principles and Applications*, ed. R. M. May, pp. 163–186. Philadelphia: W. B. Saunders.

Dickson, J. H. (1993). Scottish woodlands: their ancient past and precarious present. *Scottish Forestry*, **47**, 73–78.

Dickson, J. H., Macpherson, P. & Watson, K. J. (2000). *The Changing Flora of Glasgow: Urban and Rural Plants through the Centuries*. Edinburgh: Edinburgh University Press.

Dickson, J. H., Rodriguez, J. C. & Machado, A. (1987). Invading plants at high altitudes on Tenerife especially in the Teide National Park. *Botanical Journal of the Linnean Society*, **95**, 155–179.

Dietz, T., Ostrom, E. & Stern, P. C. (2003). The struggle to govern the commons. *Science*, **302**, 1907–912.

Dillehay, T. D., Rossen, J., Andres, T. C. & Williams, D. E. (2007). Preceramic adoption of peanut, squash and cotton in northern Peru. *Science*, **316**, 1890–1893.

Dimbleby, G. W. (1967). *Plants and Archaeology*. London: John Baker.

Dirnböck, T., Greimler, J., López, P. & Stuessy, T. F. (2003). Predicting future threats to the native vegetation of Robinson Crusoe Island, Juan Fernandez Archipelago, Chile. *Conservation Biology*, **17**, 1650–1659.

DiTomaso, J. M. (1998). Impact, biology, and ecology of saltcedar (*Tamarix* spp.) in southwestern United States. *Weed Technology*, **12**, 326–336.

Dixon, P. M. & May, B. (1990). Genetic diversity and population structure of a rare plant, Northern Monkshood (*Aconitum noveboracense*). *New York State Museum Bulletin*, **471**, 167–175.

Dobzansky, T. (1935). A critique of the species concept in biology. *Philosophy of Science*, **2**, 344–355.
Dolan, R. W. (1994). Patterns of isozyme variation in relation to population size, isolation, and phyogeographic history in royal catchfly *Silene regia* (Caryphyllaceae). *American Journal of Botany*, **81**, 965–972.
Dolek, M. & Geyer, A. (2002). Conserving biodiversity on calcareous grasslands in the Franconian Jura by grazing: a comprehensive approach. *Biological Conservation*, **104**, 351–360.
Donlan, C. J. (2007). Restoring America's big, wild animals. *Scientific American*, June 2007, pp. 48–55.
Donlan, J. *et al.* (2005). Re-wilding North America. *Nature*, **436**, 913–914.
Donlan, C. J. *et al.* (2006). Pleistocene rewilding: an optimistic agenda for twenty-first century conservation. *American Naturalist*, **168**, 660–681.
Dosmann, M. & Del Tredici, P. (2003). Plant introduction, distribution and survival: a case study of the 1980 Sino-American Botanical Expedition. *Bioscience*, **53**, 588–597.
Dosmann, M. S. (2006). Research in the garden: averting the collections crisis. *Botanical Review*, **72**, 207–234.
Dover, G. A. (1998). Neutral theory of evolution. In *The Encyclopedia of Ecology & Environmental Management*, ed. P. Calow, pp. 479–480. Oxford: Blackwell Science.
Drake, J. A., Mooney, H. A., diCastri, F. *et al.* (eds.) (1989). *Biological Invasions: A Global Perspective*. Chichester: Wiley.
Drayton, B. & Primack, R. B. (1996). Plant species lost in an isolated conservation area in Metropolitan Boston 1894 to 1993. *Conservation Biology*, **10**, 30–39.
Drayton, B. & Primack, R. B. (2000). Rates of success in the reintroduction by four methods of several perennial plant species in eastern Massachusetts. *Rhodora*, **102**, 299–331.
Druce, G. C. (1930). *The Flora of Northamptonshire*. Arbroath: T. Buncle.
Duarte, C. M., Marbá, N. & Holmer, M. (2007). Rapid domestication of marine species. *Science*, **316**, 382–383.
Dudash, M. R. (1990). Relative fitness of selfed and outcrossed progeny in a self-compatible, protandrous species, *Sabatia angularis* (Gentianaceae): a comparison of three environments. *Evolution*, **44**, 1129–1139.
Dudgeon, D. *et al.* (2006). Freshwater biodiversity: importance, threats, status and conservation challenges. *Biological Review*, **81**, 163–182.
Dudley, T. L. (2000). *Arundo donax*. In *Invasive Plants of California's Wildlands*, eds. C. C. Bossard, J. M. Randall & M. C. Hozhovsky, pp. 53–58. Berkeley, CA: University of California Press.
Dueck, T. A., Endedijk, G. J. & Klein-Ikkink, H. G. K. (1987). Soil pollution and changes in vegetation due to heavy metals in sinter-pavements. *Chemosphere*, **16**, 1021–1030.
Dukes, J. S. (2002). Species composition and diversity affect grassland susceptibility and response to invasion. *Ecological Applications*, **12**, 602–617.
Dunn, D. B. (1959). Some effects of air pollution on Lupinus in the Los Angeles area. *Ecology*, **40**, 621–625.
Durka, W., Bossdorf, O., Prati, D. & Auge, H. (2005). Molecular evidence for multiple introductions of garlic mustard (*Alliaria petiolata*, Brassicaceae) to North America. *Molecular Ecology*, **14**, 1697–1706.
Dyer, B. D. (2002). Chloroplasts, genetics of. In *Encyclopedia of Genetics*, eds. S. Brenner & J. H. Miller, Vol. 1, pp. 337–338. San Diego, CA: Academic Press.

Dyer, A. R. & Rice, K. J. (1999). Effect of competition on resource availability and growth of a Californian bunch grass. *Ecology*, **80**, 2697–2710.

Eckert, C. G. & Barrett, S. C. H. (1992). Stochastic loss of style morphs from populations of tristylous *Lythrum salicaria* and *Decodon verticillatus* (Lythraceae). *Evolution*, **46**, 1014–1029.

Eckert, C. G, Manicacci, D. & Barrett, S. C. H. (1996). Genetic drift and founder effect in native versus introduced populations of an invading plant *Lythrum salicaria* (Lythraceae). *Evolution*, **50**, 1512–1519.

Edelist, C. *et al.* (2006). Microsatellite signature of ecological selection for salt tolerance in a wild sunflower hybrid species, *Helianthus paradoxus. Molecular Ecology*, **15**, 4623–4634.

Edwards, P. J., Webb N. R., Urbanska, K. M & Bornkamm, R. (1997). Restoration ecology: science technology and society. In *Restoration Ecology and Sustainable Development*, eds. K. M. Urbanska, N. R. Webb & P. J. Edwards, pp. 381–390. Cambridge: Cambridge University Press.

Ehrlich, P. R. & Pringle, R. M. (2008). Where does biodiversity go from here? A grim business-as-usual forecast and hopeful portfolio of partial solutions. *Proceedings of the National Academy of Sciences, USA*, **105**, 11579–11586.

Ellegren, H. & Sheldon, B. C. (2008). Genetic basis of fitness differences in natural populations. *Nature*, **452**, 169–175.

Ellenberg, H. (1979). Man's influence on tropical mountain ecosystems in South America. *Journal of Ecology*, **67**, 401–416.

Ellenberg, H. (1988). *Vegetation Ecology of Central Europe*, 4th edn. Cambridge: Cambridge University Press.

Elliot, R. (1997). *Faking Nature: The Ethics of Environmental Restoration*. London: Routledge.

Ellis, R. H. & Roberts, E. H. (1980). Improved equations for the prediction of seed longevity. *Annals of Botany*, **45**, 13–30.

Ellstrand, N. C. (2003). Current knowledge of gene flow in plants: implications for transgenic flow. *Philosophical Transactions of the Royal Society of London, B*, **358**, 1163–1170.

Ellstrand, N. C. & Schierenbeck, K. (2000). Hybridization as a stimulus for the evolution of invasiveness in plants? *Proceedings of the National Academy of Sciences, USA*, **97**, 7043–7050.

Ellstrand, N. C. & Schierenbeck, K. A. (2006). Hybridization as a stimulus for the evolution of invasiveness in plants? *Euphytica*, **48**, 35–46.

Ellstrand, N. C., Whitkus, R. & Rieseberg, L. H. (1996). Distribution of spontaneous plant hybrids. *Proceedings of the National Academy of Sciences, USA*, **93**, 5090–5093.

Ellstrand, N. C., Prentice, H. C. & Hancock, J. F. (1999). Gene flow and introgression from domesticated plants into their wild relatives. *Annual Review of Ecology & Systematics*, **30**, 539–563.

Elmqvist, T. (2000). Pollinator extinction in the Pacific Islands. *Conservation Biology*, **14**, 1237–1239.

Eloff, J. N. (1985). *Botanic Gardens: Victorian Relic or 21st Century Challenge*. Inaugural Lecture, Cape Town: University of Cape Town.

Elton, C. S. (1958). *The Ecology of Invasions by Animals and Plants*. London: Methuen.

Emanuelsson, U. (1988). A model for describing the development of the cultural landscape. In *The Cultural Landscape: Past, Present and Future*, eds. H. H. Birks,

H. J. B. Birks, P. E. Kaland & D. Moe, pp. 111–121. Cambridge: Cambridge University Press.

Emms, S. K. & Arnold, M. L. (1997). The effect of habitat on parental and hybrid fitness: transplant experiments with Louisiana Irises. *Evolution*, **51**, 1112–1119.

Endler, J. A. (1986). *Natural Selection in the Wild*. Princeton, NJ: Princeton University Press.

Epling, C. & Lewis, H. (1952). Increase of the adaptive range of the genus Delphinium. *Evolution*, **6**, 253–267.

Ernst, W. H. O. (1990). Mine vegetation in Europe. In *Heavy Metal Tolerance in Plants: Evolutionary Aspects*, ed. A. J. Shaw, pp. 21–37. Boca Raton, FL: CRC Press.

Ernst, W. H. O. (1998). Evolution of plants on soils anthropogenically contaminated by heavy metals. In *Plant Evolution in Man-made Habitats*, eds. L. W. D. van Raamsdonk & J. C. M. den Nijs, pp. 13–27. Amsterdam: Hugo de Vries Laboratory.

Erwin, T. L. (1983). Beetles and other insects of tropical forest canopy at Manaus, Brazil sampled by insecticidal fogging. In *Tropical Rain Forest: Ecology and Management*, ed. S. L. Sutton, pp. 59–75. Oxford: Blackwell.

Erwin, D. H. & Anstey, R. L. (1995). *New Approaches to Speciation in the Fossil Record*. New York: Columbia University Press.

Etkin, N. L. (1994). The cull of the wild. In *Eating on the Wild Side*, ed. N. L. Etkin, pp. 1–21. Tucson, AZ & London: University of Arizona Press.

Ettl, G. J. & Cottone, N. (2004). Whitebark pine (*Pinus albicaulis*) in Mt. Rainier National Park, Washington, USA: response to blister rust infection. In *Species Conservation and Management*, eds. H. R. Akçakaya *et al.*, pp. 36–47. New York: Oxford University Press.

Evans, D. (1992). *A History of Nature Conservation in Britain*. London & New York: Routledge.

Evans, D. (1997). *A History of Nature Conservation in Britain*, 2nd edn. London & New York: Routledge.

Evans, L. T. (1998). *Feeding the Ten Billion: Plants and Population Growth*. Cambridge: Cambridge University Press.

Evelyn, J. (1661). *Fumifugium: Or the Inconvenience of the Aer and Smoake of London Dissipated*. London: W. Godbid.

Ewel, J. J. (1986). Invasibility: lessons from southern California. In *Ecology of Biological Invasions of North America and Hawaii*, eds. H. A. Mooney & J. A. Drake, pp. 214–239. New York: Springer Verlag.

Ewers, R. M. & Didham, R. K. (2006). Confounding factors in the detection of species responses to habitat fragmentation. *Biological Reviews*, **81**, 117–142.

Excoffier, L., Smouse, P. E. & Quattro, J. M. (1992). Analysis of molecular variance inferred from metric distances among DNA haplotypes: application to human mitochondrial DNA restriction data. *Genetics*, **131**, 479–491.

Faegri, K. (1988). Preface to *The Cultural Landscape: Past, Present and Future*. eds. H. H. Birks, H. J. B. Birks, P. E. Kaland & D. Moe, pp. 1–4. Cambridge: Cambridge University Press.

Fairhead, J. & Leach, M. (1998). *Misreading the African Landscape*. Cambridge: Cambridge University Press.

Falk, D. A., Millar, C. I. & Olwell, M. (1996). *Restoring Diversity*. Washington, DC & Covelo, CA: Island Press.

Falk, D. A., Palmor, M. A. & Zedler, J. B. (2006). *Foundations of Restoration Ecology*. Washington, DC, Covelo, CA & London: Island Press.

Falk, D. A. *et al.* (2006). Population and ecological genetics in restoration ecology. In *Foundations of Restoration Ecology*, eds. D. A. Falk, M. A. Palmer & J. B. Zedler, pp. 14–41. Washington, DC, Covelo, CA & London: Island Press.

Farnsworth, E. J., Klionsky, S., Brumback, W. E. & Havens, K. (2006). A set of simple decision matrices for prioritizing collections of rare plant species for ex situ conservation. *Biological Conservation*, **128**, 1–12.

Feltwell, J. (1992). *Meadows: A History and Natural History*. Stroud, UK: Sutton.

Fenner, F. & Ratcliffe, F. N. (1965). *Myxomatosis*. Cambridge: Cambridge University Press.

Fenner, M. & Thompson, K. (2005). *The Ecology of Seeds*. Cambridge: Cambridge University Press.

Ferris, C., King, R. A. & Gray, A. J. (1997). Molecular evidence for maternal parentage in the hybrid origin of *Spartina anglica* C. E. Hubbard. *Molecular Ecology*, **6**, 185–187.

Ferris-Kaan, R. (ed.) (1995). *The Ecology of Woodland Creation*. Chichester & New York: Wiley.

Finkeldy, R. & Ziehe, M. (2004). Genetic implications of silvicultural regimes. *Forest Ecology and Management*, **197**, 231–244.

Firbank, L. G. (1988). Biological Flora of the British Isles. No. 165. *Agrostemma githago* L. *Journal of Ecology*, **76**, 1232–46.

Firbank, L. G. (2005). Striking a new balance between agricultural production and biodiversity. *Annals of Applied Biology*, **146**, 163–175.

Fisher, R. A. (1929). *The Genetical Theory of Natural Selection*, 2nd edn. Reprinted 1958. London: Constable & New York: Dover Books.

Floate, K. D. & Whitham, T. G. (1993). The "hybrid bridge" hypothesis: host shifting via plant hybrid swarms. *American Naturalist*, **141**, 651–662.

Flux, J. E. C. & Fullagar, P. J. (1992). World distribution of the rabbit, *Oryctolagus cuniculus*, on islands. *Mammal Review*, **22**, 151–205.

Foose, T. J., Boer, L. de, Seal, U. S. & Lande, R. (1995). Conservation management strategies based on viable populations. In *Population Management for Survival and Recovery*, eds. J. D. Ballou, M. Gilpin & T. F. Foose, pp. 273–294. New York: Columbia University Press.

Ford, E. B. (1971). *Ecological Genetics*. London: Chapman & Hall.

Foreman, D. (2004). *Rewilding North America*. Washington, DC, Covelo, CA & London: Island Press.

Forman, R. T. T. (1995). *Land Mosaics*. Cambridge: Cambridge University Press.

Forman, R. T. T. & Deblinger, R. D. (2000). The ecological road-effect zone of a Massachusetts (USA) suburban highway. *Conservation Biology*, **14**, 36–46.

Fox, M. D. (1989). Mediterranean weeds. Exchanges of invasive plants between the five Mediterranean regions of the world. In *Biological Invasions in Europe and the Mediterranean Basin*, eds. F. Di Castri, A. J. Hanson & M. Debussche, pp. 179–200. Dordrecht: Kluwer.

Frangmeier, A., Bender, J., Weigel, H. J. & Jäger, H. J. (2002). Effects of pollutant mixtures. In *Air Pollution and Plant Life*, eds. J. N. B. Bell & M. Treshow, 2nd edn., pp. 251–272. Chichester & New York: Wiley.

Frankel, O. H. & Soulé, M. E. (1981). *Conservation and Evolution*. Cambridge: Cambridge University Press.

Frankel. O. H., Brown, A. D. H. & Burdon, J. (1995). *The Conservation of Plant Biodiversity*. Cambridge: Cambridge University Press.

Frankham, R. (1996). Relationship of genetic variation to population size in wildlife. *Conservation Biology*, **10**, 1500–1508.

Frankham, R. (2005a). Genetics and extinction. *Biological Conservation*, **126**, 131–140.

Frankham, R. (2005b). Stress and adaptation in conservation genetics. *Journal of Evolutionary Biology*, **18**, 750–755.

Frankham, R. K. & Loebel, D. A. (1992). Modelling problems in conservation genetics using captive Drosophila populations: rapid genetic adaptation to captivity. *Zoo Biology*, **11**, 333–342.

Frankham, R., Ballou, J. D. & Briscoe, D. A. (2002). *Introduction to Conservation Genetics*. Cambridge: Cambridge University Press.

Franklin, I. A. (1980). Evolutionary change in small populations. In *Conservation: An Evolutionary-Ecological Perspective*, eds. M. E. Soulé & B. A. Wilcox, pp. 135–150. Sunderland, MA: Sinauer.

Franks, S. J., Sim, S. & Weis, A. E. (2007). Rapid response of flowering time by an annual plant in response to a climate fluctuation. *Proceedings of the National Academy of Sciences, USA*, **104**, 1278–1282.

Freckleton, R. P. & Watkinson, A. R. (2003). Are all plant populations metapopulations? *Journal of Ecology*, **91**, 321–324.

Friday, L. (1997). *Wicken Fen: The Making of a Wetland Nature Reserve*. Colchester, UK: Harley.

Fridlender, A. & Boisselier-Dubayle, M.-C. (2000). Comparison de la diversité génétique (RAPD) de collections ex situ et populations naturelles de *Naufraga balearica* Constance & Cannon. *C. R. Acad. Paris, Sciences de la vie*, **323**, 399–406.

Friedland, A. J. (1990). The movement of metals through soils and ecosystems. In *Heavy Metal Tolerance in Plants: Evolutionary Aspects*, ed. A. J. Shaw, pp. 7–19. Boca Raton, FL: CRC Press.

Frye, R. J. (1996). Population viability analysis of *Pediocactus paradenei*. In *Southwest Rare and Endangered Plants*, eds. J. Maschinski, H. D. Hammond & L. Holter, pp. 39–46. Flagstaff, AZ: USDA Forest Service General Technical Report.

Fryer, J. D. & Chancellor, R. J. (1979). Evidence of changing weed populations in arable land. *Proceedings of the 14th British Weed Control Conference*, pp. 958–964.

Galton, F. (1871). Experiments in pangenesis, by breeding from rabbits of a pure variety, into whose circulation blood taken from other varieties had previously been largely infused. *Proceedings of the Royal Society of London*, **19**, 393–410.

Gartside, D. W. & McNeilly, T. (1974). The potential for evolution of metal tolerance in plants. III. Copper tolerance in normal populations of different species. *Heredity*, **32**, 335–348.

Gascon, C., Laurance, W. F. & Lovejoy, T. E. (2003). Forest fragmentation and biodiversity in central Amazonia. In *How Landscapes Change: Human Disturbance and Ecosytem Fragmentation in the Americas*, eds. G. A. Bradshaw & P. A. Marquet, pp. 33–48. Berlin: Springer Verlag.

Gaskin, J. F., Zhang, D.-Y. & Bon, M.-C. (2005). Invasion of *Lepidium draba* (Brassicaceae) in the western United States: distribution and origins of chloroplast DNA haplotypes. *Molecular Ecology*, **14**, 2331–2341.

Gaston, K. J. & May, R. M. (1992). Patterns in the numbers and distribution of taxonomists. *Nature*, **356**, 281–282.

Gauch, H. G. Jr. (2003). *Scientific Method in Practice*. Cambridge: Cambridge University Press.

Gawthrop, D. (1999). *Vanishing Halo: Saving the Boreal Forest*. Vancouver: Greystone.

Ge, S., Wang, K.-Q., Hong, D.-Y., Zhang, W.-H. & Zu, Y.-G. (1998). Comparisons of genetic diversity in the endangered *Adenophora lobophylla* and its widespread congener, *A. potaninii*. *Conservation Biology*, **13**, 509–513.

Ge, Y. Z., Cheng, X. F., Hopkins, A. & Wang, Z. Y. (2007). Generation of transgenic *Lolium temulentum* plants by *Agrobacterium tumefaciens*-mediated transformation. *Plant Cell Reports*, **26**, 783–789.

Geburek, T. (1997). Isozymes and DNA markers in gene conservation of forest trees. *Biodiversity and Conservation*, **6**, 1639–1654.

Gee, H. (1999). *In Search of Deep Time*. New York: The Free Press.

Gent, G. & Wilson, R. J. (1995). *The Flora of Northamptonshire and the Soke of Peterborough*. Kettering and District Natural History Society, Northants, UK.

George, K., Ziska, L. H., Bunce, J. A. & Quebedeaux, B. (2007). Elevated atmospheric CO_2 concentration and temperature across an urban-rural transect. *Atmospheric Environment*, **41**, 7654–7665.

Gepts, P. (2004). Crop domestication as a long-term selection experiment. *Plant Breeding Reviews*, **24**, 1–44.

Ghazoul, J. (2005a). Buzziness as usual? Questioning the global pollination crisis. *Trends in Ecology & Evolution*, **20**, 367–373.

Ghazoul, J. (2005b). Response to Steffan-Dewenter *et al.* questioning the global pollination crisis. *Trends in Ecology & Eolution*, **20**, 652–653.

Ghazoul, J. (2005c). Pollen and seed dispersal among dispersed plants. *Biological Reviews*, **80**, 413–443.

Ghersa, C. M., de la Fuente, E., Suarez, S. & Leon, R. J. C. (2002). Woody species invasion in the Rolling Pampa grasslands, Argentina. *Agriculture Ecosystems & Environment*, **88**, 271–278.

Gibbs, A. J. & Meischke, H. R. C. (1985). *Pests and Parasites as Migrants*. Cambridge: Cambridge University Press.

Giddings, G., Allison, G., Brooks, D. & Carter, A. (2000). Transgenic plants as factories for biopharmaceuticals. *Nature Biotechnology*, **18**, 1151–1155.

Gilbert, O. L. & Anderson, P. (1998). *Habitat Creation and Repair*. Oxford & New York: Oxford University Press.

Giles, J. (2007). From words to action. *Nature*, **445**, 578.

Giles, B. E. & Goudet, J. (1997). Genetic differentiation in *Silene dioica* metapopulations: estimation of spatiotemporal effects in a successional plant species. *American Naturalist*, **149**, 507–526.

Giles, B. E., Lundqvist, E. & Goudet, J. (1998). Restricted gene flow and subpopulation differentiation in *Silene dioica*. *Heredity*, **80**, 715–723.

Gill, T. E. (1996). Eolian sediments generated by anthropogenic disturbance of playas: human impacts on the geomorphic system and geomorphic impacts on the human system. *Geomorphology*, **17**, 207–228.

Gilmour, J. S. L. & Walters, S. M. (1963). Philosophy and classification. *Vistas in Botany*, **4**, 1–22.

Gilpin, M. E. & Soulé, M. E. (1986). Minimum viable populations: processes of species extinctions. In *Conservation Biology: The Science of Scarcity and Diversity*, ed. M. E. Soulé, pp. 19–34. Sunderland, MA: Sinauer Associates.

Given, D. R. (1994). *Principles and Practice of Plant Conservation*. London: Chapman & Hall.

Glasscock, R. (ed.) (1992). *Historic Landscape of Britain from the Air*. Cambridge: Cambridge University Press.

Glassman, S. F. (1971). A new hybrid palm from the Fairchild Tropical Garden. *Principes*, **15**, 79–88.

Glaubitz, J. C. Murrell, J. C. & Moran, G. F. (2003). Effects of native forest regeneration practices on genetic diversity in *Eucalyptus consideniana. Theoretical and Applied Genetics*, **107**, 422–431.

Gleick, P. H. (2003). Global freshwater resources; soft-path solutions for the 21st century. *Science*, **302**, 1524–1527.

Glen, W.. (ed.) (1994). *The Mass Extinction Debates*. Stanford, CA: Stanford University Press.

Glut, D. F. (2000). *Dinosaurs: The Encyclopedia*. Supplement 1. Jefferson, NC: McFarland & Co.

Godt, M. J. W., Hamrick, J. L. and Bratton, S. (1994). Genetic diversity in a threatened wetland species, *Helonias bullata* (Liliaceae). *Conservation Biology*, **9**, 596–604.

Gohre, V. & Paszkowski, U. (2006). Contribution of arbuscular mycorrhizal symbiosis to heavy metal phytoremediation. *Planta*, **223**, 1115–1122.

Goldblatt, P. (1980). Polyploidy in angiosperms: monocotyledons. In *Polyploidy*, ed. W. H. Lewis, pp. 219–239. New York & London: Plenum Press.

Gonzales-Martinez, S. C. *et al.* (2002). Seed gene flow and fine-scale structure in a Mediterranean pine (*Pinus pinaster* Ait.) using nuclear microsatellite markers. *Theoretical and Applied Genetics*, **104**, 1290–1297.

Goodall-Copestake, W. P. *et al.* (2005). Molecular markers and ex situ conservation of European elms (*Ulmus* spp.). *Biological Conservation*, **122**, 537–546.

Goodman, P. J. (1969). Intra-specific variation in mineral nutrition of plants from different habitats. In *Ecological Aspects of Mineral Nutrition of Plants*, ed. I. H. Rorison, pp. 237–253. Oxford: Blackwell.

Goodman. J. (1997). A collection of computer programs for calculating estimates of genetic differentiation from microsatellite data and determining their significance. *Molecular Ecology*, **6**, 881–885.

Gordon, D. R. & Rice, K. J. (1998). Patterns of differentiation in wire grass (*Aristida beyrichiana*): implications for restoration efforts. *Restoration Ecology*, **6**, 166–174.

Gornitz, V., Rosenzweig, D. & Hillel, D. (1997). Effects of anthropogenic intervention in the land hydrologic cycle on global sea level rise. *Global and Planetary Change*, **14**, 147–161.

Goudie, A. (1981). *The Human Impact: Man's Role in Environmental Change*. Oxford: Blackwell.

Goudriaan, J. *et al.* (1999). Use of models in global change studies. In *The Terrestrial Biosphere and Global Change*, eds. B. Walker, W. Steffen, J. Canadell & J. Ingram, pp. 106–140. Cambridge: Cambridge University Press.

Gould, S. J. & Eldredge, N. (1977). Punctuated equilibria: the tempo and mode of evolution reconsidered. *Palaeobiology*, **3**, 115–151.

Gould, S. J. & Eldredge, N. (1993). Punctuated equilibrium comes of age. *Nature*, **366**, 223–227.

Govindaraju, D. R. & Cullis, C. A. (1992). Ribosomal DNA variation among populations of a *Pinus rigida* Mill (Pitch pine) ecosystem. 1. Distribution of copy number. *Heredity*, **69**, 133–140.

Grabherr, G., Gottfried, M. & Pauli, H. (1994). Climate effects on mountain plants. *Nature*, **369**, 448.

Grant, V. (1981). *Plant Speciation*, 2nd edn. New York: Columbia University Press.

Grant, W. F. (1994). The current status of higher plant bioassays for the detection of environmental mutagens. *Mutation Research*, **310**. 175–185.
Grant, P. R. & Grant, B. R. (2002). Unpredictable evolution in a 30-year study of Darwin's finches. *Science*, **296**, 707–711.
Gray, A. J. (2002). The evolutionary context: a species perspective. In *Handbook of Ecological Restoration*, eds. M. R. Perrow & A. J. Davy, pp. 66–80. Cambridge: Cambridge University Press.
Grayson, D. K. (2008). Holocene overkill. *Proceedings of the National Academy of Sciences, USA*, **105**, 4077–4078.
Greally, J. M. (2007). Encyclopaedia of humble DNA. *Nature*, **447**, 782–783.
Green, S. (2003). A review of the potential for the use of bioherbicides to control forest weeds in the UK. *Forestry*, **76**, 285–298.
Gregor, J. W. & Sansome, F. W. (1927). Experiments on the genetics of wild populations. I. Grasses. *Journal of Genetics*, **17**, 349–364.
Greig, J. (1988). Some evidence of the development of grassland plant communities. In *Archaeology and the Flora of the British Isles. Human Influences on the Evolution of Plant Communities*, ed. M. Jones, Oxford University Committee for Archaeology Monograph 14. Botanical Society of the British Isles Conference Report 19, pp. 39–54. Oxford: Oxford University Committee for Archaeology.
Grenyer, R. *et al.* (2006). Global distribution and conservation of rare and threatened vertebrates. *Nature*, **444**, 93–96.
Gressel, J. (1991). Why get resistance? It can be prevented or delayed. In *Herbicide Resistance in Weeds and Crops*, eds. J. Caseley, G. W. Cussans & R. K. Atkin, pp. 1–25. Oxford: Butterworth Heinemann.
Gressel, J. (2005a). Introduction: the challenges of ferality. In *Crop Ferality and Volunteerism*, ed. J. Gressel, pp. 1–7. Boca Raton, FL: Taylor & Francis.
Gressel, J. (ed.) (2005b). *Crop Ferality and Volunteerism*. Boca Raton, FL: Taylor & Francis.
Greuter, W. (1994). Extinctions in the Mediterranean areas. *Philosophical Transactions of the Royal Society of London B*, **344**, 41–46.
Greuter, W., Burdet, H. & Long, G. (1986). *MedChecklist*, Vol. 3. Berlin & Geneva: OPTIMA.
Griffiths, M. (1994). *Index of Garden Plants*. London: Macmillan.
Grigg, D. B. (1974). *The Agricultural Systems of the World: An Evolutionary Approach*. Cambridge: Cambridge University Press.
Grignac P. (1978). The evolution of resistance to herbicides in weedy species. *Agro-ecosystems*, **4**, 377–385.
Grime, J. P., Hodgson, J. G. & Hunt, R. (1989). *Comparative Plant Ecology: A Functional Approach to Common British Species*. London: Unwin Hyman.
Groenendael, J. M. van (1986). Life history characteristics of two ecotypes of *Plantago lanceolata*. *Acta Botanica Neerlandica*, **35**, 71–86.
Grove, R. H. (1995). *Green Imperialism: Colonial Expansion, Tropical Island Edens, and the Origins of Environmentalism, 1600–1860*. Cambridge: Cambridge University Press.
Grove, A. T. & Rackham, O. (2001). *The Nature of Mediterranean Europe: An Ecological History*. New Haven, CT & London: Yale University Press.
Groves, R. H. (2006). Are some weeds sleeping? Some concepts and reasons. *Euphytica*, **148**, 111–120.

Grubb, P. J. (1977). Maintenance of species-richness in plant communities: importance of the regeneration niche. *Biological Reviews of the Cambridge Philosophical Society*, **52**, 107–145.

Gruber, S., Pekrun, C. & Claupein, W. (2004). Population dynamics of volunteer oilseed rape (*Brassica napus* L.) affected by tillage. *European Journal of Agronomy*, **20**, 351–361.

Gruen, L. & Jamieson, D. (1994). *Reflecting on Nature: Readings in Environmental Philosophy*. New York: Oxford University Press.

Guarino, L., Ramanatha Rao & Reid, R. (1995). *Collecting Plant Genetic Diversity: Technical Guidelines*. Wallingford, UK: CAB International.

Guerrant, E. O. Jr. (1992). Genetic and demographic considerations in the sampling and reintroduction of rare plants. In *Conservation Biology*, eds. P. L Fiedler & S. K. Jain, pp. 321–344. New York & London: Chapman & Hall.

Guerrant, E. O. & Fiedler, P. L. (2004). Accounting for sample decline during ex situ storage and reintroduction. In *Ex situ Plant Conservation: Supporting Species Survival in the Wild*, eds. E. O. Guerrant, K. Havens & M. Maunder, pp. 365–386. Washington, DC, Covelo, CA & London: Island Press.

Guerrant, E. O. Jr., Fiedler, P. L., Havens, K. & Maunder, M. (2004a). Revised genetic sampling guidelines for conservation collections of rare and endangered plants. In *Ex Situ Plant Conservation*, eds. E. O. Guerrant, K. Havens & M. Maunder, pp. 419–441. Washington, DC, Covelo, CA & London: Island Press.

Guerrant, E. O., Havens, K. & Maunder, M. (2004b). *Ex Situ Plant Conservation, Supporting Species Survival in the Wild*. Washington, DC, Covelo, CA & London: Island Press.

Gullison, R. E. (1998). Will bigleaf mahogany be conserved through sustainable use? In *Conservation of Biological Resources*, eds. E. J. Milner-Gulland & R. Mace, pp. 193–205. Oxford: Blackwell.

Gurevitch, J. & Padilla, D. K. (2004). Are invasive species a major cause of extinctions? *Trends in Ecology and Evolution*, **19**, 470–474.

Gurnell, A., Trockner, K., Edwards, P. & Petts, G. (2005). Effects of deposited wood on biocomplexity of river corridors. *Frontiers in Ecology and the Environment*, **3**, 377–382.

Gurney, M. (2000). Population genetics and conservation biology of *Primula elatior*. Ph.D. thesis, University of Cambridge.

Gustafsson, L. & Gustafsson, P. (1994). Low genetic variation in Swedish populations of the rare species *Vicia pisiformis* (Fabaceae) revealed with RFLP (rDNA) and RAPD. *Plant Systematics and Evolution*, **189**, 133–148.

Haddad, N. M. *et al.* (2003). Corridor use by diverse taxa. *Ecology*, **84**, 609–615.

Haeupler, H. & Schoenfelder, P. (eds.) (1989). *Atlas der Farn-, und Blutenpfanzen der Bundesrepublik Deutschland*. Stuttgard: E. Ulmer.

Hails, R. S. & Morley, K. (2005). Genes invading new populations: a risk assessment perspective. *Trends in Ecology & Evolution*, **20**, 245–252.

Hakam, N. & Simon, J. P. (2000). Molecular forms and thermal and kinetic properties of purified glutathione reductase from two populations of barnyard grass (*Echinochloa crus-galli* (L.) Beauv.: Poaceae) from contrasting climatic regions of North America. *Canadian Journal of Botany*, **78**, 969–980.

Håkansson, S. (1983). Competition and production in short-lived crop-weed stands: density effects. Uppsala: Department of Plant Husbandry. Report No. 127. Swedish University of Agricultural Sciences.

Hald, A. (1998). *A History of Mathematical Statistics from 1750 to 1930*. New York & Chichester: Wiley.

Haldane, J. B. S. (1932). *The Causes of Evolution*. London: Longmans.

Halka, O. & Halka, L. (1974). Polymorphic balance in small island populations of *Lythrum salicaria*. *Annales Botanici Fennici*, **11**, 267–270.

Hall, M. (2000). Comparing damages: Italian and American concepts of restoration. In *Methods and Approaches in Forest History*, eds. M. Agnoletti & S. Anderson, pp. 165–172. Wallingford, UK: CAB International.

Hallam, A. (1973). *A Revolution in the Earth Sciences: From Continental Drift to Plate Tectonics*. Oxford: Clarendon Press.

Hallam, A. (2002). Mass extinctions. In *Encyclopedia of Evolution*, ed. M. Pagel, pp. 661–668. Oxford: Oxford University Press.

Halpern, B. S. *et al.* (2008). A global map of human impact on marine ecosystems. *Science*, **319**, 948–952.

Hamann, O. (1984). Changes and threats to vegetation. In *Key Environments: Galapagos*, ed. R. Perry, pp. 115–131. Oxford: Pergamon Press.

Hamilton, M. B. (1994). Ex situ conservation of wild plant species: time to reassess the genetic assumptions and implications of seed banks. *Conservation Biology*, **8**, 39–49.

Hammer, K. (2003). A paradigm shift in the discipline of plant genetic resources. *Genetic Resources & Crop Evolution*, **50**, 3–10.

Hammond, P. M. (1992). Species inventory. In *Global Biodiversity. Status of the Earth's Living Resources*, ed. B. Groombridge, pp. 17–39. London: Chapman & Hall.

Hamrick, J. L. & Godt, M. J. W. (1989). Allozyme diversity in plant species. In *Plant Population Genetics, Breeding & Genetic Resources*, eds. A. H. D. Brown, M. T. Clegg, A. L. Kahler & B. S. Weir, pp. 43–63. Sunderland, MA: Sinaur.

Hancocks, H. (1994). Conservation genetics and the role of botanic gardens. In *Conservation Genetics*, eds. V. Loesche, J. Tomiuk & S. K. Jain, pp. 371–380. Basel: Birkhäuser-Verlag.

Hancocks, D. (2001). *A Different World: The Paradoxical World of Zoos and their Uncertain Future*. Los Angeles & London: University of California Press, Berkeley.

Hannah, L. *et al.* (2002). Conservation of biodiversity in a changing climate. *Conservation Biology*, **16**, 264–268.

Hannah, L., Lovejoy, T. E. & Schneider, S. H. (2005). Biodiversity and climate change in context. In *Climate Change and Biodiversity*, eds. T. E. Lovejoy & L. Hannah, pp. 3–14. New Haven, CT & London: Yale University Press.

Hansson, L., Fahrig, L. & Merriam, G. (eds.) (1995). *Mosaic Landscapes and Ecological Processes*. London: Chapman & Hall.

Hardin, G. (1968). The Tragedy of the Commons. *Science*, **162**, 1243–1248.

Harding, M. (1993). Redgrave and Lopham Fens, East Anglia: a case study of change in Flora and Fauna due to groundwater abstraction. *Biological Conservation*, **66**, 35–45.

Harlan, J. R. (1975). *Crops and Man*. Madison, WI: American Society of Agronomy and Crop Science.

Harlan, J. R. (1995). *The Living Fields: Our Agricultural Heritage*. Cambridge: Cambridge University Press.

Harlan, J. R. & de Wet, J. M. J. (1965). Some thoughts about weeds. *Economic Botany*, **19**, 16–24.

Harms, B., Knaapen, J. P. & Rademakers, J. G. (1993). Landscape planning for nature restoration: comparing regional scenarios. In *Landscape Ecology of a Stressed Environment*, eds. C. C. Vos & P. Opdam, pp. 197–218. London: Chapman & Hall.

Harper, J. L. (1956). The evolution of weeds in relation to resistance to herbicides. *Proceedings of the British Weed Control Conference*, **3**, 179.

Harper, J. L. (1977). *Population Biology of Plants*. London & New York: Academic Press.

Harper, J. L. (1981). The meanings of rarity. In *The Biological Aspects of Rare Plant Conservation*, ed. H. Synge, pp. 189–203. London: Wiley.

Harper, J. L. (1983). A Darwinian plant ecology. In *Evolution from Molecules to Man*, ed. D. S. Bendall, pp. 323–345. Cambridge: Cambridge University Press.

Harper, J. L. (1987). The heuristic value of ecological restoration. In *Restoration Ecology*, eds. W. R. Jordan III, M. E. Gilpin & J. D. Aber, pp. 35–52. Cambridge: Cambridge University Press.

Harris, T. M. (1946). Zinc poisoning of wild plants from wire netting. *New Phytologist*, **45**, 50–55.

Harris, D. R. (1989). An evolutionary continuum of people-plant interaction. In *Foraging and Farming: The Evolution of Plant Exploitation*, eds. D. R. Harris & G. C. Hillman, pp. 9–26. London & Boston: Unwin Hyman.

Harris, D. R. (1996). Domesticatory relationships of people, plants and animals. In *Redefining Nature: Ecology, Culture and Domestication*, pp. 437–463. Oxford & Washington, DC: Berg.

Harris, E. E. & Meyer, D. (2006). The molecular signature of selection underlying human adaptations. *Yearbook of Physical Anthropology*, **49**, 89–130.

Harrison, R. G. (1993). *Hybrid Zones and the Evolutionary Process*. Oxford & New York: Oxford University Press.

Harrison, S. (1994). Metapopulations and conservation. In *Large-scale Ecology and Conservation Biology*, eds. P. J. Edwards, R. M. May & N. R. Webb, pp. 111–128. Oxford: Blackwell Scientific.

Harry, I. B. & Clarke, D. D. (1986). Race-specific resistance in groundsel (*Senecio vulgaris*) infected with rust (*Erysiphe fischeri*). *New Phytologist*, **103**, 167–175.

Havens, K., Guerrant, E. O. Jr., Maunder, M. & Vitt, P. (2004). Guidelines for ex situ conservation management: minimizing risks, In *Ex Situ Plant Conservation*, eds. E. O. Guerrant, K. Havens & M. Maunder, pp. 454–473. Washington, DC, Covelo, CA & London: Island Press.

Havens, K. *et al.* (2006). Ex situ plant conservation and beyond. *Bioscience*, **56**, 525–531.

Hawkes, J. G. (1987). A strategy for seed banking in botanic gardens. In *Botanic Gardens and the World Conservation Strategy*, pp. 131–149. London: Academic Press for IUCN.

Hawkesworth, D. L. (1991). The fungal dimension of biodiversity: magnitude, significance, and conservation. *Mycological Research*, **95**, 441–456.

Hayward, I. M. & Druce, G. C. (1919). *Adventive Flora of Tweedside*. Arbroath, UK: Buncle.

Heap, I. & LeBaron, H. (2001). Introduction and overview of resistance. In *Herbicide Resistance and World Grains*, eds. S. B. Powles & D. L. Shaner, pp. 1–22. Boca Raton, FL: CRC Press.

Hedge, P., Kriwoken L. K. & Patten, K. (2003). A review of *Spartina* management in Washington State, US. *Journal of Aquatic Plant Management*, **41**, 82–90.

Hegarty, M. J. & Hiscock, S. J. (2005). Hybrid speciation in plants: new insights from molecular studies. *New Phytologist*, **165**, 411–423.

Hegarty, M. J. *et al.* (2006). Transcription shock after interspecific hybridization in Senecio is ameliorated by genome duplication. *Current Biology*, **16**, 1652–1659.

Heggestad, H. E. & Middleton, J. T. (1959). Ozone in high concentrations as a cause of tobacco injury. *Science*, **129**, 208–210.
Heller, N. E. & Zavaleta, E. S. (2009). Biodiversity management in the face of climate change: a review of 22 years of recommendations. *Biological Conservation*, **142**, 14–32.
Hendry, G. W. (1931) The adobe brick as a historical source. *Agricultural History*, **5**, 125.
Hendry, A. P., Nosil, P. & Rieseberg, L. H. (2007). The speed of ecological speciation. *Functional Ecology*, **21**, 455–464.
Hepper, F. N. (1989). *Plant Hunting for Kew*. London: HMSO.
Heslop-Harrison, J. & Lucas, G. (1978). Plant genetic resource conservation and ecosystem rehabilitation. In *The Breakdown and Restoration of Ecosystems*, eds. M. W. Holdgate & M. J. Woodman, pp. 297–306. New York & London: Plenum Press.
Hewitt, G. (2000). The genetic legacy of the Quaternary ice ages. *Nature*, **405**, 907–913.
Hewitt, G. M. & Nichols, R. A. (2005). Genetic and evolutionary impacts of climate change. In *Climate Change & Biodiversity*, eds. T. E. Lovejoy & L. Hannah, pp. 176–192. New Haven, CT & London: Yale University Press.
Heywood, V. H. (1976). The role of seed lists in Botanic gardens today. In *Conservation of Threatened Plants*, eds. J. B. Simmons, R. I. Beyer, P. E. Brandham, G. L. Lucas & V. T. H. Parry, pp. 225–231. New York & London: Plenum Press.
Heywood, V. H. (1978). *Flowering Plants of the World.* Oxford: Oxford University Press.
Heywood, V. H. (1983). Botanic gardens and taxonomy: their economic role. *Bulletin of the Botanical Survey of India*, **25**, 134–147.
Heywood, V. H. (1987). The changing rôle of the botanic garden. In *Botanic Gardens and the World Conservation Strategy*, pp. 3–18. London: Academic Press for IUCN.
Heywood, V. H. (1989). Patterns, extents and mode of invasion by terrestrial plants. In *Biological Invasions: A Global Perspective*, eds. J. A. Drake *et al.*, SCOPE 37, pp. 31–55. Chichester: Wiley.
Heywood, V. H. (1991). Developing a strategy for germplasm conservation in botanic gardens. In *Tropical Botanic Gardens: Their Role in Conservation and Development*, eds. V. H. Heywood & P. S. Wyse Jackson, pp. 11–23. London: Academic Press.
Heywood, V. H. & Stuart, S. N. (1992). Species extinctions in tropical forests. In *Tropical Deforestation*, eds. T. C. Whitmore & J. Sayer, pp. 91–117. London: Chapman & Hall.
Hickey, D. A. & McNeilly, T. (1975). Competition between metal tolerant and normal plant populations: a field experiment. *Evolution*, **29**, 458–464.
Hickling, R., Roy, D. B., Hill, J. K., Fox, R. & Thomas, C. D. (2006). The distributions of a wide range of taxonomic groups are expanding polewards. *Global Change Biology*, **12**, 450–455.
Hicks, C. R. & Turner, K. V. (1999). *Fundamental Concepts in the Design of Experiments*, 5th edn. Oxford & New York: Oxford University Press.
Hierro, J. L., Maron, J. L. & Callaway, R. M. (2005). A biogeographical approach to plant introductions: the importance of studying exotics in their introduced and native range. *Journal of Ecology*, **93**, 5–15.
Higgs, E. S. (1995). What is good ecological restoration? *Conservation Biology*, **11**, 338–348.
Hill, K. & Kaplan, H. (1989). Population and dry-season subsistence strategies of recently contracted Yora of Peru. *National Geographic Research*, **5**, 317–334.
Hillman, F. H. & Henry, H. H. (1928). The incidental seeds found in commercial seed of alfalfa and red clover. *Prodceedings of the International Seed Testing Association*, **6**, 1–20.

Hobbs, R. J., Saunders, D. A. & Arnold, G. W. (1993). Integrated landscape ecology: a western Australian perspective. *Biological Conservation*, **64**, 231–238.
Hodge, W. H. & Erlanson, C. O. (1956). Federal plant introductions: a review. *Economic Botany*, **10**, 299–334.
Hodkinson, D. J. & Thompson, K. (1997). Plant dispersal: the role of man. *Journal of Applied Ecology*, **34**, 1484–1496.
Holden, C. (2006). Report warns of looming pollination crisis in North America. *Science*, **314**, 397.
Holdgate, M. (1996). *From Care to Action*. London: Earthspan.
Holdgate, M. W. (1979). *A Perspective of Environmental Pollution*. Cambridge: Cambridge University Press.
Holdgate, M. & Philips, A. (1999). Protected areas in context. In *Integrated Protected Area Management*, eds. M. Walkey & I. Swingland, pp. 1–24. Boston, Dordrecht & London: Kluwer.
Hollingsworth, M. L. & Bailey, J. P. (2000). Evidence for massive clonal growth in the invasive weed *Fallopia japonica* (Japanese Knotweed). *Botanical Journal of the Linnean Society*, **133**, 463–472.
Holm, L., Pancho, J. V., Herberger, J. P. & Plucknett, D. L. (1977a). *A Geographic Atlas of World Weeds*. New York: Wiley.
Holm, L. G., Pluncknett, D. L., Pancho, J. V. & Herberger, J. P. (1977b). *The World's Worst Weeds: Distribution and Biology*. Hawaii: University of Honolulu Press.
Holsinger, K. E. (2000). Demography and extinction in small populations. In *Genetics, Demography and Viability of Fragmented Populations*, eds. A. G. Young & G. M. Clarke, pp. 55–74. Cambridge: Cambridge University Press.
Holsinger, R. F., Lewis, P. O. & Dey, D. K. (2002). A Bayesian approach to inferring population structure from dominant markers. *Molecular Ecology*, **11**, 1157–1164.
Holt, J. S. & LeBaron, H. M. (1990). Significance and distribution of herbicide resistance. *Weed Technology*, **4**, 141–149.
Holt, J. S., Powles, S. B. & Holtum, J. A. M. (1993). Mechanisms and agronomic aspects of herbicide resistance. *Annual Review of Plant Physiology and Plant Molecular Biology*, **44**, 203–229.
Holttum, R. E. (1970). The historical significance of botanic gardens in S.E. Asia. *Taxon*, **19**, 707–714.
Holzner, W. (1982). Concepts, categories and characteristics of weeds. In *Biology and Ecology of Weeds*, eds. W. Holzner & N. Numata, pp. 3–20. The Hague: Junk.
Holzner, W. & Numata, M. (1982). *Biology & Ecology of Weeds*. The Hague: Junk.
Hondelmann, W. (1976). Seed banks. In *Conservation of Threatened Plants*, eds. J. B. Simmons, R. I. Beyer, P. E. Brandham, G. L. Lucas & V. T. H. Parry, pp. 213–224. New York & London: Plenum Press.
Hong, T. D. & Ellis, R. H. (1996). *A Protocol to Determine Seed Storage Behaviour*. Rome: International Plant Genetic Resources Institute.
Hong, T. D., Linington, S. & Ellis, R. H. (1998). *Compendium of Information on Seed Storage Behaviour*, 2 volumes. Kew: Royal Botanic Garden.
Hood, W. G. & Naiman, R. J. (2000). Vulnerability of riparian zones to invasion by exotic vascular plants. *Plant Ecology*, **148**, 105–114.
Hopkins, I. (1914). History of the bumblebee in New Zealand: its introduction and results. *Bulletin of the New Zealand Department of Agriculture (New Series)*, **46**, 1–28.

Hopper, S. D. (1996). The use of genetic information in establishing reserves for nature conservation. In *Biodiversity in Managed Landscapes*, eds. R. C. Szaro & D. W. Johnston, pp. 253–260. Oxford & New York: Oxford University Press.

Horsman, D. C., Roberts, T. M., Lambert, M. & Bradshaw, A. D. (1979a). Studies on the effect of sulphur dioxide on perennial rye grass (*Lolium perenne* L). 1. Characteristics of fumigation system and preliminary experiments. *Journal of Experimental Botany*, **30**, 485–493.

Horsman, D. C., Roberts, T. M. & Bradshaw, A. D. (1979b). Studies on the effect of sulphur dioxide on perennial rye grass (*Lolium perenne* L). II. Evolution of sulphur dioxide tolerance. *Journal of Experimental Botany*, **30**, 495–501.

Hosius, B., Leinemann, L., Konnert, M. & Bergmann, F. (2006). Genetic aspects of forestry in the central Europe. *European Journal of Forest Research*, **125**, 407–417.

Hoskin, G. W. (1977). *The Making of the English Landscape*: London: Hodder & Stoughton.

Hufford, K. M. & Mazer, S. J. (2003). Plant ecotypes: genetic differentiation in the age of ecological restoration. *Trends in Ecology & Evolution*, **18**, 147–155.

Hughes, L. (2000). Biological consequences of global warming: is the signal already apparent? *Trends in Ecology and Evolution*, **15**, 56–61.

Hughes, C. E. & Styles, B. T. (1989). The benefits and risks of woody legume introductions. *Advances in Legume Biology. Systematic Monograph of the Botanic Garden of Missouri*, **29**, 505–531.

Hulbert, L. C. (1955). Ecological studies of *Bromus tectorum* and other annual bromegrasses. *Ecological Monographs*, **25**, 181–213.

Hulme, P. E. (2005). Adapting to climate change: is there scope for ecological management in the face of a global threat? *Journal of Applied Ecology*, **42**, 784–794.

Humborg, C., Ittekkot, V., Cociasu, A. & von Bodungen, B. (1997). Effect of Danube river dam on Black Sea biogeochemistry and ecosystem structure. *Nature*, **386**, 385.

Hunt, D. R. (1974). The role of reserve collections. In *Succulents in Peril*, ed. D. R. Hunt, pp. 17–20. Kew: XII Congress of the International Organisation for Succulent Plant Study.

Hunter, R. (1890). The preservation and enjoyment of open spaces. *Nature Note*, **1**, 101–104.

Hunter, J. (1995). *On the Other Side of Sorrow: Nature and Peoples in the Scottish Highlands*. Edinburgh: Mainstream.

Hunter, J. (2000). *The Making of the Crofting Community*, 2nd edn. Edinburgh: John Donald.

Huntley, B. (1991). How plants respond to climate change: migration rates, individualism and the consequences for plant communities. *Annals of Botany*, **67** (suppl.), 15–22.

Hurka, H. (1994). Conservation genetics and the role of botanical gardens. In *Conservation Genetics*, eds. V. Loeschcke, J. Tomiuk & S. K. Jain, pp. 371–380. Basel, Boston & Berlin: Birkhäuser Verlag.

Husband, B. C. & Schemske, D. W. (1996). Evolution of the magnitude and timing of inbreeding depression in plants. *Evolution*, **50**, 54–70.

Husheer, S. W. & Frampton, C. M. (2005). Fallow deer impacts on Wakatipu beech forest. *New Zealand Journal of Ecology*, **29**, 83–94.

Hutchings, M. J. (1989). Population biology and conservation of *Ophrys sphegodes*. In *Modern Methods of Orchid Conservation*, ed. H. W. Pritchard, pp. 101–115. Cambridge: Cambridge University Press.

Huxley, T. H. (1906). The conditions of existence as affecting the perpetuation of living beings. In *Man's Place in Nature and Other Essays*, pp. 225–244. London: Dent & Sons.

Huxley, J. S. (1942). *Evolution: The Modern Synthesis*. Oxford: Clarendon Press.

Hyam, R. (1998). Field collection: plants. In *Molecular Tools for Screening Biodiversity: Plants and Animals*, eds. A. Karp, P. G. Isaac & D. S. Ingram, pp. 49–50. London: Chapman & Hall.

Hymowitz, T. (1984). Dorsett-Morse soybean collection trip to east Asia: a 50 year retrospective. *Economic Botany*, **38**, 378–388.

Ibáñez, O., Calero, C., Mayol, M. & Rosselló, J. A. (1999). Isozymic uniformity in a wild extinct insular plant, *Lysimachia minoricensis* J. J. Rodr. (Primulaceae). *Molecular Ecology*, **8**, 813–817.

Imam, A. G. & Allard, R. W. (1965). Population studies in predominantly self-pollinated species. VI. Genetic variability between and within natural populations of wild oats from differing habitats in California. *Genetics*, **51**, 49–62.

Imper, D. K. (1997). Ecology and conservation of Wolf's evening primrose in northwestern California. In *Conservation and Management of Native Plants and Fungi*, eds. T. N. Kaye *et al.*, pp. 34–40. Corvallis, OR: Native Plant Society of Oregon.

Imrie, B. C., Kirkman, C. J. & Ross, D. R. (1972). Computer simulation of a sporophytic self-incompatible breeding system. *Australian Journal of Biological Sciences*, **25**, 343–349.

Ingram, R. & Noltie, H. J. (1995). *Senecio cambrensis* Rosser. *Journal of Ecology*, **83**, 537–546.

Ingvarsson, P. K. & Giles, B. E. (1999). Kin-structured colonization and small-scale genetic differentiation in *Silene dioica*. *Evolution*, **53**, 605–611.

IPCC (2007a). *Fourth Assessment Report, February 2007. Working Group I. The Physical Science Basis*. Contribution to the Intergovernmental Panel on Climate Change (IPCC). Summary for Policymakers available on the internet, www.ipcc.ch/index.html.

IPCC (2007b). *Fourth Assessment Report, April 2007. Working Group III. Mitigation of Climate Change*. Contribution to the Intergovernmental Panel on Climate Change (IPCC). Summary for Policymakers available on the internet, www.ipcc.ch/index.html.

IPCC (2007c). *Fourth Assessment Report, 4 May 2007. Working Group III. Mitigation of Climate Change*. Contribution to the Intergovernmental Panel on Climate Change (IPCC). Summary for Policymakers available on the internet, www.ipcc.ch/index.html.

IPCC (2007d). *Fourth Assessment Report, 16 November 2007. Final Assessment*. Draft summary for policy makers available on the internet, www.ipcc.ch/index.html.

Jachmann, H., Berry, P. S. M. & Imae, H. (1995). Tusklessness in African elephants: a future trend. *African Journal of Ecology*, **33**, 230–235.

Jackson, J. B. C. (2001). What was natural in the coastal oceans? *Proceedings of the National Academy of Sciences, USA*, **98**, 5411–5418.

Jacobsen, T. & Adams, R. M. (1958). Salt and silt in ancient Mesopotamian agriculture. *Science*, **128**, 1251–1258.

Jaeger, P. (1963). Premièries observations sur les ecotypes de l'*Heracleum sphondylium* L. (Ombellifère). *Comptes Rendues de l'Academie des Sciences, Paris*, **257**, 1147–1149.

Jaeger, K. E., Graf, A. & Wigge, P. A. (2006). The control of flowering in time and space. *Journal of Experimental Botany*, **57**, 3415–3418.
Jahodova, S. *et al.* (2007). Invasive species of *Heracleum* in Europe: an insight into genetic relationships and invasion history. *Diversity & Distributions*, **13**, 99–114.
Jain, S. K. & Martins, P. S. (1979). Ecological genetics of the colonizing ability of rose clover (*Trifolium hirtum* All.). *American Journal of Botany*, **66**, 361–366.
Jäkäläniemi, A., Tuomi, J., Siikamäki, P. & Kilpiä, A. (2005). Colonization-extinction and patch dynamic of the perennial riparian plant, *Silene tatarica*. *Journal of Ecology*, **93**, 670–680.
James, J. K. & Abbott, R. J. (2005). Recent, allopatric, homoploid hybrid speciation: the origin of *Senecio squalidus* (Asteraceae) in the British Isles from a hybrid zone on Mount Etna, Sicily. *Evolution*, **59**, 2533–2547.
Jana, S. & Thai, K. M. (1987). Patterns of changes of dormant genotypes in *Avena fatua* populations under different cultural conditions. *Canadian Journal of Botany*, **65**, 1741–1745.
Janzen, D. H. (2001). Latent extinction: the Living Dead. In *Encyclopedia of Biodiversity*, Vol. 3, ed. S. A. Levin, pp. 689–699. London: Academic Press.
Jasieniuk, M., Brule-Babel, A. L. & Morrison, I. N. (1996). The evolution and genetics of herbicide resistance in weeds. *Weed Science*, **44**, 176–193.
Jennersten, O. (1988). Pollination in *Dianthus deltoides* (Caryophyllaceae): effects of habitat fragmentation on visitation and seed set. *Conservation Biology*, **2**, 359–366.
Jewgenow, K., Dehnhard, M., Hildebrandt, T. B. & Goritz, F. (2006). Contraception for population control in exotic carnivores. *Theriogenolgy*, **66**, 1525–1529.
Johannsen, W. (1909). *Elemente der exakten Erblichkeitlehre*. Jena: Fischer.
John, D. M. (1994). Biodiversity and conservation: an algal perspective. *The Phycologist*, **38**, 3–15.
Johnstone, I. M. (1986). Plant invasion windows: a time-based classification of invasion potential. *Biological Reviews*, **61**, 369–394.
Jones, M. E. (1971a). The population genetics of *Arabidopsis thaliana*. I. The breeding system. *Heredity*, **27**, 39–50.
Jones, M. E. (1971b). The population genetics of *Arabidopsis thaliana*. II. Population structure, *Heredity*, **27**, 51–58.
Jones, M. E. (1971c). The population genetics of *Arabidopsis thaliana*. III. The effect of vernalisation. *Heredity*, **27**, 59–72.
Jones, S. (1999). *Almost Like a Whale: The Origin of Species Updated*. London & New York: Doubleday.
Jones, C. J. *et al.* (1997). Reproducibility testing of RAPD, AFLP and SSR markers in plants by a network of European laboratories. *Molecular Breeding*, **3**, 381–390.
Jones, C. J. *et al.* (1998). Reproducibility testing of RAPDs by a network of European laboratories. In *Molecular Tools for Screening Biodiversity*. eds. A. Karp, P. G. Isaac & D.S. Ingram. London: Chapman Hall.
Jones, T. L. *et al.* (2008). The protracted Holocene extinction of California's flightless sea duck (*Chendytes lawi*) and its implications for the Pleistocene overkill hypothesis. *Proceedings of the National Academy of Sciences, USA*, **105**, 4105–4108.
Jongman, R. H. & Pungetti, G. (2004). *Ecological Networks and Green Ways: Concept, Design, Implementation*. Cambridge: Cambridge University Press.
Jordan, C. F. & Miller, C. (1996). Scientific uncertainty as a constraint to environmental problem solving: large-scale ecosystems. In *Scientific Uncertainty and*

Environmental Problem Solving, ed. J. Lemons, pp. 91–117. Cambridge, MA: Blackwell.

Jordan III, W. R., Gilpin, M. E. & Aber, J. D. (1987). *Restoration Ecology*. Cambridge: Cambridge University Press.

Jørgensen, R. B. & Andersen, B. (1994). Spontaneous hybridization between oilseed rape (*Brassica napus*) and weedy *Brassica campestris* (Brassicaceae): a risk of growing genetically-modified oilseed rape. *American Journal of Botany*, **81**, 1620–1626.

Jump, A. S. & Peñuelas, J. (2005). Running to stand still: adaptation and the response of plants to rapid climate change. *Ecology Letters*, **8**, 1010–1020.

Jump, A. S., Hunt, J. M., Martinez-Izquierdo, J. A. & Peñuelas, J. (2006). Natural selection and climate change: temperature-linked spatial and temporal trends in gene frequency in *Fagus sylvatica. Molecular Ecology*, **15**, 3469–3480.

Kadereit, J. W. (1984). Studies on the biology of *Senecio vulgaris* L. ssp. denticulatus (OF-Muell) PD Sell. *New Phytologist*, **97**, 681–689.

Kadereit, J. W. & Briggs, D. (1985). Speed of development of radiate and non-radiate plants of *Senecio vulgaris* L. from habitats subject to different degrees of weeding pressure. *New Phytologist*, **99**, 155–169.

Kalisz, S., Horth, L. & McPeek, M. A. (2000). Fragmentation and the role of seed banks in promoting persistence in isolated populations of *Collinsia verna*. In *Conservation in Highly Fragmented Landscapes*, ed. M. W. Schwartz, pp. 286–312. New York: Chapman & Hall.

Kareiva, P. *et al.* (2007). Domesticated nature: shaping landscapes and ecosystems for human welfare. *Science*, **316**, 1866–1869.

Karlsson, T. (1976). *Euphrasia* in Sweden: hybridization, parallelism and species concept. *Botaniska Notiser*, **129**, 49–60.

Karlsson, T. (1984). Early flowering taxa of *Euphrasia* (Scrophulariaeae) on Gotland, Sweden. *Nordic Journal of Botany*, **4**, 303–326.

Karp, A., Isaac, P. G. & Ingram, D. S. (1998). *Molecular Tools for Screening Biodiversity*. London: Chapman Hall.

Kautz, R. *et al.* (2006). How much is enough? Landscape-scale conservation for the Florida panther. *Biological Conservation*, **130**, 118–133.

Keane, R. M. & Crawley, M. J. (2002). Exotic plant invasions and the enemy release hypothesis. *Trends in Ecology and Evolution*, **17**, 164–170.

Keith, D. A. (2004). Australian Heath Shrub (*Epacris barbata*): viability under management options for fire and disease. In *Species Conservation and Management*, ed. H. R. Akçakaya. *et al.*, pp. 90–103. New York: Oxford University Press.

Keitt, T. H. (2003). Spatial autocorrelation, dispersal and the maintenance of source-sink populations. In *How Landscapes Change: Human Disturbance and Ecosystem Fragmentation in the Americas*, eds. G. A. Bradshaw & P. A. Marquet, pp. 225–238. Berlin: Springer Verlag.

Keller, M., Kollmann, J. & Edwards, P. J. (2000). Genetic introgression from distant provenances reduces fitness in local weed populations. *Journal of Applied Ecology*, **37**, 647–659.

Kellogg, V. L. (1907). *Darwinism Today*. London: Bell.

Kelly, M. (2000). *Cynara cardunculus* L. In *Invasive Plants of California's Wildlands*, eds. C. C. Bossard, J. M. Randall & M. C. Hozhovsky, pp. 139–145. Berkeley, CA: University of California Press.

Kemp, W. B. (1937). Natural selection within plant species as exemplified in a permanent pasture. *Journal of Heredity*, **28**, 329–333.

Kendrew, J. (ed.) (1994). Fitness. In *The Encyclopedia of Molecular Biology*, ed. J. Kendrew, p. 377. Oxford: Blackwell Science.

Kennington, W. J., Waycott, M. & James, S. H. (1996). DNA fingerprinting supports notions of clonality in a rare mallee, *Eucalyptus argutifolia. Molecular Ecology*, **5**, 693–696.

Kent, D. H. (1975). *The Historical Flora of Middlesex*. London: Ray Society.

Kerr, R. A. (2005). Confronting the bogeyman of the climate system. *Science*, **310**, 432–433.

Kiang, Y. T. (1982). Local differentiation of *Anthoxanthum odoratum* L. populations on roadsides. *American Midland Naturalist*, **107**, 340–350.

Kiang, Y. T., Antonovics, J. & Wu, L. (1979). The extinction of wild rice (*Oryza perennis formosana*) in Taiwan. *Journal of Asian Ecology*, **1**, 1–9.

Kimura, M. (1983). *The Neutral Theory of Molecular Evolution*. Cambridge: Cambridge University Press.

King, L. J. (1966). *Weeds of the World: Biology and Control*. New York: Interscience Publishers; London: Leonard Hill.

Kirby, C. (1980). *The Hormone Weedkillers*. Croydon, UK: BCPC Publications.

Kirch, P. V. & Hunt, T. L. (eds.) (1997). *Historical Ecology in the Pacific Islands*. New Haven, CT & London: Yale University Press.

Kirkpatrick, J. B. (1989). *A Continent Transformed: Human Impact on the Natural Vegetation of Australia*, 2nd edn. Melbourne: Oxford University Press.

Kirkpatrick, J. & Gilfedder, L. (1995). Maintaining integrity compared with maintaining rare and threatened taxa in remnant bushland in subhumid Tasmania. *Biological Conservation*, **74**, 1–8.

Kirkpatrick, J. B., McDougall, K. & Hyde, M. (1995). *Australia's Most Threatened Ecosystem: The Southeastern Lowland Native Grasslands*. Chipping Norton, NSW: Surrey Beatty.

Knobloch, I. W. (1971). Intergeneric hybridization in flowering plants. *Taxon*, **21**, 97–103.

Kondrashov, A. S. (2005). Fruitfly genome is not junk. *Nature*, **437**, 1106.

Koopowitz, H. & Kaye, H. (1990). *Plant Extinction: A Global Crisis*. London: Christopher Helm.

Kornás, J. (1961). The extinction of the association Sperguleto–Lolietum remoti in flax cultures in the Gorce (Polish Western Carpathian Mountains). *Bulletin de l'Academie Polonaise des Sciences. Class II*, **IX**, 37–40.

Kovalchuk, I., Kovalchuk, O., Arkhipov, A., Hohn, B. & Dubrova, Y. E. (2003). Extremely complex pattern of microsatellite mutation in the germline of wheat exposed to the post-Chernobyl radioactive contamination. *Mutation Research*, **525**, 93–101.

Kovarik, A. *et al.* (2005). Rapid concerted evolution of nuclear ribosomal DNA in two tragopogon allopolyploids of recent and recurrent origin. *Genetics*, **169**, 931–944.

Kowarik, I. (2005). Urban ornamentals escaped from cultivation. In *Crop Ferality and Volunteerism*, ed. J. Gressel, pp. 97–121. Boca Raton, FL: Taylor & Francis.

Kraus, G. K. M. (1894). Gesichte der Pflanzeneinführung in der europäishen Gärten. *Leipzig*.

Krause, J. (1944). Studien über den Saisondimorphismus bei Pflanzen. *Beiträge zur Biologie der Pflanzen*, **27**, 1–91.

Krauss, S. L., Dixon, B. & Dixon, K. W. (2002). Rapid genetic decline in a translocated population of the endangered plant *Grevillea scapigera. Conservation Biology*, **16**, 986–994.

Krech, S. (1999). *The Ecological Indian: Myth & History*. New York & London: Norton.

Kreitman, M. & Di Rienzo, A. (2004). Balancing claims for balancing selection. *Trends in Genetics*, **20**, 300–304.

Kremen, C. & Ricketts, T. (2000). Global perspectives on pollination disruptions. *Conservation Biology*, **14**, 1226–1228.

Kremen, C. *et al.* (2007). Pollination and other ecosytem services produced by mobile organisms: a conceptual framework for the effects of land-use change. *Ecology Letters*, **10**, 299–314.

Kuhn, T. (1970). *The Structure of Scientific Revolutions*. Chicago: University of Chicago Press.

Kuparinen, A. & Schurr, F. M. (2007). A flexible modelling framework linking the spatiotemporal dynamics of plant genotypes and populations: application to gene flow from transgenic forests. *Ecological Modelling*, **202**, 476–486.

Kutner, L. S. & Morse, L. E. (1996). Reintroduction in a changing climate. In *Restoring Diversity*, eds. D. A. Falk, C. I. Millar & M. Olwell, pp. 23–48. Washington, DC: Island Press.

Ladizinsky, G. (1998). *Plant Evolution under Domestication*. Dordrecht, Boston & London: Kluwer Academic Publishers.

Lahaye, R. *et al.* (2008). DNA barcoding the floras of biodiversity hotspots. *Proceedings of the National Academy of Sciences, USA*, **105**, 2923–2928.

LaHaye, W. S., Gutierrez, R. J. & Akçakaya, H. R. (1994). Spotted owl metapopulation dynamics in southern California. *Journal of Animal Ecology*, **63**, 775–785.

Laimer, M. *et al.* (2005). Biotechnology of temperate fruit trees and grapevines. *Acta Biochimica Polonica*, **52**, 673–678.

Laland, K. N. (2002). Niche construction. In *Encyclopedia of Evolution*, ed. M. Patel, pp. 821–823. Oxford & New York: Oxford University Press.

Laland, K. N., Odling-Smee, F. J. & Feldman, M. W. (1999). Evolutionary consequences of niche construction and their implications for ecology. *Proceedings of the National Academy of Sciences, USA*, **96**, 10242–10247.

Lambrecht, S. C., Loik, M. E., Inouve, D. W. & Harte, J. (2007). Reproduction and physiological responses to simulated climate warming for four subalpine species. *New Phytologist*, **173**, 121–134.

Lampkin, N. (1990). *Organic Farming*. Ipswich, UK: Farming Press.

Lande, R. (1988). Genetics and demography in biological conservation. *Science*, **241**, 1455–1460.

Langmead, C. (1995). *A Passion for Plants. From the Rainforests of Brazil to Kew Gardens. The Life and Vision of Ghillean Prance*. Oxford: Lion Publishing.

Laurance, W. F. (1991). Edge effects in tropical forest fragments: application of a model for the design of nature reserves. *Biological Conservation*, **57**, 205–219.

Laurance, W. F. (1998). A crisis in the making: responses of Amazonian forest to land use and climate change. *Trends in Ecology and Evolution*, **13**, 411–415.

Lavergne, S. & Molofsky, J. (2007). Increased genetic variation and evolutionary potential drive the success of an invasive grass. *Proceedings of the National Academy of Sciences, USA*, **104**, 3883–3888.

Lawrence, M. J. (2002). A comprehensive collection and regeneration strategy for ex situ conservation. *Genetic Resources & Crop Evolution*, **49**, 199–209.

Lebaron, H. M. & Gressel, J. (eds.) (1982). *Herbicide Resistance in Plants*. New York: Wiley.

Leck, M. A., Parker, V. T. & Simpson, R. L. (1989). *Ecology of Soil Seed Banks*. San Diego, CA: Academic Press.

Ledig, F. T. (1992). Human impacts on genetic diversity in forest ecosystems. *Oikos*, **63**, 87–108.

Ledwig, F. T. (1996). *Pinus torreyana* at the Torrey Pines State Reserve, California. In *Restoring Diversity*, eds. D. A. Falk, C. I. Millar & M. Olwell, pp. 265–271. Washington, DC, Covelo, CA & London: Island Press.

Lee, R. B. (1984). *The Dobe !Kung*. Chicago: Aldine.

Lee, C. E. (2002a). Evolutionary genetics of invasive species. *Trends in Ecology and Evolution*, **17**, 386–392.

Lee, W. G. (2002b). Negative effects of introduced plants. *Encyclopedia of Biodiversity*, **3**, 501–515.

Lee, C. T., Wickneswari, R., Mahani, M. C. & Zakri, A. H. (2002). Effect of selective logging on the genetic diversity of *Scaphium macropodum*. *Biological Conservation*, **104**, 107–118.

Lehmann, E. (1944). *Veronica filiformis* Sm. eine selbststerile Planzen. *Jahrbücher für wissenschaftliche Botanik*, **91**, 395–403.

Leopold, A. (1933). *Game Management*. New York: Charles Scriber's Sons.

Les, D. H., Reinartz, J. A. & Esselman, E. J. (1991). Genetic consequences of rarity in *Aster furcatus* (Asteraceae), a threatened, self-incompatible plant. *Evolution*, **45**, 1641–1650.

Lesica, P. & Allendorf, F. W. (1999). Ecological genetics and the restoration of plant communities: mix or match? *Restoration Ecology*, **7**, 42–50.

Lever, C. (1992). *They Dined on Eland: The Story of Acclimatisation Societies*. London: Quiller.

Levey, D. J. *et al.* (2005). Effects of landscape corridors on seed dispersal by birds. *Science*, **309**, 146–148.

Levin, D. A. (1976). Consequences of long-term artificial selection, inbreeding and isolation in Phlox. II. The organization of allozymic variability. *Evolution*, **30**, 463–472.

Levin, D. A. (2000). *The Origin, Expansion, and Demise of Plant Species*. New York & Oxford: Oxford University Press.

Levin, D. A. & Kerster, H. W. (1974). Gene flow in plants. *Evolutionary Biology*, **7**, 139–220.

Levin, J. M., Vilà, M., D'Antonia, C. M., Dukes, J. S., Grigulis, K. & Lavorel, S. (2003). Mechanisms underlying the impacts of exotic plant invasions. *Proceedings of the Royal Society of London, B*, **270**, 775–781.

Levins, D. (1969). Some demographic and genetic consequences of environmental heterogeneity for biological control. *Bulletin of the Entomological Society of America*, **7**, 237–240.

Lewis, W. H. (1980). Polyploidy in Angiosperms: dicotyledons. In *Polyploidy*, ed. W. H. Lewis, pp. 241–268. New York & London: Plenum Press.

Li, Q., Xu, Z. & He, T. (2002). Ex situ conservation of endangered *Vatica guangxiensis* (Dipterocarpaceae) in China. *Biological Conservation*, **106**, 151–156.

Lichtenberger, E. (1988). The succession of an agricultural society to a leisure society: the high mountains of Europe. In *Human Impact on Mountains*, eds. N. J. R. Allan, G. W. Knapp & C. Stadel, pp. 218–227. Lanham, MD: Rowman & Littlefield.

Liebman, M., Mohler, C. L. & Staver, C. P. (2001). *Ecological Management of Agricultural Weeds*. Cambridge: Cambridge University Press.

Linder, C. R. *et al.* (1998). Long-term introgression of crop genes into wild sunflower populations. *Theoretical and Applied Genetics*, **96**, 339–347.

Ling Hwa, T. & Morishima, H. (1997). Genetic characteristics of weedy rices and the inference on their origins. *Breeding Science*, **47**, 153–160.
Linington, S. (2001). The Millennium seed bank project. In *Biological Collections and Biodiversity*, eds. B. S. Rushton, P. Hackney & C. R. Tyrie, pp. 121–125. Otley, UK: Linnean Society of London, Westbury Publishing.
Linington, S. H. & Pritchard, H. W. (2001). Gene banks. In *Encyclopedia of Biodiversity*, Vol. 3, ed. S. A. Levin, pp. 165–181. San Diego & London: Academic Press.
Lock, J. M., Friday, L. E. & Bennett, T. J. (1997). The management of the Fen. In *Wicken Fen. The Making of a Wetland Nature Reserve*, ed. L. E. Friday, pp. 213–254. Colchester: Harley Books.
Lodge, R. W. (1964). Autecology of *Cynosurus cristatus* (L.) IV. Germinability of *Cynosurus cristatus*. *Journal of Ecology*, **52**, 43–52.
Lofflin, D. L. & Kephart, S. R. (2005). Outbreeding, seedling establishment and maladaption in natural reintroduced populations of rare and common *Silene douglasii* (Caryophyllaceae). *American Journal of Botany*, **92**, 1691–1700.
Lonsdale, W. M. (1993). Rates of spread of an invading species: *Mimosa pigra* in northern Australia. *Journal of Ecology*, **81**, 513–521.
Lonsdale, W. M. (1994). Inviting trouble: introduced pasture species in northern Australia. *Australian Journal of Ecology*, **19**, 345–354.
Lourmas, M., Kjellberg, F., Dessard, H., Joly, H. I. & Chevallier, M.-H. (2007). Reduced density due to logging and its consequences on mating system and pollen flow in the African mahogany *Entandrophragma cylindricum*. *Heredity*, **99**, 151–160.
Lovejoy, T. E. & Hannah, L. (eds.) (2005). *Climate Change & Biodiversity.* New Haven & London: Yale University Press.
Lovelock, J. (2006). *The Revenge of Gaia*. London: Allen Lane.
Low, T. (2002a). *Feral Future: The Untold Story of Australia's Exotic Invaders*. Chicago: University of Chicago Press.
Low, T. (2002b). *The New Nature: Winners and Losers in Wild Australia*. Victoria, Australia: Viking.
Lowe, A. J. & Abbott, R. J. (1996). Origins of the new allopolyploid species *Senecio cambrensis* (Asteraceae) and its relationship to the Canary Islands endemic *Senecio teneriffae*. *American Journal of Botany*, **83**, 1365–1372.
Lowe, A. J. & Abbott, R. J. (2004). Reproductive isolation of a new hybrid species, *Senecio eboracensis* Abbott & Lowe (Asteraceae). *Heredity*, **92**, 386–395.
Lowenthal, D. (1985). *The Past is a Foreign Country*. Cambridge: Cambridge University Press.
Lucas, G. & Synge, H. (1978). *The IUCN Plant Red Data Book*. Morges, Switzerland: Threatened Plant Committee, Kew & IUCN.
Lumaret, R. & Ouazzani, N. (2001). Ancient wild olives in Mediterranean forests. *Nature*, **413**, 700.
Lush, W. M. (1988a). Biology of *Poa annua* in a temperate zone golf putting green (*Agrostis stolonifera/Poa annua*). I. The above-ground population. *Journal of Applied Ecology*, **25**, 977–988.
Lush, W. M. (1988b). Biology of *Poa annua* in a temperate zone golf putting green (*Agrostis stolonifera/Poa annua*). II. The seed bank. *Journal of Applied Ecology*, **25**, 989–997.
Lynas, M. (2007). *Six Degrees*. London: Fourth Estate.
Lyons, T. M., Barnes, J. D. & Davison, A. W. (1997). Relationships between ozone resistance and climate in European populations of *Plantago major*. *New Phytologist*, **136**, 503–510.

Mabey, R. (1980). *The Common Ground. A Place for Nature in Britain's Future*. London: Hutchinson.
MacArthur, R. & Wilson, E. O. (1967). *The Theory of Island Biogeography*. Princeton, NJ: Princeton University Press.
Macdonald, J. A. W. (1988). The history, impacts and control of introduced species to the Kruger National Park, South Africa. *Transactions of the Royal Society of South Africa*, **46**, 252–276.
Macdonald, J. A. W., Ortiz, L. & Lawesson, J. E. (1988). The invasion of the highlands of the Galápagos Islands by the red quinine tree *Cinchona succirubra*. *Environmental Conservation*, **15**, 215–220.
Mace, R. (2002). Demographic transition. In *Encyclopedia of Evolution*, ed. M. Pagel, Vol. 1, pp. 235–238. Oxford: Oxford University Press.
MacEachern, A. (2001). *A Natural Selection. National Parks in Atlantic Canada 1935–1970*. Montreal, London & Ithaca: McGill-Queen's University Press.
Macgregor, C. J. & Warren, C. R. (2006). Adopting sustainable farm management practices within a nitrate vulnerable zone in Scotland. *Agriculture Ecosystems & Environment*, **113**, 108–119.
Mack, R. N. (2000). Assessing the extent, status, and dynamism of plant invasions: current and emerging approaches. In *Invasive Species in a Changing World*, eds. H. A. Mooney & R. J. Hobbs, pp. 141–168. Washington, DC & Covelo, CA: Island Press.
Mack, R.N. (1984). Invaders at home on the range. *Natural History*, **93**, 40–47.
Mack, R. N. (1985). Invading plants: their potential contribution to population biology. In *Studies in Plant Demography*, ed. J. White, pp. 127–142. London: Academic Press.
Mack, R. N. (1991). The commercial seed trade: an early disperser of weeds in the United States. *Economic Botany*, **45**, 257–273.
Mack, R. N. (1996). Predicting the identity and fate of plant invaders: emergent and emerging approaches. *Biological Conservation*, **78**, 107–121.
Mack, R. N. (2000). Assessing the extent, status, and dynamism of plant invasions: current and emerging approaches. In *Invasive Species in a Changing World*, eds. H. A. Mooney & R. J. Hobbs, pp. 141–168. Washington, DC & Covelo, CA: Island Press.
Mack, R. N. & Lonsdale, W. M. (2001). Humans as global dispersers: getting more than we bargained for. *BioScience*, **51**, 95–102.
Mack, R. N., Simberloff, D., Lonsdale, W. M., Evans, H., Clout, M. & Bazzaz, F. A. (2000). Biotic invasions: causes, epidemiology, global consequences and control. *Ecological Applications*, **10**, 689–710.
MacKenzie, J. M. (1987). Chivalry, social Darwinism and ritualised killing: the hunting ethos in Central Africa up to 1914. In *Conservation in Africa, People, Policies and Practice*, eds. D. Anderson & R. Grove, pp. 41–61. Cambridge: Cambridge University Press.
Mackinnen, K. (1998). Sustainable use as a conservation tool in the forests of south-east Asia. In *Conservation of Biological Resources*, eds. E. J. Milner-Gulland & R. Mace, pp. 174–192. Oxford: Blackwell.
MacKinnon, J., MacKinnon, K., Child, G. & Thorsell, J. (1986). *Managing Protected Areas in the Tropics*. Gland, Switzerland: IUCN/UNEP.
Mackworth-Praed, H. (1991). *Conservation Piece*. Chichester: Packard Publishing.
Macnair, M. R. (1981). The tolerance of higher plants to toxic materials. In *Genetic Consequences of Man-made Change*, eds. J. A. Bishop & L. M. Cook, pp. 177–207. London: Academic Press.

Macnair, M. R. (1990). The genetics of metal tolerance in natural populations. In *Heavy Metal Tolerance in Plants*, ed. J. A. Shaw, pp. 235–255, Boca Raton, FL: CRC Press.
Macnair, M. R. (1993). Tansley Review. No. 49. The genetics of metal tolerance in vascular plants. *New Phytologist*, **124**, 541–559.
Macnair, M. R. (1997). The evolution of plants in metal-contaminated environments. In *Environmental Stress, Adaptation and Evolution*, eds. R. Bijlsma & V. Loeschke, pp. 3–24. Basel: Birkhauser Verlag.
Macnair, M. R. (2000). The genetics of metal tolerance in natural populations. In *Heavy Metal Tolerance in Plants: Evolutionary Aspects*, ed. A. J. Shaw, pp. 235–253. Boca Raton, FL: CRC Press.
Madgwick, F. J. (1999). Restoring nutrient-enriched shallow lakes: integration of theory and practice in the Norfolk Broads, UK. *Hydrobiologia*, **409**, 1–12.
Magome, H. & Murombedzi, J. (2003). Sharing South African National Parks: community land and conservation in a democratic South Africa. In *Decolonizing Nature, Strategies for Conservation in a Post-colonial Era*, eds. W. M. Adams & M. Mulligan, pp. 108–134. London & Sterling, VA: Earthspan Publications.
Majerus, M. E. N. (1998). *Melanism: Evolution in Action*. Oxford: Oxford University Press.
Malhi, Y. & Phillips, O. L. (2005). *Tropical Forests and Global Atmospheric Change*. Oxford: Oxford University Press.
Malthus, T. R. (1826). *An Essay on the Principle of Population, as It Affects the Future Improvement of Society*. London: Murray.
Mann, C. C. (1991). Extinction: are ecologists crying wolf? *Science*, **235**, 736–738.
Mann, C. C. & Plummer, M. L. (1996). *Noah's Choice*. New York: Knopf.
Mannion, A. M. (1998). Island biogeography. In *Encyclopedia of Ecology & Environmental Management*, ed. P. Calow, pp. 217–219. Oxford: Blackwell.
Marchant, C. J. (1967). Evolution in *Spartina* (Gramineae). 1. The history and morphology of the genus in Britain. *Journal of the Linnean Society (Botany)*, **60**, 1–24.
Marchant C. J. (1968). Evolution in *Spartina* (Gramineae). 2. Chromosomes, basic relationships and the problem of *S.* × *townsendii* agg. *Journal of the Linnean Society (Botany)*, **60**, 381–409.
Maron, J. L., Vilà, M., Bommarco, R., Elmendorf, S. & Beardsley, P. (2004). Rapid evolution of an invasive plant. *Ecological Monographs*, **74**, 261–280.
Marren, P. (1999). *Britain's Rare Flowers*. London: Poyser in association with Plantlife & English Nature.
Marren, P. (2005). The wolf at your door. *Independent*, 22 August 2005.
Marrero-Gómez, M. V. *et al.* (2000). Study of the establishment of the endangered *Echium acanthocarpum* (Boraginaceae) in the Canary Islands. *Biological Conservation*, **94**, 183–190.
Marsh, G. P. (1864). *Man and Nature: or, Physical Geography as Modified by Human Action*. New York: Scribners; London: Sampson Low.
Marshall, F. M. (2002). Effect of air pollution in developing countries. In *Air Pollution and Plant Life*, eds. J. N. B. Bell & M. Treshow, 2nd edn., pp. 407–416. London: Wiley.
Marshall, D. R. & Brown, A. H. D. (1975). Optimum sampling strategies in genetic conservation. In *Crop Genetic Resources for Today and Tomorrow*, eds. O. H. Frankel & J. G. Hawkes, Chapter 4. Cambridge: Cambridge University Press.

Marshall, D. R. & Weiss, P. W. (1982). Isozyme variation within and among Australian populations of *Emex spinosa* (L.) Campd. *Australian Journal of Biological Sciences*, **35**, 327–332.

Martin, P. S. (1984). Prehistoric overkill: the global model. In *Quaternary Extinctions: A Prehistoric Revolution*, eds. P. S. Martin & R. G. Klein, pp. 354–823. Tucson: University of Arizona Press.

Martin, P. S. & Klein, R. G. (eds.) (1984). *Quaternary Extinctions: A Prehistoric Revolution*. Tucson: University of Arizona Press.

Martin, J., Waldren, S., O'Sullivan, A. & Curtis, T. G. F. (2001). The establishment of the Threatened Irish Plant Seed Bank. In *Biological Collections and Biodiversity*, eds. B. S. Rushton, P. Hackney & C. R. Tyrie, pp. 127–138. Otley, UK: Linnean Society of London, Westbury Publishing.

Maschinski, J. & Duquesnel, J. (2006). Successful reintroductions of the endangered long-lived Sargent's cherry palm, *Pseudophoenix sargentii*, in the Florida Keys. *Biological Conservation*, **134**, 122–129.

Maschinski, J., Frye, R. & Rutman, S. (1997). Demography and population viability of an endangered plant species before and after protection from trampling. *Biological Conservation*, **11**, 990–999.

Massart, J. (1912). *Pour la protection de la Nature en Belgique*. Bruxelles: Lamartin.

Masson, G. (1966). *Italian Gardens*. London: Thames & Hudson.

Mather, K. (1953). The genetical structure of populations. *Symposium of the Society for Experimental Biology*, **7**, 66–95.

Mathiasen, P., Rovere, A. E. & Premoli, A. C. (2007). Genetic structure and early effects of inbreeding in fragmented temperate forests of a self-incompatible tree *Embothrium coccineum*. *Conservation Biology*, **21**, 232–240.

Matocq, M. D. & Villablanca, F. X. (2001). Low genetic diversity in an endangered species: recent or historic pattern. *Biological Conservation*, **98**, 61–68.

Matson, P. A., Parton, W. J., Power, A. G. & Swift, M. J. (1997). Agricultural intensification and ecosystem properties. *Science*, **277**, 504–509.

Matsuoka, Y. *et al.* (2002). A single domestication for maize shown by multilocus microsatellite genotyping. *Proceedings of the National Academy of Sciences, USA*, **99**, 6080–6084.

Matthes, M. C., Daly, A. & Edwards, K. J. (1998). Amplified fragment length polymorphism (AFLP). In *Molecular Tools for Screening Biodiversity*, eds. A. Karp, P. G. Isaac & D. S. Ingram, pp. 183–190. London: Chapman & Hall.

Mattiangelli, V. *et al.* (2006). A genome-wide approach to identify genetic loci with a signature of natural selection in the Irish population. *Genome Biology*, **7**, article No. 74.

Matyasek, R. *et al.* (2007). Concerted evolution of rDNA in recently formed tragopogon allotetraploids is typically associated with an inverse correlation between gene copy number and expression. *Genetics*, **176**, 2509–2519.

Maughen, G. L. (1984). Survey of weed beet in sugar beet in England, 1978–81. *Crop Protection*, **3**, 315–325.

Maunder, M. (1992). Plant reintroduction: an overview. *Biodiversity and Conservation*, **1**, 51–61.

Maunder, M. (1997). Botanic garden response to the biodiversity crisis: implications for threatened species management. Ph.D. thesis, University of Reading, UK.

Maunder, M. *et al.* (1999). Genetic diversity and pedigree for *Sophora toromiro* (Leguminosae): a tree extinct in the wild. *Molecular Ecology*, **8**, 725–738.

Maunder, M., Higgens, S. & Culham, A. (2001a). The effectiveness of botanic garden collections in supporting plant conservation: a European case history. *Biodiversity and Conservation*, **10**, 383–401.

Maunder, M., Lyte, B., Dransfield, J. & Baker, W. (2001b). The conservation value of botanic garden palm collections. *Biological Conservation*, **98**, 259–271.

Maunder, M., Hughes, C., Hawkins, J. A & Culham, A. (2004). Hybridization in ex situ plant collections; conservation concerns, liabilities and opportunities. In *Ex Situ Conservation: Supporting Species Survival in the Wild*, eds. E. O. Guerrant, K. Havens & M. Maunder, pp. 325–364. Washington, DC, Covelo, CA & London: Island Press.

Maxted, N. (2001). Ex situ, in situ conservation. In *Encyclopedia of Biodiversity*, Vol. 2, ed. S. A. Levin, pp. 683–695. San Diego, CA & London: Academic Press.

Maxwell, B. D. & Mortimer, A. M. (1994). Selection for herbicide resistance. In *Herbicide Resistance in Plants: Biology and Biochemistry*, eds. S. B. Powles & J. A. M. Holtum, pp. 1–26. Boca Raton & London: Lewis Publishers.

May, R. M. (1990). How many species? *Philosophical Transactions of the Royal Society of London, B*, **330**, 293–304.

May, R. M., Lawton, J. H. & Stork, N. E. (1995). Assessing extinction rates. In *Extinction Rates*, eds. J. H. Lawton & R. M. May, pp. 1–24. Oxford: Oxford University Press.

Mayr, E. (1942). *Systematics and the Origin of Species*. New York: Columbia University Press.

Mayr, E. (1963). *Animal Species and Evolution*. Cambridge, MA: Harvard University Press.

Mayr. E. (1991). *One Long Argument: Charles Darwin and the Genesis of Modern Evolutionary Thought*. London: Allen Lane, The Penguin Press.

McCaskill, L. W. (1973). *Hold This Land. A History of Soil Conservation in New Zealand*. Wellington, Sydney & London: A. H. & A. W. Reed.

McCauley, D. E. & Wade, M. J. (1981). The population effect of inbreeding in *Trilobium*. *Heredity*, **46**, 59–67.

McCollin, D., Moore, L. & Sparks, T. (2000). The flora of a cultural landscape: environmental determinants of change revealed using archival sources. *Biological Conservation*, **92**, 249–263.

McCracken, D. P. (1997). *Gardens of Empire: Botanical Institutions of the Victorian British Empire*. London: Leicester University Press.

McCracken, A. R. (2001). Plant pathogens: importance, spread and correct identification. In *Biological Collections and Biodiversity*, eds. B. S. Rushton, P. Hackney & C. R. Tyrie, pp. 199–207. Otley, UK: Linnean Society of London, Westbury Publishing.

McDonald *et al.* (2009). An alternative to climate change for explaining species loss in Thoreau's woods. *Proceedings of the National Academy of Sciences of the USA*, **106**. On line E28 and E29.

McEachern, A. K., Bowles, M. L. & Pavlovic, N. B. (1994). A metapopulation approach to Pitcher's thistle (*Cirsium pitcheri*) recovery in southern Lake Michigan dunes. In *Restoration of Endangered Species*, eds. M. L. Bowles & C. J. Whelan, pp. 194–218. Cambridge: Cambridge University Press.

McGraw, J. B. & Furedi, M. A. (2005). Deer browsing and population viability of a forest understory plant. *Science*, **307**, 920–922.

McKay, J. K., Christian, C. E., Harrison, S. & Rice, K. J. (2005). How local is local? Review of practical and conceptual issues in the genetics of restoration. *Restoration Ecology*, **13**, 432–440.

McKechnie, S. W. & Geer, B. W. (1993). Microevolution in a wine cellar population: an historical perspective. *Genetica*, **90**, 201–215.
McKibben, B. (1990). *The End of Nature*. London: Viking.
McKinney, M. L. & Lockwood, J. L. (1999). Biotic homogenization: a few winners replacing many losers in the next mass extinction. *Trends in Ecology & Evolution*, **14**, 450–453.
McMillan, C. (1969). Survival patterns in four prairie grasses transplanted to central Texas. *American Journal of Botany*, **56**, 108–115.
McNeely, J. A. (1989). Protected areas and human ecology: how National Parks can contribute to sustaining societies of the twenty-first century. In *Conservation for the Twenty-first Century*, eds. D. Western & M. Pearl, pp. 150–157. New York: Oxford University Press.
McNeely, J. A. (1999). The great reshuffling: how alien species help feed the global economy. In *Invasive Species and Biodiversity Management*, eds. O. T. Sundlund, P. J. Schei & Å. Viken, pp. 11–31. Dordrecht, Boston & London: Kluwer.
McNeill, J. R. (2000). *Something New Under the Sun*. London & New York: Allen Lane, The Penguin Press.
McNeilly, T. (1968). Evolution in closely adjacent populations. III. *Agrostis tenuis* on a small copper mine. *Heredity*, **23**, 99–108.
McNeilly, T. & Antonovics, J. (1968). Evolution in closely adjacent plant populations. III. Barriers to gene flow. *Heredity*, **23**, 99–108.
McVean, G. (2002). Chromosomes. In *Encyclopedia of Evolution*, ed. M. Patel, pp. 151–154. Oxford & New York: Oxford University Press.
Meerts, P. (1995). Phenotypic plasticity in the annula weed *Polygonum aviculare*. *Botanica Acta*, **108**, 414–424.
Meffe, G. K. (1995). Genetic and ecological guidelines for species reintroduction programs: application to Great Lakes fishes. *Journal of Great Lakes Research*, **21**, 3–9.
Meffe, G. K. & Carroll, C. R. (1994). *Principles of Conservation Biology* (1997, 2nd edn.). Sunderland, MA: Sinauer.
Meharg, A. A. (2003). The mechanistic basis of interactions between mycorrhizal associations and toxic metal cations. *Mycological Research*, **107**, 1253–1265.
Meharg, A. A., Cumbes, Q. J. & Macnair, M. R. (1993). Pre-adaptation of Yorkshire Fog, *Holcus lanatus* L. (Poaceae) to arsenate tolerance. *Evolution*, **47**, 313–316.
Mehrhoff, L. A. (1996). Reintroducing endangered Hawaiian plants. In *Restoring Diversity*, eds. D. A. Falk, C. I. Millar & M. Olwell, pp. 101–120. Washington, DC, Covelo, CA & London: Island Press.
Meilan, R. *et al.* (2002). The cp4 transgene provides high levels of tolerance to Roundup® herbicide in field-grown hybrid poplars. *Canadian Journal of Forest Research*, **32**, 967–976.
Meine, C. (2001). Conservation movement, historical. In *Encyclopedia of Biodiversity*, Vol. 1, ed. S. A. Levin, pp. 883–896. London: Academic Press.
Melville, E. G. K. (1994). *A Plague of Sheep: Environmental Consequences of the Conquest of Mexico* (Paperback 1997). Cambridge: Cambridge University Press.
Menchari, Y. *et al.* (2006). Weed response to herbicides: regional-scale distribution of herbicide resistance alleles in the grass weed *Alopecurus myosuroides*. *New Phytologist*, **171**, 861–874.
Mendel, G. (1866). Versuche über Planzenhybriden. *Verhandlungen des Naturforschenden Vereins in Brünn*, **4**, 3–44.

Menges, E. S. (1990). Population viability analysis for a rare plant. *Conservation Biology*, **5**, 158–164.
Menges, E. S. (1991). The application of minimum viable population theory to plants. In *Genetics and Conservation of Rare Plants*, eds. D. A. Falk & K. E. Holsinger, pp. 45–61. Oxford: Oxford University Press.
Menges, E. S. (2000). Population viability analysis for an endangered plant. *Conservation Biology*, **4**, 52–61.
Menninger, H. L. & Palmer, M. A. (2006). Restoring ecological communities: from theory to practice. In *Foundations of Restoration Ecology*, eds. D. A. Falk, M. A. Palmer & J. B. Zedler, pp. 88–112. Washington, DC, Covelo, CA & London: Island Press.
Menzel, A., Sparks, T. H., Estrella, N. & Eckhardt, S. (2005). 'SSW to NNE': North Atlantic Oscillation affects the progress of seasons across Europe. *Global Change Biology*, **11**, 909–918.
Menzel, A. *et al.* (2006). European phenological response to climate change matches the warming pattern. *Global Change Biology*, **12**, 1969–1976.
Mercer, D. (1995). *'A Question of Balance': Natural Resources Conflict Issues in Australia*. Sydney: Federation Press.
Mercer, K. L., *et al.* (2007). Stress and domestication traits increase the relative fitness of crop-wild hybrids in sunflower. *Ecology Letters*, **10**, 383–393.
Merkle, S. A. *et al.* (2007). Restoration of threatened species: a noble cause for transgenic trees. *Tree Genetics and Genomes*, **3**, 111–118.
Meyer, S. M. (2006). *The End of the Wild.* Cambridge, MA: Boston Review, MIT Press.
Michaels, S. D., He, Y. H., Scortecci, K. G. & Amasino, R. M. (2003). Attenuation of flowering locus C activity as a mechanism for the evolution of summer annual flowering behaviour in Arabidopsis. *Proceedings of the National Academy of Sciences, USA*, **100**, 10102–10107.
Midgley, G. F. & Thuiller, W. (2005). Global environmental change and the uncertain fate of biodiversity. *New Phytologist*, **167**, 638–641.
Miettinen, J., Langner, A. & Siegert, F. (2007). Burnt area estimation for the year 2005 in Borneo using multi-resolution satellite imagery. *International Journal of Wildland Fire*, **16**, 45–53.
Mikkelsen, T. R., Anderson, B. & Jørgensen, R. B. (1996). The risk of transgenic spread. *Nature*, **380**, 31.
Millar. D. I. & Libby, W. J. (1989). Disneyland or native ecosystem: genetics and the restorationist. *Restoration and Management Notes*, **7**, 18–24.
Miller, J. R. (2006). Restoration, reconciliation, and reconnecting with nature nearby. *Biological Conservation*, **127**, 356–361.
Miller, R. P. & Nair, P. K. R. (2006). Indigenous agroforestry systems in Amazonia: from prehistory to today. *Agroforestry Systems*, **66**, 151–164.
Miller, W. P., McFee, W. W. & Kelly, J. M. (1983). Mobility and retention of heavy-metals in sandy soils. *Journal of Environmental Quality*, **12**, 579–584.
Miller, G. H. *et al.* (2005). Ecosystem collapse in Pleistocene Australia and a human role in megafaunal extinction. *Science*, **309**, 287–290.
Milne, R. I. & Abbott, R. J. (2000). Origin and evolution of invasive naturalized material of *Rhododendron ponticum* L. in the British Isles. *Molecular Ecology*, **9**, 541–556.
Milner-Gulland, E. J. & Mace, R. (1998). *Conservation of Biological Resources*. Oxford: Blackwell.

Mittermeier, A. R. *et al.* (2003). *Wilderness: Earth's Last Wild Places*. CEMEX, S.A. de C.V.: Conservation International.

Mohan, J. E. *et al.* (2006). Biomass and toxicity responses of poison ivy (*Toxicodendron radicans*) to elevated atmospheric CO_2. *Proceedings of the National Academy of Sciences, USA*, **103**, 9086–9089.

Montalvo, A. M. & Ellstrand, N. C. (2000). Transplantation of the subshrub *Lotus scoparius*: testing the home-site advantage hypothesis. *Conservation Biology*, **14**, 1034–1045.

Montgomery, D. C. (1991). *Design and Analysis of Experiments*, 3rd edn. New York & Chichester: Wiley.

Mooney, H. A. & Cleland, E. E. (2001). The evolutionary impact of invasive species. *Proceedings of the National Academy of Sciences, USA*, **98**, 5446–5451.

Mooney, H. A. & Drake, J. A. (eds.) (1986). *Ecology of Biological Invasions of North America and Hawaii*. New York: Springer-Verlag.

Mooney, H. A. & Hobbs, R. J. (eds.) (2000). *Invasive Species in a Changing World*. Washington, DC & Covelo, CA: Island Press.

Moran, G. F. & Hopper, S. D. (1983). Genetic diversity and the insular population structure of the rare granite rock species, *Eucalyptus caesia* Benth. *Australian Journal of Botany*, **31**, 161–172.

Moran, G. F. & Marshall, D. R. (1978). Allozyme uniformity within and between races of the colonizing species *Xanthium strumarium* L. (Noogoora Burr). *Australian Journal of Biological Sciences*, **31**, 283–292.

Morgan, T. H. (1910). Sex limited inheritance in *Drosophila*. *Science, NY*, **32**, 120–122.

Morgan, J. W. (1997). The effect of grassland gap size on establishment, growth and flowering of the endangered *Rutidosis leptorrhynchoides* (Asteraceae). *Journal of Applied Ecology*, **34**, 566–576.

Morgan, C. S. (2001). Systematics and the National Pinetum, Bedgebury. In *Biological Collections and Biodiversity*, eds. B. S. Rushton, P. Hackney & C. R. Tyrie, pp. 183–189. Otley, UK: Linnean Society of London, Westbury Publishing.

Morris, D. W. & Heidinga, L. (1997). Balancing the books on biodiversity. *Conservation Biology*, **11**, 287–289.

Morrison, D. A. *et al.* (1996). Conservation conflicts over burning bush in south-eastern Australia. *Biological Conservation*, **76**, 167–175.

Moss, S. R., Storkey, J., Cussans, J. W., Perryman, S. A. M. & Hewitt, M. V. (2004). The Broadbalk long-term experiment at Rothhamsted: what has it told us about weeds? *Weed Science*, **52**, 864–873.

Mount, C. (1992). The incidence of simazine resistant weeds in the streets and courts of Cambridge. *University of Cambridge BA project.*

Mousseau, T. A. & Fox, C. W. (eds.) (1998). *Maternal Effects as Adaptations*. New York & Oxford: Oxford University Press.

Muhly, J. D. (1997). Artifacts of the Neolithic, Bronze and Iron Ages. In *The Oxford Encyclopedia of Archaeology in the Near East*, ed. E. M. Meyers, Vol. 4, pp. 5–15. New York & Oxford: Oxford University Press.

Müller-Stark, G. (1998). Isozymes. In *Molecular Tools for Screening Biodiversity*, eds. A. Karp, P. G. Isaac & D. S. Ingram, pp. 75–81. London: Chapman & Hall.

Müllerova, J., Pyšek, P., Jarošík, V. & Pergl, J. (2005). Aerial photographs as a tool for assessing the regional dynamics of the invasive plant species *Heracleum mantegazzianum*. *Journal of Applied Ecology*, **42**, 1042–1053.

Mulligan, G. A & Frankton, C. (1962). Taxonomy of the genus *Cardaria* with particular reference to the species introduced into North America. *Canadian Journal of Botany*, **40**, 1411–1425.

Munne-Bosch, S. & Alegre, L. (2002). Plant aging increases oxidative stress in chloroplasts. *Planta*, **214**, 608–615.

Murdy, W. H. (1979). Effect of SO_2 on sexual reproduction in *Lepidium virginicum* L. originating from regions with different SO_2 concentrations. *Botanical Gazette*, **140**, 299–303.

Murphy, C. E. & Lemerle, D. (2006). Continuous cropping systems and weed selection. *Euphytica*, **148**, 61–73.

Murphy, S. D., Clements, D. R., Belaoussoff, S., Kevan, P. G. & Swanton, C. (2006). Promotion of weed species diversity and reduction of weed seedbanks with conservation tillage and crop rotation. *Weed Science*, **54**, 69–77.

Murray, J. A. H. *et al.* (1961). *Oxford English Dictionary*. Oxford: Oxford University Press.

Musil, C. F., Schmiedel, U. & Midgley, G. F. (2005). Lethal effects of experimental warming approximating a future climate scenario on southern African quartz-field succulents: a pilot study. *New Phytologist*, **165**, 539–547.

Myers, N. (1979). *The Sinking Ark: A New Look at the Problem of Disappearing Species*. Oxford: Pergamon Press.

Myers, N. (1988). Tropical forests and their species: going, going . . . ? In *Biodiversity*, ed. E. O. Wilson, pp. 28–35. Washington, DC: National Academy of Sciences.

Myers, N. (1997). The rich diversity of biodiversity issues. In *Biodiversity II*, eds. M. L. Reaka-Kudla, D. E. Wilson & E. O. Wilson, pp. 125–138. Washington, DC: Joseph Henry Press.

Myers, N. & Knoll, A. H. (2001). The biotic crisis and the future of evolution. *Proceedings of the National Academy of Sciences, USA*, **98**, 5389–5391.

Nabhan, G. P. (1987). *The Desert Smells Like Rain: A Naturalist in Papago Indian Country*. San Francisco: Northpoint Press.

Nadel, H., Frank, J. H. & Knight, R. J. (1992). Escapes and accomplices: the naturalization of exotic Ficus and their associated faunas in Florida. *Florida Naturalist*, **75**, 29–38.

Nash, R. (1982). *Wilderness and the American Mind*, 3rd edn. New Haven: Yale University Press.

Nebel, B. & Fuhrer, J. (1994). Ozone sensitivity of species in semi-natural communities. In *Critical Levels for Ozone*, UN-ECE Workshop Report Number 16, eds. J. Fuhrer & B. Achemann, pp. 264–268. Siebfeld-Berm, Switzerland: Federal Research Station for Agricultural Chemistry and Environmental Hygiene.

Nei, M. (1987). *Molecular Evolutionary Biology*. New York: Columbia University Press.

Nelson, J. C. (1917). The introduction of foreign weeds in ballast, as illustrated by the ballast plants at Linnton, Oregon. *Torreya*, **17**, 151–160.

Nelson, A. P. (1965). Taxonomic and eolutionary implications of lawn races in *Prunella vulgaris* (Labiatae). *Brittonia*, **17**, 160–174.

Nelson, M. P. (1996). Rethinking wilderness: the need for a new idea of wilderness. *Philosophy in the Contemporary World*, **3**, 6–9.

Nepstad, D. C. *et al.* (2002). The effects of partial throughfall exclusion on canopy processes above ground production, and biogeochemistry of the Amazon forest. *Journal of Geophysical Research: Atmospheres*, **107** (D20), Art. No. 8085.

Nethercott, P. J. M. (1998). In *The Conservation Status of Sorbus in the UK*, eds. A. Jackson & M. Flanagan, pp. 40–43. Kew: Royal Botanic Gardens.

Nettencourt, D. de (1977). *Incompatibility in Angiosperms*. New York: Springer-Verlag.
Netting, R. M. (1981). *Balancing an Alp: Ecological Change and Continuity in a Swiss Mountain Community*. Cambridge: Cambridge University Press.
Neuffer, B. & Hurka, H. (1999). Colonization history and introduction dynamics of *Capsella bursa-pastoris* (Brassicaceae) in North America: isozymes and quantitative traits. *Molecular Ecology*, **8**, 1667–1681.
Neuffer, B. & Linde, M. (1999). *Capsella bursa-pastoris*: colonisation and adaptation; a globetrotter conquers the world. In *Plant Evolution in Man-made Habitats*, eds. L. W. D. van Raamsdonk & J. C. M. den Nijs, pp. 49–72. Amsterdam: Hugo de Vries Laboratory.
Neumann, R. P. (1998). *Imposing Wilderness*. Berkeley, CA & London: University of California Press.
Newmark, W. D. (1995). Extinction of mammal populations in western North-American National Parks. *Conservation Biology*, **9**, 512–526.
Newton, A. C. (2007). *Forest Ecology & Conservation*. Oxford: Oxford University Press.
Nicholson, M. (1970). *The Environmental Revolution. A Guide for the New Masters of the World*. London: Hodder & Stoughton.
Nilan, R. A. (1964). *The Cytology and Genetics of Barley, 1951–1962*. Publication of Washington State University, USA.
Nisbet, E. G. (1991). *Leaving Eden*: *To Protect and Manage the Earth.* Cambridge: Cambridge University Press.
Noble, I. R. & Dirzo, R. (1997). Forests as human-dominated ecosystems. *Science*, **277**, 522–525.
Noblick, L. (1992). Reassessment of the garden's *Syagrus* collection. *Fairchild Tropical Garden Bulletin*, **47**, 31–35.
Norton, D. A., Lord, J. M., Given, D. R. & De Lange, P. J. (1994). Over-collecting: an overlooked factor in the decline of plant taxa. *Taxon*, **43**, 181–185.
Noss, R. F. & Daly, K. M. (2006). Incorporating connectivity into broad-scale conservation planning. In *Connectivity Conservation*, eds. K. R. Crooks & M. Sanjayan, pp. 587–619. Cambridge: Cambridge University Press.
Novak, S. J. (2007). The role of evolution in the invasion process. *Proceedings of the National Academy of Sciences, USA*, **104**, 3671–3672.
Novak, S. J. & Mack, R. N. (1993). Genetic variation in *Bromus tectorum* (Poaceae): comparison between native and introduced populations. *Heredity*, **71**, 167–176.
Novak, S. J. & Mack, R. N. (2001). Tracing plant introduction and spread: genetic evidence from *Bromus tectorum* (cheatgrass). *Bioscience*, **51**, 114–122.
Novak, S. J., Mack, R. N. & Soltis, P. S. (1993). Genetic variation in *Bromus tectorum* (Poaceae): introduction dynamics in North America. *Canadian Journal of Botany*, **71**, 1441–1448.
Nunn, P. D. (1994). *Oceanic Islands*. Oxford: Blackwell.
O'Dea, K. (1991). Traditional diet and food preferences of Australian Aboriginal hunter-gatherers. *Philosophical Transactions of the Royal Society of London, B*, **334**, 233–241.
O'Donnell, K. & Cigelnik, E. (1997). Two divergent intergenomic rDNA ITS2 types within a monophyletic lineage of the fungus *Fusarium* are nonorthologous. *Molecular Phylogenetics and Evolution*, **7**, 103–116.
Odling-Smee, F. J., Laland, K. N. & Feldman, M. W. (2003). *Niche Construction*: *The Neglected Process in Evolution*. Princeton, NJ: Princeton University Press.

Odum, E. P (2001). Concept of ecosystem. In *Encyclopedia of Biodiversity*, Vol. 2, ed. S. A. Levin, pp. 305–310. San Diego & London: Academic Press.
Oelschlaeger, M. (1991). *The Idea of Wilderness: From Prehistory to the Age of Ecology*. New Haven: Yale University Press.
Ohlemüller, R., Gritti, E. S., Sykes, M. T. & Thomas, C. D. (2006). Quantifying components of risk for European woody species under climate change. *Global Change Biology*, **12**, 1788–1799.
Oka, H. I. & Chang, W. T. (1959). The impact of cultivation on populations of wild rice *Oryza sativa* f. *spontanea*. *Phyton*, **13**, 105–117.
Okubo, N., Taniguchi, H. & Motokawa, T. (2005). Successful methods for transplanting fragments of *Acropora formosa* and *Acropora hyacinthus*. *Coral Reefs*, **24**, 333–342.
Oliver, F. W. (1913). Report of Meeting of the British Ecological Society at Manchester University, December 1913. *Journal of Ecology*, **1**, 55–56.
Oostermeijer, J. G. B., den Nijs, J. C. M., Raijmann, L. E. L. & Menken, S. B. J. (1992). Population biology and management of the marsh gentian (*Gentiana pneumonanthe* L.), a rare species in the Netherlands. *Botanical Journal of the Linnean Society*, **108**, 117–130.
Oostermeijer, J. G. B., Luijten, S. H., Ellis-Adam, A. C. & den Nijs, J. C. M. (2002). Future prospects for the rare, late-flowering *Gentianella germanica* and *Gentianopsis ciliata* in Dutch nutrient-poor calcareous grasslands. *Biological Conservation*, **104**, 39–350.
Oreskes, N. (2004). The scientific consensus on climate change. *Science*, **306**, 1686.
Oreskes, N., Shrader-Frechette, K. S. & Belitz, K. (1994). Verification, validation, and confirmation of numerical models in the earth sciences. *Science*, **263**, 641–646.
Orr, J. C. *et al.* (2005). Anthropogenic ocean acidification over the twenty-first century and its impact on calcifying organisms. *Nature*, **437**, 681–686.
OTA (Office of Technology Assessment, US Congress) (1993). *Harmful Non-indigenous Species in the USA. OTA-F-565*. Washington, DC: US Government Printing Office.
Ownbey, M. (1950). Natural hybridization and amphidiploidy in the genus *Tragopogon*. *American Journal of Botany*, **37**, 487–496.
Ownbey, M. & McCollum, G. D. (1954). Cytoplasmic inheritance and reciprocal amphidiploidy in *Tragopogon*. *American Journal of Botany*, **40**, 788–796.
Pacala, S. & Socolow, R. (2004). Stabilization wedges: solving the climate problem for the next 50 years with current technology. *Science*, **305**, 968–972.
Page, C. N. & Gardner, M. F. (1994). Conservation of rare temperate rainforest conifer tree species: a fast-growing role for arboreta in Britain and Ireland. In *The Common Ground of Wild and Cultivated Plants*, eds. A. R. Perry & R. G. Ellis, pp. 119–143. Cardiff: National Museum of Wales.
Paine, R. (2002). Origins of agriculture. In *Encyclopedia of Evolution*, ed. M. Pagel, Vol. 1, pp. 15–19. Oxford: Oxford University Press.
Paland, S. & Schmidt, B. (2003). Population size and the nature of genetic load in *Gentianella germanica*. *Evolution*, **57**, 2242–2251.
Palmer, M. *et al.* (2004). Ecology for a crowded planet. *Science*, **304**, 1251–1252.
Palumbi, S. R. (2001). *The Evolution Explosion: How Humans Cause Rapid Evolutionary Change*. New York & London: Norton.
Pammenter, N. W. & Berjak, P. (2000). Evolutionary and ecological aspects of recalcitrant seed biology. *Seed Science Research*, **10**, 301–306.

Pappert, R. A., Hamrick, J. L. & Donovan, L. A. (2000). Genetic variation in *Pueraria lobata* (Fabaceae), an introduced, clonal, invasive plant of the southeastern United States. *American Journal of Botany*, **87**, 1240–1245.

Parisod, C., Trippi, C. & Galland, N. (2005). Genetic variability and founder effect in the Pitcher Plant *Sarracenia purpurea* (Sarraceniaceae) in populations introduced into Switzerland: from inbreeding to invasion. *Annals of Botany*, **95**, 277–286.

Parker, D. M. (1982). The conservation, by restocking, of *Saxifraga cespitosa* in North Wales. *Watsonia*, **14**, 104–105.

Parker, I. M., Rodriguez, J. & Loik, M. E. (2002). An evolutionary approach to understanding the biology of invasions: local adaptation and general-purpose genotypes in the weed *Verbascum thapsus*. *Conservation Biology*, **17**, 59–72.

Parmesan, C. (2005). Detection at multiple levels: *Euphydryas editha* and climate change. In *Climate Change & Biodiversity*, eds. T. E. Lovejoy & L. Hannah, pp. 56–60. New Haven & London: Yale University Press.

Parmesan, C. (2006). Ecological and evolutionary responses to recent climate change. *Annual Review of Ecology, Evolution and Systematics*, **37**, 637–669.

Parsons, J. J. (1970). The Africanization of the new world tropical grasslands. *Tubinger Geographische Studien*, **34**, 141–153.

Paterniani, E. (1969). Selection for reproductive isolation between two populations of Maize, *Zea mays L. Evolution*, **23**, 534–47.

Paterson, A. H. *et al.* (2005). Ancient duplication of cereal genomes. *New Phytologist*, **165**, 658–661.

Paton, D. C. (1997). Honeybees and the disruption of plant-pollinator systems in Australia. *Victorian Naturalist*, **114**, 23–29.

Paton, D. C. (2000). Disruption of bird-plant pollination systems in southern Australia. *Conservation Biology*, **14**, 1232–1234.

Paul, N. D. & Ayres, P. G. (1986). The impact of a pathogen (*Puccinia lagenophorae*) on populations of groundsel (*Senecio vulgaris*) overwintering in the field. *Journal of Ecology*, **74**, 1085–1094.

Pauwels, M., Saumitou-Laprade, P., Holl, A. C., Petit, D. & Bonnin, I. (2005). Multiple origin of metallicolous populations of the pseudometallophyte *Arabidopsis halleri* (Brassicaceae) in central Europe: the cpDNA testimony. *Molecular Ecology*, **14**, 4403–4414.

Pavlik, B. M. (1994). Demographic monitoring and the recovery of endangered plants. In *Restoration of Endangered Species*, eds. M. L. Bowles & C. J. Whelan, pp. 322–350. Cambridge: Cambridge University Press.

Pearce, F. (2001). Global green belt. *New Scientist*, 15 September 2001, 15.

Pellmyr, O. (2002). Microevolution. In *Encyclopedia of Evolution.* ed. M. Patel, pp. 731–732. Oxford & New York: Oxford University Press.

Perring, F. & Walters, S. M. (1962). *Atlas of the British Flora*. London: BSBI, Nelson. 2nd edn. (1976) Wakefield: EP Publishing for the BSBI; 3rd edn. (1982) London: BSBI.

Perring, F. H., Sell, P. D. & Walters, S. M. (1964). *A Flora of Cambridgeshire*. Cambridge: Cambridge University Press.

Perrow, M. R. & Davy, A. J. (2002). *Handbook of Ecological Restoration*. Cambridge: Cambridge University Press.

Peterken, G. F. & Montford, E. P. (1998). Long-term change in an unmanaged population of wych elm subjected to Dutch elm disease. *Journal of Ecology*, **86**, 205–218.

Peters, R. L. (1992). Conservation of biological diversity in the face of climate change. In *Global Warming and Biological Diversity*, eds. R. L. Peters & T. E. Lovejoy, pp. 15–30. New Haven & London: Yale University Press.

Peters, G. B., Lonie, J. S. & Moran, G. F. (1980). The breeding system, genetic diversity and pollen sterility in *Eucalyptus pulverulenta*, a rare species with small disjunct populations. *Australian Journal of Botany*, **38**, 559–570.

Peterson, D. L. (1994). Recent changes in the growth and establishment of subalpine conifers in western North America. In *Mountain Environments in Changing Climates*, ed. M. Beniston, pp. 234–243. London: Routledge.

Peterson, G. D., Cumming, G. S. & Carpenter, S. R. (2003). Scenario planning: a tool for conservation in an uncertain world. *Conservation Biology*, **17**, 358–366.

Petty, A. M. *et al.* (2007). Savanna responses to feral buffalo in Kakadu National Park. *Ecological Monographs*, **77**, 441–463.

Phillips, M. & Mighall, T. (2000). *Society and Exploitation Through Nature*. London: Prentice Hall.

Pickett, S. T. A., Parker, V. T. & Fiedler, P. L. (1992). The new paradigm in ecology: implications for conservation biology above the species level. In *Conservation Biology*, eds. P. L. Fiedler & S. K. Jain, pp. 65–88. New York & London: Chapman & Hall.

Pigott, C. D. & Huntley, J. P. (1981). Factors controlling the distribution of *Tilia cordata* at the northern limit of its geographical range. III. Nature and cause of seed sterility. *New Phytologist*, **87**, 817–839.

Pimentel, D. & Pimentel, M. (2002). Agricultural production. In *Encyclopedia of Life Sciences*, Vol. 1, pp. 260–272. London & New York: Macmillan.

Pimm, S. L. (1998). Extinction. In *Conservation Science and Action*, ed. W. J. Sutherland, pp. 20–38. Oxford: Blackwell Science.

Piniak, G. A. & Brown, E. K. (2008). Growth and mortality of coral transplants (*Pocillopora damicornis*) along a range of sediment influence in Maui, Hawai'i. *Pacific Science*, **62**, 39–55.

Pires, J. C. *et al.* (2004). Molecular cytogenetic analysis of recently evolved *Tragopogon* (Asteraceae) allopolyploids reveal a karyotype that is additive of the diploid parents. *American Journal of Botany*, **91**, 1022–1035.

Pivard, S. *et al.* (2008). Where do the feral oilseed rape populations come from? A large-scale study of their possible origin in a farmland area. *Journal of Applied Ecology*, **45**, 476–485.

Plate, L. (1913). *Selektionsprinzip und Probleme der Artbildung: ein Handbuch des Darwinismus*. Leipzig & Berlin: Engelmann.

Podolsky, R. H. (2001). Genetic variation for morphological and allozyme variation in relation to population size in *Clarkia dudleyana*, an endemic annual. *Conservation Biology*, **15**, 412–423.

Pollard, A. J. (1992). The importance of deterrence: responses of grazing animals to plant variation. In *Plant Resistance to Herbivores and Pathogens*, eds. R. S. Fitz & E. L. Simms, pp. 216–239. Chicago: University of Chicago Press.

Pollard, E., Hooper, M. D. & Moore, N. W. (1974). *Hedges*. London: Collins New Naturalist.

Pond, W. G. & Pond, K. R. (2002). *Introduction to Animal Science*. New York & Chichester, UK: Wiley.

Pons, T. L. (1991). Induction of dark dormancy in seeds: its importance for the seed bank in the soil. *Functional Ecology*, **5**, 669–675.

Pontin, C. (1993). *A Green History of the World*. London: Penguin Books.
Poschlod, P. & Jackel, A. K. (1993). Untersuchungen zur Dynamik von generativen Diasporenbanken von Samenpfanzen auf beweideten, gemähten, brachgefallenen und aufgeforsteten Kalkmagerrasenstandorten. *Verhandlungen Geselleschaft für Ökologie*, **20**, 893–904.
Poschlod, P. & WallisDeVries, M. F. (2002). The historical and socioeconomic perspective of calcareous grasslands: lessons from the distant and recent past. *Biological Conservation*, **104**, 361–376.
Possingham, H. P. (1996). Decision theory and biodiversity management: how to manage a metapopulation. In *The Ecological Basis for Conservation: Heterogeneity, Ecosystems and Biodiversity*, eds. S. T. A. Pickett, R. S. Ostfeld, M. Shachak & G. E. Likens, pp. 298–304. New York: Chapman & Hall.
Potvin, C. & Tousignant, D. (1996). Evolutionary consequences of simulated global change: genetic adaptation or adaptive phenotypic plasticity. *Oecologia*, **108**, 683–693.
Pounds, J. A., Fogden, M. P. L. & Campbell, J. H. (1999). Biological response to climate change on a tropical mountain. *Nature*, **398**, 611–615.
Powles, S. B. & Shaner, D. L. (2001). *Herbicide Resistance and World Grains*. Boca Raton, FL & London: CRC Press.
Prance, G. T. (2004). Introduction. In *Ex Situ Plant Conservation*, eds. E. O. Guerrant, K. Havens & M. Maunder, pp. xxiii–xxix. Washington, DC, Covelo, CA & London: Island Press.
Prat, S. (1934). Die Erblichkeit der Resistenz gegen Kupfer. *Berichte der Deutschen Botanischen Gesellschaft*, **102**, 65–67.
Pressey, R. L., Possingham, H. P. & Margules, C. R. (1996). Optimality in reserve selection algorithms: when does it matter and how much? *Biological Conservation*, **76**, 259–267.
Preston, C. D. (1986). An additional criterion for assessing native status. *Watsonia*, **16**, 83.
Preston, C. D. (2002). 'Babingtonia pestifera': the explosive spread of *Elodea canadensis* and its intellectual reverberations. *Nature in Cambridgeshire*, **44**, 40–49.
Preston, C. & Mallory-Smith, C. A. (2001). Biochemical mechanisms, inheritance, and molecular genetics of herbicide resistance in weeds. In *Herbicide Resistance and World Grains*, eds. S. B. Powles & D. L. Shaner, pp. 23–60. Boca Raton, FL: CRC Press.
Preston, C. D. & Sheail, J. (2007). The transformation of the riparian commons in Cambridge from undrained pastures to level recreation areas, 1833–1932. *Nature in Cambridgeshire*, **4**, 70–84.
Preston, C. D., Pearman, D. A. & Dines, T. D. (2002). *New Atlas of the British & Irish Flora*. Oxford: Oxford University Press.
Primack, R. B. (1993). *Essentials of Conservtion Biology*. Sunderland, MA: Sinauer.
Prince, S. D. & Carter, R. N. (1985). The geographical distribution of prickly lettuce (*Lactuca serriola*). III. Its performance in transplant sites beyond its distribution limit in Britain. *Journal of Ecology*, **73**, 49–64.
Pringle, C. M. (2000). Threats to the U.S. Public Lands from cumulative hydrologic alterations outside of their boundaries. *Ecological Applications*, **10**, 971–989.
Prins, H. H. T. (1998). Origins and development of grassland communities in northwestern Europe. In *Grazing and Conservation Management*, eds. M. F. WallisDeVries, J. P. Bakker & S. E. Van Wieren, pp. 55–106. Dordrecht: Kluwer Academic Press.

Pritchard, T. (1960). Race formation in weedy species with special reference to *Euphorbia cyparissias* L. and *Hypericum perforatum* L. In *The Biology of Weeds*, ed. J. L. Harper, pp. 61–66. Oxford: Blackwell.
Pritchard, H. W. (1989). *Modern Methods in Orchid Conservation: The Role of Physiology, Ecology and Management*. London & New York: Cambridge University Press.
Prober, S. M. & Brown, A. D. H. (1994). Conservation of the grassy white box woodlands: population genetics and fragmentation of *Eucalyptus albens*. *Conservation Biology*, **8**, 1003–1013.
Proctor, M. C. F., Yeo, P. F. & Lack, A. J. (1996). *The Natural History of Pollination*. London: Harper Collins.
Provine, W. B. (1986). *Sewall Wright and Evolutionary Biology*, 2nd edn. Chicago: University of Chicago Press.
Provine, W. B. (1987). *The Origins of Theoretical Population Genetics*. Chicago: Chicago University Press.
Purseglove, J. W. (1959). History and functions of botanic gardens with special reference to Singapore. *Gardeners' Bulletin, Singapore*, **17**, 53–72.
Pyne, S. J. (1982). *Fire in America: A Cultural History of Wildland and Rural Fire*. Princeton, NJ: Princeton University Press.
Quintana-Ascencio, P. F. & Menges, E. S. (1996). Inferring metapopulation dynamics from patch-level incidence of Florida scrub plants. *Conservation Biology*, **10**, 1210–1219.
Rabinowitz, D., Cairns, S. & Dillon, T. (1986). Seven forms of rarity and their frequency in the flora of the British Isles. In *Conservation Biology: The Science of Scarcity and Diversity*, ed. M. E. Soulé, pp. 182–204. Sunderland, MA: Sinauer Associates.
Rackham, O. (1975). *Hayley Wood: Its History and Ecology*. Cambridge: Cambridge & Isle of Ely Naturalists' Trust.
Rackham, O. (1980). *Ancient Woodland: Its History, Vegetation and Uses in England*. London: Edward Arnold. Second revised edition, 2003. Colvend: Castlepoint Press.
Rackham, O. (1986). *The History of the Countryside*. London: Dent.
Rackham, O. (1987). *The History of the Countryside*, Paperback edition. London: Dent.
Rackham. O. (1990). The greening of Myrtos. In *Man's Role in the Shaping of the Eastern Mediterranean Landscape*, eds. S. Bottema, G. Entjes-Nieborg & W. van Zeist, pp. 341–343. Rotterdam: Balkema.
Rackham, O. (1998). Implications of historical ecology for conservation. In *Conservation Science and Action*, ed. W. J. Sutherland, pp. 152–175. Oxford: Blackwell.
Rackham, O. (2001). Land-use patterns, historic. In *Encyclopedia of Biodiversity*, Vol. 3, ed. S. A. Levin, pp. 675–687. San Diego, CA & London: Academic Press.
Rackham, O. (2006). *Woodlands*. London: New Naturalist, HarperCollins.
Rackham, O. & Moody, J. (1996). *The Making of the Cretan Landscape*. Manchester: Manchester University Press.
Radosevich, S., Holt, J. & Ghersa, G. (1997). *Weed Ecology: Implications for Management*, 2nd edn. New York & Chichester, UK: John Wiley & Sons.
Rafinski, J. N. (1979). Geographic variation of flower colour in *Crocus scepusiensis* (Iridaceae). *Plant Systematics and Evolution*, **131**, 107–125.
Raijmann, L. E. L., Leeuwan, N. C., Kersten, R., Oostermeijer, J. R. B., den Nijs H. C. M. & Menken, S. B. J. (1994). Genetic variation and outcrossing rate in relation to population size in *Gentiana pneumonanthe* L. *Conservation Biology*, **8**, 1014–1026.

Raimondo, D. C. & Donaldson, J. S. (2003). Responses of cycads with different life histories to the impact of plant collecting: simulation models to determine important life history stages and population recovery times. *Biological Conservation*, **111**, 345–358.

Ralls, K. & Meadows. R. (1993). Conservation genetics: breeding like flies. *Nature*, **361**, 689–690.

Ramakrishnan, P. S. (1996). Conserving the sacred: from species to landscapes. *Nature & Resources*, **32**, 11–19.

Ramsey, M. M. & Stewart, J. (1998). Re-establishment of the lady's slipper orchid (*Cypripedium calceolus* L.) in Britain. *Botanical Journal of the Linnean Society*, **126**, 173–181.

Randi, E. (2000). Mitochondrial DNA. In *Molecular Methods in Ecology*, ed. E. A. J. Baker, pp. 136–167. Oxford: Blackwell Science.

Randles, J. W. (1985). Exotic plant pathogens. In *Pests and Parasites as Migrants*, eds. A. Gibbs & R. Meischke, pp. 40–42. Cambridge: Cambridge University Press.

Randolph, L. F., Nelson, I. S. & Plaisted, R. L. (1967). Negative evidence of introgression affecting the stability of Louisiana Iris species. *Cornell University Agriculture Experimental Station Memoir* no. 398.

Raup, D. M. (1991). *Extinction. Bad Genes or Bad Luck?* Oxford: Oxford University Press.

Raven, P. H. (1981). Research in botanical gardens. *Botanische Jahrbücher für Systematik*, **102**, 53–72.

Raven, P. H. (1988). Our diminishing tropical forests. In *Biodiversity*, ed. E. O. Wilson, pp. 119–122. Washington, DC: National Academic Press.

Raven, P. H. (2004). Foreword. In *Ex Situ Plant Conservation*, eds. E. O. Guerrant, K. Havens & M. Maunder, pp. xiii–xv. Washington, DC, Covelo, CA & London: Island Press.

Ray, J. (1660). *Catalogus Plantarum circa Cantabrigiam nascentium*, Cambridge.

Raybould, A. F. (1995). Wild crops. In *Encyclopedia of Environmental Biology*, Vol. 3, ed. W. A. Nierenberg, pp. 551–565. San Diego, CA & New York: Academic Press.

Raybould, A. F., Gray, A. J., Lawrence, M. J. & Marshall, D. F. (1990). The origin and taxonomy of *Spartina* × *neyrautii* Foucaud. *Watsonia*, **18**, 207–209.

Raybould, A. F., Gray, A. J., Lawrence, M. J. & Marshall, D. F. (1991a). The evolution of *Spartina anglica* C. E. Hubbard (Gramineae): origin and genetic variability. *Biological Journal of the Linnean Society*, **43**, 111–126.

Raybould, A. F., Gray, A. J., Lawrence, M. J. & Marshall, D. F. (1991b). The evolution of *Spartina anglica* C. E. Hubbard (Gramineae): genetic variation and status of the parental species in Britain. *Biological Journal of the Linnean Society*, **44**, 369–380.

Read, M. (1989). *Grown in Holland?* Brighton: Flora & Fauna Preservation Society.

Read, M. (1993). The indigenous propagation project: a search for co-operation and long-term solutions. In *Species Endangered by Trade: A Role for Horticulture?*, eds. M. Groves, M. Read & B. A. Thomas, pp. 54–62. London: Fauna and Flora Preservation Society.

Read, M. I. & Thomas, B. A. (2001). Propagation not collection. In *Biological Collections and Biodiversity*, eds. B. S. Rushton, P. Hackney & C. R. Tyrie, pp. 101–110. Otley, UK: Linnean Society of London, Westbury Publishing.

Réale, D., McAdam, A. G., Boutin, S. & Berteaux, D. (2003). Genetic and plastic responses of a northern mammal to climate change. *Proceedings of the Royal Society of London, B*, **270**, 591–596.

Reed, C. F. (1977). Economically important foreign weeds: potential problems in the United States. *United States Department of Agriculture Handbook*, No. 498.

Reed, D. H. (2005). Relationship between population size and fitness. *Conservation Biology*, **19**, 563–568.

Reed, M., Mills, I. S., Dunning, J. B. *et al.* (2001). Emerging issues in population viability analysis. *Conservation Biology*, **16**, 7–19.

Reed, D. H., Lowe, E., Briscoe, D. A. & Frankham, R. (2002). Inbreeding and extinction: effects of environmental stress and lineages. *Conservation Genetics*, **3**, 301–307.

Rees, W. E. (2001). Ecological footprint, concept of. In *Encyclopedia of Biodiversity*, ed. S. A. Levin, Vol. 2, pp. 229–244. London: Academic Press.

Regan, H. M. & Auld, T. D. (2004). Australian shrub *Grevillea caleyi:* recovery through management of fire and predation. In *Species Conservation and Management*, ed. H. R. Akçakaya *et al.*, pp. 23–35. New York: Oxford University Press.

Reid, W. V. (1992). How many species will there be? In *Tropical Deforestation and Species Extinction*, eds. T. C. Whitmore & J. A Sayer, pp. 55–73. The World Conservation Union, London & New York: Chapman & Hall.

Reid, L. M. & Steyn, J. N. (1990). South Africa. In *International Handbook of National Parks and Nature Reserves*, ed. C. W. Allin, pp. 337–371. New York: Greenwood Press.

Reidy, M. M., Campbell, T. A. & Hewitt, D. G. (2008). Evaluation of electric fencing to inhibit feral pig movements. *Journal of Wildlife Management*, **72**, 1012–1018.

Reiling, K. & Davison, A. W. (1992a). The response of native, herbaceous species to ozone: growth and fluorescence screening. *New Phytologist*, **122**, 29–37.

Reiling, K. & Davison, A. W. (1992b). Spatial variation in ozone resistance of British populations of *Plantago major* L. *New Phytologist*, **122**, 699–708.

Reiling, K. & Davison, A. W. (1994). Effects of exposure to ozone resistance at different stages of development of *Plantago major* L. on chorophyll fluorescence and gas exchange. *New Phytologist*, **129**, 509–514.

Reiling, K. & Davison, A. W. (1995). Effects of ozone on stomatal conductance and photosynthesis in populations of *Plantago major* L. *New Phytologist*, **129**, 587–594.

Reilly, M. (2007). Alien vine is public enemy number one. *New Scientist*, 11 August 2007, 13.

Reinhartz, J. A. & Les, D. H. (1994). Bottle-neck-induced dissolution of self-incompatibility and breeding systems consequences in *Aster furcatus* (Asteraceae). *American Journal of Botany*, **81**, 446–455.

Reisch, C. & Poschlod, P. (2003). Intraspecific variation, land use and habitat quality: a phenologic and morphometric analysis of *Sesleria albicans* (Poaceae). *Flora*, **198**, 321–328.

Rejmánek, M. (1996). A theory of seed plant invasiveness: the first sketch. *Biological Conservation*, **78**, 171–181.

Remison, S. U. (1976). A study of root interactions among grass species. Ph.D. thesis, University of Reading, England.

Renfrew, J. M. (1973). *Paleoethnobotany: The Prehistoric Food Plants of the Near East and Europe*. London: Methuen.

Rhymer, J. M. & Simberloff, D. (1996). Extinction by hybridization and introgression. *Annual Review of Ecology & Systematics*, **27**, 83–109.

Rich, T. C. G. & Houston, L. (2004). The distribution and population sizes of the rare English endemic *Sorbus wilmottiana* E. F. Warburg, Wilmott's Whitebeam (Rosaceae). *Watsonia*, **25**, 185–191.

Rich, T. C. G. & Woodruff, E. R. (1992). Recording bias in botanical surveys. *Watsonia*, **19**, 73–95.

Richards, A. J. (1986). *Plant Breeding Systems*. London: Allen & Unwin.

Richards, A. J. (1997). *Plant Breeding Systems*, 2nd edn. London: Chapman & Hall.

Richards, C. M. *et al.* (2007). Capturing genetic diversity of wild populations for ex situ conservation: Texas wild rice (*Zizania texana*) as a model. *Genetic Resources and Crop Evolution*, **54**, 837–848.

Richardson, D. M. (1999). Commercial forestry and agroforestry as sources of alien invasive trees and shrubs. In *Invasive Species and Biodiversity Management*, eds. O. T. Sandlund, P. J. Schei & A. Viken, pp. 237–257. Dordrecht & Boston: Kluwer.

Richardson, D. M. & Bond, W. J. (1991). Determinant of plant distribution: evidence from pine invasions. *American Naturalist*, **137**, 639–668.

Richardson, D. M. & Higgins, S. I. (1999). Pines as invaders in the southern hemisphere. In *Ecology and Biogeography of Pinus*, ed. D. M. Richardson, pp. 450–473. Cambridge: Cambridge University Press.

Richardson, D. M., Allsopp, N., D'Antonio, C. M., Milton, S. J. & Rejmánek, M. (2000). Plant invasions: the role of mutualisms. *Biological Reviews*, **75**, 65–93.

Richens, R. H. (1947). Biological Flora of the British Isles. *Allium vineale L. Journal of Ecology*, **34**, 209–226.

Ridley, H. N. (1930). *The Dispersal of Plants Throughout the World*. Ashford, Kent: Reeve.

Ridley, M. (2002). Natural selection. In *Encyclopedia of Evolution*, ed. M. Patel, pp. 797–804. Oxford & New York: Oxford University Press.

Rieger, R., Michaelis, A. & Green, M. M. (1968). *Glossary of Genetics and Cytogenetics*, 3rd edn., Heidelberg & New York: Springer Verlag.

Rieseberg, L. H. & Gerber, D. (1995). Hybridization in the Catalina Island mountain Mahogony (*Cerocarpus traskiae*): RAPD evidence. *Conservation Biology*, **9**, 199–203.

Rieseberg, L. H. & Wendel, J. F. (1993). Introgression and its consequences. In *Hybrid Zones and the Evolutionary Process*, ed. R. G. Harrison, pp. 70–109. New York & Oxford: Oxford University Press.

Rieseberg, L. H. & Willis, J. H. (2007). Plant speciation. *Science*, **17**, 910–914.

Rieseberg, L. H., Zona, S., Aberbom, L. & Martin, T. D. (1989). Hybridization in the island endemic Catalina Mahogany. *Conservation Biology*, **3**, 52–58.

Rieseberg, L. H. *et al.* (2007). Hybridization and the colonization of novel habitats by annual sunflowers. *Genetica*, **129**, 149–165.

Riley, H. P. (1938). A character analysis of colonies of *Iris fulva* and *Iris hexagona* var. *giganti-caerulea* and natural hybrids. *American Journal of Botany*, **25**, 727–738.

Rindos, D. (1989). Darwinism and its role in the explanation of domestication. In *Foraging and Farming: The Evolution of Plant Exploitation*, eds. D. R. Harris & G. C. Hillman, pp. 27–41. London & Boston: Unwin Hyman.

Rinschede, G. (1988). Transhumance in European and American mountains. In *Human Impact on Mountains*, eds. N. J. R. Allan, G. W. Knapp & C. Stadel, pp. 96–115. Lanham, MD: Rowman & Littlefield.

Roach, D. A. & Wulff, R. D. (1987). Maternal effects in plants. *Annual Review of Ecology & Systematics*, **18**, 209–235.

Roberton, A. W., Kelly, D., Ladley, J. J. & Sparrow, A. D. (1998). Effects of pollinator loss on endemic New Zealand mistletoes (Loranthaceae). *Conservation Biology*, **13**, 499–508.
Robertson, K. R., Anderson, R. C. & Schwartz, M. W. (1997). The tallgrass prairie mosaic. In *Conservation in Highly Fragmented Landscapes*, ed. M. W. Schwartz, pp. 53–87. New York: Chapman & Hall.
Robertson, A., Newton, A. C. & Ennos, R. A. (2004). Multiple hybrid origins, genetic diversity and population structure of two endemic *Sorbus* taxa on the Isle of Arran, Scotland. *Molecular Ecology*, **13**, 123–134.
Robichaux, R. H., Friar, E. A. & Mount, D. W. (1997). Molecular genetic consequences of a population bottleneck associated with reintroduction of the Mauna Kea silversword (*Argyroxiphium sandwicense* ssp. *sandwicense Asteraceae*). *Conservation Biology*, **11**, 1140–1146.
Robinson, W. (1894). *The Wild Garden*. London: The Scolar Press.
Robson, G. C. & Richards, O. W. (1936). *The Variation of Animals in Nature*. London & New York: Longmans.
Rojas, M. (1992). The species problem and conservation: what are we protecting. *Conservation Biology*, **6**, 170–178.
Rolston, H. III (2004). In situ and ex situ conservation: philosophical and ethical concerns. In *Ex Situ Plant Conservation*, eds. E. O. Guerrant, K. Havens & M. Maunder, pp. 21–39. Washington, DC, Covelo, CA & London: Island Press.
Roos, M. C. (1993). State of affairs regarding Flora Malesiana: progress in revision work and publication schedule. *Flora Malesiana Bulletin*, **11**(2), 248–252.
Roos, M. C. (2000). Charting tropical plant diversity: Europe's contribution. In *Systematics Agenda 2000: The Challenge for Europe*, eds. S. Blackmore & D. Cutler, pp. 55–88. Cardigan, Wales: Published for the Linnean Society of London by Samara Publishing.
Roose, M. L. & Gottlieb, L. D. (1976). Genetic and biochemical consequences of polyploidy in *Tragopogon*. *Evolution*, **30**, 818–830.
Root, T. L. MacMynowski, D. P., Mastrandrea, M. D. & Schneider, S. H. (2005). Human-modified temperatures induce species changes: joint attribution. *Proceeding of the National Academy of Sciences, USA*, **102**, 7465–7469.
Rosenzweig, M. L (2001). Loss of speciation rate will impoverish future diversity. *Proceedings of the National Academy of Sciences, USA*, **98**, 5404–5410.
Ross, M. A. & Lembi, C. A. (1985). *Applied Weed Science*. Minneapolis, MN: Burgess Publishing Company.
Rosser, E. M. (1955). A new British species of *Senecio*. *Watsonia*, **3**, 228–232.
Ross-Ibarra, J., Morrell, P. L. & Gaut, B. S. (2007). Plant domestication, a unique opportunity to identify the genetic basis of adaptation. *Proceedings of the National Academy of Sciences, USA*, **104**, 8641–8648.
Rothschild, M. & Marren, P (1997). *Rothschild's Reserves: Time and Fragile Nature*. Jerusalem: Balaban Press & Colchester: Harley Books.
Roughgarden, J. (1979). *Theory of Population Genetics and Evolutionary Ecology: An Introduction*. New York & London: Macmillan.
Roux, F., Touzet, P., Cuguen, J. & Le Corre, V. (2006). How to be early flowering: an evolutionary perspective. *Trends in Plant Science*, **11**, 375–381.
Rowell, T. A. (1997). The history of Wicken Fen. In *Wicken Fen: The Making of a Wetland Nature Reserve*, pp. 187–212. Colchester: Harley Books.

Ruddiman, W. F. (2005). How did humans first alter global climate? *Scientific American*, March 2005, 34–41.
Rueda, J., Linacero, R. & Vázquez, A. M. (1998). Plant total DNA extraction. In *Molecular Tools for Screening Biodiversity*, eds. A. Karp, P. G. Isaac & D. S. Ingram, pp. 10–14. London: Chapman Hall.
Runte, A. (1997). *National Parks: The American Experience*, 3rd edn. University of Lincoln & London: Nebraska Press.
Russell-Smith, J. *et al.* (2007). Bushfires 'down under': patterns and implications of contemporary Australian landscape burning. *International Journal of Wildland Fire*, **16**, 361–377.
Rustad, L. E. *et al.* (2001). A meta-analysis of the response of soil respiration, net nitrogen mineralization, and aboveground plant growth to experimental ecosystem warming. *Oecologia*, **126**, 543–562.
Ryan, G. F. (1970). Resistance of common groundsel to simazine and atrazine. *Weed Science*, **18**, 614–616.
Rymer, J. M. & Simberloff, D. (1996). Extinction by hybridization and introgression. *Annual Review of Ecology & Systematics*, **27**, 83–109.
Sage, R. F. (1995). Was low atmospheric CO_2 during the Pleistocene a limiting factor for the origin of agriculture? *Global Climate Change*, **1**, 93–106.
Sakai, A. K. *et al.* (2001). The population biology of invasive species. *Annual Review of Ecology and Systematics*, **32**, 305–332.
Sale, K. (1990). *The Conquest of Paradise: Christopher Columbus and the Columbian Legacy*. New York: Knopf.
Salisbury, E. J. (1943). The flora of bombed areas. *Proceedings of the Royal Institution of Great Britain*, **32**, 435–455.
Salisbury, E. J. (1964). *Weeds and Aliens*, 2nd edn. London: Collins.
Salmon, A., Ainouche, M. L. & Wendel, J. F. (2005). Genetic and epigenetic consequences of recent hybridization and polyploidy in *Spartina* (Poaceae). *Molecular Ecology*, **14**, 1163–1175.
Saltonstall, K. (2003). Cryptic invasion by a non-native genotype of the common reed, *Phragmites australis*, into North America. *Proceedings of the National Academy of Sciences, USA*, **99**, 2445–2449.
Sanderson, E. W. *et al.* (2002). The human footprint and the last of the wild. *Bioscience*, **52**, 891–904.
Sandlund, O. T., Schei., P. J. & Viken, Å. (eds.) (2001). *Invasive Species and Biodiversity Management.* London: Kluwer Academic Publishers.
Sanz-Elorza, M., Dana, E. D., González, A. & Sobrino, E. (2003). Changes in the high-mountain vegetation of the Central Iberian Peninsula as a probable sign of global warming. *Annals of Botany (London)*, **92**, 273–280.
Sarasan, V. *et al.* (2006). Conservation in vitro of threatened plants: progress in the past decade. *In Vitro Cellular & Developmental Biology-Plant*, **42**, 206–214.
Sauer, C. O. (1950). Grassland climax, fire, and man. *Journal of Range Management*, **3**, 16–21.
Sauer, C. O. (1958). Man in the ecology of tropical America. *Proceedings of the Ninth Pacific Science Congress*, 1957, 104–110.
Sauer, C. O. (1975). Man's dominance by use of fire. *Geoscience and Man*, **10**, 1–13.
Saunders, D. A., Smith, G. T., Ingram, J. A. & Forrester, R. I. (2003). Changes in a remnant of salmon gum *Eucalyptus salmonophloia* and York gum *E. loxophleba*

woodland, 1978 to 1997. Implications for woodland conservation in the wheat-sheep regions of Australia. *Biological Conservation*, **110**, 245–256.

Sax, D. F., Stachowicz, J. J. & Gaines, S. D. (2006). *Species Invasions: Insights into Ecology, Evolution and Biogeography*. Sunderland, MA: Sinauer.

Scarre, C. (1996). Mines and quarries. In *Oxford Companion to Archaeology*, ed. B. M. Fagan, pp. 469–470. Oxford & New York: Oxford University Press.

Scavia, D. & Bricker, S. B. (2006). Coastal eutrophication assessment in the United States. *Biogeochemistry*, **79**, 187–208.

Schama, S. (1995). *Landscape and Memory*. London: Fontana Press, HarperCollins.

Schat, H. & ten Bookum, W. M. (1992). Genetic control of copper tolerance in *Silene vulgaris*. *Heredity*, **68**, 219–229.

Schat, H., Vooijs, R. & Kuiper, E. (1996). Identical major gene loci for heavy metal tolerances that have independently evolved in different local populations and subspecies of *Silene vulgaris*. *Evolution*, **50**, 1888–1895.

Scherer-Lorenzen, M., Elend, A., Nöllert, S. & Schulze, E.-D. (2000). Plant invasions in Germany: general aspects and impact of nitrogen deposition. In *Invasive Species in a Changing World*, eds. H. A. Mooney & R. J. Hobbs, pp. 351–368. Washington, DC & Covelo, CA: Island Press.

Schindler, D. W. & Donahue, W. F. (2006). An impending water crisis in Canada's western prairie provinces. *Proceeding of the National Academy of Sciences*, **103**, 7210–7216.

Schmid, E. & Sinabell, F. (2007). On the choice of farm management practices after the reform of the Common Agricultural Policy in 2003. *Journal of Environmental Management*, **82**, 332–340.

Schnoor, J. L., Licht, L. A., McCutcheon, S. C., Wolfe, N. L. & Carreira, L. H. (1995). Phytoremediation of organic and nutrient contaminants. *Environmental Science & Technology*, **29**, 318–323.

Schoch-Bodmer, H. (1938). The proportion of long-, mid- and short-styled plants in natural populations of *Lythrum salicaria*. *Journal of Genetics*, **36**, 39–43.

Schoner, C. A., Norris, R. F. & Chilcote, W. (1978). Yellow foxtail (*Setaria lutescens*) biotype studies: growth and morphological characteristics. *Weed Science*, **26**, 632–636.

Schroder, D. (2004). Use of set-aside land according to the EU regulation and use of biotope sites according to the Federal German Nature Conservation Act for protecting the environment, soils, landscapes, nature, species and biotopes. *Berichte über Landwirtschaft*, **82**, 518–528.

Schrödinger, E. (1944). *What is Life?* Cambridge: Cambridge University Press.

Schullery, P. (1997). *Searching for Yellowstone*. Boston & New York: Houghton Mifflin.

Schwaegerle, K. E. & Schaal, B. A. (1979). Genetic variability and founder effect in the Pitcher Plant *Sarracenia purpurea*. *Evolution*, **33**, 1210–1218.

Schwanitz, F. (1966). *The Origin of Cultivated Plants*. Cambridge, MA: Harvard University Press.

Schwartz, M. W. (1997). Introduction. In *Conservation in Highly Fragmented Landscapes*, ed. M. W. Schwartz, pp. xiii–xvi. New York: Chapman & Hall.

Schwartz, M. W. & van Mantgem, P. J. (1997). The value of small preserves in chronically fragmented landscapes. In *Conservation in Highly Fragmented Landscapes*, ed. M. W. Schwartz, pp. 379–394. New York: Chapman & Hall.

Schwartz, M. D., Ahas, R. & Aasa, A. (2006). Onset of spring starting earlier across the Northern Hemisphere. *Global Change Biology*, **12**, 343–351.

Scott, R. H. (1944). Life history of the wild onion and its bearing on control. *Agriculture*, **LI**, 162–170.
Scott, J. W. (1999). *The incidence of triazine resistant Senecio vulgaris L.* (common groundsel) in Cambridge. University of Cambridge BA project.
Scott, D. (2005). Integrating climate change into Canada's National Park System. In *Climate Change & Biodiversity*, eds. T. E. Lovejoy & L. Hannah, pp. 342–345. New Haven & London: Yale University Press.
Scott, N. E. & Davison, A. W. (1985). The distribution and ecology of coastal species on roadsides. *Vegetatio*, **62**, 43–440.
Scott, S. E. & Wilkinson, M. J. (1998). Transgenic risk is low. *Nature*, **393**, 320.
Scribner, T. K. & Pearce, J. M. (2000), Microsatellites: evolutionary and morphological background and empirical applications at individual, population and phylogenetic levels. In *Molecular Methods in Ecology*, ed. A. J. Baker, pp. 235–273. Oxford: Blackwell Science.
Searcy, K. B & Mulcahy, D. L. (1985a). Pollen-tube competition and selection for metal tolerance in *Silene dioica* (Caryophyllaceae), and *Mimulus guttatus* (Scrophulariaceae). *American Journal of Botany*, **72**, 1695–1699.
Searcy, K. B & Mulcahy, D. L. (1985b). Pollen selection and gametophytic expression of metal tolerance in *Silene dioica* (Caryophyllaceae), and *Mimulus guttatus* (Scrophulariaceae). *American Journal of Botany*, **72**, 1700–1706.
Searcy, K. B & Mulcahy, D. L. (1985c). The parallel expression of metal tolerance in pollen and sporophytes of *Silene dioica* (L.) Clairv., *Silene alba* (Mill) Krause and *Mimulus guttatus* (DC). *Theoretical and Applied Genetics*, **69**, 597–602.
Seaward, M. R. D. & Richardson, D. H. S. (1990). Atmopheric sources of metal pollution and effects on vegetation. In *Heavy Metal Tolerance in Plants: Evolutionary Aspects*, ed. A. J. Shaw, pp. 75–92. Boca Raton, FL: CRC Press.
Segarra-Moragues, J. G., Iriondo, J. M. & Catalán, P. (2005). Genetic fingerprinting of germplasm accessions as an aid for species conservation: a case study with *Borderea chouardii* (Discoreaceae), one of the most critically endangered Iberian plants. *Annals of Botany*, **96**, 1283–1292.
Sellars, R. W. (1997). *Preserving Nature in the National Parks: A History*. New Haven & London: Yale Univerity Press.
Seymour, J. & Girardet, H. (1986). *Far from Paradise. The Story of Human Impact on the Environment*. Basingstoke, UK: Greenprint.
Shafer, C. L. (1990). *Nature Reserves: Island Theory and Conservation Practice*. Washington, DC & London: Smithsonian Institute Press.
Shafer, C. L. (1997). Terrestrial nature reserve design at the urban/rural interface. In *Conservation In Highly Fragmented Landscapes*, ed. M. W. Schwartz, pp. 345–378. New York: Chapman Hall.
Shafer, C. L. (1999a). US national park buffer zones: historic, scientific, social, and legal aspects. *Environmental Management*, **23**, 49–79.
Shafer, C. L. (1999b). National park and reserve planning to protect biological diversity: some basic elements. *Landscape and Urban Planning*, **44**, 123–153.
Shafer, C. L. (2001). Inter-reserve distance. *Biological Conservation*, **100**, 215–227.
Shah, J. *et al.* (2000). Integrated analysis for acid rain in Asia: policy implications and results of RAINS-ASIA model. *Annual Review of Energy and the Environment*, **25**, 339–375.

Shapcott, A. (1994). Genetic and ecological variation in *Atherosperma moschatum* and the implications for conservation of its biodiversity. *Australian Journal of Botany*, **42**, 663–686.

Shapcott, A. (1998). The genetics of *Ptychosperma bleeseri*, a rare palm from the Northern Territory, Australia. *Biological Conservation*, **85**, 203–209.

Sharma, C. B. S. R. & Panneerselvam, N. (1990). Genetic toxicity of pesticides in higher plant systems. *Critical Reviews in Plant Sciences*, **9**, 409–442.

Shaw, A. K. (ed.) (2001). *Heavy Metal Tolerance in Plants: Evolutionary Aspects*. Boca Raton, FL: CRC Press.

Sheail, J. (1976). *Nature in Trust*. Glasgow & London: Blackie.

Sheail, J. (1981). *Rural Conservation in Inter-war Britain*. Oxford: Clarendon Press.

Sheail, J. (1987). *Seventy-five Years in Ecology: The British Ecological Society*. Oxford: Blackwell.

Sheail, J. (1998). *Nature Conservation in Britain: The Formative Years*. London: The Stationery Office.

Sheail, J., Treweek, J. R. & Mountford, J. O. (1997). The UK transition from nature preservation to 'creative conservation'. *Environmental Conservation*, **24**, 224–235.

Sheppard, P. M. (1975). *Natural Selection & Heredity*, 4th edn. London: Hutchinson University Library.

Sherratt, A. (1996). Central and Eastern European copper mines. In *Oxford Companion to Archaeology*, ed. B. M. Fagan, p. 471. Oxford & New York: Oxford University Press.

Shindo, C., Bernasconi, G. & Hardtke, C. S. (2007). Natural genetic variation in *Arabidopsis*: tools, traits and prospects. *Annals of Botany*, **99**, 1043–1054.

Shoard, M. (1980). *The Theft of the Countryside*. London: Temple Smith.

Shoard, M. (1997). *This Land is Our Land*. London: Gaia Books.

Shrader-Frechette, K. S. & McCoy, E. D. (1993). *Method in Ecology*. Cambridge: Cambridge University Press.

Shugart, H. H. (2001). Phenomenon of succession. In *Encyclopedia of Biodiversity*, ed. S. A. Levin, Vol. 5, pp. 541–552. San Diego, CA & London: Academic Press.

Simberloff, D. (1998). Flagships, umbrellas, and keystones: is single-species management passé in the landscape era. *Biological Conservation*, **83**, 247–257.

Simberloff, D. (2001). Introduced species, effects and distribution of. In *Encyclopedia of Biodiversity*, ed. S. A. Levin, Vol. 3, pp. 517–529. San Diego, CA: Academic Press.

Simmonds, N. W. (ed.) (1976). *Evolution of Crop Plants*. London: Longmans.

Simon, J., Bosch, M., Molero, J. & Blanché, C. (2001). Conservation biology of Pyrenean larkspur (*Delphinium montanum*): a case of conflict of plant versus animal conservation? *Biological Conservation*, **98**, 305–314.

Simpson, G. G. (1961). *Principles of Animal Taxonomy. The Species and Lower Categories*. New York: Columbia University Press.

Simpson, D. A. (1984). A short history of the introduction and spread of *Elodea Michx* in the British Isles. *Watsonia*, **17**, 121–132.

Simpson, M. (2001). Plant cataloguing in the National Trust. In *Biological Collections and Biodiversity*, eds. B. S. Rushton, P. Hackney & C. R. Tyrie, pp. 117–120. Otley, UK: Linnean Society of London, Westbury Publishing.

Singer, M. C., Thomas, C. D. & Parmesan, C. (1993). Rapid human-induced evolution of insect-host associations. *Nature*, **366**, 681–683.

Sinskaia, E. N. & Beztuzheva, A. A. (1931). The forms of *Camelina sativa* in connection with climate, flax and man. *Trudy po Prikladnoi. Botanika Genetikei Selektsi*, **25**, 98–200.

Skolmen, R. G. (1979). *Plantings on Forest Reserves of Hawaii, 1910–1960*. Honolulu: Institute of Pacific Island Forestry.
Silvertown, J. *et al.* (1994). Short-term effects and long-term after-effects of fertilizer application on the flowering population of the Green-winged orchid *Orchis morio*. *Biological Concervation*, **69**, 191–197.
Slotte, T., Holm, K., McIntyre, L. M., Lagercrantz, U. & Lascoux, M. (2007). Differential expression of genes important for adaptation in *Capsella bursa-pastoris* (Brassicaceae). *Plant Physiology*, **145**, 160–173.
Small, E. (1984). Hybridization in the domesticated-weed-wild complex. In *Plant Biosciences*, ed. W. F. Grant, pp. 195–210. Toronto: Academic Press.
Smirnov, Y. S. & Tkachenko, K. G. (2001). The Botanical Garden of the Komarov Botanical Institute of the Russian Academy of Science celebrates 285 years. In *Biological Collections and Biodiversity*, eds. B. S. Rushton, P. Hackney & C. R. Tyrie, pp. 109–115. Otley, UK: Linnean Society of London, Westbury Publishing.
Smith, R. I. L. (1994). Vascular plants as bioindicators of regional warming in the Antarctic. *Oecologia*, **99**, 322–328.
Smith, H. (2007). WWF despair over Greek fire damage. *The Guardian*. 28 September 2007.
Smith, E. V. & Mayton, E. L. (1938). Nut grass eradication studies. II. The eradication of nut grass *Cyperus rotundus* L. by certain tillage treatments. *Journal of the American Society of Agronomy*, **30**, 18–21.
Smith, F. D. M., May, R. M., Pellew, R., Johnson, T. H. & Walter, K. R. (1993). How much do we know about the current extinction rate? *Trends in Ecology & Evolution*, **8**, 375–378.
Smith, B. M., Diaz, A., Winder, L. & Daniels, R. (2005a). The effect of provenance on the establishment and performance of *Lotus corniculatus* L. in the re-creation environment. *Biological Conservation*, **125**, 37–46.
Smith, G. C., Henderson, I. S. & Robertson, P. A. (2005b). A model of ruddy duck *Oxyura jamaicensis* eradication for the UK. *Journal of Applied Ecology*, **42**, 546–555.
Snaydon, R. W. (1970). Rapid differentiation in a mosaic environment. I. The response of *Anthoxanthum odoratum* populations to soils. *Evolution*, **24**, 257–269.
Snaydon, R. W. (1976). Genetic changes within species. An appendix to: The Park Grass Experiment on the effect of fertilisers and liming on the botanical composition of permanent grassland and the yield of hay by J. M. Thurston, G. V. Dyke & E. D. Williams. Publication of Rothamsted Experimental Station.
Snaydon, R. W. (1978). Genetic changes in pasture populations. In *Plant Relations in Pastures*, ed. J. R. Wilson, pp. 253–269. Melbourne: CSIRO.
Snaydon, R. W. & Davies, M. S. (1972). Rapid differentiation in a mosaic environment. II. Morphological variation in *Anthoxanthum odoratum*. *Evolution*, **26**, 390–405.
Sniegowski, P. D. (2002). Mutation. In *Encyclopedia of Evolution*, ed. M. Patel, pp. 777–783. Oxford & New York: Oxford University Press.
Snogerup, S. (1979). Cultivation and continued holding of Aegean endemics in an artificial environment. In *Survival or Extinction*, eds. H. Synge & H. Townsend, pp. 85–90. Kew, UK: Bentham-Moxon Trust.
Snow, A. A. *et al.* (2003). A Bt transgene reduces herbivory and enhances fecundity in wild sunflowers. *Ecological Applications*, **13**, 279–286.
Sobey, D. G. (1987). Differences in seed production between *Stellaria media* populations from different habitat types. *Annals of Botany*, **59**, 543–549.

Sokal, R. R. & Rohlf, F. J. (1993). *Biometry: The Principles and Practice of Statistics in Biological Research*. New York: Freeman.
Solbrig, O. T. (1994). Biodiversity: an introduction. In *Biodiversity and Global Change*, eds. O. T. Solbrig, H. M. van Emden & P. G. W. J. van Oordt, pp. 13–20. Wallingford, UK: CAB International.
Solecki, M. K. (1993). Cut-leaved and common teasel (*Dipsacus laciniatus* L. and *D. sylvestris* Huds.): profile of two invasive aliens. In *Biological Pollution: The Control and Impact of Invasive Exotic Species*, ed. B. N. McKnight, pp. 85–92. Indianapolis: Indiana Academy of Science.
Soltis, P. S. (2005). Ancient and recent polyploidy in angiosperms. *New Phytologist*, **166**, 5–8.
Soltis, D. E. & Soltis, P. S. (1989). Allopolyploid speciation in *Tragopogon*: insights from chloroplast DNA. *American Journal of Botany*, **76**, 1119–1124.
Soltis, P. S. & Soltis, D. E. (2000). The role of genetic and genomic attributes in the success of polyploids. *Proceedings of the National Academy of Sciences, USA*, **97**, 7051–7057.
Soltis, P. S., Plunckett, G. M., Novak, S. J. & Soltis, D. E. (1995). Genetic variation in Tragopogon species: additional origins of the allopolyploids *T. mirus* and *T. miscellus* (Compositae). *American Journal of Botany*, **82**, 1329–1341.
Soltis, D. E. *et al.* (2004). Recent and recurrent polyploidy in *Tragopogon* (Asteraceae): cytogentic, genomic and genetic comparisons. *Biological Journal of the Linnean Society*, **82**, 485–501.
Somers, C. M., Yauk, C. L., White, P. A., Parfett, C. L. J. & Quinn, J. S. (2002). Air pollution induces heritable DNA mutations. *Proceeding of the National Academy of Sciences, USA*, **99**, 15904–15907.
Sonneveld, A. (1955). Photoperiodic adaptations of grassland plants. *Proceedings Xth Internatinal Grassland Conference*, 711–714.
Soó, R. v. (1927). Systematische Monographie der Gattung *Melampyrum*. *Feddes Repert*, **23**, 159–176, 385–397.
Sørenson, T. (1954). Adaptation of small plants to deficient nutrition and a short growing season. Illustrated by cultivation experiments with *Capsella bursa-pastoris* (L.) Med. *Botanisk Tidsskrift*, **51**, 339–361.
Sorenson, J. C. (1991). On the relationship of phonological strategy to ecological success: the case of broomsedge in Hawaii. *Vegetatio*, **95**, 137–147.
Sork, V. L., Nason, J., Campbell, D. R. & Fernandez, J. F. (1999). Landscape approaches to historical and contemporary gene flow in plants. *Trends in Ecology & Evolution*, **14**, 219–224.
Soukup, J. & Holec, J. (2004). Crop-wild interaction within the *Beta vulgaris* complex: agronomic aspects of weed beet in the Czech Republic. In *Introgression from Genetically Modified Plants into Wild Relatives*, eds. H. C. M. den Nijs, D. Bartsch & J. Sweet, pp. 203–218. Wallingford, UK: CAB International.
Soulé, M. E. (1980). Thresholds for survival: maintaining fitness and evolutionary potential. In *Conservation: An Evolutionary-Ecological Perspective*, eds. M. E. Soulé & B. A. Wilcox, pp. 151–170, Sunderland, MA: Sinauer.
Soulé, M. E. (ed.) (1987). *Viable Populations for Conservation*. Cambridge: Cambridge University Press.
Soulé, M. E. & Simberloff, D. (1986). What do genetics and ecology tell us about the design of nature reserves? *Biological Conservation*, **35**, 19–40.

Soulé, M., Gilpin, M., Conway, W. & Foose, T. (1986). The millennium ark: how long a voyage, how many staterooms, how many passengers? *Zoo Biology*, **5**, 101–113.
Spence, M. D. (1999). *Dispossessing the Wilderness: Indian Removal and the Making of the National Parks*. New York & Oxford: Oxford University Press.
Spiers, A. G. & Hopcroft, D. H. (1994). Comparative studies of the poplar rusts *Melampsora medusae, M. laricipopulina* and their interspecific hybrid *M. medusae-populina*. *Mycological Research*, **98**, 889–903.
Spurway, H. (1952). Can wild animals be bred in captivity? *New Biology*, **13**, 11–30.
Spurway, H. (1955). The causes of domestication: an attempt to integrate some ideas of Konrad Lorenz with evolutionary theory. *Journal of Genetics*, **53**, 325–362.
Stace, C. A. (1975). *Hybridization and the Flora of the British Isles*. London: Academic Press.
Stace, C. A. (1980). *Plant Taxonomy and Biosystematics*. London: Arnold (2nd edn., 1989).
Stace, C. A. (1991). *New Flora of the British Isles*. Cambridge: Cambridge University Press.
Stachowicz, J. J., Fried, H., Osman, R. W. & Whitlatch, R. B. (2002). Biodiversity, invasion resistance, and marine ecosystem function: reconciling pattern and process. *Ecology*, **83**, 2575–2590.
Stapledon, R. G. (1928). Cocksfoot grass (*Dactylis glomerata*) ecotypes in relation to the biotic factor. *Journal of Ecology*, **16**, 72–104.
Stearn, W. T. (1984). The introduction of plants into the gardens of Western Europe during 2000 years. *Supplement to The Australian Garden Journal*, April 1984, pp. 1–7.
Stearns, S. C. (1976). Life history tactics: review of ideas. *Quarterly Journal of Biology*, **51**, 3–47.
Stearns, S. C. (1992). *The Evolution of Life Histories*. Oxford: Oxford University Press.
Stebbins, G. L. (1950). *Variation and Evolution in Plants*. New York & London: Columbia University Press.
Stebbins, G. L. (1957). Self-fertilisation and population variability in higher plants. *American Naturalist*, **41**, 337–354.
Stebbins, G. L. (1966). *Processes of Organic Evolution*. Englewood Cliffs, NJ: Prentice Hall.
Stebbins, G. L. (1971). *Chromosomal Evolution in Higher Plants*. London: Arnold.
Steinberg, E. K. & Jordon, C. E. (1998). Using molecular genetics to learn about the ecology of threatened species: the allure and the illusion of measuring genetic structure in natural populations. In *Conservation Biology: For the Coming Decade*, 2nd edn., eds. P. L. Fiedler & P. M. Karieva, pp. 440–460. New York: Chapman & Hall.
Steinmeyer, B., Wöhrmann, K. & Hurka, H. (1985). Phänotypenvariabilität und Umwelt bei *Capsella bursa-pastoris* (Cruciferae). *Flora*, **177**, 323–334.
Stenstrom, A. & Jonsdottir, I. S. (2006). Effects of simulated climate change on phenology and life history traits in *Carex bigelowii*. *Nordic Journal of Botany*, **24**, 355–371.
Stephens, P. A. & Sutherland, W. J. (1999). Consequences of the Allee effect for behaviour, ecology and conservation. *Trends in Ecology and Evolution*, **14**, 401–405.
Stern, N. H. (2007). *The Economics of Climate Change: The Stern Review*. Cambridge: Cambridge University Press.
Sterneck, J. v. (1895). Beitrag zur Kenntnis der Gattung *Alectorolophus* All. *Österreichischer Botanische Zeitschrift*, **63**, 195–303.

Stevens, C. J. *et al.* (2004). Impact of nitrogen deposition on the species richness of grasslands. *Science*, **303**, 1876–1879.
Steward, J. M. (1934). Ethnography of the Owens Valley Paiute. *University of California Publications of American Archaeology and Ethnology*, **33**, 233–240.
Stiling, P., Rossi, A. & Gordon, D. (2000). The difficulties of single factor thinking in restoration: replanting a rare cactus in the Florida Keys. *Biological Conservation*, **94**, 327–333.
Stockwell, C. A., Hendry, A. P. & Kinnison, M. T. (2003). Contemporary evolution meets conservation biology. *Trends in Ecology & Evolution*, **18**, 94–101.
Stockwell, C. A., Kinnison, M. T. & Hendry, A. P. (2006). Evolutionary restoration ecology. In *Foundations of Restoration Ecology*, eds. D. A. Falk, M. A. Palmer & J. B. Zedler, pp. 113–137. Washington, DC, Covelo, CA & London: Island Press.
Stokstad, E. (2005a). Experimental drought predicts grim future for rainforest. *Science*, **308**, 346–347.
Stokstad, E. (2005b). Ecology – flying on the edge: bluebirds make use of habitat corridors. *Science*, **309**, 35.
Stokstad, E. (2006). The case of the empty hives. *Science*, **316**, 970–972.
Stork, N. E. (1997). Measuring global diversity and its decline. In *Biodiversity II. Understanding and Protecting our Biological Resources*, eds. M. L. Reaka-Kudla, D. E. Wilson & E. O. Wilson, pp. 41–68. Washington, DC: Joseph Henry Press.
Stork, N. E. (2007). World of insects. *Nature*, **448**, 657–658.
Storme, V. *et al.* (2004). Ex situ conservation of Black poplar in Europe: genetic diversity in nine gene bank collections and their value for nature development. *Theoretical and Applied Genetics*, **108**, 969–981.
Stott, P. (2001). Jungles of the mind. The invention of the tropical rain forest (and emergence of myths). *History Today*, **51**, 38–44.
Stout, A. B. (1923). Studies of *Lythrum salicaria*. I. The efficiency of self-pollination. *American Journal of Botany*, **10**, 440–449.
Strid, A. (1971). Past and present distribution of *Nigella arvensis* L. *Botaniska Notiser*, **124**, 231–236.
Stuart, S. N. *et al.* (2004). Status and trends of amphibium declines and extinctions worldwide. *Science*, **306**, 1783–1786.
Sugii, N. & Lamoureux, C. (2004). Tissue culture as a conservation method: an empirical view from Hawaii. In *Ex situ Plant Conservation*, eds. E. O. Guerrant Jr., K. Havens & M. Maunder, pp. 189–205. Washington, DC, Covelo, CA & London: Island Press.
Sullivan, J. J., Timmins, S. M. & Williams, P. A. (2005). Movement of exotic plants into coastal native forests from gardens in northern New Zealand. *New Zealand Journal of Ecology*, **29**, 1–10.
Sultan, S. E. (1987). Evolutionary implications of phenotypic plasticity in plants. *Evolutionary Biology*, **21**, 127–178.
Sundlund, O. T., Schei, P. J. & Viken, Å. (1999). *Invasive Species and Biodiversity Management*. Dordrecht, Boston & London: Kluwer.
Susaria, S., Medina, V. F. & McCutcheon, S. C. (2002). Phytoremediation: an ecological solution to organic chemical contamination. *Ecological Engineering*, **18**, 647–658.
Sutherland, W. J. (1998). Managing habitats and species. In *Conservation Science and Action*, ed. W. J. Sutherland, pp. 202–219. Oxford: Blackwell.
Sutherland, W. J. & Hill, D. A. (eds.) (1995). *Managing Habitats for Conservation*. Cambridge: Cambridge University Press.

Sutherland, W. J., Pullin, A. S., Dolman, P. M. & Knight, T. M. (2004). The need for evidence-based conservation. *Trends in Ecology & Evolution*, **19**, 305–308.

Sutton, W. S. (1903). The chromosomes in heredity. *Biological Bulletin, Marine Biological Laboratory, Woods Hole, MA*, **4**, 231–248.

Svensson, R. & Wigren, M. (1983). A survey of the history, biology and preservation of some retreating synanthropic plants. *Acta Universitatis Upsaliensis Symbolae Botanicae Upsaliensis*, **25**, 1–73.

Swensen, S. M., Allen, G. J., Howe, M., Elisens, W. J., Junak, S. A. & Rieseberg, L. H. (1995). Genetic analysis of the endangered island endemic *Malacothamnus fasciculatus* (Nutt.) Greene var. *nesioticus* (Rob.) Kearn. (Malvaceae). *Conservation Biology*, **9**, 404–415.

Szaro, R. C. (1996). Biodiversity in managed landscapes: principles, practice and policy. In *Biodiversity in Managed Landscapes. Theory and Practice*, eds. R. C. Szaro & D. W. Johnston, pp. 727–770. New York & Oxford: Oxford University Press.

Taggart, J. B., McNally, S. F. & Sharp, P. M. (1989). Genetic variability and differentiation among founder populations of the pitcher plant (*Sarracenia purpurea* L.) in Ireland. *Heredity*, **64**, 177–183.

Tansley, A. G. (1935). The use and abuse of vegetational concept terms. *Ecology*, **16**, 284–307.

Tansley, A. G. (1945). *Our Heritage of Wild Nature: A Plea for Organized Nature Conservation*. Cambridge: Cambridge University Press.

Tansley, S. A. & Brown, C. R. (2000). RAPD variation in the rare and endangered *Leucadendron elimense* (Proteaceae): implications for their conservation. *Biological Conservation*, **95**, 39–48.

Tattersall, F. & Manley, W. (eds.) (2003). *Conservation and Conflict: Mammals and Farming in Britain*. Otley, UK: Westbury Publishing.

Taylor, P. (2005). *Beyond Conservation: A Wildland Strategy*. London: Earthspan.

Taylor, G. E. Jr. & Murdy, W. H. (1975). Population differentiation of an annual plant species, *Geranium carolinianum*, in response to sulfur dioxide. *Botanical Gazette*, **136**, 212–215.

Taylor, G. E. Jr., Pitelka, L. F. & Clegg, M. T. (eds.) (1991). *Ecological Genetics and Air Pollution*. New York & Heidelberg: Springer-Verlag.

Tedin, O. (1925). Vererbung, Variation, und Systematik in der Gattung *Camelia*. *Hereditas*, **6**, 275–386.

Ter Borg, S. J. (1972). Variability of *Rhinanthus serotinus* (Schönh.) Oborny in relation to environment. Thesis, Rijksuniversiteit, Groningen.

Terborgh, J. (1999). *Requiem for Nature*. Washington, DC & Covelo, CA: Island Press.

Tewkesbury, J. J. *et al.* (2002). Corridors affect plants, animals, and their interactions in fragmented landscapes. *Proceedings of the National Academy of Sciences, USA*, **99**, 12923–12926.

Teyssedre, A. & Couvet, D. (2007). Expected impact of agricultural expansion on the world avifauna. *Comtes Rendues Biologies*, **330**, 247–254.

Thacker, C. (1979). *The History of Gardens*. London: Croom Helm.

Theaker, A. J. (1990). Life history variation in *Senecio vulgaris* L. Ph.D. dissertation, University of Cambridge.

Theaker, A. J. & Briggs, D. (1992). Genecological studies of groundsel (*Senecio vulgaris* L.). III. Population variation and its maintenance in the University Botanic Garden, Cambridge. *New Phytologist*, **121**, 281–291.

Theaker, A. J. & Briggs, D. (1993). Genecological studies of groundsel (*Senecio vulgaris* L.). IV. Rate of development in plants from different habitat types. *New Phytologist*, **123**, 185–194.

Thomas, W. L. (1956). *Man's Role in Changing the Face of the Earth*. Chicago: University of Chicago Press.

Thomas, D. H. (1974). *Predicting the Past: An Introduction to Anthropological Archaeology*. New York: Holt, Rinehart & Winston.

Thomas, M. B. & Willis, A. J. (1998). Biocontrol – risky but necessary? *Trends in Ecology & Evolution*, **13**, 325–329.

Thomas, C. D. *et al.* (2004). Extinction risk from climate change. *Nature*, **427**, 145–148.

Thompson, P. A. (1973a). Effects of cultivation on the germination character of the corncockle (*Agrostemma githago* L.). *Annals of Botany*, **37**, 133–154.

Thompson, P.A. (1973b). The effects of geographic dispersal by man on the evolution of physiological races of the corn cockle (*Agrostemma githago* L.). *Annals of Botany*, **37**, 413–421.

Thompson, J. N. (2002). Coevolution. In *Encyclopedia of Evolution*, ed. M. Patel, pp. 178–183. Oxford & New York: Oxford University Press.

Thompson, K. & Grime, J. P. (1979). Seasonal variation in the seed banks of herbaceous species in ten contrasting habitats. *Journal of Ecology*, **67**, 893–921.

Thompson, D. Q., Stuckey, R. L. & Thompson, E. B. (1987). Spread, impact and control of purple loosestrife (*Lythrum salicaria*) in North America. *Fish and Wildlife Research*, **2**, 1–55.

Thompson, K., Bakker, J. P. & Bekker, R. M. (1997). *The Soil Seed Banks of North West Europe: Methodology, Density and Longevity*. Cambridge: Cambridge University Press.

Thomson, J. D., Herre, E. A., Hamrick, J. L. & Stone, J. L. (1991). Genetic mosaics in Strangler Fig trees: implications for tropical conservation. *Science*, **254**, 1214–1216.

Thrall, P. H., Burdon, J. J. & Murray, B. R. (2000). The metapopulation paradigm: a fragmented view of conservation biology. In *Genetics, Demography and Viability of Fragmented Populations*, eds. A. G. Young & G. M. Clarke, pp. 75–95. Cambridge: Cambridge University Press.

Thuiller, W., Lavorel, S., Araújo, M. B., Sykes, M. T. & Prentice, C. (2005). Climate change threats to plant diversity in Europe. *Proceedings of the National Academy of Sciences, USA*, **102**, 8245–8250.

Thurston, J. M., Williams, E. D. & Johnston, A. E. (1976). Modern developments in an experiment on permanent grassland started in 1856: effects of fertilizers and lime on botanical composition and crop and soil analyses. *Annales Agronomique*, **27**, 1043–1082.

Tienderen, P. H. van & Toorn, J. van der (1991a). Genetic differentiation between populations of *Plantago lanceolata* L. I. Local adaptation in three contrasting habitats. *Journal of Ecology*, **79**, 27–42.

Tienderen, P. H. van & Toorn, J. van der (1991b). Genetic differentiation between populations of *Plantago lanceolata* L. II. Phenotypic selection in a transplant experiment in three contrasting habitats. *Journal of Ecology*, **79**, 43–59.

Tollefson, J. (2008). Not your father's biofuels. *Nature*, **451**, 880–883.

Towns, D. & Atkinson, I. (1991). New Zealand restoration ecology. *New Scientist*, **130**, 36–39.

Townsend, D. W. H. (1977). Policies of micropropagation unit, progress and potential. In *Technical Advances in Modern Botanic Gardens*, Abstract of papers for meeting 23

June 1977 at Royal Botanic Garden, Kew, pp. 14–15. London: Ministry of Agriculture, Fisheries & Food.

Townsend, H. (1979). The potential and progress of the technical propagation unit at the Royal Botanic Gardens, Kew. In *Survival or Extinction*, eds. H. Synge & H. Townsend, pp. 189–193. Kew: Bentham-Moxon Trust.

Trebst, A. (1996). Molecular genetics and evolution of pesticide resistance. *ACS Symposium Series*, pp. 44–51. Washington, DC: American Chemical Society.

Trimen, H. & Thiselton-Dyer, W. T. (1869). *Flora of Middlesex*. London: Robert Hardwicke.

Troup, R. S. (1952). *Silvicultural Systems*. Oxford: Clarendon Press.

Tudge, C. (1998). *Neanderthals, Bandits and Farmers. How Agriculture Really Began*. London: Weidenfeld & Nicolson.

Tull, J. (1733). *Horse-Hoeing Husbandry*, Dublin: Gunn, Risk, Ewing, Smith, Smith & Bruce.

Turelli, M., Barton N. H. & Coyne, J. A. (2001). Theory and speciation. *Trends in Ecology and Evolution*, **16**, 330–343.

Turesson, G. (1930). The selective effect of climate upon plant species. *Hereditas*, **14**, 99–152.

Turesson, G. (1943). Variation in the apomictic microspecies of *Alchemilla vulgaris* L. *Botaniska Notiser*, **1943**, 413–427.

Turker, M. (2002). Ageing. In *Encyclopedia of Life Sciences*. London: Macmillan.

Turnbull, P. C. B. *et al.* (2004). Vaccine-induced protection against anthrax in cheetah (*Acinonyx jubatus*) and black rhinoceros (*Diceros bicornis*). *Vaccine*, **22**, 3340–3347.

Turner, I. M. (1996). Species loss in fragments of tropical rain forest. *Journal of Applied Ecology*, **33**, 200–209.

Turner, I. M. *et al.* (1994). A study of plant species extinction in Singapore: lessons for the conservation of tropical biodiversity. *Conservation Biology*, **8**, 705–712.

Turrill, W. B. (1959). *The Royal Botanic Gardens Kew: Past and Present*. London: Jenkins.

Tutin, T. G. *et al.* (1964–1980). Flora Europaea. Cambridge: Cambridge University Press.

Tyler, C., Pullin, A. S. & Stewart, G. B. (2006). Effectiveness of management interventions to control invasion by *Rhododendron ponticum*. *Environmental Management*, **37**, 513–522.

Uhl, C. *et al.* (1990). Studies of ecosystem response to natural and anthropogenic disturbances provide guidelines for designing sustainable land-use systems in Amazonia. In *Alternatives to Deforestation: Steps Towards Sustainable Use of Amazon Rain Forest*, ed. A. B. Anderson, pp. 24–42. New York: Columbia University Press.

Umina, P. A., Weeks, A. R., Kearney, M. R., McKechnie, S. W. & Hoffmann, A. A. (2005). A rapid shift in a classic clinal pattern in *Drosophila* reflecting climate change. *Science*, **308**, 691–693.

Underwood, E. C., Klinger, R. & Moore, P. E. (2004). Predicting patterns of non-native plant invasions in Yosemite National Park, California, USA. *Diversity & Distributions*, **10**, 447–459.

Urbanska, K. M. (1994). Restoration ecology research above the timberline: demographic monitoring of whole trial plots in the Swiss Alps. *Botanica Helvetica*, **104**, 141–156.

Urbanska, K. M. (1997). Safe sites: interface of plant ecology and restoration ecology. In *Restoration Ecology & Sustainable Development*, eds. K. M. Urbanska, N. R. Webb & P. J. Edwards, pp. 81–110, Cambridge: Cambridge University Press.

Usher, M. B. (1986). Invasibility and wildlife conservation – invasive species on nature reserves. *Philosophical Transactions of the Royal Society of London Series B, Biological Sciences*, **314**, 695–710.

Vacher, C. *et al.* (2004). Impact of ecological factors on the initial invasion of Bt transgenes into wild populations of birdseed rape (*Brassica rapa*). *Theoretical and Applied Genetics*, **109**, 806–814.

Van Andel, J. & Aronson, J. (2006). *Restoration Ecology: The New Frontier*. Oxford: Blackwell.

Van Der Meijden, R., Plate, C. L. & Weeda, E. J. (1989). *Atlas van de Nederlandse Flora*, Vol. 3. Leiden: Rijksherbarium/Hortus Botanicus.

van Dijk, G. E. (1955). The influence of sward age and management on the type of timothy and cocksfoot. *Euphytica*, **4**, 83–93.

Van Dijk, P. J. (2003). Ecological and evolutionary opportunities of apomixis: insights from *Taraxacum* and *Chondrilla*. *Philosophical Transactions of the Royal Society of London, Series B, Biological Sciences*, **358**, 1113–1120.

van Dijk, H. (2004). Gene exchange between wild and crop in *Beta vulgaris*: how easy is hybridization and what will happen in later generations? In *Introgression from Genetically Modified Plants into Wild Relatives*, eds. H. C. M. den Nijs, D. Bartsch & J. Sweet, pp. 53–61, Wallingford, UK: CAB International.

van Gemerden, B. S., Olff, H., Parren, M. P. E. & Bongers, F. (2003). The pristine rain forest? Remnants of historical human impact on current tree species composition and diversity. *Journal of Biogeography*, **30**, 1381–1390.

van Keuren, R. W. & Davis, R. R. (1968). Persistence of birdsfoot trefoil, *Lotus corniculatus* L. *Agronomy Journal*, **60**, 92–95.

Van Slageren, M. W. (2003). The Millennium Seed Bank: building partnerships in arid regions for the conservation of wild species. *Journal of Arid Environments*, **54**, 195–201.

Van Treuren, R., Bijlmsa, R., Ouborg, N. J. & van Delden, W. (1993). The effects of population size and plant density on outcrossing rates in locally endangered *Salvia pratensis*. *Evolution*, **47**, 1094–1104.

van Valen, L. (1976). Ecological species, multispecies, and oaks. *Taxon*, **25**, 223–239.

Varley, J. A. (1979). Physical and chemical soil factors affecting the growth and cultivation of endemic plants. In *Survival or Extinction*, eds. H. Synge & H. Townsend, pp. 199–205. Kew, UK: Bentham-Moxon Trust.

Vavilov, N. I. (1951). *The Origin, Variation and Breeding of Cultivated Plants*. *Chronica Botanica*, **13**, 1–366.

Vegte, F. W. van der (1978). Population differentiation and germination ecology in *Stellaria media* (L.) Vill. *Oecologia*, **37**, 231–245.

Vekemans, X, Schierup, M. H. & Christiansen, F. B. (1998). Mate availability and fecundity selection in multi-allelic self-incompatibility systems in plants. *Evolution*, **52**, 19–29.

Velkov, V. V., Medvinsky, A. B., Sokolov, M. S. & Marchenko, A. I. (2005). Will transgenic plants adversely affect the environment? *Journal of Biosciences*, **30**, 515–548.

Vera, F. W. M. (2000). *Grazing Ecology and Forest History*. Wallingford, UK: CABI Publishing.

Vergeer, P., Rengelink, R., Copal, A. & Ouborg, N. J. (2003). Interacting effects of genetic variation, habitat quality and population size on performance of *Succisa pratensis*. *Journal of Ecology*, **91**, 18–26.

Vietmeyer, N. (1995). Applying biodiversity. *Journal of the Federation of American Scientists*, **48**, 1–8.
Vighi, M. & Funari, E. (eds.) (1995). *Pesticide Risk in Groundwater*. Boca Raton, FL & New York: Lewis Publishers.
Vincent, C., Goettel, M. S. & Lazarovitis, G. (2007). *Biological Control: A Global Perspective: Case Histories from Around the World*. Cambridge, MA: CABI.
Viosca, P. Jr. (1935). The Irises of southwestern Louisiana: a taxonomic and ecological interpretation. *Bulletin of the American Iris Society*, **57**, 3–56.
Virginia, R. A. & Wall, D. H. (2001). Ecosystem function, principles of. In *Encyclopedia of Biodiversity*, Vol. 2, ed. S. A. Levin, pp. 345–352. San Diego, CA & London: Academic Press.
Vision, T., Brown, D. & Tanksley, S. (2000). The origins of genomic duplications in *Arabidopsis*. *Science*, **152**, 2114–2117.
Visser, M. E., van Noordwijk, A. J., Tinbergen, J. M. & Lessels, C. M. (1998). Warmer springs lead to mistimed reproduction in great tits (*Parus major*). *Proceedings of the Royal Society of London, B*, **264**, 1867–1870.
Vitousek, P. M. (1990). Biological invasions and ecosystem processes: towards an integration of population biology and ecosystem studies. *Oikos*, **57**, 7–13.
Vitousek, P. M., Loope, L. L. & Adsersen, H. (eds.) (1995). *Islands: Biological Diversity and Ecosystem Function*. Berlin: Springer Verlag.
Vitousek, P. M., Mooney, H. A., Lubchenco, J. & Melillo, J. M. (1997). Human domination of the earth's ecosystems. *Science*, **277**, 494–499.
Vogt, K. *et al.* (2001). Conservation efforts, contemporary. In *Encyclopedia of Biodiversity*, ed. S. A. Levin, Vol. 1, pp. 865–881. San Diego, CA: Academic Press.
Vos, C. C. & Opdam, P. (1993). *Landscape Ecology of a Stressed Environment*. London: Chapman & Hall.
Wagner, M. (1868). *Die Darwinische Theorie und das Migrationsgesetz der Organismen*. Leipzig: Leopold Voss.
Wagner, M. (1889). *Die Entstehung der Arten durch räumliche Sonderung*. Basel: Benno Schwable.
Walker, N. F., Hulme, P. E. & Hoelzel, A. R. (2003). Population genetics of an invasive species *Heracleum mantegazzianum*: implications for the role of life history, demographics and independent introductions. *Molecular Ecology*, **12**, 1747–1756.
Wallace, A. R. (1876). *The Geographical Distribution of Animals, with a Study of the Relations of Living Faunas as Elucidating Past Changes of the Earth's Surface*, Vol. 1. New York: Harper & Brothers.
Wallace, A. R. (1895). *The Geographical Distribution of Animals*, Vol. 1. London: MacMillan.
Wallace, A. R. (1910). *The World of Life*. London: Chapman & Hall.
Wallace, A. R. (1911). *The World of Life*. New York: Moffat, Yard.
Waller, D. M., O'Malley, D. M. & Gawler, S. C. (1987). Genetic variation in the extreme endemic *Pedicularis furbishiae* (Scrophulariaceae). *Conservation Biology*, **1**, 335–340.
Walley, K. A., Khan, M. S. I. & Bradshaw, A. D. (1974). The potential for evolution of heavy metal tolerance in plants. I. Copper and zinc tolerance in *Agrostis tenuis*. *Heredity*, **32**, 309–319.
WallisDeVries, M. F., Poschlod, P. & Willems, J. H. (2002). Challenges for the conservation of calcareous grasslands in northwestern Europe; integrating the requirements of flora and fauna. *Biological Conservation*, **104**, 265–273.

Wallstedt, T. & Borg, H. (2003). Effects of experimental acidification on mobilisation of metals from sediments of limed and unlimed lakes. *Environmental Pollution*, **126**, 381–391.

Walmsley, C. A. & Davy, A. J. (1997a). Germination characteristics of shingle beach species, effects of seed ageing and their implications for vegetation restoration. *Journal of Applied Ecology*, **34**, 131–142.

Walmsley, C. A. & Davy, A. J. (1997b). The restoration of coastal shingle vegetation: effects of substrate composition on the establishment of seedlings. *Journal of Applied Ecology*, **34**, 143–153.

Walmsley, C. A. & Davy, A. J. (1997c). The restoration of coastal shingle vegetation: effects of substrate composition on the establishment of container grown plants. *Journal of Applied Ecology*, **34**, 154–165.

Walters, S. M. (1957). Distribution maps of plants: an historical survey. *Conference Report 1957 of BSBI*, pp. 89–96. British Museum (Natural History), London: Botanical Society of the British Isles.

Walters, S. M. (1970). Dwarf variants of *Alchemilla* L. *Fragmenta Floristica et Geobotanica*, 91–98.

Walters, S. M. (1986). *Alchemilla*: a challenge to biosystematists. *Acta Universitatus Upsaliensis, Symbolae Botanicae Upsaliensis*, **XXVII**, 193–198.

Walters, I. (1989). Intensified fishery production at Morton Bay, southeast Queensland, in the late Holocene. *Antiquity*, **63**, 215–224.

Walters, S. M. (1993). *Wild and Garden Plants*. London: HarperCollins.

Walters, S. M. (1995). The taxonomy of European vascular plants: a review of the past half-century and the influence of the Flora Europea project. *Biological Reviews*, **70**, 361–374.

Walters, C. (2004). Principles for preserving germplasm in gene banks. In *Ex Situ Plant Conservation*, eds. E. O. Guerrant, K. Havens & M. Maunder, pp. 113–138. Washington, DC, Covelo, CA & London: Island Press.

Walters, S. M. *et al.* (1984–2000). *European Garden Flora*. Cambridge: Cambridge University Press.

Walther, G.-C. *et al.* (2002). Ecological responses to recent climate change. *Nature*, **416**, 389–395.

Walton, D. W. & Bridgewater, P. (1996). Of gardens and gardeners. *Nature & Resources*, **32**, 15–19.

Wang, R. L., Wendel, J. F. & Dekker, J. H. (1995). Weedy adaptation in *Setaria* spp. II. Genetic diversity and population structure in *Setaria glauca, Setaria geniculata* and *Setaria faberii* (Poaceae). *American Journal of Botany*, **82**, 1031–1039.

Wang, Z. Y., Hopkins, A. & Mian, R. (2001). Forage and turf grass biotechnology. *Critical Reviews in Plant Sciences*, **20**, 573–619.

Wang, X. Y. *et al.* (2005). Duplication and DNA segmental loss in the rice genome: implications for diploidization. *New Phytologist*, **165**, 937–946.

Wanless, R. M. *et al.* (2009). From both sides: dire demographic consequences of carnivorous mice and longlining for the critically endangered Tristan albatrosses on Gough Island. *Biological Conservation*, **142**, 1710–1718.

Warren, C. (2002). *Managing Scotland's Environment*. Edinburgh: Edinburgh University Press.

Warren, M. S. *et al.* (2001). Rapid responses of British butterflies to opposing forces of climate and habitat change. *Nature*, **414**, 65–69.

Warwick, S. I. (1980). The genecology of lawn weeds. VII. The response of different growth forms of *Plantago major* L., and *Poa annua* L. to simulated trampling. *New Phytologist*, **85**, 461–469.

Warwick, S. I. (1990). Allozyme and life history variation in five northwardly colonizing North American weed species. *Plant Systematics and Evolution*, **169**, 41–54.

Warwick, S. I. & Black, L. D. (1986). Genecological variation in recently established populations of *Abutilon theophrasti* (velvetleaf). *Canadian Journal of Botany*, **64**, 1632–1643.

Warwick, S. I. & Black, L. D. (1994). Relative fitness of herbicide-resistant and susceptible biotypes of weeds. *Phytoprotection*, **75**, 37–49.

Warwick, S. I. & Briggs, D. (1978a). The genecology of lawn weeds. I. Population differentiation in *Poa annua* L. in a mosaic environment of bowling green lawns and flowerbeds. *New Phytologist*, **81**, 711–721.

Warwick, S. I. & Briggs, D. (1978b). The genecology of lawn weeds. II. Evidence for disruptive selection in *Poa annua* L. in a mosaic environment of bowling green lawns and flowerbeds. *New Phytologist*, **81**, 725–737.

Warwick, S. I. & Briggs, D. (1979). The genecology of lawn weeds. III. Experiments with *Achillea millefolia* L., *Bellis perennis* L., *Plantago lanceolata*, L., *Plantago major*, L. and *Prunella vulgaris* L. collected from lawns and contrasting grassland habitats. *New Phytologist*, **83**, 509–536.

Warwick, S. I. & Briggs, D. (1980a). The genecology of lawn weeds. IV. Adaptive significance of variation in *Bellis perennis* L. as revealed in a transplant experiment. *New Phytologist*, **85**, 275–288.

Warwick, S. I. & Briggs, D. (1980b). The genecology of lawn weeds. V. The adaptive significance of different growth habit in lawn and roadside populations of *Plantago major* L. *New Phytologist*, **85**, 289–300.

Warwick, S. I. & Briggs, D. (1980c). The genecology of lawn weeds. VI. The adaptive significance of variation in *Achillea millefolium* L. as investigated by transplant experiments. *New Phytologist*, **85**, 451–460.

Warwick, S. I. & Small, E. (1999). Invasive plant species: evolutionary risks from transgenic crops. In *Plant Evolution in Man-made Habitats*, eds. L. W. D. van Raamsdonk & J. C. M. den Nijs, pp. 235–256. Amsterdam: Hugo de Vries Laboratory.

Warwick, S. I., Thompson, B. K. & Black, L. D. (1984). Population variation in *Sorghum halepense*, Johnson grass, at the northern limits of its range. *Canadian Journal of Botany*, **62**, 1781–1790.

Warwick, S. I., Thompson, B. L. & Black, L. D. (1987). Genetic variation in Canadian and European populations of the colonizing weed species, *Apera spica-venti*. *New Phytologist*, **106**, 301–317.

Warwick, S. I., Beckie, H. & Small, E. (1999). Transgenic crops: new weed problems for Canada? *Phytoprotection*, **80**, 71–84.

Warwick, S. I. *et al.* (2003). Hybridisation between transgenic *Brassica napus* L. and its wild relatives: *Brassica rapa* L., *Rhaphanus raphanistrum* L., *Sinapis arvensis* L., and *Erucastrum gallicum* (Willd.) OE Schultz. *Theoretical and Applied Genetics*, **107**, 528–539.

Watkins, C. (1998). *European Woods and Forests*. Wallingford, UK: CAB International.

Watson P. J. (1970). Evolution in closely adjacent plant populations. VII. An entomophilous species, *Potentilla erecta*, in two contrasting habitats. *Heredity*, **24**, 407–422.

Watson, J. D. & Crick, F. H. C. (1953). A structure of deoxyribose nucleic acid. *Nature*, **171**, 737–738.

Weaver, S. E. & Warwick, S. I. (1983). Comparative relationships between atrazine resistant and susceptible populations of *Amaranthus retroflexus* and *A. powellii* from southern Ontario. *New Phytologist*, **92**, 131–139.

Webb, D. A. (1985). What are the criteria for presuming native status? *Watsonia*, **15**, 231–236.

Weber, B. H. (2005). The origins of Darwinism. *Nature*, **438**, 287–287.

Weber, E. & Schmidt, B. (1998). Latitudinal population differentiation in two species of *Solidago* (Asteraceae) introduced into Europe. *American Journal of Botany*, **85**, 1110–1121.

Webster, P. J., Holland, G. J., Curry, J. A. & Chang, H. R. (2005). Changes in tropical cyclone number, duration, and intensity in a warming environment. *Science*, **309**, 1844–1846.

Weekes, R. *et al.* (2005). Crop-to-crop gene flow using farm scale sites of oilseed rape (*Brassica napus*) in the UK. *Transgenic Research*, **14**, 749–759.

Weekley, C. W. & Race, T. (2001). The breeding system of *Ziziphus celata* Judd and D. W. Hall (Rhamnaceae), a rare endemic plant of the Lake Wales Ridge, Florida, USA: implications for recovery. *Biological Conservation*, **100**, 207–213.

Wegener, A. (1915). *Die Entstehung der Kontinente und Ozeane.* Vieneg, Braunschweig. (English translation 1925, *The Origin of Continents and Oceans*. London: Methuen).

Weir, J. & Ingram, R. (1980). Ray morphology and cytological investigations of *Senecio cambrensis* Rosser. *New Phytologist*, **86**, 237–241.

Welch, R., Remillard, M. & Doren, R. F. (1995). GIS database development for South Florida's national parks and preserves. *Photogrammetric Engineering and Remote Sensing*, **61**, 1371–1381.

Werth, C. R., Riopel, J. L. & Gillespie, N. W. (1984). Genetic uniformity in an introduced population of Witchweed (*Striga asiatica*) in the United States. *Weed Science*, **32**, 645–648.

Western, D. (1989). Conservation without Parks: wildlife in the rural landscape. In *Conservation for the Twenty-first Century*, eds. D. Western & M. Pearl, pp. 158–165. New York: Oxford University Press.

Western, D. (2001). Human modified ecosystems and future evolution. *Proceedings of the National Academy of Sciences, USA*, **98**, 5458–5465.

Wettstein, R.v. (1900). Descendenztheoretische Untersuchungen. I. Untersuchen über den Saison-Dimorphismus im Pflanzenreich. *Denkschriften der Kaiserlichen Akademie der Wissenschaften Wien, Mathematische: Naturwissenschaftliche Klasse*, **70**, 305–346.

Whitburn, T. (1898). My sanctuary. *Nature Notes*, **9**, 24–25.

White, J. (1985). The census of vegetation. In *The Population Structure of Vegetation*, ed. J. White, pp. 33–88. Dordrecht: Junk.

White, P. S. (1996). Spatial and biological scales of reintroduction. In *Restoring Diversity, Strategies for Reintroduction of Endangered Species*, eds. D. A. Falk, C. I. Millar & M. Olwell, pp. 49–86. Washington, DC & Corvelo, CA: Island Press.

White, M. A., Diffenbaugh, N. S., Jones, G. V., Pal, J. S. & Giorgi, F. (2006). Extreme heat reduces and shifts United States premium wine production in the 21st century. *Proceeding of the National Academy of Sciences, USA*, **103**, 11217–11222.

Whitehouse, H. L. K. (1973). *Towards an Understanding of the Mechanism of Heredity*, 3rd edn. London: Arnold.
Whitehouse, A. M. (2002). Tusklessness in the elephant population of the Addo Elephant National Park, South Africa. *Journal of Zoology*, **257**, 249–254.
Whitfield, C. P., Davison, A. W. & Ashenden, T. W. (1997). Artificial selection and heritability of ozone resistance in two populations of *Plantago major*. *New Phytologist*, **137**, 645–655.
Whitlock, M. C. & Michalakis, Y. (2002). Metapopulation. In *Encyclopedia of Evolution*, ed. M. Patel, pp. 724–727. Oxford & New York: Oxford University Press.
Whitmore, T. C. & Sayer, J. A. (1992). *Tropical Deforestation and Species Extinction*. London: Chapman & Hall.
Whitney, G, G. (1994). *From Coastal Wilderness to Fruited Plain: A History of Environmental Change in Temperate North America, 1500 to the Present*. Cambridge: Cambridge University Press.
Whitton, J. *et al.* (1997). The persistence of cultivar alleles in wild populations of sunflowers five generations after hybridization. *Theoretical and Applied Genetics*, **95**, 33–40.
Wickler, W. (1968). *Mimicry in Plants and Animals*. New York: McGraw-Hill.
Wiedemann, A. M. & Pickart, A. (1996). The *Ammophila* problem on the northwest coast of North America. *Landscape and Urban Planning*, **34**, 287–299.
Wieland, G. D. (1993). *Guidelines for the Management of Orthodox Seeds*. St. Louis: Center for Plant Conservation.
Wilcock, C. C. & Jennings, S. B. (1999). Partner limitation and restoration of sexual reproduction in the clonal dwarf shrub *Linnaea borealis* L. (Caprifoliaceae). *Protoplasma*, **208**, 76–86.
Wilcox, B. A. & Murphy, D. D. (1985). Migration and control of purple loosestrife (*Lythrum salicaria*) along highway corridors. *Environmental Management*, **13**, 365–370.
Wiley, E. O. (1981). *Phylogenetics: The Theory and Practice of Phylogenetic Systematic*. New York: Wiley.
Wilkes, H. G. (1977). Hybridization of maize and teosinte in Mexico and Gutemala and improvement of maize. *Economic Botany*, **31**, 254–293.
Wilkins, D. A. (1957). A technique for the measurement of lead tolerance in plants. *Nature*, **180**, 37–38.
Wilkins, D. A. (1960). *The Measurement and Ecological Genetics of Lead Tolerance in* Festuca Ovina. Report of the Scottish Plant Breeding Station, pp. 85–98.
Wilkins, D. A. (1978). The measurement of tolerance to edaphic factors by means of root growth. *New Phytologist*, **80**, 623–634.
Willems, J. H. (1995). Soil seed bank, seedling recruitment and actual species composition in an old and isolated chalk grassland site. *Folia Geobotanica et Phytotaxonomica, Praha*, **30**, 141–156.
Williams, R. (1976). *Keywords: A Vocabulary of Culture and Society*. London: Fontana, Croom Helm.
Williams, R. (1980). *Problems in Materialism and Culture*. London: Verso.
Williams, M. (1989). *Americans and their Forests*. Cambridge: Cambridge University Press.
Williams, M. (2006). *Deforesting the Earth: From Prehistory to Global Crisis. An Abridgement*. Chicago & London: University of Chicago Press.

Williams, S. L. & Davis, C. A. (1996). Population genetic analyses of transplanted eelgrass (*Zostera marina*) beds reveal reduced genetic diversity in southern California. *Restoration Ecology*, **4**, 163–180.
Williams, D. A., Wang, Y. Q., Borchetta, M. & Gaines, M. S. (2007). Genetic diversity and spatial structure of a keystone species in fragmented pine rockland habitat. *Biological Conservation*, **138**, 256–268.
Williamson, M. H. (1996). *Biological Invasions*. London: Chapman & Hall.
Williamson, T. (1997). *The Norfolk Broads. A Landscape History*. Manchester & New York: Manchester University Press.
Williamson, M. H. & Brown, K. C. (1986). The analysis and modelling of British invasions. *Philosophical Transactions of the Royal Society of London, Series B*, **314**, 505–521.
Williamson, M. H. & Fitter, A. (1996). The characteristics of successful invaders. *Biological Conservation*, **78**, 163–170.
Willis, E. O. (1974). Populations and local extinctions of birds on Barro Colorado Island, Panama. *Ecological Monographs*, **44**, 153–169.
Willis, K. J. & McElwain, J. C. (2002). *The Evolution of Plants*. Oxford: Oxford University Press.
Willis, K. J., Gillson, L. & Brncic, T. M. (2004). How "virgin" is virgin rainforest? *Science*, **304**, 402–403.
Willis, C. G. *et al.* (2008). Phylogenetic patterns of species loss in Thoreau's woods are driven by climate change. *Proceedings of the National Academy of Sciences of the USA*, **105**, 17029–17033.
Wilson, E. H. (1919). The romance of our trees: II. The ginkgo. *Garden Magazine*, **30**, 144–148.
Wilson, E. O. (1988). The current state of biological diversity. In *Biodiversity*, ed. E. O. Wilson, pp. 3–18. Washington, DC: National Academic Press.
Wilson, E. O. (1992). *The Diversity of Life*. Cambridge, MA: Belknap Press.
Wilson, G. B & Bell, J. N. B. (1985). Studies on the tolerance to SO_2 of grass populations in polluted areas. III. Investigations on the rate of development of tolerance. *New Phytologist*, **100**, 63–77.
Wilson, G. B. & Bell, J. N. B. (1986). Studies of the tolerance to sulphur dioxide of grass populations in polluted areas. IV. The spatial relationship between tolerance and a point source of pollution. *New Phytologist*, **102**, 563–574.
Winston, M. L. (1997). *Nature Wars: People vs. Pest*. Cambridge, MA: Harvard University Press.
Wiser, S. K., Allen, R. B., Clinton, P. W. & Platt, K. H. (1998). Community structure and forest invasion by an exotic herb over 23 years. *Ecology*, **79**, 2071–2081.
Wolff, K., Morgan-Richards, M. & Davison, A. W. (2000). Patterns of molecular genetic variation in *Plantago major* and *P. intermedia* in relation to ozone resistance. *New Phytologist*, **145**, 501–509.
Woodland, D. J. (1991). *Contemporary Plant Systematics*. Engelwood Cliffs, NJ: Prentice Hall.
Woods, M. (2001). Wilderness. In *A Companion to Environmental Philosophy*, ed. D. Jamieson, pp. 349–361. Maiden, MA & Oxford, UK: Blackwell.
Woodward, F. I. (1987). *Climate and Plant Distribution*. Cambridge: Cambridge University Press.
World Conservation Monitoring Centre (WCMC) (1992). *Global Diversity: Status of the Earth's Living Resources*. Cambridge: WCMC.

Worster, D. (ed.) (1988). *The Ends of the Earth*. Cambridge: Cambridge University Press.
Worster, D. (1992). *Under Western Skies: Nature and History in the American West*. New York & Oxford: Oxford University Press.
Wright, R. G. (1999). Wildlife management in the national parks: questions in search of answers. *Ecological Applications*, **9**, 30–36.
Wright, S. (1931). Evolution in Mendelian populations. *Genetics*, **16**, 97–159.
Wright, S. (1951). The genetical structure of populations. *Annals of Eugenics*, **15**, 323–354.
Wright, S. J., Zeballos, H., Domínguez, I., Gallardo, M. M., Moreno, M. C. & Ibáñnez, R. (2000). Poachers alter mammal abundance, seed dispesal, and seed predation in a neotropical forest. *Conservation Biology*, **14**, 227–239.
Wu, L. (1990). Colonization and establishment of plants in contaminated sites. In *Heavy Metal Tolerance in Plants: Evolutionary Aspects*, ed. A. J. Shaw, pp. 269–284. Boca Raton, FL: CRC Press.
Wu, L. & Antonovics, J. (1976). Experimental ecological genetics in *Plantago*. II. Lead tolerance in *Plantago lanceolata* and *Cynodon dactylon* from a roadside. *Ecology*, **57**, 205–208.
Wu., L., Bradshaw, A. D. & Thurman, D. A. (1975). The potential for evolution of heavy metal tolerance in plants. III. The rapid evolution of copper tolerance in *Agrostis stolonifera*. *Heredity*, **34**, 165–187.
Wu, L., Till-Bottraud, I. & Torres, A. (1987). Genetic differentiation in temperature-enforced seed dormancy among golf course populations of *Poa annua* L. *New Phytologist*, **107**, 623–631.
Wulff, E. V. (1943). *An Introduction to Historical Plant Geography*. Waltham, MA: Chronica Botanica Company.
Wyse Jackson, P. S. & Sutherland, L. A. (2000). *International Agenda for Botanic Gardens in Conservation*. Kew: Botanic Gardens Conservation International.
Yannic, G., Baumel, A. & Ainouche, M. (2004). Uniformity of the nuclear and chloroplast genomes of *Spartina maritima* (Poaceae), a salt-marsh species in decline along the Western European coast. *Heredity*, **93**, 182–188.
Yates, C. J. & Ladd, P. G. (2005). Relative importance of reproductive biology and establishment ecology for persistence of a rare shrub in a fragmented landscape. *Conservation Biology*, **19**, 239–249.
Young, A. (1995). Landscape structure and genetic variation in plants: empirical evidence. In *Mosaic Landscapes and Ecological Processes*, eds. L. Hansson, L. Fahrig & G. Merriam, pp. 153–177. London: Chapman & Hall.
Young, A. (1998). *Land Resources: Now and for the Future*. Cambridge: Cambridge University Press.
Young, A. G. & Clarke, G. M. (2000). *Genetics, Demography and Viability of Fragmented Populations*. Cambridge: Cambridge University Press.
Young, A. G., Brown, A. H. D., Murray, B .G., Thrall, P. H. & Miller, C. (2000). Genetic erosion, restricted mating and reduced viability in fragmented populations of the endangered grassland herb *Rutidosis leptorrhynchoides*. In *Genetics, Demography and Viability of Fragmented Populations*, eds. A. G. Young & G. M. Clarke, pp. 335–359. Cambridge: Cambridge University Press.
Zabinsi, C. & Davis, M. B. (1989). Hard times ahead for the Great Lakes forest: a climate threshold model predicts responses to CO_2-induced climate change. In *The Potential*

Effects of Global Climate Change on the United States, eds. J. B. Smith & D. Tirpak, pp. 5.1–5.19, Appendix D. Washington, DC: US Environmental Protection Agency.
Zapiola, M. L. *et al.* (2008). Escape and establishment of transgenic glyphosate-resistant creeping bentgrass *Agrostis stolonifera* in Oregon, USA: a 4-year study. *Journal of Applied Ecology*, **45**, 486–494.
Zar, J. H. (1999). *Biostatistical Analysis*, 4th edn. London: Prentice-Hall.
Zavaleta, E. S., Hobbs, R. J. & Mooney, H. A. (2001). Viewing invasive species removal in a whole-ecosystem context. *Trends in Ecology & Evolution*, **16**, 454–459.
Zeder, M. A. (2006). Central questions in the domestication of plants and animals. *Evolutionary Anthropology*, **15**, 105–117.
Zhang, D.-X. & Hewitt, G. M. (1998). Extraction of DNA from preserved specimens. In *Molecular Tools for Screening Biodiversity*, eds. A. Karp, P. G. Isaac & D. S. Ingram, pp. 41–45. London: Chapman Hall.
Zimdahl, R. C. (1999). *Fundamentals of Weed Science*. San Diego, CA and London: Academic Press.
Zimmerman, C. A. (1976). Growth characteristics of weediness of *Portulaca oleracea* L. *Ecology*, **57**, 964–974.
Zinger, H. B. (1909). On the species of *Camelina* and *Spergularia* occurring as weeds in sowings of flax and their origin. *Trudy Botanicheskoy Imperatorskoi Akademii Nauk*, **6**, 1–303.
Ziska, L. H. (2002). Influence of rising atmospheric CO_2 since 1900 on early growth and photosynthetic response of a noxious weed Canada thistle (*Cirsium arvense*). *Functional Plant Biology*, **29**, 1387–1392.
Ziska, L. H. & Bunce, J. A. (2006). Plant responses to rising atmospheric carbon dioxide. In *Plant Growth and Climate Change*, eds. J. I. L. Morison & M. D. Morecroft, pp. 17–47. Oxford: Blackwell.
Ziska, L. H., Faulkner, S. & Lydon, J. (2004). Changes in biomass and root: shoot ratio of field grown Canada thistle (*Cirsium arvense*), a noxious invasive weed, with elevated CO_2: implications for control with glyphosate. *Weed Science*, **52**, 584–588.
Ziska, L. H., Reeves, J. B. & Blank, B. (2005). The impact of recent increases in atmospheric CO_2 on biomass production and vegetative retention of Cheatgrass (*Bromus tectorum*): implications for fire disturbance. *Global Change Biology*, **11**, 1325–1332.
Zoller, H. (1954). Die Arten der *Bromus erectus* Wiesen der Schweizer Juras. Veröffentlichungen des Geobotanischen Institute ETH. *Stiftung Rübel*, **28**, 1–283.
Zolman, J. F. (1993). *Biostatistics: Experimental Design and Statistical Inference*. Oxford & New York: Oxford University Press.
Zopfi, H.-J. (1991). Aestival and autumnal vicariads of *Gentianella* (Gentianaceae): a myth? *Plant Systematics and Evolution*, **174**, 139–158.
Zopfi, H.-J. (1993a). Ecotypic variation in *Rhinanthus alectorolophus* (Scopoli) Pollich in relation to grassland management. I. Morphological delimitations and habitats of seasonal ecotypes. *Flora*, **188**, 15–39.
Zopfi, H.-J. (1993b). Ecotypic variation in *Rhinanthus alectorolophus* (Scopoli) Pollich in relation to grassland management. II. The genetic basis of seasonal ecotypes. *Flora*, **188**, 153–173.
Zopfi, H.-J. (1995). Life history variation and infraspecific heterochrony in *Rhinanthus glacialis* (Scrophulariaceae). *Plant Systematics and Evolution*, **198**, 209–233.

Index

The index includes the latin names of plants and animals involved in the 'case studies' analysed in the text. In the space available, it has not been possible to include all the names used in the book: in general, common names of organisms are not listed.